Logic Overview III — The Seven Stages
The Scientific Method (see Chapter 13)

1. Identify the problem
2. Devise preliminary hypotheses
3. Collect additional facts
4. Formulate a refined explanatory hypothesis
5. Deduce consequences from the refined hypothesis
6. Test the consequences deduced
7. Apply the theory

Logic Overview IV — Mill's Methods of Inductive Inference
(see Chapter 12)

1. **The Method of Agreement:** The one factor or circumstance that is *common* to all the cases of the phenomenon under investigation is likely to be the cause (or effect) of that phenomenon.
   A B C D occur together with w x y z.
   A E F G occur together with w t u v.
   Therefore A is the cause (or the effect) of w.

2. **The Method of Difference:** The one factor or circumstance whose *absence* or *presence distinguishes* all cases in which the phenomenon under investigation occurs from those cases in which it does not occur, is likely to be the cause, or part of the cause, of that phenomenon.
   A B C D occur together with w x y z.
   B C D occur together with x y z.
   Therefore A is the cause, or the effect, or an indispensable part of the cause of w.

3. **The Joint Method of Agreement and Difference:** The *combination*, in the same investigation, of the Method of Agreement and the Method of Difference.
   A B C — x y z.   A B C — x y z.
   A D E — x t w.   B C — y z.
   Therefore A is the effect, or the cause, or an indispensable part of the cause, of x.

4. **The Method of Residues:** When some portion of the phenomenon under examination is known to be the consequence of well-understood antecedent circumstances, we may infer that the *remainder* of that phenomenon is the effect of the remaining antecedents.
   A B C — x y z.
   B is known to be the cause of y.
   C is known to be the cause of z.
   Therefore A is the cause of x.

5. **The Method of Concomitant Variation:** When the variations in one phenomenon are highly *correlated* with the variation in another phenomenon, one of the two is likely to be the cause of the other, or they may be related as the products of some third factor causing both.
   A B C — x y z.
   A+B C — x+y z.
   Therefore A and x are causally connected.

INTRODUCTION TO
LOGIC

INTRODUCTION TO
LOGIC

FOURTEENTH EDITION

Irving M. Copi
University of Hawaii

Carl Cohen
University of Michigan

Kenneth McMahon
Hawaii Pacific University

Prentice Hall
Boston Columbus Indianapolis New York San Francisco Upper Saddle River
Amsterdam Cape Town Dubai London Madrid Milan Munich Paris Montreal Toronto
Delhi Mexico City Sao Paulo Sydney Hong Kong Seoul Singapore Taipei Tokyo

Editorial Director: Craig Campanella
Editor in Chief: Dickson Musslewhite
Publisher: Nancy Roberts
Editorial Project Manager: Maggie Barbieri
Editorial Assistant: Nart Varoqua
Director of Marketing: Brandy Dawson
Senior Marketing Manager: Laura Lee Manley
Managing Editor: Maureen Richardson
Project Manager: Marianne Peters-Riordan
AV/Project Manager: Maria Piper
Senior Operations Supervisor: Sherry Lewis
Operations Specialist: Christina Amato
Senior Art Director: Nancy Wells
Text and Cover Design: Ilze Lemesis
Manager, Visual Research: Beth Brenzel
Photo Researcher: Billy Ray
Manager, Rights and Permissions: Zina Arabia
Manager, Cover Visual Research & Permissions: Karen Sanatar
Cover Art: Vladimir Caplinskij/Shutterstock.Images
Director, Digital Media: Brian Hyland
Digital Media Editor: Rachel Comerford
Full-Service Project Management and Composition: Prepare
Printer/Binder and Insert Printer: R.R. Donnelley & Sons
Cover Printer: Lehigh/Phoenix
Text Font: 10.5/14 Palatino

Credits and acknowledgments borrowed from other sources and reproduced, with permission, in this textbook appear on appropriate page within text (or on page 607).

Copyright © 2011, 2009, 2005, 2002, 1998 by Pearson Education, Inc., publishing as Prentice Hall, 1 Lake St. Upper Saddle River, NJ 07458. All rights reserved. Manufactured in the United States of America. This publication is protected by Copyright, and permission should be obtained from the publisher prior to any prohibited reproduction, storage in a retrieval system, or transmission in any form or by any means, electronic, mechanical, photocopying, recording, or likewise. To obtain permission(s) to use material from this work, please submit a written request to Pearson Education, Inc., Permissions Department, 1 Lake St. Upper Saddle River, NJ07458.

Many of the designations by manufacturers and seller to distinguish their products are claimed as trademarks. Where those designations appear in this book, and the publisher was aware of a trademark claim, the designations have been printed in initial caps or all caps.

Library of Congress Cataloging-in-Publication Data

Copi, Irving M.
  Introduction to logic. — 14th ed./Irving M. Copi, Carl Cohen, K. D. McMahon.
    p. cm.
    Includes index.
    ISBN 0-205-82037-9 (alk. paper)
 1. Logic—Textbooks.   I. Cohen, Carl, 1931-   II. McMahon, K. D. (Kenneth D.)   III. Title.
BC108.C69 2011
160—dc22
            2010036272

10 9 8 7 6 5 4 3 2 1

Prentice Hall
is an imprint of

PEARSON

Student Edition	ISBN-13:	978-0-205-82037-5
	ISBN-10:	0-205-82037-9
Exam Edition	ISBN-13:	978-0-205-82048-1
	ISBN-10:	0-205-82048-4
A La Carte Edition	ISBN-13:	978-0-205-82865-4
	ISBN-10:	0-205-82865-5

www.pearsonhighered.com

We dedicate this fourteenth edition of *Introduction to Logic* to the many thousands of students and their teachers, at hundreds of universities in the United States and around the world, who have relied on earlier editions of the book, and have greatly helped to improve it, over five decades.

Brief Contents

Part I LOGIC AND LANGUAGE 1

SECTION A REASONING 1

chapter 1 Basic Logical Concepts 2

chapter 2 Analyzing Arguments 34

SECTION B INFORMAL LOGIC 64

chapter 3 Language and Definitions 64

chapter 4 Fallacies 105

Part II DEDUCTION 163

SECTION A CLASSICAL LOGIC 163

chapter 5 Categorical Propositions 164

hapter 6 Categorical Syllogisms 205

chapter 7 Syllogisms in Ordinary Language 245

SECTION B MODERN LOGIC 287

chapter 8 Symbolic Logic 287

chapter 9 Methods of Deduction 337

chapter 10 Quantification Theory 397

Part III INDUCTION 443

SECTION A ANALOGY AND CAUSATION 443

chapter 11 Analogical Reasoning 444

chapter 12 Causal Reasoning 470

SECTION B SCIENCE AND PROBABILITY 513

chapter 13 Science and Hypothesis 513

chapter 14 Probability 539

Contents

Foreword xiii
Preface xv
Acknowledgments xvii
A Very Brief History of Logic xxi

Part I LOGIC AND LANGUAGE 1

SECTION A REASONING 1

chapter 1   Basic Logical Concepts 2

- **1.1** What Logic Is 2
- **1.2** Propositions and Arguments 2
- **1.3** Recognizing Arguments 11
- **1.4** Arguments and Explanations 18
- **1.5** Deductive and Inductive Arguments 24
- **1.6** Validity and Truth 27

chapter 2   Analyzing Arguments 34

- **2.1** Paraphrasing Arguments 34
- **2.2** Diagramming Arguments 38
- **2.3** Complex Argumentative Passages 49
- **2.4** Problems in Reasoning 54

SECTION B INFORMAL LOGIC 64

chapter 3   Language and Definitions 64

- **3.1** Language Functions 64
- **3.2** Emotive Language, Neutral Language, and Disputes 71
- **3.3** Disputes and Ambiguity 75
- **3.4** Definitions and Their Uses 79
- **3.5** The Structure of Definitions: Extension and Intension 86
- **3.6** Definition by Genus and Difference 94

chapter 4   Fallacies 105

- **4.1** What Is a Fallacy? 105
- **4.2** Classification of Fallacies 106
- **4.3** Fallacies of Relevance 108
- **4.4** Fallacies of Defective Induction 126
- **4.5** Fallacies of Presumption 134
- **4.6** Fallacies of Ambiguity 140
       Logic in the Real World 156

Part II DEDUCTION 163

SECTION A CLASSICAL LOGIC 163

chapter 5   Categorical Propositions 164

- **5.1** The Theory of Deduction 164
- **5.2** Classes and Categorical Propositions 165

ix

- **5.3** The Four Kinds of Categorical Propositions 166
- **5.4** Quality, Quantity, and Distribution 170
- **5.5** The Traditional Square of Opposition 176
- **5.6** Further Immediate Inferences 180
- **5.7** Existential Import and the Interpretation of Categorical Propositions 189
- **5.8** Symbolism and Diagrams for Categorical Propositions 197

chapter 6   Categorical Syllogisms 205

- **6.1** Standard-Form Categorical Syllogisms 205
- **6.2** The Formal Nature of Syllogistic Argument 210
- **6.3** Venn Diagram Technique for Testing Syllogisms 213
- **6.4** Syllogistic Rules and Syllogistic Fallacies 224
- **6.5** Exposition of the Fifteen Valid Forms of the Categorical Syllogism 234
  Appendix: Deduction of the Fifteen Valid Forms of the Categorical Syllogism 239

chapter 7   Syllogisms in Ordinary Language 245

- **7.1** Syllogistic Arguments 245
- **7.2** Reducing the Number of Terms to Three 246
- **7.3** Translating Categorical Propositions into Standard Form 249
- **7.4** Uniform Translation 258
- **7.5** Enthymemes 264
- **7.6** Sorites 269
- **7.7** Disjunctive and Hypothetical Syllogisms 272
- **7.8** The Dilemma 278

SECTION B   MODERN LOGIC 287

chapter 8   Symbolic Logic 287

- **8.1** Modern Logic and Its Symbolic Language 287
- **8.2** The Symbols for Conjunction, Negation, and Disjunction 288
- **8.3** Conditional Statements and Material Implication 300
- **8.4** Argument Forms and Refutation by Logical Analogy 310
- **8.5** The Precise Meaning of "Invalid" and "Valid" 314
- **8.6** Testing Argument Validity Using Truth Tables 314
- **8.7** Some Common Argument Forms 316
- **8.8** Statement Forms and Material Equivalence 323
- **8.9** Logical Equivalence 329
- **8.10** The Three "Laws of Thought" 333

chapter 9   Methods of Deduction 337

- **9.1** Formal Proof of Validity 337
- **9.2** The Elementary Valid Argument Forms 340
- **9.3** Formal Proofs of Validity Exhibited 345
- **9.4** Constructing Formal Proofs of Validity 348
- **9.5** Constructing More Extended Formal Proofs 350
- **9.6** Expanding the Rules of Inference: Replacement Rules 357
- **9.7** The System of Natural Deduction 364
- **9.8** Constructing Formal Proofs Using the Nineteen Rules of Inference 368
- **9.9** Proof of Invalidity 383

- 9.10 Inconsistency 386
- 9.11 Indirect Proof of Validity 392
- 9.12 Shorter Truth-Table Technique 394

chapter 10 Quantification Theory 397

- 10.1 The Need for Quantification 397
- 10.2 Singular Propositions 399
- 10.3 Universal and Existential Quantifiers 401
- 10.4 Traditional Subject–Predicate Propositions 406
- 10.5 Proving Validity 415
- 10.6 Proving Invalidity 424
- 10.7 Asyllogistic Inference 429
- Logic in the Real World 438

Part III INDUCTION 443

SECTION A ANALOGY AND CAUSATION 443

chapter 11 Analogical Reasoning 444

- 11.1 Induction and Deduction Revisited 444
- 11.2 Argument by Analogy 445
- 11.3 Appraising Analogical Arguments 452
- 11.4 Refutation by Logical Analogy 463

chapter 12 Causal Reasoning 470

- 12.1 Cause and Effect 470
- 12.2 Causal Laws and the Uniformity of Nature 473
- 12.3 Induction by Simple Enumeration 474
- 12.4 Methods of Causal Analysis 476
- 12.5 Limitations of Inductive Techniques 501

SECTION B SCIENCE AND PROBABILITY 513

chapter 13 Science and Hypothesis 513

- 13.1 Scientific Explanation 513
- 13.2 Scientific Inquiry: Hypothesis and Confirmation 516
- 13.3 Evaluating Scientific Explanations 521
- 13.4 Classification as Hypothesis 529

chapter 14 Probability 539

- 14.1 Alternative Conceptions of Probability 539
- 14.2 The Probability Calculus 542
- 14.3 Probability in Everyday Life 553
- Logic in the Real World 561

Appendix: Graduate-Level Admission Tests 565

Solutions to Selected Exercises 573

Photocredits 607

Glossary/Index 608

Foreword

> In a republican nation, whose citizens are to be led by persuasion and not by force, the art of reasoning becomes of the first importance.
>
> —*Thomas Jefferson*

Logic has sometimes been defined as the science of the laws of thought. This is inaccurate. Thinking is one of the processes studied by psychologists. If thought refers to any process that occurs in people's minds, not all thought is an object of study for the logician. Thus, one may think of a number between one and ten without doing any reasoning about it. One may also remember, imagine, free-associate, or perform any of a number of mental processes. All reasoning is thinking, but not all thinking is reasoning. The laws that describe the movements of the mind are psychological laws rather than logical principles. To define logic in this way is to include too much.

Logic has also been called the science of reasoning. This is better, but reasoning is a kind of thinking in which inference takes place and conclusions are drawn from premises. This process is extremely complex, characterized by a combination of trial and error, occasionally illuminated by flashes of insight. Logicians are not concerned with the ways in which the mind arrives at its conclusions in the process of reasoning; they are concerned only with the correctness of the completed process: Does the conclusion reached *follow* by necessity from the premises? The study of the methods and principles used to distinguish correct from incorrect reasoning is the central issue with which logic deals.

Reason is the instrument on which one must depend when reliable judgments are needed. Nonrational instruments—feelings, beliefs, habits, hunches, and the like—are commonly employed, but when a great deal depends on the judgments one makes—when one must decide how to act in complicated circumstances, or determine what is true in matters that have a serious impact—reason is our best recourse; nothing can replace it.

There are rational methods, methods well tested and confirmed, for determining what is true. There are well-established, rational techniques, for drawing new inferences from what is already known to be true. Our ignorance is vast, and therefore people often resort to some authority in reaching judgment—but the need for reason cannot be escaped even then, because one must decide which authorities deserve respect. Every serious intellectual pursuit comes ultimately to rely on reasoning, because there is nothing that can successfully replace it.

By nature humans are endowed with powers of reasoning. Logic is the study of the uses of those powers. Intuitively, we may have long acted on sound principles, only partly grasped. With care, these principles can be brought to the surface, formulated precisely, and applied with confidence to problems solvable by reason. Through studying logic, people come first to recognize their own native capacities, then to learn to strengthen them through practice. The study of logic helps one to reason well by illuminating the principles of *correct* reasoning.

Whatever the sphere in which knowledge is sought—whether in science, politics, or in the conduct of one's private life—logic is used to reach warranted conclusions. In the formal study of logic, with which this book is concerned, the student will learn how to acquire truths and how to evaluate competing claims for truth, to critique arguments, recognize inconsistencies, detect logical fallacies, and construct formal arguments according to demonstrably valid inference patterns. In sum, the study of logic will help the student to reason more carefully and, in general, to think and act more rationally.

Ideally, every college course should contribute to this end, yet many do not. Much that is taught in college classes soon grows out of date. But the skills of accurate thinking never become obsolete, and the development of these skills lies squarely within the province of the study of logic. The study of logic helps us to identify arguments that are good and to understand why they are good. The study of logic helps us to identify arguments that are bad and to understand why they are bad. No study is more useful or more widely relevant to serious concerns than this.

This considered assurance we give to our readers: A command of the fundamental principles of correct reasoning, which the study of this book promotes, will make a deeply satisfying, significant, and permanent contribution to one's intellectual life.

Preface

Earlier editions of *Introduction to Logic* have been warmly received by our philosophical colleagues around the world. James Druley from Reedley College, Madera, CA, who was one of the reviewers of the twelfth edition, wrote: "Several times, after reading a part of the text I have thought, 'That could not have been written any more insightfully or elegantly; that could not have been explained better.'" We are gratified by such kind words, of course, but we are not content. As a consequence, a number of alterations have been made in an effort to refine, polish, and improve an already excellent text. Certain fallacies have been reclassified in Chapter 4, and the flowchart in Chapter 6 has been restructured. In Chapter 8 we have provided and explained alternative symbols that are used in other texts; in addition, much of the material in Chapters 8–10 has been reorganized. Some arguments, definitions, explanations, and formulations have been tightened up and clarified, while the essential structure and substance of the book have been retained.

In Part One, the basic concepts of logic are presented. We explore the difficulties encountered in everyday uses of language, and the different functions of language. We explain the nature and functions of definitions in ordinary discourse, and then identify and exhibit the many informal fallacies that are commonly encountered. In Part Two the methods of deductive reasoning are presented and analyzed. Here we first give an account of traditional or classical syllogistics, and then introduce the more powerful notation and techniques of modern symbolic logic. Finally, in Part Three, the techniques of inductive logic are presented, beginning with the appraisal of simple analogical arguments and proceeding to an analysis of causal reasoning, and the canons of scientific method. We conclude with an account of alternative theories of probability, and the uses of the theorems in the elementary calculus of probability.

New to This Edition

- A short history of logic appears just before the beginning of Part One.
- Brief biographies of a number of the great logicians have been included throughout the chapters, both in order to historically contextualize logic, as well as to indicate how contributions to logic are not something remote and detached from life, but an expression of the lives and passions of very human beings.
- A new Appendix describes the practical applications of logic to the taking of admissions tests, such as the GRE (Graduate Record Examination), LSAT (Law School Admission Test), GMAT (Graduate Management Admission Test), and MAT (Miller Analogies Test).

Finally, *Introduction to Logic* in all of its editions has been rich with illustrative materials and exercises taken from events and controversies in real life, from history and some classical sources, but mainly from contemporary periodicals and books. We take pride in the fact that, as our reviewers have noted, those studying *Introduction to Logic* are inescapably introduced to a very wide range of intellectual concerns and thus learn much more than logic. Exhibiting arguments and theories (good and bad) by illustrating them with genuine controversies in the world of college students, rather than with illustrations artificially devised for the purpose, has been our continuing concern. Logical theory is most fully grasped when it is vividly applied to contemporary human affairs. In this edition of *Introduction to Logic* we have added many fresh illustrations, replacing those more dated, along with some new arguments arising in connection with the lively issues of the first decade of the twenty-first century. In the selection of these illustrations and exercises we seek scrupulously to avoid partisanship. On all sides of controversial issues, good arguments, and bad ones, may appear. Support for one view or another in current controversy is not our proper business; the understanding and analysis of arguments is.

In all editions of *Introduction to Logic* we strove to achieve that combination of accuracy, clarity, and penetration that has always been our objective. To this end we have relied on the support and advice of students and instructors who use the book and who are sensitive to its shortcomings. We conclude, therefore, with an earnest invitation to our readers to join us in advancing this never-ending project. Send us corrections as needed, and suggestions of every kind. Your contributions, warmly welcome, may be most conveniently addressed to Carl Cohen at ccohen@umich.edu or Kenneth D. McMahon at kmcmahon@campus.hpu.edu. The feedback from students and instructors who rely on *Introduction to Logic* has helped to make it the world's most widely used book in the study of logic. We will receive your responses to it with respect and heartfelt gratitude.

Carl Cohen
The University of Michigan, Ann Arbor
Kenneth McMahon
Hawaii Pacific University, Honolulu

A Note on Color

Logic is often perceived as a "black-and-white" subject. However, in this edition of *Introduction to Logic* we have introduced color. Color plays a role in guiding the eye more readily to sections, subsections, and text boxes, but it also serves important pedagogical functions. In Chapter 6, for instance, color has been used to strengthen the application of the Venn diagram technique for the evaluation of categorical syllogisms. Mastery of this technique is more readily achieved if the circles representing the subject, predicate, and middle terms of categorical syllogisms are consistently positioned relative to each other. Accordingly, previous editions maintained a consistent and predictable placement. This edition strengthens those spatial clues by consistently assigning a specific color to each circle. This provides an additional visual reminder of the nature of each circle.

Acknowledgments

For accurate corrections, penetrating observations, and wise suggestions in very great number, two persons deserve our heartfelt thanks:

Prof. Victor Cosculluela, Polk State College, Winter Haven, Florida
Mr. Manuel Sánchez Nazario, Chicago, Illinois

Contributors to this edition have been very numerous. College students, as well as instructors, have written to suggest improvements, to point out ambiguities or inaccuracies, to note typographical errors, to suggest useful illustrative materials. All receive our direct response, of course; but we take satisfaction in listing here also the names of some of those to whom we are indebted for contributions large and small to this fourteenth edition of *Introduction to Logic*:

Prof. John M. Abbarno, D'Youville College, Buffalo, New York
Prof. Benjamin Abellera, DCF Foundation, Silver Spring, Maryland
Dr. Gerald Abrams, University of Michigan, Ann Arbor, Michigan
Mr. Russell Alfonso, University of Hawaii, Honolulu, Hawaii
Mr. Wyatt Dean Ammon, Hamline University, St. Paul, Minnesota
Ms. Tamara Andrade, Ann Arbor, Michigan
Emil Badici, University of Florida, Gainesville, Florida
Stephen Barnes, Northwest Vista College, San Antonio, Texas
Mr. Jason Bates, University of Michigan, Ann Arbor, Michigan
Mr. Maximilian Bauer, Ann Arbor, Michigan
Mr. George Beals, Michigan Bible School, Plymouth, Michigan
Drew Berkowitz, Bridgewater State University, Bridgewater, Massachusetts
Ms. Amelia Bischof, Ithaca College, Ithaca, New York
Mr. Evan Blanchard, Ann Arbor, Michigan
Mr. Benjamin Block, Ann Arbor, Michigan
Prof. Jeffery Borrowdale, Cuesta College, San Luis Obispo, California
Mr. John Bransfield, University of Connecticut, Storrs, Connecticut
Mr. Nicholas Bratton, Seattle, Washington
Teresa Britton, Eastern Illinois University, Charleston, Illinois
Prof. Keith Burgess-Jackson, University of Texas at Arlington, Texas
Mr. Bryan Campbell, Vanderbilt University, Nashville, Tennessee
Prof. Rebecca Carr, George Washington University, Washington, D.C.
Prof. Sidney Chapman, Richland College, Dallas, Texas
Mr. Kun-Hung Chen, National Taiwan University, Taipei
Prof. Zoe Close, Grossmont College, El Cajon, California
Jennifer Caseldine-Bracht, Indiana University/Purdue University, Ft. Wayne, Indiana
Prof. William S. Cobb, University of Michigan, Ann Arbor, Michigan
Prof. Malcolm S. Cohen, University of Michigan, Ann Arbor, Michigan
Mr. Keith Coleman, University of Kansas, Lawrence, Kansas
Ms. Meredith Crimp, Ann Arbor, Michigan
Mr. Dennis A. De Vera, Dept. of Social Sciences, CAS, CLSU, Philippines
Mr. Joshua De Young, University of Michigan, Ann Arbor, Michigan
James Druly, Reedley College, Madera Center, Madera, California
Prof. Elmer H. Duncan, Baylor University, Waco, Texas
R. Valentine Dusek, University of New Hampshire, Durham, New Hampshire
Mr. Eric Dyer, University of Michigan, Ann Arbor, Michigan
Mr. Kumar Eswaran, Temple University, Philadelphia, Pennsylvania
Mr. Joshua Fay
William Ferraiolo, San Joaquin Delta College, Stockton, California

Ms. Morgan Fett, Ann Arbor, Michigan
Prof. Daniel E. Flage, James Madison University, Harrisonburg, Virginia
Jason Flato, Georgia Perimeter College, Clarkston, Georgia
Prof. Kevin Funchion, Salem State College, Salem, Massachusetts
Ms. Elizabeth Gartner, University of Michigan, Ann Arbor, Michigan
Prof. Faith Gielow, Villanova University, Villanova, Pennsylvania
Prof. Joseph Gilbert, State University of New York at Brockport, New York
Mr. Anand Giridharadas, Mumbai, India
Prof. Sidney Gospe, University of Washington, Seattle, Washington
Mr. Michael Graubert, London, England
Mr. Joseph Grcic, Indiana State University, Terre Haute, Indiana
Dr. Robert A. Greene, University of Michigan, Ann Arbor, Michigan
Ms. Janice Grzankowski, Cheektowaga, New York
Mr. Matthew Hampel, University of Michigan, Ann Arbor, Michigan
Prof. Allan Hancock, Cuesta College, San Luis Obispo, California
Prof. Warren Harbison, Boise State University, Boise, Idaho
Mr. Abdul Halim B. Abdul Karim, of the National University of Singapore
Prof. Clare Swift Heiller, of Bakersfield College, Bakersfield, California
Prof. Jeremiah Joaquin, De La Salle University, Manila, Philippines
Prof. Royce Jones, Illinois College, Jacksonville, Illinois
Prof. Gale Justin, California State University at Sacramento
Mr. Rory Kraft, Jr., Michigan State University, East Lansing, Michigan
Prof. Richard T. Lambert, Carroll College, Helena, Montana
Mr. Charles Lambros, State University of New York at Buffalo
Mr. Andrew LaZella, Hamline University, St. Paul, Minnesota
Prof. Gerald W. Lilje, Washington State University, Pullman, Washington
Mr. James Lipscomb, Tarrytown, New York
Ms. Linda Lorenz, Ann Arbor, Michigan
Prof. E. M. Macierowski, Benedictine College, Atchison, Kansas
Ms. Erika Malinoski, University of Michigan, Ann Arbor, Michigan
Prof. Krishna Mallik, Bentley College, Waltham, Massachusetts
Mr. Neil Manson, University of Aberdeen, United Kingdom
Prof. Edwin Martin, North Carolina State University, Raleigh, North Carolina
Prof. Michael J. Matthis, Kutztown University, Kutztown, Pennsylvania
Prof. George Mavrodes, University of Michigan, Ann Arbor, Michigan
Prof. Leemon McHenry, Wittenberg University, Springfield, Ohio
Mr. Christopher Melley, Univ. of Maryland, Asian Division, Okinawa, Japan
Ms. Medeline Metzger, Ann Arbor, Michigan
Mr. David A. Mihaila, Honolulu, Hawaii
Prof. Richard W. Miller, University of Missouri at Rolla, Missouri
Prof. Masato Mitsuda, San Francisco State University, San Francisco, California
Ms. Erin Moore, Ohio State University, Columbus, Ohio
Ms. Susan Moore, Fairgrove, Michigan
Prof. Kippy Myers, Freed-Hardeman University, Henderson, Tennessee
Mr. Michael North, University of Michigan, Ann Arbor, Michigan
David O'Connor, Seton Hall University, South Orange, New Jersey
Mr. John Oltean, Ann Arbor, Michigan
Prof. Sumer Pek, University of Michigan, Ann Arbor, Michigan
Prof. Ray Perkins, Plymouth State College, Plymouth, New Hampshire
Mr. Robert Picciotto, Gastonia, North Carolina
Prof. Howard Pospesel, University of Miami, Coral Gables, Florida
Mr. Wayne Praeder, of the U.S. Chess Federation
Ms. Deborah Pugh, Stanford, California
Prof. Dennis P. Quinn, St. Vincent College, Latrobe, Pennsylvania
Mr. Nicholas Quiring, University of Michigan, Ann Arbor, Michigan
Mr. Chris Raabe, of Yakatut, Alaska
Mr. Jay Rapaport, University of Michigan, Ann Arbor, Michigan

Dr. Patrick Rarden, Appalachian State University, Boone, North Carolina
Prof. Lee C. Rice, Marquette University, Milwaukee, Wisconsin
Dr. Thomas Riggins, New York University, New York City
David C. Ring, Orange Coast College, Costa Mesa, California
Prof. Blaine B. Robinson, South Dakota School of Mines, Rapid City, South Dakota
Eric Saidel, George Washington University, Washington, D.C.
Rudy Saldana, Citrus College, Glendora, California
Mr. Milton Schwartz, Esq., New York City
Mr. Amit Sharma, of V. S. Niketan College, Kathmandu, Nepal
Prof. Emeritus Albert C. Shaw, Rowan College, Glassboro, New Jersey
Prof. Edward Sherline, University of Wyoming, Laramie, Wyoming
Mr. Amjol Shrestha, Hawaii Pacific University, Honolulu, Hawaii
Ms. Lauren Shubow, University of Michigan, Ann Arbor, Michigan
Mr. Jason A. Sickler, University of North Dakota, Grand Forks, North Dakota
Ms. Stefanie Silverman, University of Michigan, Ann Arbor, Michigan
Prof. Michael Slattery, Villanova University, Villanova, Pennsylvania
Dr. Barbara M. Sloat, University of Michigan, Ann Arbor, Michigan
Mahadevan Srinivasan, University of Tennessee, Knoxville, Tennessee
Prof. James Stewart, Bowling Green State University, Bowling Green, Ohio
Mr. Paul Tang, California State University, Long Beach, California
Mr. Andrew Tardiff, North Kingstown, Rhode Island
Mark L. Thomas, Blinn College, Bryan, Texas
David A. Truncellito, George Washington University, Washington, DC
Ms. Meghan Urisko, Ann Arbor, Michigan
Mr. J. A. Van de Mortel, Cerritos College, Norwalk, California
David Vessey, Grand Valley State University, Allendale, Michigan
Mr. Chris Viger, University of Western Ontario
Mr. Roy Weatherford, University of South Florida, Tampa, Florida
Prof. Allen Weingarten, Morristown, New Jersey
Prof. Warren Weinstein, California State University at Long Beach, California
Ms. Jessica Wheeler, Springfield, Missouri
Prof. Philip H. Wiebe, Trinity Western University, British Columbia, Canada
Mr. Michael Wingfield, Lake Dallas, Texas
Mr. Isaiah Wunsch, University of Michigan, Ann Arbor, Michigan
Ms. Cynthia Yuen, Ann Arbor, Michigan
Maria Zaccaria, Georgia Perimeter College, Dunwoody, Georgia

A Very Brief History of Logic

Philosophy begins with wonder. What is the world made of? Where does it come from? Why are we here? The speculations of primitive peoples were often imaginative, but were unfounded, irrational. Philosophy as we think of it today did not arise until the Greek philosophers of the sixth century BCE sought some overriding theories about the world. Is there one stuff of which the world is made? One principle that is fundamental throughout?

We think of Socrates and Plato as the great figures in the birth of Western philosophy, and we study them still today. Their greatness lies in part in their efforts to bring things into intellectual order—to provide, or at least to seek, some coherent system that can explain why things are the way they are. But even before Socrates there had been deep thinkers—Thales, Parmenides, Heraclitus, Democritus and others who had proposed assorted accounts of the fundamental stuff of the world, or of the fundamental principle by which all is governed.

They were theorizing, not merely guessing—but there was no real science in these early speculations. Dogmatic suppositions, supernatural forces, the gods, ancient myths and legends had always to be called upon. As philosophy gradually matured there grew the drive to *know*, to discover principles that could be relied upon in giving explanations.

Thus logic begins. Judgments are sought that can be tested and confirmed. The *methods* with which we discover and confirm whatever we really know need to be identified and refined. We must *reason* about things, and we hunger to understand the principles of right reasoning.

That first climb from chaotic thought into some well-ordered system of reasoning was an enterprise of extraordinary difficulty. Its first master, Aristotle (see p. 3), having developed a system within which the principles of reasoning could be precisely formulated, was rightly held in awe by rational thinkers from his day to ours. He was the first great logician.

Aristotle approached reasoning as an activity in which we first identify *classes* of things. We then recognize the *relations* among these classes. Then we can manipulate the propositions in which these relations are specified. The fundamental elements of reasoning are, he thought, the groups themselves, the categories into which we can put things. He therefore distinguished types of *categorical* propositions (e.g., "All Xs are Ys"—a universal affirmative proposition; "Some Ys are not Xs"—a particular negative proposition; and so on) and with those understood we can reason immediately to conclusions about the relations among these propositions (e.g., "If some Xs are Ys, then it cannot be true that no Ys are Xs"). More importantly, by combining categorical propositions involving three terms (say, Xs, Ys, and Zs) in various ways, we can reason accurately by constructing *categorical syllogisms* (e.g., "If all Xs are Ys, and some Xs are Zs, it must be that some Zs are Ys"). Using such techniques, a great system of deductive logic can be built, as will be shown in Chapters 5, 6 and 7 of this book.

A century after Aristotle the work of the Stoic philosopher, Chrysippus (see p. 7), carried logical analysis to a higher level. The fundamental elements of reasoning were taken to be not the Aristotelian categories, but *propositions*, the units with which we can affirm or deny some states of affairs (e.g., "X is in Athens," or "X is in Sparta"). We can then discover the logical relations among propositions: "If X is in Athens then X is not in Sparta." We can then identify elementary arguments that depend upon these various relations: "If X is in Athens then X is not in Sparta. X is in Athens. Therefore X is not in Sparta." The form of this simple argument, called *modus ponens*, is common and useful; many other such elementary forms may be identified and applied in rational discourse, as we will see in later portions of this book.

With these advances it soon becomes clear that the validity of a deductive argument, the solidity with which a conclusion may be inferred if the premises are true, depends upon the *form* of the argument, its shape rather than its content—or as logicians say, its syntactic features rather than its semantic content. *Modus ponens*, and every such argument form, can have an unlimited number of realizations, or instances. The consequences of this formal nature of validity remained to be investigated. With the decline of the Roman Empire, the work of the Greek logicians had been preserved by Muslim scholars, most notably Al-Farabi (c. 872–c. 950), who wrote, in Baghdad, a commentary on the works of Aristotle, and came to be called "the Second Teacher," second only to Aristotle in breadth and depth of learning. He was followed by the great Muslim polymath, Ibn Sina, known by his Latinized name, Avicenna. Their scholarship eventually penetrated and refreshed Western thought. Syntactic forms came again to be of central interest in logic in the twelfth century, in France, with the work of the monk, Peter Abelard (1079–1142).

In England the great logical figure of those early modern years was William of Ockham (1287–1348). He identified some of the theorems more precisely formulated many years later by the mathematical logician, Augustus De Morgan; De Morgan's theorems we will encounter and apply in the second part of this book. Ockham sought to rid metaphysics, in which he was chiefly interested, of useless concepts. He urged that when a term or notion has been shown fruitless it should be simply cut out and discarded. This imperative principle, "Ockham's razor," remains a common guideline: In all rational thinking, entities must not be multiplied beyond necessity.

Deductive logic had largely begun with Aristotle's compiled treatises, *The Organon*. That logic allowed and encouraged the powerful manipulation of what is already known, and that is indeed extremely useful. However, the long-studied analysis of propositions and their relations did not provide the stuff of new knowledge, desperately needed and widely sought in the early modern centuries. What the intellectual world required, many thought, was a *new Organon*. That *Novum Organum* was published by Francis Bacon (1561–1626) in England in 1620. The Baconian method aimed to codify the procedures used by scientists when investigating all natural things. Called "the father of empiricism", Bacon, with other pioneers of the scientific revolution in astronomy and medicine, did not reject the work of classical logicians, but supplemented that work by formulating the methods that make possible the *acquisition* of empirical truths. Facts—what we learn about the world—constitute the premises upon which deductive arguments can be built. These were the first great steps in formulating the principles of *inductive logic*.

It was time to gather the threads of logical analysis, deductive and inductive, into one coherent fabric. The first textbook of logic (*Logic, or the Art of Thinking*), was published anonymously in 1662 by a group known as the Port-Royal logicians. The principal authors, Antoine Arnauld (famous for his published disputes with Descartes) and Pierre Nicole, were joined by Blaise Pascal (1623–1662), a great French mathematician who had invented, while a teenager, a functioning mechanical calculator. Pascal was also one of the originators of the theory of probability—a sphere of logic that we will enter in the final chapter of this book. Other textbooks followed, including *Logick, or the Right Use of Reason* (1725) by Isaac Watts; then *Logic* (1826) by Richard Whately. Then, in 1843, there was published in England one of the greatest of all logic textbooks: *A System of Logic*, by John Stuart Mill (1806–1873). In this work the techniques with which we uncover and confirm causal connections in the real world were for the first time set forth in accurate detail. Mill's methods, his still relevant contributions to the study of inductive logic, we discuss at length in Part III of this book.

In deductive logic much creative work remained to be done. Reasoning was known to be burdened by the ambiguities and imprecision of ordinary language. One of the greatest of early modern thinkers, Gottfried Wilhelm Leibniz (1646–1716), set himself the task of overcoming these deficiencies by developing a mathematically exact symbolic language, one in which concepts might be expressed with unambiguous clarity. Leibniz (also one of the independent inventors of the infinitesimal calculus) had envisioned a sort of logic machine—one with which operations of a logical nature might be performed efficiently and accurately, as can be done in the algebra that he knew well. That great logic machine he never produced, but his dream of it may be seen as the foreshadowing of the modern electronic computer.

A major advance toward Leibniz's goal was made by the English logician George Boole (see p. 189), who devised, in his *Investigation into the Laws of Thought* (1854), a general system for the accurate expression and thus manipulation of propositions. Propositions had played a central role in logic since the time of Aristotle and Chrysippus. But it was only with Boole's deep analysis of propositions—the *Boolean interpretation* discussed in great detail in Chapter 5 of this book—that a fully consistent system of the logic of propositions was at last possible.

Other mathematicians and logicians made significant advances that brought greater precision and efficiency to the realm of deductive logic. One of these was Augustus De Morgan (1806–1871), alluded to above in connection with the work of William of Ockham. The theorems that still carry his name remain to this day critical logical tools in proving the validity of deductive arguments. Another English logician, John Venn (1834–1923), contributed brilliantly to the process of determining deductive validity by designing a system, as beautiful as it is simple, for the iconic exhibition of the relations of the terms in categorical propositions. Venn diagrams, consisting of interlocking circles, are now very widely used. They serve as an easily applied device with which the sense of propositions can be given visual force, and with which the validity or invalidity of categorical syllogisms can be established. We use Venn diagrams extensively in Part II of this book.

One of the greatest American philosophers, Charles Sanders Peirce (1839–1914), best known as the founder of the movement known as *pragmatism*, thought of himself primarily as a logician. Logic was for him a very broad study, involving the methods of all inquiry; formal deductive logic, to which he made some notable contributions, he took to be one of its branches. We think with signs, said Peirce, and logic is the formal theory of signs. He introduced some new concepts, such as inclusion and logical sum; he devised symbols for the expression of novel logical operations; he explored the logic of relations—and he anticipated work later done in expressing Boolean operations using the features of electrical switching circuits, a key step toward the actual development of the all-conquering logic machine that had been envisioned by Gottfried Leibniz.

A rigorous, formal system of propositional logic was produced by the German logician Gottlob Frege (1848–1925). That system, and his invention of the concept of *quantification*, establish him as one of the greatest of modern logicians. With quantification—as we explain in detail in Chapter 10 of this book—it is possible to deal accurately with a huge body of deductive argument that cannot otherwise be readily penetrated by the machinery of modern symbolic logic.

Bertrand Russell (1872–1970) and Alfred North Whitehead (1861–1947) sought to integrate all this modern work on deductive logic in one great and remarkable treatise: *Principia Mathematica*, published in segments from 1910 to 1913. Using (with some adjustments) the notation that had been devised by the Italian logician Giuseppe Peano (1858–1932), as well as the logical system earlier developed by Frege, Russell and Whitehead attempted to show that the whole of mathematics could be derived from a few basic logical axioms. Much of what appears in chapters 8, 9, and 10 of this book is derived from their work.

Deductive logic continued to develop. The completeness of axiomatic systems became a matter of great interest in the twentieth century. Kurt Gödel (1906–1978) was able to demonstrate that any formal axiomatic system, if it is consistent, must in fact be incomplete, and from Gödel's incompleteness theorems it follows that within any formal system there will be some formulas that must remain undecidable. Other aspects of deductive logic have been more recently investigated: the distinction between "fuzzy" and "crisp" logic has been explored; modal logic, in which the concepts of possibility and necessity are manipulated, has been highly developed.

But perhaps nothing that modern logicians have accomplished has had more profound impact than the development—by John von Neumann (1903–1957) and others—of the intellectual architecture of the circuits of digital computers. Not long thereafter, with the actual construction and gradual perfection of the electronic digital computer during the twentieth century, Leibniz's great vision was at last made real.

The account above sketches the history of logic in the West, mainly in Europe and North America. Elsewhere on the planet logic was also studied, of course—but we do not have accessible and accurate records of the discoveries made long ago in China and India. We know that in India much work had been done on the principles of logic. Augustus De Morgan was influenced by that work; the theorems that bear his name, explained in Chapter 9 of this book, were developed independently in India. George Boole was influenced by Indian thinkers as well. The rules of immediate inference, discussed in this book in chapter 5, appear also to have been articulated in India, but logic there emphasized effective philosophical argumentation, including both deductive and inductive elements, rather than formal systems. In China, at the time of the philosopher Mozi (470–391 BCE), the principles of analogical reasoning, discussed in chapter 11 of this book, were developed. But of that history we cannot be sure, because in the years 213–206 BCE the Qin dynasty, to erase all marks of preceding dynasties, burned many books and killed many scholars. Much work done in earlier periods was thus permanently lost.

From the time of Aristotle's *Organon* to the twenty-first century more people have studied logic from one book than from any other; that book, now in your hands, is *Introduction to Logic*, originally conceived and written by one of the most powerful and incisive thinkers of the twentieth century, the late Irving Copi (1917–2002).

INTRODUCTION TO
LOGIC

Waterfall, by M. C. Escher, is all wrong. The water flows away, and going away comes closer; it flows downward, and going down it comes up, returning to the point from which it began. What can account for the plausible appearance of what we know to be impossible? The artist plays with the normal assumptions of our vision. The corner points of the central cube in this picture are connected in ways that cause us to perceive what is farthest away on the structure as closest, and to perceive its highest points as also its lowest. We are duped by Escher's artistry.

As perception may be tricked by a clever picture, our thinking may be tricked by a clever argument. There are principles that underlie good reasoning, but when we violate them we are likely to be misled—or by carelessness to mislead ourselves. In *Waterfall* we confront disorder in seeing, and then with scrutiny detect its cause. In the study of logic we confront many bad arguments, and then with scrutiny learn what makes them bad.

M.C. Escher's *Waterfall* © 2005 The M.C. Escher Company—Holland. All rights reserved. *www.mcescher.com*

part I

Logic and Language

SECTION A	REASONING
chapter 1	Basic Logical Concepts
chapter 2	Analyzing Arguments
SECTION B	INFORMAL LOGIC
chapter 3	Language and Definitions
chapter 4	Fallacies

"
Come now, and let us reason together.

—Isaiah 1:18

All our lives we are giving and accepting reasons. Reasons are the coin we pay for the beliefs we hold.

—*Edith Watson schipper*
"

Basic Logical Concepts

1.1 What Logic Is
1.2 Propositions and Arguments
1.3 Recognizing Arguments
1.4 Arguments and Explanations
1.5 Deductive and Inductive Arguments
1.6 Validity and Truth

1.1 What Logic Is

Logic is the study of the methods and principles used to distinguish correct from incorrect reasoning.

When we reason about any matter, we produce arguments to support our conclusions. Our arguments include reasons that we think justify our beliefs. However, not all reasons are good reasons. Therefore we may always ask, when we confront an argument: Does the conclusion reached *follow* from the premises assumed? To answer this question there are objective criteria; in the study of logic we seek to discover and apply those criteria.

Reasoning is not the only way in which people support assertions they make or accept. They may appeal to authority or to emotion, which can be very persuasive, or they may rely, without reflection, simply on habits. However, when someone wants to make judgments that can be completely relied upon, their only solid foundation will be correct reasoning. Using the methods and techniques of logic—the subject matter of this book—one can distinguish reliably between sound and faulty reasoning.

1.2 Propositions and Arguments

We begin by examining more closely the most fundamental concepts in the study of logic, concepts presupposed in the paragraphs just above. In reasoning we construct and evaluate *arguments*; arguments are built with *propositions*. Although these concepts are apparently simple, they require careful analysis.

A. Propositions

Propositions are the building blocks of our reasoning. A **proposition** asserts that something is the case or it asserts that something is not. We may affirm a proposition, or deny it—but every proposition either asserts what really is the case, or it asserts something that is not. Therefore every proposition is either true or false.

Logic
The study of the methods and principles used to distinguish correct from incorrect reasoning.

Proposition
A statement; what is typically asserted using a declarative sentence, and hence always either true or false—although its truth or falsity may be unknown.

2

Biography

Aristotle

Of all the great philosophers and logicians, ancient and modern, none is greater than Aristotle (384–322 BCE), whose works and influence largely ruled the world of intellect for two millennia. He was often referred to as "The Philosopher"; his authority (even when he was mistaken!) was rarely questioned.

Born in Macedonia, in the city of Stagira, where his father was physician to the king, he was viewed from birth as a member of the aristocracy, and was a friend of the king's son, Philip. When Philip became king of Macedonia, he summoned Aristotle, who had for many years been studying in Athens at Plato's school, The Academy, to return to Macedonia as tutor to his son Alexander (who later would be known as Alexander the Great). As he advanced on his subsequent conquests in Asia, Alexander remained in contact with his respected teacher, sending back, at Aristotle's request, specimens and artifacts that contributed to the early growth of the sciences.

Aristotle—one of the trio, with Plato and Socrates, who largely founded Western philosophy—had a truly encyclopedic mind. He investigated, contributed to, wrote about, and taught virtually all subjects on which some knowledge had been accumulated at his time: the natural sciences (biology, zoology, embryology, anatomy, astronomy, meteorology, physics, and optics); the arts (poetry, music, theater, and rhetoric); government and politics; psychology and education; economics; ethics; metaphysics—and of course logic, of which he alone was the systematic founder. His treatises on logic, later combined into one great work entitled *The Organon* ("The Instrument"), constitute the earliest formal study of our subject. The penetration and coherence of his logical analyses, and the comprehensiveness and general accuracy of his scientific studies, justify his acknowledged status as one of the finest thinkers ever to have graced our planet.

At the age of 49 Aristotle returned to Athens and established his own highly influential school, the Lyceum, where he taught for twelve years. He died of natural causes in 322 BCE. In his will, he asked to be buried next to his wife, Pythias.

In logic Aristotle grasped the overriding necessity of determining the rules of correct reasoning. He explained validity and characterized the four fundamental types of categorical propositions and their relations. In the *Prior*

Analytics, one of the six books of *The Organon*, he developed a sophisticated theoretical account of categorical syllogisms, an account that long dominated the realm of deductive logic and that remains today an effective tool of sound reasoning.

It is said of Aristotle that he was probably the last person to know everything there was to be known in his own time. ■

There are many propositions about whose truth we are uncertain. "There is life on some other planet in our galaxy," for example, is a proposition that, so far as we now know, may be true or may be false. Its "truth value" is unknown, but this proposition, like every proposition, must be either true or false.

A question *asserts* nothing, and therefore it is not a proposition. "Do you know how to play chess?" is indeed a sentence, but that sentence makes no claim about the world. Neither is a command a proposition ("Come quickly!"), nor is an exclamation a proposition ("Oh my gosh!"). Questions, commands, and exclamations—unlike propositions—are neither true nor false.

When we assert some proposition, we do so using a sentence in some language. However, the proposition we assert is not identical to that sentence. This is evident because two different sentences, consisting of different words differently arranged, may have the same meaning and may be used to assert the very same proposition. For example, "Leslie won the election" and "The election was won by Leslie" are plainly two different sentences that make the same assertion.

Sentences are always parts of some language, but propositions are not tied to English or to any given language. The four sentences

It is raining.	(English)
Está lloviendo.	(Spanish)
Il pleut.	(French)
Es regnet.	(German)

are in different languages, but they have a single meaning: all four, using different words, may be uttered to assert the very same proposition. *Proposition* is the term we use to refer to what it is that declarative sentences are typically used to assert.

The term **statement** is not an exact synonym of *proposition*, but it is often used in logic in much the same sense. Some logicians prefer *statement* to *proposition*, although the latter has been more commonly used in the history of logic. Other logicians eschew both terms as metaphysical, using only the term *sentence*.

Statement
A proposition; what is typically asserted by a declarative sentence, but not the sentence itself. Every statement must be either true or false, although the truth or falsity of a given statement may be unknown.

However, the concept of a proposition is seen by many as making a useful distinction between a sentence and what the sentence asserts. Consequently, in this book we use both terms.

The very same sentence can be used to make very different statements (or to assert very different propositions), depending on the context in which it is expressed. For example, the sentence, "The largest state in the United States was once an independent republic," once expressed a true statement or proposition (about Texas), but if asserted today would express a false statement or proposition (about Alaska). The same words assert different propositions at different times.

Propositions may be *simple*, like those used in the preceding illustrations, but they may also be *compound*, containing other propositions within themselves. Consider the following proposition, from a recent account of the exploitation of the Amazon Basin in Brazil:

> The Amazon Basin produces roughly 20 percent of the Earth's oxygen, creates much of its own rainfall, and harbors many unknown species.[1]

This sentence simultaneously asserts three propositions, concerning what the Amazon Basin produces and what it creates and what it harbors. The passage thus constitutes a *conjunctive* proposition. Asserting a conjunctive proposition is equivalent to asserting each of its component propositions separately.

Some compound propositions do not assert the truth of their components. In *disjunctive* (or *alternative*) *propositions*, no one of the components is asserted. Abraham Lincoln (in a message to Congress in December 1861) said, "Circuit courts are useful, or they are not useful." This disjunctive proposition is plainly true, but either one of its components might be false.

Other compound propositions that do not assert their components are *hypothetical* (or *conditional*) *propositions*. The eighteenth-century freethinker, Voltaire, said, "If God did not exist, it would be necessary to invent him." Here, again, neither of the two components is asserted. The proposition "God does not exist," is not asserted, nor is the proposition, "it is necessary to invent him." Only the "if–then" proposition is asserted by the hypothetical or conditional statement, and that compound statement might be true even if both of its components were false.

In logic, the internal structure of propositions is important. To evaluate an argument we need a full understanding of the propositions that appear in that argument. Propositions of many different kinds will be analyzed in this book.

B. Arguments

With propositions as building blocks, we construct *arguments*. In any argument we affirm one proposition on the basis of some other propositions. In doing this, an *inference* is drawn. **Inference** is a process that may tie together a cluster of propositions. Some inferences are *warranted* (or correct); others are not. The logician analyzes these clusters, examining the propositions with which the process begins and with which it ends, as well as the relations among these propositions.

Inference
A process by which one proposition is arrived at and affirmed on the basis of some other proposition or propositions.

6   CHAPTER 1 Basic Logical Concepts

Such a cluster of propositions constitutes an *argument*. Arguments are the chief concern of logic.

Argument is a technical term in logic. It need not involve disagreement, or controversy. In logic, **argument** refers strictly to any group of propositions of which one is claimed to follow from the others, which are regarded as providing support for the truth of that one. For every possible inference there is a corresponding argument.

In writing or in speech, a passage will often contain several related propositions and yet contain no argument. An argument is not merely a collection of propositions; it is a cluster with a structure that captures or exhibits some inference. We describe this structure with the terms *conclusion* and *premise*. The **conclusion** of an argument is the proposition that is affirmed on the basis of the other propositions of the argument. Those other propositions, which are affirmed (or assumed) as providing support for the conclusion, are the **premises** of the argument.

We will encounter a vast range of arguments in this book—arguments of many different kinds, on many different topics. We will analyze arguments in politics, in ethics, in sports, in religion, in science, in law, and in everyday life. Those who defend these arguments, or who attack them, are usually aiming to establish the truth (or the falsehood) of the conclusions drawn. As logicians, however, our interest is in the arguments as such. As agents or as citizens we may be deeply concerned about the truth or falsity of the conclusions drawn. However, as logicians we put those interests aside. Our concerns will be chiefly two. First, we will be concerned about the *form* of an argument under consideration, to determine if that argument is *of a kind* that is likely to yield a warranted conclusion. Second, we will be concerned about the *quality* of the argument, to determine whether it *does in fact* yield a warranted conclusion.

Arguments vary greatly in the degree of their complexity. Some are very simple. Other arguments, as we will see, are quite intricate, sometimes because of the structure or formulation of the propositions they contain, sometimes because of the relations among the premises, and sometimes because of the relations between premises and conclusion.

The simplest kind of argument consists of one premise and a conclusion that is claimed to follow from it. Each may be stated in a separate sentence, as in the following argument that appears on a sticker affixed to biology textbooks in the state of Alabama:

> No one was present when life first appeared on earth. Therefore any statement about life's origins should be considered as theory, not fact.

Both premise and conclusion may be stated within the same sentence, as in this argument arising out of recent advances in the science of human genetics:

> Since it turns out that all humans are descended from a small number of African ancestors in our recent evolutionary past, believing in profound differences between the races is as ridiculous as believing in a flat earth.[2]

Argument
Any group of propositions of which one is claimed to follow from the others, which are regarded as providing support or grounds for the truth of that one.

Conclusion
In any argument, the proposition to which the other propositions in the argument are claimed to give support, or for which they are given as reasons.

Premises
In an argument, the propositions upon which inference is based; the propositions that are claimed to provide grounds or reasons for the conclusion.

Biography

Chrysippus

Of all the logicians of ancient times, Aristotle and Chrysippus stand out as the two greatest. The enormous influence of Aristotle, who first systematized logic and was its principal authority for two thousand years, has already been recognized. Born a century later, Chrysippus (c. 279–c. 206 BCE) developed a conceptual scheme whose influence has only more recently been appreciated.

The logic of Aristotle was one of classes. In the Aristotelian argument "All men are mortal; Greeks are men; therefore Greeks are mortal," the fundamental elements are the categories, or terms ("men," "mortal things," and "Greeks"). In contrast, the logic of Chrysippus was one built of *propositions* and the connections between them (e.g., "If it is now day, it is now light. It is now day. Therefore it is now light."). This simple argument form (now called *modus ponens*) and many other fundamental argument forms, Chrysippus analyzed and classified. His logical insights were creative and profound.

Born in Asia Minor, in Soli, Chrysippus studied the philosophy of the Stoics —most famous among them Zeno and Cleanthes—and eventually became head of the Stoic school in Athens. In that capacity he taught the need to control one's emotions, which he thought to be disorders or diseases. He urged the patient acceptance of the outcomes of a fate one cannot control, and the recognition that the one God (of which the traditional Greek gods are but aspects) is the universe itself.

But it is as a logician that his influence has been greatest. He grasped, as Aristotle did not, the central role of the proposition—"*that which is, in itself, capable of being denied or affirmed.*" From this base he developed the first coherent system of propositional logic. ∎

The order in which premises and conclusion appear can also vary, but it is not critical in determining the quality of the argument. It is common for the conclusion of an argument to *precede* the statement of its premise or premises. On the day Babe Ruth hit his 700th home run (13 July 1934), the following argument appeared in *The New York Times*:

> A record that promises to endure for all time was attained on Navin Field today when Babe Ruth smashed his seven-hundredth home run in a lifetime career. It promises to live, first because few players in history have enjoyed the longevity on the diamond of the immortal Bambino, and, second, because only two other players in the history of baseball have hit more than 300 home runs.

This is an example of an argument whose two premises, each numbered, appear after the conclusion is stated. It is also an example of a very plausible argument whose conclusion is false, given that Hank Aaron hit his 700th home run on 21 July 1973, thirty-nine years later.

Even when premise and conclusion are united in one sentence, the conclusion of the argument may come first. The English utilitarian philosopher, Jeremy Bentham, presented this crisp argument in his *Principles of Legislation* (1802):

> Every law is an evil, for every law is an infraction of liberty.

Although this is only one short sentence, it is an argument because it contains two propositions, of which the first (every law is an evil) is the conclusion and the second (every law is an infraction of liberty) is the premise. However, no single proposition can be an argument, because an argument is made up of a group of propositions. Yet some propositions, because they are compound, do sound like arguments, and care must be taken to distinguish them from the arguments they resemble. Consider the following hypothetical proposition:

> If a state aims to be a society composed of equals, then a state that is based on the middle class is bound to be the best constituted.

Neither the first nor the second component of this proposition is asserted. All that is asserted is that the former implies the latter, and both might well be false. No inference is drawn, no conclusion is claimed to be true. Aristotle, who studied the constitution and quality of actual states in Greece more than two thousand years ago, wrote confidently in *Politics*, Book IV, Chapter 11:

> A state aims at being a society composed of equals, and therefore a state that is based on the middle class is bound to be the best constituted.

In this case we *do* have an argument. This argument of Aristotle is short and simple; most arguments are longer and more complicated. Every argument, however—short or long, simple or complex—consists of a group of propositions of which one is the conclusion and the other(s) are the premises offered to support it.

Although every argument is a structured cluster of propositions, not every structured cluster of propositions is an argument. Consider this very recent account of global inequality:

> In the same world in which more than a billion people live at a level of affluence never previously known, roughly a billion other people struggle to survive on the purchasing power equivalent of less than one U.S. dollar per day. Most of the world's poorest people are undernourished—lack access to safe drinking water or even the most basic health services and cannot send their children to school. According to UNICEF, more than 10 million children die every year—about 30,000 per day—from avoidable, poverty-related causes.[3]

This report is deeply troubling—but there is no argument here.

Reasoning is an art, as well as a science. It is something we do, as well as something we understand. Giving reasons for our beliefs comes naturally, but skill in the art of building arguments, and testing them, requires practice. One who has practiced and strengthened these skills is more likely to reason correctly than one who has never thought about the principles involved. Therefore we provide in this book very many opportunities for practice in the analysis of arguments.

EXERCISES

Identify the premises and conclusions in the following passages. Some premises do support the conclusion, others do not. Note that premises may support conclusions directly or indirectly and that even simple passages may contain more than one argument.

EXAMPLE

1. A well-regulated militia being necessary to the security of a free state, the right of the people to keep and bear arms shall not be infringed.
—*The Constitution of the United States*, Amendment 2

SOLUTION

Premise: A well-regulated militia is necessary for the security of a free state.
Conclusion: The right of the people to keep and bear arms shall not be infringed.

2. What stops many people from photocopying a book and giving it to a pal is not integrity but logistics; it's easier and inexpensive to buy your friend a paperback copy.
—Randy Cohen, *The New York Times Magazine,* 26 March 2000

3. Thomas Aquinas argued that human intelligence is a gift from God and therefore "to apply human intelligence to understand the world is not an affront to God, but is pleasing to him."
—Recounted by Charles Murray in *Human Accomplishment* (New York: HarperCollins, 2003)

4. Sir Edmund Hillary is a hero, not because he was the first to climb Mount Everest, but because he never forgot the Sherpas who helped him achieve this impossible feat. He dedicated his life to helping build schools and hospitals for them.
—Patre S. Rajashekhar, "Mount Everest," *National Geographic,* September 2003

5. Standardized tests have a disparate racial and ethnic impact; white and Asian students score, on average, markedly higher than their black and Hispanic peers. This is true for fourth-grade tests, college entrance exams, and every other assessment on the books. If a racial gap is evidence of discrimination, then all tests discriminate.
—Abigail Thernstrom, "Testing, the Easy Target," *The New York Times,* 15 January 2000

6. Good sense is, of all things in the world, the most equally distributed, for everybody thinks himself so abundantly provided with it that even

those most difficult to please in all other matters do not commonly desire more of it than they already possess.
—René Descartes, *A Discourse on Method*, 1637

7. When Noah Webster proposed a *Dictionary of the American Language*, his early 19th-century critics presented the following argument against it: "Because any words new to the United States are either stupid or foreign, there is no such thing as the American language; there's just bad English."
—Jill Lepore, "Noah's Mark," *The New Yorker*, 6 November 2006

8. The death penalty is too costly. In New York State alone taxpayers spent more than $200 million in our state's failed death penalty experiment, with no one executed.

    In addition to being too costly, capital punishment is unfair in its application. The strongest reason remains the epidemic of exonerations of death row inmates upon post-conviction investigation, including ten New York inmates freed in the last 18 months from long sentences being served for murders or rapes they did not commit.
—L. Porter, "Costly, Flawed Justice," *The New York Times*, 26 March 2007

9. Houses are built to live in, not to look on; therefore, let use be preferred before uniformity.
—Francis Bacon, "Of Building," in *Essays*, 1597

10. To boycott a business or a city [as a protest] is not an act of violence, but it can cause economic harm to many people. The greater the economic impact of a boycott, the more impressive the statement it makes. At the same time, the economic consequences are likely to be shared by people who are innocent of any wrongdoing, and who can ill afford the loss of income: hotel workers, cab drivers, restaurateurs, and merchants. The boycott weapon ought to be used sparingly, if for no other reason than the harm it can cause such bystanders.
—Alan Wolfe, "The Risky Power of the Academic Boycott," *The Chronicle of Higher Education*, 17 March 2000

11. Ethnic cleansing was viewed not so long ago as a legitimate tool of foreign policy. In the early part of the 20th century forced population shifts were not uncommon; multicultural empires crumbled and nationalism drove the formation of new, ethnically homogenous countries.
—Belinda Cooper, "Trading Places," *The New York Times Book Review*, 17 September 2006

12. If a jury is sufficiently unhappy with the government's case or the government's conduct, it can simply refuse to convict. This possibility puts powerful pressure on the state to behave properly. For this reason a jury is one of the most important protections of a democracy.
—Robert Precht, "Japan, the Jury," *The New York Times*, 1 December 2006

13. Without forests, orangutans cannot survive. They spend more than 95 percent of their time in the trees, which, along with vines and termites,

provide more than 99 percent of their food. Their only habitat is formed by the tropical rain forests of Borneo and Sumatra.

—Birute Galdikas, "The Vanishing Man of the Forest," *The New York Times*, 6 January 2007

14. Omniscience and omnipotence are mutually incompatible. If God is omniscient, he must already know how he is going to intervene to change the course of history using his omnipotence. But that means he can't change his mind about his intervention, which means he is not omnipotent.

—Richard Dawkins, *The God Delusion* (New York: Houghton Mifflin, 2006)

15. Reason is the greatest enemy that faith has; it never comes to the aid of spiritual things, but more frequently than not struggles against the divine Word, treating with contempt all that emanates from God.

—Martin Luther, *Last Sermon in Wittenberg*, 17 January 1546

1.3 Recognizing Arguments

Before we can evaluate an argument, we must *recognize* it. We must be able to distinguish argumentative passages in writing or speech. Doing this assumes, of course, an understanding of the language of the passage. However, even with a thorough comprehension of the language, the identification of an argument can be problematic because of the peculiarities of its formulation. Even when we are confident that an argument is intended in some context, we may be unsure about which propositions are serving as its premises and which as its conclusion. As we have seen, that judgment cannot be made on the basis of the order in which the propositions appear. How then shall we proceed?

A. Conclusion Indicators and Premise Indicators

One useful method depends on the appearance of certain common indicators, certain words or phrases that typically serve to signal the appearance of an argument's conclusion or of its premises. Here is a partial list of **conclusion indicators**:

therefore	for these reasons
hence	it follows that
so	I conclude that
accordingly	which shows that
in consequence	which means that
consequently	which entails that
proves that	which implies that
as a result	which allows us to infer that
for this reason	which points to the conclusion that
thus	we may infer

Conclusion indicator A word or phrase (such as "therefore" or "thus") appearing in an argument and usually indicating that what follows it is the conclusion of that argument.

Other words or phrases typically serve to mark the premises of an argument and hence are called **premise indicators**. Usually, but not always, what follows any one of these will be the premise of some argument. Here is a partial list of premise indicators:

since	as indicated by
because	the reason is that
for	for the reason that
as	may be inferred from
follows from	may be derived from
as shown by	may be deduced from
inasmuch as	in view of the fact that

B. Arguments in Context

The words and phrases we have listed may help to indicate the presence of an argument or identify its premises or conclusion, but such indicators do not necessarily appear. Sometimes it is just the meaning of the passage, or its setting, that indicates the presence of an argument. For example, during the intense controversy over the deployment of additional U.S. troops to Iraq in 2007, one critic of that deployment wrote:

> As we send our young men and women abroad to bring order to Iraq, many of its so-called leaders have abandoned their posts. We have given the Iraqis an opportunity to iron out their differences and they throw it back in our faces. Iraq does not deserve our help.[4]

No premise indicators or conclusion indicators are used here, yet the argument is clear. Indicators are also absent in the following argument in Sam Harris's *Letter to a Christian Nation*, whose premises and conclusions are unmistakable:

> Half the American population believes that the universe is 6,000 years old. They are wrong about this. Declaring them so is not "irreligious intolerance." It is intellectual honesty.[5]

Often, however, the force of an argument can be appreciated only when one understands the *context* in which that argument is presented. For example, the undergraduate admission system of the University of Michigan that gave a fixed number of extra points to all members of certain minority groups was held unconstitutional by the U.S. Supreme Court in *Gratz v. Bollinger* in 2003. Justice Ruth Bader Ginsburg dissented, defending the Michigan system with the following argument:

> Universities will seek to maintain their minority enrollment . . . whether or not they can do so in full candor. . . . [They] may resort to camouflage. If honesty is the best policy, surely Michigan's accurately described, fully disclosed College affirmative action program is preferable to achieving similar numbers through winks, nods, and disguises.[6]

This argument derives its force from the realization that universities had in fact long disguised their preferential admission programs to avoid attacks based

---

Premise indicator
In an argument, a word or phrase (like "because" and "since") that normally signals that what follows it are statements serving as premises.

on the equal protection clause of the 14th Amendment to the U.S. Constitution. Chief Justice William Rehnquist's response to Justice Ginsburg's argument is also intelligible only in the context of her defense of the preferential admission system. Rehnquist wrote the following:

> These observations are remarkable for two reasons. First, they suggest that universities—to whose academic judgment we are told we should defer—will pursue their affirmative action programs whether or not they violate the United States Constitution. Second, they recommend that these violations should be dealt with, not by requiring the Universities to obey the Constitution, but by changing the Constitution so that it conforms to the conduct of the universities.[7]

Rehnquist's reference to "changing the Constitution" must be understood in light of the fact that the Michigan undergraduate admission system had been held unconstitutional. His reference to the pursuit of affirmative action programs "whether or not they violate the United States Constitution" can best be understood in light of Ginsburg's earlier reference to the possible use of "winks, nods, and disguises."

The full force of argument and counterargument can be grasped, in most circumstances, only with an understanding of the *context* in which those arguments are presented. In real life, context is critical. For example, if you are told that I am bringing a lobster home for dinner, you will have little doubt that I intend to eat it, not feed it.

C. Premises or Conclusions Not in Declarative Form

It is not uncommon for the premises of an argument to be presented in the form of questions. However, if questions assert nothing, and do not express propositions, how is this possible? On the surface they make no assertions; beneath the surface an interrogative sentence can serve as a premise when its question is **rhetorical**—that is, when it suggests or assumes an answer that is made to serve as the premise of an argument. The sentence may be interrogative even though its meaning is declarative.

This use of questions is sometimes obvious, as in a letter dated 7 January 2007 to *The New York Times*, objecting to a new series of U.S. coins that will honor former presidential wives. Irit R. Rasooly wrote:

> I am irked by the new set of coins being issued. While some first ladies have influenced our country, should we bestow this honor on people who are unelected, whose only credential is having a prominent spouse?

Plainly, the critic means to affirm the proposition that we should not bestow this honor on such people. He continues:

> Wouldn't honoring women who have served as governors, Supreme Court justices or legislators be a more fitting tribute to this nation's women than coins featuring "First Spouses"?

This critic obviously believes that honoring such achievements would be a more fitting tribute, but he again expresses that proposition with a question. His

Rhetorical question An utterance used to make a statement, but which, because it is in interrogative form and is therefore neither true nor false, does not literally assert anything.

letter also provides an illustration of the need to rely on context to interpret declarative statements that are actually made. The writer's report that he is "irked" by the new set of coins is no doubt true, but this statement is more than a mere description of his state of mind; he means to express the judgment that such a set of coins *ought not* be issued.

Using questions to express a premise is sometimes counterproductive, however, because it may invite answers (by the listener, or silently by the reader) that threaten the conclusion at which the argument aims. For example, the archbishop of the Anglican Church in Nigeria, who is an ardent opponent of homosexuality and views it as deeply sinful, argues thus:

> Why didn't God make a lion to be a man's companion? Why didn't He make a tree to be a man's companion? Or better still, why didn't He make another man to be a man's companion? So even from the creation story you can see that the mind of God, God's intention, is for man and woman to be together.[8]

Conclusions drawn about God's intentions, using as premises questions that invite a myriad of different responses, may be undermined by the answers they elicit.

Questions can serve most effectively as premises when the answers assumed really do seem to be clear and inescapable. In such cases the readers (or hearers) are led to provide the apparently evident answers for themselves, thus augmenting the persuasiveness of the argument. Here is an example: Some who find euthanasia morally unacceptable reject the defense of that practice as grounded in the right to self-determination possessed by the terminally ill patient. They argue as follows:

> If a right to euthanasia is grounded in self-determination, it cannot reasonably be limited to the terminally ill. If people have a right to die, why must they wait until they are actually dying before they are permitted to exercise that right?[9]

The question is forceful because its answer appears to be undeniable. It seems obvious that there is no good reason why, if people have a right to die grounded in self-determination, they must wait until they are dying to exercise that right. Hence (this critique concludes) the right to euthanasia, if there is one, cannot be limited to the terminally ill. The argument has much merit, but from the perspective of its religious advocates, it may prove to be a two-edged sword.

Arguments that depend on rhetorical questions are always suspect. Because the question is neither true nor false, it may be serving as a device to suggest the truth of some proposition while avoiding responsibility for asserting it. That proposition is likely to be dubious, and it may in fact be false. To illustrate: In 2007 Arab leaders in Jerusalem expressed great anxiety about the safety of the Al-Aqsa mosque when the Israeli government began construction of a ramp leading to the platform (also sacred to the Jews) on which that very holy mosque is situated. In reviewing the situation, David Gelernter, an Israeli partisan asked: "Is it possible that Arab leaders are more interested in attacking Israel than protecting religious and cultural monuments?"[10] Well, yes, that is possible, of course—but it may not be true, and the question framed in this way is plainly intended to cause the reader to believe that Arab leaders were being duplicitous in voicing their concerns. Did the author assert that such duplicity lay behind the Arab objections? No, he didn't say that!

Gossip columnists thrive on suggestive questions. Celebrity tidbits commonly appear in the form, "Does Paris Hilton have any talent as an actress?" Similarly, in discussing social issues, rhetorical questions can be an effective method of covert assertion. When riots in France spread through Islamic neighborhoods, many wondered what motivated those rioters. Journalist Christopher Caldwell wrote:

> Were they admirers of France's majority culture, frustrated at not being able to join it on equal terms? Or did they simply aspire to burn to the ground a society they despised?[11]

Accusers who protect themselves by framing their accusations in interrogative sentences may shield themselves from the indignant complaints of their target. "No," they may insist, "that is not what I said!"

It is wise policy to refrain from arguing with questions.

In some arguments the conclusion appears in the form of an imperative. The reason, or reasons, we ought to perform a given act are set forth as premises, and we are then directed to act in that way. Thus in Proverbs 4:7 we read:

> Wisdom is the principal thing; therefore get wisdom.

Here the second clause is a command, and a command, like a question, is neither true nor false and cannot express a proposition. Therefore, strictly speaking, it cannot be the conclusion of an argument. Nonetheless, it surely is meant to be the conclusion of an argument in this passage from Proverbs. How can we explain this apparent inconsistency? It is useful in many contexts to regard a command as no different from a proposition in which hearers (or readers) are told that they would be wise to act, or ought to act, in the manner specified in the command. Thus the conclusion of the argument in Proverbs may be rephrased as "Getting wisdom is what you should do." Assertions of this kind may be true or false, as most will agree. What difference there is between a command to do something and a statement that it should be done is an issue that need not be explored here. By ignoring that difference (if there really is one), we are able to deal uniformly with arguments whose conclusions are expressed in this form.

Reformulations of this kind can clarify the roles of an argument's constituent propositions. It is necessary to grasp the *substance* of what is being asserted, to understand which claims are serving to support which inferences, whatever their external forms. Some needed reformulations are merely grammatical. A proposition that functions as a premise may take the form of a phrase rather than a declarative sentence. This is well illustrated in the following argumentative passage, whose conclusion is a very sharp criticism of the United States:

> What is a failed state? It is one that fails to provide security for the population, to guarantee rights at home or abroad, or to maintain functioning democratic institutions. On this definition the United States is the world's biggest failed state.[12]

The second and third premises of this argument are compressed into phrases, but the propositions for which these phrases are shorthand are clear enough, and their critical role in the author's reasoning is evident.

D. Unstated Propositions

Arguments are sometimes obscure because one (or more) of their constituent propositions is not stated but is assumed to be understood. An illustration will be helpful here. The chair of the Department of Sociology at City College, CUNY, presented two strong but controversial arguments, in parallel, regarding the justifiability of the death penalty. The first premise of each argument is the hypothesis that the factual belief (of the proponent, or of the opponent, of the penalty) about what does in fact deter homicide is mistaken. The second premise of each argument, although entirely plausible, is not stated, leaving the reader the task of reconstructing it.

The first argument went like this:

> If the proponent of the death penalty is incorrect in his belief that the [death] penalty deters homicide, then he is responsible for the execution of murderers who should not be executed.[13]

This argument relies on the unstated second premise: "No one should be executed to advance an objective that is not promoted by execution." Hence one who *mistakenly* believes that the objective (deterring murders) is achieved by executing those convicted is responsible for the execution of murderers who should not be executed.

The second argument went like this:

> If the opponent of the death penalty is incorrect in his belief that the death penalty doesn't deter, he is responsible for the murder of innocent individuals who would not have been murdered if the death penalty had been invoked.[14]

This argument relies on the unstated second premise: "Protecting the lives of innocent individuals from murder justifies the execution of murderers if other murderers are then deterred by the fear of execution." Hence one who *mistakenly* believes that the death penalty does not deter murderers is responsible for the lives of innocents who are subsequently murdered.

In each of these arguments the assumed but unstated second premise is plausible. One might find both arguments persuasive—leaving open for empirical investigation the question of whether, in fact, the death penalty does deter murder.

However, the force of each of the arguments depends on the truth of the unstated premise on which it relies.

A premise may be left unstated because the arguer supposes that it is unquestioned common knowledge. In the controversy over the cloning of human beings, one angry critic wrote:

> Human cloning—like abortion, contraception, pornography and euthanasia—is intrinsically evil and thus should never be allowed.[15]

This is plainly an argument, but part of it is missing. The argument relies on the very plausible but unstated premise that "what is intrinsically evil should never be allowed." Arguments in everyday discourse very often rely on some proposition that is understood but not stated. Such arguments are called **enthymemes**. We will examine them more closely later in this book.

The unstated premise on which an enthymeme relies may not be universally accepted; it may be uncertain or controversial. An arguer may deliberately refrain from formulating that critical premise, believing that by allowing it to remain tacit, the premise is shielded from attack. For example, medical research using embryonic stem cells (cells found in the human embryo that can develop into other types of cells and into most types of tissue) is highly controversial. One U.S. senator used the following enthymeme in attacking legislation that would permit government financing of such research:

> This research [involving the use of embryonic stem cells] is illegal, for this reason: The deliberate killing of a human embryo is an essential component of the contemplated research.[16]

The stated premise is true: Research of this kind is not possible without destroying the embryo. However, the conclusion that such research is illegal depends on the unstated premise that the killing of a human embryo is illegal—and *that* claim is very much in dispute.

The effectiveness of an enthymeme may depend on the hearer's knowledge that some proposition is false. To emphasize the falsity of some proposition, a speaker may construct an argument in which the first premise is a hypothetical proposition of which the antecedent (the "if" component), is the proposition whose falsity the speaker wishes to show, and the consequent (the "then" component) is a proposition known by everyone to be false. The unstated falsehood of this second component is the second premise of the enthymematic argument. The unstated falsehood of the first component is the conclusion of the argument. To illustrate: The distinguished political philosopher John Rawls admired Abraham Lincoln as the president who most appreciated the moral equality of human beings. Rawls frequently quoted Lincoln's enthymematic argument, "If slavery is not wrong, nothing is wrong."[17] It is of course wildly false to say that nothing is wrong—from which it follows that it is equally false to say that slavery is not wrong. Similarly, distinguished psychiatrist Bruno Bettelheim, survivor of both Dachau and Buchenwald, wrote: "If all men are good, then there never was an Auschwitz."

Enthymeme
An argument that is stated incompletely, the unstated part of it being taken for granted.

1.4 Arguments and Explanations

Passages that appear to be arguments are sometimes not arguments but *explanations*. The appearance of words that are common indicators—such as "because," "for," "since," and "therefore"—cannot settle the matter, because those words are used both in explanations and in arguments (although "since" can sometimes refer to temporal succession). We need to know the intention of the author. Compare the following two passages:

> 1. Lay up for yourselves treasures in heaven, where neither moth nor rust consumes and where thieves do not break in and steal. For where your treasure is, there will your heart be also.
> —Matt. 7:19
>
> 2. Therefore is the name of it [the tower] called Babel; because the Lord did there confound the language of all the earth.
> —Gen. 11:19

The first passage is clearly an argument. Its conclusion, that one ought to lay up treasures in heaven, is supported by the premise (here marked by the word "for") that one's heart will be where one's treasure is laid up. The second passage, which uses the word "therefore" quite appropriately, is not an argument. It *explains* why the tower (whose construction is recounted in Genesis) is called Babel. The tower was given this name, we are told, because it was the place where humankind, formerly speaking one language, became confounded by many languages—the name is derived from a Hebrew word meaning "to confound." The passage assumes that the reader knows that the tower had that name; the intention is to explain why that name was given to it. The phrase, "Therefore is the name of it called Babel," is not a conclusion but a completion of the explanation of the naming. In addition, the clause, "because the Lord did there confound the language of all the earth," is not a premise; it could not serve as a reason for believing that Babel was the name of the tower, because the fact that that *was* the name is known by those to whom the passage is addressed. In this context, "because" indicates that what follows will *explain* the giving of that name, Babel, to that tower.

These two passages illustrate the fact that superficially similar passages may have very different functions. Whether some passage is an argument or an explanation depends on the *purpose* to be served by it. If our aim is to establish the truth of some proposition, Q, and we offer some evidence, P, in support of Q, we may appropriately say "Q because P." In this case we are giving an argument *for* Q, and P is our premise. Alternatively, suppose that Q is known to be true. In that case we don't have to give any reasons to support its truth, but we may wish to give an account of *why* it is true. Here also we may say "Q because P"—but in this case we are giving not an argument *for* Q, but an explanation *of* Q.

In responding to a query about the apparent color of quasars (celestial objects lying far beyond our galaxy), one scientist wrote:

> The most distant quasars look like intense points of infrared radiation. This is because space is scattered with hydrogen atoms (about two per cubic meter) that absorb blue light, and if you filter the blue from visible white light, red is what's left. On its multibillion-light-year journey to earth quasar light loses so much blue that only infrared remains.[18]

The author is not seeking to convince his reader that quasars have the apparent color they do, but rather giving the causes of this fact; he is explaining, not arguing.

However, it may be difficult at times to determine whether an author intends to be explaining some state of affairs, or to be arguing for some conclusion that is critical in that explanation. Here, for example, is a passage that may be interpreted in either way.

> I would like to highlight another property of water, unique but also vital to making life on Earth possible. As water cools, approaching its freezing point, its density suddenly decreases, reversing the usual "natural convection" patterns in which colder fluids sink. This reversal causes the coldest strata of water to rise to the top of an ocean or lake. These large bodies of water now freeze from the top down. Were it not for this unique property of water, the oceans and lakes would have long and completely frozen over from the bottom up with dire consequences for any life-sustaining liquid water on Earth.[19]

More than one conclusion may be inferred from the same premise, thus presenting two arguments. Similarly, more than one thing may be accounted for by the same fact, thus presenting two explanations. Here is an illustration:

> The *Oxford English Dictionary* is a historical dictionary, providing citations meant to show the evolution of every word, beginning with the earliest known usage. Therefore, a key task, and a popular sport for thousands of volunteer word aficionados, is antedating: finding earlier citations than those already known.[20]

That antedating is a key task for the makers of that dictionary is accounted for by the fact that the *Oxford English Dictionary* is a *historical* dictionary. This fact about the dictionary also explains why, for word aficionados, antedating is a popular sport.

If an author writes "*Q* because *P*," how can we tell whether he intends to explain or to persuade? We can ask: What is the status of *Q* in that context? Is *Q* a proposition whose truth needs to be established or confirmed? In that case, "because *P*" is probably offering a premise in its support; "*Q* because *P*" is in that instance an argument. Or is *Q* a proposition whose truth is known, or at least not in doubt in that context? In that case, "because *P*" is probably offering some account of why *Q* has come to be true; "*Q* because *P*" is in that instance an explanation.

In an explanation, one must distinguish *what* is being explained from what the explanation *is*. In the explanation from Genesis given at the beginning of this section, what is being explained is how the tower of Babel came to have that name; the explanation is that it was there that the Lord did confound the language of all the Earth. In the astronomical example given subsequently, what is being explained is the fact that quasars appear to be red; the explanation is that as light travels from the very distant quasar to Earth all the blue in that light is filtered out.

If we are sensitive to the context, we will usually be able to distinguish an explanation from an argument. However, there will always be some passages whose purpose is uncertain, and such passages may deserve to be given alternative, equally plausible "readings"—viewed as arguments when interpreted in one way and as explanations when interpreted in another.

EXERCISES

Some of the following passages contain explanations, some contain arguments, and some may be interpreted as either an argument or an explanation. What is your judgment about the chief function of each passage? What would have to be the case for the passage in question to be an argument? To be an explanation? Where you find an argument, identify its premises and conclusion. Where you find an explanation, indicate what is being explained and what the explanation is.

EXAMPLE

1. Humans have varying skin colors as a consequence of the distance our ancestors lived from the Equator. It's all about sun. Skin color is what regulates our body's reaction to the sun and its rays. Dark skin evolved to protect the body from excessive sun rays. Light skin evolved when people migrated away from the Equator and needed to make vitamin D in their skin. To do that they had to lose pigment. Repeatedly over history, many people moved dark to light and light to dark. That shows that color is not a permanent trait.

   —Nina Jablonski, "The Story of Skin," *The New York Times*, 9 January 2007

SOLUTION

This is essentially an explanation. *What* is being explained is the fact that humans have varying skin colors. The explanation is that different skin colors evolved as humans came to live at different distances from the Equator and hence needed different degrees of protection from the rays of the sun. One might interpret the passage as an argument whose conclusion is that skin color is not a permanent trait of all humans. Under this interpretation, all the propositions preceding the final sentence of the passage serve as premises.

2. David Bernstein [in *Only One Place of Redress: African Americans, Labor Regulations, and the Courts from Reconstruction to the New Deal*, 2001] places labor laws at the center of the contemporary plight of black Americans.

Many of these ostensibly neutral laws (e.g., licensing laws, minimum-wage laws, and collective bargaining laws) were either directly aimed at stymieing black economic and social advancement or, if not so aimed, were quickly turned to that use. A huge swath of the American labor market was handed over to labor unions from which blacks, with few exceptions, were totally excluded. The now longstanding gap between black and white unemployment rates dates precisely from the moment of government intervention on labor's behalf. In short (Bernstein argues) the victories of American labor were the undoing of American blacks.
—Ken I. Kirsch, "Blacks and Labor—the Untold Story," *The Public Interest*, Summer 2002

3. Animals born without traits that led to reproduction died out, whereas the ones that reproduced the most succeeded in conveying their genes to posterity. Crudely speaking, sex feels good because over evolutionary time the animals that liked having sex created more offspring than the animals that didn't.
—R. Thornhill and C. T. Palmer, "Why Men Rape," *The Sciences*, February 2000

4. Changes are real. Now, changes are only possible in time, and therefore time must be something real.
—Immanuel Kant, *Critique of Pure Reason* (1781), "Transcendental Aesthetic," section II

5. The nursing shortage in the United States has turned into a full-blown crisis. Because fewer young people go into nursing, one-third of registered nurses in the United States are now over 50 years of age, and that proportion is expected to rise to 40 percent over the next decade. Nurses currently practicing report high rates of job dissatisfaction, with one in five seriously considering leaving the profession within the next five years.... Hospitals routinely cancel or delay surgical cases because of a lack of nursing staff.
—Ronald Dworkin, "Where Have All the Nurses Gone?," *The Public Interest*, Summer 2002

6. To name causes for a state of affairs is not to excuse it. Things are justified or condemned by their consequences, not by their antecedents.
—John Dewey, "The Liberal College and Its Enemies," *The Independent*, 1924

7. One may be subject to laws made by another, but it is impossible to bind oneself in any matter which is the subject of one's own free exercise of will.... It follows of necessity that the king cannot be subject to his own laws. For this reason [royal] edicts and ordinances conclude with the formula, "for such is our good pleasure."
—Jean Bodin, *Six Books of the Commonwealth*, 1576

8. I like Wagner's music better than anybody's. It is so loud that one can talk the whole time without people hearing what one says.
—Oscar Wilde, *The Picture of Dorian Gray*, 1891

9. Three aspects of American society in recent decades make cheating more likely.

    First, there is the rise of a market-drenched society, where monetary success is lauded above all else. Second, there is the decline of religious, communal, and family bonds and norms that encourage honesty. Finally, there is the absence of shame by those public figures who are caught in dishonest or immoral activities. No wonder so many young people see nothing wrong with cutting corners or worse.
    —Howard Gardner, "More Likely to Cheat," *The New York Times*, 9 October 2003

10. Love looks not with the eyes, but with the mind;
    And therefore is wing'd Cupid painted blind.
    —William Shakespeare, *A Midsummer Night's Dream*, act 1, scene 1

11. An article in *The New York Times*, "Why Humans and Their Fur Parted Ways," suggested that the fact that women have less body hair than men is somehow related to greater sexual selection pressure on women. A reader responded with the following letter:

    > Here is an elaboration for which I have no evidence but it is consistent with what we think we know: sexual selection has probably strongly influenced numerous traits of both sexes.
    >
    > Youthful appearance is more important to men when selecting a mate than it is to women. The longer a woman can look young, the longer she will be sexually attractive and the more opportunities she will have to bear offspring with desirable men. Hairlessness advertises youth.
    >
    > Hence a greater sexual selection pressure on women to lose body hair.
    —T. Doyle, "Less Is More," *The New York Times*, 26 August 2003

12. MAD, mutually assured destruction, was effective in deterring nuclear attack right through the cold war. Both sides had nuclear weapons. Neither side used them, because both sides knew the other would retaliate in kind. This will not work with a religious fanatic [like Mahmoud Ahmadinejad, President of the Islamic Republic of Iran]. For him, mutual assured destruction is not a deterrent, it is an inducement. We know already that Iran's leaders do not give a damn about killing their own people in great numbers. We have seen it again and again. In the final scenario, and this applies all the more strongly if they kill large numbers of their own people, they are doing them a favor. They are giving them a quick free pass to heaven and all its delights.
    —Bernard Lewis, quoted in *Commentary*, June 2007

13. About a century ago, we discovered that planetary orbits are not stable in four or more dimensions, so if there were more than three space dimensions, planets would not orbit a sun long enough for life to originate. And in one or two space dimensions, neither blood flow nor large numbers of neuron connections can exist. Thus, interesting life can exist only in three dimensions.
    —Gordon Kane, "Anthropic Questions," *Phi Kappa Phi Journal*, Fall 2002

14. Translators and interpreters who have helped United States troops and diplomats now want to resettle in the United States. They speak many strategically important languages of their region. The United States does not have an adequate number of interpreters and translators who are proficient in these languages. Therefore, we need them. Q.E.D.
—Oswald Werner, "Welcome the Translators," *The New York Times*, 3 November 2007

15. The Treasury Department's failure to design and issue paper currency that is readily distinguishable to blind and visually impaired individuals violates Section 504 of the Rehabilitation Act, which provides that no disabled person shall be "subjected to discrimination under any program or activity conducted by any Executive agency."
—Judge James Robertson, Federal District Court for the District of Columbia, *American Council of the Blind v. Sec. of the Treasury*, No. 02-0864 (2006)

16. Rightness [that is, acting so as to fulfill one's duty] never guarantees moral goodness. For an act may be the act which the agent thinks to be his duty, and yet be done from an indifferent or bad motive, and therefore be morally indifferent or bad.
—Sir W. David Ross, *Foundations of Ethics* (Oxford: Oxford University Press, 1939)

17. Man did not invent the circle or the square or mathematics or the laws of physics. He discovered them. They are immutable and eternal laws that could only have been created by a supreme mind: God. And since we have the ability to make such discoveries, man's mind must possess an innate particle of the mind of God. To believe in God is not "beyond reason."
—J. Lenzi, "Darwin's God," *The New York Times Magazine*, 18 March 2007

18. Many of the celebratory rituals [of Christmas], as well as the timing of the holiday, have their origins outside of, and may predate, the Christian commemoration of the birth of Jesus. Those traditions, at their best, have much to do with celebrating human relationships and the enjoyment of the goods that this life has to offer. As an atheist I have no hesitation in embracing the holiday and joining with believers and nonbelievers alike to celebrate what we have in common.
—John Teehan, "A Holiday Season for Atheists, Too," *The New York Times*, 24 December 2006

19. All ethnic movements are two-edged swords. Beginning benignly, and sometimes necessary to repair injured collective psyches, they often end in tragedy, especially when they turn political, as illustrated by German history.
—Orlando Patterson, "A Meeting with Gerald Ford," *The New York Times*, 6 January 2007

20. That all who are happy, are equally happy, is not true. A peasant and a philosopher may be equally *satisfied*, but not equally *happy*. Happiness consists in the multiplicity of agreeable consciousness. A peasant has not the capacity for having equal happiness with a philosopher.

—Samuel Johnson, in Boswell's *Life of Johnson*, 1766

1.5 Deductive and Inductive Arguments

Every argument makes the claim that its premises provide grounds for the truth of its conclusion; that claim is the mark of an argument. However, there are two very different ways in which a conclusion may be supported by its premises, and thus there are two great classes of arguments: the *deductive* and the *inductive*. Understanding this distinction is essential in the study of logic.

A *deductive argument* makes the claim that its conclusion is supported by its premises *conclusively*. An *inductive argument*, in contrast, does not make such a claim. Therefore, if we judge that in some passage a claim for conclusiveness is being made, we treat the argument as deductive; if we judge that such a claim is not being made, we treat it as inductive. Because every argument either makes this claim of conclusiveness (explicitly or implicitly) or does not make it, every argument is either deductive or inductive.

When the claim is made that the premises of an argument (if true) provide incontrovertible grounds for the truth of its conclusion, that claim will be either correct or not correct. If it is correct, that argument is *valid*. If it is not correct (that is, if the premises when true fail to establish the conclusion irrefutably although claiming to do so), that argument is *invalid*.

For logicians the term *validity* is applicable only to deductive arguments. To say that a deductive argument is valid is to say that it is not possible for its conclusion to be false if its premises are true. Thus we define **validity** as follows: A deductive argument is *valid* when, if its premises are true, its conclusion *must* be true. In everyday speech, of course, the term *valid* is used much more loosely.

Although every deductive argument makes the claim that its premises guarantee the truth of its conclusion, not all deductive arguments live up to that claim. Deductive arguments that fail to do so are invalid.

Because every deductive argument either succeeds or does not succeed in achieving its objective, every deductive argument is either valid or invalid. This point is important: If a deductive argument is not valid, it must be invalid; if it is not invalid, it must be valid.

The central task of deductive logic (treated at length in Part II of this book) is to discriminate valid arguments from invalid ones. Over centuries, logicians have devised powerful techniques to do this—but the traditional techniques for determining validity differ from those used by most modern logicians. The former, collectively known as *classical logic* and rooted in the analytical works of Aristotle, are explained in Chapters 5, 6, and 7 of this book. The techniques of *modern symbolic logic* are presented in detail in Chapters 8, 9, and 10. Logicians of

Validity
A characteristic of any deductive argument whose premises, if they were all true, would provide conclusive grounds for the truth of its conclusion. Such an argument is said to be *valid*. Validity is a formal characteristic; it applies only to arguments, as distinguished from truth, which applies to propositions.

the two schools differ in their methods and in their interpretations of some arguments, but ancients and moderns agree that the fundamental task of deductive logic is to develop the tools that enable us to distinguish arguments that are valid from those that are not.

In contrast, the central task of inductive arguments is to ascertain the facts by which conduct may be guided directly, or on which other arguments may be built. Empirical investigations are undertaken—as in medicine, or social science, or astronomy—leading, when inductive techniques are applied appropriately, to factual conclusions, most often concerning cause-and-effect relationships of some importance.

A great variety of inductive techniques are examined in detail in Part III of this book, but an illustration of the inductive process will be helpful at this point to contrast induction with deduction. Medical investigators, using inductive methods, are eager to learn the causes of disease, or the causes of the transmission of infectious diseases. Sexually transmitted diseases (STDs), such as acquired immune deficiency syndrome (AIDS), are of special concern because of their great seriousness and worldwide spread. Can we learn inductively how to reduce the spread of STDs? Yes, we can.

In 2006 the National Institutes of Health announced that large-scale studies of the spread of STDs in Kenya and Uganda (African countries in which the risk of HIV infection, commonly resulting in AIDS, is very high) was sharply lower among circumcised men than among those who were not circumcised. Circumcision is not a "magic bullet" for the treatment of disease, of course. However, we did learn, by examining the experience of very many voluntary subjects (3,000 in Uganda, 5,000 in Kenya, divided into circumcised and uncircumcised groups) that a man's risk of contracting HIV from heterosexual sex is *reduced by half* as a result of circumcision. The risk to women is also reduced by about 30 percent.*

These are discoveries (using the inductive method called *concomitant variation*, which is discussed in detail in Chapter 12) of very great importance. The causal connection between the absence of circumcision and the spread of HIV is not known with certainty, the way the conclusion of a deductive argument is known, but it is now known with a very high degree of probability.

Inductive arguments make weaker claims than those made by deductive arguments. Because their conclusions are never certain, the terms *validity* and *invalidity* do not apply to inductive arguments. We can evaluate inductive arguments, of course; appraising such arguments is a central task of scientists in every sphere. The higher the level of probability conferred on its conclusion by the premises of an inductive argument, the greater is the merit of that argument. We can say that inductive arguments may be "better" or "worse," "weaker" or "stronger," and so on. The argument constituted by the circumcision study is very strong, the probability of its conclusion very high. Even when the premises are all true, however, and provide strong support for the conclusion, that conclusion is

---

*So great is the advantage of circumcision shown by these studies that they were stopped, on 13 December 2006, by the Data Safety and Monitoring Board of the National Institutes of Health, to be fair to all participants by announcing the probable risks of the two patterns of conduct.

not established with certainty. The entire theory of induction, techniques of analogical and causal reasoning, and methods for appraising inductive arguments and for quantifying and calculating probabilities are presented at length in Chapters 11, 12, 13, and 14.

Because an inductive argument can yield no more than some degree of probability for its conclusion, it is always possible that additional information will strengthen or weaken it. Newly discovered facts may cause us to change our estimate of the probabilities, and thus may lead us to judge the argument to be better (or worse) than we had previously thought. In the world of inductive argument—even when the conclusion is judged to be very highly probable—*all the evidence is never in.* New discoveries may eventually disconfirm what was earlier believed, and therefore we never assert that the conclusion of an inductive argument is absolutely certain.

Deductive arguments, on the other hand, cannot become better or worse. They either succeed or they do not succeed in exhibiting a compelling relation between premises and conclusion. If a deductive argument is valid, no additional premises can possibly add to the strength of that argument. For example, if all humans are mortal and Socrates is human, we may conclude without reservation that Socrates is mortal—*and that conclusion will follow from those premises no matter what else may be true in the world, and no matter what other information may be discovered or added.* If we come to learn that Socrates is ugly, or that immortality is a burden, or that cows give milk, none of those findings nor any other findings can have any bearing on the validity of the original argument. The conclusion that follows with certainty from the premises of a deductive argument follows from any enlarged set of premises with the same certainty, regardless of the nature of the premises added. If an argument is valid, nothing in the world can make it more valid; if a conclusion is validly inferred from some set of premises, nothing can be added to that set to make that conclusion follow more strictly, or more validly.

This is not true of inductive arguments, however, for which the relationship claimed between premises and conclusion is much less strict and very different in kind. Consider the following inductive argument:

> Most corporation lawyers are conservatives.
> Miriam Graf is a corporation lawyer.
> Therefore Miriam Graf is probably a conservative.

This is a fairly good inductive argument; its first premise is true, and if its second premise also is true, its conclusion is more likely to be true than false. But in this case (in contrast to the argument about Socrates' mortality), new premises added to the original pair might weaken or (depending on the content of those new premises) strengthen the original argument. Suppose we also learn that

> Miriam Graf is an officer of the American Civil Liberties Union (ACLU).

and suppose we add the (true) premise that

> Most officers of the ACLU are not conservatives.

Now the conclusion (that Miriam Graf is a conservative) no longer seems very probable; the original inductive argument has been greatly weakened by the presence of this additional information about Miriam Graf. Indeed, if the final premise were to be transformed into the universal proposition

> No officers of the ACLU are conservatives.

the opposite of the original conclusion would then follow deductively—and validly—from the full set of premises affirmed.

On the other hand, suppose we enlarge the original set of premises by adding the following additional premise:

> Miriam Graf has long been an officer of the National Rifle Association (NRA).

The original conclusion (that she is a conservative) would be supported by this enlarged set of premises with even greater likelihood than it was by the original set.

Inductive arguments do not always acknowledge explicitly that their conclusions are supported only with some degree of probability. On the other hand, the mere presence of the word "probability" in an argument gives no assurance that the argument is inductive. There are some strictly deductive arguments *about* probabilities themselves, in which the probability of a certain combination of events is deduced from the probabilities of other events. For example, if the probability of three successive heads in three tosses of a coin is $1/8$, one may infer deductively that the probability of getting at least one tail in three tosses of a coin is $7/8$. Other illustrations of such arguments are given in Chapter 14.

In sum, the distinction between induction and deduction rests on the nature of the claims made by the two types of arguments about the *relations* between their premises and their conclusions. Thus we characterize the two types of arguments as follows: A **deductive argument** is one whose conclusion is claimed to follow from its premises with absolute necessity, this necessity not being a matter of degree and not depending in any way on whatever else may be the case. In sharp contrast, an **inductive argument** is one whose conclusion is claimed to follow from its premises only with probability, this probability being a matter of degree and dependent on what else may be the case.

1.6 Validity and Truth

A deductive argument is valid when it succeeds in linking, with logical necessity, the conclusion to its premises. Its validity refers to the relation between its propositions—between the set of propositions that serve as the premises and the one proposition that serves as the conclusion of that argument. If the conclusion follows with logical necessity from the premises, we say that the argument is valid. Therefore *validity can never apply to any single proposition by itself*, because the needed relation cannot possibly be found within any one proposition.

Truth and falsehood, on the other hand, are attributes of individual propositions. A single statement that serves as a premise in an argument may be true; the statement that serves as its conclusion may be false. This conclusion might have been validly inferred, but to say that any conclusion (or any single premise) is itself valid or invalid makes no sense.

Deductive argument
One of the two major types of argument traditionally distinguished, the other being the inductive argument. A deductive argument claims to provide conclusive grounds for its conclusion. If it does provide such grounds, it is valid; if it does not, it is invalid.

Inductive argument
One of the two major types of argument traditionally distinguished, the other being the deductive argument. An inductive argument claims that its premises give only some degree of probability, but not certainty, to its conclusion.

Truth is the attribute of those propositions that assert what really is the case. When I assert that Lake Superior is the largest of the five Great Lakes, I assert what really is the case, what is true. If I had claimed that Lake Michigan is the largest of the Great Lakes my assertion would not be in accord with the real world; therefore it would be false. This contrast between validity and truth is important: *Truth and falsity are attributes of individual propositions or statements; validity and invalidity are attributes of arguments.*

Just as the concept of validity cannot apply to single propositions, the concept of truth cannot apply to arguments. Of the several propositions in an argument, some (or all) may be true and some (or all) may be false. However, the argument as a whole is neither true nor false. Propositions, which are statements about the world, may be true or false; deductive arguments, which consist of inferences from one set of propositions to other propositions, may be valid or invalid.

The relations *between* true (or false) propositions and valid (or invalid) arguments are critical and complicated. Those relations lie at the heart of deductive logic. Part II of this book is devoted largely to the examination of those complex relations, but a preliminary discussion of the relation between validity and truth is in order here.

We begin by emphasizing that an argument may be valid even if one or more of its premises is not true. Every argument makes a claim about the relation between its premises and the conclusion drawn from them; that relation may hold even if the premises turn out to be false or the truth of the premises is in dispute. This point was made dramatically by Abraham Lincoln in 1858 in one of his debates with Stephen Douglas. Lincoln was attacking the *Dred Scott* decision of the Supreme Court, which had held that slaves who had escaped into Northern states must be returned to their owners in the South. Lincoln said:

> I think it follows [from the *Dred Scott* decision], and I submit to the consideration of men capable of arguing, whether as I state it, in syllogistic form, the argument has any fault in it:
>
> > Nothing in the Constitution or laws of any State can destroy a right distinctly and expressly affirmed in the Constitution of the United States.
> >
> > The right of property in a slave is distinctly and expressly affirmed in the Constitution of the United States.
> >
> > Therefore, nothing in the Constitution or laws of any State can destroy the right of property in a slave.
>
> I believe that no fault can be pointed out in that argument; assuming the truth of the premises, the conclusion, so far as I have capacity at all to understand it, follows inevitably. There is a fault in it as I think, but the fault is not in the reasoning; the falsehood in fact is a fault of the premises. I believe that the right of property in a slave is not distinctly and expressly affirmed in the Constitution, and Judge Douglas thinks it is. I believe that the Supreme Court and the advocates of that decision [the *Dred Scott* decision] may search in vain for the place in the Constitution where the right of property in a slave is distinctly and expressly affirmed. I say, therefore, that I think one of the premises is not true in fact.[21]

The reasoning in the argument that Lincoln recapitulates and attacks is not faulty—but its second premise (that "the right of property in a slave is . . . affirmed in the Constitution") is plainly false. The conclusion has therefore not been established. Lincoln's logical point is correct and important: *An argument may be valid even when its conclusion and one or more of its premises are false*. The validity of an argument, we emphasize once again, depends only on the *relation* of the premises to the conclusion.

There are many possible combinations of true and false premises and conclusions in both valid and invalid arguments. Here follow seven illustrative arguments, each prefaced by the statement of the combination (of truth and validity) that it represents. With these illustrations (whose content is deliberately trivial) before us, we will be in a position to formulate some important principles concerning the relations between truth and validity.

I. Some *valid* arguments contain *only true propositions*—true premises and a true conclusion:

> All mammals have lungs.
> All whales are mammals.
> Therefore all whales have lungs.

II. Some *valid* arguments contain *only false propositions*—false premises and a false conclusion:

> All four-legged creatures have wings.
> All spiders have exactly four legs.
> Therefore all spiders have wings.

This argument is valid because, if its premises were true, its conclusion would have to be true also—even though we know that in fact both the premises *and* the conclusion of this argument are false.

III. Some *invalid* arguments contain *only true propositions*—all their premises are true, and their conclusions are true as well:

> If I owned all the gold in Fort Knox, then I would be wealthy.
> I do not own all the gold in Fort Knox.
> Therefore I am not wealthy.

The true conclusion of this argument does not follow from its true premises. This will be seen more clearly when the immediately following illustration is considered.

IV. Some *invalid* arguments contain *only true premises* and have a *false conclusion*. This is illustrated by an argument exactly like the previous one (III) in form, changed only enough to make the conclusion false.

> If Bill Gates owned all the gold in Fort Knox, then Bill Gates would be wealthy.
> Bill Gates does not own all the gold in Fort Knox.
> Therefore Bill Gates is not wealthy.

The premises of this argument are true, but its conclusion is false. Such an argument cannot be valid because it is impossible for the premises of a valid argument to be true and its conclusion to be false.

V. Some *valid* arguments have *false premises* and a *true conclusion*:

> All fishes are mammals.
> All whales are fishes.
> Therefore all whales are mammals.

The conclusion of this argument is true, as we know; moreover, it may be validly inferred from these two premises, both of which are wildly false.

VI. Some *invalid* arguments also have *false premises* and a *true conclusion*:

> All mammals have wings.
> All whales have wings.
> Therefore all whales are mammals.

From Examples V and VI taken together, it is clear that we cannot tell from the fact that an argument has false premises and a true conclusion whether it is valid or invalid.

VII. Some *invalid* arguments, of course, contain *all false propositions*—false premises and a false conclusion:

> All mammals have wings.
> All whales have wings.
> Therefore all mammals are whales.

These seven examples make it clear that there are valid arguments with false conclusions (Example II), as well as invalid arguments with true conclusions (Examples III and VI). Hence it is clear that *the truth or falsity of an argument's conclusion does not by itself determine the validity or invalidity of that argument*. Moreover, *the fact that an argument is valid does not guarantee the truth of its conclusion* (Example II).

Two tables (referring to the seven preceding examples) will make very clear the variety of possible combinations. The first table shows that *invalid* arguments can have every possible combination of true and false premises and conclusions:

	Invalid Arguments	
	True Conclusion	**False Conclusion**
True Premises	Example III	Example IV
False Premises	Example VI	Example VII

The second table shows that *valid* arguments can have only three of those combinations of true and false premises and conclusions:

Valid Arguments	True Conclusion	False Conclusion
True Premises	Example I	—
False Premises	Example V	Example II

The one blank position in the second table exhibits a fundamental point: *If an argument is valid and its premises are true, we may be certain that its conclusion is true also.* To put it another way: *If an argument is valid and its conclusion is false, not all of its premises can be true.* Some perfectly valid arguments do have false conclusions, but any such argument must have at least one false premise.

When an argument is valid *and* all of its premises are true, we call it *sound*. The conclusion of a sound argument obviously must be true—and only a sound argument can establish the truth of its conclusion. If a deductive argument is not sound—that is, if the argument is not valid or if not all of its premises are true—it fails to establish the truth of its conclusion even if in fact the conclusion is true.

To test the truth or falsehood of premises is the task of science in general, because premises may deal with any subject matter at all. The logician is not (professionally) interested in the truth or falsehood of propositions so much as in the logical relations between them. By *logical relations between propositions* we mean those relations that determine the correctness or incorrectness of the arguments in which they occur. The task of determining the correctness or incorrectness of arguments falls squarely within the province of logic. The logician is interested in the correctness even of arguments whose premises may be false.

Why do we not confine ourselves to arguments with true premises, ignoring all others? Because the correctness of arguments whose premises are not known to be true may be of great importance. In science, for example, we verify theories by *deducing* testable consequences from uncertain theoretical premises—but we cannot know beforehand which theories are true. In everyday life also, we must often choose between alternative courses of action, first seeking to deduce the consequences of each. To avoid deceiving ourselves, we must reason correctly about the consequences of the alternatives, taking each as a premise. If we were interested only in arguments with true premises, we would not know which set of consequences to trace out until we knew which of the alternative premises was true. But if we knew which of the alternative premises was true, we would not need to reason about it at all, because our purpose was to help us decide which alternative premise to *make* true. To confine our attention to arguments with premises known to be true would therefore be self-defeating.

Effective methods for establishing the validity or invalidity of deductive arguments are presented and explained at length in Part II of this book.

EXERCISES

For each of the argument descriptions provided below, construct a deductive argument (on any subject of your choosing) having only two premises.

1. A valid argument with one true premise, one false premise, and a false conclusion
2. A valid argument with one true premise, one false premise, and a true conclusion
3. An invalid argument with two true premises and a false conclusion
4. An invalid argument with two true premises and a true conclusion
5. A valid argument with two false premises and a true conclusion
6. An invalid argument with two false premises and a true conclusion
7. An invalid argument with one true premise, one false premise, and a true conclusion
8. A valid argument with two true premises and a true conclusion

chapter 1 Summary

The most fundamental concepts of logic are introduced in this chapter.

In Section 1.1 we explained what logic is and why it is necessary, and we defined it as the study of the methods and principles used to distinguish correct from incorrect reasoning.

In Section 1.2 we gave an account of propositions, which may be asserted or denied, and which are either true or false, and of arguments, which are clusters of propositions of which one is the conclusion and the others are the premises offered in its support. Arguments are the central concern of logicians.

In Section 1.3 we discussed difficulties in the recognition of arguments, arising from the variety of ways in which the propositions they contain may be expressed, and sometimes even from the absence of their express statement in arguments called enthymemes.

In Section 1.4 we discussed the differences between arguments and explanations, showing why this distinction often depends on the context and on the intent of the passage in that context.

In Section 1.5 we explained the fundamental difference between deductive arguments, whose conclusions may be certain (if the premises are true and the reasoning valid), and inductive arguments, aiming to establish matters of fact, whose conclusions may be very probable but are never certain.

In Section 1.6 we discussed validity and invalidity (which apply to deductive arguments) as contrasted with truth and falsity (which apply to propositions). We explored some of the key relations between validity and truth.

END NOTES

[1] Scott Wallace, "Last of the Amazon," *National Geographic*, January 2007.

[2] David Hayden, "Thy Neighbor, Thy Self," *The New York Times*, 9 May 2000.

[3] Peter Singer, "What Should a Billionaire Give—and What Should You?" *The New York Times Magazine*, 17 December, 2006.

[4] Roger Woody, "Why Iraq's a Mess," *The New York Times*, 26 January 2007.

[5] Harris, Sam, *Letter to a Christian Nation* (New York: Alfred A. Knopf, 2006).

[6] *Gratz v. Bollinger*, 539 U.S. 244 (2003).

[7] Ibid.

[8] Archbishop Peter Akinola, quoted in *The New York Times*, 25 December 2006.

[9] Ramsey Colloquium of the Institute on Religion and Public Life, "Always to Care, Never to Kill," *The Wall Street Journal*, 17 November 1991.

[10] David Gelernter, "Ramping Up the Violence," *The Weekly Standard*, 26 February 2007.

[11] Christopher Caldwell, *Reflections on the Revolution in Europe: Immigration, Islam, and the West* (Doubleday, 2009).

[12] Noam Chomsky, *Failed States* (New York: Henry Holt, 2006).

[13] Steven Goldberg, "The Death Penalty," *The New York Times*, 20 December 2004.

[14] Ibid.

[15] "The Vote to Ban Human Cloning," *The New York Times*, 2 August 2001.

[16] Senator Sam Brownback, of Kansas, at a Senate hearing in April 2000.

[17] Samuel Freeman, "John Rawls, Friend and Teacher," *Chronicle of Higher Education*, 13 December 2002.

[18] Jeff Greenwald, "Brightness Visible," *The New York Times Magazine*, 14 May 2000.

[19] Joseph Bamberger, "Water's Inimitable Qualities," *The New York Times*, 17 July 2007.

[20] James Gleick, "Cyber-Neologisms," *The New York Times Magazine*, 5 November 2006.

[21] From *The Collected Works of Abraham Lincoln*, vol. 3, Roy P. Basler, editor (New Brunswick, NJ: Rutgers University Press, 1953).

PEARSON mylogiclab

For additional exercises and tutorials about concepts covered in this chapter, log in to MyLogicLab at *www.mylogiclab.com* and select your current textbook.

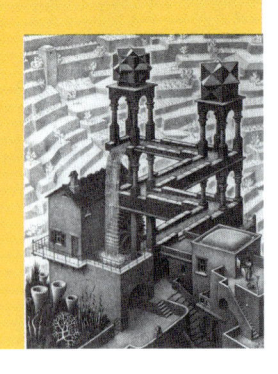

Analyzing Arguments

2.1 Paraphrasing Arguments

2.2 Diagramming Arguments

2.3 Complex Argumentative Passages

2.4 Problems in Reasoning

2.1 Paraphrasing Arguments

Arguments in everyday life are often more complex—more tangled and less precise—than those given as illustrations in Chapter 1. Premises may be numerous and in topsy-turvy order; they may be formulated awkwardly, and they may be repeated using different words; even the meaning of premises may be unclear. To sort out the connections of premises and conclusions so as to evaluate an argument fairly, we need some analytical techniques.

The most common, and perhaps the most useful technique for analysis is *paraphrase*. We paraphrase an argument by setting forth its propositions in clear language and in logical order. This may require the reformulation of sentences, and therefore great care must be taken to ensure that the paraphrase put forward captures correctly and completely the argument that was to be analyzed.

The following passage, whose premises are confusingly intertwined, was part of the majority decision of the U.S. Supreme Court when, in 2003, it struck down as unconstitutional a Texas statute that had made it a crime for persons of the same sex to engage in certain forms of intimate sexual conduct. Justice Anthony Kennedy, writing for the majority, said this:

> The [present] case does involve two adults who, with full and mutual consent from each other, engaged in sexual practices common to a homosexual life style. The petitioners are entitled to respect for their private lives. The state cannot demean their existence or control their destiny by making their private sexual conduct a crime. Their right to liberty under the Due Process Clause [of the 14th Amendment to the U.S. Constitution] gives them the full right to engage in their conduct without intervention of the government. It is a premise of the Constitution that there is a realm of personal liberty which the government may not enter. The Texas statute furthers no legitimate state interest which can justify its intrusion into the personal and private life of the individual.[1]

Although the general thrust of this decision is clear, the structure of the argument, which is really a complex of distinct arguments, is not. We can clarify the whole by paraphrasing the decision of the Court as follows:

1. The Constitution of the United States guarantees a realm of personal liberty that includes the private, consensual sexual activity of adults.

2. The conduct of these petitioners was within that realm of liberty and they therefore had a full right, under the Constitution, to engage in the sexual conduct in question without government intervention.

Biography

Peter Abelard

Peter Abelard was born near Nantes, in Brittany, in 1079 CE, to a noble family. He could have become a wealthy knight, but rejected such a life, instead choosing an academic career. He left home for Paris, and studied with William of Champeaux, with whom he quarreled acrimoniously, resulting in his opening a school of his own. Eventually he was elected to the faculty of the school of Notre Dame, where he was extremely popular, attracting students from all over Europe. He was primarily interested in logic, which was then called dialectic, and in metaphysics. He confronted the deep metaphysical problem of universals, or abstract objects. General terms (e.g., *justice, yellow, smooth*) plainly do exist, but are there abstract *objects* that actually exist, beneath or behind those terms, in some non-physical world? Abelard held that there are no such entities, but that we are sometimes misled by the words we use for the common properties of things. His position came to be known as *nominalism*.

While working as a tutor to Heloise, the seventeen-year-old niece of a Parisian named Fulbert, a relationship developed that resulted in her becoming pregnant. The couple ran away to his home in Brittany, where she gave birth to a child. Abelard eventually married Heloise, later sending her to become a nun. Her uncle, outraged by the scandal, hired thugs to assault Abelard and castrate him. Abelard then became a monk and lecturer, unpopular among colleagues because of his intellectual arrogance. He was obliged to move from abbey to abbey, became embroiled in theological controversies, and died in Paris in 1142.

In logic, Abelard explored the relations of premises and conclusions in deductive arguments. He was one of the first to emphasize the *syntactic* nature of validity. An argument is valid, he pointed out, not because of the semantic content of its propositions, but because of the *formal relations* among those propositions.

PEARSON mylogiclab

---

3. The Texas statute intrudes, without justification, into the private lives of these petitioners, and demeans them, by making their protected, private sexual conduct a crime.

4. The Texas statute that criminalizes such conduct therefore wrongly denies the rights of these petitioners and must be struck down as unconstitutional.

In this case the paraphrase does no more than set forth clearly what the premises indubitably assert. Sometimes, however, paraphrasing can bring to the surface what was assumed in an argument but was not fully or clearly stated. For example, the great English mathematician, G. H. Hardy, in *A Mathematician's Apology* (Cambridge University Press, 1940), argued thus: "Archimedes will be remembered when Aeschylus is forgotten, because languages die and mathematical ideas do not." We may paraphrase this argument by spelling out its claims:

1. Languages die.
2. The plays of Aeschylus are written in a language.
3. So the work of Aeschylus will eventually die.
4. Mathematical ideas never die.
5. The work of Archimedes was with mathematical ideas.
6. So the work of Archimedes will never die.
7. Therefore Archimedes will be remembered when Aeschylus is forgotten.

This paraphrase enables us to distinguish and examine the premises and inferences compressed into Hardy's single sentence.

EXERCISES

Paraphrase each of the following passages, which may contain more than one argument.

1. The [Detroit] Pistons did not lose because of the lack of ability. They are an all-around better team. They lost because of the law of averages. They will beat the [San Antonio] Spurs every two times out of three. When you examine the NBA finals [of 2005], that is exactly how they lost the seventh (last game) because that would have been three out of three. The Spurs will beat the Pistons one out of three. It just so happens that, that one time was the final game, because the Pistons had already won two in a row.
    —Maurice Williams, "Law of Averages Worked Against Detroit Pistons," *The Ann Arbor* (Michigan) *News*, 8 July 2005

2. Hundreds of thousands of recent college graduates today cannot express themselves with the written word. Why? Because universities have shortchanged them, offering strange literary theories, Marxism, feminism, deconstruction, and other oddities in the guise of writing courses.
    —Stanley Ridgeley, "College Students Can't Write?" *National Review Online*, 19 February 2003

3. Racially diverse nations tend to have lower levels of social support than homogenous ones. People don't feel as bound together when they are divided on ethnic lines and are less likely to embrace mutual support programs. You can have diversity or a big welfare state. It's hard to have both.
    —David Brooks (presenting the views of Seymour Lipset), "The American Way of Equality," *The New York Times*, 14 January 2007

4. Orlando Patterson claims that "freedom is a natural part of the human condition." Nothing could be further from the truth. If it were true, we

could expect to find free societies spread throughout human history. We do not. Instead what we find are every sort of tyrannical government from time immemorial.

—John Taylor, "Can Freedom Be Exported?" *The New York Times*, 22 December 2006

5. *The New York Times* reported, on 30 May 2000, that some scientists were seeking a way to signal back in time. A critical reader responded thus:

> It seems obvious to me that scientists in the future will never find a way to signal back in time. If they were to do so, wouldn't we have heard from them by now?

—Ken Grunstra, "Reaching Back in Time," *The New York Times*, 6 June 2000

6. Nicholas Kristof equates the hunting of whales by Eskimos with the whaling habits of Japanese, Norwegians, and Icelanders. The harsh environment of the Inupiat [Eskimos] dictates their diet, so not even the most rabid antiwhaling activist can deny their inalienable right to survive. The Japanese and the European whale-hunting countries can choose the food they consume; they have no need to eat whales. It is not hypocritical to give a pass to the relatively primitive society of the Inupiat to hunt a strictly controlled number of whales for survival while chastising the modern societies that continue to hunt these magnificent mammals for no good reason.

—Joseph Turner, "Their Whale Meat, and Our Piety," *The New York Times*, 18 September 2003

7. Space contains such a huge supply of atoms that all eternity would not be enough time to count them and count the forces which drive the atoms into various places just as they have been driven together in this world. So we must realize that there are other worlds in other parts of the universe with races of different men and different animals.

—Lucretius, *De Rerum Natura*, First Century BCE

8. If you marry without love, it does not mean you will not later come to love the person you marry. And if you marry the person you love, it does not mean that you will always love that person or have a successful marriage. The divorce rate is very low in many countries that have prearranged marriage. The divorce rate is very high in countries where people base their marriage decisions on love.

—Alex Hammoud, "I Take This Man, for Richer Only," *The New York Times*, 18 February 2000

9. Our entire tax system depends upon the vast majority of taxpayers who attempt to pay the taxes they owe having confidence that they're being treated fairly and that their competitors and neighbors are also paying what is due. If the public concludes that the IRS cannot meet these basic expectations, the risk to the tax system will become very high, and the effects very difficult to reverse.

—David Cay Johnston, "Adding Auditors to Help IRS Catch Tax Cheaters," *The New York Times*, 13 February 2000

10. People and governments want to talk, talk, talk about racism and other forms of intolerance; we are obsessed with racial and ethnic issues. But we come to these issues wearing earplugs and blinders, and in a state of denial that absolves us of complicity in any of these hateful matters. Thus, the other guy is always wrong.

—Bob Herbert, "Doomed to Irrelevance," *The New York Times*, 6 November 2001

2.2 Diagramming Arguments

A second technique for the analysis of arguments is *diagramming*. With a diagram we can represent the structure of an argument graphically; the flow of premises and conclusions is displayed in a two-dimensional chart, or picture, on the page. A diagram is not needed for a simple argument, even though drawing one can enhance our understanding. When an argument is complex, with many premises entwined in various ways, a diagram can be exceedingly helpful.

To construct the diagram of an argument we must first number all the propositions it contains, in the order in which they appear, circling each number. Using arrows between the circled numbers, we can then construct a diagram that shows the relations of premises and conclusions without having to restate them. To convey the process of inference on the two-dimensional page, we adopt this convention: A conclusion always appears in the space *below* the premises that give it support; coordinate premises are put on the same horizontal level. In this way, an argument whose wording may be confusing can be set forth vividly in iconic form. The structure of the argument is displayed visually.[2]

Here follows a straightforward argument that may be readily diagrammed:

① There is no consensus among biologists that a fertilized cell is alive in a sense that an unfertilized egg or unused sperm is not. ② Nor is there a consensus about whether a group of cells without even a rudimentary nervous system is in any sense human. ③ Hence there are no compelling experimental data to decide the nebulous issue of when "human" life begins.[3]

The circled numbers serve to represent the propositions, so we can diagram the argument as follows:

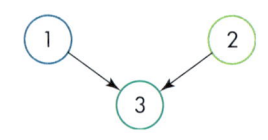

When the several premises of an argument are not all coordinate—that is, when some premises give direct support not to the conclusion but to other premises that support the conclusion—the diagram can show this quite clearly. Here is an argument illustrating this feature of diagramming:

① Football analysis is trickier than the baseball kind because ② Football really is a team sport. ③ Unlike in baseball, all eleven guys on the field are involved in every play. ④ Who deserves the credit or blame is harder to know than it looks.[4]

The diagram looks like this:

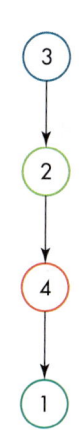

```
    3
    ↓
    2
    ↓
    4
    ↓
    1
```

An alternative plausible interpretation of this argument can be represented by a different diagram:

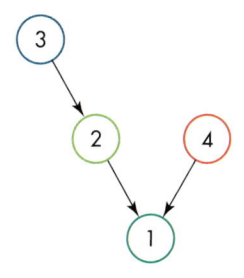

```
  3
   ↘
    2   4
     ↘ ↙
      1
```

Another strength of diagrams is their ability to exhibit relations between the premises—relations that may be critical to the argument. Each premise of an argument may support its conclusion separately, as in the arguments above. In some arguments, however, the premises support the conclusion only when they are considered *jointly*—and this is a feature of the reasoning that a diagram is well suited to display, by providing a visual representation of that connection. The following argument illustrates this:

① General Motors makes money (when it does) on new cars and on the financing of loans. ② Car dealers, by contrast, make most of their money on servicing old cars and selling used ones. ③ So car dealers can thrive even when the automaker languishes.[5]

By bracketing the premises in the diagram of this argument, we show that its premises give support only because they are joined, thus:

```
    1           2
    └─────┬─────┘
          ↓
          3
```

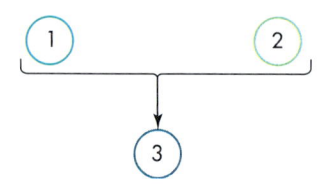

In this argument, neither premise supports the conclusion independently. It is the combination of the facts that General Motors makes most of its money in one way, while car dealers make most of their money in another way, that supports the conclusion that the latter may thrive while the former languishes.

Often we can *show* what we cannot as conveniently say. Diagrams are particularly useful when an argument's structure is complicated. Consider the following argument:

> ① Desert mountaintops make good sites for astronomy. ② Being high, they sit above a portion of the atmosphere, enabling a star's light to reach a telescope without having to swim through the entire depths of the atmosphere. ③ Being dry, the desert is also relatively cloud-free. ④ The merest veil of haze or cloud can render a sky useless for many astronomical measures.[6]

Proposition ① is plainly the conclusion of this argument, and the other three provide support for it—but they function differently in giving that support. Statement ② supports, by itself, the claim that mountaintops are good sites for telescopes. But statements ③ and ④ must work together to support the claim that desert mountaintops are good sites for telescopes. A diagram shows this neatly:

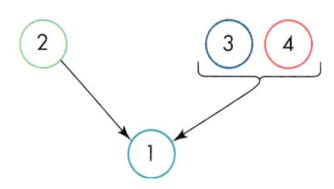

```
    2            3   4
                 └─┬─┘
     \             /
      \           /
       ↘         ↙
            1
```

Some complications may be revealed more clearly using paraphrase. When an argument has a premise that is not stated explicitly, a paraphrase allows us to formulate the tacit premise and then add it to the list explicitly. A diagram requires the representation of the tacit premise in some way that indicates visually that it has been added (a broken circle around a number is commonly used), but even then the added premise remains to be precisely formulated. Thus the argument

> Since there are no certainties in the realm of politics, politics must be the arena for negotiation between different perspectives, with cautious moderation likely to be the best policy.[7]

is best clarified by a paraphrase in which its tacit premise and internal complexity is made explicit, thus:

1. There are no certainties in the realm of politics.
2. Where there are no certainties, those with different perspectives must negotiate their differences.
3. The best policy likely to emerge from such negotiation is one of cautious moderation.
4. Therefore politics is the realm for negotiation between different perspectives, with cautious moderation likely to be the best policy.

The number of arguments in a passage is determined, most logicians agree, by the number of conclusions it contains. If a passage contains two or more arguments, and a number of propositions whose relations are not obvious, a diagram may prove particularly useful in sorting things out. A passage in a letter from Karl Marx to Friedrich Engels illustrates this nicely:

> ① To hasten the social revolution in England is the most important object of the International Workingman's Association. ② The sole means of hastening it is to make Ireland independent. Hence ③ the task of the "International" is everywhere to put the conflict between England and Ireland in the foreground, and ④ everywhere to side openly with Ireland.[8]

There are two conclusions in this passage and hence two arguments. But both conclusions are inferred from the same two premises. A diagram exhibits this structure:

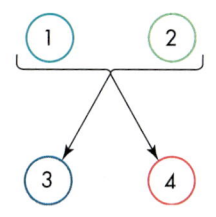

Two conclusions (and hence two arguments) may have a single stated premise. For example,

> Older women have less freedom to fight sexual harassment at their jobs or to leave a battering husband, because age discrimination means they won't easily find other ways of supporting themselves.[9]

The single premise here is that older women cannot easily find alternative ways to support themselves. The two conclusions supported by that premise are (a) that older women have less freedom to fight sexual harassment at their jobs, and (b) that older married women have less freedom to leave a battering husband. A *single argument* ordinarily means an argument with a single conclusion, regardless of how many premises are adduced in its support.

When there are two or more premises in an argument, or two or more arguments in a passage, the order of appearance of premises and conclusions may need to be clarified. The conclusion may be stated last, or first; it may sometimes be sandwiched between the premises offered in its support, as in the following passage:

> The real and original source of inspiration for the Muslim thinkers was the Quran and the sayings of the Holy Prophet. It is therefore clear that the Muslim philosophy was not a carbon copy of Greek thought, as it concerned itself primarily and specifically with those problems which originated from and had relevance to Muslims.[10]

Here the conclusion, that "Muslim philosophy was not a carbon copy of Greek thought," appears after the first premise of the argument and before the second.

The same proposition that serves as a conclusion in one argument may serve as premise in a different argument, just as the same person may be a commander in one context and a subordinate in another. This is well illustrated by a passage from the work of Thomas Aquinas. He argues:

> Human law is framed for the multitude of human beings.
> The majority of human beings are not perfect in virtue.
> Therefore human laws do not forbid all vices.[11]

The conclusion of this argument is used immediately thereafter as a premise in another, quite different argument:

> Vicious acts are contrary to acts of virtue.
> But human law does not prohibit all vices. . . .
> Therefore neither does it prescribe all acts of virtue.[12]

No special techniques are needed to grasp these arguments of St. Thomas. However, when the cascade of arguments is compressed, a paraphrase is helpful in showing the flow of reasoning. Consider the following passage:

> Because ① the greatest mitochondrial variations occurred in African people, scientists concluded that ② they had the longest evolutionary history, indicating ③ a probable African origin for modern humans.[13]

We might diagram the passage thus:

```
    1
    ↓
    2
    ↓
    3
```

A paraphrase of this passage, although perhaps more clumsy, exhibits more fully the cascade of the two arguments that are compressed in it:

1. The more mitochondrial variation in a people, the longer its evolutionary history.
2. The greatest mitochondrial variations occur in African people.

   Therefore African people have had the longest evolutionary history.

1. African people have had the longest evolutionary history.
2. Modern humans probably originated where people have had the longest evolutionary history.

   Therefore modern humans probably originated in Africa.

These examples make it evident that the same proposition can serve as a premise where it occurs as an assumption in an argument; or as a conclusion where it is claimed to follow from other propositions assumed in an argument. "Premise" and "conclusion" are always *relative* terms.

Multiple arguments may be interwoven in patterns more complicated than cascades, and these will require careful analysis. The diagramming technique then becomes particularly useful. In John Locke's *Second Treatise of Government*, for example, two arguments are combined in the following passage:

> It is not necessary—no, nor so much as convenient—that the legislative should be always in being; but absolutely necessary that the executive power should, because there is not always need of new laws to be made, but always need of execution of the laws that are made.

The component propositions here may be numbered thus: ① It is not necessary or convenient that the legislative [branch of government] should be always in being; ② it is absolutely necessary that the executive power should be always in being; ③ there is not always need of new laws to be made; ④ there is always need of execution of the laws that are made. The diagram for this passage is

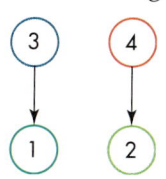

```
    3       4
    │       │
    ▼       ▼
    1       2
```

which shows that the conclusion of the second argument is stated between the conclusion and the premise of the first argument, and that the premise of the first argument is stated between the conclusion and the premise of the second argument. The diagram also shows that both conclusions are stated before their premises.

That very same diagram shows the logical structure of two related arguments of the Roman philosopher Seneca, in support of the deterrence theory of punishment. He wrote:

> ① No one punishes because a sin has been committed, ② but in order that a sin will not be committed. [For] ③ what has passed cannot be recalled, but ④ what lies in the future may be prevented.

That "no one punishes because a sin has been committed" is the conclusion of one argument; its premise is that "what has passed cannot be recalled." That "[we do punish] in order that a sin will not be committed" is the conclusion of a second argument, whose premise is that "what lies in the future may be prevented."

Diagramming and paraphrasing are both very useful tools with which we can analyze arguments so as to understand more fully the relations of premises to conclusions.

EXERCISES

A. Diagram each of the following passages, which may contain more than one argument.

EXAMPLE

1. In a recent attack upon the evils of suburban sprawl, the authors argue as follows:

   The dominant characteristic of sprawl is that each component of a community—housing, shopping centers, office parks, and civic institutions—is segregated, physically separated from the others, causing the residents of suburbia to spend an inordinate amount of time and money moving from one place to the next. And since nearly everyone drives alone, even a sparsely populated area can generate the traffic of a much larger traditional town.[14]

SOLUTION

① The dominant characteristic of sprawl is that each component of a community—housing, shopping centers, office parks, and civic institutions—is segregated, physically separated from the others, causing ② the residents of suburbia to spend an inordinate amount of time and money moving from one place to the next. And since ③ nearly everyone drives alone, ④ even a sparsely populated area can generate the traffic of a much larger traditional town.

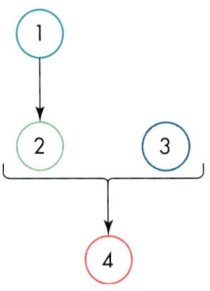

```
        1
        ↓
    ┌───┴───┐
    2       3
    └───┬───┘
        ↓
        4
```

2. At any cost we must have filters on our Ypsilanti Township library computers. Pornography is a scourge on society at every level. Our public library must not be used to channel this filth to the people of the area.
   —Rob. J. and Joan D. Pelkey, *The Ann Arbor* (Michigan) *News*, 3 February 2004

3. At his best, Lyndon Johnson was one of the greatest of all American presidents. He did more for racial justice than any president since Abraham Lincoln. He built more social protections than anyone since Franklin Roosevelt. He was probably the greatest legislative politician in American history. He was also one of the most ambitious idealists. Johnson sought power to use it to accomplish great things.
   —Alan Brinkley, "The Making of a War President," *The New York Times Book Review*, 20 August 2006

4. Married people are healthier and more economically stable than single people, and children of married people do better on a variety of indicators. Marriage is thus a socially responsible act. There ought to be some way of spreading the principle of support for marriage throughout the tax code.
   —Anya Bernstein, "Marriage, Fairness and Taxes," *The New York Times*, 15 February 2000

5. The distinguished economist J. K. Galbraith long fought to expose and improve a society exhibiting "private opulence and public squalor." In

his classic work, *The Affluent Society* (Boston: Houghton Mifflin, 1960), he argued as follows:

> Vacuum cleaners to insure clean houses are praiseworthy and essential in our standard of living. Street cleaners to insure clean streets are an unfortunate expense. Partly as a result, our houses are generally clean and our streets generally filthy.

6. Defending the adoption of the euro in place of the pound as the monetary unit of the United Kingdom, Prime Minister Tony Blair said this: "The argument is simple. We are part of Europe. It affects us directly and deeply. Therefore we should exercise leadership in order to change Europe in the direction we want."

   —Reported by Alan Cowell in the *The New York Times*, 9 December 2001

7. California's "three strikes and you're out" law was enacted 10 years ago this month (March, 2004). Between 1994 and 2002, California's prison population grew by 34,724, while that of New York, a state without a "three strikes" law, grew by 315. Yet during that time period New York's violent crime rate dropped 20 percent more than California's. No better example exists of how the drop in crime cannot be attributed to draconian laws with catchy names.

   —Vincent Schiraldi, "Punitive Crime Laws,"
   The New York Times, 19 March 2004

8. No one means all he says, and yet very few say all they mean, for words are slippery and thought is viscous.

   —Henry Adams, *The Education of Henry Adams* (1907)

9. The first impression becomes a self-fulfilling prophesy: we hear what we expect to hear. The interview is hopelessly biased in favor of the nice.

   —Malcom Gladwell, "The New-Boy Network," *The New Yorker*, 29 May 2000

10. No government can ever guarantee that the small investor has an equal chance of winning. It is beyond dishonest to pretend that rules can be written to prevent future financial scandals. No set of regulations can insure fairness and transparency in the [securities] markets.

    —Lester Thurow, "Government Can't Make the Market Fair,"
    The New York Times, 23 July 2002

B. There may be one argument or more than one argument in each of the following passages. Paraphrase the premises and conclusions (or use diagrams if that is helpful) to analyze the arguments found in each passage.

EXAMPLE

1. An outstanding advantage of nuclear over fossil fuel energy is how easy it is to deal with the waste it produces. Burning fossil fuels produces 27,000 million tons of carbon dioxide yearly, enough to make, if

solidified, a mountain nearly one mile high with a base twelve miles in circumference. The same quantity of energy produced from nuclear fission reactions would generate two million times less waste, and it would occupy a sixteen-meter cube. All of the high-level waste produced in a year from a nuclear power station would occupy a space about a cubic meter in size and would fit safely in a concrete pit.

—James Lovelock, *The Revenge of Gaia: Earth's Climate Crisis and the Fate of Humanity* (New York: Basic Books, 2006)

SOLUTION

① An outstanding advantage of nuclear over fossil fuel energy is how easy it is to deal with the waste it produces. ② Burning fossil fuels produces 27,000 million tons of carbon dioxide yearly, enough to make, if solidified, a mountain nearly one mile high with a base twelve miles in circumference. ③ The same quantity of energy produced from nuclear fission reactions would generate two million times less waste, and it would occupy a sixteen-meter cube. ④ All of the high level waste produced in a year from a nuclear power station would occupy a space about a cubic meter in size and would fit safely in a concrete pit.

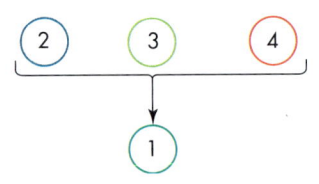

```
    2    3    4
    └────┼────┘
         ↓
         1
```

2. Why decry the wealth gap? First, inequality is correlated with political instability. Second, inequality is correlated with violent crime. Third, economic inequality is correlated with reduced life expectancy. A fourth reason? Simple justice. There is no moral justification for chief executives being paid hundreds of times more than ordinary employees.

    —Richard Hutchinsons, "When the Rich Get Even Richer," *The New York Times*, 26 January 2000

3. Genes and proteins are discovered, not invented. Inventions are patentable, discoveries are not. Thus, protein patents are intrinsically flawed.

    —Daniel Alroy, "Invention vs. Discovery," *The New York Times*, 29 March 2000

4. Ultimately, whaling's demise in Japan may have little to do with how majestic, smart, or endangered the mammals are, but a good deal to do with simple economics. A Japanese newspaper conducted a survey in Japan regarding the consumption of whale meat, and reported that of all the thousands of respondents, only 4 percent said that they actually ate whale meat at least sometimes. The newspaper then wrote this: "A growing number of Japanese don't want to eat whale meat. And if they won't eat it, they won't buy it. And if they won't buy it, say goodbye to Japanese whaling."

    —Reported in *Asahi Shimbun*, April 2002

5. On the 18th of July, 2002, the *Consejo Juvenil Sionista Argentino* (Young Zionists of Argentina) held a mass demonstration to promote

widespread remembrance of the horror of the bombing of the Jewish Community Center in Buenos Aires, exactly eight years earlier. At this demonstration the Young Zionists carried a huge banner, which read: *"Sin memoria, no hay justicia. Sin justicia, no hay futuro."* ("Without remembrance, there is no justice. Without justice, there is no future.")

6. Back in 1884, Democratic nominee Grover Cleveland was confronted by the charge that he had fathered an out-of-wedlock child. While Republicans chanted, "Ma, Ma, where's my Pa," Cleveland conceded that he had been supporting the child. No excuses, no evasions. One of his supporters—one of the first spin doctors—gave this advice to voters:

   Since Grover Cleveland has a terrific public record, but a blemished private life, and since his opponent, James G. Blaine, has a storybook private life but a checkered public record, why not put both where they perform best—return Blaine to private life, keep Cleveland in public life.

7. "Wars don't solve problems; they create them," said an October 8 letter about Iraq.

   World War II solved problems called Nazi Germany and militaristic Japan, and created alliances with the nations we crushed. The Revolutionary War solved the problem of taxation without representation, and created the United States of America. The Persian Gulf War solved the problem of the Iraqi invasion of Kuwait. The Civil War solved the problem of slavery.

   These wars created a better world. War is the only way to defeat evil enemies with whom there is no reasoning. It's either us or them. What creates true peace is victory.

   —Keith Kraska, "Necessary Wars,"
   The New York Times, 15 October 2002

8. In the Crito, Plato presents the position of the Athenian community, personified as "the Laws," speaking to Socrates or to any citizen of the community who may contemplate deliberate disobedience to the state:

   He who disobeys us is, as we maintain, thrice wrong; first, because in disobeying us he is disobeying his parents; secondly, because we are the authors of his education; thirdly, because he has made an agreement with us that he will duly obey our commands.

9. The reality is that money talks. Court officers, judges and juries treat private lawyers and their clients differently from those who cannot pay for representation. Just as better-dressed diners get prime tables at a restaurant, human nature dictates better results for those who appear to have money.

   —Desiree Buenzle, "Free Counsel and Fairness," *The New York Times*,
   15 January 2007

10. The town of Kennesaw, GA passed a *mandatory gun ownership* law, in 1982, in response to a *handgun ban* passed in Morton Grove, IL. Kennesaw's crime rate dropped sharply, while Morton Grove's did not. Criminals, unsurprisingly, would rather break into a house where they aren't at risk of being shot. . . . Criminals are likely to suspect that towns with

laws like these on the books will be unsympathetic to malefactors in general, and to conclude that they will do better elsewhere. To the extent that's true, we're likely to see other communities adopting similar laws so that criminals won't see them as attractive alternatives.

—Glenn Reynolds, "A Rifle in Every Pot," *The New York Times*, 16 January 2007

Biography

William of Ockham

William of Ockham, sometimes spelled Occam, (c. 1288–c. 1348) was an influential Franciscan friar, born in the village in Surrey, England, after which he was named. Sent while young to a monastery, he went on to study theology and philosophy at Oxford, and then at the University of Paris, where he eventually taught.

The great intellectual theme of William's life was *simplification*. This was manifested most famously in what came to be known as "Ockham's Razor"—the drive for parsimony in the construction of theories. If any phenomenon can be explained without the assumption of this or that hypothetical entity, we ought not assume that entity; one should not multiply entities beyond necessity. In metaphysics this drive for simplification led him to the position known as *nominalism*: what exists in the universe are only individuals. The universals, or Platonic forms, of which some philosophers write, he believed to be no more than the products of abstraction by the human mind.

William became deeply involved in the theological controversies of those medieval days. He was summoned to the Papal court in Avignon in 1324, apparently under charges of heresy. While there a dispute arose concerning the poverty of Christ, many zealous Franciscans insisting that Jesus and his apostles owned no personal property. William came to share that view; he asserted that Pope John XXII, unwilling to accept the poverty of Jesus, was himself an heretic. William was then obliged to take refuge in the court of the Holy Roman Emperor, Ludwig of Bavaria, and while there, not surprisingly, was himself excommunicated. He died in Munich in 1348.

William of Ockham was an inventive logician, suggesting that we might better rely upon a logical system that did not force us to view all propositions as either true or false (a so-called *two-valued logic*), but that a *three-valued logic*, developed more fully many centuries later, would permit a better reflection of the state of our knowledge. Some central logical equivalences, which came later to be known as De Morgan's theorems, he well understood and actually wrote out in words, not having at his disposal the modern notation with which we now express them.

A powerful and widely respected mind, William of Ockham was referred to by many as *Doctor Invincibilis*—"unconquerable teacher."

PEARSON **mylogiclab**

2.3 Complex Argumentative Passages

Some arguments are exceedingly complicated. Analyzing passages in which several arguments are interwoven, with some propositions serving as both premises and subconclusions while other propositions serve only as premises, and still others are repeated in different words, can be a challenge. The diagramming technique is certainly helpful, but there is no mechanical way to determine whether the diagram actually does represent the author's intent accurately. More than one plausible interpretation may be offered, and in that case more than one diagram can reasonably be used to show the logical structure of that passage.

To analyze fairly, we must strive to understand the flow of the author's reasoning, and to identify the role of each element in the passage as part of that flow. The examples that follow (in which component propositions have been numbered for purposes of analysis) show the ways in which we can set forth the connections between premises and conclusions. Only after that is done, when we have identified the arguments within a passage and the relations of those arguments, can we go about deciding whether the conclusions do indeed follow from the premises affirmed.

In the following set of arguments, the final conclusion of the passage appears in the very first statement, which is not unusual. Four premises directly support this conclusion; two of these are subconclusions, which in turn are supported, in different ways, by other premises affirmed in the passage:

> ① It is very unlikely that research using animals will be unnecessary or poorly done. ② Before an experiment using a vertebrate animal is carried out, the protocol for that experiment must be reviewed by an institutional committee that includes a veterinarian and a member of the public, and ③ during the research the animal's health and care are monitored regularly. ④ Researchers need healthy animals for study in science and medicine, because ⑤ unhealthy animals could lead to erroneous results. This is a powerful incentive for ⑥ scientists to make certain that any animals they use are healthy and well nourished. Furthermore, ⑦ research involving animals is expensive, and because ⑧ funding is limited in science, ⑨ only high-quality research is able to compete effectively for support.[15]

The following diagram shows the logical structure of this passage. To "read" the diagram we replace the numbers with the indicated propositions, beginning with those highest on the page and therefore earliest in the logical cascade. We thus follow each of the several paths of reasoning to the final conclusion.

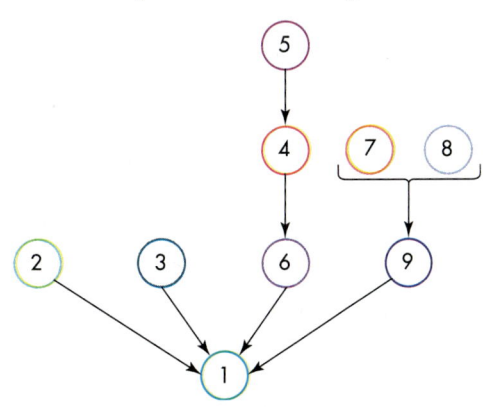

50   CHAPTER 2 Analyzing Arguments

Repetition complicates the task of analysis. Individual propositions are sometimes repeated within an argument in differently worded sentences, sometimes for emphasis and at other times by oversight. The diagram reveals this because we can assign the same number to different formulations of the same proposition. The following passage, comprising three distinct arguments, exhibits this confusing duplication of propositions:

> ① The Big Bang theory is crumbling. . . . ② According to orthodox wisdom, the cosmos began with the Big Bang—an immense, perfectly symmetrical explosion 20 billion years ago. The problem is that ③ astronomers have confirmed by observation the existence of huge conglomerations of galaxies that are simply too big to have been formed in a mere 20 billion years. . . . Studies based on new data collected by satellite, and backed up by earlier ground surveys, show that ④ galaxies are clustered into vast ribbons that stretch billions of light years, and ⑤ are separated by voids hundreds of millions of light years across. Because ⑥ galaxies are observed to travel at only a small fraction of the speed of light, mathematics shows that ⑦ such large clumps of matter must have taken at least one hundred billion years to come together—five times as long as the time since the hypothetical Big Bang. . . . ③ Structures as big as those now seen can't be made in 20 billion years. . . . ② The Big Bang theorizes that matter was spread evenly through the universe. From this perfection, ③ there is no way for such vast clumps to have formed so quickly.[16]

In this passage the premises that report observational evidence, ④, ⑤, and ⑥, give reasons for ⑦, the great length of time that would have had to elapse since the Big Bang. This passage of time is used to support the subconclusion (formulated in three slightly different ways) that ③ structures as big as those now seen are too big to have been formed in that period of time. From that subconclusion, combined with ②, a short statement (formulated in two slightly different ways) of the original symmetry and spread that the Big Bang theory supposes, we infer the final conclusion of the passage, ①: that the Big Bang theory is crumbling—the proposition with which the passage begins. The following diagram shows this set of logical relations:

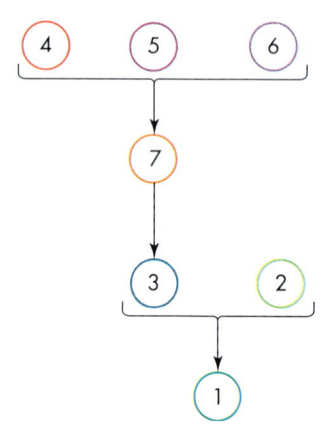

```
    4      5      6
     \     |     /
      \    |    /
           ▼
           7
           │
           ▼
        3     2
         \   /
          ▼
          1
```

The fact that a premise may appear in compressed form, sometimes as a short noun phrase, must be borne in mind. In the following argument the phrase "the scattering in the atmosphere" serves as a premise, ④, that may be reformulated as "the sun's energy is scattered in the atmosphere." This compression, along with repetition, makes it more difficult to analyze this argument:

> ① Solar-powered cars can never be anything but experimental devices. ② Solar power is too weak to power even a mini-car for daily use. ③ The solar power entering the atmosphere is about 1 kilowatt per square yard. Because of ④ the scattering in the atmosphere, and because ⑤ the sun shines half a day on the average at any place on earth, ⑥ average solar power received is 1/6 kilowatt, or 4 kilowatt hours a day.... Tests on full-size cars indicate that ⑦ 300,000 watt hours are required in a battery for an electric car to perform marginally satisfactorily. So, ⑧ 40 square yards of cells would be needed to charge the car batteries, about the size of the roof of a tractor-trailer. ① It is not undeveloped technologies that put solar power out of the running to be anything but a magnificently designed experimental car. It is cosmology.[17]

The first proposition in this passage, asserting that "solar powered cars can never be more than experimental," is the final conclusion. It is repeated in more elaborate form at the end of the passage, as a diagram of the passage shows:

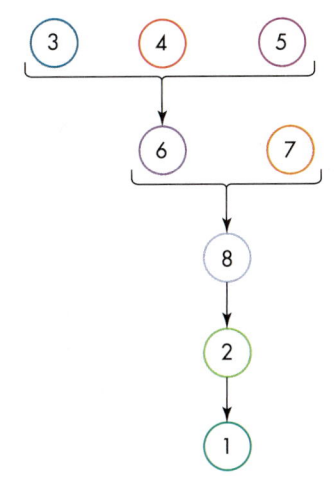

Complex argumentative passages can be entirely cogent. The following complex argument, for example, was offered by a distinguished editor in defense of her highly controversial editorial policy:

> The Journal [the New England Journal of Medicine] . . . has taken the position that ① it will not publish reports of unethical research, regardless of their scientific merit. . . . There are three reasons for our position. First, ② the policy of publishing only ethical research, if generally applied, would deter unethical work. ③ Publication is an important part of the reward system in medical research, and ④ investigators would not undertake unethical studies if they knew the results would not be published. Furthermore, ⑤ any other policy would tend to lead to more unethical work, because, as I have indicated, ⑥ such studies may be easier to carry out and thus ⑦ may give their practitioners a competitive edge. Second, ⑧ denying publication even when the ethical violations are minor protects the principle of the primacy of the research subject. ⑨ If small lapses were permitted we would become inured to them, and ⑩ this would lead to larger violations. And finally, ⑪ refusal to publish unethical work serves notice to society at large that even scientists do not consider science the primary measure of a civilization. ⑫ Knowledge, although important, may be less important to a decent society than the way it is obtained.[18]

Again, the final conclusion appears at the beginning of the passage, and the three major premises that support it directly, ②, ⑧, and ⑪, are themselves supported by various other premises arranged differently. However, each of the many propositions in the passage has a clear logical role in leading to the conclusion that the passage aims to justify: Reports of research done in unethical ways will not be published in the *New England Journal of Medicine*, regardless of their scientific merit. The following diagram shows the logical structure of this complicated but carefully reasoned passage:

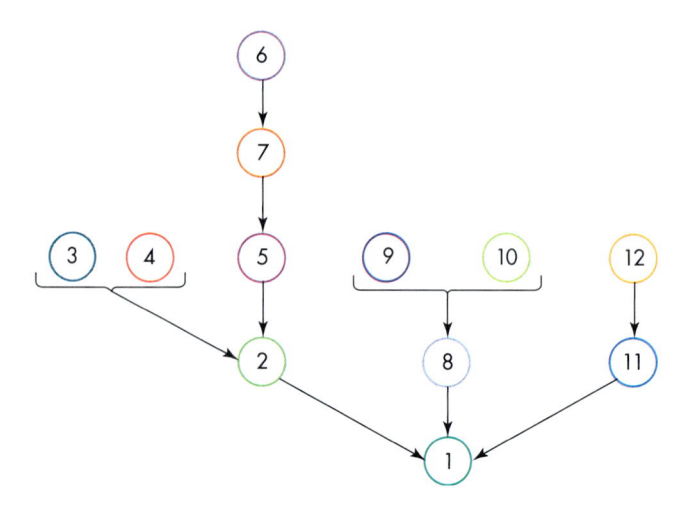

Arguments in newspaper editorials and letters-to-the-editor columns often fall short of this standard. They may include statements whose role is unclear; connections among the statements in the argument may be tangled or misstated; the flow of argument may be confused even in the mind of the author. Logical analysis, paraphrase supported by diagrams, can expose such deficiencies. By exhibiting the *structure* of a reasoning process, we can better see what its strengths and weaknesses may be. The aim and special province of logic is the evaluation of arguments, but successful evaluation presupposes a clear grasp of the structure of the argument in question.

EXERCISES

Each of the following famous passages, taken from classical literature and philosophy, comprises a set of arguments whose complicated interrelations are critical for the force of the whole. Construct for each the diagram that you would find most helpful in analyzing the flow of argument in that passage. More than one interpretation will be defensible.

   1. A question arises: whether it be better [for a prince] to be loved than feared or feared than loved? One should wish to be both, but, because it is difficult to unite them in one person, it is much safer to be feared than loved, when, of the two, one must be dispensed with. Because this is to

be asserted in general of men, that they are ungrateful, fickle, false, cowards, covetous. . . . and that prince who, relying entirely on their promises, has neglected other precautions, is ruined, because friendships that are obtained by payments may indeed be earned but they are not secured, and in time of need cannot be relied upon. Men have less scruple in offending one who is beloved than one who is feared, for love is preserved by the link of obligation which, owing to the baseness of men, is broken at every opportunity for their advantage; but fear preserves you by a dread of punishment which never fails.

—Niccolò Machiavelli, *The Prince*, 1515

2. Democratic laws generally tend to promote the welfare of the greatest possible number; for they emanate from the majority of the citizens, who are subject to error, but who cannot have an interest opposed to their own advantage. The laws of an aristocracy tend, on the contrary, to concentrate wealth and power in the hands of the minority; because an aristocracy, by its very nature, constitutes a minority. It may therefore be asserted, as a general proposition, that the purpose of a democracy in its legislation is more useful to humanity than that of an aristocracy.

—Alexis de Tocqueville, *Democracy in America*, 1835

3. ". . . You appeared to be surprised when I told you, on our first meeting, that you had come from Afghanistan."
"You were told, no doubt."
"Nothing of the sort. I *knew* you came from Afghanistan. From long habit the train of thoughts ran so swiftly through my mind that I arrived at the conclusion without being conscious of intermediate steps. There were such steps, however. The train of reasoning ran, 'Here is a gentleman of medical type, but with the air of a military man. Clearly an army doctor, then. He has just come from the tropics, for his face is dark, and that is not the natural tint of his skin, for his wrists are fair. He has undergone hardship and sickness, as his haggard face says clearly. His left arm has been injured. He holds it in a stiff and unnatural manner. Where in the tropics could an English army doctor have seen much hardship and got his arm wounded? Clearly in Afghanistan.' The whole train of thought did not occupy a second. I then remarked that you came from Afghanistan, and you were astonished."
"It is simple enough as you explain it," I said, smiling.

—A. Conan Doyle, *A Study in Scarlet*, 1887

4. Nothing is demonstrable unless the contrary implies a contradiction. Nothing that is distinctly conceivable implies a contradiction. Whatever we conceive as existent, we can also conceive as nonexistent. There is no being, therefore, whose nonexistence implies a contradiction. Consequently there is no being whose existence is demonstrable.

—David Hume, *Dialogues Concerning Natural Religion, Part IX*, 1779

Challenge to the Reader

In the *Ethics* (1677), Baruch Spinoza, one of the most influential of all modern thinkers, presents a deductive philosophical system in which the central conclusions—about God, about nature, and about human life and human freedom—are demonstrated in "geometrical" fashion. Here follows an example. Proposition 29 of the first book of the *Ethics* (there are five books in all) reads:

> In nature there is nothing contingent, but all things are determined from the necessity of the divine nature to exist and act in a certain manner.

Immediately after the statement of each proposition in the *Ethics* appears its proof. The proof of Prop. 29 (from which internal references to proofs given earlier in the same work have been omitted for the sake of clarity) appears immediately below. Analyze this proof, by constructing a diagram that shows the structure of the argument, or by paraphrasing it in a way that makes it clear and persuasive to a modern reader.

> Whatever is, is in God. But God cannot be called a contingent thing, for He exists necessarily and not contingently. Moreover, the modes of the divine nature [the creations which depend on, or have been created by, God immediately] have followed from it necessarily and not contingently. . . . But God is the cause of these modes not only in so far as they simply exist, but also in so far as they are considered as determined to any action. If they are not determined by God it is an impossibility and not a contingency that they should determine themselves; and, on the other hand, if they are determined by God it is an impossibility and not a contingency that they should render themselves indeterminate. Wherefore all things are determined from a necessity of the divine nature, not only to exist, but to exist and act in a certain manner, and there is nothing contingent.

2.4 Problems in Reasoning

In reasoning we advance from premises known (or affirmed for the purpose) to conclusions. We construct arguments of our own every day, in deciding how we shall act, in judging the conduct of others, in defending our moral or political convictions, and so on. Skill in devising good arguments (and in deciding whether a proffered argument is good) is of enormous value, and this skill can be improved with practice. Ancient games of reasoning, such as chess and go, exercise that skill, and there are some widely known commercial games (Clue and Mastermind are examples) that also have this merit.

Problems may be contrived which are designed to test and strengthen logical skills; some of these are presented in this section. Such problems are far neater than those that arise in real life, of course. But solving them may require extended reasoning in patterns not very different from those employed by a detective, a journalist, or a juror. Chains of inferences will be needed, in which subconclusions are used as premises in subsequent arguments. Finding the solution may require the creative recombination of information given earlier or discovered. Contrived problems can prove frustrating—but solving them, like every successful application of reasoning, is quite satisfying. In addition to being models for the employment of reason, logical games and puzzles are good fun. "The enjoyment of the doubtful," wrote the philosopher John Dewey, "is a mark of the educated mind."

One type of reasoning problem is the common brainteaser in which, using only the clues provided, we must determine the names or other facts about certain specified characters. Here is a simple example:

> In a certain flight crew, the positions of pilot, copilot, and flight engineer are held by three persons, Allen, Brown, and Carr, though not necessarily in that order. The copilot, who is an only child, earns the least. Carr, who married Brown's sister, earns more than the pilot. What position does each of the three persons hold?

To solve such problems we look first for a sphere in which we have enough information to reach some conclusions going beyond what is given in the premises. In this case we know most about Carr: he is not the pilot, because he earns more than the pilot; and he is not the copilot because the copilot earns the least. By elimination we may infer that Carr must be the flight engineer. Using that subconclusion we can determine Brown's position. Brown is not the copilot because he has a sister and the copilot is an only child; he is not the flight engineer because Carr is. Brown must therefore be the pilot. Allen, the only one left, must therefore be the copilot.

When problems of this type become more complex, it is useful to construct a graphic display of the alternatives, called a *matrix*, which we fill in as we accumulate new information. The helpfulness of such a matrix will be seen in solving the following problem:

> Alonzo, Kurt, Rudolf, and Willard are four creative artists of great talent. One is a dancer, one is a painter, one is a singer, and one is a writer, though not necessarily in that order.
>
> 1. Alonzo and Rudolf were in the audience the night the singer made his debut on the concert stage.
> 2. Both Kurt and the writer have had their portraits painted from life by the painter.
> 3. The writer, whose biography of Willard was a best-seller, is planning to write a biography of Alonzo.
> 4. Alonzo has never heard of Rudolf.
>
> What is each man's artistic field?

To remember the facts asserted in these premises, as well as the subconclusions that may be inferred from them, would be a demanding task. Written notes could become a confusing clutter. We need a method for storing and exhibiting the information given and the intermediate conclusions drawn, keeping it all available for use as the number of inferences increases and the chain of arguments lengthens. The matrix we construct allows us to represent all the relevant possibilities and to record each inference drawn.

For this problem the matrix must display an array of the four persons (in four rows) and the four artistic professions (in four columns) that they hold. It would look like this:

	Dancer	Painter	Singer	Writer
Alonzo				
Kurt				
Rudolf				
Willard				

When we conclude that one of those individuals (named at the left of one of the rows) cannot be the artist whose profession is at the top of one of the columns, we write an N (for "No") in the box to the right of that person's name and in the column headed by that profession. We can immediately infer, from premise (1), that neither Alonzo nor Rudolph is the singer, so we place an N to the right of their names, in the third (singer) column. We can infer from premise (2) that Kurt is neither the painter nor the writer, so we enter an N to the right of his name in the second (painter) and the fourth (writer) columns. From premise (3) we see that the writer is neither Alonzo nor Willard, so we enter an N to the right of their names in the fourth column. The entries we have made thus far are all justified by the information given originally, and our matrix now looks like this:

	Dancer	Painter	Singer	Writer
Alonzo			N	N
Kurt		N		N
Rudolf			N	
Willard				N

From the information now clearly exhibited, we can conclude by elimination that Rudolf must be the writer, so we enter a Y (for "Yes") in the box to the right of Rudolf's name in the fourth (writer) column, and we place an N in the other boxes to the right of his name. The array now makes it evident that the painter must be either Alonzo or Willard, and we can eliminate Alonzo in this way: Rudolf had his portrait painted by the painter (from premise 2), and Alonzo has never heard of Rudolf (from premise 4)—therefore Alonzo cannot be the painter. So we enter an N to the right of Alonzo's name under column 2 (painter). We may conclude that Alonzo must be the dancer, so we enter a Y to the right of Alonzo's name in the first (dancer) column. In that same column we can now enter an N for both Kurt and Willard. The only possible category remaining for Kurt is singer, and therefore we enter a Y in that box for him, and an N in the singer column for Willard. By elimination, we conclude that Willard must be the painter and put a Y in the last empty box in the matrix. Our completed graphic display looks like this:

	Dancer	Painter	Singer	Writer
Alonzo	Y	N	N	N
Kurt	N	N	Y	N
Rudolf	N	N	N	Y
Willard	N	Y	N	N

Our matrix now filled in, the full solution is evident: Alonzo is the dancer; Kurt is the singer; Rudolf is the writer; Willard is the painter.

Some brainteasers of this kind, requiring solutions on several dimensions, are very challenging and almost impossible to solve without using a matrix.

In the real world, we are often called upon to reason from some present state of affairs to its causes, from what is to what was. Scientists—especially archeologists, geologists, astronomers, and physicians—commonly confront events or conditions whose origins are problematic. Reasoning that seeks to explain how things must have developed from what went before is called **retrograde analysis**. For example, to the amazement of astronomers, comet Hyakutake, streaking by the earth in 1996, was found to be emitting variable X-rays a hundred times stronger than anyone had ever predicted a comet might emit. A comet expert at the Max Planck Institute in Germany remarked, "We have our work cut out for us in explaining these data—but that's the kind of problem you love to have."

We do love to have them, and for that reason problems in retrograde analysis are often devised for amusement. In the real world, logical problems arise within a theoretical framework that is supplied by scientific or historical knowledge; but in contrived problems that framework must be provided by the problem itself. Some rules or laws must be set forth within which logical analysis can proceed. The chessboard is the setting for the most famous of all problems in retrograde analysis; the rules of chess provide the needed theoretical context. No skill in playing chess is required, but readers who are not familiar with the rules of chess may skip the illustration that follows.

Retrograde problems in chess commonly take this form: An arrangement of pieces on the chessboard is given; it was reached in a game of chess in which all the rules of the game were obeyed. What move, or series of moves, has just been completed? An example of such a problem follows. The diagram presents a position reached in an actual game of chess, all moves in that game having been made in accordance with the rules of chess. The black king has just moved.

Retrograde analysis
Reasoning that seeks to explain how things must have developed from what went before.

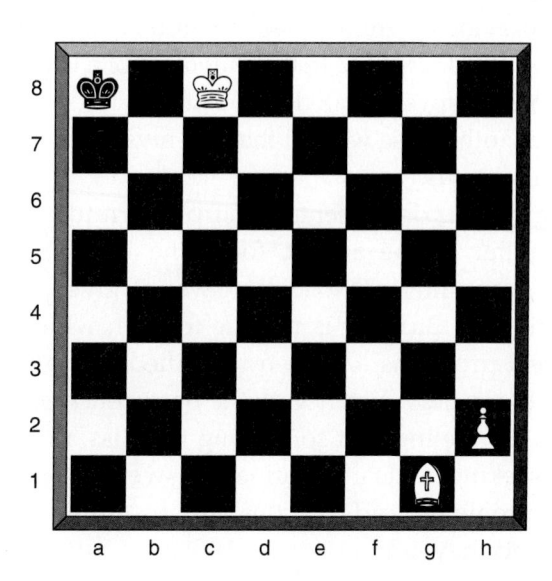

For the purpose of analysis, the rows are numbered from bottom to top, 1 to 8, and the columns are lettered from left to right, a to h. Each square on the board can then be identified by a unique letter-number combination: The black king is on a8, the white pawn on h2, and so on. The problem is this: The last move was made by black. What was that move? And what was white's move just before that? Can you reason out the solution before reading the next paragraph?

Solution: Because the two kings may never rest on adjacent squares, the black king could not have moved to its present position from b7 or from b8; therefore we may be certain that the black king has moved from a7, where it was in check.

That much is easily deduced. But what preceding white move could have put the black king in check? No move by the white bishop (on g1) could have done it, because there would have been no way for that bishop to move to that square, g1, without the black king having been in check with white to move. Therefore it must be that the check was *discovered* by the movement of a white piece that had been blocking the bishop's attack and was captured by the black king on its move to a8. What white piece could have been on that black diagonal and moved from there to the white square in the corner? Only a knight that had been on b6. We may therefore be certain that before black's last move (the black king from a7 to a8), white's last move was that of a white knight from b6 to a8.[19]

Problems of reasoning that confront us in the real world are rarely this tidy. Many real problems are not described accurately, and their misdescription may prove so misleading that no solution can be reached. In cases of that kind, some part or parts of the description of the problem need to be rejected or replaced. However, we cannot do this when we are seeking to solve logical puzzles of the sort presented here.

In the real world, moreover, even when they are described accurately, problems may be incomplete in that something not originally available may be essential for the solution. The solution may depend on some additional scientific discovery, or some previously unimagined invention or equipment, or the search of some as-yet-unexplored territory. In the statement of a logical puzzle, as in the writing of a good murder mystery, all the information that is sufficient for the solution must be given; otherwise we feel that the mystery writer, or the problem maker, has been unfair to us.

Finally, the logical puzzle presents a sharply formulated question (for example, which member of the artistic foursome is the singer? What were black's and white's last moves?) whose answer, if given and proved, solves the problem definitively. But that is not the form in which many real-world problems arise. Real problems are often identified, initially at least, only by the recognition of some inconsistency or the occurrence of an unusual event, or perhaps just by the feeling that something is amiss, rather than by a well-formed question seeking a clearly defined answer. In spite of these differences, contrived problems and puzzles are useful in strengthening our reasoning skills—and they are fun.

EXERCISES

The following problems require reasoning for their solution. To prove that an answer is correct requires an argument (often containing subsidiary arguments) whose premises are contained in the statement of the problem—and whose final conclusion is the answer to it. If the answer is correct, it is possible to construct a valid argument proving it. In working these problems, readers are urged to concern themselves not merely with discovering the answers but also with formulating arguments to prove that those answers are correct.

1. In a certain mythical community, politicians never tell the truth, and non-politicians always tell the truth. A stranger meets three natives and asks the first of them, "Are you a politician?" The first native answers the question. The second native then reports that the first native denied being a politician. The third native says that the first native is a politician.

    How many of these three natives are politicians?

2. Of three prisoners in a certain jail, one had normal vision, the second had only one eye, and the third was totally blind. The jailor told the prisoners that, from three white hats and two red hats, he would select three and put them on the prisoners' heads. None could see what color hat he wore. The jailor offered freedom to the prisoner with normal vision if he could tell what color hat he wore. To prevent a lucky guess, the jailor threatened execution for any incorrect answer. The first prisoner could not tell what hat he wore. Next the jailor made the same offer to the one-eyed prisoner. The second prisoner could not tell what hat he wore either. The jailor did not bother making the offer to the blind prisoner, but he agreed to extend the same terms to that prisoner when he made the request. The blind prisoner said:

    > I do not need to have my sight;
    > From what my friends with eyes have said,
    > I clearly see my hat is _____!

    How did he know?

3. On a certain train, the crew consists of the brakeman, the fireman, and the engineer. Their names, listed alphabetically, are Jones, Robinson, and Smith. On the train are also three passengers with corresponding names, Mr. Jones, Mr. Robinson, and Mr. Smith. The following facts are known:
    a. Mr. Robinson lives in Detroit.
    b. The brakeman lives halfway between Detroit and Chicago.
    c. Mr. Jones earns exactly $40,000 a year.
    d. Smith once beat the fireman at billiards.
    e. The brakeman's next-door neighbor, one of the three passengers mentioned, earns exactly three times as much as the brakeman.
    f. The passenger living in Chicago has the same name as the brakeman.

    What is the engineer's name?

4. The employees of a small loan company are Mr. Black, Mr. White, Mrs. Coffee, Miss Ambrose, Mr. Kelly, and Miss Earnshaw. The positions they occupy are manager, assistant manager, cashier, stenographer, teller, and clerk, though not necessarily in that order. The assistant manager is the manager's grandson, the cashier is the stenographer's son-in-law, Mr. Black is a bachelor, Mr. White is twenty-two years old, Miss Ambrose is the teller's stepsister, and Mr. Kelly is the manager's neighbor.
   Who holds each position?

5. Benno Torelli, genial host at Miami's most exclusive nightclub, was shot and killed by a racketeer gang because he fell behind in his protection payments. After considerable effort on the part of the police, five suspects were brought before the district attorney, who asked them what they had to say for themselves. Each of them made three statements, two true and one false. Their statements were

   Lefty: I did not kill Torelli. I never owned a revolver in all my life. Spike did it.

   Red: I did not kill Torelli. I never owned a revolver. The others are all passing the buck.

   Dopey: I am innocent. I never saw Butch before. Spike is guilty.

   Spike: I am innocent. Butch is the guilty one. Lefty did not tell the truth when he said I did it.

   Butch: I did not kill Torelli. Red is the guilty one. Dopey and I are old pals.

   Whodunnit?

6. Mr. Short, his sister, his son, and his daughter are fond of golf and often play together. The following statements are true of their foursome:

   a. The best player's twin and the worst player are of the opposite sex.
   b. The best player and the worst player are the same age.

   Which one of the foursome is the best player?

7. Daniel Kilraine was killed on a lonely road, 2 miles from Pontiac, Michigan, at 3:30 A.M. on March 17 of last year. Otto, Curly, Slim, Mickey, and the Kid were arrested a week later in Detroit and questioned. Each of the five made four statements, three of which were true and one of which was false. One of these persons killed Kilraine.
   Their statements were

   Otto: I was in Chicago when Kilraine was murdered. I never killed anyone. The Kid is the guilty one. Mickey and I are pals.

Curly: I did not kill Kilraine. I never owned a revolver in my life. The Kid knows me. I was in Detroit the night of March 17.

Slim: Curly lied when he said he never owned a revolver. The murder was committed on St. Patrick's Day. Otto was in Chicago at this time. One of us is guilty.

Mickey: I did not kill Kilraine. The Kid has never been in Pontiac. I never saw Otto before. Curly was in Detroit with me on the night of March 17.

The Kid: I did not kill Kilraine. I have never been in Pontiac. I never saw Curly before. Otto erred when he said I am guilty.

Whodunnit?

8. Six balls confront you. Two are red; two are green; two are blue. You know that in each color pair, one ball is heavier than the other. You also know that all three of the heavier balls weigh the same, as do all three of the lighter balls. The six balls (call them R1, R2, G1, G2, B1, and B2) are otherwise indistinguishable. You have only a balance scale; if equal weights are placed on the two sides of your scale, they will balance; if unequal weights are placed on the two sides, the heavier side will go down. With no more than two weighings on that balance scale, how can you identify the heavier and the lighter balls in all three pairs?

9. In the same mythical community described in Exercise 1, a stranger meets three other natives and asks them, "How many of you are politicians?" The first native replies, "We are all politicians." The second native says, "No, just two of us are politicians." The third native then says, "That isn't true either."

    Is the third native a politician?

10. Imagine a room with four walls, with a nail placed in the center of each wall, as well as in the ceiling and floor, six nails in all. The nails are connected to each other by strings, each nail connected to every other nail by a separate string. These strings are of two colors, red or blue, and of no other color. All these strings obviously make many triangles, because any three nails may be considered the apexes of a triangle.

    Can the colors of the strings be distributed so that no one triangle has all three sides (strings) of the same color? If so, how? And if not, why not?

Challenge to the Reader

Here is a final reasoning problem whose solution requires the construction of a set of sustained arguments. It isn't easy—but solving it is well within your power and will give you great pleasure.

You are presented with a set of twelve metal balls, apparently identical in every respect: size, color, and so on. In fact, eleven of them are identical, but one of them is "odd": It differs from all the rest in weight only; it is either heavier, or lighter, than all the others. You are given a balance scale, on which the balls can be weighed against one another. If the same number of balls are put on each side of the balance, and the "odd" ball is on one side, that side will go down if the odd ball is heavier, or up if the odd ball is lighter; the two sides will balance if the odd ball is not among those weighed and the same number of balls are placed on each side. You are allowed three weighings only; any removal or addition of a ball constitutes a separate weighing.

Your challenge is this: Devise a set of three weighings that will enable you to identify the odd ball wherever it may lie in a random mixing of the twelve balls, *and* that will enable you to determine whether the odd ball is heavier or lighter than the rest. ●

chapter 2 Summary

In this chapter we have discussed techniques for the analysis of arguments, and some of the difficulties confronted in that process.

In Section 2.1 we explained the paraphrasing of an argumentative passage, in which the essential propositions may be reworded (or supplied if they are assumed but missing), and in which premises and conclusions are put into the most intelligible order.

In Section 2.2 we explained the diagramming of an argument, in which the propositions of an argument are represented by numbers, and the relations of the premises and conclusions are then exhibited graphically in two dimensions, by showing on a page the relations of those numbered propositions.

In Section 2.3 we discussed complex argumentative passages, in which the conclusions of subarguments may serve as premises for further arguments, and whose complete analysis generally requires an intricate diagram or an extensive paraphrase.

In Section 2.4 we discussed contrived problems of reasoning, which often mirror the complexities confronted by many different kinds of investigation in real life, and whose solutions require the construction of extended sets of arguments and subarguments.

END NOTES

[1] *Lawrence v Texas*, 539 Us558 (2003).

[2] This technique was first developed and perfected decades ago by several distinguished logicians: Monroe C. Beardsley, in *Practical Logic* (Englewood Cliffs, NJ: Prentice Hall, 1950); Stephen N. Thomas, in *Practical Reasoning in Natural Language* (Englewod Cliffs, NJ: Prentice Hall, 1973); and Michael Scriven, in *Reasoning* (New York: McGraw-Hill, 1976). We follow their lead.

[3] John Blass "Stem Cell Research: When Does Life Begin?" *The New York Times*, 25 January 2007.

[4] Adam Gopnick, "The Unbeautiful Game," *The New Yorker*, 8 January 2007.

[5] James Surowieki, "Dealer's Choice," *The New Yorker*, 4 September 2006.

[6] Blanchard Hiatt, *University of Michigan Research News*, September 1979.

[7] Robert Alter, "Neocon or Not?" *The New York Times Book Review*, 25 June 2006.

[8] Karl Marx, Letter #141, 9 April 1870, in *Karl Marx and Friedrich Engels Correspondence*, 1846–1895 (New York: International Publishers, 1936).

[9] Boston Women's Health Book Collective, *Our Bodies, Ourselves* (New York: Simon & Schuster, 1984).

[10] C. A. Quadir, *Philosophy and Science in the Islamic World* (London: Croom Helm, 1988).

[11] Thomas Aquinas, *Summa Theologiae*, I, Question 96, Article 2, circa 1265.

[12] Ibid., Article 3.

[13] *Science*, 26 May 1995.

[14] Paraphrased in part from Andres Duany, Elizabeth Plater-Zyberk, and Jeff Speck, *Suburban Nation: The Rise of Sprawl and the Decline of the American Dream* (New York: North Point Press, 2000).

[15] *Science, Medicine, and Animals* (Washington, DC: National Academy of Sciences, 1991).

[16] Eric J. Lerner, "For Whom the Bang Tolls," *The New York Times*, 2 June 1991.

[17] Victor Wouk, "You Can't Drive Solar Cars to Work," *The New York Times,* 15 July 1991.

[18] Marcia Angell, "The Nazi Hypothermia Experiments and Unethical Research Today," *New England Journal of Medicine*, 17 May 1990.

[19] Readers who find retrograde analysis enjoyable will take delight in a collection of such problems, compiled by the logician Raymond Smullyan, and entitled *The Chess Mysteries of Sherlock Homes* (New York: Alfred A. Knopf, 1979).

For additional exercises and tutorials about concepts covered in this chapter, log in to MyLogicLab at *www.mylogiclab.com* and select your current textbook.

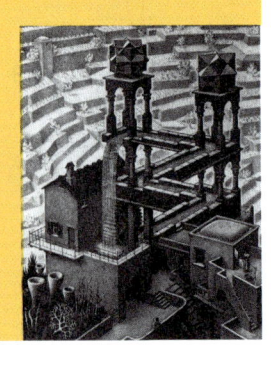

Language and Definitions

3.1 Language Functions
3.2 Emotive Language, Neutral Language, and Disputes
3.3 Disputes and Ambiguity
3.4 Definitions and Their Uses
3.5 The Structure of Definitions: Extension and Intension
3.6 Definition by Genus and Difference

3.1 Language Functions

When people reason, they typically do so using language, manipulating propositions in a logical or *informative* spirit. But language is used in a great variety of ways, only some of which are informative. Without the intention to inform, we may *express* ourselves using language: "That's really great!" we may say; and the poet, overcome by the beauty of an ancient city, channels his emotions in writing these lines:

> Match me such marvel, save in Eastern clime—
> A rose-red city—"half as old as time."[1]

Of course, some expressive discourse also has informative content, and may express attitudes as well as beliefs.

> Grow old along with me!
> The best is yet to be,
> The last of life for which the first was made.[2]

Moreover, some discourse is *directive*, with or without expressive or informative elements. It seeks to guide or to command. "Step on the scale, please," we may be told, or we may receive this good advice:

Drive defensively. The cemetery is full of law-abiding citizens who had the right of way.

A mixture of functions is a natural feature of almost all our uses of language. We can see this in our own speech and writing. *Emotive* language may be used to advance our purposes in directing others: "That conduct is utterly disgusting!" says parent to child, expressing an attitude, seeking to direct behavior, and (with those same words) probably reporting a fact. We may say that language has three major functions:

1. *Informative*
2. *Expressive*

3. *Directive*

   To these we may add less common types of use:

4. *Ceremonial* language (as when we say, "How do you do?" upon being introduced to a stranger), in which words may combine expressive and other functions; and

5. *Performative* language (as when we say, "I apologize for my foolish remark"), in which words themselves serve, when spoken or written, to perform the function they announce. Other examples are "I congratulate you, . . ." "I accept your offer, . . ." and "I promise you that. . . ."

Logicians are chiefly concerned with language used informatively–affirming or denying propositions, formulating or evaluating arguments, and so on. In reasoning it is this informative function of language that is the principal concern.

In this informative mode we can distinguish between facts a sentence formulates and facts about the speaker who formulates them. If someone says, "War is always the wrong solution to international conflict," that may indeed be true, but it is also evidence of the *beliefs* of the person who utters that remark. When someone says, "I strongly oppose our involvement in this war on moral grounds," that is a statement (very probably true) about the speaker, but it also serves to express a judgment about the morality of the war under discussion. To open an argument with a statement of one's own views is by no means deceptive; it is one of the common ways in which judgment and biographical report are appropriately integrated.

The *uses* of language must be distinguished from the *forms* of language. The several uses of language (informative, expressive, etc.) are implemented using different forms. Sentences (the units of our language that express complete thoughts) may be *declarative* in form, or *exclamatory*, or *imperative*, or *interrogative*. When we are reasoning our sentences are usually declarative. When we are expressing emotion our sentences (e.g., "That's fantastic!") are often exclamatory. When we are seeking to direct conduct our sentences (e.g., "Take off your shirt!") are likely to be imperative in form—but there is no strict correlation between function and form.

For example, we noted earlier that a premise may be affirmed by asking a rhetorical question. The devout believer asks in prayer, "Who is like unto Thee?"—but it is plain that this interrogative expresses a religious belief. When one responds in a conversation, "What can you possibly mean by that?" a skeptical attitude is very plainly being expressed. Similarly, a directive function may be served by reporting a fact in apparently declarative mode, as when we urge a companion to move more quickly by saying, "It is very late; we are running short of time." And the exclamation, "What lovely flowers!" uttered by a young woman to her gentleman friend as they pass a florist's window, may be intended to function more directively than expressively.

The combination of functions can create a kind of dissonance, even at times leading to troubling controversy. Here is a famous example: During the Vietnam War, a young man protesting the military draft was arrested in the Los Angeles County Courthouse for wearing a jacket on which a deliberate obscenity was emblazoned. He was convicted of "offensive conduct" under the California penal code. His conviction was reversed, however, by the Supreme Court of the United States, whose majority recognized that there was in this case a tension between the expressive spirit of his language and the informative function of his protest, the latter being protected by the First Amendment of the U.S. Constitution. Justice John Harlan wrote:

> [M]uch linguistic expression serves a dual communicative function: it conveys not only ideas capable of relatively precise, detached explication, but otherwise inexpressible emotions as well. In fact, words are often chosen as much for their emotive as their cognitive force. We cannot sanction the view that the Constitution, while solicitous of the cognitive content of individual speech, has little or no regard for that emotive function which, practically speaking, may often be the more important element of the message sought to be communicated. . . . and in the same vein, we cannot indulge the facile assumption that one can forbid particular words without also running a substantial risk of suppressing ideas in the process.[3]

The emotional offensiveness of some words may (in some contexts) be overridden by their more important informative function. Being sensitive to the flexibility of language, and recognizing the different functions served by language in a given context, are necessary precursors to the application of the logical analysis that is our central concern in this book.

It would be convenient if a given function were invariably executed using language in some specific grammatical form, but that is simply not the case. Language is too loose, and its uses too variable to expect that. In determining the real function of a sentence, therefore, context is always critical.

In summary, the principal uses of language are three: informative, expressive, and directive. The grammatical forms of language are essentially four: declarative, interrogative, imperative, and exclamatory. There is no sure connection between the grammatical form of a passage and the use or uses its author intends. Language that serves any one of the three principal functions may take any one of the four grammatical forms.

EXERCISES

A. Which of the various functions of language are exemplified by each of the following passages?

1. Check the box on line 6a unless your parent (or someone else) can claim you as a dependent on his or her tax return.

—U.S. Internal Revenue Service, "Instructions,"
Form 1040, 2006

2. 'Twas brillig, and the slithy toves
   Did gyre and gimble in the wabe;
   All mimsy were the borogoves,
   And the mome raths outgrabe.

   —Lewis Carroll, *Through the Looking-Glass*, 1871

3. What traveler among the ruins of Carthage, of Palmyra, Persepolis, or Rome, has not been stimulated to reflections on the transiency of kingdoms and men, and to sadness at the thought of a vigorous and rich life now departed . . . ?

   —G. W. F. Hegel, *Lectures on the Philosophy of History*, 1823

4. Moving due south from the center of Detroit, the first foreign country one encounters is not Cuba, nor is it Honduras or Nicaragua or any other Latin American nation; it is Canada.

5. I was a child and she was a child,
   In this kingdom by the sea,
   But we loved with a love that was more than love—
   I and my Annabel Lee—

   —Edgar Allan Poe, "Annabel Lee," 1849

6. Reject the weakness of missionaries who teach neither love nor brotherhood, but chiefly the virtues of private profit from capital, stolen from your land and labor. Africa awake, put on the beautiful robes of Pan-African Socialism!

   —W. E. B. Dubois, "Pan-Africa," 1958

7. If I speak in the tongues of men and of angels, but have not love, I am a noisy gong or a clanging cymbal.

   —I Cor. 13:1

8. I herewith notify you that at this date and through this document I resign the office of President of the Republic to which I was elected.

   —President Fernando Collor De Mello, in a letter to the Senate of Brazil, 29 December 1992

9. American life is a powerful solvent. It seems to neutralize every intellectual element, however tough and alien it may be, and to fuse it in the native good will, complacency, thoughtlessness, and optimism.

   —George Santayana, *Character and Opinion in the United States*, 1934

10. The easternmost point of land in the United States—as well as the northernmost point and the westernmost point—is in Alaska.

B. What language functions are most probably *intended* to be served by each of the following passages?

1. There is no caste here. Our Constitution is color-blind, and neither knows nor tolerates classes among citizens. In respect of civil rights, all citizens are equal before the law. The humblest is the peer of the most powerful.
   —Justice John Harlan, dissenting in *Plessy v. Ferguson*, 163 U.S. 537, 1896

2. Judges do not know how to rehabilitate criminals—because no one knows.
   —Andrew Von Hirsch, *Doing Justice—The Choice of Punishment* (New York: Hill & Wang, 1976)

3. When tillage begins, other arts follow. The farmers therefore are the founders of human civilization.
   —Daniel Webster, "On Agriculture," 1840

4. The only thing necessary for the triumph of evil is for good men to do nothing.
   —Edmund Burke, letter to William Smith, 1795

5. They have no lawyers among them, for they consider them as a sort of people whose profession it is to disguise matters.
   —Sir Thomas More, *Utopia*, 1516

6. White society is deeply implicated in the ghetto. White institutions created it, white institutions maintain it, and white society condones it.
   —The National Commission on Civil Disorders (Kerner Commission), 1968

7. The bad workmen who form the majority of the operatives in many branches of industry are decidedly of the opinion that bad workmen ought to receive the same wages as good.
   —John Stuart Mill, *On Liberty*, 1859

8. War is the greatest plague that can afflict humanity; it destroys religion, it destroys states, it destroys families. Any scourge is preferable to it.
   —Martin Luther, *Table Talk*, 1566

9. Human history becomes more and more a race between education and catastrophe.
   —H. G. Wells, *The Outline of History*, 1920

10. The man who insists upon seeing with perfect clearness before he decides, never decides.
    —Henri-Frédéric Amiel, *Amiel's Journal*, 1885

11. Among other evils which being unarmed brings you, it causes you to be despised.
    —Niccolò Machiavelli, *The Prince*, 1515

12. Eternal peace is a dream, and not even a beautiful one. War is a part of God's world order. In it are developed the noblest virtues of man: courage and abnegation, dutifulness and self-sacrifice. Without war the world would sink into materialism.
    —Helmuth von Moltke, 1892

13. Language! the blood of the soul, sir, into which our thoughts run, and out of which they grow.
    —Oliver Wendell Holmes, *The Autocrat of the Breakfast-Table*, 1858

14. Over the past 133 years, more than 7,500 scientists, including social scientists, have been elected to the National Academy of Sciences. It appears that only three of them have been black.
    —*The Journal of Blacks in Higher Education*, Summer 1996

15. A little philosophy inclineth man's mind to atheism; but depth in philosophy bringeth man's mind about to religion.
    —Francis Bacon, *Essays*, 1601

16. You'll never have a quiet world until you knock the patriotism out of the human race.
    —George Bernard Shaw, *O'Flaherty, V.C.*, 1915

17. If [he] does really think that there is no distinction between virtue and vice, why, sir, when he leaves our houses let us count our spoons.
    —Samuel Johnson, 1763

18. Man scans with scrupulous care the character and pedigree of his horses, cattle, and dogs before he matches them; but when he comes to his own marriage he rarely, or never, takes any such care.
    —Charles Darwin, *The Descent of Man*, 1871

19. The story of the whale swallowing Jonah, though a whale is large enough to do it, borders greatly on the marvelous; but it would have approached nearer to the idea of miracle if Jonah had swallowed the whale.
    —Thomas Paine, *The Age of Reason*, 1796

20. The notion of race is the hydra-headed monster which stifles our most beautiful dreams before they are fairly dreamt, calling us away from the challenges of normal human interaction to a dissonance of suspicion and hatred in pursuit of a fantasy that never was.
    —C. Eric Lincoln, *Coming Through the Fire*
    (Durham, NC: Duke University Press, 1996)

C. For the following passages, indicate what propositions they may be intended to assert, if any; what overt actions they may be intended to cause, if any; and what they may be regarded as providing evidence for about the speaker, if anything.

1. I will not accept if nominated and will not serve if elected.
   —William Tecumseh Sherman, message to the Republican National Convention, 1884

2. The government in its wisdom considers ice a "food product." This means that Antarctica is one of the world's foremost food producers.
   —George P. Will

3. Mankind has grown strong in eternal struggles and it will only perish through eternal peace.
   —Adolf Hitler, *Mein Kampf*, 1925

4. Without music, earth is like a barren, incomplete house with the dwellers missing. Therefore the earliest Greek history and Biblical history, nay the history of every nation, begins with music.
—Ludwig Tieck, quoted in Paul Henry Lang, *Music in Western Civilization* (New York: W. W. Norton, 1941)

5. Research is fundamentally a state of mind involving continual reexamination of doctrines and axioms upon which current thought and action are based. It is, therefore, critical of existing practices.
—Theobald Smith, *American Journal of Medical Science*, 1929

6. I have tried sedulously not to laugh at the acts of man, nor to lament them, nor to detest them, but to understand them.
—Baruch Spinoza, *Tractatus Theologico-politicus*, 1670

7. Of what use is political liberty to those who have no bread? It is of value only to ambitious theorists and politicians.
—Jean-Paul Marat, *L'Ami du peuple*, 1789

8. While there is a lower class I am in it, while there is a criminal element I am of it, and while there is a soul in prison I am not free.
—Eugene Debs, 1918

9. If there were a nation of gods they would be governed democratically, but so perfect a government is not suitable to men.
—Jean-Jacques Rousseau, *The Social Contract*, 1762

10. There are three classes of citizens. The first are the rich, who are indolent and yet always crave more. The second are the poor, who have nothing, are full of envy, hate the rich, and are easily led by demagogues. Between the two extremes lie those who make the state secure and uphold the laws.
—Euripides, *The Suppliant Women*

11. I am convinced that turbulence as well as every other evil temper of this evil age belongs not to the lower but to the middle classes—those middle classes of whom in our folly we are so wont to boast.
—Lord Robert Cecil, *Diary in Australia*, 1852

12. God will see to it that war shall always recur, as a drastic medicine for ailing humanity.
—Heinrich von Treitschke, *Politik*, 1916

13. I would rather that the people should wonder why I wasn't President than why I am.
—Salmon P. Chase, at the Republican National Convention, 1860

14. He [Benjamin Disraeli] is a self-made man, and worships his creator.
—John Bright, 1882

15. We hear about constitutional rights, free speech and the free press. Every time I hear these words I say to myself, "That man is a Red, that man is a Communist." You never heard a real American talk in that manner.
—Frank Hague, speech before the Jersey City Chamber of Commerce, 12 January 1938

16. Even a fool, when he holdeth his peace, is counted wise: And he that shutteth his lips is esteemed a man of understanding.

—Prov. 17:28

17. A word fitly spoken is like apples of gold in ornaments of silver.

—Prov. 25:11

18. I have sworn upon the altar of God eternal hostility against every form of tyranny over the mind of man.

—Thomas Jefferson, 1800

19. A free man thinks of nothing less than of death, and his wisdom is not a meditation upon death but upon life.

—Baruch Spinoza, *Ethics*, 1677

20. I have seen, and heard, much of Cockney impudence before now; but never expected to hear a coxcomb ask two hundred guineas for flinging a pot of paint in the public's face.

—John Ruskin, on Whistler's painting, *Nocturne in Black and Gold*, 1878

3.2 Emotive Language, Neutral Language, and Disputes

Because a given sentence, or passage, can serve several functions—that is, for example, it can express feelings while reporting facts—the clever use of language can be deceptive or manipulative, and the careless use of language can lead to needless misunderstanding and dispute.

The words we use to convey *beliefs* may be neutral and exact, but they may also have (by accident or by design) an impact on the *attitudes* of our listeners. A rose by any other name would smell as sweet (as Shakespeare wrote), but our response to a flower is likely to be influenced if we are told, as it is handed to us, that it is commonly called "skunkweed." The negative attitudes that are commonly evoked by some words lead to the creation of *euphemisms* to replace them—gentle words for harsh realities. Janitors become "maintenance workers," and then "custodians." "Waiters" become "waitpersons," and then "servers"—and so on.

The medical vocabulary dealing with human reproduction and elimination is neutral and not offensive, but the four-letter words that are vulgar synonyms of those medical terms are shocking to many because of the attitudes they evoke. There are "seven dirty words" that may not be used on the broadcast media in the United States—because they have unacceptable emotive meanings that are sharply distinguishable from their literal meanings.[4]

Emotionally colored language is appropriate in some contexts—in poetry for example—but it is highly inappropriate in other contexts, for example, in survey research. The responses to a survey will certainly depend in good measure on the words used in asking the questions. Whether we should avoid emotive language, or rely on it, depends on the purpose language is intended to serve in the context. If we aim to provide an unbiased report of facts, we undermine that objective if we use words that are heavily charged with emotional meaning. Sometimes, however, it is nearly impossible to avoid some emotive content—such as when those in conflict about the morality of abortion call themselves either "pro-life," or "pro-choice." In logic we generally strive for language that is, so far as possible, free of the distortion that emotive meanings introduce.

Playing on the emotions of readers and listeners is a central technique in the advertising industry. When the overriding aims are to persuade and sell, manipulating attitudes becomes a sophisticated professional art. Rhetorical tricks are also common in political campaigns, and the choice of words is critical. The best defense against trickery, for voters as for consumers, is an awareness of the real uses to which the language before us is being put. We must be on guard against those who use words to make the worse appear the better cause. "With words," said Benjamin Disraeli, "we govern men."

When parties are in dispute, the differences between them that lead to that dispute may be disagreements in beliefs about the facts, or disagreements in attitude about facts that are actually agreed upon. This uncertainty, and the confusion to which it can lead, may arise because the words being used in the dispute have very different emotive meanings. To illustrate this, imagine a dispute between X and Y about legislation authorizing the death penalty for murder. X and Y may agree or disagree about the facts: whether capital punishment really is an effective deterrent to murder. They may also agree or disagree about whether it is right for the state to execute criminals, whatever may be the facts about its deterrent effectiveness. So it is possible that they could agree about factual beliefs but disagree in their attitudes, or they might agree in their attitudes but disagree about their beliefs. It is also possible, of course, that they disagree both in attitude and in belief.

When one seeks to resolve disputes that have both factual and emotional aspects, it is important to determine what really is at issue between the disputing parties. If the disagreement truly is one about whether the death penalty deters in fact, then resolution of the dispute will require, first, an effort to determine those facts objectively—although this may not be easy to do. If, on the other hand, the disagreement arises from conflicting convictions about the rightness of state-authorized executions, whether or not the death penalty deters, coming to agreement about the facts is likely to prove insufficient to resolve the dispute.

In many cases a disagreement in attitude about some event or possible outcome is rooted in a disagreement in some belief about facts; in other cases it is not. One of the greatest of all football coaches and one of the greatest of all writers on sports differed profoundly about the importance of winning. Wrote the journalist, Grantland Rice:

> For when the One Great Scorer comes
> To write against your name,
> He marks—not that you won or lost—
> But how you played the game.

Said the coach, Vince Lombardi:

> Winning isn't everything. It's the only thing.

Do you believe that this disagreement in attitude was rooted in a disagreement in belief?

Of course, we do not reach agreement simply by recognizing the nature of the dispute. But until we recognize the real nature of a dispute, and the differing functions of the language used by the conflicting parties, it is unlikely that the resolution of differences can be achieved.

EXERCISES

Identify the kinds of agreement or disagreement most probably exhibited by the following pairs:

1. **a.** Answer a fool according to his folly,
   Lest he be wise in his own conceit.
   —Prov. 26:5

   b. Answer not a fool according to his folly,
   Lest thou also be like unto him.
   —Prov. 26:4

2. **a.** Our country: in her intercourse with foreign nations may she always be in the right; but our country, right or wrong!
   —Stephen Decatur, toast at a dinner in Norfolk, Virginia, April 1816

   b. Our country, right or wrong. When right, to be kept right; when wrong, to be put right.
   —Carl Schurz, speech in the U.S. Senate, January 1872

3. **a.** A bad peace is even worse than war.
   —Tacitus, *Annals*

   b. The most disadvantageous peace is better than the most just war.
   —Desiderius Erasmus, *Adagia*, 1539

4. **a.** A stitch in time saves nine.
   b. Better late than never.

5. **a.** Absence makes the heart grow fonder.
   b. Out of sight, out of mind.

6. **a.** The race is not to the swift, nor the battle to the strong.
   —Eccl. 9:11

   b. But that's the way to bet.
   —Jimmy the Greek

7. **a.** For that some should rule and others be ruled is a thing not only necessary, but expedient; from the hour of their birth, some are marked out for subjection, others for rule.... It is clear, then, that some men are by nature free, and others slaves, and that for these latter slavery is both expedient and right.
—Aristotle, *Politics*

   b. If there are some who are slaves by nature, the reason is that men were made slaves against nature. Force made the first slaves, and slavery, by degrading and corrupting its victims, perpetuated their bondage.
—Jean-Jacques Rousseau, *The Social Contract*, 1762

8. **a.** War alone brings up to its highest tension all human energy and puts the stamp of nobility upon the peoples who have the courage to face it.
—Benito Mussolini, *Encyclopedia Italiana*, 1932

   b. War crushes with bloody heel all justice, all happiness, all that is Godlike in man. In our age there can be no peace that is not honorable; there can be no war that is not dishonorable.
—Charles Sumner, *Addresses on War*, 1904

9. **a.** Next in importance to freedom and justice is popular education, without which neither freedom nor justice can be permanently maintained.
—James A. Garfield, 1880

   b. Education is fatal to anyone with a spark of artistic feeling. Education should be confined to clerks, and even them it drives to drink. Will the world learn that we never learn anything that we did not know before?
—George Moore, *Confessions of a Young Man*, 1888

10. **a.** Belief in the existence of god is as groundless as it is useless. The world will never be happy until atheism is universal.
—J. O. La Mettrie, *L'Homme Machine*, 1865

    b. Nearly all atheists on record have been men of extremely debauched and vile conduct.
—J. P. Smith, *Instructions on Christian Theology*

11. **a.** I know of no pursuit in which more real and important services can be rendered to any country than by improving its agriculture, its breed of useful animals, and other branches of a husbandman's cares.
—George Washington, in a letter to John Sinclair

    b. With the introduction of agriculture mankind entered upon a long period of meanness, misery, and madness, from which they are only now being freed by the beneficent operations of the machine.
—Bertrand Russell, *The Conquest of Happiness*, 1930

12. **a.** Whenever there is, in any country, uncultivated land and unemployed poor, it is clear that the laws of property have been so far extended as to violate natural right.

—Thomas Jefferson

   b. Every man has by nature the right to possess property of his own. This is one of the chief points of distinction between man and the lower animals.

—Pope Leo XIII, *Rerum Novarum*, 1891

13. **a.** The right of revolution is an inherent one. When people are oppressed by their government, it is a natural right they enjoy to relieve themselves of the oppression, if they are strong enough, either by withdrawal from it, or by overthrowing it and substituting a government more acceptable.

—Ulysses S. Grant, *Personal Memoirs*, vol. 1

   b. Inciting to revolution is treason, not only against man, but against God.

—Pope Leo XIII, *Immortale Dei*, 1885

14. **a.** Language is the armory of the human mind; and at once contains the trophies of its past, and the weapons of its future conquests.

—Samuel Taylor Coleridge

   b. Language—human language—after all, is little better than the croak and cackle of fowls, and other utterances of brute nature—sometimes not so adequate.

—Nathaniel Hawthorne, *American Notebooks*, 1835

15. **a.** How does it become a man to behave towards the American government today? I answer, that he cannot without disgrace be associated with it.

—Henry David Thoreau, *An Essay on Civil Disobedience*, 1849

   b. With all the imperfections of our present government, it is without comparison the best existing, or that ever did exist.

—Thomas Jefferson

3.3 Disputes and Ambiguity

Many disputes, whether about beliefs or about attitudes, are genuine. However, some disputes are merely verbal, arising only as a result of linguistic misunderstanding. The terms used by the disputing parties may have more than one meaning—they may be *ambiguous*—but such ambiguity may be unrecognized by the disputing parties. To uncover and to resolve verbal disagreements, ambiguities must be identified, and the alternative meanings of the critical terms in the dispute must be distinguished and clarified.

Disputes fall into three categories. The first is the *obviously genuine dispute*. If A roots for the Yankees, and B for the Red Sox, they are in genuine disagreement, although they disagree mainly in attitude. If C believes that Miami is south of

Honolulu, and D denies this, they too are in genuine disagreement, but in this dispute about geographic facts a good map can settle the matter.

A second category is disputes in which the apparent conflict is not genuine and can be resolved by coming to agreement about how some word or phrase is to be understood. These may be called *merely verbal disputes*. F may hold that a tree falling in the wilderness with no person to hear it creates no sound, while G insists that a sound really is produced by the falling tree. If a "sound" is the outcome of a human auditory sensation, then F and G may agree that there was none; or if a "sound" is simply what is produced by vibrations in the air, then they may agree that a sound was indeed produced. Getting clear about what is meant by "sound" will resolve the disagreement, which was no more than verbal.

A third category, more slippery, is disputes that are *apparently verbal but really genuine*. A misunderstanding about the use of terms may be involved in such cases, but when that misunderstanding has been cleared up there remains a disagreement that goes beyond the meanings of the words. For example, should a film in which explicit sexual activity is depicted be considered "pornography"? J holds that its explicitness makes it pornographic and offensive; K holds that its beauty and sensitivity make it art and not pornography. Plainly they disagree about what "pornography" means—but after that ambiguity has been exposed, it is likely that the parties will still disagree in their judgment of that film. Whether the film is "pornographic" may be settled by a definition of that term, but a deeper disagreement is then likely to be exposed. The word "pornographic" plainly carries pejorative associations. J, who finds the film objectionable, understands the word "pornographic" in one way, while K, who approves of the film, uses the word "pornographic" differently. Does the sexually explicit content of the film make it objectionable and thus "pornographic"? J and K differ in their uses of the word, but for both of them the emotional meaning of the word is very negative; and they also differ about the criteria for the application of that negative word, "pornography."

In summary, when confronting a dispute that arises in discourse, we must first ask whether there is some ambiguity that can be eliminated by clarifying the alternative meanings in play. If there is, then we must ask whether clearing up that linguistic issue will resolve the matter. If it does, the dispute was indeed merely verbal. If it does not, the dispute was genuine, although it may have appeared to be merely verbal.

EXERCISES

A. Identify three disagreements in current political or social controversy that are of the three types described in this section: one that is genuine, one that is merely verbal, and one that is apparently verbal but really genuine. Explain the disagreements in each case.

B. Discuss each of the following disputes. If the dispute is obviously genuine, indicate each of the disputers' positions with respect to the proposition at issue.

If it is merely verbal, resolve it by explaining the different senses attached by the disputers to the key word or phrase that is used ambiguously. If it is an apparently verbal dispute that is really genuine, locate the ambiguity and explain the real disagreement involved.

1. **Daye:** Pete Rose was the greatest hitter in the history of baseball. He got more hits than any other major-league player.

    Knight: No, Barry Bonds deserves that title. He hit more home runs than any other major-league player.

2. **Daye:** Despite their great age, the plays of Sophocles are enormously relevant today. They deal with eternally recurring problems and values such as love and sacrifice, the conflict of generations, life and death—as central today as they were over two thousand years ago.

    Knight: I don't agree with you at all. Sophocles has nothing to say about the pressing and immediate issues of our time: inflation, unemployment, the population explosion, and the energy crisis. His plays have no relevance to today.

3. **Daye:** Bob Jones is certainly a wonderful father to his children. He provides a beautiful home in a fine neighborhood, buys them everything they need or want, and has made ample provision for their education.

    Knight: I don't think Bob Jones is a good father at all. He is so busy getting and spending that he has no time to be with his children. They hardly know him except as somebody who pays the bills.

4. **Daye:** Amalgamated General Corporation's earnings were higher than ever last year, I see by reading their annual report.

    Knight: No, their earnings were really much lower than in the preceding year, and they have been cited by the Securities and Exchange Commission for issuing a false and misleading report.

5. **Daye:** Business continues to be good for National Conglomerate, Inc. Their sales so far this year are 25 percent higher than they were at this time last year.

    Knight: No, their business is not so good now. Their profits so far this year are 30 percent lower than they were last year at this time.

6. **Daye:** Ann is an excellent student. She takes a lively interest in everything and asks very intelligent questions in class.

    Knight: Ann is one of the worst students I've ever seen. She never gets her assignments in on time.

7. **Daye:** Tom did it of his own free will. No pressure was brought to bear on him; no threats were made; no inducements were offered; there was no hint of force. He deliberated about it and made up his own mind.

   Knight: That is impossible. Nobody has free will, because everything anyone does is inevitably determined by heredity and environment according to inexorable causal laws of nature.

8. **Daye:** Professor Graybeard is one of the most productive scholars at the university. The bibliography of his publications is longer than that of any of his colleagues.

   Knight: I wouldn't call him a productive scholar. He is a great teacher, but he has never produced any new ideas or discoveries in his entire career.

9. **Daye:** Betty finally got rid of that old Chevy and bought herself a new car. She's driving a Buick now.

   Knight: No, Betty didn't buy herself a new car. That Buick is a good three years old.

10. **Daye:** Dick finally got rid of that old Ford of his and bought himself a new car. He's driving a Pontiac now.

    Knight: No, Dick didn't buy himself a new car. It's his roommate's new Pontiac that he's driving.

11. **Daye:** Helen lives a long way from campus. I walked out to see her the other day, and it took me nearly two hours to get there.

    Knight: No, Helen doesn't live such a long way from campus. I drove her home last night, and we reached her place in less than ten minutes.

12. **Daye:** Senator Gray is a fine man and a genuine liberal. He votes for every progressive measure that comes before the legislature.

    Knight: He is no liberal, in my opinion. The old skinflint contributes less money to worthy causes than any other man in his income bracket.

13. **Daye:** The University of Winnemac overemphasizes athletics, for it has the largest college stadium in the world and has constructed new sports buildings instead of badly needed classroom space.

    Knight: No, the University of Winnemac does not overemphasize athletics. Its academic standards are very high, and it sponsors a wide range of extracurricular activities for students in addition to its athletic program.

14. **Daye:** It was in bad taste to serve roast beef at the banquet. There were Hindus present, and it is against their religion to eat beef.

    Knight: Bad taste, nothing! That was the tastiest meal I've had in a long time. I think it was delicious!

15. **Daye:** Don't ask your wife about it. You ought to use your own judgment.

    Knight: I will use my own judgment, and in my judgment, I should ask my wife.

3.4 Definitions and Their Uses

Good definitions are plainly very helpful in eliminating verbal disputes, but there are other uses of definition that are important in logic. Before distinguishing these uses, one feature of all definitions must be emphasized: Definitions are definitions of *symbols* (not of objects), because only symbols have the meanings that definitions may explain. To illustrate, we can define the word "chair" because it has meaning; but a chair itself we cannot define. We can sit on a chair, or paint it, or burn it, or describe it—but we cannot define it because an actual chair is not a symbol that has a meaning to be explained. Sometimes we say, misleadingly, that the thing is being defined; in fact, what we *define* are always *symbols*.

Two commonly used technical terms are useful in discussing definitions. The *definiendum* is the symbol being defined. The *definiens* is the symbol (or group of symbols) used to explain the meaning of the *definiendum*. Put otherwise, the *definiendum* is the term to be defined and the *definiens* is the definition of it. However, it would be a mistake to say that the *definiens* is the meaning of the *definiendum*—rather, it is another symbol (or group of symbols) that *has the same meaning* as the *definiendum*.

With this preface, we may say that definitions, depending on how they are used, are of five kinds: (1) stipulative, (2) lexical, (3) precising, (4) theoretical, and (5) persuasive. We shall consider each in turn:

A. Stipulative Definitions

A definition that has a meaning that is deliberately assigned to some symbol is called a **stipulative definition**. One who introduces a new symbol is free to assign to it, or *stipulate*, whatever meaning she cares to. Even an old term put into a new context may have its meaning stipulated. Definitions of this sort are sometimes called *nominal*.

Why introduce a term by stipulation? Many reasons can justify doing so. It may simply be convenient; one word may stand for many words in a message. It may protect secrecy, if the sender and the receiver are the only persons who understand the stipulation. It may advance economy of expression. In the sciences, new symbols are often defined by stipulation to mean what has been meant by a long sequence of familiar words, thus saving time and increasing clarity. Many

---

Definiendum
In any definition, the word or symbol being defined.

Definiens
In any definition, a symbol or group of symbols that is said to have the same meaning as the *definiendum*.

Stipulative definition
A definition in which a new symbol is introduced to which some meaning is arbitrarily assigned; as opposed to a lexical definition, a stipulative definition cannot be correct or incorrect.

numbers that would be cumbersome to write out, for example, have been given names by stipulation: The prefix "zetta-" has been stipulatively defined as the number equal to a billion trillions (10^{21}), and the prefix "yotta-" as the number equal to a trillion trillions (10^{24}). These were defined stipulatively in 1991 by the Conférence générale des poids et mesures (General Committee on Weights and Measures), the international body that governs in the realm of scientific units. At the other extreme, "zepto-" has been stipulatively defined as "a billionth of a trillionth," and "yocto-" as a trillionth of a trillionth. Perhaps the most famous of all stipulations was the arbitrary naming of the number 10^{100} (represented by the digit 1 followed by 100 zeros) as a "googol"—a name suggested by the 9-year-old nephew of the mathematician, Edward Kasner, when he was asked for a word that might appropriately represent a very large number. The name of the now-famous Internet search firm, Google, is a deliberate misspelling of this term.

Some stipulative definitions are introduced in science to free the investigator from the distractions of the emotive associations of more familiar terms. In modern psychology, for example, the word "intelligence" is widely replaced by Spearman's "*g* factor"—a term intended to convey the same descriptive meaning without any emotional baggage. Excitement and interest may also be provided by introducing a catchy new term, as when "black hole" was introduced to replace "gravitationally completely collapsed star." The term was introduced by Dr. John Archibald Wheeler at a 1967 meeting of the Institute for Space Studies in New York City. The word "quark," now widely used in physics, was introduced by the physicist Murray Gell-Mann in 1963 to name a type of subatomic particle about which he had been theorizing. In James Joyce's novel *Finnegan's Wake*, the word "quark" appears in the line, "Three quarks for Muster Mark," but Dr. Gell-Mann reported that he had chosen this name for the particle before he had encountered it in that novel. In philosophy, Charles Sanders Peirce had long referred to his philosophy as "pragmatism," but when that word came to be used carelessly he stipulated that his views would henceforth be known as "pragmaticism"—a word that is ugly enough, he said, that no one would want to steal it!

A stipulative definition is neither true nor false; it is neither accurate nor inaccurate. A symbol defined by a stipulative definition did not have that meaning before it was given that meaning by the definition, so the definition cannot be a report of the term's meaning. For anyone who accepts the stipulative definition, the *definiendum* and the *definiens* have the *same* meaning; that is a consequence of the definition, not a fact asserted by it. *A stipulative definition is a proposal* (or a resolution or a request or an instruction) *to use the* definiendum *to mean what is meant by the* definiens. Such a definition is therefore directive rather than informative. Proposals may be rejected, requests refused, instructions disobeyed—but they can be neither true nor false.

Stipulative definitions may be evaluated as useful in advancing some purpose, or as useless because they are too complex or unclear, but they cannot resolve genuine disagreements. By reducing the emotive role of language, however, and by simplifying discourse, they can help to prevent fruitless conflict.

B. Lexical Definitions

Most often the term being defined has some established use. When the purpose of the definition is to explain that use, or to eliminate ambiguity, the definition is called lexical. A **lexical definition** reports a meaning the *definiendum* already has. That report may be correct or incorrect—and therefore it is clear that a lexical definition may be either true or false. Thus the definition "the word 'bird' means any warm-blooded vertebrate with feathers" is true; that is a correct report of how the word "bird" is generally used by speakers of English. On the other hand, the definition "the word 'bird' means any two-footed mammal" is obviously false.

Mistakes in word usage are usually not so obvious. We may call muddy water "turgid" when we mean to say that it is "turbid"; the lexical definition of "turgid" is "swollen" or "pompous." Some mistakes are downright funny, as when Mrs. Malaprop, a comically misspeaking character of the Restoration dramatist Richard Sheridan, gives the order to "illiterate him . . . from your memory" or uses the phrase "as headstrong as an allegory on the banks of the Nile." Nor are such confusions always fictional. At a U.S. university not long ago, students defined "actuary" as "a home for birds," and the definition of "duodenum" was given as "a number system in base 2."[5] Whether they are funny or sad, these are mistakes—incorrect reports of how English-speaking people use these words.

Here lies the central difference between lexical and stipulative definitions: Truth or falsity may apply to the former but not the latter. In a stipulative definition the *definiendum* has no meaning apart from (or before) the definition that introduces it, so that the definition cannot be true or false. But the *definiendum* of a lexical definition does have a prior and independent meaning, and therefore its definition may be true, or false, depending on whether that meaning is reported correctly or incorrectly.

What we here call a *lexical* definition has been referred to by some as a "real" definition—to indicate that the *definiendum* really does have the meaning identified. However, the question of whether the *definiendum* names any real or actually existing thing has nothing to do with whether the definition is lexical or stipulative. The definition "the word 'unicorn' means an animal like a horse but having a single straight horn projecting from its forehead" surely is a lexical definition, and a correct one; its *definiendum* means exactly what is meant by the *definiens*—but the *definiendum* in this case does not name or denote any existing thing, because there are no unicorns.

A qualification must be made at this point. Some definitions are indeed simply mistaken, but some uses that depart from what is normal may be better described as unusual or unorthodox. Word usage is a statistical matter, subject to variation over time—and therefore we cannot always specify "the" correct meaning of a term, but must give an account of its various meanings, as determined by the uses it has in actual speech and writing.

Lexical definition
A definition that reports the meaning that the *definiendum* already has. A lexical definition can be true or false.

Some lexicographers try to overcome this variability by referring to "best" usage or "correct" usage. This effort cannot fully succeed, however, because "best" usage is also an inexact matter, measured by the number of prominent authors and speakers whose uses of the given term are in accord with that definition. Literary and academic uses of words lag behind changes in a living language, so definitions that report meanings accepted by some intellectual aristocracy are likely to be out of date. What is unorthodox at a given time may soon become commonplace. So lexical definitions must not ignore the ways in which a term is used by great numbers of those who speak that language, because if lexical definitions are not true to actual usage, the reports they give will not be entirely correct. To take account of language growth, good dictionaries often indicate which meanings of words are "archaic" or "obsolete," and which meanings are "colloquial" or "slang."

With this qualification understood—that is, bearing in mind the variability of a living language—lexical definitions are in essence true or false, in the sense that they may be true to actual usage, or may fail to be true to it.

C. Precising Definitions

Some terms are ambiguous; some terms are vague. A term is *ambiguous* in a given context when it has more than one distinct meaning and the context does not make clear which meaning is intended. A term is *vague* when there are borderline cases to which the term might or might not apply. A word or a phrase—for example, "libel" or "freedom of speech"—may be both ambiguous and vague. **Precising definitions** are those used to eliminate ambiguity or vagueness.

Every term is vague to some degree, but excessive vagueness causes serious practical problems. This is particularly true in the law, where acts that are forbidden by some statute need to be sharply defined. For example, as this is being written the precise meaning of the phrase "unreasonable searches," which lies at the heart of the Fourth Amendment to the U.S. Constitution, is becoming the focus of a sharp debate within, and among, appellate courts. Global Positioning Devices surreptitiously placed by police now make possible the tracking of all the movements of persons suspected of a crime. Such tracking yields evidence that sometimes results in criminal conviction. Is evidence gathered in this way permissible? Simply trailing a suspect is not a violation of the Fourth Amendment because people have no expectation of privacy for actions exposed to public view. But GPS technology permits prolonged surveillance; it reveals business practices, church-going habits, recreational interests, the identity of associates and even sexual escapades. Is this a search that requires a judicial warrant? In 2010 the U.S. Court of Appeals for the District of Columbia held that it is, overturning a conviction that had been obtained using such evidence without a warrant.[6] The supreme courts of Massachusetts, New York, Oregon and Washington agree, recently ruling that their state constitutions require police to obtain a warrant for the use of such devices. But decisions in three similar GPS-related cases (in Chicago, St. Louis, and San Francisco) have been criticized by the U.S. Supreme Court. Some judges have argued that tracing the movements of a car is not a search at all. "Unreasonable searches" is certainly a vague phrase that cries

Precising definition
A definition devised to eliminate ambiguity or vagueness by delineating a concept more sharply.

out for more precision. A precising definition of that phrase is very likely to come soon from the Supreme Court of the United States.

The vagueness of units of measurement in science is a serious problem. "Horsepower," for example, is commonly used in reporting the power of motors, but its vagueness invited commercial deception. To overcome that, a precise definition was needed. "One horsepower" is now defined precisely as "the power needed to raise a weight of 550 pounds by one foot in one second"—calculated to be equal to 745.7 watts. (The power of one real horse is much greater, estimated to be about equal to 18,000 watts! A 200-horsepower automobile, therefore, has approximately the power of ten real horses.)

A meter is the internationally accepted unit of measure for distance. Originally it was defined, by stipulation, as one ten-millionth of the distance from one of the Earth's poles to the equator, and this was represented by a pair of carefully inscribed scratches on a metal bar made of platinum-iridium, kept in a vault near Paris, France. However, scientific research required more precision. A "meter" is now defined, precisely, as "the distance light travels in vacuum in one 299,792,458th of a second." Building on this, a "liter" is defined precisely as the volume of a cube having edges of 0.1 meter.

The vagueness of terms such as "horsepower" and "meter" cannot be eliminated by appealing to ordinary usage, because ordinary usage is not sufficiently exact. If it were, the terms would not have been vague. Therefore, borderline cases can be resolved only by going beyond the report of normal usage with the definition given. Such definitions are called *precising definitions*.

A precising definition differs from both lexical and stipulative definitions. It differs from stipulative definitions in that its *definiendum* is not a new term, but one whose usage is known, although unhappily vague. In constructing a precising definition, therefore, we are not free to assign to the *definiendum* any meaning we please. Established usage must be respected as far as possible, while making the known term more precise. Neither can a precising definition be a simple report, because it must go beyond established usage if the vagueness of the *definiendum* is to be reduced. How that is done—how the gaps in ordinary language are filled in—may indeed be a matter of outright stipulation.

Appellate court judges are often obliged to define some common terms more precisely. The definitions they provide are not mere stipulations, because even when the judges go beyond established usage, they will explain their reasons for the refinements being introduced. For example, unreasonable searches and seizures are forbidden by the Fourth Amendment of the U.S. Constitution, and evidence obtained through an unreasonable seizure is generally held to be inadmissible in court. But what is a "seizure"? Suppose a suspect, running from the police, throws away a packet of drugs, which is then confiscated. Have those drugs been seized? A precising definition was formulated by the U.S. Supreme Court to resolve this matter. A seizure, the Court concluded, must involve either the use of some physical force that restrains movement, or the assertion of authority (such as an order to stop) to which a subject yields. If the subject keeps running, no seizure has occurred; the packet of drugs he throws while running from the police therefore cannot be the product of an unreasonable seizure, and will be admissible as evidence.[7]

The precise definitions of terms can be very important in the world of commerce. For example, is a sport utility vehicle (SUV) a car or a light truck? The fuel economy standards applied to "light trucks" are more lenient than those applied to "cars," and therefore auto manufacturers must know the criteria that will be used by the U.S. Department of Transportation to define these categories precisely.[8]

If a law is so vague that a citizen cannot be expected to be sure when he is disobeying it, it may be struck down by a court. U.S. Supreme Court Justice Thurgood Marshall long ago explained the need for precising definitions in law:

> It is a basic principle of due process that an enactment is void for vagueness if its prohibitions are not clearly defined. Vague laws offend several important values. First . . . we insist that laws give the person of ordinary intelligence a reasonable opportunity to know what is prohibited, so that he may act accordingly. Vague laws may trap the innocent by not providing fair warning. Second, if arbitrary and discriminatory enforcement is to be prevented, law must provide explicit standards for those who apply them. A vague law impermissibly delegates basic policy matters to policemen, judges, and juries for resolution on an ad hoc and subjective basis, with the attendant dangers of arbitrary and discriminatory application. Third . . . where a vague statute abuts upon sensitive areas of basic First Amendment freedoms, it operates to inhibit the exercise of those freedoms. Uncertain meanings inevitably lead citizens to "steer far wider of the unlawful zone" than if the boundaries of the forbidden areas were clearly marked.[9]

This principle was applied in 1996 when a federal law making it illegal to transmit "indecent" or "patently offensive" materials on the Internet was struck down as impermissibly vague.[10] To avoid such uncertainties, legislatures often preface the operative portions of a statute with a section called "definitions," in which the precise meanings of key terms in that statute are spelled out. Similarly, in labor–management contracts, the terms setting forth the agreed-upon rules of the workplace will be very carefully defined. Precising definitions are conceptual instruments of wide importance.

D. Theoretical Definitions

In science, and in philosophy, definitions often serve as a compressed summary, or recapitulation, of some theory. Such definitions, when they are faulty, are criticized not so much because they are not precise as because they are not adequate—they do not correctly encapsulate the theory in question.

How, for example, should we define the word "planet"? For many years it was believed with little controversy, and all children were taught, that planets are simply bodies in orbit around the sun and that there are nine planets in the solar system—of which the smallest is Pluto, made of unusual stuff, with an unusual orbit, and most distant from the sun. But other bodies, larger than Pluto and oddly shaped, have been recently discovered orbiting the sun. Are they also planets? Why not? Older definitions had become conceptually inadequate. An intense controversy within the International Astronomical Union (IAU), still not fully resolved, has recently resulted in a new definition of "planet," according to which there are only eight planets in our solar system. And now a new category, "dwarf planet" (for bodies such as Pluto, Ceres, and Eris), has been defined. Needed were definitions that would accommodate new discoveries as well as old, while maintaining a consistent and fully intelligible account of the entire system. Such definitions (not as simple as we might like) were adopted by the IAU in 2006. A planet is "a celestial body that, within the Solar System, (1) is in orbit around the Sun; and (2) has sufficient mass for its self-gravity to overcome rigid body forces so that it assumes a hydrostatic equilibrium (nearly round) shape; and (3) has cleared the neighborhood around its orbit. In a system other than our solar system, the new definition requires that the body (1) be in orbit around a star or stellar remnant; and (2) have a mass below the limiting mass for thermonuclear fusion of deuterium; and (3) be above the minimum mass/size requirement for planetary status in the solar system.

In such controversies it is not simply the use of some word, such as "planet," that is at issue. What is wanted is a comprehensive grasp of the theory in which that term is a key element. A definition that encapsulates this larger understanding we rightly call a **theoretical definition**.

In philosophy also, theoretical definitions are sought. When Socrates struggles to find the correct definition of "justice" in Plato's *Republic*, he is not simply seeking a set of words that can serve as a synonym for "justice." When Spinoza, in the *Ethics*, seeks to define "bondage" and "freedom," he is not examining how people use those words, nor is he merely hoping to eliminate borderline cases. Neither lexical nor precising (and certainly not stipulative) definitions are the philosophical objectives. More deeply, philosophers commonly seek to develop an account of human virtues that will help us to understand these and other forms of right conduct.

The quest for theoretical definitions remains compelling. What is a "right"? Is health care a right? Do nonhuman animals have rights? How might we best define the term? Which nations truly manifest "democracy"? Is the fact that leaders are elected by popular vote sufficient to make a government democratic? If not, what other political institutions or patterns of citizen conduct characterize democratic communities? What is the most appropriate application of that term? Theoretical definitions are the *products* of our comprehensive understanding in some sphere.

E. Persuasive Definitions

The four categories we have discussed so far are concerned chiefly with the informative use of language. But definitions are also used at times to express feelings as well, so as to influence the conduct of others. A definition put forward to resolve a dispute by influencing attitudes or stirring emotions may be called a **persuasive definition**.

Persuasive definitions are common in political argument. From the left we hear *socialism* defined as "democracy extended to the economic sphere." From the right we hear *capitalism* defined as "freedom in the economic sphere." The directive intent of the emotive language in these definitions is obvious—but emotive coloration

Theoretical definition
A definition that encapsulates an understanding of the theory in which that term is a key element.

Persuasive definition
A definition formulated and used to resolve a dispute by influencing attitudes or stirring emotions, often relying upon the use of emotive language.

may also be injected subtly into wording that purports to be a correct lexical definition, and that appears on the surface to be that. As we seek to distinguish good reasoning from bad, we must be on guard against *persuasive* definitions.

In summary, we have distinguished five ways in which definitions are used. Thus any definition may be categorized in accordance with its principal function:

Stipulative
Lexical
Precising
Theoretical
Persuasive

Of course, some definitions may serve more than one of these functions. A stipulative definition may be intended to influence hearers manipulatively. A lexical definition may be used objectively to make discussion of some matter more precise, and so on. Here, as everywhere in language, context is critical.

EXERCISES

A. Find examples of definitions that function in each of the five ways distinguished and explain, in each case, how the definition serves that purpose.

B. Discuss the following:

Federal law imposes a five-year mandatory prison sentence on anyone who "uses or carries a firearm" in connection with a narcotics crime. In 1998 the U.S. Supreme Court faced this question: Does traveling in a car with a gun in a locked glove compartment or trunk—as opposed to carrying a gun on one's person—satisfy the meaning of "carry" in that law? Justice Stephen Breyer argued that Congress intended the word in its ordinary, everyday meaning, without the artificial limitation that it be immediately accessible. Quoting *Robinson Crusoe* and *Moby Dick*, he pointed to the common use of "carry" to mean "convey in a vehicle." The mandatory sentence, he concluded, is thus properly imposed. Justice Ruth Bader Ginsburg found Breyer's literary evidence selective and unpersuasive; in response, she offered quotations from Rudyard Kipling, the TV series *M*A*S*H.*, and President Theodore Roosevelt's "Speak softly and carry a big stick" to show that "carry" is properly understood in the federal statute to mean "the gun at hand, ready for use as a weapon" [*Muscarello v. U.S.*, U.S. 96-1654 (1998)]. In this controversy, which side puts forward the better precising definition?

3.5 The Structure of Definitions: Extension and Intension

A definition states the *meaning* of a term. When we look closely at the literal (or descriptive) meaning of a term, however, we see that there are different *senses* in which that term has meaning. With those different senses distinguished (our object just below), we will also see that definitions may be grouped and understood not only on the basis of their functions (as in the preceding section), but in view of the way those definitions are built: their *structure*.

We focus on general terms—terms that are applicable to more than one object—which are of critical importance in reasoning. The word "planet" is a typical general term; it is applicable to a number of objects, and it applies in the same sense equally to Mercury, Venus, Earth, Mars, Jupiter, Saturn, Uranus, and Neptune. (But not to Pluto! As explained in the preceding section, Pluto is now classified by the International Astronomical Union as a "dwarf planet.") What is meant by the word "planet" is (in one sense) that set of objects. The collection of planets constitutes the meaning of the term, its *extensional* meaning. If I say that all planets have elliptical orbits, part of what I assert is that Mars has an elliptical orbit, and another part is that Venus has an elliptical orbit, and so on. The *extension* of the general term "planet" consists of the objects to which the term may be correctly applied. The *extensional meaning* (also called the *denotative meaning*) of a general term is the collection of the objects that constitutes the **extension** (or *denotation*) of the term.

To understand the meaning of a general term is to know how to apply it correctly; however, it is not necessary to know all the objects to which it may be applied correctly in order to apply it correctly. All the objects within the extension of a given term have some *common attributes* or characteristics that lead us to use the same term to denote them. If we know these attributes, we may know the meaning of a term in a different sense, without knowing its extension. In this second sense, *meaning* supposes some *criterion for deciding*, with respect to any given object, whether it falls within the extension of that term. This sense of meaning is called the *intensional meaning* (or, sometimes, *connotative meaning*) of the term. The set of attributes shared by all and only those objects to which a general term refers is called the **intension** (or *connotation*) of that term.

Every general term has *both* an *in*tensional (or connotative) meaning and an *ex*tensional (or denotative) meaning. Consider the general term "skyscraper." It applies correctly to *all buildings over a certain height*; that is its intension. The extension of the term "skyscraper" is the class of buildings that contains the Empire State Building in New York, the Willis Tower in Chicago, the Shanghai World Financial Center, the Petronas Twin Towers in Kuala Lumpur, and others also—that is, the collection of the objects to which the term applies.

The extension of a term (its membership) is determined by its intension. The intension of the term "equilateral triangle" is the attribute of being a plane figure enclosed by three straight lines of equal length. The extension of "equilateral triangle" is the class of all those objects, and only those objects, that have this attribute. Because any object that has this attribute must be a member of that class, we say that the term's intension *determines* its extension.

However, the reverse is not true: The extension of a term does not determine its intension. Consider "equiangular triangle," which has an intension different from that of "equilateral triangle." The intension of "equiangular triangle" is the attribute of being a plane figure enclosed by three straight lines that intersect each other to form equal angles. It is true, of course, that the extension of the term

Extension
The collection of all the objects to which a term may correctly be applied.

Intension
The attributes shared by all and only the objects in the class that a given term denotes; the connotation of the term.

"equiangular triangle" is exactly the same as the extension of the term "equilateral triangle." So if we were to identify the extension of one of these terms, that would leave the intension of the class uncertain; intension is not determined by extension. Terms may have different intensions and the same extension; but terms with different extensions cannot possibly have the same intension.

When attributes are added to the intension of a term, we say that the intension increases. Begin with a general term such as "person." Add "living." Add "over twenty years old." Add "born in Mexico." With each such addition the intension increases; the intension of the term, "Living person over twenty years old born in Mexico," is far greater than that of "person." So these terms are given here in order of *increasing* intension. However, increasing their intention *decreases* their extension. The number of living persons is much lower than that of persons, and the number of living persons over twenty years old is lower still, and so on.

One may be tempted to say that extension and intension always vary inversely, but in fact that is not the case. This is because there comes a point when increasing the intension of the term has no effect on its extension. Consider this series: "living person," "living person with a spinal column," "living person with a spinal column less than one thousand years old," "living person with a spinal column less than one thousand years old who has not read all the books in the Library of Congress." These terms are clearly in order of increasing intension, but the extension of each of them is exactly the same, not decreasing at all. So we can say that, if terms are arranged in order of increasing intension, their extensions will be in *nonincreasing* order. That is, if extensions vary, they will vary inversely with the intensions.

Note that the extensions of some terms are empty; there simply are no objects having the indicated attributes. In Greek mythology, Bellerophon killed the fire-breathing Chimera, a monster with a lion's head, a goat's body, and a serpent's tail. We fully understand the intension of the term Chimera, but it has no extension.

Some bad arguments play on the fact that meaning can refer to extension or to intension, while extension may be empty. For example:

> The word "God" is not meaningless; therefore it has a meaning. But by definition, the word "God" means a being who is all-powerful and supremely good. Therefore that all-powerful and supremely good being, God, must exist.

The word "God" is certainly not meaningless, and so there is an intension that is its meaning. However, it does not follow from the fact that a term has an intension that it denotes any existent thing. The useful distinction between intension and extension was introduced and emphasized by St. Anselm of Canterbury (1033–1109), who is best known for his "ontological argument"—to which the preceding fallacious argument has little resemblance.

A contemporary critic has argued in similar fashion:

> Kitsch is the sign of vulgarity, sleaze, schlock, sentimentality, and bad faith that mark and mar our human condition. That is why utopia can be defined as a state of affairs in which the term has disappeared because it no longer has a referent.[11]

Here the writer has failed to distinguish between *meaning* and *referent*. Many valuable terms—those naming mythological creatures, for example—have no existing referent, no extension, but we do not want or expect such terms to disappear. Terms with intension but no extension are very useful. If utopia someday comes, we may wish to express our good fortune in having eliminated "kitsch" and "sleaze," but to do that we will need to be able to use those very words meaningfully.

We now use the distinction between intension and extension to explain some techniques for constructing definitions. Some definitions approach a general term by focusing on the class of *objects* to which the term refers. Some definitions approach a general term by focusing on the *attributes* that determine the class. Each approach, as we shall see, has advantages and disadvantages.

EXERCISES

A. Arrange each of the following groups of terms in order of increasing intension:
   1. Animal, feline, lynx, mammal, vertebrate, wildcat.
   2. Alcoholic beverage, beverage, champagne, fine white wine, white wine, wine.
   3. Athlete, ball player, baseball player, fielder, infielder, shortstop.
   4. Cheese, dairy product, Limburger, milk derivative, soft cheese, strong soft cheese.
   5. Integer, number, positive integer, prime number, rational number, real number.

B. Divide the following list of terms into five groups of five terms each, arranged in order of increasing intension:

Aquatic animal, beast of burden, beverage, brandy, cognac, domestic animal, filly, fish, foal, game fish, horse, instrument, liquid, liquor, musical instrument, muskellunge, parallelogram, pike, polygon, quadrilateral, rectangle, square, Stradivarius, string instrument, violin.

A. Extension and Denotative Definitions

Denotative definitions employ techniques that identify the extension of the term being defined. The most obvious way to explain the extension of a term is to identify the objects denoted by it. This is one very effective technique, but it has serious limitations.

We saw in the preceding section that two terms with different intensions (e.g., "equilateral triangle" and "equiangular triangle") may have the same extension. Therefore, even if we could enumerate all the objects denoted by a general term, that would not distinguish it from another term that has the very same extension.

Of course it is usually impossible to enumerate all the objects in a class. The objects denoted by the term "star" are literally astronomical in number; the objects denoted by the term "number" are infinitely many. For most general terms, complete enumeration is practically out of the question. Therefore denotative

Denotative definition
A definition that identifies the extension of a term, by (for example) listing the members of the class of objects to which the term refers. An extensional definition.

definitions are restricted to partial enumerations of the objects denoted—and this limitation gives rise to serious difficulties. The core of the problem is this: Partial enumeration of a class leaves the meaning of the general term very uncertain.

Any given object has many attributes, and thus may be included in the extensions of many different general terms. Therefore, any object given as an example of a general term is likely to be an example of many general terms with very different intensions. If I give the example of the Empire State Building to explain the term "skyscraper," there are many other classes of things to which I could be referring. Even if we give two examples, or three, or four, the same problem arises. Suppose I list, along with the Empire State Building, the Chrysler Building and the Trump Tower. What is the class I have in mind? It could be skyscrapers, but all these are also "great structures of the twentieth century," or "expensive pieces of real estate in Manhattan," or "landmarks in New York City." In addition, each of these general terms denotes objects not denoted by the others. Hence, partial enumeration cannot distinguish among terms that have different extensions.

We may seek to overcome this problem by naming groups of members of the class as examples. This technique, definition by subclass, does sometimes make complete enumeration possible. Thus we might define "vertebrate" to mean "amphibians and birds and fishes and reptiles and mammals." The completeness of the list gives some psychological satisfaction—but the meaning of the term "vertebrate" has not been adequately specified by such a definition.

Instead of naming or describing the objects denoted by the term being defined, as ordinary denotative definitions do, we might try pointing at them. Such definitions are called **ostensive definitions** or *demonstrative definitions*. An example of an ostensive definition is "the word 'desk' means *this*," accompanied by a gesture such as pointing a finger in the direction of a desk.

Ostensive definitions have all the limitations mentioned earlier, as well as some limitations peculiar to themselves. Gestures have a geographic limitation; one can only indicate what is visible. We cannot ostensively define the word "ocean" in an inland valley. More seriously, gestures are invariably ambiguous. To point to a desk is also to point to a part of it, as well as to its color and its size and its shape and material, and so on—in fact, one points to everything that lies in the general direction of the desk, including the lamp or the wall behind it.

This ambiguity might sometimes be resolved by adding a descriptive phrase to the *definiens*, thus producing a **quasi-ostensive definition**—for example, "the word 'desk' means *this* article of furniture" accompanied by the appropriate gesture. However, such an addition supposes the prior understanding of the phrase "article of furniture," which defeats the purpose that ostensive definitions have been claimed to serve, having been alleged by some to be the "primary" (or primitive) definitions—the way we first learn the meanings of words. In reality, we first learn language by observing and imitating, not by relying on definitions.

Ostensive definition
A kind of denotative definition in which the objects denoted by the term being defined are referred to by means of pointing, or with some other gesture; sometimes called a demonstrative definition.

Quasi-ostensive definition
A variety of denotative definition that relies upon gesture, in conjunction with a descriptive phrase.

Beyond such difficulties, all denotative definitions have this further inadequacy: They *cannot* define words that, although perfectly meaningful, do not denote anything at all. When we say that there are no unicorns we are asserting, meaningfully, that the term "unicorn" does not denote, that its extension is empty. Terms with no extension are very important, and this shows that techniques of definition that rely on extension cannot reach the heart of the matter. "Unicorn" has no extension, but the term is certainly not meaningless. If it were meaningless, it would also be meaningless to say, "There are no unicorns." This statement we fully understand, and it is true. Meaning pertains more to intension than to extension; the real key to definition is intension.

EXERCISES

C. Define the following terms by example, enumerating three examples for each term:

1. actor
2. boxer
3. composer
4. dramatist
5. element
6. flower
7. general (officer)
8. harbor
9. inventor
10. poet

D. For each of the terms given in Exercise Set A, find a nonsynonymous general term that your three examples serve equally well to illustrate.

B. Intension and Intensional Definitions

A term that is sometimes used instead of "intension" is "connotation"; intensional definitions are connotative definitions. We avoid the use of the word "connotation" here because, in everyday English, the connotation of a term is its total significance, including especially its emotive as well as its descriptive meaning. Because we are concerned here only with informative significance, we put the term "connotation" aside; this section therefore uses the terms "intension" and "intensional."

The *intension* of a term, we have said, consists of the attributes shared by all the objects denoted by the term, and shared only by those objects. If the attributes that define the term "chair" are "being a single raised seat" and "having a back," then *every* chair is a single raised seat with a back, and *only* chairs are single raised seats with a back.

Even within this restriction, three different senses of intension must be distinguished: the subjective, the objective, and the conventional. The **subjective intension** of a word for a speaker is the set of all the attributes the speaker believes to be possessed by objects denoted by that word. This set varies from individual to individual, and even from time to time for the same individual, and thus cannot serve the purposes of definition. The public meanings of words, not their private interpretations, are the logician's concern. The **objective intension** of a word is the total set of characteristics shared by all the objects in the word's extension. Within the objective intension of the term "circle," therefore, is the attribute that a circle encloses a greater area than any other plane figure having an equal perimeter. However, this attribute of circles is one that many who use the word are completely unaware of. No one possesses the omniscience required to understand all the attributes shared by the objects denoted by general terms, and therefore objective intension cannot be the public meaning whose explanation we seek to give.

People do communicate with one another and therefore do understand the terms they use; hence there must be publicly available intensions that are neither subjective nor objective in the senses just explained. Terms have stable meanings because there is an implicit agreement to use the same criterion for deciding about any object whether it is part of a given term's extension. What makes a thing a circle, in common discourse, is its being a closed plane curve, all points of which are equidistant from a point within called the center. It is by convention that this criterion is established, and this meaning is the **conventional intension** of the term "circle." This is the important sense of intension for purposes of definition: It is public but does not require omniscience to use. The word "intension" is normally taken to mean *conventional intension*, and that is our usage here.

What are the techniques, using intension, for defining terms? Several methods are common. The simplest and most frequently used is that of providing another word, whose meaning is already understood, that has the same meaning as the word being defined. Two words with the same meaning are called synonyms, so a definition given in this way is called a **synonymous definition**. Dictionaries, especially smaller ones, rely heavily on this method of defining terms. Thus a dictionary may define *adage* as meaning "proverb"; *bashful* may be defined as "shy"; and so on. Synonymous definitions are particularly useful when it is the meanings of words in another language that call for explanation. The word *chat* means "cat" in French; *amigo* means "friend" in Spanish; and so on. One learns the vocabulary of a foreign language by studying definitions using synonyms.

This is a good method of defining terms; it is easy, efficient, and helpful; but it has very serious limitations. Many words have no exact synonym, and therefore synonymous definitions are often not fully accurate and may mislead. Translation from one language to another can never be perfectly faithful to the original, and often fails to catch its spirit or convey its depth. From this realization comes the Italian proverb, *"Traduttore, traditore"* ("Translator, traitor").

Subjective intension
The set of all attributes that the speaker believes to be possessed by objects denoted by a given term.

Objective intension
The total set of attributes shared by all the objects in the extension of a term.

Conventional intension
The commonly accepted intension of a term; the criteria generally agreed upon for deciding, with respect to any object, whether it is part of the extension of that term.

Synonymous definition
A kind of connotative definition in which a word, phrase or symbol is defined in terms of another word, phrase or symbol that has the same meaning and is already understood.

A more serious limitation of synonymous definitions is this: When the concept the word aims to convey is simply not understood, every synonym may be as puzzling to the reader or hearer as the *definiendum* itself. Synonyms are virtually useless, therefore, when the aim is to construct a precising or a theoretical definition.

One may seek to explain the intension of a term by tying the *definiendum* to some clearly describable set of actions or operations; doing that is giving the term what is called an **operational definition**.

The term *operational definition* was first used by the Nobel Prize-winning physicist P. W. Bridgeman in his 1927 book, *The Logic of Modern Physics*.

For example, in the wake of the success of Einstein's theory of relativity, space and time could no longer be defined in the abstract way that Newton had used. It was therefore proposed to define such terms "operationally"—that is, by means of the operations actually undertaken when we measure distances and durations. An operational definition of a term states that the term is applied correctly to a given case if and only if the performance of specified operations in that case yields a specified result. The numerical value given for length can be defined operationally by referring to the results of a specified measuring procedure, and so on. Only public and repeatable operations are accepted in the *definiens* of an operational definition. Social scientists have also applied this technique. Some psychologists, for example, have sought to replace abstract definitions of "mind" and "sensation" by operational definitions that refer only to behavior or to physiological observations.

Of all the kinds of definition, the one that is most widely applicable is **definition by genus and difference**. This is the most important of all uses of the intension of general terms, and it is by far the technique that is most commonly relied upon in defining terms. We therefore devote the next and final section of this chapter to a detailed examination of definition by genus and difference, and the rules that properly guide its use.

The following table summarizes the kinds of definition by function (of which there are five), and the six techniques that depend on extension (three) and intension (three).

Operational definition
A kind of connotative definition that states that the term to be defined is correctly applied to a given case if and only if the performance of specified operations in that case yields a specified result.

Definition by genus and difference
A type of connotative definition of a term that first identifies the larger class ("genus") of which the *definiendum* is a species or subclass, and then identifies the attribute ("difference") that distinguishes the members of that species from members of all other species in that genus.

Five Types of Definition

1. Stipulative
2. Lexical
3. Precising
4. Theoretical
5. Persuasive

Six Techniques for Defining Terms

A. Extensional Techniques	B. Intensional Techniques
1. Definitions by example	4. Synonymous definitions
2. Ostensive definitions	5. Operational definitions
3. Quasi-ostensive definitions	6. Definitions by genus and difference

> **EXERCISES**
>
> E. Give synonymous definitions for each of the following terms:
>
> | | |
> |---|---|
> | 1. absurd | 2. buffoon |
> | 3. cemetery | 4. dictator |
> | 5. egotism | 6. feast |
> | 7. garret | 8. hasten |
> | 9. infant | 10. jeopardy |
> | 11. kine | 12. labyrinth |
> | 13. mendicant | 14. novice |
> | 15. omen | 16. panacea |
> | 17. quack | 18. rostrum |
> | 19. scoundrel | 20. tepee |

3.6 Definition by Genus and Difference

Definition by genus and difference relies directly on the intension of the terms defined, and it does so in the most helpful way. In view of their exceedingly common use, we look very closely at definitions of this type. Definitions by genus and difference are also called *analytical* definitions, or by their Latin name, definitions *per genus et differentia*.

Earlier we referred to the attributes that define a class. Normally these attributes are complex—that is, they can be analyzed into two or more other attributes. This complexity and analyzability can be understood in terms of classes. Any class of things having members may have its membership divided into subclasses. For example, the class of all triangles can be divided into three nonempty subclasses: equilateral triangles, isosceles triangles, and scalene triangles. The class whose membership is thus divided into subclasses is called the *genus*, and the various subclasses are its *species*. As used here, the terms "genus" and "species" are *relative* terms, like "parent" and "offspring." The same persons may be parents in relation to their children, but also offspring in relation to their parents. Likewise, a class may be a genus with respect to its own subclasses, but also a species with respect to some larger class of which it is a subclass. Thus the class of all triangles is a genus relative to the species *scalene triangle* and a species relative to the genus *polygon*. The logician's use of the words "genus" and "species" as relative terms is different from the biologist's use of them as fixed or absolute terms, and the two uses should not be confused.

A *class* is a collection of entities having some common characteristic. Therefore all members of a given genus have some characteristic in common. All members of the genus *polygon* (for example) share the characteristic of being closed plane figures bounded by straight line segments. This genus may be divided into different species or subclasses, such that all the members of each subclass have some further attribute in common that is shared by no member of any other subclass. The genus *polygon* is divided into triangles, quadrilaterals, pentagons, hexagons, and so on. Each species of the genus *polygon* differs from all the rest. What differentiates members of the subclass *hexagon* from the members of all other subclasses is *having precisely six sides*. All members of all species of a given genus share some attribute that makes them members of the genus, but the members of any one species share some further attribute that differentiates them from the members of every other species of that genus. The characteristic that serves to distinguish them is called the *specific difference*. Having six sides is the specific difference between the species *hexagon* and all other species of the genus *polygon*.

Thus, we may say that the attribute of being a hexagon is analyzable into the attributes of (1) being a polygon and (2) having six sides. To someone who did not know the meaning of the word "hexagon" or of any synonym of it, but who did know the meanings of the words "polygon," "sides," and "six," the meaning of the word "hexagon" can be readily explained by means of a definition by genus and difference: The word *hexagon* means "a polygon having six sides."

Using the same technique, we can readily define "prime number": A prime number is any natural number greater than one that can be divided exactly, without remainder, only by itself or by one.

Two steps are required to define a term by genus and difference. First, a genus must be named—the genus of which the species designated by the *definiendum* is the subclass. Second, the specific difference must be named—the attribute that distinguishes the members of that species from members of all others species in that genus. In the definition of prime number just given, the genus is the class of natural numbers greater than one: 2, 3, 4, . . . and so on; the specific difference is the quality of being divisible without remainder only by itself or by one: 2, 3, 5, 7, 11, . . . and so on. Definitions by genus and difference can be very precise.

Two limitations of definitions by genus and difference deserve notice, although such definitions remain, nevertheless, exceedingly useful. First, the method is applicable only to terms whose attributes are complex in the sense indicated above. If there are any attributes that are absolutely *unanalyzable*, then the words with those intensions cannot be defined by genus and difference. The sensed qualities of the specific shades of a color have been thought by some to be simple and unanalyzable in this sense. Whether there really are such unanalyzable attributes remains an open question, but if there are, they limit the applicability of definition by genus and difference. Second, the technique is not applicable when the attributes of the term are universal. Words such as "being," "entity," "existent," and "object" cannot be defined by the method of genus and difference because the class of all entities (for example) is not a species of some broader genus. A universal class (if there is one) constitutes the very highest class, or *summum genus*, as it is called. The same limitation applies to words referring to ultimate metaphysical categories, such as "substance" or "attribute." Neither of these limitations, however, is a serious handicap in most contexts in which definitions are needed.

Constructing good definitions by genus and difference is by no means a simple task; it requires thoughtful selection of the most appropriate genus for the term in question, as well as identification of the most helpful specific difference for that term. In appraising proposed definitions by genus and difference, especially when they are intended as lexical, there are five good rules that have been traditionally laid down.

Rule 1: A definition should state the essential attributes of the species.

Earlier we distinguished the conventional intension of a term from the subjective intension and the objective intension. To define a term using, as its specific difference, some attribute that is not normally recognized as its attribute, even though it may be a part of that term's objective intension, would be a violation of the spirit of this rule. The rule itself might best be expressed, using our terminology, by saying that *a definition should state the conventional intension of the term being defined*.

The conventional intension of a term is not always an intrinsic characteristic of the things denoted by that term. It may concern the origin of those things, or relations of the members of the class defined to other things, or the uses to which the members of that class are normally put. Thus the term "Stradivarius violin," which denotes a number of violins, has as its conventional intension no actual physical characteristic but rather the attribute of being a violin made in the Cremona workshop of Antonio Stradivari. The essential attributes of "governors" or "senators" would not be any specific mental or physical features that differentiate them from other persons, but the special relations they have to other citizens. The use of shape, or material, as the specific difference of a class is usually an inferior way to construct a definition. It is not an essential attribute of a "shoe," for example, that it is made of leather; what is critical in its definition is the use to which it is put, as an outer covering for the foot.

Rule 2: A definition must not be circular.

If the *definiendum* itself appears in the *definiens*, the definition can explain the meaning of the term being defined only to those who already understand it. So if a definition is *circular* it must fail in its purpose, which is to explain the meaning of the *definiendum*.

A book on gambling contains this blatant violation of the rule: "A compulsive gambler is a person who gambles compulsively."[12] As another example, a sophisticated scientist, writing in a medical journal, lapses into definitional circularity in this passage: "This review defines stress as a specific morphological, biochemical, physiological, and/or behavioral change experienced by an organism in response to a stressful event or stressor."[13]

As applied to definitions by genus and difference, avoiding circularity rules out the use, in the *definiens*, of any synonym of the *definiendum*. For example, there is no point in defining *lexicon* as "a compilation of words in the form of a dictionary." If the synonym "dictionary" is assumed to be understood, one could as well give a straightforward synonymous definition of "lexicon" instead of resorting to the more powerful but more complicated technique of genus and difference. Similarly, antonyms of the *definiendum* are also ruled out.

Rule 3: A definition must be neither too broad nor too narrow.

This is an easy rule to understand, but it is often difficult to respect. We don't want the *definiens* to denote more things than are denoted by the *definiendum*, or fewer things either, of course, but mistakes are often made. When Plato's successors in the Academy at Athens settled on the definition of "man" as "featherless biped," their critic, Diogenes, plucked a chicken and threw it over the wall into the Academy. There was a featherless biped—but no man! The *definiens* was too broad. Legend has it that to narrow the definition of "man," the attribute "having broad nails" was added to the *definiens*.

Finding or constructing the *definiens* that has precisely the correct breadth is the task faced by the lexicographer, and it is often very challenging, but if Rule 1 has been fully observed, the essence of the *definiendum* stated in the *definiens*, this rule will have been obeyed, because the conventional intension of the term cannot be too broad or too narrow.

Rule 4: Ambiguous, obscure, or figurative language must not be used in a definition.

Ambiguous terms in the *definiens* obviously prevent the definition from performing its function of explaining the *definiendum*. Obscure terms also defeat that purpose, but obscurity is a relative matter. What is obscure to amateurs may be perfectly familiar to professionals. A "dynatron oscillator" does truly mean "a circuit that employs a negative-resistance volt-ampere curve to produce an alternating current." Although it may be obscure to the ordinary person, the language of this *definiens* is wholly intelligible to the students of electrical engineering for whom the definition was written; its technical nature is unavoidable. Obscure language in nontechnical definitions may result in an effort to explain the unknown using what is even more unknown. Dr. Samuel Johnson, in his great *Dictionary of the English Language* (1755), defined *net* as meaning "anything reticulated or decussated at equal distances with interstices between the intersections"—a good example of obscurity in definition.

Another sort of obscurity arises when the language of the *definiens* is metaphorical. Figurative language may convey a "feel" for the term being defined, but it cannot give a clear explanation of the term. We do not learn the meaning of the word "bread" if we are told only that it is "the staff of life." *The Devil's Dictionary* (1911), by Ambrose Bierce, is a collection of witty definitions, many of which have a cynical bite. Bierce defined "fib" as "a lie that has not cut its teeth," and "oratory" as "a conspiracy between speech and action to cheat the understanding." Entertaining and insightful such definitions may be, but serious explanations of the definienda they are not.

Rule 5: A definition should not be negative when it can be affirmative.

What a term *does* mean, rather than what it does *not* mean, is what the definition seeks to provide. There are far too many things that the vast majority of terms do not mean; we are unlikely to cover them all in a definition. "A piece of furniture that is not a bed or a chair or a stool or a bench" does not define a couch; neither does it define a dresser. We need to identify the attributes that the *definiendum* has, rather than those it does not have.

Of course there are some terms that are essentially negative and therefore require negative definitions. The word *baldness* means "the state of not having hair on one's head," and the word *orphan* means "a child who does not have parents." Sometimes affirmative and negative definitions are about equally useful; we may define a "drunkard" as "one who drinks excessively," but also as "one who is not temperate in drinking." In those cases in which negatives are used appropriately in specifying the essential attributes, the genus must first be mentioned affirmatively. Then, sometimes, the species can be characterized accurately by rejecting all other species of that genus. Only rarely are the species few enough to make this possible. If, for example, we define "scalene" triangle as "a triangle that is neither equilateral nor isosceles," we respect poorly the spirit of Rule 1—because it is the essential attribute that the class does possess, "having sides of unequal length," that best defines it. In general, affirmative definitions are much preferred over negative ones.

In summary, intensional definitions, and among them definitions by genus and difference especially, can serve any of the purposes for which definitions are sought. They may help to eliminate ambiguity, to reduce vagueness, to give theoretical explanation, and even to influence attitudes. They are also commonly used to increase and enrich the vocabulary of those to whom they are provided. For most purposes, intensional definitions are much superior to extensional definitions, and of all definitions that rely on intensions, those constructed by genus and difference are usually the most effective and most helpful.

EXERCISES

A. Construct definitions for the following terms (in the box on the left side) by matching the *definiendum* with an appropriate genus and difference (from the box on the right side).

Definiendum		Definiens	
		Genus	Difference
1. banquet	11. lamb	1. offspring	1. female
2. boy	12. mare	2. horse	2. male
3. brother	13. midget	3. man	3. very large
4. child	14. mother	4. meal	4. very small
5. foal	15. pony	5. parent	5. young
6. daughter	16. ram	6. sheep	
7. ewe	17. sister	7. sibling	
8. father	18. snack	8. woman	
9. giant	19. son	9. person	
10. girl	20. stallion		

B. Criticize the following in terms of the rules for definition by genus and difference. After identifying the difficulty (or difficulties), state the rule (or rules) that are being violated. If the definition is either too narrow or too broad, explain why.

1. A genius is one who, with an innate capacity, affects for good or evil the lives of others.
   —Jacqueline Du Pre, in *Jacqueline Du Pre: Her Life, Her Music, Her Legend* (Arcade Publishing, 1999)

2. Knowledge is true opinion.
   —Plato, *Theaetetus*

3. Life is the art of drawing sufficient conclusions from insufficient premises.
   —Samuel Butler, *Notebooks*

4. "Base" means that which serves as a base.
   —Ch'eng Wei-Shih Lun, quoted in Fung Yu-Lan, *A History of Chinese Philosophy*, 1959

5. Alteration is combination of contradictorily opposed determinations in the existence of one and the same thing.
   —Immanuel Kant, *Critique of Pure Reason*, 1787

6. Honesty is the habitual absence of the intent to deceive.

7. Hypocrisy is the homage that vice pays to virtue.
   —François La Rochefoucauld, *Reflections*, 1665

8. The word *body*, in the most general acceptation, signifieth that which filleth, or occupieth some certain room, or imagined place; and dependeth not on the imagination, but is a real part of that we call the universe.
   —Thomas Hobbes, *Leviathan*, 1651

9. Torture is "any act by which severe pain or suffering, whether physical or mental, is intentionally inflicted on a person for such purposes as obtaining from him or a third person information or a confession."
   —United Nations Convention Against Torture, 1984

10. "Cause" means something that produces an effect.

11. War . . . is an act of violence intended to compel our opponent to fulfill our will.
    —Carl von Clausewitz, *On War*, 1911

12. A raincoat is an outer garment of plastic that repels water.

13. A hazard is anything that is dangerous.
    —*Safety with Beef Cattle*, U.S. Occupational Safety and Health Administration, 1976

14. To sneeze [is] to emit wind audibly by the nose.
    —Samuel Johnson, *Dictionary*, 1814

15. A bore is a person who talks when you want him to listen.
    —Ambrose Bierce, 1906

16. Art is a human activity having for its purpose the transmission to others of the highest and best feelings to which men have risen.
—Leo Tolstoi, *What Is Art?*, 1897

17. Murder is when a person of sound memory and discretion unlawfully killeth any reasonable creature in being, and under the king's peace, with malice aforethought, either express or implied.
—Edward Coke, *Institutes*, 1684

18. A cloud is a large semi-transparent mass with a fleecy texture suspended in the atmosphere whose shape is subject to continual and kaleidoscopic change.
—U. T. Place, "Is Consciousness a Brain Process?" *The British Journal of Psychology*, February 1956

19. Freedom of choice: the human capacity to choose freely between two or more genuine alternatives or possibilities, such choosing being always limited both by the past and by the circumstances of the immediate present.
—Corliss Lamont, *Freedom of Choice Affirmed*, 1967

20. Health is a state of complete physical, mental, and social well-being and not merely the absence of disease or infirmity.
—Constitution of the World Health Organization, 1946

21. By analysis, we mean analyzing the contradictions in things.
—Mao Zedong, *Quotations from Chairman Mao*, 1966

22. Noise is any unwanted signal.
—Victor E. Ragosine, "Magnetic Recording," *Scientific American*, February 1970

23. To explain (explicate, *explicare*) is to strip reality of the appearances covering it like a veil, in order to see the bare reality itself.
—Pierre Duhem, *The Aim and Structure of Physical Theory*, 1991

24. The Master said, Yu, shall I teach you what knowledge is? When you know a thing, to recognize that you know it, and when you do not know a thing, to recognize that you do not know it. That is knowledge.
—Confucius, *The Analects*

25. I would define political correctness as a form of dogmatic relativism, intolerant of those, such as believers in "traditional values," whose positions are thought to depend on belief in objective truth.
—Philip E. Devine, *Proceedings of the American Philosophical Association*, June 1992

C. Discuss the following definitions:

1. Faith is the substance of things hoped for, the evidence of things not seen.
—Heb. 11:1

2. Faith is when you believe something that you know ain't true.
—Definition attributed to a schoolboy by William James in "The Will to Believe," 1897

3. Faith may be defined briefly as an illogical belief in the occurrence of the improbable.
—H. L. Mencken, *Prejudice*, 1922

4. Poetry is simply the most beautiful, impressive, and widely effective mode of saying things.
   —Matthew Arnold, 1865

5. Poetry is the record of the best and happiest moments of the happiest and best minds.
   —Percy Bysshe Shelley, *The Defence of Poetry*, 1821

6. Dog, n. A kind of additional or subsidiary Deity designed to catch the overflow and surplus of the world's worship.
   —Ambrose Bierce, *The Devil's Dictionary*, c. 1911

7. Conscience is an inner voice that warns us somebody is looking.
   —H. L. Mencken, 1949

8. A bond is a legal contract for the future delivery of money.
   —Alexandra Lebenthal, Lebenthal and Company, 2001

9. "The true," to put it very briefly, is only the expedient in the way of our thinking, just as "the right" is only the expedient in the way of our behaving.
   —William James, "Pragmatism's Conception of Truth," 1907

10. To be conceited is to tend to boast of one's own excellences, to pity or ridicule the deficiencies of others, to daydream about imaginary triumphs, to reminisce about actual triumphs, to weary quickly of conversations which reflect unfavorably upon oneself, to lavish one's society upon distinguished persons and to economize in association with the undistinguished.
    —Gilbert Ryle, *The Concept of Mind*, 1949

11. Economics is the science which treats of the phenomena arising out of the economic activities of men in society.
    —J. M. Keynes, *Scope and Methods of Political Economy*, 1891

12. Justice is doing one's own business, and not being a busybody.
    —Plato, *The Republic*

13. Legend has it that the distinguished economist, John Maynard Keynes, enjoyed referring to a university education as "the inculcation of the incomprehensible into the indifferent by the incompetent."

14. By good, I understand that which we certainly know is useful to us.
    —Baruch Spinoza, *Ethics*, 1677

15. Political power, then, I take to be a right of making laws with penalties of death, and consequently all less penalties, for the regulating and preserving of property, and of employing the force of the community in the execution of such laws, and in defense of the commonwealth from foreign injury, and all this only for the public good.
    —John Locke, *Essay Concerning Civil Government*, 1690

16. And what, then, is belief? It is the demi-cadence which closes a musical phrase in the symphony of our intellectual life.
    —Charles Sanders Peirce, "How to Make Our Ideas Clear," 1878

17. Political power, properly so called, is merely the organized power of one class for oppressing another.
—Karl Marx and Friedrich Engels, *The Communist Manifesto*, 1847

18. Grief for the calamity of another is pity; and ariseth from the imagination that the like calamity may befall himself.
—Thomas Hobbes, *Leviathan*, 1651

19. We see that all men mean by justice that kind of state of character which makes people disposed to do what is just and makes them act justly and wish for what is just.
—Aristotle, *Nicomachean Ethics*

20. Inquiry is the controlled or directed transformation of an indeterminate situation into one that is so determinate in its constituent distinctions and relations as to convert the elements of the original situation into a unified whole.
—John Dewey, *Logic: The Theory of Inquiry*, 1938

21. A fanatic is one who can't change his mind and won't change the subject.
—Winston Churchill

22. Regret is the pain people feel when they compare what is with what might have been.
—Richard Gotti, "How Not to Regret Regret," *Bottom Line Personal*, 30 September 1992

23. Happiness is the satisfaction of all our desires, *extensively*, in respect of their manifoldness, *intensively*, in respect of their degree, and *potensively*, in respect of their duration.
—Immanuel Kant, *Critique of Pure Reason*, 1787

24. A tragedy is the imitation of an action that is serious and also, as having magnitude, complete in itself; in language with pleasurable accessories, each kind brought in separately in the parts of the work; in a dramatic, not in a narrative form; with incidents arousing pity and fear, wherewith to accomplish its catharsis of such emotions.
—Aristotle, *Poetics*

25. Propaganda is manipulation designed to lead you to a simplistic conclusion rather than a carefully considered one.
—Anthony Pratkanis, *The New York Times*, 27 October 1992

26. . . . the frequently celebrated female intuition . . . is after all only a faculty for observing tiny insignificant aspects of behavior and forming an empirical conclusion which cannot be syllogistically examined.
—Germaine Greer, *The Female Eunuch*, 1971

27. A fetish is a story masquerading as an object.
—Robert Stoller, "Observing the Erotic Imagination," 1985

28. Religion is a complete system of human communication (or a "form of life") showing in primarily "commissive," "behabitive," and

"exercitive" modes how a community comports itself when it encounters an "untranscendable negation of . . . possibilities."
—Gerald James Larson, "Prolegomenon to a Theory of Religion," *Journal of the American Academy of Religion*, 1978

29. Robert Frost, the distinguished New England poet, used to define a liberal as someone who refuses to take his own side in an argument.
—"Dreaming of JFK," *The Economist*, 17 March 1984

30. The meaning of a word is what is explained by the explanation of the meaning.
—Ludwig Wittgenstein, *Philosophical Investigations*, 1953

chapter 3 Summary

In this chapter we have been concerned with the uses of language and with definitions.

In Section 3.1 we identified the three chief uses of language—the *informative*, the *expressive*, and the *directive*—and two less common uses—the *ceremonial* and the *performative*.

In Section 3.2 we discussed the emotive and the neutral meanings of words. Disputes, we explained, may arise from conflicting beliefs about facts, or from conflicting attitudes about facts whose truth may (or may not) be agreed on, and we emphasized the importance of the neutral uses of language in logical discourse.

In Section 3.3 we explained that ambiguous terms are those that have more than one distinct meaning in a given context. We distinguished three different kinds of disputes: those that are *genuine*, whether the conflict be about beliefs or attitudes; those that are *merely verbal*, arising from the unrecognized use of ambiguous terms, and those that are *genuine but appear on the surface to be verbal*, in which a real difference remains even after apparent ambiguity has been eliminated.

In Section 3.4 we began the discussion of definitions, distinguishing the *definiendum* (the symbol that is to be defined) from the *definiens* (the symbol or group of symbols used to explain the meaning of the *definiendum*). We distinguished five different kinds of definition based on their functions: (1) *stipulative definitions*, with which a meaning is assigned to a term (and hence which cannot be true or false); (2) *lexical definitions*, which report the meaning that the term already has (and hence can be true or false); (3) *precising definitions*, which aim to eliminate vagueness or ambiguity; (4) *theoretical definitions*, which aim to encapsulate our understanding of some intellectual sphere; and (5) *persuasive definitions*, which aim to influence conduct.

In Section 3.5 we explained the structure of definitions, first distinguishing the *extension* of a general term, the objects denoted by it, from its *intension*, the attributes shared by all and only the members of the class designated by that term. We explained three varieties of extensional definition: *definitions by example*, in which we list or give examples of the objects denoted by the term; *ostensive definitions*, in which we point to, or indicate by gesture the extension of the term being defined; and *semi-ostensive definitions*, in which the pointing or gesture is accompanied by a descriptive phrase whose meaning is assumed known.

We also distinguished three varieties of intensional definition: *synonymous definitions*, in which we provide another word whose meaning is already understood that has the same meaning as the word being defined; *operational definitions*, which state that a term is applied correctly to a given case if and only if the performance of specified operations in that case yields a specified result; and *definitions by genus and difference*, of which a full account was given in Section 3.6.

In Section 3.6 we closely examined definitions by genus and difference, in which we first name the genus of which the species designated by the *definiendum* is a subclass, and then name the attribute (or specific difference) that distinguishes the members of that species from members of all other species of that genus. We formulated and explained five rules for the construction of good definitions by genus and difference: (1) A definition should state essential attributes; (2) a definition must not be circular; (3) a definition must not be too broad or too narrow; (4) definitions should not rely on ambiguous, obscure, or figurative language; and (5) when possible, definitions should not be negative.

END NOTES

[1] John Burgon, "Petra" (1845), on the ruins of Petra, now in Jordan.

[2] Robert Browning, "Rabbi Ben Ezra," 1864.

[3] *Cohen v. California*, 403 U.S. 15, at p. 26 (1971).

[4] By provision of the Federal Communications Decency Act. The reason those seven words are held not fit to broadcast is the reason they are not listed here. The letters with which they begin are: S, P, F, C, C, M, and T.

[5] See *The Chronicle of Higher Education*, 30 May 1993.

[6] *The Washinton Post*, Washington D.C., 7 August 2010.

[7] *California v. Hoary D.*, 499 U.S. 621 (1991).

[8] D. Hakim, "Government May Alter Line Between a Car and Truck," *The New York Times*, 25 March 2003.

[9] *Greyned v. City of Rockford*, 408 U.S. 104 (1972).

[10] *American Civil Liberties Union v. Reno*, 929 Fed. Supp. 824 (1996).

[11] John P. Sisk, "Art, Kitsch and Politics," *Commentary*, May 1988.

[12] Jay Livingston, *Compulsive Gamblers* (New York: Harper & Row, 1974), p. 2.

[13] W. H. Voge, "Stress—The Neglected Variable in Experimental Pharmacology and Toxicology," *Trends in Pharmacological Science*, January 1987.

For additional exercises and tutorials about concepts covered in this chapter, log in to MyLogicLab at *www.mylogiclab.com* and select your current textbook.

Fallacies

- **4.1** What Is a Fallacy?
- **4.2** Classification of Fallacies
- **4.3** Fallacies of Relevance
- **4.4** Fallacies of Defective Induction
- **4.5** Fallacies of Presumption
- **4.6** Fallacies of Ambiguity

4.1 What Is a Fallacy?

When we reason, we (presumably) strive to reason correctly, so one of the central tasks of logic is to identify the ways in which we are tempted to reason *in*correctly. One reasons incorrectly when the premises of an argument fail to support its conclusion, and arguments of that sort may be called fallacious. So in a very general sense, any error in reasoning is a **fallacy**. Similarly, any mistaken idea or false belief may sometimes be labeled "fallacious."

Logicians, however, commonly use the term "fallacy" more narrowly, to designate not just *any* error in reasoning, but *typical* errors—mistakes in reasoning that exhibit a pattern that can be identified and named. The great logician Gottlob Frege observed that it is one of the logician's tasks to "indicate the *pitfalls* laid by language in the way of the thinker." In this book we will use the term in this way.

In this narrower sense, each fallacy is a *type* of incorrect argument. Of course, many different arguments may make an error of some given type; that is, it may exhibit the *same kind of mistake* in reasoning. Any argument that does exhibit that kind of mistake is said to *commit* that fallacy. The particular argument that commits some known fallacy is commonly said to *be* a fallacy, because it is an individual example of that typical mistake.

To illustrate: If one accepts the premise that all science is essentially materialistic and then goes on to argue that Karl Marx, a very influential philosopher of the nineteenth century who was certainly a materialist, must therefore have been scientific, one reasons badly. It may indeed be true that Marx was scientific (as he claimed to be), but it does not *follow* from the fact that he was a materialist (which he certainly was) that he was scientific. The bad reasoning here is fallacious. If every P is a Q, it does not follow from the fact that one is a Q that one is a P. All dogs are mammals, but not every mammal is a dog. What is identified here is a pattern of mistake; it is a very common mistake that we will explore in detail in Chapter 8. Because that pattern of error, or fallacy, appears in many different contexts, it is flagged, and labeled: "the fallacy of affirming the consequent." The argument concerning Karl Marx is a fallacy because it commits that fallacy, and the fallacy it commits

chapter 4

Fallacy
A type of argument that seems to be correct, but contains a mistake in reasoning.

is the fallacy of affirming the consequent. This is independent of the equivocation in the use of the term "materialist," which means different things in science and in Marxism.

In this illustration the mistake that has been made is called a *formal* fallacy; it is a pattern of mistake that appears in deductive arguments of a certain specifiable form. There are other formal fallacies, and we shall examine them in Chapter 8. Most fallacies, however, are not formal but *in*formal: They are patterns of mistake that are made in the everyday uses of language. Informal fallacies, which we examine very closely in this chapter, arise from confusions concerning the *content* of the language used. There is no limit to the variety of forms in which that content may appear, and thus informal fallacies are often more difficult to detect than formal ones. It is language that deceives us here; we may be tricked by inferences that seem plausible on the surface but that are in reality not warranted. Such traps, the "pitfalls" that language sets, can be avoided if the *patterns* of those mistakes are well understood. Considerable attention will be devoted to these informal fallacies—the kinds of mistakes made in everyday speaking and writing, and commonly encountered, for example, in the "letters to the editor" in daily newspapers. These are the logical mistakes that we will name and explain.

Because language is slippery and imprecise, we must be cautious in this enterprise. Of course we must be careful not to make the mistakes in question, but we must also be careful to refrain from accusing others of making mistakes when they do not really do so. If we encounter an argument that appears to be fallacious, we must ask ourselves what really was meant by terms being used. The accusation of fallacy is sometimes unjustly leveled at a passage intended by its author to make a point that the critic has missed—perhaps even to make a joke. As patterns of mistakes in spoken and written language, are identified, the type of language used needs to be unerdstood. Our logical standards should be high, but our application of those standards to arguments in ordinary life should also be generous and fair.

4.2 Classification of Fallacies

Informal fallacies are numerous and can therefore be best understood if they are grouped into categories, each with clearly identifiable features. This classification of fallacies is a controversial matter in logic. There is no one correct taxonomy of fallacies. Logicians have proposed lists of fallacies that vary greatly in length; different sets have been specified, and different names have been given to both the sets and the individual fallacies. Any classification of the kind that will follow here is bound to be arbitrary in some degree. Our aim is to provide a comprehensive scheme within which the most common informal fallacies can be helpfully identified—and avoided.

The outline of this classification appears immediately below. After presenting it, we will examine each group, and each individual fallacy, in detail.

- **Fallacies of relevance.** Fallacies of relevance are the most numerous and the most frequently encountered. In these fallacies, *the premises of the argument are simply not relevant to the conclusion*. However, because they are made to appear to be relevant, they may deceive. We will distinguish and discuss:
  - R1: The appeal to the populace
  - R2: The appeal to emotion
  - R3: The red herring
  - R4: The straw man
  - R5: The attack on the person
  - R6: The appeal to force
  - R7: Missing the point (irrelevant conclusion)

- **Fallacies of defective induction.** In fallacies of defective induction, which are also common, the mistake arises from the fact that *the premises of the argument*, although relevant to the conclusion, *are so weak and ineffective* that relying on them is a blunder. We will distinguish and discuss:
  - D1: The argument from ignorance
  - D2: The appeal to inappropriate authority
  - D3: False cause
  - D4: Hasty generalization

- **Fallacies of presumption.** In fallacies of presumption, *too much is assumed* in the premises. The inference to the conclusion depends mistakenly on these unwarranted assumptions. We will distinguish and discuss:
  - P1: Accident
  - P2: Complex question
  - P3: Begging the question

- **Fallacies of ambiguity.** The incorrect reasoning in fallacies of ambiguity arises from the *equivocal use of words or phrases*. Some word or phrase in one part of the argument has a meaning different from that of the same word or phrase in another part of the argument. We will distinguish and discuss:
  - A1: Equivocation
  - A2: Amphiboly
  - A3: Accent
  - A4: Composition
  - A5: Division

Which of all these fallacies is actually committed by a specific passage is often disputable. The mistake that is made in a given argument might be construed in different ways and thus might reasonably be viewed as an instance of more than one fallacy. Once again, in the realm of natural language, context is critical, and much depends on reasonable interpretation.[1]

4.3 Fallacies of Relevance

Fallacies of relevance are bald mistakes; they might better be called fallacies of *irrelevance*, because they arise when there is no real connection between the premises and the conclusion of an argument. Because that connection is missing, the premises offered cannot possibly establish the truth of the conclusion drawn. Of course, the premises may still be *psychologically* relevant, in that they may evoke attitudes likely to cause the acceptance of the conclusion. The mistake arises when some emotive features of language are used to support the truth of a claim for which no objective reasons have been given. The modern names of these fallacies are used here, but many of them have traditional names as well (usually in Latin), which will also be included. Seven fallacies of relevance are of principal interest.

R1. The Appeal to the Populace (*Argumentum ad Populum*)

This fallacy is sometimes defined as the fallacy committed in making an emotional appeal; but this definition is so broad as to include most of the fallacies of relevance. It is defined more narrowly as the attempt to win popular assent to a conclusion by arousing the feelings of the multitude. The **argument *ad populum*** ("**to the populace**") is the baldest of all fallacies, and yet it is one of the most common. It is the instrument on which every demagogue and propagandist relies when faced with the task of mobilizing public sentiment. It is a fallacy because, instead of evidence and rational argument, the speaker (or writer) relies on expressive language and other devices calculated to excite enthusiasm for or against some cause. Patriotism is one common cause about which it is easy to stir emotions, and we know that terrible abuses and injustices have been perpetrated in the name of patriotism. The oratory of Adolf Hitler, whipping up the racist enthusiasms of his German listeners, is a classic example. Love of country is an honorable emotion, but the appeal to that emotion in order to manipulate and mislead one's audience is intellectually disreputable. "Patriotism," Samuel Johnson observed, "is the last refuge of a scoundrel."

The patriotic argument may be used when the national cause is good and the argument's author is no scoundrel. An emotional defense of belief lacks intellectual merit, but the conclusion of that bad argument may be supportable by other premises of a more rational sort. Still, offered as the premises of an argument, sheer emotion is fallacious. On 23 March 1775 the Virginia House of Burgesses passed a resolution delivering Virginia's troops to the Revolutionary War. The House was spurred to adopt this resolution by an oration whose emotional content has rarely been exceeded. Patrick Henry concluded this famous speech with the following appeal:

> . . . if we mean not basely to abandon the noble struggle in which we have been so long engaged, and which we have pledged ourselves never to abandon until the glorious object of our contest shall be obtained—we must fight! I repeat it, sir, we must fight! An appeal to arms and to the God of hosts is all that is left to us. . . . There is no retreat but in submission and slavery! Our chains are forged! Their clanking may be

---

Fallacy of relevance A fallacy in which the premises are irrelevant to the conclusion.

Appeal to the populace An informal fallacy in which the support given for some conclusion is an appeal to popular belief. Also known as argument *ad populum*.

heard on the plains of Boston!. . . . Is life so dear, or peace so sweet, as to be purchased at the price of chains and slavery? Forbid it, Almighty God! I know not what course others may take; but as for me, give me liberty or give me death!

It is reported that the crowd, upon hearing his speech, jumped up and shouted: "To arms! To arms!"

A qualification may be in order here. If the passions of the speaker are used to convince his listeners that some beliefs are true, the argument is indeed fallacious. However, if the speaker and his listener are in complete agreement in their beliefs, and the speaker aims only to spur his listeners to act in support of those mutual beliefs, the emotion he exhibits may serve a useful purpose. There is a distinction to be drawn between emotions used improperly as premises in argument and emotions used reasonably as triggers for appropriate conduct. However, this distinction will always be problematic because, when the speaker succeeds in spurring to action, it may be said that he has relied on emotion to convince his audience of the truth of some claim—the claim that now is the time to act, or the claim that the way to act in pursuit of the common goal is his way. In controversy, in deciding what conduct is appropriate, the appeal to emotion is unavoidably troubling.

The heaviest reliance on arguments *ad populum* is to be found in commercial advertising, where its use has been elevated almost to the status of a fine art. The products advertised are associated, explicitly or slyly, with things that we yearn for or that excite us favorably. Breakfast cereal is associated with trim youthfulness, athletic prowess, and vibrant good health; whiskey is associated with luxury and achievement, and beer with high adventure; the automobile is associated with romance, riches, and sex. The men depicted using the advertised product are generally handsome and distinguished, the women are sophisticated and charming, very well-dressed or hardly dressed at all. So clever and persistent are the ballyhoo artists of our time that we are all influenced to some degree, in spite of our resolution to resist. Almost every imaginable device may be used to command our attention, even to penetrate our subconscious thoughts. We are manipulated by relentless appeals to emotion of every kind.

Of course, the mere association of some product with an agreeable feeling or satisfying emotion is by itself no argument at all, but when such associations are systematically impressed on us, there usually is an argument *ad populum* lurking not far below the surface. It is suggested that the product—some beer perhaps, or some perfume, or some brand of jeans—is sexy, or is associated with wealth, or power, or some other admired characteristic, and therefore we, in purchasing it, will acquire some of that same merit.

One variety of this bad argument is particularly crass because it suggests no more than that one is well advised to buy (or join, or support, etc.) simply because that is what everyone else is doing. Some call this the "bandwagon fallacy," from the known phenomenon that, in an exciting campaign, many will be anxious to "jump on the bandwagon"—to do what others do because so many others are doing it. Brazen examples of this bandwagon fallacy are common in the public media; here, for example, are the exact words of a recent advertisement on ABC TV:

Why are so many people attracted to the Pontiac Grand Prix? It could be that so many people are attracted to the Grand Prix because—so many people are attracted to the Grand Prix!

This is the essence of an appeal to the populace.

Playing on the emotions of the general population is pernicious in the context of public polling. Those who are conducting the poll, if they are unscrupulous, may frame questions in ways designed to get the responses they seek, by using words or phrases with known emotive impact. Alternatively, if used without design but carelessly, some words may have an impact that will vitiate the poll results. In serious survey research, therefore, questions will be worded with the very greatest care, avoiding terms that are emotionally loaded, to preserve the integrity of the poll results. It is sometimes difficult to avoid all emotional taint. Many Americans support "affirmative action," viewing it as a policy designed to treat minorities fairly. But many Americans also oppose "racial preferences" in college admissions or in employment. The outcome of any random poll on this topic will depend critically on which set of words—"affirmative action" or "racial preference"—is used in the questions asked.

When results using different words conflict, it may be said that importantly different questions have been asked. Perhaps. This is a perennial problem in survey research. In *argument*, however, the logical point remains very important: A conclusion defended with premises that are directed mainly at emotions is a fallacious argument *ad populum*.

R2. Appeals to Emotion

Appeal to Pity (*ad Misericordiam*)

One variety of the appeal to emotion that appears with great frequency is the argument *ad misericordiam*. The Latin word *misericordiam* literally means "merciful heart"; this fallacy is the emotional **appeal to pity**.

Pity is often an admirable human response. Justice, it is wisely said, should be tempered with mercy. Surely there are many situations in which leniency in punishment is justified by the special circumstances of the offender. In such situations—in the sentencing phase of a trial, for example—the identification of those circumstances and the reasons they might apply to a criminal already convicted are appropriately put before the court. That is no fallacy. It would be a fallacy, however, if such considerations were registered in the effort to cause a jury to acquit a defendant who is indeed guilty of the acts with which he or she is charged. When the premises (or intimated premises) of an argument boil down to no more than an appeal to the merciful heart, the argument is plainly *ad misericordiam*, and fallacious. What is special about this variety is only that the emotions appealed to are of a particular kind: generosity and mercy.

In civil suits, when attorneys are seeking compensatory damages for the injuries suffered by their clients, there is often an effort to rely implicitly on the appeal to pity. The cause of the injury may be described as a faceless and unfeeling corporate juggernaut; or the injured party may be presented as the helpless victim

Appeal to pity
A fallacy in which the argument relies on generosity, altruism, or mercy, rather than on reason. Also known as argument *ad misericordiam*.

of an uncaring bureaucracy or an incompetent professional. The miseries of the client's continuing disability may be depicted in some heart-rending way. The injured plaintiff may make it a point to limp painfully into the courtroom. A study by the Harvard School of Public Health has demonstrated that the appeal to pity really works. When doctors are sued for malpractice, this study shows, the size of the monetary award to successful plaintiffs depends much more on the nature of the disability they suffered than on whether it could be shown that the doctor accused had in fact done anything wrong.[2]

In criminal trials, the sympathies of the jury plainly have no bearing on the guilt or innocence of the accused, but an appeal to those sympathies may nevertheless be made. Such an appeal may be made obliquely. At his trial in Athens, Socrates referred with disdain to other defendants who had appeared before their juries accompanied by their children and families, seeking acquittal by evoking pity. Socrates continued:

> I, who am probably in danger of my life, will do none of these things. The contrast may occur to [each juror's] mind, and he may be set against me, and vote in anger because he is displeased at me on this account. Now if there be such a person among you—mind, I do not say that there is—to him I may fairly reply: My friend, I am a man, and like other men, a creature of flesh and blood, and not "of wood or stone" as Homer says; and I have a family, yes, and sons, O Athenians, three in number, one almost a man, and two others who are still young; and yet I will not bring any of them here to petition you for acquittal.[3]

There are many ways to pull heartstrings. Although it is often successful, the appeal to pity is an obvious fallacy, ridiculed in the story of the trial of a youth accused of the murder of his mother and father with an ax. Confronted with overwhelming proof of his guilt, his attorney pleads for leniency on the grounds that his client is now an orphan!

Logicians give special names to other clusters of fallacious emotional appeals. Thus one might also distinguish the appeal to envy (*ad invidiam*), the appeal to fear (*ad metum*), the appeal to hatred (*ad odium*), and the appeal to pride (*ad superbium*). In all of these, the underlying mistake is the argument's reliance on feelings as premises.

R3. The Red Herring

The **red herring** is a fallacious argument whose effectiveness lies in *distraction*. Attention is deflected; readers or listeners are drawn to some aspect of the topic under discussion by which they are led away from the issue that had been the focus of the discussion. They are urged to attend to some observation or some claim that may be associated with the topic, but that is not relevant to the truth of what had originally been in dispute. A red herring has been drawn across the track.

This fallacy has a fascinating history. The phrase is believed to have been derived from the practice of those who tried to save a fox being hunted by leaving a misleading trail of scent (a smoked herring is very smelly and does become dark red) that would be likely to distract or confuse the dogs in hot pursuit. In

Red herring
A fallacy in which attention is deliberately deflected away from the issue under discussion.

many contexts, any *deliberately misleading trail* is commonly called a red herring. Especially in literature, and above all in suspense or detective stories, it is not rare for some character or event to be introduced deliberately to mislead the investigators (and the readers) and thus to add to the excitement and complexity of the plot. An ulterior political motivation may be suggested, a sexual scandal may be intimated—whatever can put the reader off the track may serve as a red herring. In the very popular novel and film, *The Da Vinci Code*,[4] one of the characters, a Catholic bishop, enters the plot in ways that very cleverly mislead. His name is the author's joke: Bishop Aringarosa—meaning "red herring" in Italian. In the world of finance, a prospectus issued to attract investors in a company about to go public, which tells much about the company but not the price of its shares, is also called a red herring.

Fallacious arguments use this technique in various ways. The opponents of an appropriate tax measure may call attention to a new and appealing way in which funds can be raised by state-sponsored gambling. A defense of the prosperity produced by an economic system may be deflected by vigorously condemning the economic inequality that system permits. Economic inequality may well be excessive or unfair, but if most of the members of a community are reasonably well off, that fact is not disproved by the reality of the enormous gap between the moderate wealth of most and the great wealth of some.

The distinguished political columnist David Broder has observed that in recent discussions of U.S. foreign policy in the Middle East, it has been the policy of some to urge that a show of military strength is a necessary element of our international posture. As Broder points out, however, it is a "rhetorical trick" to respond, whenever there is criticism of military expansion, that "its critics are soft on terror."[5] This is a classic red herring.

Another recent example arose during debate in Congress over legislation originally designed to oblige corporations to protect the accumulated funds that had been set aside for the pensions of their employees. One legislator, apparently seeking to protect his corporate donors, entered the debate with the irrelevant point that there is a serious need for the provision of better advice to retired persons on the investment of their pensions. No doubt there is. But one commentator astutely observed, "What does this have to do with employers squandering their workers' retirement? It's a red herring. . . . Mr. Smith's herring replaces a major national scandal with a minor scandal, in an attractive rhetorical wrapping."[6]

Again: At Duke University in 2006, three student athletes were indicted for rape; the indictments were plainly unfounded and soon withdrawn. When the prosecutor was charged with misconduct in office, feelings at the university grew intense. One member of the Duke faculty, writing in the local newspaper, defended the prosecutor and some other faculty members who had supported him. In the course of this defense she argued that the real "social disaster" in the Duke rape case was that "18 percent of the American population lives below the poverty line," and that we do not have "national health care or affordable childcare." That herring was bright red.[7]

R4. The Straw Man

It is very much easier to win a fight against a person made of straw than against one made of flesh and blood. If one argues against some view by presenting an opponent's position as one that is easily torn apart, the argument is fallacious, of course. Such an argument commits the fallacy of the **straw man**.

One may view this fallacy as a variety of the red herring, because it also introduces a distraction from the real dispute. In this case, however, the distraction is of a particular kind: It is an effort to shift the conflict from its original complexity into a different conflict, between parties other than those originally in dispute. So common is this variety of distraction that the pattern of argument that relies on it has long carried its own name: the *straw man argument*.

In controversies of a moral or a political nature, a successful argument almost invariably requires some reasonable and nuanced distinctions, and perhaps also some narrowly described exceptions. The extreme position in any dispute—the claim that conduct of a certain kind is *always* wrong, or *always* justified—is likely to be difficult if not impossible to defend. Therefore it is often a fallacious device to contend that what one aims to defeat is indefensible because it is categorical or absolute. Victory may be achieved over this fictitious opponent, but one will have destroyed only a straw man.

One who urges the enlargement of the authority of some central administration may be fallaciously accused of seeking to transform the state into a "big brother" whose reach will extend into every corner of citizens' private lives. Such a "big brother" is likely to be no more than a straw man. One who urges the devolution of authority from central to more local governments may be portrayed, with similar fallacy, as the enemy of an efficient and effective administration—and that, too, is likely to be a straw man. Straw man arguments often take the form of supposing that the position under attack adopts the most extreme view possible—that *every* act or policy of a certain kind is to be rejected. The argument is easy to win, but its premises are not relevant to the conclusion that was originally proposed. The straw man argument often presents a genuine objection or criticism, and the objection may be sound, but it is aimed at a new and irrelevant target.

Straw man arguments present a special risk to their proponents. If, in controversy, a critic depicts his or her opponents in a way that is clearly more extreme and more unreasonable than is justifiable given what they had written or said, readers or members of the audience are likely to recognize the exaggeration and to respond in a way quite opposite to what was hoped for. Readers (or listeners) may sense the unreasonableness of the portrayal and be offended by the unfairness. Even further, the readers or listeners, recognizing the distortion, may be caused by its unfairness to move intellectually to the side of the party that has been misrepresented, formulating in their own minds the response that may justly be made to the fallacious attack. Neutral persons who were to be persuaded may be thus transformed, by unfair fallacious argument, into adversaries. Every fallacious argument presents some risk of this kind; the fallacy of the straw man invites it with special force.

Straw man
A fallacy in which an opponent's position is depicted as being more extreme or unreasonable than is justified by what was actually asserted.

R5. Argument Against the Person (*Argumentum ad Hominem*)

Of all the fallacies of irrelevance, the **argument against the person**, or *ad hominem*, is among the most pernicious. Such arguments are common, as many fallacies are. These, in addition to being unfair to the adversary (as straw man arguments are also), are hurtful, often inflicting serious personal damage without any opportunity for the fallacy to be exposed or its author chastised.

The phrase *ad hominem* translates as "against the person." An *ad hominem* argument is one in which the thrust is directed, not at a conclusion, but at some person who defends the conclusion in dispute. This personalized attack might be conducted in either of two different ways, for which reason we distinguish two major forms of the argument *ad hominem*: the *abusive* and the *circumstantial*.

A. *Argumentum ad hominem*, Abusive

One is tempted, in heated argument, to disparage the character of one's opponents, to deny their intelligence or reasonableness, to question their understanding, or their seriousness, or even their integrity. However, the character of an adversary is logically irrelevant to the truth or falsity of what that person asserts, or to the correctness of the reasoning employed. A proposal may be attacked as unworthy because it is supported by "radicals," or by "reactionaries," but such allegations, even when plausible, are not relevant to the merit of the proposal itself.

Personal abuse can be psychologically persuasive, however, because it may induce strong disapproval of some advocate, and by unjustifiable extension in the mind of the hearer, disapproval of what had been advocated. For example, Judge Constance Baker Motley, long active in the civil rights movement, defends affirmative action with an *ad hominem* attack on its critics. She writes:

> Those who resist [affirmative action programs] deny that they are racists, but the truth is that their real motivation is racism, a belief in the inherent inferiority of African-Americans and people of mixed racial backgrounds.[8]

However, the merits (or demerits) of arguments about affirmative action are not illuminated by denigrating the character of those who take the side one rejects.

Ad hominem abusive has many variations. The opponent may be reviled (and his claims held unworthy) because he is of a certain religious or political persuasion: a "Papist" or an "atheist," a member of the "radical right" or the "loony left," or the like. A conclusion may be condemned because it has been defended by persons believed to be of bad character, or because its advocate has been closely associated with those of bad character. Socrates was convicted of impiety partly because of his long association with persons known to have been disloyal to Athens and rapacious in conduct. Very recently, when Clyde Collins Snow was called a racist because of the conclusions he reached as a forensic scientist, he replied as follows:

> My work devoted to the investigation of the disappearance, torture, and extrajudicial execution of human rights victims in many countries has often made me the target of public criticism and official outrage. To date, however, none of my critics has called me a racist. Among my detractors have been apologists for the brutal military junta in Argentina, representatives of General Pinochet's military in Chile, the Guatemalan

Argument against the person A fallacy in which the argument relies upon an attack against the person taking a position. This fallacy is also known as "argument *ad hominem*."

Defense Minister, and Serbian government spokesmen. Thus Mr. Goodman [Snow's accuser] finds himself in interesting company.[9]

The accusation of *guilt by association* is a common form of *ad hominem* abuse.

B. *Argumentum ad hominem*, Circumstantial

The circumstances of one who makes (or rejects) some claim have no more bearing on the truth of what is claimed than does his character. The mistake made in the *circumstantial* form of the *ad hominem* fallacy is to treat those personal circumstances as the premise of an opposing argument.

Thus it may be argued fallaciously that an opponent should accept (or reject) some conclusion merely because of that person's employment, or nationality, or political affiliation, or other circumstances. It may be unfairly suggested that a member of the clergy must accept a given proposition because its denial would be incompatible with the Scriptures; or it may be claimed that political candidates must support a given policy because it is explicitly propounded in the platform of their party. Such argument is irrelevant to the *truth* of the proposition in question; it simply urges that some persons' circumstances require its acceptance. Hunters, accused of the needless slaughter of unoffending animals, sometimes reply by noting that their critics eat the flesh of harmless cattle. Such a reply is plainly *ad hominem*: The fact that the critic eats meat does not even begin to prove that it is right for the hunter to kill animals for amusement.

When the circumstances of the speaker are used not merely as grounds for attack—suggesting a foolish inconsistency or the like—but used rather in a plainly negative spirit, a special name is given to such *ad hominem* arguments. They are called by their traditional Latin name, *tu quoque*. This Latin expression does not translate simply, but it means, in essence, "You're another," or more loosely, "Look who's talking." The substance of the fallacy is to contend that you (the first party) are just as bad as I am, just as guilty of whatever it is that you complained about. But of course, that response is not a refutation of the original complaint. It may be true that the first party is guilty of the conduct in question, but calling that guilt to attention does not support the innocence of the second party, which is the issue in the argument at hand.

An illustration will be helpful. A correspondent for CNN interviewed Osama bin Laden, leader of the terrorist organization Al Qaeda, some years ago in Afghanistan. The exchange went like this:

Cliff Arnett (CNN): The United States government says that you are still funding military training camps here in Afghanistan for militant Islamic fighters and that you're a sponsor of international terrorism. . . . Are these accusations true?

Osama bin Laden: . . . At the time that they condemn any Muslim leader who calls for his rights, they receive the highest official of the Irish Republican Army at the White House as a political leader. Wherever we look we find the

U.S. as the leader of terrorism and crime in the world. The U.S. does not consider it a terrorist act to throw atomic bombs at nations thousands of miles away.... The U.S. does not consider it terrorism when hundreds of thousands of our sons and brothers in Iraq die for lack of food or medicine. So there is no basis for what the U.S. says.[10]

How the United States conducts its international relations is rightly open to criticism—but whatever may be true about that behavior, attacking it is no response to the allegation of Al Qaeda terrorism. This is a classic *tu quoque*.

The circumstances of an opponent are not properly the issue in serious argument. It is the substance of what is claimed, or denied, that must be addressed. It is true that highlighting one's opponent's circumstances may prove rhetorically effective in winning assent, or in persuading others, but the effectiveness of this device does not make up for its error. Arguments of this kind are fallacious.

Circumstantial *ad hominem* arguments are sometimes used to suggest that the opponents' conclusion should be rejected because their judgment is warped, dictated by their special situation rather than by reasoning or evidence. However, an argument that is favorable to some group deserves discussion on its merits; it is fallacious to attack it simply on the ground that it is presented by a member of that group and is therefore self-serving. The arguments in favor of a protective tariff (for example) may be bad, but they are not bad because they are presented by a manufacturer who benefits from such tariffs.

One argument of this kind, called **poisoning the well**, is particularly perverse. The incident that gave rise to the name illustrates the argument forcefully. The British novelist and Protestant clergyman Charles Kingsley, attacking the famous Catholic intellectual John Henry Cardinal Newman, argued thus: Cardinal Newman's claims were not to be trusted because, as a Roman Catholic priest (Kingsley alleged), Newman's first loyalty was not to the truth. Newman countered that this *ad hominem* attack made it impossible for him, and indeed for all Catholics, to advance their arguments, because anything they might say to defend themselves would then be undermined by others' alleging that, after all, truth was not their first concern. Kingsley, said Cardinal Newman, had "poisoned the well of discourse."

Between the abusive and the circumstantial varieties of argument *ad hominem* there is a clear connection: The circumstantial may be regarded as a special case of the abusive. When a circumstantial *ad hominem* argument explicitly or implicitly charges the opponents with *inconsistency* (among their beliefs, or between what they profess and what they practice, not logical inconsistency), that is clearly one kind of abuse. When a circumstantial *ad hominem* argument charges the opponents with a lack of trustworthiness in virtue of membership in a group, that is an accusation of *prejudice* in defense of self-interest and is clearly also an abuse.

An important qualification is called for at this point. *Ad hominem* arguments are fallacious (and often unfair to the adversary) because an attack against some

Poisoning the well
A variety of abusive *ad hominem* argument in which continued rational exchange is undermined by attacking the good faith or intellectual honesty of the opponent.

person is generally not relevant to the objective merits of the argument that person has put forward. However, there are some circumstances in which it is indeed reasonable to raise doubts about some conclusion by impeaching the testimony of one who makes a claim that would (if true) support the conclusion in question. In courtroom proceedings, for example, it is acceptable, and often effective, to call a jury's attention to the unreliability of a witness, and by so doing to undermine the claims upheld by the testimony of that witness. This may be done by exhibiting contradictions in the testimony given, showing that at least some of what has been asserted must be false. It may be done by showing (not merely asserting) that the witness lied—an abusive but in this context appropriate counterargument. Testimony may also be undermined by exhibiting the great benefits that would accrue to the witness from the acceptance of his testimony—impeaching by circumstance. These are, strictly speaking, *ad hominem* considerations, and yet they are not fallacious because of the special context in which those assertions are being put forward, and because of the agreed-upon rules for the evaluation of conflicting witnesses.

Even in these special circumstances, an attack on the person of the witness does not establish the falsehood of what had been asserted. Revealing a pattern of past dishonesty or duplicity, or showing an inconsistency with testimony earlier given, may cast justifiable doubt on the reliability of the speaker, but the truth or falsity of the factual claim made can be established only with evidence that bears directly on that claim, and not merely on some person who denies or asserts it. In each case we must ask: Is the attack on the person relevant to the truth of what is at issue? When, as commonly occurs, the attack is not relevant to the merits of the claim, the *ad hominem* argument is indeed fallacious.

R6. The Appeal to Force (*Argumentum ad Baculum*)

It seems odd to suppose that one could hope to establish some proposition as true, or persuade some other person of its truth, by resorting to force. Threats or strong-arm methods to coerce one's opponents can hardly be considered arguments at all. Traditionally, a category of fallacies of this kind has been identified as the **appeal to force** or the argument *ad baculum* (*appeal ad baculum* means literally "appeal to the stick"!), and it surely is clear that however expedient force may prove to be, it cannot replace rational methods of argument. "Might makes right" is not a subtle principle, and we all reject it.

The force threatened need not be physical, of course. In 2000, two professors of law at Boise State University published (in a law journal of the University of Denver) an article that was harshly critical of the Boise Cascade Corporation, one of the world's largest producers of paper and wood products. Subsequently, the university issued a formal "errata" notice that "this article has been retracted for its lack of scholarship and false content."

Why did the university retract the article? Did Boise Cascade threaten the university with a lawsuit? "Well," said the university's general counsel, "'threaten' is an interesting word. Let's just say they pointed out that the objections they

Appeal to force
A fallacy in which the argument relies upon an open or veiled threat of force. Also known as "argument *ad baculum*."

raised did rise to the level of being actionable." The university, it turns out, had received a highlighted copy of the article in question from the general counsel of Boise Cascade, together with a letter saying, "I have been advised to proceed with litigation against Denver University if any of these highlighted areas are republished by Denver University in any form."[11]

There are some circumstances in which threats may be introduced with more subtlety, and in such circumstances we may say that something like an argument—a plainly fallacious argument, to be sure—has been presented. What is put forward may be a veiled threat, or a proposition that suggests some danger if the proposition in question is not given full assent. It may be that certain *behaviors* are of importance, whatever may be doubted or believed. To illustrate, when the U.S. attorney general in the administration of President Ronald Reagan was under strong attack in the press for misconduct, the White House chief of staff at the time, Howard Baker, opened one meeting of his staff by saying:

> The President continues to have confidence in the Attorney General and I have confidence in the Attorney General and you ought to have confidence in the Attorney General, because we work for the President and because that's the way things are. And if anyone has a different view of that, or any different motive, ambition, or intention, he can tell me about it because we're going to have to discuss your status.[12]

One might say that nobody is fooled by an argument of this sort; the threatened party may *behave* appropriately but need not, in the end, accept the *truth* of the conclusion being insisted on. To this it was answered, by representatives of twentieth-century Italian fascism, that real persuasion can come through many different instruments, of which reason is one and the blackjack is another. Once the opponent is truly persuaded, they held, the instrument of persuasion may be forgotten. That fascist view appears to guide many of the governments of the globe to this day; but the argument *ad baculum*—reliance on the club, or on the threat of force in any form—is by reason unacceptable. The appeal to force is the abandonment of reason.

R7. Missing the Point (*Ignoratio Elenchi*)

Among the fallacies of relevance, the final category to be identified is perhaps the most difficult to describe with precision. A variety of alternative names have been applied to this category, including irrelevant conclusion and mistaken refutation. It arises when the argument goes awry—when, on close examination, there is a "disconnect" between the premises and the conclusion. The twist may on occasion be an instrument of deliberate deception, but more often the fallacy is the product of sloppy thinking, a confusion in reasoning that the author of the argument herself does not fully recognize, or grasp.

Aristotle, the first to give a systematic classification of the informal fallacies, explains the fallacy we call **missing the point**, or *ignoratio elenchi*, as a *mistake* that is made in seeking to refute another's argument. The Latin word *elenchi* is derived from a Greek word that means a "disproof," or a "refutation." An *ignoratio elenchi* is a *mistaken refutation*, one that goes haywire because the person present-

Missing the point
A fallacy in which the premises support a different conclusion from the one that is proposed. Also known as "irrelevant conclusion" and "*ignoratio elenchi*."

ing it does not fully understand the proposition in dispute. He refutes, or tries to refute, a claim other than that which was originally at issue. *He misses the point.*

As an example, suppose that one person emphasizes how important it is to increase funding for the public schools. His opponent responds by insisting that a child's education involves much more than schooling and gets underway long before her formal schooling begins. That assertion is entirely reasonable, of course, but it misses the point of what was said earlier. One party presents an argument for *P*, to alleviate the need for funds; his interlocutor counters with an irrelevant *Q*, about the importance of pre-school education.

Suppose that some very controversial amendment to the tax code is proposed—say, the elimination of inheritance taxes. Such taxes, it is argued, are not fair, because the money in the estate of a deceased person was already taxed at the time it was earned—and therefore to tax it again upon the person's death is to tax the same funds twice. But, responds the supporter of the tax, inheritance taxes are imposed only on large estates that can well afford the tax; and furthermore (the advocate of the tax continues), our government needs that money. This response is an *ignoratio elenchi*. The inheritance tax may certainly be defended, but the size of estates taxed and the need for the resulting funds misses the point of the argument that had been put forward: the claim of unfair double taxation. Similarly, in a controversy over a new and very expensive weapons system for the military, criticized (let us suppose) for its doubtful practicality and enormous expense, the premises of an argument offered in support of the new weapons will miss the point if they do no more than underscore the pressing need for strong national defense. Objectives stated in general terms—national security, a balanced budget—are easy to endorse; the difficult questions in dispute are likely to be whether some particular proposed measure (a particular weapon system, a particular tax) will in fact promote the end sought, and whether it is likely to do so as effectively and efficiently as its alternatives. Bypassing the hard questions by emphasizing our agreement on easy generalizations about larger objectives commits the *ignoratio elenchi*: It misses the point.

There is a sense in which every fallacy of irrelevance is an *ignoratio elenchi*, because in all these fallacies there is a gap between the premises and the conclusion. Premises that are not relevant—red herrings, straw men, personal attacks—all miss the point; that is true. But we reserve this name for those fallacies of irrelevance that do not fit into other categories. The *ignoratio elenchi* is, we may say, a catchall class of fallacies: fallacies in which the premises simply fail to connect to the intended conclusion with the coherence that rational argument requires.

There is another expression with similar breadth and flexibility, the widely used phrase *non sequitur*. Its meaning is "does not follow": A *non sequitur* is an argument in which the conclusion simply does not follow from the premises. Thus every fallacy is, in that general sense, also a *non sequitur*. As a candidate for the presidency of the United States in 2000, George W. Bush indicated that he was planning to grant a reprieve (under his authority as the governor of Texas) to a man who had been convicted of murder and was scheduled for execution. Why, he was asked, did he telegraph his intention before announcing his formal decision? He replied:

I believe this is a case where it's important for me to send a signal about what I may do because it's a case where we're dealing with a man's innocence or guilt.[13]

The term *non sequitur* is most commonly applied when the failure of the argument is obvious, when the gap between the premises and the conclusion is painfully wide. "A great, rough *non sequitur*," Abraham Lincoln observed in a speech in 1854, "was sometimes twice as dangerous as a well polished fallacy."[14]

Yet, there are times when what appears at first to be a *non sequitur* will be seen upon reflection not to be one. Consider this report of a historic "legal fiasco."

> The prisoner pleaded guilty. He then said he had made a mistake, and the judge allowed him to change his plea to not guilty. The case was tried. The jury acquitted. "Prisoner," said Mr. Justice Hawkins, "a few minutes ago you said you were a thief. Now the jury say you are a liar. Consequently you are discharged."[15]

overview

Fallacies of Relevance

R1. The Appeal to the Populace (*ad Populum*)

An informal fallacy committed when the support offered for some conclusion is an inappropriate appeal to the multitude.

R2. The Appeal to Emotion

An informal fallacy committed when the support offered for some conclusion is emotions—fear, envy, pity, or the like—of the listeners.

R3. The Red Herring

An informal fallacy committed when some distraction is used to mislead and confuse.

R4. The Straw Man

An informal fallacy committed when the position of one's opponent is misrepresented and that distorted position is made the object of attack.

R5. Argument Against the Person (*ad Hominem*)

An informal fallacy committed when, rather than attacking the substance of some position, one attacks the person of its advocate, either abusively or as a consequence of his or her special circumstances.

R6. Appeal to Force (*ad Baculum*)

An informal fallacy committed when force, or the threat of force, is relied on to win consent.

R7. Missing the Point (*Ignoratio Elenchi*)

An informal fallacy committed when one refutes, not the thesis one's interlocutor is advancing, but some different thesis that one mistakenly imputes to him or her.

EXERCISES

A. Identify and explain the fallacies of relevance in the following passages:

1. If you can't blame the English language and your own is unforgivingly precise, blame the microphone. That was the route Jacques Chirac took after his nuclear remark about a nuclear Iran. "Having one or perhaps a second bomb a little later, well, that's not very dangerous," Mr. Chirac said with a shrug. The press was summoned back for a retake. "I should rather have paid attention to what I was saying and understood that perhaps I was on the record," Mr. Chirac offered, as if the record rather than the remark were the issue.
   —Stacy Schiff, "Slip Sliding Away," *The New York Times*, 2 February 2007

2. Nietzsche was personally more philosophical than his philosophy. His talk about power, harshness, and superb immorality was the hobby of a harmless young scholar and constitutional invalid.
   —George Santayana, *Egotism in German Philosophy*, 1915

3. Like an armed warrior, like a plumed knight, James G. Blaine marched down the halls of the American Congress and threw his shining lances full and fair against the brazen foreheads of every defamer of his country and maligner of its honor.
   For the Republican party to desert this gallant man now is worse than if an army should desert their general upon the field of battle.
   —Robert G. Ingersoll, nominating speech at the Republican National Convention, 1876

4. However, it matters very little now what the king of England either says or does; he hath wickedly broken through every moral and human obligation, trampled nature and conscience beneath his feet, and by a steady and constitutional spirit of insolence and cruelty procured for himself an universal hatred.
   —Thomas Paine, *Common Sense*, 1776

5. This embarrassing volume is an out-and-out partisan screed made up of illogical arguments, distorted and cherry-picked information, ridiculous generalizations and nutty asides. It's a nasty stewpot of intellectually untenable premises and irresponsible speculation that frequently reads like a "Saturday Night Live" parody of the crackpot right.
   —Michiko Kakutani, "Dispatch from Gomorrah, Savaging the Cultural Left," *The New York Times*, 6 February 2007.

6. I was seven years old when the first election campaign which I can remember took place in my district. At that time we still had no political parties, so the announcement of this campaign was received with very little interest. But popular feeling ran high when it was disclosed that one of the candidates was "the Prince." There was no need to add Christian and surname to realize which Prince was meant. He was the owner

of the great estate formed by the arbitrary occupation of the vast tracts of land reclaimed in the previous century from the Lake of Fucino. About eight thousand families (that is, the majority of the local population) are still employed today in cultivating the estate's fourteen thousand hectares. The Prince was deigning to solicit "his" families for their vote so that he could become their deputy in parliament. The agents of the estate, who were working for the Prince, talked in impeccably liberal phrases: "Naturally," said they, "naturally, no one will be forced to vote for the Prince, that's understood; in the same way that no one, naturally, can force the Prince to allow people who don't vote for him to work on his land. This is the period of real liberty for everybody; you're free, and so is the Prince." The announcement of these "liberal" principles produced general and understandable consternation among the peasants. For, as may easily be guessed, the Prince was the most hated person in our part of the country.

—Ignazio Silone, *The God That Failed*, 1949

7. According to R. Grunberger, author of *A Social History of the Third Reich*, Nazi publishers used to send the following notice to German readers who let their subscriptions lapse: "Our paper certainly deserves the support of every German. We shall continue to forward copies of it to you, and hope that you will not want to expose yourself to unfortunate consequences in the case of cancellation."

8. In *While Europe Slept: How Radical Islam Is Destroying the West from Within* (2006), Bruce Bawer argues that "by appeasing a totalitarian [Muslim] ideology Europe is "imperiling its liberty." Political correctness, he writes, is keeping Europeans from defending themselves, resulting in "its self-destructive passivity, softness toward tyranny, its reflexive inclination to appease." A review of the book in *The Economist* observes that Mr. Bawer "weakens his argument by casting too wide a net," and another reviewer, Imam Fatih Alev, says of Bawer's view that "it is a constructed idea that there is this very severe difference between Western values and Muslim values."

—"Clash Between European and Islamic Views," in Books, *The New York Times*, 8 February 2007.

9. To know absolutely that there is no God one must have infinite knowledge. But to have infinite knowledge one would have to be God. It is impossible to be God and an atheist at the same time. Atheists cannot prove that God doesn't exist.

—"Argument Against Atheism," http://aaron_mp.tripod.com/id2.html (2007)

10. When we had got to this point in the argument, and everyone saw that the definition of justice had been completely upset, Thrasymachus, instead of replying to me, said: "Tell me, Socrates, have you got a nurse?"

"Why do you ask such a question," I said, "when you ought rather to be answering?"

"Because she leaves you to snivel, and never wipes your nose; she has not even taught you to know the shepherd from the sheep."

—Plato, *The Republic*

11. I also admit that there are people for whom even the reality of the external world [is] a grave problem. My answer is that I do not address *them*, but that I presuppose a minimum of reason in my readers.

—Paul Feyerabend, "Materialism and the Mind-Body Problem," *The Review of Metaphysics*, 1963

12. Clarence Darrow, renowned criminal trial lawyer, began one shrewd plea to a jury thus:

> You folks think we city people are all crooked, but we city people think you farmers are all crooked. There isn't one of you I'd trust in a horse trade, because you'd be sure to skin me. But when it comes to having sympathy with a person in trouble, I'd sooner trust you folks than city folks, because you come to know people better and get to be closer friends.

—Irving Stone, *Clarence Darrow for the Defense*, 1943

13. A national organization called In Defense of Animals registered protest, in 1996, against alleged cruelty to animals being sold live or slaughtered in Chinese markets in San Francisco. Patricia Briggs, who brought the complaint to the city's Animal Welfare Commission, said: "The time of the crustaceans is coming. You'd think people wouldn't care about lobsters, because they aren't cuddly and fuzzy and they have these vacant looks and they don't vocalize. But you'd be surprised how many people care." To which response was given by Astella Kung, proprietor of Ming Kee Game Birds, where fowl are sold live: "How about the homeless people? Why don't the animal people use their energy to care for those people? They have no homes! They are hungry!"

—"Cuisine Raises Debate on Cruelty and Culture," *The New York Times*, 26 August 1996

14. The U.S. Department of Agriculture operates a price support program for the benefit of tobacco producers; its regulations limit the amount of tobacco that can be grown, and thus keep the price of tobacco high. Those same producers fight against consumer health regulations. On what ground? One analyst observed:

> For the proponent of price support regulations to turn around and fight consumer-health regulations on the grounds that government regulation is unwarranted interference by big brother and bad for the economy is the kind of argument that makes rational people wince.

—A. L. Fritschler, *Smoking and Politics* (Englewood Cliffs, NJ: Prentice-Hall, 1983)

15. During World War I, the British government deliberately inflamed the anti-German sentiments of the people with cartoons. One of these cartoons appears on the next page.

RED CROSS OR IRON CROSS?

WOUNDED AND A PRISONER
OUR SOLDIER CRIES FOR WATER.
THE GERMAN "SISTER"
POURS IT ON THE GROUND BEFORE HIS EYES.
THERE IS NO WOMAN IN BRITAIN
WHO WOULD DO IT.
THERE IS NO WOMAN IN BRITAIN
WHO WILL FORGET IT.

Source: Wilson, David (20th century). The Bridgeman Art Library International. Private Collection/The Bridgeman Art Library.

B. Each of the following passages may be plausibly criticized by some who conclude that it contains a fallacy, but each will be defended by some who deny that the argument is fallacious. Discuss the merits of each argument and explain why you conclude that it does or does not contain a fallacy of relevance.

1. The chairman of General Electric, Jack Welch, was challenged at a stockholder's meeting recently by a nun who argued that GE was responsible for the cleanup of the Hudson River where pollutants from GE's plants had for many years been allowed to collect. Welch flatly denied the company's responsibility, saying, "Sister, you have to stop this conversation. You owe it to God to be on the side of truth here."

    —Elizabeth Kolbert, "The River," *The New Yorker*, 4 December 2000

2. Gender feminism is notoriously impossible to falsify: it chews up and digests all counterevidence, transmuting it into confirming evidence. The fact that most people, including most women, do not see the pervasive and tenacious system of male power only shows how thoroughly they have been socialized to perpetuate it. The more women who reject the gender feminist perspective, the more this proves them in thrall to the androcentric system. Nothing and no one can refute the hypothesis of the sex-gender system for those who . . . see it so clearly "everywhere."

    —Christina Sommers, *Proceedings of the American Philosophical Association*, June 1992

3. As the American Revolution began to appear likely, some Americans sought reconciliation with England; Thomas Paine opposed reconciliation bitterly. In *Common Sense* (1776), he wrote:

   > . . . all those who espouse the doctrine of reconciliation may be included within the following descriptions. Interested men, who are not to be trusted, weak men who cannot see, prejudiced men who will not see, and a certain set of moderate men who think better of the European world than it deserves; and this last class, by an ill-judged deliberation, will be the cause of more calamities to this Continent than all the other three.

4. "But I observe," says Cleanthes, "with regard to you, Philo, and all speculative sceptics, that your doctrine and practice are as much at variance in the most abstruse points of theory as in the conduct of common life."
   —David Hume, *Dialogues Concerning Natural Religion*, 1779

5. A press release from the National Education Association (NEA) begins with the following statement: "America's teachers see smaller classes as the most critical element in doing a better job, a survey by the NEA indicates." . . . But the NEA, of course, is interested in having as many teachers in the schools as possible. For example, in a 3,000-pupil school system with 30 pupils assigned to each class, the teaching staff would be approximately 100. But if class size were changed to 25 the total number of teachers would rise to 120. And in a time of shrinking enrollments, that is a way to keep teachers on the public payroll. . . .

   It is unfortunate that an organization with the professional reputation the National Education Association enjoys should be so self-serving.
   —Cynthia Parsons, *Christian Science Monitor Service*

6. I testify unto every man that heareth the words of the prophecy of this book. If any man shall add unto these things, God shall add unto him the plagues that are written in this book: And if any man shall take away from the words of the book of this prophecy, God shall take away his part out of the book of life, and out of the holy city and *from* the things which are written in this book.
   —Rev. 22: 18–19

7. Anytus: "Socrates, I think that you are too ready to speak evil of men: and, if you will take my advice, I would recommend you to be careful. Perhaps there is no city in which it is not easier to do men harm than to do them good, and this is certainly the case at Athens, as I believe that you know."
   —Plato, *Meno*

8. The Greek historian Thucydides, in his *History of the Peloponnesian War*, gave the following account of an Athenian's appeal to representatives of the small island of Melos, to join Athens in its war against Sparta:

   > You know as well as we do that, in the logic of human nature, right only comes into question where there is a balance of power, while it is might that determines what the strong exhort and the weak concede. . . . Your strongest weapons are hopes yet unrealized, while the weapons in your

hands are somewhat inadequate for holding out against the forces already arranged against you. . . . Reflect that you are taking a decision for your country, a country whose fate hangs upon a single decision right or wrong.

9. In that melancholy book, *The Future of an Illusion*, Dr. Freud, himself one of the last great theorists of the European capitalist class, has stated with simple clarity the impossibility of religious belief for the educated man of today.
—John Strachey, *The Coming Struggle for Power*, 1933

10. The classic trap for any revolutionary is always "What's your alternative?" But even if you could provide the interrogator with a blueprint, this does not mean he would use it; in most cases he is not sincere in wanting to know.
—Shulamith Firestone, *The Dialectic of Sex: The Case for Feminist Revolution*, 1970

4.4 Fallacies of Defective Induction

The premises of the fallacious arguments described in the preceding section are not relevant to the conclusions drawn. However, there are many fallacious arguments in which the premises are relevant and yet are wholly inadequate. These we call **fallacies of defective induction**. What are asserted as premises simply do not serve as good reasons to reach the conclusion drawn.

D1. The Argument from Ignorance (*Argumentum ad Ignorantiam*)

Someone commits the fallacy *argumentum ad ignorantiam* if he or she argues that something is true because it has not been proved false, or false because it has not been proved true. Just because some proposition has not yet been proved false, we are not entitled to conclude that it is true. The same point can be made in reverse: If some proposition has not yet been proved true, we are not entitled to conclude that it is false. Many true propositions have not yet been proved true, of course, just as many false propositions have not yet been proved false. The fact that we cannot now be confident rarely serves as a good reason to assert knowledge of falsity, or of truth. Such an inference is defective; the fallacy is called the **argument from ignorance**, or the argument *ad ignorantiam*. Ignorance sometimes obliges us to suspend judgment, assigning neither truth nor falsity to the proposition in doubt.

As a current illustration, the great abolitionist, Frederick Douglass, will soon have a memorial, now being built at the northwest corner of Central Park in New York City. Beneath an 8-foot statue of Douglass himself is planned a quilt in granite, an array of squares that are supposed, in legend, to be part of a secret code used along the Underground Railroad to aid slaves escaping from their southern owners. However, prominent historians now agree that there never was such a code. There is no surviving example of such a quilt, and there is not a single mention of quilting codes in any diaries or memoirs from that period. The designer of

---

Fallacy of defective induction
A fallacy in which the premises are too weak or ineffective to warrant the conclusion.

Argument from ignorance
A fallacy in which a proposition is held to be true just because it has not been proven false, or false because it has not been proven true. Also known as "argument *ad ignorantiam*."

the memorial, Algernon Miller, nevertheless insists that the quilt remain part of the memorial project. "No matter what anyone has to say," argues Miller, "they [his scholarly critics] weren't there in that particular moment." Not knowing that the legend is false, he concludes that we are justified in presuming it true.

The fallacious appeal to ignorance crops up in science when plausible claims are held to be false because evidence of their truth cannot be provided. There may be good reason for its absence: In archeology or in paleontology, for instance, that evidence may have been destroyed over time. In astronomy or in physics, the evidence desired may be so distant in space or in time that it is physically unobtainable. The fact that some desired evidence has not been gathered does not justify the conclusion that an otherwise plausible claim is false.

The argument from ignorance is particularly attractive to those who defend propositions that are very doubtful, even far-fetched. Pseudo-scientists who make unverifiable claims about psychic phenomena (for example, about telepathy, or about contact with the dead) may insist that the truth of their claims is supported by the fact that their critics have been unable to prove their falsehood.

An argument from ignorance was confronted by Galileo, whose newly invented telescope, early in the seventeenth century, plainly revealed the mountains and valleys of the moon. In his day, the "truth" that the moon was a perfect crystalline sphere was unquestioned; it had to be perfect because that was what Aristotle had taught. Confronted by the evidence the telescope revealed, Galileo's Aristotelian opponents responded with an argument that seemed irrefutable: Any apparent irregularities on the moon's surface are filled in with a crystalline substance that is, of course, invisible! This hypothesis saved the moon's perfection, was in accord with what Aristotle had taught—and could not be proved false. This fallacy deserved ridicule. Galileo answered with an *argumentum ad ignorantiam* of his own, absurd enough to expose his critics: The moon is not a perfect sphere, he replied, because there are surely crystal mountains—invisible!—rising high from its surface. Because my theological critics cannot prove the claim false, we cannot conclude that such mountains are not there!

Whenever some great change is proposed, within an institution, or in society at large, those threatened by it are likely to attack with an argument from ignorance. How do we know it will work? How do we know that it is safe? We do not know; and without the knowledge that it is workable and safe, we must not adopt the change proposed. To prove workability or safety in advance, however, is often impossible. The objection sometimes takes the form of questions that suggest (but do not assert) the most horrific outcomes.

The fallacy can be a serious hindrance to progress. When the recombination of DNA, now an invaluable tool in medical science, first became possible in the 1970s, objections to further experimentation in that field were based largely on ignorance. All experiments with recombinant DNA should be stopped immediately, said one prominent scientist, who asked: "If Dr. Frankenstein must go on producing his little biological monsters . . . how can we be sure what would happen once the little beasts escaped from the laboratory?"[16] Another fearful scientist who sought to block these investigations made the appeal to ignorance explicitly:

> Can we predict the consequences? We are ignorant of the broad principles of evolution. . . . We simply do not know. We are ignorant of the various factors we currently perceive to participate in the evolutionary process. We are ignorant of the depth of security of our own environmental niche. . . . We do not know.[17]

What we do not know does not justify condemning the effort to learn. Fortunately, these appeals to ignorance were not successful in halting experimentation in a scientific realm whose value in saving and improving lives has proved, in the years since, to be incalculable.

Policy changes may be supported, as well as opposed, by an appeal to ignorance. When the federal government issued a waiver allowing Wisconsin to reduce the additional benefits it had been giving to welfare mothers for having more than one child, the governor of Wisconsin was asked if there was any evidence that unwed mothers were having additional children simply to gain the added income. His reply, *ad ignorantiam*, was this: "No, there isn't. There really isn't, but there is no evidence to the contrary, either."[18]

In some circumstances, of course, the fact that certain evidence or results have not been obtained, even after they have been actively sought in ways calculated to reveal them, may have substantial argumentative force. New drugs being tested for safety, for example, are commonly given to rodents or other animal subjects for prolonged periods; the absence of any toxic effect on the animals is taken to be evidence (although not conclusive evidence) that the drug is probably not toxic to humans. Consumer protection often relies on evidence of this kind. In circumstances like these we rely, not on ignorance, but on our knowledge, or conviction, that if the result we are concerned about were likely to arise, it would have arisen in some of the test cases. This use of the inability to prove something true supposes that investigators are highly skilled, and that they very probably would have uncovered the evidence sought had that been possible. Tragic mistakes sometimes are made in this sphere, but if the standard is set too high—if what is required is a conclusive proof of harmlessness that cannot ever be given—consumers will be denied what may prove to be valuable, even lifesaving, medical therapies.

Similarly, when a security investigation yields no evidence of improper conduct by the persons investigated, it would be wrong to conclude that the investigation has left us ignorant. A thorough investigation will properly result in the persons being "cleared." *Not* to draw a conclusion, in some cases, is as much a breach of correct reasoning as it would be to draw a mistaken conclusion.

The appeal to ignorance is common and often appropriate in a criminal court, where an accused person, in U.S. jurisprudence and British common law, is presumed innocent until proved guilty. We adopt this principle because we recognize that the error of convicting the innocent is far more grave than that of acquitting the guilty—and thus the defense in a criminal case may legitimately claim that if the prosecution has not proved guilt beyond a reasonable doubt, the only verdict possible is "not guilty." The U.S. Supreme Court strongly reaffirmed this standard of proof in these words:

> The reasonable-doubt standard . . . is a prime instrument for reducing the risk of convictions resting on factual error. The standard provides concrete substance for the presumption of innocence—that bedrock axiomatic and elementary principle whose enforcement lies at the foundation of the administration of our criminal law.[19]

However, *this* appeal to ignorance succeeds only when innocence must be assumed in the absence of proof to the contrary; in other contexts, such an appeal is indeed an argument *ad ignorantiam*.

D2. The Appeal to Inappropriate Authority (*Argumentum ad Verecundiam*)

The argument *ad verecundiam* is committed when someone argues that a proposition is true because an expert in a given field has said that it is true. This fallacy is predicated upon the feeling of respect that people have for the famous. An expert's judgment constitutes no conclusive proof; experts disagree, and even when they are in agreement they may be wrong. However, reference to an authority in an area of competence may carry some weight, but it doesn't prove a conclusion. Ultimately, even experts need to rely upon empirical evidence and rational inference.

The fallacy of the appeal to inappropriate authority arises when the appeal is made to parties who have no *legitimate* claim to authority in the matter at hand. Thus, in an argument about morality, an appeal to the opinions of Darwin, a towering authority in biology, would be fallacious, as would be an appeal to the opinions of a great artist such as Picasso to settle an economic dispute. Care must be taken in determining whose authority it is reasonable to rely on, and whose to reject. Although Picasso was not an economist, his judgment might plausibly be given some weight in a dispute pertaining to the economic value of an artistic masterpiece; and if the role of biology in moral questions were in dispute, Darwin might indeed be an appropriate authority. This is not to say that an authority in one field might not be correct when speaking outside his or her area of expertise—to allege that would constitute a species of *argumentum ad hominem* circumstantial. In every instance, an argument must be judged upon its own merits.

The most blatant examples of misplaced appeals to inappropriate authority appear in advertising "testimonials." We are urged to drive an automobile of a particular make because a famous golfer or tennis player affirms its superiority; we are urged to drink a beverage of a certain brand because some movie star or football coach expresses enthusiasm about it. Whenever the truth of some proposition is asserted on the basis of the authority of one who has no special competence in that sphere, the appeal to inappropriate authority is the fallacy committed.

This appears to be a simple-minded mistake that is easy to avoid, but there are circumstances in which the fallacious appeal is tempting, and therefore

Appeal to inappropriate authority A fallacy in which a conclusion is accepted as true simply because an expert has said that it is true. This is a fallacy whether or not the expert's area of expertise is relevant to the conclusion. Also known as "argument *ad verecundiam*."

intellectually dangerous. Here are two examples: In the sphere of international relations, in which weapons and war unhappily play a major role, one opinion or another is commonly supported by appealing to those whose special competence lies in the technical design or construction of weapons. Physicists such as Robert Oppenheimer and Edward Teller, for example, may indeed have been competent to give authoritative judgments regarding how certain weapons can (or cannot) function, but their knowledge in this sphere did not give them special wisdom in determining broad political goals. An appeal to the strong judgment of a distinguished physicist as to the wisdom of ratifying some international treaty would be an argument *ad verecundiam*. Similarly, we admire the depth and insight of great fiction—say, in the novels of Alexander Solzhenitsyn or Saul Bellow—but to resort to their judgment in determining the real culprit in some political dispute would be an appeal *ad verecundiam*. The Latin name was originated by John Locke, whose criticism was directed chiefly at those who think that citing learned authorities is enough to win any argument, who think it "a breach of modesty for others to derogate any way from it, and question authority," and who "style it impudence in anyone who shall stand out against them." That argument Locke named *ad verecundiam*—literally, an appeal to the modesty of those who might be so bold as to oppose authority.[20]

This species of the *argumentum ad verecundiam* is an appeal to one who has no legitimate claim to authority. Even one who does have a legitimate claim to authority may well prove mistaken, of course, and we may later regret our choice of experts. However, if the experts we chose deserved their reputation for knowledge, it was no fallacy to consult them even if they erred. Our mistake becomes a fallacy when our conclusion is based exclusively upon the verdict of an authority.

D3. False Cause (Argument *non Causa pro Causa*)

It is obvious that any reasoning that relies on treating as the cause of some thing or event what is not really its cause must be seriously mistaken. Often we are tempted to suppose, or led to suppose, that we understand some specific cause-and-effect relation when in fact we do not. The nature of the connection between cause and effect, and how we determine whether such a connection is present, are central problems of inductive logic and scientific method, discussed in detail in Part III of this book. Presuming the reality of a causal connection that does not really exist is a common mistake; in Latin the mistake is called the fallacy of *non causa pro causa*; we call it simply the fallacy of **false cause**.

Whether the causal connection alleged is indeed mistaken may sometimes be a matter for dispute. Some college faculty members, it has been argued, grade leniently because they fear that rigorous grading will cause lowered evaluations of them by their students and damage to their careers. Gradual "grade inflation" is said to be the result of this fear. One college professor wrote this:

False cause
A fallacy in which something that is not really the cause of something else is treated as its cause. Also known as *non causa pro causa*.

> Course evaluation forms [completed by students] are now required in many institutions, and salaries are influenced by the results. When I joined the University of Michigan 30 years ago, my salary was higher than that of any member of the anthropology department who is still active today. My standards for grading have not followed the trend toward inflation. Student complaints about grades have increased, and now my salary is at the bottom of the professorial list.[21]

Do you think the author of this passage commits the fallacy of false cause?

We sometimes mistakenly presume that one event is caused by another because it follows that other closely in time. In primitive cultures such mistakes were common; the sun would invariably reappear after an eclipse if the drums had been beaten in the darkness, but we know that it is absurd to suppose that the beating of the drums was the cause of the sun's reappearance. Mere temporal succession does not establish a causal connection. This variety of false cause is called the fallacy of ***post hoc ergo propter hoc***—"after this, therefore because of this."

Even very sophisticated people sometimes commit this fallacy. A few years ago, a critic ridiculed the reasoning of a U.S. congressman this way:

> I'm getting tired of assertions like those of Rep. Ernest Istook, Jr.—"As prayer has gone out of the schools, guns, knives, drugs, and gangs have come in"—with the unsupported implication that there is some causal connection between these events.... We could just as well say, "After we threw God out of the schools, we put a man on the moon." Students may or may not need more faith, but Congress could certainly use more reason.[22]

Some suggested that the 1954 insertion of the words "under God" into the Pledge of Allegiance was the cause of the host of social ills that followed![23]

Mistakes of this kind are widespread. Unusual weather conditions are blamed on some unrelated celestial phenomenon that happened to precede them; an infection really caused by a virus is thought to be caused by a chill wind, or wet feet, and so on. Perhaps no sphere is more vulnerable to this sort of argument than that of crimes and punishments. Typical is this remark in a letter to *The New York Times*:

> The death penalty in the United States has given us the highest crime rate and greatest number of prisoners per 100,000 population in the industrialized world.[24]

Post hoc ergo propter hoc is an easy fallacy to detect when it is blatant, but even the best of scientists and statesmen are occasionally misled by it.

False cause is also the fallacy committed when one mistakenly argues against some proposal on the ground that any change in a given direction is sure to lead to further changes in the same direction—and thus to grave consequences. Taking this step, it may be said, will put us on a slippery slope to disaster—and such reasoning is therefore called the fallacy of the **slippery slope**. Whether the feared consequences will indeed arise is not determined by the first step in a given direction; the suggestion that a change in that direction will trigger a catastrophic chain reaction is not generally warranted, although such argument is commonly invoked in defense of the status quo. What needs to be determined is what, in fact, probably will (or will not) cause the results feared.

Post hoc ergo propter hoc A fallacy in which an event is presumed to have been caused by a closely preceding event. Literally, "After this; therefore, because of this."

Slippery slope A fallacy in which change in a particular direction is asserted to lead inevitably to further changes (usually undesirable) in the same direction.

Consider the following illustration: One common objection to the legalization of assisted suicide is that once formal permission has been given to medical doctors to act in a way that is of disputable morality, doctors will be led to engage in more and greater immorality of the same or similar type. The first leniency ought to be avoided, according to this argument, because it will leave us insecure on a slope so slippery that our first step down cannot be our last. To this argument one keen critic responded:

> The slippery slope argument, although influential, is hard to deal with rationally. It suggests that once we allow doctors to shorten the life of patients who request it, doctors could and would wantonly kill burdensome patients who do not want to die. This suggestion is not justified. . . .
>
> Physicians often prescribe drugs which, in doses greater than prescribed, would kill the patient. No one fears that the actual doses prescribed will lead to their use of lethal doses. No one objects to such prescriptions in fear of a "slippery slope." Authorizing physicians to assist in shortening the life of patients who request this assistance no more implies authority to shorten the life of patients who want to prolong it, than authority for surgery to remove the gall bladder implies authority to remove the patient's heart.[25]

The supposition that moving in a given direction, however prudently, is sure to produce the dreadful result of moving in the same direction to excess, is the fallacy of the slippery slope.

There are circumstances, of course, in which the first step in a new direction does establish a precedent that makes additional movement in that direction easier to achieve. This may be good or bad. Opposing new legislation that would punish crimes more severely if they were motivated by racial hatred, one critic writes:

> There should not be a separate category for hate crimes. A murder is a murder; a beating is a beating. We should prosecute people for the crimes they commit, not why they commit them. If we start to categorize crimes by their motivation, we start down a very slippery slope.[26]

Some arguments of this kind have merit, because precedent can affect subsequent decision making. The slippery slope is indeed a fallacy—but the mere allegation that that fallacy has been committed does not prove the argument in question faulty.

D4. Hasty Generalization

Hasty generalization
A fallacy of defective induction in which one moves carelessly from a single case, or a very few cases, to a large-scale generalization about all or most cases. Also known as "converse accident."

Throughout our lives, we rely on statements about how things generally are and how people generally behave. Nonetheless, general claims, although critical in reasoning, must be carefully scrutinized: The universality of their application ought never be accepted or assumed without justification. **Hasty generalization** is the fallacy we commit when we draw conclusions about *all* the persons or things in a given class on the basis of our knowledge about only one (or only a very few) of the members of that class. We all know of persons who have generalized mistakenly about certain companies or governments because of a single experience. Stereotypes about people who come from certain countries, or cultures, are widespread and commonly mistaken; hasty generalizations about foreign cultures can be downright nasty, and are good illustrations of the fallacious leap to broad generalization on the basis of very little evidence.

An anecdote or single instance may indeed be relevant support for a general rule or theory; but when it is treated as proof of that theory, the generalization is not well founded—the induction is defective. Here is an example: Eating deep-fried foods tends to raise one's cholesterol level. A single instance in which it does not do so is hardly sufficient to show that such foods are healthy. The owner of a "fish and chips" shop in England fallaciously defended the healthfulness of his deep-fried cookery with this argument:

> Take my son, Martyn. He's been eating fish and chips his whole life, and he just had a cholesterol test, and his level is below the national average. What better proof could there be than a fryer's son?[27]

Foods or drugs that are harmless in one context may be harmful in another. To move from a single case, or a very few cases, to a large-scale generalization about all or most cases, is fallacious reasoning, but it is common and often tempting. It is also called the *fallacy of converse accident* because it is the reverse of another common mistake, known as the *fallacy of accident*, in which generalizations are misused in another way. We turn to it next.

overview

Fallacies of Defective Induction

D1. The Argument from Ignorance (*ad Ignorantiam*)

An informal fallacy in which a conclusion is supported by an illegitimate appeal to ignorance, as when it is supposed that something is likely to be true because we cannot prove that it is false.

D2. The Appeal to Inappropriate Authority (*ad Verecundiam*)

An informal fallacy in which the appeal to authority is illegitimate, either because the authority appealed to has no special claim to expertise on the topic at issue, or, more generally, because no authority is assured to be reliable.

D3. False Cause (*non Causa pro Causa*)

An informal fallacy in which the mistake arises from accepting as the cause of an event what is not really its cause.

D4. Hasty Generalization

An informal fallacy in which a principle that is true of a particular case is applied, carelessly or deliberately, to the great run of cases.

4.5 Fallacies of Presumption

Some mistakes in everyday reasoning are the consequence of an *unjustified assumption*, often suggested by the way in which the argument is formulated. That suggestion may be deliberate, or the assumption may be only an oversight. In either case, the upshot is that the reader, the listener, and even the author of the passage may be led to assume the truth of some unproved and unwarranted proposition. When such dubious propositions, buried in the argument, are crucial for the support of the conclusion, the argument is bad and can be very misleading. Arguments that depend on such unwarranted leaps are called **fallacies of presumption**.

In fallacious arguments of this kind the premises may indeed be relevant to the conclusion drawn, but that relevance is likely to flow from the tacit supposition of what has not been given support and may even be unsupportable. The presumption often goes unnoticed. To expose such a fallacy it is therefore usually sufficient to call attention to the smuggled assumption, or supposition, and to its doubtfulness or its falsity. Three common fallacies are included in this category.

P1. Accident

Circumstances alter cases. A generalization that is largely true may not apply in a given case (or to some subcategory of cases) for good reasons. The reasons the generalization does not apply in those cases have to do with the special circumstances, also called the "accidental" circumstances, of that case or those cases. If these accidental circumstances are ignored, and we assume that the generalization applies universally, we commit the **fallacy of accident**.

In the preceding section we explained the *fallacy of converse accident*, or *hasty generalization*, the mistake of moving carelessly or too quickly to a generalization that the evidence does not support. *Accident* is the fallacy that arises when we move carelessly or unjustifiably *from* a generalization to some particulars that it does not in fact cover.

Experience teaches us that even generalizations that are widely applicable and very useful are likely to have exceptions for which we must be on guard. For example, there is a general principle in law that hearsay evidence—statements made by a third party outside court—may not be accepted as evidence in court; this is the "hearsay rule," and it is a good rule. However, when the person whose oral communications are reported is dead, or when the party reporting the hearsay in court does so in conflict with his own best interest, that rule may not apply. Indeed, there is hardly any rule or general principle that does not have plausible exceptions, and we are likely to argue fallaciously if we reason on the supposition that some rule applies universally.

P2. Complex Question (*Plurium Interrogationum*)

One of the most common fallacies of presumption is to ask a question in such a way as to presuppose the truth of some conclusion that is buried in the question. The question itself is likely to be rhetorical, with no answer actually being sought. But putting the question seriously, thereby introducing its presupposition surreptitiously, often achieves the questioner's purpose—fallaciously.

Fallacy of presumption
Any fallacy in which the conclusion depends on a tacit assumption that is dubious, unwarranted, or false.

Fallacy of accident
A fallacy in which a generalization is mistakenly applied to a particular case to which the generalization does not apply.

Thus an essayist recently asked:

> With all of the hysteria, all of the fear, all of the phony science, could it be that man-made global warming is the greatest hoax ever perpetrated on the American people?[28]

Such a statement assumes that much of the evidence supporting global warming is unreliable or "phony." Or a homeowner might ask, regarding a proposed increase in the property tax, "How can you expect the majority of the voters, who rent but don't own property and don't have to pay the tax, to care if the tax burden of others is made even more unfair?"—assuming both that the burden of the proposed tax is unfair and that those who rent rather than own their own homes are not affected by tax increases on property. Because assumptions like these are not asserted openly, the questioners evade the need to defend them forthrightly.

The **complex question** is often a deceitful device. The speaker may pose some question, then answer it or strongly suggest the answer with the truth of the premise that had been buried in the question simply assumed. A letter writer asks, "If America's booming economy depends on people's using consumer credit beyond their means, thus creating poverty, do we really have a healthy economy?"[29] But the role and the results of consumer credit remain to be addressed.

One critic of research in genetics hides his assumptions in this question: "What are the consequences of reducing the world's gene pool to patented intellectual property, controlled by a handful of life-science corporations?"[30] The "consequences" asked about are never actually discussed; they are only a device with which the reader may be frightened by the assumptions of the question—that the world's gene pool is soon likely to be reduced to patented intellectual property, and that a handful of corporations will soon control that gene pool. Establishing the plausibility of such threats requires much more than asking questions designed to presuppose them.

The appearance of a question in an editorial or headline often has the purpose of suggesting the truth of the unstated assumptions on which it is built: "Judge Took Bribe?" This technique is a common mark of what is called "yellow journalism." In debate, whenever a question is accompanied by the aggressive demand that it be answered "yes or no," there is reason to suspect that the question is "loaded"—that it is unfairly complex.

> Does the distinguished senator believe that the American public is really so naïve that they will endorse just any stopgap measure?

This "question," of course, cannot be answered "Yes." It conceals several unchallenged assumptions: that what is proposed is a "stopgap" measure, that it is inadequate, and that the American public would reject it.

The mistake that underlies the fallacy of complex question also underlies a common problem in parliamentary procedure. Deliberative bodies sometimes confront a motion that, although not intended deceptively, is covertly complex. In such circumstances there is a need, before discussion, to simplify the issues confronting the body. This accounts for the privileged position, in parliamentary procedure governed by *Robert's Rules of Order* or similar manuals, of the motion to *divide the question*. For example, a motion that the body "postpone for one year" action on

Complex question
An informal fallacy in which a question is asked in such a way as to presuppose the truth of some conclusion buried in that question.

some controversial matter may wisely be divided into the questions of whether to postpone action, and *if* that is done, then to determine the length of the postponement. Some members may support the postponement itself yet find the one-year period intolerably long; if the opportunity to divide the question were not given priority, the body might be maneuvered into taking action on a motion that, because of its complexity, cannot be decided in a way that captures the true will of the body. A presiding officer, having the duty to promote a fully rational debate, may solicit the motion to divide the question before beginning the substantive discussion.

Egregious examples of the fallacy of the complex question arise in dialogue or cross-examination in which one party poses a question that is complex, a second party answers the question, and the first party then draws a fallacious inference for which that answer was the ground. For example:

Lawyer: The figures seem to indicate that your sales increased as a result of these misleading advertisements. Is that correct?

Witness: They did not!

Lawyer: But you do admit, then, that your advertising was misleading. How long have you been engaging in practices like these?

When a question is complex, and all of its presuppositions are to be denied, they must be denied individually. The denial of only one presupposition may lead to the assumption of the truth of the other. In law, this has been called "the negative pregnant." Here is an illustration from a notorious murder trial:

Q: Lizzie, did you not take an axe and whack your mother forty times, and then whack your father forty-one times when faced with the prospect of cold mutton stew?

A: Not true. We were to eat Brussels sprouts fondue that day.

P3. Begging the Question (*Petitio Principii*)

The fallacy called **begging the question** is widely misunderstood, partly because its name is misleading. It is the mistake of assuming the truth of what one seeks to prove. The "question" in a formal debate is the issue that is in dispute; to "beg" the question is to ask, or to suppose, that the very matter in controversy be conceded. This is an argument with no merit at all, of course, and one who makes such an assumption commits a gross fallacy.

The Latin name of the fallacy, for which "begging the question" is the translation, is *petitio principii*, so each instance of it is called a *petitio*. One might think the fallacy would be so obvious that no one would ever commit it, but that is not the case. The logical mistake arises because it is obscured, even from its author, by the language used. Logician Richard Whately used this classic example of a deceptive *petitio*:

> To allow every man unbounded freedom of speech must always be, on the whole, advantageous to the state; for it is highly conducive to the interests of the community that each individual should enjoy a liberty, perfectly unlimited, of expressing his sentiments.[31]

Begging the question
An informal fallacy in which the conclusion of an argument is stated or assumed in any one of the premises. Also known as "circular argument" and *petitio principii*.

This statement says only that freedom of speech is a good thing because it is a good thing—which is not much of an argument.

In the effort to establish the desired conclusion, an author may cast about, searching for premises that will do the trick. Of course, the conclusion itself, reformulated in other words, will do the trick very nicely. Another illustration, equally fallacious, is found in this claim by a sixteenth-century Chinese philosopher:

> There is no such thing as knowledge which cannot be carried into practice, for such knowledge is really no knowledge at all.[32]

This fallacy, like the fallacy of missing the point, is often a mistake that is not recognized by the author of the passage. The presumption that is the heart of the fallacy is buried in the verbiage of the premises, sometimes obscured by confusing or unrecognized synonyms. The arguments are *circular*—every *petitio* is a circular argument—but the circle that has been constructed may be large and confusing, and thus the logical mistake goes unseen.

It would be wrong to suppose that only silly authors make this mistake. Even powerful minds are on occasion snared by this fallacy, as is illustrated by a highly controversial issue in the history of philosophy. Logicians have long sought to establish the reliability of inductive procedures by establishing the truth of what is called the *principle of induction*. This is the principle that the laws of nature will operate tomorrow as they operate today, that in basic ways nature is essentially uniform, and that therefore we may rely on past experience to guide our conduct in the future. "That the future will be essentially like the past" is the claim at issue, but this claim, never doubted in ordinary life, turns out to be very difficult to prove. Some thinkers have claimed that they could prove it by showing that, when we have in the past relied on the inductive principle, we have always found that this method has helped us to achieve our objectives. They ask, "Why conclude that the future will be like the past?" and answer, "Because it always has been like the past."

As David Hume pointed out, however, this common argument is a *petitio*—it begs the question. The point at issue is whether nature *will continue* to behave regularly. That it *has* done so in the past cannot serve as proof that it will do so in the future, unless one assumes the very principle that is here in question: that the future will be like the past. Hence Hume, granting that in the past the future has been like the past, asked the telling question with which philosophers still tussle: How can we know that future futures will be like past futures? They *may* be so, of course, but we cannot *assume* that they will be for the sake of *proving* that they will.[33]

Because the name of this fallacy is widely misunderstood, that name is sometimes wrongly used to refer to a linguistic device that is not a fallacy, not even an argument of any kind, but merely a provocative observation. A claim "begs" the question (in this sense) when it *raises* some question or opens the door to some controversy. Thus a magazine headline may mistakenly read, "The President's decision to invade Iraq begs the question: What are the limits of the President's war-making authority?" This use of the phrase is simply a linguistic mistake. To "beg the question" is not to raise the issue, but to *assume* the truth of the conclusion sought.

Circular arguments are certainly fallacious, but the premises are not irrelevant to the conclusions drawn. They are relevant; indeed, they prove the conclusion, but they do so trivially—they end where they began. A *petitio principii* is always technically valid, but always worthless.

> ### overview
>
> ### Fallacies of Presumption
>
> **P1. Accident**
>
> An informal fallacy in which a generalization is applied to individual cases that it does not govern.
>
> **P2. Complex Question (*Plurium Interrogationum*)**
>
> An informal fallacy in which a question is asked in such a way as to presuppose the truth of some proposition buried in the question.
>
> **P3. Begging the Question (*Petitio Principii*)**
>
> An informal fallacy in which the conclusion of an argument is stated or assumed in one of the premises.

EXERCISES

Identify and explain any fallacies of defective induction or of presumption in the following passages:

1. My generation was taught about the dangers of social diseases, how they were contracted, and the value of abstinence. Our schools did not teach us about contraception. They did not pass out condoms, as many of today's schools do. And not one of the girls in any of my classes, not even in college, became pregnant out of wedlock. It wasn't until people began teaching the children about contraceptives that our problems with pregnancy began.
   —Frank Webster, "No Sex Education, No Sex," *Insight*, 17 November 1997

2. A national mailing soliciting funds, by People for the Ethical Treatment of Animals (PETA), included a survey in which questions were to be answered "yes" or "no." Two of the questions asked were these:

   "Do you realize that the vast majority of painful animal experimentation has no relation at all to human survival or the elimination of disease?"

   "Are you aware that product testing on animals does not keep unsafe products off the market?"

3. If you want a life full of sexual pleasures, don't graduate from college. A study to be published next month in *American Demographics* magazine shows that people with the most education have the least amount of sex.
   —*The Chronicle of Higher Education*, 23 January 1998

4. There is no surprise in discovering that acupuncture can relieve pain and nausea. It will probably also be found to work on anxiety, insomnia, and itching, because these are all conditions in which placebos work. Acupuncture works by suggestion, a mechanism whose effects on humans are well known.

   The danger in using such placebo methods is that they will be applied by people inadequately trained in medicine in cases where essential preliminary work has not been done and where a correct diagnosis has not been established.
   —Fred Levit, M.D., "Acupuncture Is Alchemy, Not Medicine," *The New York Times*, 12 November 1997

5. In a motion picture featuring the famous French comedian Sacha Guitry, three thieves are arguing over division of seven pearls worth a king's ransom. One of them hands two to the man on his right, then two to the man on his left. "I," he says, "will keep three." The man on his right says, "How come you keep three?" "Because I am the leader." "Oh. But how come you are the leader?" "Because I have more pearls."

6. ". . . I've always reckoned that looking at the new moon over your left shoulder is one of the carelessest and foolishest things a body can do. Old Hank Bunker done it once, and bragged about it; and in less than two years he got drunk and fell off of the shot tower, and spread himself out so that he was just a kind of a layer, as you may say; and they slid him edgeways between two barn doors for a coffin, and buried him so, so they say, but I didn't see it. Pap told me. But anyway it all come of looking at the moon that way, like a fool."
   —Mark Twain, *The Adventures of Huckleberry Finn*, 1885

7. Former Senator Robert Packwood of Oregon became so angry at the state's leading newspaper, the Portland Oregonian, that in response to a request from that paper for a quote, he offered this: "Since I quit talking to the *Oregonian*, my business has prospered beyond all measure. I assume that my business has prospered because I don't talk to the *Oregonian*. Therefore I will continue that policy. Thanks."
   —*The New York Times*, 7 February 1999

8. Mr. Farrakhan, the Black Muslim leader, citing the example of Israel, said black Americans should also be able to form a country of their own on the African continent, and said he plans to ask African leaders to "carve out a territory for all people in the diaspora." He said black Americans should also be granted dual citizenship by all African countries. "We want dual citizenship," he said, "and because we don't know where we came from, we want dual citizenship everywhere."
   —Kenneth Noble, "U.S. Blacks and Africans Meet to Forge Stronger Ties," *The New York Times*, 27 May 1993

9. The French claim to be a nation of rebels. In fact their heyday of revolution is over. Twenty-first century France rebels against change, not for it.

What typically happens is that a French government decides to do something radical like, say, enable companies to fire service-sector workers who assault their customers. The unions see this as the first step on the road to slavery and call a national strike. After a week of posturing the government backs down and waiters and sales clerks go back to insulting customers just as they have done since time immemorial.

—S. Clarke, "No Sex, Please, We're French,"
The New York Times, 23 March 2007

10. Hiroyuki Suzuki was formerly a member of the Sakaume gumi, an independent crime family in Japan known for its role in gambling. Mr. Suzuki's wife Mariko broke her kneecap, and when Mariko went to church the next Sunday, the minister put his hands on her broken knee and pronounced it healed. She walked away from church that day. Mr. Suzuki regarded her religion as a silly waste of time—but he was fascinated by the recovery of her knee. "In gambling," he said, "you use dice. Dice are made from bone. If God could heal her bone, I figured he could probably assist my dice and make me the best dice thrower in all of Japan." Mr. Suzuki's gambling skills did improve, enabling him to pay off his debts. He now says his allegiance is to Jesus.

—Stephanie Strom, "He Watched over His Rackets,"
The New York Times, 22 June 1999

4.6 Fallacies of Ambiguity

The meaning of words or phrases may shift as a result of inattention, or may be deliberately manipulated within the course of an argument. A term may have one sense in a premise but quite a different sense in the conclusion. When the inference drawn depends on such changes it is, of course, fallacious. Mistakes of this kind are called **fallacies of ambiguity** or sometimes "sophisms." The deliberate use of such devices is usually crude and readily detected—but at times the ambiguity may be obscure, the error accidental, the fallacy subtle. Five varieties are distinguished here.

A1. Equivocation

Most words have more than one literal meaning, and most of the time we have no difficulty keeping those meanings separate by noting the context and using our good sense when reading and listening. Yet when we confuse the several meanings of a word or phrase—accidentally or deliberately—we are using the word equivocally. If we do that in the context of an argument, we commit the **fallacy of equivocation**.

Sometimes the equivocation is obvious and absurd and is used in a joking line or passage. Lewis Carroll's account of the adventures of Alice in *Through the Looking-Glass* is replete with clever and amusing equivocations. One of them goes like this:

Fallacy of ambiguity
An informal fallacy caused by a shift or a confusion in the meanings of words or phrases within an argument. Also known as a "sophism."

Fallacy of equivocation
A fallacy in which two or more meanings of a word or phrase are used, accidentally or deliberately, in different parts of an argument.

"Who did you pass on the road?" the King went on, holding his hand out to the messenger for some hay.

"Nobody," said the messenger.

"Quite right," said the King; "this young lady saw him too. So of course Nobody walks slower than you."

The equivocation in this passage is rather subtle. As it is first used here, the word "nobody" simply means "no person." Reference is then made using a pronoun ("him"), as though that word ("nobody") had *named* a person. When subsequently the same word is capitalized and plainly used as a name ("Nobody"), it putatively names a person having a characteristic (being passed on the road) derived from the first use of the word. Equivocation is sometimes the tool of wit—and Lewis Carroll was a very witty logician.*

Equivocal arguments are always fallacious, but they are not always silly or comical, as in the example discussed in the following excerpt:

> There is an ambiguity in the phrase "have faith in" that helps to make faith look respectable. When a man says that he has faith in the president he is assuming that it is obvious and known to everybody that there is a president, that the president exists, and he is asserting his confidence that the president will do good work on the whole. But, if a man says he has faith in telepathy, he does not mean that he is confident that telepathy will do good work on the whole, but that he believes that telepathy really occurs sometimes, that telepathy exists. Thus the phrase "to have faith in x" sometimes means to be confident that good work will be done by x, who is assumed or known to exist, but at other times means to believe that x exists. Which does it mean in the phrase "have faith in God"? It means ambiguously both; and the self-evidence of what it means in the one sense recommends what it means in the other sense. If there is a perfectly powerful and good god it is self-evidently reasonable to believe that he will do good. In this sense "have faith in God" is a reasonable exhortation. But it insinuates the other sense, namely "believe that there is a perfectly powerful and good god, no matter what the evidence." Thus the reasonableness of trusting God if he exists is used to make it seem also reasonable to believe that he exists.[34]

One kind of equivocation deserves special mention. This is the mistake that arises from the misuse of "relative" terms, which have different meanings in different contexts. For example, the word "tall" is a relative word; a tall man and a tall building are in quite different categories. A tall man is one who is taller than most men, a tall building is one that is taller than most buildings. Certain forms of argument that are valid for nonrelative terms break down when relative terms are substituted for them. The argument "an elephant is an animal; therefore a gray elephant is a gray animal" is perfectly valid. The word "gray" is a nonrelative term. In contrast, the argument "an elephant is an animal; therefore a small

---

*This passage very probably inspired David Powers, who formally changed his name to Absolutely Nobody and ran as an independent candidate for lieutenant governor of the state of Oregon. His campaign slogan was "Hi, I'm Absolutely Nobody. Vote for me." In the general election of 1992, he drew 7 percent of the vote.

elephant is a small animal" is ridiculous. The point here is that "small" is a relative term: A small elephant is a very large animal. The fallacy is one of equivocation with respect to the relative term "small." Not all equivocation on relative terms is so obvious, however. The word "good" is a relative term and is frequently equivocated on when it is argued, for example, that so-and-so is a good general and would therefore be a good president, or that someone is a good scholar and is therefore likely to be a good teacher.

A2. Amphiboly

The **fallacy of amphiboly** occurs when one is arguing from premises whose formulations are ambiguous because of their grammatical construction. The word "amphiboly" is derived from the Greek, its meaning in essence being "two in a lump," or the "doubleness" of a lump. A statement is amphibolous when its meaning is indeterminate because of the loose or awkward way in which its words are combined. An amphibolous statement may be true in one interpretation and false in another. When it is stated as premise with the interpretation that makes it true, and a conclusion is drawn from it on the interpretation that makes it false, then the fallacy of amphiboly has been committed.

In guiding electoral politics, amphiboly can mislead as well as confuse. During the 1990s, while he sat in the U.S. House of Representatives as a Democrat from California, Tony Coelho is reported to have said: "Women prefer Democrats to men." Amphibolous statements make dangerous premises—but they are seldom encountered in serious discourse.

What grammarians call "dangling" participles and phrases often present amphiboly of a striking sort, as in "The farmer blew out his brains after taking affectionate farewell of his family with a shotgun." Some tidbits in *The New Yorker* make acid fun of writers and editors who overlook careless amphiboly:

> Dr. Salick donated, along with his wife, Gloria, $4.5 million to Queens College for the center.
>
> *Gloria is tax-deductible.*[35]

Fallacy of amphiboly
A fallacy in which a loose or awkward combination of words can be interpreted in more than one way; the argument contains a premise based upon one interpretation, while the conclusion relies on a different interpretation.

Fallacy of Accent
A fallacy of ambiguity that occurs when an argument contains a premise that relies on one possible emphasis of certain words, but the conclusion relies on a different emphasis that gives those same words a different meaning.

A3. Accent

We have seen that shifting the meaning of some term in an argument may result in a fallacy of ambiguity. Most commonly that shift is an equivocation, as noted earlier. Sometimes, however, the shift is the result of a change in *emphasis* on a single word or phrase, whose meaning does not change. When the premise of an argument relies on one possible emphasis, but a conclusion drawn from it relies on the meaning of the same words emphasized differently, the **fallacy of accent** has been committed.

This fallacy can be very serious, and in argument it can be very damaging. Its name seems innocuous. This is due, in part, to the origin of the name in the classification of fallacies first presented by Aristotle.[36] It happens that in the Greek language of Aristotle's day, some words spelled identically had different meanings depending on the way in which they were pronounced, or accented. Those different meanings could result in a deceptive argument, appropriately called a fallacy of accent. In English today there are not very many cases in which changing the accent in a word changes the meaning of the word. Three of the most common are *in*crease and in*crease,* *in*sult and in*sult,* *re*cord and re*cord.* These pairs of words accented differently mean different parts of speech—one member of each pair is a noun, the other a verb—and thus it is unlikely that fallacious argument would now arise from those differently accented words.

Over the centuries, however, while the Aristotelian name has been retained, it has come to be applied to a much wider category, which includes the misleading uses of *emphasis* in various forms and the use of meanings deliberately taken out of context. We are greatly stretching the name "accent" that Aristotle used. If we could overcome the weight of tradition, we might wisely rename the argument that misleads in this way "the fallacy of emphasis."

Consider, as an illustration, the different meanings that can be given to the statement

> We should not speak ill of our friends.

When the sentence is read without any special stress on one of its words, this injunction is surely one with which we would all agree. But, if the sentence is read with stress on the word "friends," we might understand it to suggest that speaking ill of those who are not our friends is not precluded. Such an injunction is no longer acceptable as a moral rule. Suppose we stress the word "speak" in this sentence. Then it might suggest that whereas nasty speech is to be avoided, one may work ill even on one's friends—a very troubling conclusion. If the word "we" is emphasized, the suggestion arises that the injunction applies to *us* but not to *others*, and so on. The various arguments that emerge are plainly the outcome of the deliberate manipulation of emphasis; the sentence can be used to achieve assorted fallacious ambiguities. How is the sentence to be rightly understood? That depends on its context, of course. Often, a phrase or a passage can be understood correctly only when its context is known, because that context makes clear the sense in which the words are intended.

Therefore the fallacy of accent may be construed broadly to include the distortion produced by pulling a quoted passage out of its context, putting it in another context, and there drawing a conclusion that could never have been drawn in the original context. Quoting out of context is sometimes done with deliberate craftiness. In the presidential election campaign of 1996 the Democratic vice-presidential candidate, Al Gore, was quoted by a Republican press aide as having said that "there is no proven link between smoking and lung cancer." Those were indeed Mr. Gore's exact words, uttered during a television interview in 1992. But they were only part of a sentence. In that interview, Mr. Gore's full statement was that some tobacco company scientists *"will claim with a straight face that* there is no proven link between smoking and lung cancer. . . . But the weight of the evidence accepted by the overwhelming preponderance of scientists is, yes, smoking does cause lung cancer."[37]

The omission of the words "will claim with a straight face"—and of Gore's express conviction that cancer is caused by smoking—unfairly reversed the sense of the passage from which the quotation was pulled. The argument suggested by the abbreviated quotation, having the apparent conclusion that Mr. Gore seriously doubts the causal link between smoking and cancer, is an egregious example of the fallacy of accent.

Deliberate distortion of this kind is not rare. A biography by Thomas DiLorenzo, purporting to show that Abraham Lincoln was not the advocate of human equality he is widely thought to have been, quotes words of Lincoln that appear to mock the principle that "all men are created equal." Lincoln is quoted thus: "I am sorry to say that I have never seen two men of whom it is true. But I must admit I never saw the Siamese Twins, and therefore will not dogmatically say that no man ever saw a proof of this sage aphorism." DiLorenzo then remarks that such mockery contrasts sharply with the "seductive words of the Gettysburg Address, eleven years later, in which he purported to rededicate the nation to the notion that all men are created equal."[38] However, DiLorenzo fails to report that those quoted words were in fact Lincoln's account of the view of an unnamed Virginia clergyman, a view he goes on immediately to reject, saying that it "sounds strangely in republican America." DiLorenzo's failure to report the context of the words quoted renders his argument fallacious and disreputable.

Advertising often relies on the same device. A theater critic who says of a new play that it is far from the funniest appearing on Broadway this year may find herself quoted in an ad for the play: "Funniest appearing on Broadway this year!" To avoid such distortions, and the fallacies of accent that are built on them, the responsible writer must be scrupulously accurate in quotation, always indicating whether italics were in the original, indicating (with dots) whether passages have been omitted, and so on.

Physical manipulation of print or pictures is commonly used to mislead deliberately through accent. Sensational words appear in large letters in the headlines of newspaper reports, deliberately suggesting mistaken conclusions to those who glance hastily at them. Later in the report the headline is likely to be qualified by other words in much smaller letters. To avoid being tricked, by news reports or in contracts, one is well advised to give careful attention to "the small print." In political propaganda the misleading choice of a sensational heading or the use of a clipped photograph, in what purports to be a factual report, will use accent shrewdly to encourage the drawing of conclusions known by the propagandist to be false. An account that may not be an outright lie may yet distort by accent in ways that are deliberately manipulative or dishonest.

Such practices are hardly rare in advertising. A remarkably low price often appears in very large letters, followed by "and up" in tiny print. Wonderful bargains in airplane fares are followed by an asterisk, with a distant footnote explaining that the price is available only three months in advance for flights on Thursdays following a full moon, or that there may be other "applicable restrictions." Costly items with well-known brand names are advertised at very low prices, with a small note elsewhere in the ad that "prices listed are for limited quantities in stock." Readers are drawn into the store but are likely to be unable to make the purchase at the advertised price. Accented passages, by themselves, are not strictly fallacies; they become embedded in fallacies when one interpretation of a phrase, flowing from its accent, is relied on to suggest a conclusion (for example, that the plane ticket or brand item can be purchased at the listed price) that is very doubtful when account is taken of the misleading accent.

Even the literal truth can be used, by manipulating its placement, so as to deceive with accent. Disgusted with his first mate, who was repeatedly inebriated while on duty, the captain of a ship noted in the ship's log, almost every day, "The mate was drunk today." The angry mate took his revenge. Keeping the log himself on a day when the captain was ill, the mate recorded, "The captain was sober today."

A4. Composition

The term **fallacy of composition** is applied to both of two closely related types of mistaken argument. The first may be described as *reasoning fallaciously from the attributes of the parts of a whole to the attributes of the whole itself*. A flagrant example is to argue that, because every part of a certain machine is light in weight, the machine "as a whole" is light in weight. The error here is manifest when we recognize that a very heavy machine may consist of a very large number of lightweight parts. Not all examples of fallacious composition are so obvious, however. Some are misleading. One may hear it seriously argued that, because each scene of a certain play is a model of artistic perfection, the play as a whole is artistically perfect. This is as much a fallacy of composition as to argue that, because every ship is ready for battle, the whole fleet must be ready for battle.

The other type of composition fallacy is strictly parallel to that just described. Here, the fallacy is *reasoning from attributes of the individual elements or members of a collection to attributes of the collection or totality of those elements*. For example, it would be fallacious to argue that because a bus uses more gasoline than an automobile, all buses use more gasoline than all automobiles. This version of the fallacy of composition turns on a confusion between the "distributive" and the "collective" use of general terms. Thus, although college students may enroll in no more than six different classes each semester, it is also true that college students enroll in hundreds of different classes each semester. This verbal conflict is easily resolved. It may be true of college students, distributively, that each may enroll in no more than six classes each semester. We call this a *distributive* use of the term "college students," because we are speaking of college students taken *singly*. But it is true of college students, taken collectively, that they enroll in hundreds of different classes each semester. This is a *collective* use of the term "college students," in that we are speaking of college students all together, as a totality. Thus, buses, distributively, use more gasoline than automobiles, but collectively, automobiles use more gasoline than buses, because there are so many more of them.

This second kind of composition fallacy may be defined as the invalid inference that what may truly be predicated of a term distributively may also be truly predicated of the term collectively. Thus, the nuclear bombs dropped during World War II did more damage than did the ordinary bombs dropped—but only

Fallacy of composition
A fallacy of ambiguity in which an argument erroneously assigns attributes to a whole (or to a collection) based on the fact that parts of that whole (or members of that collection) have those attributes.

distributively. The matter is exactly reversed when the two kinds of bombs are considered collectively, because so many more conventional bombs were dropped than nuclear ones. Ignoring this distinction in an argument permits the fallacy of composition.

These two varieties of composition, though parallel, are really distinct because of the difference between a mere collection of elements and a whole constructed out of those elements. Thus, a mere collection of parts is no machine; a mere collection of bricks is neither a house nor a wall. A whole, such as a machine, a house, or a wall, has its parts organized or arranged in certain definite ways. Because organized wholes and mere collections are distinct, so are the two versions of the composition fallacy, one proceeding invalidly to wholes from their parts, the other proceeding invalidly to collections from their members or elements.

A5. Division

The **fallacy of division** is simply the reverse of the fallacy of composition. In it the same confusion is present, but the inference proceeds in the opposite direction. As in the case of composition, two varieties of the fallacy of division may be distinguished. The first kind of division consists of *arguing fallaciously that what is true of a whole must also be true of its parts*. To argue that, because a certain corporation is very important and Mr. Doe is an official of that corporation, therefore Mr. Doe is very important, is to commit the fallacy of division. This first variety of the division fallacy is committed in any such argument, as in moving from the premise that a certain machine is heavy, or complicated, or valuable, to the conclusion that this or any other part of the machine must be heavy, or complicated, or valuable. To argue that a student must have a large room because the room is located in a large dormitory would be still another instance of the first kind of fallacy of division.

The second type of division fallacy is committed *when one argues from the attributes of a collection of elements to the attributes of the elements themselves*. To argue that, because university students study medicine, law, engineering, dentistry, and architecture, therefore each, or even any, university student studies medicine, law, engineering, dentistry, and architecture is to commit the second kind of division fallacy. It is true that university students, collectively, study all these various subjects, but it is false that university students, distributively, do so. Instances of this fallacy of division often look like valid arguments, for what is true of a class distributively is certainly true of each and every member. Thus the argument

Fallacy of division
A fallacy of ambiguity in which an argument erroneously assigns attributes to parts of a whole (or to members of a collection) based on the fact that the whole (or the collection) has those attributes.

Dogs are carnivorous.
Afghan hounds are dogs.
Therefore Afghan hounds are carnivorous.

is perfectly valid. Closely resembling this argument is another,

Dogs are frequently encountered in the streets.
Afghan hounds are dogs.
Therefore Afghan hounds are frequently encountered in the streets.

which is invalid, committing the fallacy of division. Some instances of division are obviously jokes, as when the classical example of valid argumentation,

> Humans are mortal.
> Socrates is a human.
> Therefore Socrates is mortal.

is parodied by the fallacious

> American Indians are disappearing.
> That man is an American Indian.
> Therefore that man is disappearing.

The old riddle, "Why do white sheep eat more than black ones?" turns on the confusion involved in the fallacy of division, for the answer ("Because there are more of them") treats collectively what seemed to be referred to distributively in the question.

The fallacy of division, which springs from a kind of ambiguity, resembles the fallacy of accident (discussed in Section 4.5), which springs from unwarranted presumption. Likewise, the fallacy of composition, also flowing from ambiguity, resembles the hasty generalization we call "converse accident." These likenesses are superficial. An explanation of the differences between the two pairs of fallacies will be helpful in grasping the errors committed in all four.

If we infer, from looking at one or two parts of a large machine, that because they happen to be well designed, every one of the machine's many parts is well designed, we commit the fallacy of converse accident or hasty generalization, for what is true about one or two parts may not be true of all. If we examine every part and find that each is carefully made, and from that finding infer that the entire machine is carefully made, we also reason fallaciously, because however carefully the parts were produced, they may have been *assembled* awkwardly or carelessly. Here the fallacy is one of composition. In converse accident, one argues that some atypical members of a class have a specified attribute, and therefore that all members of the class, distributively, have that attribute; in composition, one argues that, because each and every member of the class has that attribute, the class *itself* (collectively) has that attribute. The difference is great. In converse accident, all predications are distributive, whereas in the composition fallacy, the mistaken inference is from distributive to collective predication.

Similarly, division and accident are two distinct fallacies; their superficial resemblance hides the same kind of underlying difference. In division, we argue (mistakenly) that, because the class itself has a given attribute, each of its members also has it. Thus, it is the fallacy of division to conclude that, because an army as a whole is nearly invincible, each of its units is nearly invincible. In accident, we argue (also mistakenly) that, because some rule applies in general, there are no special circumstances in which it might not apply. Thus, we commit the fallacy of accident when we insist that a person should be fined for ignoring a "No Swimming" sign when jumping into the water to rescue someone from drowning.

CHAPTER 4 Fallacies

> ## overview
>
> ### Fallacies of Ambiguity
>
> **A1. Equivocation**
>
> An informal fallacy in which two or more meanings of the same word or phrase have been confused.
>
> **A2. Amphiboly**
>
> An informal fallacy arising from the loose, awkward, or mistaken way in which words are combined, leading to alternative possible meanings of a statement.
>
> **A3. Accent**
>
> An informal fallacy committed when a term or phrase has a meaning in the conclusion of an argument different from its meaning in one of the premises, the difference arising chiefly from a change in emphasis given to the words used.
>
> **A4. Composition**
>
> An informal fallacy in which an inference is mistakenly drawn from the attributes of the parts of a whole to the attributes of the whole itself.
>
> **A5. Division**
>
> An informal fallacy in which a mistaken inference is drawn from the attributes of a whole to the attributes of the parts of the whole.
>
> Unlike accident and converse accident, composition and division are fallacies of *ambiguity*, resulting from the multiple meanings of terms. Wherever the words or phrases used may mean one thing in one part of the argument and another thing in another part, and those different meanings are deliberately or accidentally confounded, we can expect the argument to be fallacious.

EXERCISES

A. Identify and explain the fallacies of ambiguity that appear in the following passages:

1. . . . the universe is spherical in form . . . because all the constituent parts of the universe, that is the sun, moon, and the planets, appear in this form.
   —Nicolaus Copernicus, *The New Idea of the Universe*, 1514

2. Robert Toombs is reputed to have said, just before the Civil War, "We could lick those Yankees with cornstalks." When he was asked after the war what had gone wrong, he is reputed to have said, "It's very simple. Those damn Yankees refused to fight with cornstalks."
   —E. J. Kahn, Jr., "Profiles (Georgia)," *The New Yorker*, 13 February 1978

3. To press forward with a properly ordered wage structure in each industry is the first condition for curbing competitive bargaining; but there is no reason why the process should stop there. What is good for each industry can hardly be bad for the economy as a whole.
   —Edmond Kelly, *Twentieth Century Socialism*, 1910

4. No man will take counsel, but every man will take money: therefore money is better than counsel.
   —Jonathan Swift

5. I've looked everywhere in this area for an instruction book on how to play the concertina without success. (Mrs. F. M., Myrtle Beach, S.C., *Charlotte Observer*)
   You need no instructions. Just plunge ahead boldly.
   —*The New Yorker*, 21 February 1977

6. . . . each person's happiness is a good to that person, and the general happiness, therefore, a good to the aggregate of all persons.
   —John Stuart Mill, *Utilitarianism*, 1861

7. If the man who "turnips!" cries
   Cry not when his father dies,
   'Tis a proof that he had rather
   Have a turnip than his father.
   —Hester L. Piozzi, *Anecdotes of Samuel Johnson*, 1932

8. Fallaci wrote her: "You are a bad journalist because you are a bad woman."
   —Elizabeth Peer, "The Fallaci Papers," *Newsweek*, 1 December 1980

9. A Worm-eating Warbler was discovered by Hazel Miller in Concord, while walking along the branch of a tree, singing, and in good view. (*New Hampshire Audubon Quarterly*)
   That's our Hazel—surefooted, happy, and with just a touch of the exhibitionist.
   —*The New Yorker*, 2 July 1979

10. The basis of logic is the syllogism, consisting of a major and a minor premise and a conclusion—thus:
    Major Premise: Sixty men can do a piece of work sixty times as quickly as one man;
    Minor Premise: One man can dig a post-hole in sixty seconds; therefore—
    Conclusion: Sixty men can dig a post-hole in one second.
    This may be called the syllogism arithmetical, in which, by combining logic and mathematics, we obtain a double certainty and are twice blessed.
    —Ambrose Bierce, *The Devil's Dictionary*, 1911

B. Each of the following passages may be plausibly criticized by some who conclude that it contains a fallacy, but each may be defended by some who deny that the argument is fallacious. Discuss the merits of the argument in each passage, and explain why you conclude that it does (or does not) contain a fallacy.

1. Seeing that eye and hand and foot and every one of our members has some obvious function, must we not believe that in like manner a human being has a function over and above these particular functions?
   —Aristotle, *Nicomachean Ethics*

2. All phenomena in the universe are saturated with moral values. And, therefore, we can come to assert that the universe for the Chinese is a moral universe.
   —T. H. Fang, *The Chinese View of Life*, 1956

3. The only proof capable of being given that an object is visible, is that people actually see it. The only proof that a sound is audible, is that people hear it: and so of the other sources of our experience. In like manner, I apprehend, the sole evidence it is possible to produce that anything is desirable, is that people actually desire it.
   —John Stuart Mill, *Utilitarianism*, 1863

4. Thomas Carlyle said of Walt Whitman that he thinks he is a big poet because he comes from a big country.
   —Alfred Kazin, "The Haunted Chamber," *The New Republic*, 23 June 1986

5. Mr. Levy boasts many excellent bona fides for the job [of Chancellor of the New York City Public Schools]. But there is one bothersome fact: His two children attend an elite private school on Manhattan's Upper East Side. Mr. Levy . . . should put his daughter and son in the public schools. I do not begrudge any parent the right to enroll a child in a private school. My wife and I considered several private schools before sending our children to a public school in Manhattan. Mr. Levy is essentially declaring the public schools unfit for his own children.
   —Samuel G. Freedman, "Public Leaders, Private Schools," *The New York Times*, 15 April 2000

C. Identify and explain the fallacies of relevance or defective induction, or presumption, or ambiguity as they occur in the following passages. Explain why, in the case of some, it may be plausibly argued that what appears at first to be a fallacy is not, when the argument is interpreted correctly.

1. John Angus Smith, approaching an undercover agent, offered to trade his firearm, an automatic, for two ounces of cocaine that he planned to sell at a profit. Upon being apprehended, Smith was charged with "using" a firearm "during and in relation to . . . a drug trafficking crime." Ordinarily conviction under this statute would result in a prison sentence of five years; however, if the firearm, as in this case, is "a machine gun or other automatic weapon," the mandatory sentence is 30 years. Smith was convicted and sentenced to 30 years in prison. The case was appealed to the U.S. Supreme Court.

Justice Antonin Scalia argued that, although Smith certainly did intend to trade his gun for drugs, that was not the sense of "using" intended by the statute. "In the search for statutory meaning we give nontechnical terms their ordinary meanings . . . to speak of 'using a firearm' is to speak of using it for its distinctive purpose, as a weapon." If asked whether you use a cane, he pointed out, the question asks whether you walk with a cane, not whether you display "your grandfather's silver-handled walking stick in the hall."

Justice Sandra Day O'Connor retorted that we may do more than walk with a cane. "The most infamous use of a cane in American history had nothing to do with walking at all—the caning (in 1856) of Senator Charles Sumner in the United States Senate."

Justice Scalia rejoined that the majority of the Court "does not appear to grasp the distinction between how a word can be used and how it is ordinarily used. . . . I think it perfectly obvious, for example, that the falsity requirement for a perjury conviction would not be satisfied if a witness answered 'No' to a prosecutor's enquiry whether he had ever 'used a firearm' even though he had once sold his grandfather's Enfield rifle to a collector."

Justice O'Connor prevailed; Smith's conviction was affirmed.
—*John Angus Smith v. United States*, 508 U.S. 223, 1 June 1993

2. *Time Magazine* book critic Lev Grossman was "quite taken aback" in the summer of 2006 when he saw a full-page newspaper advertisement for Charles Frazier's novel, *Thirteen Moons*, that included a one-word quotation attributed to *Time*. Grossman had written, "Frazier works on an epic scale, but his genius is in the detail." The one-word quotation by which he was struck was "Genius."
—Henry Alford, "Genius!," *The New York Times Review of Books*, 29 April 2007

3. In the Miss Universe Contest of 1994, Miss Alabama was asked: If you could live forever, would you? And why? She answered:

> I would not live forever, because we should not live forever, because if we were supposed to live forever, then we would live forever, but we cannot live forever, which is why I would not live forever.

4. Order is indispensable to justice because justice can be achieved only by means of a social and legal order.
—Ernest van den Haag, Punishing Criminals, 1975

5. The Inquisition must have been justified and beneficial, if whole peoples invoked and defended it, if men of the loftiest souls founded and created it severally and impartially, and its very adversaries applied it on their own account, pyre answering to pyre.
—Benedetto Croce, *Philosophy of the Practical*, 1935

6. The following advertisement for a great metropolitan newspaper appears very widely in Pennsylvania:

> In Philadelphia nearly everybody reads the *Bulletin*.

7. . . . since it is impossible for an animal or plant to be indefinitely big or small, neither can its parts be such, or the whole will be the same.
   —Aristotle, *Physics*

8. For the benefit of those representatives who have not been here before this year, it may be useful to explain that the item before the General Assembly is that hardy perennial called the "Soviet item." It is purely a propaganda proposition, not introduced with a serious purpose of serious action, but solely as a peg on which to hang a number of speeches with a view to getting them into the press of the world. This is considered by some to be very clever politics. Others, among whom the present speaker wishes to be included, consider it an inadequate response to the challenge of the hour.
   —Henry Cabot Lodge, speech to the United Nations General Assembly, 30 November 1953

9. The war-mongering character of all this flood of propaganda in the United States is admitted even by the American press. Such provocative and slanderous aims clearly inspired today's speech by the United States Representative, consisting only of impudent slander against the Soviet Union, to answer which would be beneath our dignity. The heroic epic of Stalingrad is impervious to libel. The Soviet people in the battles at Stalingrad saved the world from the fascist plague and that great victory which decided the fate of the world is remembered with recognition and gratitude by all humanity. Only men dead to all shame could try to cast aspersions on the shining memory of the heroes of that battle.
   —Anatole M. Baranovsky, speech to the United Nations General Assembly, 30 November 1953

10. Prof. Leon Kass reports a notable response to an assignment he had given students at the University of Chicago. Compose an essay, he asked, about a memorable meal you have eaten. One student wrote as follows:

    > I had once eaten lunch with my uncle and my uncle's friend. His friend had once eaten lunch with Albert Einstein. Albert Einstein was once a man of great spirituality. Therefore, by the law of the syllogism, I had once eaten lunch with God.

    —Leon Kass, *The Hungry Soul: Eating and the Perfecting of Our Nature* (New York: The Free Press, 1995)

11. Consider genetically engineered fish. Scientists hope that fish that contain new growth hormones will grow bigger and faster than normal fish. Other scientists are developing fish that could be introduced into cold, northern waters, where they cannot now survive. The intention is to boost fish production for food. The economic benefits may be obvious, but not the risks. Does this make the risks reasonable?
    —Edward Bruggemann, "Genetic Engineering Needs Strict Regulation," *The New York Times*, 24 March 1992

12. The multiverse theory actually injects the concept of a transcendent Creator at almost every level of its logical structure. Gods and worlds, creators and creatures, lie embedded in each other, forming an infinite regress in unbounded space.

    This *reductio ad absurdum* of the multiverse theory reveals what a very slippery slope it is indeed. Since Copernicus, our view of the universe has enlarged by a factor of a billion billion. The cosmic vista stretches one hundred billion trillion miles in all directions—that's a 1 with 23 zeros. Now we are being urged to accept that even this vast region is just a miniscule fragment of the whole.
    —Paul Davies, "A Brief History of the Multiverse,"
    The New York Times, 12 April 2003

13. When Copernicus argued that the Ptolemaic astronomy (holding that the celestial bodies all revolved around the Earth) should be replaced by a theory holding that the Earth (along with all the other planets) revolved around the sun, he was ridiculed by many of the scientists of his day, including one of the greatest astronomers of that time, Clavius, who wrote in 1581:

    > Both [Copernicus and Ptolemy] are in agreement with the observed phenomena. But Copernicus's arguments contain a great many principles that are absurd. He assumed, for instance, that the earth is moving with a triple motion . . . [but] according to the philosophers a simple body like the earth can have only a simple motion. . . . Therefore it seems to me that Ptolemy's geocentric doctrine must be preferred to Copernicus's doctrine.

14. All of us cannot be famous, because all of us cannot be well known.
    —Jesse Jackson, quoted in *The New Yorker*, 12 March 1984

15. The God that holds you over the pit of hell, much as one holds a spider or some loathsome insect over the fire, abhors you, and is dreadfully provoked; his wrath towards you burns like fire; he looks upon you as worthy of nothing else but to be cast into the fire; you are ten thousand times so abominable in his eyes as the most hateful and venomous serpent is in ours. You have offended him infinitely more than a stubborn rebel did his prince; and yet it is nothing but his hand that holds you from falling into the fire every moment.
    —Jonathan Edwards, "The Pit of Hell," 1741

16. Mysticism is one of the great forces of the world's history. For religion is nearly the most important thing in the world, and religion never remains for long altogether untouched by mysticism.
    —John McTaggart, Ellis McTaggart, "Mysticism," *Philosophical Studies*, 1934

17. If science wishes to argue that we cannot know what was going on in [the gorilla] Binti's head when she acted as she did, science must also acknowledge that it cannot prove that nothing was going on. It is because of our irresolvable ignorance, as much as fellow-feeling, that we

should give animals the benefit of doubt and treat them with the respect we accord ourselves.

—Martin Rowe and Mia Macdonald, "Let's Give Animals Respect They Deserve," *The New York Times*, 26 August 1996

18. If we want to know whether a state is brave we must look to its army, not because the soldiers are the only brave people in the community, but because it is only through their conduct that the courage or cowardice of the community can be manifested.

—Richard L. Nettleship, *Lectures on the Republic of Plato*, 1937

19. Whether we are to live in a future state, as it is the most important question which can possibly be asked, so it is the most intelligible one which can be expressed in language.

—Joseph Butler, "Of Personal Identity," 1736

20. Which is more useful, the Sun or the Moon? The Moon is more useful since it gives us light during the night, when it is dark, whereas the Sun shines only in the daytime, when it is light anyway.

—George Gamow (inscribed in the entry hall of the Hayden Planetarium, New York City)

Chapter 4 Summary

A fallacy is a type of argument that may seem to be correct, but that proves on examination not to be so. In this chapter we have grouped the major informal fallacies under four headings: (1) fallacies of relevance, (2) fallacies of defective induction, (3) fallacies of presumption, and (4) fallacies of ambiguity. Within each group we have named, explained, and illustrated the most common kinds of reasoning mistakes.

1. Fallacies of Relevance

R1. The appeal to the populace (*ad populum*): When correct reasoning is replaced by devices calculated to elicit emotional and nonrational support for the conclusion urged.

R2. The appeal to emotion: When correct reasoning is replaced by appeals to specific emotions, such as pity, pride, or envy.

R3. The red herring: When correct reasoning is manipulated by the introduction of some event or character that *deliberately misleads* the audience and thus hinders rational inference.

R4. The straw man: When correct reasoning is undermined by the *deliberate misrepresentation* of the opponent's position.

R5. The attack on the person (*ad hominem*): When correct reasoning about some issue is replaced by an attack upon the *character or special circumstances* of the opponent.

R6. The appeal to force (*ad baculum*): When reasoning is replaced by *threats* in the effort to win support or assent.

R7. Missing the point (*ignoratio elenchi*): When correct reasoning is replaced by the mistaken refutation of a position that was not really at issue.

2. Fallacies of Defective Induction

In fallacies of defective induction, the premises may be relevant to the conclusion, but they are far too weak to support the conclusion. Four major fallacies are as follows:

D1. Appeal to ignorance (*ad ignorantiam*): When it is argued that a proposition is true on the ground that it has not been proved false, or when it is argued that a proposition is false because it has not been proved true.

D2. Appeal to inappropriate authority (*ad verecundiam*): When the premises of an argument appeal to the judgment of some person or persons who have no legitimate claim to authority in the matter at hand.

D3. False cause (*non causa pro causa*): When one treats as the cause of a thing that which is not really the cause of that thing, often relying (as in the subtype *post hoc ergo propter hoc*) merely on the close temporal succession of two events.

D4. Hasty generalization (converse accident): When one moves carelessly or too quickly from one or a very few instances to a broad or universal claim.

3. Fallacies of Presumption

In fallacies of presumption, the mistake in argument arises from relying on some proposition that is assumed to be true but is without warrant and is false or dubious. Three major fallacies are as follows:

P1. Accident: When one mistakenly applies a generalization to an individual case that it does not properly govern.

P2. Complex question (*plurium interrogationum*): When one argues by asking a question in such a way as to presuppose the truth of some assumption buried in that question.

P3. Begging the question (*petitio principii*): When one assumes in the premises of an argument the truth of what one seeks to establish in the conclusion of that same argument.

4. Fallacies of Ambiguity

In fallacies of ambiguity, the mistakes in argument arise as a result of the shift in the meaning of words or phrases, from the meanings that they have in the premises to different meanings that they have in the conclusion. Five major fallacies are as follows:

A1. Equivocation: When the same word or phrase is used with two or more meanings, deliberately or accidentally, in formulating an argument.

A2. Amphiboly: When one of the statements in an argument has more than one plausible meaning, because of the loose or awkward way in which the words in that statement have been combined.

A3. Accent: When a shift of meaning arises within an argument as a consequence of changes in the emphasis given to its words or parts.

A4. Composition: This fallacy is committed (a) when one reasons mistakenly from the attributes of a part to the attributes of the whole, or (b) when one reasons mistakenly from the attributes of an individual member of some collection to the attributes of the totality of that collection.

A5. Division: This fallacy is committed (a) when one reasons mistakenly from the attributes of a whole to the attributes of one of its parts, or (b) when one reasons mistakenly from the attributes of a totality of some collection of entities to the attributes of the individual entities within that collection.

LOGIC IN THE REAL WORLD

SUPERFASHIONALITY

The following is an excerpt from fashion and celebrity magazine *Superfashionality*. Browse this thrilling and edifying publication, and then answer the questions that follow.

How to Tell if He Likes You

Have you been wondering whether that special guy likes you back? Here are some ways to find out!

1. In a study, 90 percent of men said that they like most people. So, he probably likes you!
2. Ask your friends. If they think he likes you, he probably likes you!
3. Statistically speaking, women who purchase mixing bowls are more likely to be in committed relationships. So buy a mixing bowl right away!
4. Any boy who doesn't like you is obviously stupid.

Trend Watch

If you love leather and you love socks, you'll love leather socks!

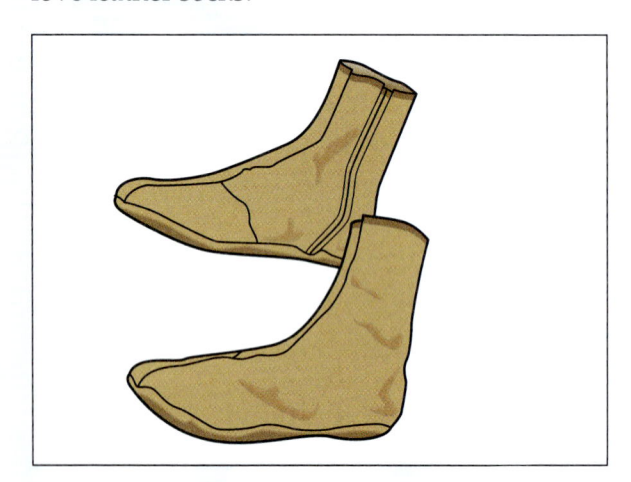

Self-Esteem Corner

Do you have self-esteem? If so, people will naturally like you. Improve your self-esteem in three easy steps:

1. Go out and make friends by getting people to like you.
2. Tell yourself, "These people like me!"
3. Feel the self-esteem!

Capybara Controversy

A hot new trend is capybara fur! Capybara vests and jackets have recently been seen on numerous C-list celebrities at events where free fur vests are given away.

What is a capybara? Dr. M. Hoffenstephen of the University of Nova Scotia explains, "A capybara is an animal that has qualities particular to the species *Hydrochoerus hydrochaeris* (i.e., the capybara)."

However, not everyone's hip to capybara fur. Martha Cupfeld, of People for the Moral Treatment of Mammals, commented, "If we accept the making of vests from capybaras, it's one small step on the path towards making vests from hirsute human beings."

Reality star Whitney Hudson, winner of this season's *Design It or Die!*, commented, "Capybara fur is what *everyone* is wearing. If your clothes are made of regular fabric, you're totally being left behind. Gross."

What is a capybara?

It's this!

Baby for Jen?

Jennifer Amberton was spotted wearing a big shirt! *Superfashionality* waited outside her house with a camera pointed at her front door for eighteen hours. When she came out, we shouted, "Jen, are you going to name your baby Beryl, after your mother?"

Before entering a chauffeur-driven black car and slamming the door behind her, she replied, "My mother's name is Susan, idiots."

There you have it! If she didn't deny that there's a baby on board, she must be pregnant! Keep reading *Superfashionality* to stay on top of this breaking news story!

Celebrity Quotes of the Week

"Haters say that my last movie, "The Last Machine Gun Kickboxer," was too violent, but you can't be against all violence. Smushing bugs is violence."

—*Buck Chatham*

"I am not just a handbag designer. I make satchels, totes. I am an engineer of containers. If it has an inside and an outside, I make it and put sequins on it."

—*Donatella Flaviatore*

"I am in favor of an amendment against Lady Cha-Cha's flag underpants because the most important thing is that millions of Americans cannot afford flags."

—*Johan Colbare*

Questions

1. **How to Tell if He Likes You:** Match each of the suggestions given (1, 2, 3, and 4) with the following four fallacy names:

   A. Appeal to Inappropriate Authority
   B. Fallacy of Equivocation
   C. Argument *ad Hominem*
   D. False Cause

LOGIC IN THE REAL WORLD

2. **Trend Watch:** The statement "If you love leather and you love socks, you'll love leather socks!" is an example of what fallacy?
   A. Fallacy of Division
   B. Accent
   C. Fallacy of Composition
   D. Begging the Question

3. **Self-Esteem Corner:** The idea that self-esteem will cause people to like you, and that you can build self-esteem by taking notice of those who like you, is an example of what fallacy?
   A. Red Herring
   B. Accident
   C. Begging the Question
   D. Straw Man

4. **Capybara Controversy:** What type of definition does Dr. M. Hoffenstephen give for the capybara?
   A. Stipulative
   B. Precising
   C. Lexical
   D. Circular

5. **Capybara Controversy:** What type of definition does the box containing the capybara photo give for the capybara?
   A. Precising
   B. Ostensive
   C. Theoretical
   D. Persuasive

6. **Capybara Controversy:** What type of fallacy does Whitney Hudson commit when urging us to wear capybara fur?
   A. Appeal to the Populace
   B. *Ad Hominem*
   C. Appeal to Ignorance
   D. Hasty Generalization

7. **Baby For Jen?:** Jennifer Amberton was asked a complex question—"Are you going to name your baby Beryl, after your mother?" The question contained two presuppositions, that Jennifer is pregnant and that her mother's name is Beryl. When Jennifer denied the second presupposition but not the first, the first was presumed to be true (that is, because the starlet didn't deny being pregnant, *Superfashionality* assumed that she was). What is the name of this error? [See page 136 of your textbook to find it).
   A. The Auspicious Bump
   B. The Negative Pregnant
   C. The Nosy Interlocutor
   D. The Conceptive Pause

8. **Celebrity Quotes of the Week:** What fallacy does Buck Chatham commit?
   A. False Cause
   B. Amphiboly
   C. Straw Man
   D. Fallacy of Division

9. **Celebrity Quotes of the Week:** Is Donatella Flaviatore's quote decreasing or increasing in intension?
   A. Increasing
   B. Decreasing
   C. Neither

10. **Celebrity Quotes of the Week:** What fallacy does Johan Colbare commit?
    A. Red Herring
    B. Appeal to the Populace
    C. Accent
    D. Equivocation

Solutions

1. B. The word "like" is being used in two different ways here.
   A. Your friends are highly biased.
   D. Maybe women who already have boyfriends are more likely to buy mixing bowls.
   C. Obviously an "against the man" attack.

2. C. This is the Fallacy of Composition: Just because we love leather and socks individually does not mean we will like them together. (The same could be said of ketchup ice cream and other questionable combinations of individually popular items).

3. C. The argument "self-esteem will make people like you and you can get self-esteem by getting people to like you" is circular (see discussion of circular definitions on page 137).

4. D. Dr. M. Hoffenstephen essentially defines a capybara as that which has the characteristics of a capybara—that's not much of a definition at all.

5. B. An ostensive definition is one which "points" at an example of the *definiendum*.

6. A. This is also called the "Bandwagon" fallacy. As in, "Hop on this bandwagon—it's taking us to the fur store at the mall! Everybody's doing it!"

7. B. Believe it or not, the name of that fallacy is "The Negative Pregnant." It was this fairly incredible name that inspired this entire fashion magazine-based exercise. Of course, the Negative Pregnant need not only be about pregnancy; for instance, if someone asks, "Did you fail to do your homework because you were robbing a bank?" and you say, "I did my homework," the (fallacious) implication is that you robbed a bank.

8. C. Rather than arguing against those who say that a particular movie was too violent, Buck argues against a Straw Man: the simplistic argument that all violence is wrong.

9. B. Donatella has moved from defining the items she makes as handbags to defining them in increasingly broad ways: finally, anything with an inside and an outside. This is decreasing in intension (see page 87).

10. A. Rather than tell us why Lady Cha-Cha's flag underpants must be outlawed, Colbare attempts to distract us by pointing out the terrible poverty of people who cannot afford to express their patriotism. Sad, but quite irrelevant to a flag-underpants amendment.

END NOTES

[1] For a discussion of the methods of classifying fallacies, see Howard Kahane, "The Nature and Classification of Fallacies," in *Informal Logic*, edited by J. A. Blair and R. J. Johnson (Inverness, CA: Edgepress, 1980). For a more extensive theoretical treatment of fallacies, see C. L. Hamblin, *Fallacies* (London: Methuen, 1970); and J. Woods and D. Walton, *Argument: The Logic of the Fallacies* (Scarborough, ONT.: McGraw-Hill Ryerson, 1982). For a more detailed listing of the varieties of fallacies, see W. W. Fernside and W. B. Holther, *Fallacy: The Counterfeit of Argument* (Englewood Cliffs, NJ: Prentice-Hall, 1959), who name and illustrate 51 different fallacies; or D. H. Fischer, *Historians' Fallacies* (New York: Harper & Row, 1979), who distinguishes 112 different fallacies.

[2] Reported in *The New England Journal of Medicine*, 26 December 1996.

[3] Plato, *Apology*, 34; Jowett translation.

[4] Dan Brown, *The Da Vinci Code* (New York: Random House, 2003).

[5] David Broder, "Deciding What to Do in Iraq Requires Thought, Not Gut Instinct," *The Washington Post*, 12 January 2007.

[6] www.Figarospeech.com, 19 March 2006.

[7] *The News & Observer*, Raleigh, NC, 5 January 2007.

[8] Constance Baker Motley, *Equal Justice Under Law* (New York: Farrar, Straus & Giroux, 2001).

[9] "'Kind' Racism," *The Sciences*, June 1997.

[10] Interview with Osama bin Laden, CNN, March 1997.

[11] Peter Monaghan, "A Journal Article Is Expunged and Its Authors Cry Foul," *The Chronicle of Higher Education*, 8 December 2000.

[12] "White House Orders Silence on Meese," *The Washington Post*, 29 April 1988.

[13] "Bush Expected to Grant a Stay of Execution," *The New York Times*, 1 June 2000.

[14] *The Collected Works of Abraham Lincoln*, R. P. Basler, editor (New Brunswick, NJ: Rutgers University Press, 1953), Vol. 2, p. 283.

[15] Stephen Tumim, *Great Legal Fiascos* (London, Arthur Barker, 1985).

[16] Erwin Chargaff, in a famous letter to the editor of *Science*, *Science*, vol. 192, p. 938, 1976.

[17] Robert Sinsheimer, "Troubled Dawn for Genetic Engineering," *New Scientist*, vol. 168, p. 148, 1975.

[18] "Wisconsin to Cut Welfare," *Ann Arbor News*, 11 April 1992.

[19] Justice William Brennan, writing for the Court, *In re Winship*, 397 U.S. 358, 1970.

[20] J. Locke, *An Essay Concerning Human Understanding*, 1690.

[21] C. Loring Brace, "Faculty Is Powerless," *The New York Times*, 24 February 1998.

[22] Douglas E. McNeil, "School Prayer Fallacy," *The New York Times*, 10 June 1998.

[23] Peter Steinfels, "Beliefs," *The New York Times*, 13 February 1996.

[24] I. Harvey, "Death Penalty Ethics," *The New York Times*, 13 February 1996.

[25] Ernest van den Haag, "Make Mine Hemlock," *National Review*, 12 June 1995.

[26] Zev Simpser, "A Murder Is a Murder," *The New York Times*, 3 May 2002.

[27] John Bedder, reported in "Fried and Salty, Yessir, Matey, but Truly English," *The New York Times*, 9 March 1993.

[28] Elizabeth Kolbert, "Talk of the Town," *The New Yorker*, 17 November 2003.

[29] Barbara Commins, "The Slide into Poverty," *The New York Times*, 10 September 2000.

[30] Jeremy Rifkin, "Issues in Genetic Research," *The Chronicle of Higher Education*, 3 July 1998.

[31] In his treatise, "Logic," which appeared in the *Encyclopaedia Metropolitana* (London, 1828).

[32] Weng Shou-Jen, *Record of Instructions* (c. 1518).

[33] See David Hume, "Sceptical Doubts Concerning the Operations of the Understanding," in *An Enquiry Concerning Human Understanding*, sec. 4 (1747).

[34] Richard Robinson, *An Atheist's Values* (Oxford: Oxford University Press, 1964), p. 121.

[35] *The New Yorker*, 3 March 2003.

[36] In the work whose title is generally given as *On Sophistical Refutations*.

[37] *The New York Times*, 18 June 1996.

[38] Thomas DiLorenzo, *The Real Lincoln: A New Look at Abraham Lincoln, His Agenda, and an Unnecessary War* (Prima Publishing, 2002).

PEARSON mylogiclab™

For additional exercises and tutorials about concepts covered in this chapter, log in to MyLogicLab at *www.mylogiclab.com* and select your current textbook.

Belvedere, by M.C. Escher, depicts a structure in which the relations of the base to the middle and upper portions are not rational; the pillars seem to connect the parts, but do so in ways that make no sense when closely examined. One pillar, resting on the railing at the rear, appears to support the upper story in front; two other pillars, which rise from the balustrade at the top of the front staircase, appear to support the upper portion of the building at its very rear! No such structure could ever stand.

A deductive argument rests upon premises that serve as its foundation. To succeed, its parts must be held firmly in place by the reasoning that connects those premises to all that is built upon them. If the deductive inferences are solid and reliable at every point, the argument may stand. But if any proposition in the argument is asserted on the basis of other propositions that cannot bear its weight, the argument will collapse as *Belvedere* would collapse. The architect studies the links that can make a building secure; the logician studies the links that can make a deductive argument valid.

M.C. Escher's *Belvedere* © 2004 The M.C. Escher Company. Baam, Holland. All rights reserved.

part II

Deduction

SECTION A		**CLASSICAL LOGIC**
chapter 5		Categorical Propositions
chapter 6		Categorical Syllogisms
chapter 7		Syllogisms in Ordinary Language
SECTION B		**MODERN LOGIC**
chapter 8		Symbolic Logic
chapter 9		Methods of Deduction
chapter 10		Quantification Theory

> For as one may feel sure that a chain will hold when he is assured that each separate link is of good material and that it clasps the two neighboring links, namely, the one preceding and the one following it, so we may be sure of the accuracy of the reasoning when the matter is good, that is to say, when nothing doubtful enters into it, and when the form consists in a perpetual concatenation of truths which allows of no gap.
>
> —*Gottfried Wilhelm Leibniz*

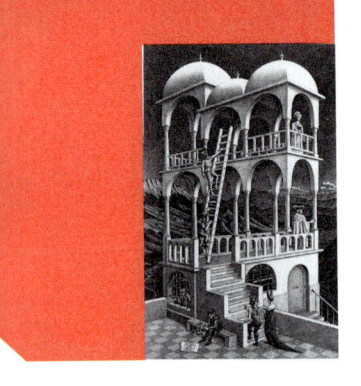

Categorical Propositions

5.1 The Theory of Deduction
5.2 Classes and Categorical Propositions
5.3 The Four Kinds of Categorical Propositions
5.4 Quality, Quantity, and Distribution
5.5 The Traditional Square of Opposition
5.6 Further Immediate Inferences
5.7 Existential Import and the Interpretation of Categorical Propositions
5.8 Symbolism and Diagrams for Categorical Propositions

5.1 The Theory of Deduction

We turn now to the analysis of the structure of arguments. Preceding chapters have dealt mainly with the language in which arguments are formulated. In this and succeeding chapters we explore and explain the relations between the premises of an argument and its conclusion.

All of Part II of this book is devoted to **deductive arguments**. A deductive argument is one whose premises are claimed to provide conclusive grounds for the truth of its conclusion. If that claim is correct—that is, if the premises of the argument really do assure the truth of its conclusion with necessity—that deductive argument is **valid**. Every deductive argument either does what it claims, or it does not; therefore, every deductive argument is either valid or invalid. If it is valid, it is impossible for its premises to be true without its conclusion also being true.

The theory of deduction aims to explain the relations of premises and conclusion in valid arguments. It also aims to provide techniques for the appraisal of deductive arguments—that is, for discriminating between valid and invalid deductions. To accomplish this, two large bodies of theory have been developed. The first is called **classical logic** (or **Aristotelian logic**, after the Greek philosopher who initiated this study). The second is called **modern logic** or **modern symbolic logic**, developed mainly during the nineteenth and twentieth centuries. Classical logic is the topic of this and the following two chapters (Chapters 5, 6, and 7); modern symbolic logic is the topic of Chapters 8, 9, and 10.

Aristotle (384–322 BCE) was one of the towering intellects of the ancient world. After studying for twenty years in Plato's Academy, he became tutor to Alexander the Great; later he founded his own school, the Lyceum, where he contributed substantially to nearly every field of human knowledge. His great treatises on reasoning were collected after his death and came to be called the *Organon*, meaning literally the "instrument," the fundamental tool of knowledge.

The word *logic* did not acquire its modern meaning until the second century CE, but the subject matter of logic was long understood to be the matters treated in

---

chapter 5

Deductive argument
An argument whose premises are claimed to provide conclusive grounds for the truth of its conclusion.

Validity
A characteristic of any deductive argument whose premises, if they were all true, would provide conclusive grounds for the truth of its conclusion. Such an argument is said to be *valid*.

Classical or **Aristotelian logic**
The traditional account of syllogistic reasoning, in which certain interpretations of categorical propositions are presupposed.

Modern or **modern symbolic logic**
The account of syllogistic reasoning accepted today. It differs in important ways from the traditional account.

Aristotle's seminal *Organon*. Aristotelian logic has been the foundation of rational analysis for thousands of years. Over the course of those centuries it has been very greatly refined: its notation has been much improved, its principles have been carefully formulated, its intricate structure has been completed. This great system of classical logic, set forth in this and the next two chapters, remains an intellectual tool of enormous power, as beautiful as it is penetrating.

5.2 Classes and Categorical Propositions

Classical logic deals mainly with arguments based on the relations of classes of objects to one another. By a **class** we mean a collection of all objects that have some specified characteristic in common. (The concept of classes was introduced briefly in Chapter 3, in explaining definitions based on the intension of terms.) Everyone can see immediately that two classes can be related in at least the following three ways:

1. All of one class may be included in all of another class. Thus the class of all dogs is *wholly included* (or wholly contained) in the class of all mammals.
2. Some, but not all, of the members of one class may be included in another class. Thus the class of all athletes is *partially included* (or partially contained) in the class of all females.
3. Two classes may have no members in common. Thus the class of all triangles and the class of all circles may be said to *exclude* one another.

These three relations may be applied to classes, or categories, of every sort. In a deductive argument we present propositions that state the relations between one category and some other category. The propositions with which such arguments are formulated are therefore called **categorical propositions.** Categorical propositions are the fundamental elements, the building blocks of argument, in the classical account of deductive logic. Consider the argument

No athletes are vegetarians.
All football players are athletes.
Therefore no football players are vegetarians.

This argument contains three categorical propositions. We may dispute the truth of its premises, of course, but the relations of the classes expressed in these propositions yield an argument that is certainly valid: If those premises are true, that conclusion must be true. It is plain that each of the premises is indeed categorical; that is, *each premise affirms, or denies, that some class S is included in some other class P, in whole or in part.* In this illustrative argument the three categorical propositions are about the class of all athletes, the class of all vegetarians, and the class of all football players.

The critical first step in developing a theory of deduction based on classes, therefore, is to identify the kinds of categorical propositions and to explore the relations among them.

Class
The collection of all objects that have some specified characteristic in common.

Categorical proposition
A proposition that can be analyzed as being about classes, or categories, affirming or denying that one class, S, is included in some other class, P, in whole or in part.

5.3 The Four Kinds of Categorical Propositions

There are four and only four kinds of **standard-form categorical propositions**. Here are examples of each of the four kinds:

1. All politicians are liars.
2. No politicians are liars.
3. Some politicians are liars.
4. Some politicians are not liars.

We will examine each of these kinds in turn.

1. **Universal affirmative propositions.** In these we assert that *the whole of one class is included or contained in another class*. "All politicians are liars" is an example; it asserts that every member of one class, the class of politicians, is a member of another class, the class of liars. Any universal affirmative proposition can be written schematically as

    All S is P.

    where the letters S and P represent the *subject* and *predicate* terms, respectively. Such a proposition *affirms* that the relation of class *inclusion* holds between the two classes and says that the inclusion is complete, or *universal*. All members of S are said to be also members of P. Propositions in this standard form are called *universal affirmative propositions*. They are also called **A** propositions.

    Categorical propositions are often represented with diagrams, using two interlocking circles to stand for the two classes involved. These are called **Venn diagrams**, named after the English logician and mathematician, John Venn (1834–1923), who invented them. Later we will explore these diagrams more fully, and we will find that such diagrams are exceedingly helpful in appraising the validity of deductive arguments. For the present we use these diagrams only to exhibit graphically the sense of each categorical proposition.

    We label one circle S, for "subject class," and the other circle P, for "predicate class." The diagram for the **A** proposition, which asserts that all S is P, shows that portion of S which is outside of P shaded out, indicating that there are no members of S that are not members of P. So the **A** proposition is diagrammed thus:

All S is P.

Standard-form categorical proposition
Any categorical proposition of the form "All S is P" (universal affirmative), "No S is P" (universal negative), "Some S is P" (particular affirmative), or "Some S is not P" (particular negative). Respectively, these four types are known as **A**, **E**, **I**, and **O** propositions.

Venn diagram
Iconic representation of a categorical proposition or of an argument, used to display their logical forms by means of overlapping circles.

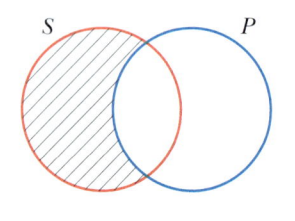

2. **Universal negative propositions.** The second example above, "No politicians are liars," is a proposition in which it is denied, universally, that any member of the class of politicians is a member of the class of liars. It asserts that the subject class, *S*, is wholly excluded from the predicate class, *P*. Schematically, categorical propositions of this kind can be written as

No *S* is *P*.

where again *S* and *P* represent the subject and predicate terms. This kind of proposition *denies* the relation of *inclusion* between the two terms, and denies it *universally*. It tells us that no members of *S* are members of *P*. Propositions in this standard form are called *universal negative propositions*. They are also called **E** propositions.

The diagram for the **E** proposition will exhibit this mutual exclusion by having the overlapping portion of the two circles representing the classes *S* and *P* shaded out. So the **E** proposition is diagrammed thus:

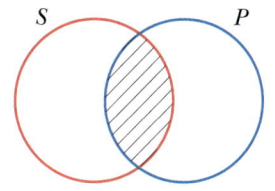

No *S* is *P*.

3. **Particular affirmative propositions.** The third example above, "Some politicians are liars," affirms that some members of the class of all politicians are members of the class of all liars. But it does not affirm this of politicians universally. Only some particular politician or politicians are said to be liars. This proposition does not affirm or deny anything about the class of all politicians; it makes no pronouncements about that entire class. Nor does it say that some politicians are not liars, although in some contexts it may be taken to suggest that. The literal and exact interpretation of this proposition is the assertion that the class of politicians and the class of liars *have some member or members in common*. That is what we understand this standard-form proposition to mean.

"Some" is an indefinite term. Does it mean "at least one," "at least two," or "at least several"? How many does it mean? Context might affect our understanding of the term as it is used in everyday speech, but logicians, for the sake of definiteness, interpret "some" to mean *"at least one."* A particular affirmative proposition may be written schematically as

Some *S* is *P*.

which says that at least one member of the class designated by the subject term *S* is also a member of the class designated by the predicate term *P*. The proposition *affirms* that the relation of class *inclusion* holds, but does not affirm it of the first class universally—it affirms it only partially; that is, it is affirmed of some *particular* member, or members, of the first class. Propositions in this standard form are called *particular affirmative propositions*. They are also called **I** propositions.

The diagram for the **I** proposition indicates that there is at least one member of *S* that is also a member of *P* by placing an *x* in the region in which the two circles overlap. So the **I** proposition is diagrammed thus:

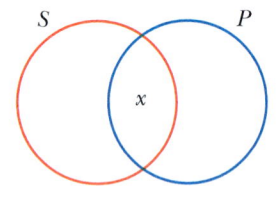

Some *S* is *P*.

4. **Particular negative propositions.** The fourth example above, "Some politicians are not liars," like the third, does not refer to politicians universally, but only to *some* member or members of that class; it is *particular*. Unlike the third example, however, it does not affirm the inclusion of some member or members of the first class in the second class; this is precisely what is *denied*. It is written schematically as

Some *S* is not *P*.

which says that at least one member of the class designated by the subject term *S* is excluded from the whole of the class designated by the predicate term *P*. The denial is not universal. Propositions in this standard form are called *particular negative propositions*. They are also called **O** propositions.

The diagram for the **O** proposition indicates that there is at least one member of *S* that is not a member of *P* by placing an *x* in the region of *S* that is outside of *P*. So the **O** proposition is diagrammed thus:

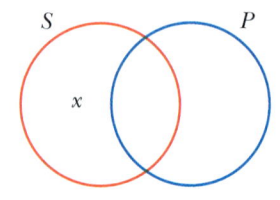

Some *S* is not *P*.

The examples we have used in this section employ classes that are simply named: politicians, liars, vegetarians, athletes, and so on. But subject and predicate terms in standard-form propositions can be more complicated. Thus, for example, the proposition "All candidates for the position are persons of honor and integrity" has the phrase "candidates for the position" as its subject term and the phrase "persons of honor and integrity" as its predicate term. Subject and predicate terms can become more intricate still, but in each of the four standard forms a relation is expressed between a subject class and a predicate class. These four—**A**, **E**, **I**, and **O** propositions—are the building blocks of deductive arguments.

This analysis of categorical propositions appears to be simple and straightforward, but the discovery of the fundamental role of these propositions, and the exhibition of their relations to one another, was a great step in the systematic development of logic. It was one of Aristotle's permanent contributions to human knowledge. Its apparent simplicity is deceptive. On this foundation—classes of objects and the relations among those classes—logicians have erected, over the course of centuries, a highly sophisticated system for the analysis of deductive argument. This system, whose subtlety and penetration mark it as one of the greatest of intellectual achievements, we now explore in the following three steps:

A. In the remainder of this chapter we will examine the features of standard-form categorical propositions more deeply, explaining their relations to one another, and what inferences may be drawn *directly* from these categorical propositions. Much of deductive reasoning can be mastered with no more than a thorough grasp of **A**, **E**, **I**, and **O** propositions and their interconnections.

B. The next chapter will examine *syllogisms*, the arguments that are commonly constructed using standard-form categorical propositions. We will there explore the nature of syllogisms, and show that every valid syllogistic form is uniquely characterized and is therefore given its own name. We will then develop powerful techniques for determining the validity (or invalidity) of syllogisms.

C. In Chapter 7 we integrate syllogistic reasoning and the language of argument in everyday life. Some limitations of reasoning based on this foundation will be identified, but the wide applicability that this foundation makes possible will be demonstrated.

overview

Standard-Form Categorical Propositions

Proposition Form	Name and Type	Example
All *S* is *P*.	**A** Universal affirmative	All lawyers are wealthy people.
No *S* is *P*.	**E** Universal negative	No criminals are good citizens.
Some *S* is *P*.	**I** Particular affirmative	Some chemicals are poisons.
Some *S* is not *P*.	**O** Particular negative	Some insects are not pests.

> **EXERCISES**
>
> Identify the subject and predicate terms in, and name the form of, each of the following propositions:
>
> *1. Some historians are extremely gifted writers whose works read like first-rate novels.
>
> 2. No athletes who have ever accepted pay for participating in sports are amateurs.
>
> 3. No dogs that are without pedigrees are candidates for blue ribbons in official dog shows sponsored by the American Kennel Club.
>
> 4. All satellites that are currently in orbit less than ten thousand miles high are very delicate devices that cost many thousands of dollars to manufacture.
>
> *5. Some members of families that are rich and famous are not persons of either wealth or distinction.
>
> 6. Some paintings produced by artists who are universally recognized as masters are not works of genuine merit that either are or deserve to be preserved in museums and made available to the public.
>
> 7. All drivers of automobiles that are not safe are desperadoes who threaten the lives of their fellows.
>
> 8. Some politicians who could not be elected to the most minor positions are appointed officials in our government today.
>
> 9. Some drugs that are very effective when properly administered are not safe remedies that all medicine cabinets should contain.
>
> *10. No people who have not themselves done creative work in the arts are responsible critics on whose judgment we can rely.

5.4 Quality, Quantity, and Distribution

A. Quality

Quality
An attribute of every categorical proposition, determined by whether the proposition affirms or denies class inclusion. Thus every categorical proposition is either universal in quality or particular in quality.

Every standard-form categorical proposition either affirms, or denies, some class relation, as we have seen. If the proposition affirms some class inclusion, whether complete or partial, its quality is *affirmative*. So the **A** proposition, "All S is P," and the **I** proposition, "Some S is P," are both affirmative in quality. Their letter names, **A** and **I**, are thought to come from the Latin word, "**AffIrmo**," meaning "I affirm." If the proposition denies class inclusion, whether complete or partial, its quality is *negative*. So the **E** proposition, "No S is P," and the **O** proposition, "Some S is not P," are both negative in quality. Their letter names, **E** and **O**, are thought to come from the Latin word, "**nEgO**," meaning "I deny." Every categorical proposition has one quality or the other, affirmative or negative.

B. Quantity

Every standard-form categorical proposition has some class as its subject. If the proposition refers to all members of the class designated by its subject term, its **quantity** is *universal*. So the **A** proposition, "All *S* is *P*," and the **E** proposition, "No *S* is *P*," are both universal in quantity. If the proposition refers only to some members of the class designated by its subject term, its quantity is *particular*. So the **I** proposition, "Some *S* is *P*," and the **O** proposition, "Some *S* is not *P*," are both particular in quantity.

The quantity of a standard-form categorical proposition is revealed by the word with which it begins—"all," "no," or "some." "All" and "no" indicate that the proposition is universal; "some" indicates that the proposition is particular. The word "no" serves also, in the case of the **E** proposition, to indicate its negative quality, as we have seen.

Because every standard-form categorical proposition must be either affirmative or negative, and must be either universal or particular, the four names uniquely describe each one of the four standard forms by indicating its quantity and its quality: universal affirmative (**A**), particular affirmative (**I**), universal negative (**E**), particular negative (**O**).

C. General Schema of Standard-Form Categorical Propositions

Between the subject and predicate terms of every standard-form categorical proposition occurs some form of the verb "to be." This verb (accompanied by "not" in the case of the **O** proposition) serves to connect the subject and predicate terms and is called the **copula**. Writing the four propositions schematically, as we did earlier (All *S* is *P*, Some *S* is *P*, etc.), only the words "is" and "is not" appear; but (depending on context) other forms of the verb "to be" may be appropriate. We may change the tense (for example, "Some Roman emperors were monsters" or "Some soldiers will not be heroes"), or change to the plural form of the verb (for example, "All squares are rectangles"). In these examples, "were," "are," and "will not be" serve as copulas. However, the general skeleton of a standard-form categorical proposition always consists of just four parts: first the quantifier, then the subject term, next the copula, and finally the predicate term. The schema may be written as

Quantifier (subject term) copula (predicate term).

D. Distribution

Categorical propositions are regarded as being about classes, the classes of objects designated by the subject and predicate terms. We have seen that a proposition may refer to classes in different ways; it may refer to *all* members of a class or refer to only *some* members of that class. Thus the proposition, "All senators are citizens," refers to, or is about, *all* senators, but it does not refer to all citizens. That proposition does not affirm that every citizen is a senator, but it does not deny it either. Every **A** proposition is thus seen to refer to all members of the class

Quantity An attribute of every categorical proposition, determined by whether the proposition refers to *all* members or only to *some* members of the class designated by its subject term. Thus every categorical proposition is either universal in quantity or particular in quantity.

Copula Any form of the verb "to be" that serves to connect the subject term and the predicate term of a categorical proposition.

designated by its subject term, *S*, but does *not* refer to all members of the class designated by its predicate term, *P*.

To characterize the ways in which terms can occur in categorical propositions, we introduce the technical term **distribution**. A proposition *distributes* a term if it refers to all members of the class designated by that term. In **A**, **E**, **I**, and **O** propositions, the terms that are distributed vary, as follows:

In the A proposition (e.g., "All senators are citizens"): In this proposition, "senators" is distributed, but "citizens" is not. In **A** propositions (universal affirmatives) the subject term is distributed, but the predicate term is undistributed.

In the E proposition (e.g., "No athletes are vegetarians"): The subject term, "athletes," is distributed, because the whole class of athletes is said to be excluded from the class of vegetarians. However, in asserting that the whole class of athletes is excluded from the class of vegetarians, it is also asserted that the whole class of vegetarians is excluded from the class of athletes. Of each and every vegetarian, the proposition says that he or she is not an athlete. Unlike an **A** proposition, therefore, an **E** proposition refers to all members of the class designated by its predicate term, and therefore also distributes its predicate term. **E** propositions (universal negatives) distribute both their subject and their predicate terms.

In the I proposition (e.g., "Some soldiers are cowards"): No assertion is made about all soldiers in this proposition, and no assertion is made about all cowards either. It says nothing about each and every soldier, and nothing about each and every coward. Neither class is wholly included, or wholly excluded, from the other. In **I** propositions (particular affirmatives) both subject and predicate terms are undistributed.

In the O proposition (e.g., "Some horses are not thoroughbreds"): Nothing is said about all horses. The proposition refers to some members of the class designated by the subject term: it says, of this part of the class of horses, that it is excluded from the class of all thoroughbreds. But they are excluded from the *whole* of the latter class. Given the particular horses referred to, the proposition says that each and every member of the class of thoroughbreds is *not* one of those particular horses. When something is said to be excluded from a class, the whole of the class is referred to, just as, when a person is excluded from a country, all parts of that country are forbidden to that person. In **O** propositions (particular negatives) the subject term is not distributed, but the predicate term is distributed.

We thus see that universal propositions, both affirmative and negative, distribute their subject terms, whereas particular propositions, whether affirmative or negative, do not distribute their subject terms. Thus the *quantity* of any standard-form categorical proposition determines whether its *subject* term is distributed or undistributed. We likewise see that affirmative propositions, whether universal or particular, do not distribute their predicate terms, whereas negative

Distribution
An attribute that describes the relationship between a categorical proposition and each one of its terms, indicating whether or not the proposition makes a statement about every member of the class represented by a given term.

propositions, both universal and particular, do distribute their predicate terms. Thus the *quality* of a standard-form categorical proposition determines whether its *predicate* term is distributed or undistributed.

In summary: the **A** proposition distributes only its subject term; the **E** proposition distributes both its subject and predicate terms; the **I** proposition distributes neither its subject nor its predicate term; and the **O** proposition distributes only its predicate term.

Which terms are distributed by which standard-form categorical propositions will become very important when we turn to the evaluation of syllogisms. The following diagram presents all these distributions graphically and may be useful in helping you to remember which propositions distribute which of their terms:

	Predicate term undistributed	Predicate term distributed
Subject term distributed	**A**: All *S* is *P*.	**E**: No *S* is *P*.
Subject term undistributed	**I**: Some *S* is *P*.	**O**: Some *S* is not *P*.

Visual Logic

The **A** proposition: All bananas are fruits

This **A** proposition asserts that *every* member of the class of bananas (the subject class) is also a member of the class of fruits (the predicate class). When a term refers to every member of a class, we say that the term is *distributed*. *In an **A** proposition, the subject term is always distributed.* But the **A** proposition does not refer to every member of the predicate class; this example does not assert that all fruits are bananas; it says nothing about every fruit. *In an **A** proposition, the predicate term is not distributed.*

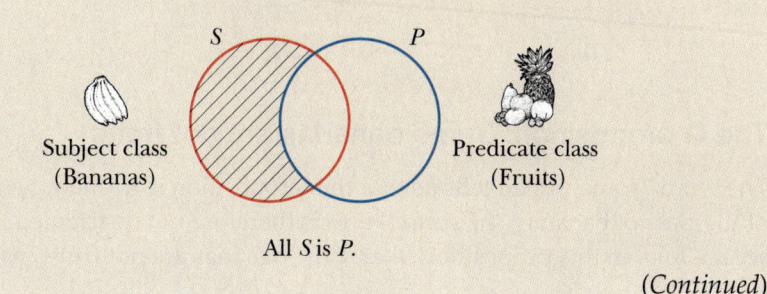

Subject class (Bananas)   Predicate class (Fruits)

All *S* is *P*.

(Continued)

The E proposition: No bananas are fruits

This E proposition asserts that *every* member of the class of bananas is *outside* the class of fruits. The *subject* term, "bananas," is plainly distributed. Because bananas are excluded from the entire class of fruits, this proposition refers to every member of the *predicate* class as well, because it plainly says that *no* fruit is a banana. *In an E proposition, both the subject term and the predicate term are distributed.*

Note that the concept of distribution has nothing to do with truth or falsity. This example proposition is certainly false—but, as in every E proposition, both of its terms are distributed.

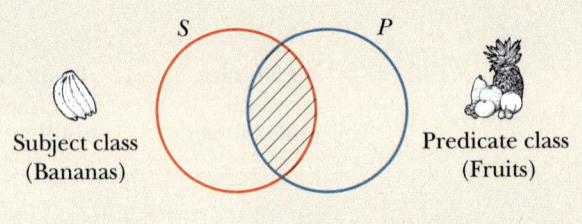

Subject class (Bananas) — Predicate class (Fruits)

No *S* is *P*.

The I proposition: Some bananas are fruits

The word "some" in this I proposition tells us that at least one member of the class designated by the subject term, "bananas," is also a member of the class designated by the predicate term, "fruits"—but this proposition makes no claim about the subject class as a whole. Therefore, in this proposition, as in every I proposition, the subject term is not distributed. Nor does this proposition say anything about every member of the class of fruits (we are told only that there is at least one member of the class of bananas in it), so the predicate is not distributed either. *In an I proposition, neither the subject term nor the predicate term is distributed.*

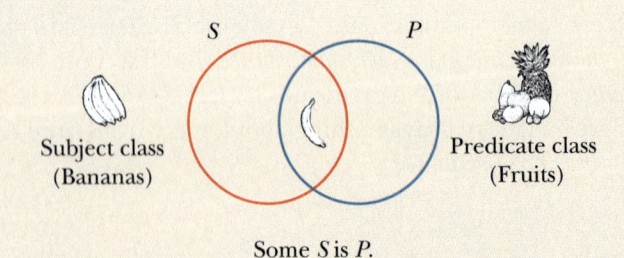

Subject class (Bananas) — Predicate class (Fruits)

Some *S* is *P*.

The O proposition: Some bananas are not fruits

The word "some" again tells us that this proposition is not about all members of the class of bananas; the *subject* term is therefore not distributed. Because we are told, in this proposition, that some bananas are not fruits, we are told something about the entire predicate class—namely, that the entire class of fruits does not have one of those subject bananas among them. *In an **O** proposition, the predicate term is distributed but the subject term is not distributed.*

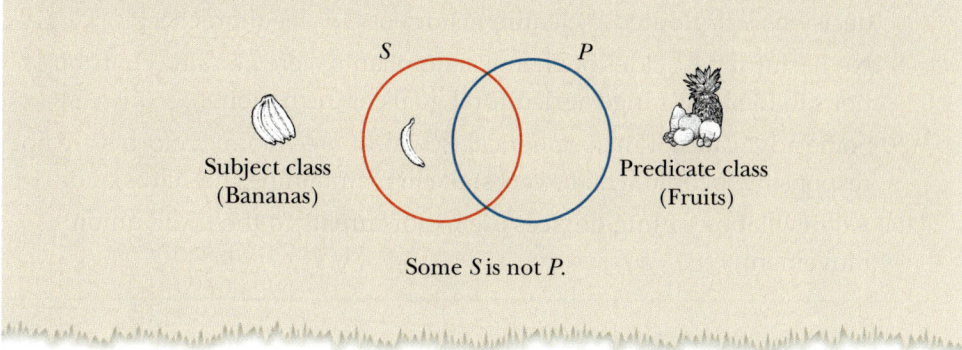

Subject class (Bananas)   Predicate class (Fruits)

Some *S* is not *P*.

We conclude this section with a table that presents all the critical information about each of the four standard-form categorical propositions:

overview

Quantity, Quality, and Distribution

Proposition	Letter Name	Quantity	Quality	Distributes
All *S* is *P*.	A	Universal	Affirmative	*S* only
No *S* is *P*.	E	Universal	Negative	*S* and *P*
Some *S* is *P*.	I	Particular	Affirmative	Neither
Some *S* is not *P*.	O	Particular	Negative	*P* only

EXERCISES

Name the quality and quantity of each of the following propositions, and state whether their subject and predicate terms are distributed or undistributed:

*1. Some presidential candidates will be sadly disappointed people.

2. All those who died in Nazi concentration camps were victims of a cruel and irrational tyranny.

3. Some recently identified unstable elements were not entirely accidental discoveries.

4. Some members of the military-industrial complex are mild-mannered people to whom violence is abhorrent.

*5. No leader of the feminist movement is a major business executive.

6. All hard-line advocates of law and order at any cost are people who will be remembered, if at all, only for having failed to understand the major social pressures of the twenty-first century.

7. Some recent rulings of the Supreme Court were politically motivated decisions that flouted the entire history of U.S. legal practice.

8. No harmful pesticides or chemical defoliants were genuine contributions to the long-range agricultural goals of the nation.

9. Some advocates of major political, social, and economic reforms are not responsible people who have a stake in maintaining the status quo.

*10. All new labor-saving devices are major threats to the trade union movement.

5.5 The Traditional Square of Opposition

The preceding analysis of categorical propositions enables us to exhibit the relations among those propositions, which in turn provide solid grounds for a great deal of the reasoning we do in everyday life. We need one more technical term: *opposition*. Standard-form categorical propositions having the same subject terms and the same predicate terms may (obviously) differ from each other in quality, or in quantity, or in both. Any such kind of differing has been traditionally called **opposition**. This term is used even when there is no apparent disagreement between the propositions. The various kinds of opposition are correlated with some very important truth relations, as follows:

A. Contradictories

Two propositions are **contradictories** if one is the denial or negation of the other—that is, if they cannot both be true and cannot both be false. Two standard-form categorical propositions that have the same subject and predicate terms but differ from each other in *both* quantity and quality are contradictories.

Thus the **A** proposition, "All judges are lawyers," and the **O** proposition, "Some judges are not lawyers," are clearly contradictories. They are opposed in both quality (one affirms, the other denies) and quantity (one refers to all, and the other to some). Of the pair, exactly one is true and exactly one is false. They cannot both be true; they cannot both be false.

Similarly, the **E** proposition, "No politicians are idealists," and the **I** proposition, "Some politicians are idealists," are opposed in both quantity and quality, and they too are contradictories.

In summary: **A** and **O** propositions are contradictories ("All S is P" is contradicted by "Some S is not P"). **E** and **I** propositions are also contradictories ("No S is P" is contradicted by "Some S is P").

Opposition
The logical relation that exists between two contradictories, between two contraries, or in general between any two categorical propositions that differ in quantity, quality, or other respects. These relations are displayed on the square of opposition.

Contradictories
Two propositions so related that one is the denial or negation of the other. On the traditional square of opposition, the two pairs of contradictories are indicated by the diagonals of the square: **A** and **E** propositions are the contradictories of **O** and **I**, respectively.

B. Contraries

Two propositions are said to be **contraries** if they cannot both be true—that is, if the truth of one entails the falsity of the other—but both can be false. Thus, "Texas will win the coming game with Oklahoma," and "Oklahoma will win the coming game with Texas," are contraries. If either of these propositions (referring to the same game, of course) is true, then the other must be false. But these two propositions are not contradictories, because the game could be a draw and then both would be false. Contraries cannot both be true, but, unlike contradictories, they can both be false.

The traditional account of categorical propositions held that universal propositions (**A** and **E**) having the same subject and predicate terms but differing in quality (one affirming, the other denying) were contraries. Thus it was said that an **A** proposition, "All poets are dreamers," and its corresponding **E** proposition, "No poets are dreamers," cannot both be true—but they can both be false and may be regarded as contraries. This Aristotelian interpretation has some troubling consequences that will be discussed in Section 5.7.

One difficulty with this Aristotelian account arises if either the **A** proposition or the **E** proposition is necessarily true—that is, if either is a logical or mathematical truth, such as "All squares are rectangles," or "No squares are circles." In such a case, the claim that the **A** proposition and the **E** proposition are contraries cannot be correct, because a necessarily true proposition cannot possibly be false and so cannot have a contrary, because two propositions can only be contraries if they can both be false. Propositions that are neither necessarily true nor necessarily false are said to be **contingent**. So the reply to this difficulty is that the present interpretation assumes (not unreasonably) that the propositions in question are contingent, in which case the claim that **A** and **E** propositions having the same subject and predicate terms are contraries may be correct. For the remainder of this chapter, we therefore make the assumption that the propositions involved are contingent.

C. Subcontraries

Two propositions are said to be **subcontraries** if they cannot both be false, although they may both be true.

The traditional account held that particular propositions (**I** and **O**) having the same subject and predicate terms but differing in quality (one affirming, the other denying) are subcontraries. It was said that the **I** proposition, "Some diamonds are precious stones," and the **O** proposition, "Some diamonds are not precious stones," could both be true—but they could not both be false and therefore must be regarded as subcontraries.

A difficulty similar to the one noted above arises here too. If either the **I** or the **O** proposition is necessarily false (for example, "Some squares are circles" or "Some squares are not rectangles"), it cannot have a subcontrary, because two propositions that are subcontraries can both be true. But if both the **I** and the **O**

Contraries
Two propositions so related that they cannot both be true, although both may be false.

Contingent
Being neither tautologous nor self-contradictory. A contingent statement may be true or false.

Subcontraries
Two propositions so related that they cannot both be false, although they may both be true.

are contingent propositions, they can both be true, and as we noted in connection with contraries just above, we shall assume for the remainder of this chapter that they are contingent.

D. Subalternation

When two propositions have the same subject and the same predicate terms, and agree in quality (both affirming or both denying) but differ in quantity (one universal, the other particular), they are called *corresponding propositions*. This is also a form of "opposition" as that term has traditionally been used. Thus the **A** proposition, "All spiders are eight-legged animals," has a corresponding **I** proposition, "Some spiders are eight-legged animals." Likewise, the **E** proposition, "No whales are fishes," has a corresponding **O** proposition, "Some whales are not fishes." This opposition between a universal proposition and its corresponding particular proposition is known as **subalternation**. In any such pair of corresponding propositions, the universal proposition is called the *superaltern*, and the particular is called the *subaltern*.

In subalternation (in the classical analysis), the superaltern implies the truth of the subaltern. Thus, from the universal affirmative, "All birds have feathers," the corresponding particular affirmative, "Some birds have feathers," was held to follow. From the universal negative, "No whales are fishes," the corresponding particular, "Some whales are not fishes," was likewise held to follow. But of course the implication does not hold from the particular to the universal, from the subaltern to the superaltern. From the proposition, "Some animals are cats," it is obvious that we cannot infer that "All animals are cats." And it would be absurd to infer from "Some animals are not cats" that "No animals are cats."

E. The Square of Opposition

There are thus four ways in which propositions may be "opposed"—as *contradictories, contraries, subcontraries*, and as *sub-* and *superalterns*. These are represented using an important and widely used diagram called the **square of opposition**, which is reproduced as Figure 5-1.

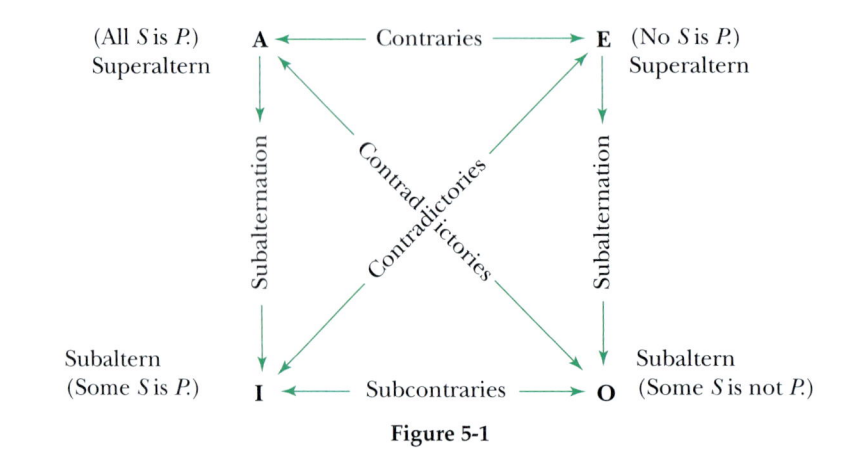

Subalternation
The relation on the square of opposition between a universal proposition (an **A** or an **E** proposition) and its corresponding particular proposition (an **I** or an **O** proposition, respectively). In this relation, the particular proposition (**I** or **O**) is called the "subaltern," and the universal proposition (**A** or **E**) is called the "superaltern."

Square of opposition
A diagram in the form of a square in which the four types of categorical propositions (**A**, **E**, **I**, and **O**) are situated at the corners, exhibiting the logical relations (called "oppositions") among these propositions.

(All *S* is *P*.) **A** ← Contraries → **E** (No *S* is *P*.)
Superaltern   Superaltern

Subalternation — Contradictories — Subalternation

Subaltern   Subaltern
(Some *S* is *P*.) **I** ← Subcontraries → **O** (Some *S* is not *P*.)

Figure 5-1

Relations exhibited by this square of opposition were believed to provide the logical basis for validating certain elementary forms of argument. To explain these, we must first distinguish between **immediate inferences** and **mediate inferences**.

When we draw a conclusion from one or more premises, some inference must be involved. That inference is said to be *mediate* when more than one premise is relied on—as is the case with syllogisms, where the conclusion is drawn from the first premise through the mediation of the second. However, when a conclusion is drawn from only one premise there is no such mediation, and the inference is said to be *immediate*.

A number of very useful immediate inferences may be readily drawn from the information embedded in the traditional square of opposition. Here are some examples:

- If an **A** proposition is the premise, then (according to the square of opposition) one can validly infer that the corresponding **O** proposition (that is, the **O** proposition with the same subject and predicate terms) is false.
- If an **A** proposition is the premise, then the corresponding **I** proposition is true.
- If an **I** proposition is the premise, its corresponding **E** proposition, which contradicts it, must be false.

Given the truth, or the falsehood, of any one of the four standard-form categorical propositions, the truth or falsehood of some or all of the others can be inferred immediately. A considerable number of immediate inferences are based on the traditional square of opposition; we list them here:

A is given as true:	**E** is false; **I** is true; **O** is false.
E is given as true:	**A** is false; **I** is false; **O** is true.
I is given as true:	**E** is false; **A** and **O** are undetermined.
O is given as true:	**A** is false; **E** and **I** are undetermined.
A is given as false:	**O** is true; **E** and **I** are undetermined.
E is given as false:	**I** is true; **A** and **O** are undetermined.
I is given as false:	**A** is false; **E** is true; **O** is true.
O is given as false:	**A** is true; **E** is false; **I** is true.*

Immediate inference
An inference that is drawn directly from one premise without the mediation of any other premise. Various kinds of immediate inferences may be distinguished, traditionally including *conversion*, *obversion*, and *contraposition*.

Mediate inference
Any inference drawn from more than one premise.

---

*A proposition is *undetermined* if its truth or falsity is not determined—fixed—by the truth or falsity of any other proposition. In another sense, a proposition is undetermined if one does not know that it is true and one also does not know that it is false. If it is given that an **A** proposition is undetermined, in either sense, we may infer that its contradictory **O** proposition must be undetermined in that same sense. For if that **O** proposition were known to be true, the **A** proposition contradicting it would be known to be false; and if that **O** proposition were known to be false, the **A** proposition contradicting it would be known to be true. The same reasoning applies to the other standard-form propositions. If any of the four categorical propositions is given as undetermined in either sense, its contradictory must be undetermined in the same sense.

EXERCISES

A. If we assume that the first proposition in each of the following sets is true, what can we affirm about the truth or falsehood of the remaining propositions in each set? B. If we assume that the first proposition in each set is false, what can we affirm?

*1. a. All successful executives are intelligent people.
    b. No successful executives are intelligent people.
    c. Some successful executives are intelligent people.
    d. Some successful executives are not intelligent people.

2. a. No animals with horns are carnivores.
    b. Some animals with horns are carnivores.
    c. Some animals with horns are not carnivores.
    d. All animals with horns are carnivores.

3. a. Some uranium isotopes are highly unstable substances.
    b. Some uranium isotopes are not highly unstable substances.
    c. All uranium isotopes are highly unstable substances.
    d. No uranium isotopes are highly unstable substances.

4. a. Some college professors are not entertaining lecturers.
    b. All college professors are entertaining lecturers.
    c. No college professors are entertaining lecturers.
    d. Some college professors are entertaining lecturers.

5.6 Further Immediate Inferences

There are three other important kinds of immediate inference: *conversion*, *obversion*, and *contraposition*. These are not associated directly with the square of opposition. Each is explained below:

A. Conversion

Conversion is an inference that proceeds by interchanging the subject and predicate terms of a proposition. "No men are angels" converts to "No angels are men," and these propositions may be validly inferred from one another. Similarly, "Some women are writers" and "Some writers are women" are logically equivalent, and by conversion either can be validly inferred from the other. Conversion is perfectly valid for all **E** propositions and for all **I** propositions. One standard-form categorical proposition is said to be the *converse* of another when we derive it by simply interchanging the subject and predicate terms of that other proposition. The proposition from which it is derived is called the *convertend*. Thus, "No idealists are politicians" is the converse of "No politicians are idealists," which is its convertend.

Conversion A valid form of immediate inference for some but not all types of propositions. To form the converse of a proposition the subject and predicate terms are simply interchanged. Thus, applied to the proposition "No circles are squares," conversion yields "No squares are circles," which is called the "converse" of the original proposition. The original proposition is called the "convertend."

The conversion of an **O** proposition is not valid. The **O** proposition, "Some animals are not dogs," is plainly true; its converse is the proposition, "Some dogs are not animals," which is plainly false. An **O** proposition and its converse are not logically equivalent.

The **A** proposition presents a special problem here. Of course, the converse of an **A** proposition does not follow from its convertend. From "All dogs are animals" we certainly may not infer that "All animals are dogs." Traditional logic recognized this, of course, but asserted, nevertheless, that something *like* conversion was valid for **A** propositions. On the traditional square of opposition, one could validly infer from the **A** proposition, "All dogs are animals," its subaltern **I** proposition, "Some dogs are animals." The **A** proposition says something about all members of the subject class (dogs); the **I** proposition makes a more limited claim, about only some of the members of that class. It was held that one could infer "Some *S* is *P*" from "All *S* is *P*." And, as we saw earlier, an **I** proposition may be converted validly; if some dogs are animals, then some animals are dogs.

So, if we are given the **A** proposition, "All dogs are animals," we first infer that "Some dogs are animals" by subalternation, and from that subaltern we can by conversion validly infer that "Some animals are dogs." Hence, by a combination of subalternation and conversion, we advance validly from "All *S* is *P*" to "Some *P* is *S*." This pattern of inference, called *conversion by limitation* (or *conversion per accidens*), proceeds by interchanging subject and predicate terms and changing the quantity of the proposition from universal to particular. This type of conversion will be considered further in the next section.

In all conversions, the converse of a given proposition contains exactly the same subject and predicate terms as the convertend, their order being reversed, and always has the same quality (of affirmation or denial). A complete picture of this immediate inference as traditionally understood is given by the following table:

overview

Valid Conversions

Convertend	Converse
A: All *S* is *P*.	**I:** Some *P* is *S*. (by limitation)
E: No *S* is *P*.	**E:** No *P* is *S*.
I: Some *S* is *P*.	**I:** Some *P* is *S*.
O: Some *S* is not *P*.	(conversion not valid)

B. Classes and Class Complements

To explain other types of immediate inference we must examine more closely the concept of a "class" and explain what is meant by the *complement of a class*. Any class, we have said, is the collection of all objects that have a certain common attribute, which we may refer to as the "class-defining characteristic." The class of all humans is the collection of all things that have the characteristic of being human; its class-defining characteristic is the attribute of being human. The class-defining characteristic need not be a "simple" attribute; any attribute may determine a class. For example, the complex attribute of being left-handed and red-headed and a student determines a class—the class of all left-handed, red-headed students.

Every class has, associated with it, a **complementary class**, or **complement**, which is the collection of all things that do not belong to the original class. The complement of the class of all people is the class of all things that are *not* people. The class-defining characteristic of that complementary class is the (negative) attribute of not being a person. The complement of the class of all people contains no people, but it contains everything else: shoes and ships and sealing wax and cabbages—but no kings, because kings are people. It is often convenient to speak of the complement of the class of all persons as the "class of all nonpersons." The complement of the class designated by the term *S* is then designated by the term *non-S*; we may speak of the term *non-S* as being the complement of the term *S*. Sometimes in reasoning one uses what is called the *relative complement of a class*, its complement within some other class. For example, within the class of "children of mine," there is a subclass, "daughters of mine," whose relative complement is another subclass, "children of mine who are not daughters," or "sons of mine." But obversions, and other immediate inferences, rely on the absolute complement of classes, as defined above.

The word *complement* is thus used in two senses. In one sense it is the complement of a class; in the other it is the complement of a term. These are different but very closely connected. One term is the (term) complement of another just in case the first term designates the (class) complement of the class designated by the second term.

Note that a class is the (class) complement of its own complement. Likewise, a term is the (term) complement of its own complement. A sort of "double negative" rule is involved here, to avoid strings of "non's" prefixed to a term. Thus, the complement of the term "voter" is "nonvoter," but the complement of "nonvoter" should be written simply as "voter" rather than as "nonnonvoter."

One must be careful not to mistake contrary terms for complementary terms. "Coward" and "hero" are contraries, because no person can be both a coward and a hero. We must not identify "cowards" with "nonheroes" because not everyone, and certainly not everything, need be one or the other. Likewise, the complement of the term "winner" is not "loser" but "nonwinner," for although not everything, or even everyone, is either a winner or a loser, absolutely everything is either a winner or a nonwinner.

C. Obversion

Obversion is an immediate inference that is easy to explain once the concept of a term complement is understood. To obvert a proposition, we change its quality (affirmative to negative or negative to affirmative) and replace the predicate term with its complement. However, the subject term remains unchanged,

---

Complement, or **complementary class**
The collection of all things that do not belong to a given class.

Obversion
A valid form of immediate inference for every standard-form categorical proposition. To obvert a proposition we change its quality (from affirmative to negative, or from negative to affirmative) and replace the predicate term with its complement. Thus, applied to the proposition "All dogs are mammals," obversion yields "No dogs are nonmammals," which is called the "obverse" of the original proposition. The original proposition is called the "obvertend."

and so does the quantity of the proposition being obverted. For example, the **A** proposition, "All residents are voters," has as its obverse the **E** proposition, "No residents are nonvoters." These two are logically equivalent propositions, and either may be validly inferred from the other.

Obversion is a valid immediate inference when applied to *any* standard-form categorical proposition:

- The **E** proposition, "No umpires are partisans," has as its obverse the logically equivalent **A** proposition, "All umpires are nonpartisans."
- The **I** proposition, "Some metals are conductors," has as its obverse the **O** proposition, "Some metals are not nonconductors."
- The **O** proposition, "Some nations were not belligerents," has as its obverse the **I** proposition, "Some nations were nonbelligerents."

The proposition serving as premise for the obversion is called the *obvertend*; the conclusion of the inference is called the *obverse*. Every standard-form categorical proposition is logically equivalent to its obverse, so obversion is a valid form of immediate inference for all standard-form categorical propositions. To obtain the obverse of any proposition, we leave the quantity (universal or particular) and the subject term unchanged; we change the quality of the proposition and replace the predicate term with its complement. The following table gives a complete picture of all valid obversions:

overview

Obversions

Obvertend	Obverse
A: All S is P.	**E:** No S is non-P.
E: No S is P.	**A:** All S is non-P.
I: Some S is P.	**O:** Some S is not non-P.
O: Some S is not P.	**I:** Some S is non-P.

D. Contraposition

Another type of immediate inference, **contraposition**, can be reduced to the first two, conversion and obversion. To form the *contrapositive* of a given proposition, we replace its subject term with the complement of its predicate term, and we replace its predicate term with the complement of its subject term. Neither the quality nor the quantity of the original proposition is changed, so the contrapositive of an **A** proposition is an **A** proposition, the contrapositive of an **O** proposition is an **O** proposition, and so forth.

For example, the contrapositive of the **A** proposition, "All members are voters," is the **A** proposition, "All nonvoters are nonmembers." These are logically

Contraposition
A valid form of immediate inference for some, but not for all types of propositions. To form the contrapositive of a given proposition, its subject term is replaced by the complement of its predicate term, and its predicate term is replaced by the complement of its subject term. Thus the contrapositive of the proposition "All humans are mammals" is the proposition "All nonmammals are nonhumans."

equivalent propositions, as will be evident on reflection. Contraposition is plainly a valid form of immediate inference when applied to **A** propositions. It really introduces nothing new, because we can get from any **A** proposition to its contrapositive by first obverting it, next applying conversion, and then applying obversion again. Beginning with "All *S* is *P*," we obvert it to obtain "No *S* is non-*P*," which converts validly to "No non-*P* is *S*," whose obverse is "All non-*P* is non-*S*." The contrapositive of any **A** proposition is the obverse of the converse of the obverse of that proposition.

Contraposition is a valid form of immediate inference when applied to **O** propositions also, although its conclusion may be awkward to express. The contrapositive of the **O** proposition, "Some students are not idealists," is the somewhat cumbersome **O** proposition, "Some nonidealists are not nonstudents," which is logically equivalent to its premise. This also can be shown to be the outcome of first obverting, then converting, then obverting again. "Some *S* is not *P*" obverts to "Some *S* is non-*P*," which converts to "Some non-*P* is *S*," which obverts to "Some non-*P* is not non-*S*."

For **I** propositions, however, contraposition is not a valid form of inference. The true **I** proposition, "Some citizens are nonlegislators," has as its contrapositive the false proposition, "Some legislators are noncitizens." The reason for this invalidity becomes evident when we try to derive the contrapositive of the **I** proposition by successively obverting, converting, and obverting. The obverse of the original **I** proposition, "Some *S* is *P*," is the **O** proposition, "Some *S* is not non-*P*," but (as we saw earlier) the converse of an **O** proposition does not follow validly from it.

In the case of **E** propositions, the contrapositive does not follow validly from the original, as can be seen when, if we begin with the true proposition, "No wrestlers are weaklings," we get, as its contrapositive, the obviously false proposition, "No nonweaklings are nonwrestlers." The reason for this invalidity we will see, again, if we attempt to derive it by successive obversion, conversion, and obversion. If we begin with the **E** proposition, "No *S* is *P*," and obvert it, we obtain the **A** proposition, "All *S* is non-*P*"—which in general cannot be validly converted *except by limitation*. If we do then convert it by limitation to obtain "Some non-*P* is *S*," we can obvert this to obtain "Some non-*P* is not non-*S*." This outcome we may call the *contrapositive by limitation*—and this too we will consider further in the next section.

Contraposition by limitation, in which we infer an **O** proposition from an **E** proposition (for example, we infer "Some non-*P* is not non-*S*" from "No *S* is *P*"), has the same peculiarity as conversion by limitation, on which it depends. Because a particular proposition is inferred from a universal proposition, the resulting contrapositive cannot have the *same* meaning and cannot be logically equivalent to the proposition that was the original premise. On the other hand, the contrapositive of an **A** proposition is an **A** proposition, and the contrapositive of an **O** proposition is an **O** proposition, and in each of these cases the contrapositive and the premise from which it is derived are equivalent.

Contraposition is thus seen to be valid only when applied to **A** and **O** propositions. It is not valid at all for **I** propositions, and it is valid for **E** propositions only by limitation. The complete picture is exhibited in the following table:

overview

Contraposition

Premise	Contrapositive
A: All S is P.	**A:** All non-P is non-S.
E: No S is P.	**O:** Some non-P is not non-S. (by limitation)
I: Some S is P.	(contraposition not valid)
O: Some S is not P.	**O:** Some non-P is not non-S.

Questions about the relations between propositions can often be answered by exploring the various immediate inferences that can be drawn from one or the other of them. For example, given that the proposition, "All surgeons are physicians," is true, what can we know about the truth or falsehood of the proposition, "No nonsurgeons are nonphysicians"? Does this problematic proposition—or its contradictory or contrary—follow validly from the one given as true? To answer we may proceed as follows: From what we are given, "All surgeons are physicians," we can validly infer its contrapositive, "All nonphysicians are nonsurgeons." From this, using conversion by limitation (valid according to the traditional view), we can derive "Some nonsurgeons are nonphysicians." But this is the contradictory of the proposition in question ("No nonsurgeons are nonphysicians"), which is thus no longer problematic but known to be false.

In the very first chapter of this book we noted that a valid argument whose premises are true *must* have a true conclusion, but also that a valid argument whose premises are false *can* have a true conclusion. Thus, from the false premise, "All animals are cats," the true proposition, "Some animals are cats," follows by subalternation. Then from the false proposition, "All parents are students," conversion by limitation yields the true proposition, "Some students are parents." Therefore, if a proposition is given to be false, and the question is raised about the truth or falsehood of some *other*, related proposition, the recommended procedure is to begin drawing immediate inferences from either (1) the contradictory of the proposition known to be false, or (2) the problematic proposition itself. The contradictory of a false proposition must be true, and all valid inferences from that will also be true propositions. If we follow the other course and are able to show that the problematic proposition implies the proposition that is given as false, we know that it must itself be false.

Here follows a table in which the forms of immediate inference—conversion, obversion, and contraposition—are fully displayed:

CHAPTER 5 Categorical Propositions

overview

Immediate Inferences: Conversion, Obversion, Contraposition

CONVERSION

Convertend	Converse
A: All *S* is *P*.	**I:** Some *P* is *S*. (by limitation)
E: No *S* is *P*.	**E:** No *P* is *S*.
I: Some *S* is *P*.	**I:** Some *P* is *S*.
O: Some *S* is not *P*.	(conversion not valid)

OBVERSION

Obvertend	Obverse
A: All *S* is *P*.	**E:** No *S* is non-*P*.
E: No *S* is *P*.	**A:** All *S* is non-*P*.
I: Some *S* is *P*.	**O:** Some *S* is not non-*P*.
O: Some *S* is not *P*.	**I:** Some *S* is non-*P*.

CONTRAPOSITION

Premise	Contrapositive
A: All *S* is *P*.	**A:** All non-*P* is non-*S*.
E: No *S* is *P*.	**O:** Some non-*P* is not non-*S*. (by limitation)
I: Some *S* is *P*.	(contraposition not valid)
O: Some *S* is not *P*.	**O:** Some non-*P* is not non-*S*.

EXERCISES

A. State the converses of the following propositions, and indicate which of them are equivalent to the given propositions:

*1. No people who are considerate of others are reckless drivers who pay no attention to traffic regulations.
2. All graduates of West Point are commissioned officers in the U.S. Army.
3. Some European cars are overpriced and underpowered automobiles.
4. No reptiles are warm-blooded animals.
5. Some professional wrestlers are elderly persons who are incapable of doing an honest day's work.

B. State the obverses of the following propositions:
   *1. Some college athletes are professionals.
   2. No organic compounds are metals.
   3. Some clergy are not abstainers.
   4. No geniuses are conformists.
   *5. All objects suitable for boat anchors are objects that weigh at least fifteen pounds.

C. State the contrapositives of the following propositions and indicate which of them are equivalent to the given propositions.
   *1. All journalists are pessimists.
   2. Some soldiers are not officers.
   3. All scholars are nondegenerates.
   4. All things weighing less than fifty pounds are objects not more than four feet high.
   *5. Some noncitizens are not nonresidents.

D. If "All socialists are pacifists" is true, what may be inferred about the truth or falsehood of the following propositions? That is, which can be known to be true, which can be known to be false, and which are undetermined?
   *1. Some nonpacifists are not nonsocialists.
   2. No socialists are nonpacifists.
   3. All nonsocialists are nonpacifists.
   4. No nonpacifists are socialists.
   *5. No nonsocialists are nonpacifists.
   6. All nonpacifists are nonsocialists.
   7. No pacifists are nonsocialists.
   8. Some socialists are not pacifists.
   9. All pacifists are socialists.
   *10. Some nonpacifists are socialists.

E. If "No scientists are philosophers" is true, what may be inferred about the truth or falsehood of the following propositions? That is, which can be known to be true, which can be known to be false, and which are undetermined?
   *1. No nonphilosophers are scientists.
   2. Some nonphilosophers are not nonscientists.
   3. All nonscientists are nonphilosophers.
   4. No scientists are nonphilosophers.
   *5. No nonscientists are nonphilosophers.
   6. All philosophers are scientists.
   7. Some nonphilosophers are scientists.
   8. All nonphilosophers are nonscientists.

9. Some scientists are not philosophers.
*10. No philosophers are nonscientists.

F. If "Some saints were martyrs" is true, what may be inferred about the truth or falsehood of the following propositions? That is, which can be known to be true, which can be known to be false, and which are undetermined?

*1. All saints were martyrs.
2. All saints were nonmartyrs.
3. Some martyrs were saints.
4. No saints were martyrs.
*5. All martyrs were nonsaints.
6. Some nonmartyrs were saints.
7. Some saints were not nonmartyrs.
8. No martyrs were saints.
9. Some nonsaints were martyrs.
*10. Some martyrs were nonsaints.
11. Some saints were not martyrs.
12. Some martyrs were not saints.
13. No saints were nonmartyrs.
14. No nonsaints were martyrs.
*15. Some martyrs were not nonsaints.

G. If "Some merchants are not pirates" is true, what may be inferred about the truth or falsehood of the following propositions? That is, which can be known to be true, which can be known to be false, and which are undetermined?

*1. No pirates are merchants.
2. No merchants are nonpirates.
3. Some merchants are nonpirates.
4. All nonmerchants are pirates.
*5. Some nonmerchants are nonpirates.
6. All merchants are pirates.
7. No nonmerchants are pirates.
8. No pirates are nonmerchants.
9. All nonpirates are nonmerchants.
*10. Some nonpirates are not nonmerchants.
11. Some nonpirates are merchants.
12. No nonpirates are merchants.
13. Some pirates are merchants.
14. No merchants are nonpirates.
*15. No merchants are pirates.

5.7 Existential Import and the Interpretation of Categorical Propositions

Categorical propositions are the building blocks of arguments, and our aim throughout is to analyze and evaluate arguments. To do this we must be able to diagram and symbolize the **A**, **E**, **I**, and **O** propositions. But before we can do that we must confront and resolve a deep logical problem—one that has been a source of controversy for literally thousands of years. In this section we explain this problem, and we provide a resolution on which a coherent analysis of syllogisms may be developed.

The issues here, as we shall see, are far from simple, but the analysis of syllogisms in succeeding chapters does not require that the complications of this controversy be mastered. It does require that the interpretation of categorical propositions that emerges from the resolution of the controversy be understood. This is commonly called the **Boolean interpretation** of categorical propositions—named after George Boole (1815–1864), an English mathematician whose contributions to logical theory played a key role in the later development of the

Boolean interpretation
The modern interpretation of categorical propositions, adopted in this book and named after the English logician George Boole. In the Boolean interpretation, often contrasted with the Aristotelian interpretation, universal propositions (**A** and **E** propositions) do not have existential import.

Biography

George Boole

George Boole was born in Lincolnshire, England, in 1815, becoming by mid-century one of the great mathematicians of his time. His family was very poor; he was self-taught in the classical languages and in mathematics. When his father, a shoemaker, was unable to support the family, George became an assistant teacher at the age of 16—and then eventually the director of a boarding school. A gold medal from the Royal Society for his mathematical research, and then a paper entitled "The Mathematical Analysis of Logic," led to his appointment, in 1849, as Professor of Mathematics at Queen's College in Cork, Ireland.

George Boole was a penetrating thinker with a great talent for synthesis. The later development of his work by others came to be called "Boolean algebra," which, combined with the properties of electrical switches with which logic can be processed, was critical in the development of modern electronic digital computers. In his great book, *An Investigation into the Laws of Thought, on Which are Founded the Mathematical Theories of Logic and Probabilities* (1854), Boole presented a fully developed system for the symbolic representation of propositions, and also for the general method of logical inference. He showed that with the correct representation of premises, however many terms they may include, it is possible with purely symbolic manipulation to draw any conclusion that is already embedded in those propositions. A modest man and creative scholar, Boole died in 1864 at the age of 49. We continue to rely upon his analyses, seminal in the development of modern symbolic logic.

PEARSON mylogiclab

modern computer. So if the outcome of the following discussion—summarized in the final paragraph of this section, on page 195—is fully grasped, the intervening pages of this section may be safely bypassed.

To understand the problem, and the Boolean outcome with which we emerge, it must be seen that some propositions have existential import, and some do not. A proposition is said to have **existential import** if it typically is uttered to assert the existence of objects of some kind. Why should this seemingly abstruse matter be of concern to the student of logic? Because the correctness of the reasoning in many arguments is directly affected by whether the propositions of which those arguments are built do, or do not, have existential import. We must arrive at a clear and consistent interpretation of categorical propositions in order to determine with confidence what may be rightly inferred from them, and to guard against incorrect inferences that are sometimes drawn from them.

We begin with **I** and **O** propositions, which surely do have existential import. Thus the **I** proposition, "Some soldiers are heroes," says that there exists at least one soldier who is a hero. The **O** proposition, "Some dogs are not companions," says that there exists at least one dog that is not a companion. Particular propositions, **I** and **O** propositions, plainly *do* assert that the classes designated by their subject terms (for example, soldiers and dogs) are not empty—the class of soldiers, and the class of dogs (if the examples given here are true), each has at least one member.*

If this is so, however—if **I** and **O** propositions have existential import (as no one would wish to deny)—wherein lies the problem? The problem arises from the *consequences* of this fact, which are very awkward. Earlier we supposed that an **I** proposition follows validly from its corresponding **A** proposition by subalternation. That is, from "All spiders are eight-legged animals" we infer validly that "Some spiders are eight-legged animals." Similarly, we supposed that an **O** proposition follows validly from its corresponding **E** proposition. But if **I** and **O** propositions have existential import, and they follow validly from their corresponding **A** and **E** propositions, then **A** and **E** propositions must *also* have existential import, because a proposition with existential import cannot be derived validly from another that does not have such import.†

Existential import
An attribute of those propositions that normally assert the existence of objects of some specified kind. Particular propositions (**I** and **O** propositions) always have existential import; thus the proposition "Some dogs are obedient" asserts that there are dogs. Whether universal propositions (**A** and **E** propositions) have existential import is an issue on which the Aristotelian and Boolean interpretations of propositions differ.

---

*A few propositions appear to be exceptions: "Some ghosts appear in Shakespeare's plays" and "Some Greek gods are described in the *Iliad*" are particular propositions that are certainly true even though there are (presumably) neither ghosts nor Greek gods. However, it is the formulation that misleads in such cases. These statements do not themselves affirm the existence of ghosts or Greek gods; they say only that there are certain other propositions that are affirmed or implied in Shakespeare's plays and in the *Iliad*. The intended meaning is "Some passages in Shakespeare's plays are about ghosts" and "Some descriptions in the *Iliad* are of Greek gods." The propositions of Shakespeare and Homer may not be true, but it is certainly true that their writings contain or imply those propositions. That is all that is affirmed by these apparent exceptions, which arise chiefly in literary or mythological contexts. **I** and **O** propositions do have existential import.

†There is another way to show that the existential import of **A** and **E** propositions must follow from that of **I** and **O** propositions, on the traditional square of opposition. In the case of the **A** proposition, we could show it by relying on the (traditionally assumed) validity of conversion by limitation; in the case of the **E** proposition, we could show it by relying on the (traditionally assumed) validity of contraposition by limitation. The result is always the same as that reached above: On the traditional square of opposition, if **I** and **O** propositions have existential import, **A** and **E** propositions must also have existential import.

Visual Logic

Aristotle v. Boole on Interpreting Categorical Propositions

There are two rival interpretations of categorical propositions: the Aristotelian, which is traditional, and the Boolean, which is modern.

In the interpretation of the ancient Greek philosopher Aristotle, the truth of a universal proposition ("All leprechauns wear little green hats," or "No frogs are poisonous") implies the truth of its corresponding particular proposition ("Some leprechauns wear little green hats" or "Some frogs are not poisonous").

In contrast, George Boole, a nineteenth-century English mathematician, argued that we cannot infer the truth of the particular proposition from the truth of its corresponding universal proposition, because (as both sides agree) every particular proposition asserts the existence of its subject class; if some frogs are not poisonous, there must be at least one frog. But if the universal proposition permits us to infer the corresponding particular proposition, then "All leprechauns wear little green hats" would permit us to infer that some leprechauns do, and that would imply that there really are leprechauns!

So, in the modern or Boolean interpretation, a universal proposition (an **A** or an **E** proposition) must be understood to assert only that "If there is such a thing as a leprechaun, it wears a little green hat," and "If there is such a thing as a frog, it is not poisonous."

Source: © Topham/The Image Works

Source: © Topham/The Image Works

This consequence creates a very serious problem. We know that **A** and **O** propositions, on the traditional square of opposition, are contradictories. "All Danes speak English" is contradicted by "Some Danes do not speak English." Contradictories cannot both be true, because one of the pair must be false; nor can they both be false, because one of the pair must be true. But *if* corresponding **A** and **O** propositions do have existential import, as we concluded in the paragraph just above, then both contradictories *could* be false! To illustrate, the **A** proposition, "All inhabitants of Mars are blond," and its corresponding

O proposition, "Some inhabitants of Mars are not blond," are contradictories; if they have existential import—that is, if we interpret them as asserting that there *are* inhabitants of Mars—then both these propositions are false if Mars has no inhabitants. Of course, we do know that Mars has no inhabitants; the class of its inhabitants is empty, so both of the propositions in the example are false. But if they are both false, they *cannot be contradictories!*

Something seems to have gone wrong with the traditional square of opposition in cases of this kind. If the traditional square is correct when it tells us that **A** and **E** propositions validly imply their corresponding **I** and **O** propositions, then the square is not correct when it tells us that corresponding **A** and **O** propositions are contradictories. In that case, the square is also mistaken in holding that the corresponding **I** and **O** propositions are subcontraries.

What is to be done? Can the traditional square of opposition be rescued? Yes, it can, but the price is high. We could rehabilitate the traditional square of opposition by introducing the notion of a *presupposition*. Much earlier (in Section 4.5), we observed that some complex questions are properly answered "yes" or "no" only if the answer to a prior question has been presupposed. "Did you spend the money you stole?" can be reasonably answered "yes" or "no" only if the presupposition that you stole some money is granted. Now, to rescue the square of opposition, we might insist that *all* propositions—that is, the four standard-form categorical propositions **A**, **E**, **I**, and **O**—presuppose (in the sense indicated above) that the classes to which they refer do have members; they are not empty. That is, questions about the truth or falsehood of propositions, and about the logical relations holding among them, are admissible and may be reasonably answered (in this interpretation) only if we presuppose that they never refer to empty classes. In this way, we may save all of the relationships set forth in the traditional square of opposition: **A** and **E** will remain contraries, **I** and **O** will remain subcontraries, subalterns will follow validly from their superalterns, and **A** and **O** will remain contradictories, as will **I** and **E**. To achieve this result, however, we must pay by accepting the blanket presupposition that all classes designated by our terms do have members—are not empty.*

Well, why not do just that? This existential presupposition is both necessary and sufficient to rescue Aristotelian logic. It is, moreover, a presupposition in full accord with the ordinary use of modern languages such as English in very many cases. If you are told, "All the apples in the barrel are Delicious," and you find when you look into the barrel that it is empty, what can you say? You would probably not say that the claim is false, or true, but would instead point out that there *are* no apples in the barrel. You would thus be explaining that the speaker

---

*Phillip H. Wiebe argues that Aristotelian logic does not require the assumption that the class designated by the complement of the subject term be nonempty. See "Existential Assumptions for Aristotelian Logic," *Journal of Philosophical Research* 16 (1990–1991): 321–328. But Aristotelian logic certainly does require the assumption that at least the classes designated by the other three terms (the subject term, the predicate term, and the complement of the predicate term) are not empty—and this existential assumption gives rise to all the difficulties noted in the remarks that follow.

had made a mistake, that in this case the existential presupposition (that there exist apples in the barrel) was false. The fact that we would respond in this corrective fashion shows that we do understand, and do generally accept, the existential presupposition of propositions that are ordinarily uttered.

Unfortunately, this blanket existential presupposition, introduced to rescue the traditional square of opposition, imposes intellectual penalties that are too heavy to bear. There are very good reasons not to do it. Here are three such reasons.

First, this rescue preserves the traditional relations among **A**, **E**, **I**, and **O** propositions, but only at the cost of reducing their power to formulate assertions that we may need to formulate. If we invariably presuppose that the class designated has members, *we will never be able to formulate the proposition that denies that the class has members!* Such denials may sometimes be very important and must surely be made intelligible. We would never be able to formulate the proposition, "No unicorns are creatures that exist."

Second, even ordinary usage of language is not in complete accord with this blanket presupposition. "All" may refer to possibly empty classes. If a property owner were to say, "All trespassers will be prosecuted," far from presupposing that the class of trespassers has members, he would be intending to ensure that the class will become and remain empty. This statement can be true even if no one is ever prosecuted and the word "all" in that statement refers to an empty class. Consider, as another example, the checklist on an IRS return envelope. There is an item that reads, "All necessary schedules have been attached." A taxpayer who did not need to attach any schedules would certainly not hesitate to check the box next to this statement, essentially declaring it to be true even though the class of necessary schedules is, in his case, empty. On the other hand, consider the **I** proposition ("Some *S* is *P*"). Going back to the case of the property owner, suppose he had asserted that "Some trespassers will be prosecuted." If there were no trespassers, then we would call his statement false. This is because, unlike "all," the word "some" in an **I** proposition makes a clear commitment that is incompatible with an empty subject class. The word "some" is interpreted to mean "at least one"—never "zero"—and that concreteness commits particular propositions, if they are to be true, to a state of affairs in which the subject class is not empty.

Third, in science, and in other theoretical spheres, *we often wish to reason without making any presuppositions about existence*. Newton's first law of motion, for example, asserts that certain things are true about bodies that are not acted on by any external forces: They remain at rest, or they continue their straight-line motion. The law may be true; a physicist may wish to express and defend it without wanting to presuppose that there actually are any bodies that are not acted on by external forces.

Objections of this kind make the blanket existential presupposition unacceptable for modern logicians. The Aristotelian interpretation of categorical propositions, long thought to be correct, must be abandoned, and a more modern interpretation employed.

In modern logic it is not assumed that the classes to which categorical propositions refer always have members. The modern interpretation that explicitly rejects this assumption is called, as we noted earlier, Boolean.*

We adopt the Boolean interpretation of categorical propositions in all that follows. This has important logical consequences. Therefore we set forth now what this Boolean interpretation of categorical propositions entails.

1. In some respects, the traditional interpretation is not upset. **I** *and* **O** *propositions continue to have existential import in the Boolean interpretation*, so the proposition "Some *S* is *P*" is false if the class *S* is empty, and the proposition "Some *S* is not *P*" is likewise false if the class *S* is empty.

2. It also remains true in this interpretation that *the universal propositions,* **A** *and* **E**, *are the contradictories of the particular propositions,* **O** *and* **I**. That is, the proposition, "All men are mortal," does contradict the proposition, "Some men are not mortal," and the proposition "No gods are mortal," does contradict the proposition, "Some gods are mortal."

3. All this is entirely coherent because, in the Boolean interpretation, *universal propositions are interpreted as having no existential import*. So even when the *S* class is empty, the proposition, "All *S* is *P*" can be true, as can the proposition, "No *S* is *P*." For example, the propositions, "All unicorns have horns" and "No unicorns have wings" may both be true, even if there are no unicorns. But if there are no unicorns, the **I** proposition, "Some unicorns have horns," is false, as is the **O** proposition, "Some unicorns do not have wings."

4. Sometimes, in ordinary discourse, we utter a universal proposition with which we do intend to assert existence. *The Boolean interpretation permits this to be expressed*, but doing so requires two propositions, one existential in force but particular, the other universal but not existential in force. For example, "All planets in our solar system revolve around the Sun." This is a universal proposition that has no existential import—it says only that if there is a planet in our solar system, then it revolves around the sun. However, if we express the proposition intending also to assert the existence of planets in our solar system that do so revolve, we would need to add: "Mars is a planet in our solar system." This proposition has that desired existential force, referring as it does to actually existing planets.

5. Some very important changes result from our adoption of the Boolean interpretation. *Corresponding* **A** *and* **E** *propositions can both be true and are therefore not contraries*. This may seem paradoxical, but the force of this claim will be understood if we think carefully about the Boolean inter-

---

*Bertrand Russell, another of the founders of modern symbolic logic, also advanced this approach in a famous essay entitled "The Existential Import of Propositions," in *Mind*, July 1905, and referred to it there as "Peano's interpretation" of propositions, after Giuseppe Peano, a great Italian mathematician of the early twentieth century.

pretation of the following two propositions: "All unicorns have wings" and "No unicorns have wings." The first of these asserts only that if there is a unicorn then it has wings, and the second asserts only that if there is a unicorn then it does not have wings—and both of these "if . . . then" propositions, which are corresponding **A** and **E** propositions, can indeed be true if there are no unicorns.

6. In like manner, in the Boolean interpretation, corresponding **I** and **O** propositions, because they do have existential import, can both be false if the subject class is empty. So *corresponding* **I** *and* **O** *propositions are not subcontraries*. If there are no unicorns (that is, if the subject class is empty) it is simply false to assert that some unicorns have horns, and in that case it is also false to assert that some unicorns do not have horns. These corresponding **I** and **O** propositions, which have existential import, are plainly false if there are no unicorns. Since, in this case, they can both be false, they are not subcontraries.

7. In the Boolean interpretation, *subalternation*—inferring an **I** proposition from its corresponding **A**, and an **O** proposition from its corresponding **E**—*is not valid*. This is because, plainly, one may not validly infer a proposition that has existential import from one that does not.

8. The Boolean interpretation *preserves* most immediate inferences: *conversion for* **E** *and for* **I** *propositions* is preserved; *contraposition for* **A** *and for* **O** *propositions* is preserved; *obversion for any proposition* is preserved. But conversion by limitation, and contraposition by limitation, are not valid.

9. The traditional square of opposition, in the Boolean interpretation, is transformed in the following general way: *Relations along the sides of the square are undone, but the diagonal, contradictory relations remain in force*.

In short, the blanket existential presupposition is rejected by modern logicians. It is a mistake, we hold, to *assume* that a class has members if it is not asserted explicitly that it does. Any argument that relies on this mistaken assumption is said to commit the fallacy of existential assumption, or more briefly, the **existential fallacy**.* With this Boolean interpretation clearly in mind, we are now in a position to set forth a powerful system for the symbolizing and diagramming of standard-form categorical syllogisms.

Existential fallacy Any mistake in reasoning that arises from assuming illegitimately that some class has members.

---

*The following exchange from *Alice in Wonderland* might serve as an example of the existential fallacy. The confusion arises because Alice, but not the March Hare or the Mad Hatter, attaches existential import to the word "more":

"Take some more tea," the March Hare said to Alice, very earnestly.

"I've had nothing yet," Alice replied in an offended tone, "so I can't take more."

"You mean you can't take less," said the Hatter: "it's very easy to take more than nothing."

All of this may seem strange at first; however, it must be borne in mind that logical formulations require greater precision than do natural languages, and sometimes meanings are assigned to words and symbols that are not in accord with ordinary usage.

EXERCISES

In the preceding discussion of existential import, it was shown why, in the Boolean interpretation of propositions adopted in this book, most of the inferences that traditionally were thought to be valid are not valid. These inferences mistakenly assume that certain classes have members, thereby committing the existential fallacy. This fallacy is committed in each of the arguments presented below. Explain the point or points at which, in each argument, the mistaken existential assumption is made.

EXAMPLE

A.   (1) No mathematician is one who has squared the circle.
therefore, (2) No one who has squared the circle is a mathematician;
therefore, (3) All who have squared the circle are nonmathematicians;
therefore, (4) Some nonmathematician is one who has squared the circle.

SOLUTION

Step (3) to step (4) is invalid. The inference at this point is conversion by limitation (that is, from "All *S* is *P*" to "Some *P* is *S*"), which was acceptable in the traditional interpretation but is invalid in the Boolean interpretation. This step relies on an inference from a universal proposition to a particular proposition, but the preceding discussion has shown that the classes in a universal proposition cannot be assumed to have members, whereas the classes in a particular proposition do have members. Thus the invalid passage from (3) to (4) permits the inference that the predicate class in (4) is not empty, and therefore that there *is* someone who has squared the circle! In inferring (4) from (3), one commits the existential fallacy.

B.   (1) No citizen is one who has succeeded in accomplishing the impossible;
therefore, (2) No one who has succeeded in accomplishing the impossible is a citizen;
therefore, (3) All who have succeeded in accomplishing the impossible are noncitizens;
therefore, (4) Some who have succeeded in accomplishing the impossible are noncitizens;
therefore, (5) Some noncitizen is one who has succeeded in accomplishing the impossible.

C.   (1) No acrobat is one who can lift himself by his own bootstraps;
therefore, (2) No one who can lift himself by his own bootstraps is an acrobat;
therefore, (3) Someone who can lift himself by his own bootstraps is not an acrobat. (From which it follows that there is at least one being who can lift himself by his own bootstraps.)

D.   (1) It is true that: No unicorns are animals found in the Bronx Zoo;
therefore, (2) It is false that: All unicorns are animals found in the Bronx Zoo;
therefore (3) It is true that: Some unicorns are not animals found in the Bronx Zoo. (From which it follows that there exists at least one unicorn.)

*E.   (1) It is false that: Some mermaids are members of college sororities;
therefore (2) It is true that: Some mermaids are not members of college sororities. (From which it follows that there exists at least one mermaid.)

5.8 Symbolism and Diagrams for Categorical Propositions

Because the Boolean interpretation of categorical propositions depends heavily on the notion of an empty class, it is convenient to have a special symbol to represent it. The zero symbol, 0, is used for this purpose. To say that the class designated by the term S has no members, we write an equals sign between S and 0. Thus the equation $S = 0$ says that there are no S's, or that S has no members.

To say that the class designated by S does have members is to deny that S is empty. To assert that there are S's is to deny the proposition symbolized by $S = 0$. We symbolize that denial by drawing a slanting line through the equals sign. Thus the inequality $S \neq 0$ says that there are S's, by denying that S is empty.

Standard-form categorical propositions refer to two classes, so the equations that represent them are somewhat more complicated. Where each of two classes is already designated by a symbol, the class of all things that belong to both of them can be represented by juxtaposing the symbols for the two original classes. For example, if the letter S designates the class of all satires and the letter P designates the class of all poems, then the class of all things that are both satires and poems is represented by the symbol SP, which thus designates the class of all satirical poems (or poetic satires). The common part or common membership of two classes is called the *product* or *intersection* of the two classes. The *product* of two classes is the class of all things that belong to both of them. The product of the class of all Americans and the class of all composers is the class of all American composers. (One must be on one's guard against certain oddities of the English language here. For example, the product of the class of all Spaniards and the class of all dancers is not the class of all Spanish dancers, for a Spanish dancer is not necessarily a dancer who is Spanish, but any person who performs Spanish dances. Similarly, with abstract painters, English majors, antique dealers, and so on.)

This new notation permits us to symbolize **E** and **I** propositions as equations and inequalities. The **E** proposition, "No S is P," says that no members of the class S are members of the class P; that is, there are no things that belong to both classes. This can be rephrased by saying that the product of the two classes is empty, which is symbolized by the equation $SP = 0$. The **I** proposition, "Some S is P," says that at least one member of S is also a member of P. This means that the product of the classes S and P is not empty and is symbolized by the inequality $SP \neq 0$.

To symbolize **A** and **O** propositions, it is convenient to introduce a new method of representing class complements. The complement of a class is the collection or class of all things that do not belong to the original class, as explained in Section 5.6. The complement of the class of all soldiers is the class of all things that are not soldiers—the class of all nonsoldiers. Where the letter S symbolizes the class of all soldiers, we symbolize the class of all nonsoldiers by $\bar{S}$ (read "S bar"), the symbol for the original class with a bar above it. The **A** proposition, "All S is P," says that all members of the class S are also members of the class P; that is, that there are no members of the class S that are not members of P or (by obversion) that "No S is non-P." This, like any other **E** proposition, says that the product of the classes designated by its subject and predicate terms is empty. It is symbolized by the equation $S\bar{P} = 0$. The **O** proposition, "Some S is not P," obverts to the logically equivalent **I** proposition, "Some S is non-P," which is symbolized by the inequality $S\bar{P} \neq 0$.

In their symbolic formulations, the interrelations among the four standard-form categorical propositions appear very clearly. It is obvious that the **A** and **O** propositions are contradictories when they are symbolized as $S\bar{P} = 0$ and $S\bar{P} \neq 0$, and it is equally obvious that the **E** and **I** propositions, $SP = 0$ and $SP \neq 0$, are contradictories. The *Boolean square of opposition* may be represented as shown in Figure 5-2.

A: $S\bar{P}=0$ **E**: $SP=0$

Contradictories

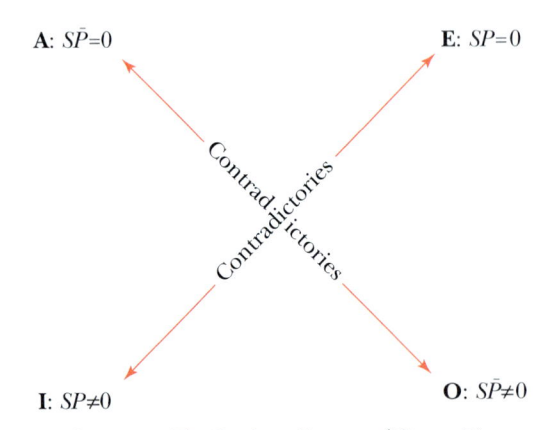

I: $SP \neq 0$ **O**: $S\bar{P} \neq 0$

Figure 5-2 The Boolean Square of Opposition

The notation shown in the table is useful, for example, in representing the relationship among contradictories in the Boolean square of opposition.

When first explaining the four types of standard-form categorical propositions, in Section 5.3, we represented the relations of the classes in those propositions graphically with intersecting circles, labeled S and P. Now we carry that process of diagramming categorical propositions somewhat further, enriching our notation in ways that will facilitate the analysis to follow. We begin by repre-

5.8 Symbolism and Diagrams for Categorical Propositions

overview

Symbolic Representation of Categorical Propositions

Form	Proposition	Symbolic Representation	Explanation
A	All S is P.	$S\bar{P} = 0$	The class of things that are both S and non-P is empty.
E	No S is P.	$SP = 0$	The class of things that are both S and P is empty.
I	Some S is P.	$SP \neq 0$	The class of things that are both S and P is not empty. (SP has at least one member.)
O	Some S is not P.	$S\bar{P} \neq 0$	The class of things that are both S and non-P is not empty. ($S\bar{P}$ has at least one member.)

senting any class with an unmarked circle, labeled with the term that designates that class. The class S is diagrammed with a simple circle, as shown in Figure 5-3.

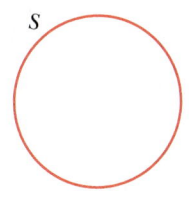

Figure 5-3

The diagram in Figure 5-3 is of a class, not a proposition. It represents the class S, but it says nothing about it. To diagram the proposition that S has no members, or that there are no S's, we shade all of the interior of the circle representing S, indicating in this way that it contains nothing and is empty. To diagram the proposition that there are S's, which we interpret as saying that there is at least one member of S, we place an *x* anywhere in the interior of the circle representing S, indicating in this way that there is something inside it, that it is not empty. Thus the two propositions, "There are no S's," and "There are S's," are represented by the two diagrams in Figure 5-4.

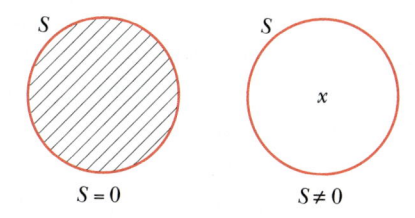

$S = 0$ $S \neq 0$

Figure 5-4

Note that the circle that diagrams the class *S* will also, in effect, diagram the class $\bar{S}$, for just as the interior of the circle represents all members of *S*, so the exterior of the circle represents all members of $\bar{S}$.

To diagram a standard-form categorical proposition, as explained in Section 5.3, two circles are required. Figure 5-5 shows a pair of intersecting circles, which we may use as the skeleton, or framework, for diagramming any standard-form categorical proposition whose subject and predicate terms are symbolized by *S* and *P*.

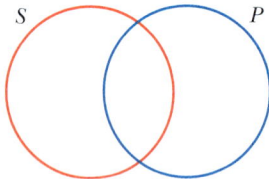

Figure 5-5

Figure 5-5 diagrams the two classes, *S* and *P*, but diagrams no proposition concerning them. It does not affirm that either or both have members, nor does it deny that they have. In fact, there are more than two classes diagrammed by the two intersecting circles. The part of the circle labeled *S* that does not overlap the circle labeled *P* diagrams all *S*'s that are not *P*'s and can be thought of as representing the product of the classes *S* and $\bar{P}$. We may label it $S\bar{P}$. The overlapping part of the two circles represents the product of the classes *S* and *P*, and diagrams all things belonging to both of them. It is labeled *SP*. The part of the circle labeled *P* that does not overlap the circle labeled *S* diagrams all *P*'s that are not *S*'s, and represents the product of the class $\bar{S}$ and *P*. It is labeled $\bar{S}P$. Finally, the part of the diagram external to both circles represents all things that are neither in *S* nor in *P*; it diagrams the fourth class $\bar{S}\bar{P}$ so labeled.

With these labels inserted, Figure 5-5 becomes Figure 5-6.

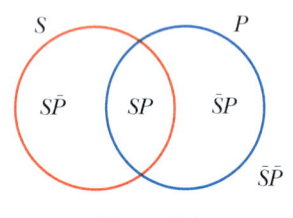

Figure 5-6

Figure 5-6 can be interpreted in terms of the several different classes determined by the class of all Spaniards (*S*) and the class of all painters (*P*). *SP* is the product of these two classes, containing all those things and only those things that belong to both of them. Every member of *SP* must be a member of both *S* and *P*; every member must be both a Spaniard and a painter. This product class *SP* is the class of all Spanish painters, which contains, among others, Velázquez and Goya. $S\bar{P}$ is the product of the first class and the complement of the second, containing all those things and only those things that belong to the class *S* but not to the class *P*. It is the class of all Spaniards who are not painters, all Spanish nonpainters, and it will contain neither Velázquez nor Goya, but it will include both the novelist Cervantes and the dictator Franco, among many others. $\bar{S}P$ is the product of the second class and the complement of the first, and is the class of all painters who are not Spaniards. This class $\bar{S}P$ of all non-Spanish painters includes, among others, both the Dutch painter Rembrandt and the American painter Georgia O'Keeffe. Finally, $\bar{S}\bar{P}$ is the product of the complements of the two original classes. It contains all those things and only those things that are neither Spaniards nor painters. It is a very large class indeed, containing not merely English admirals and Swiss mountain climbers, but such things as the Mississippi River and Mount Everest. All these classes are diagrammed in Figure 5-6, where the letters S and P are interpreted as in this paragraph.

Diagrams of this kind, as noted earlier, are called *Venn diagrams* after John Venn, the English logician who introduced this notation. When, in such diagrams, the several areas are labeled, but not marked in any other way, they represent *classes* only. Figure 5-6 illustrates this. It does not represent any proposition. In such a diagram, if a circle or part of a circle is blank, that signifies nothing—neither that there are, nor that there are not, members of the class represented by that space.

With certain additions, however, Venn diagrams can be used to represent *propositions* as well as classes. By shading out some spaces, or by inserting x's in various parts of the picture, we can accurately diagram any one of the four standard-form categorical propositions. Because Venn diagrams (with appropriate markings) represent categorical propositions so fully and so graphically, these diagrams have become one of the most powerful and most widely used instruments for the appraisal of syllogistic arguments. Let us consider how each of the four basic categorical propositions can be represented using this technique.

To diagram the **A** proposition, "All S is P," symbolized by $S\bar{P} = 0$, we simply shade out the part of the diagram that represents the class $S\bar{P}$, thus indicating that it has no members or is empty. To diagram the **E** proposition, "No S is P," symbolized by $SP = 0$, we shade out the part of the diagram that represents the class SP, to indicate that it is empty. To diagram the **I** proposition, "Some S is P," symbolized by $SP \neq 0$, we insert an x into the part of the diagram that represents the class SP. This insertion indicates that the class product is not empty but has at least one member. Finally, for the **O** proposition, "Some S is not P," symbolized by $S\bar{P} \neq 0$, we insert an x into the part of the diagram that represents the class $S\bar{P}$, to indicate that it is not empty but has at least one member. Placed side by side, diagrams for the four standard-form categorical propositions display their different meanings very clearly, as shown in Figure 5-7.

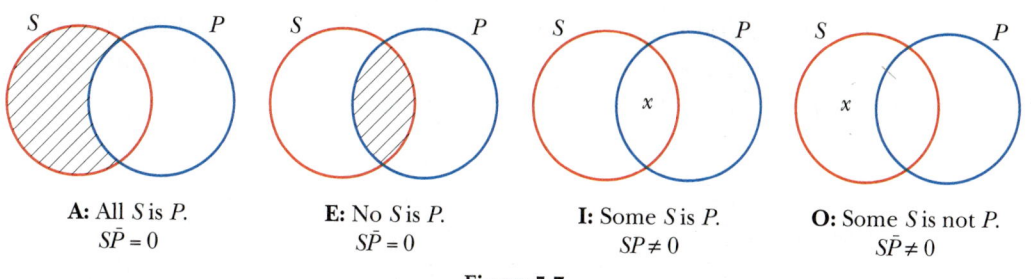

A: All S is P.
$S\bar{P} = 0$

E: No S is P.
$SP = 0$

I: Some S is P.
$SP \neq 0$

O: Some S is not P.
$S\bar{P} \neq 0$

Figure 5-7

We have constructed diagrammatic representations for "No S is P" and "Some S is P," and because these are logically equivalent to their converses, "No P is S" and "Some P is S," the diagrams for the latter have already been shown. To diagram the **A** proposition, "All P is S" (symbolized by $P\bar{S} = 0$) within the same framework, we must shade out the part of the diagram that represents the class $P\bar{S}$. It should be obvious that the class $P\bar{S}$ is the same as the class $\bar{S}P$—if not immediately, then by recognizing that every object that belongs to the class of all painters and the class of all non-Spaniards must (also) belong to the class of all non-Spaniards and the class of all painters—all painting non-Spaniards are non-Spanish painters, and vice versa. To diagram the **O** proposition, "Some P is not S," symbolized by $P\bar{S} \neq 0$ we insert an x into the part of the diagram that represents the class $P\bar{S}\,(=\bar{S}P)$. Diagrams for these propositions then appear as shown in Figure 5-8.

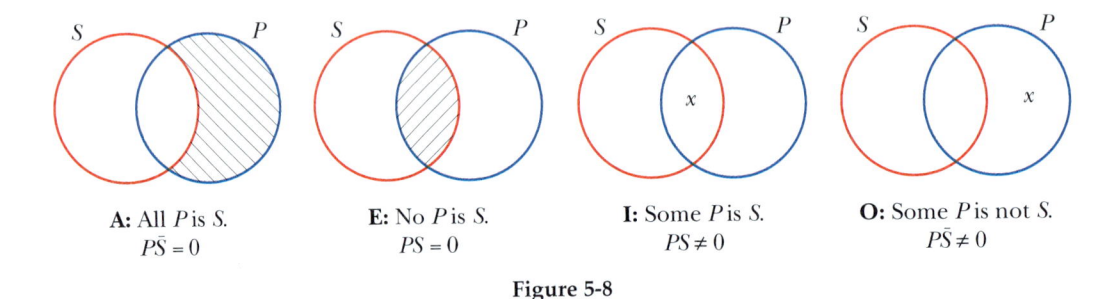

A: All P is S.
$P\bar{S} = 0$

E: No P is S.
$PS = 0$

I: Some P is S.
$PS \neq 0$

O: Some P is not S.
$P\bar{S} \neq 0$

Figure 5-8

This further adequacy of the two-circle diagrams is mentioned because in the next chapter it will be important to be able to use a given pair of overlapping circles with given labels—say, S and M—to diagram any standard-form categorical proposition containing S and M as its terms, regardless of the order in which they occur in it.

The Venn diagrams constitute an *iconic* representation of the standard-form categorical propositions, in which spatial inclusions and exclusions correspond to the nonspatial inclusions and exclusions of classes. They provide an exceptionally clear method of notation. They also provide the basis for the simplest and most direct method of testing the validity of categorical syllogisms, as will be explained in Chapter 6.

EXERCISES

Express each of the following propositions as equalities or inequalities, representing each class by the first letter of the English term designating it, and symbolizing the proposition by means of a Venn diagram.

EXAMPLE

1. Some sculptors are painters.

SOLUTION

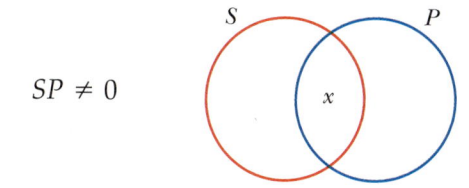

$SP \neq 0$

[Venn diagram with circles labeled S and P, with an x in the overlapping region]

2. No peddlers are millionaires.
3. All merchants are speculators.
4. Some musicians are not pianists.
*5. No shopkeepers are members.
6. Some political leaders of high reputation are scoundrels.
7. All physicians licensed to practice in this state are medical school graduates who have passed special qualifying examinations.
8. Some stockbrokers who advise their customers about making investments are not partners in companies whose securities they recommend.
9. All puritans who reject all useless pleasure are strangers to much that makes life worth living.
*10. No modern paintings are photographic likenesses of their objects.
11. Some student activists are middle-aged men and women striving to recapture their lost youth.
12. All medieval scholars were pious monks living in monasteries.
13. Some state employees are not public-spirited citizens.
14. No magistrates subject to election and recall will be punitive tyrants.
*15. Some patients exhibiting all the symptoms of schizophrenia have bipolar disorder.
16. Some passengers on large jet airplanes are not satisfied customers.
17. Some priests are militant advocates of radical social change.
18. Some stalwart defenders of the existing order are not members of a political party.
19. No pipelines laid across foreign territories are safe investments.
*20. All pornographic films are menaces to civilization and decency.

chapter 5 Summary

This chapter has introduced and explained the basic elements of classical, or Aristotelian, deductive logic, as distinguished from modern symbolic logic. (To review this distinction, see Section 5.1.)

In Section 5.2 we introduced the concept of classes, on which traditional logic is built, and the categorical propositions that express relations between classes.

In Section 5.3 we explained the four basic standard-form categorical propositions:

- **A:** universal affirmative
- **E:** universal negative
- **I:** particular affirmative
- **O:** particular negative

In Section 5.4 we discussed various features of these standard-form categorical propositions: their quality (affirmative or negative) and their quantity (universal or particular). We also explained why different terms are distributed or undistributed, in each of the four basic kinds of propositions.

In Section 5.5 we explored the kinds of opposition arising among the several standard-form categorical propositions: what it means for a proposition to have a contradictory, a contrary, a subcontrary, or a sub- or superaltern. We showed how these relations are exhibited on the traditional square of opposition, and explained the immediate inferences that can be drawn from them.

In Section 5.6 we examined other kinds of immediate inferences that are based on categorical propositions: conversion, obversion, and contraposition.

In Section 5.7 we explored the controversial issue of existential import, showing that the traditional square of opposition can be retained only if we make a blanket assumption that the classes to which propositions refer always do have some members—an assumption that modern logicians are unwilling to make. We then explained the interpretation of propositions to be adopted throughout this book: the interpretation called Boolean, which retains much, but not all, of the traditional square of opposition while rejecting the blanket assumption of nonempty classes. In this Boolean interpretation, we explained that particular propositions (**I** and **O** propositions) are interpreted as having existential import, whereas universal propositions (**A** and **E** propositions) are interpreted as not having such import. We carefully detailed the consequences of adopting this interpretation of propositions.

In Section 5.8 we returned to the use of Venn diagrams, using intersecting circles to represent classes. We showed how, with additional markings, Venn diagrams may also be used to represent categorical propositions.

This chapter has provided the tools we will need to analyze categorical syllogisms, of which standard-form propositions are the essential building blocks.

For additional exercises and tutorials about concepts covered in this chapter, log in to MyLogicLab at *www.mylogiclab.com* and select your current textbook.

Categorical Syllogisms

6.1 Standard-Form Categorical Syllogisms
6.2 The Formal Nature of Syllogistic Argument
6.3 Venn Diagram Technique for Testing Syllogisms
6.4 Syllogistic Rules and Syllogistic Fallacies
6.5 Exposition of the Fifteen Valid Forms of the Categorical Syllogism

Appendix: Deduction of the Fifteen Valid Forms of the Categorical Syllogism

6.1 Standard-Form Categorical Syllogisms

Categorical propositions can now be used in more extended reasoning. Arguments that rely on **A**, **E**, **I**, and **O** propositions commonly have two categorical propositions as premises and one categorical proposition as a conclusion. Such arguments are called *syllogisms*; a **syllogism** is a deductive argument in which a conclusion is inferred from two premises.

The syllogisms with which we are concerned here are called *categorical* because they are arguments based on the relations of classes, or categories—relations that are expressed by the categorical propositions with which we are familiar. More formally, we define a **categorical syllogism** as a deductive argument consisting of three categorical propositions that together contain exactly three terms, each of which occurs in exactly two of the constituent propositions.

Syllogisms are very common, very clear, and readily testable. The system of categorical syllogisms that we will explore is powerful and deep. The seventeenth-century philosopher and mathematician Gottfried Leibniz said, of the invention of the form of syllogisms, that it was "one of the most beautiful and also one of the most important made by the human mind." Syllogisms are the workhorse arguments with which deductive logic, as traditionally practiced, has been made effective in writing and in controversy.

It will be convenient to have an example to use as we discuss the parts and features of the syllogism. Here is a valid standard-form categorical syllogism that we shall use as an illustration:

> No heroes are cowards.
> Some soldiers are cowards.
> Therefore some soldiers are not heroes.

To analyze such an argument accurately, it needs to be in *standard form*. A categorical syllogism is said to be in **standard form** (as the above example is) when two things are true of it: (1) its premises and its conclusion are all standard-form categorical propositions (**A**, **E**, **I**, or **O**); and (2) those propositions are arranged in

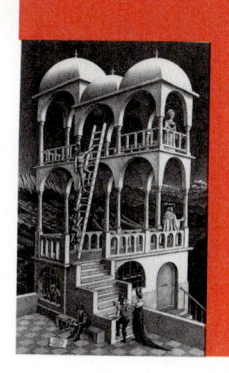

Syllogism
Any deductive argument in which a conclusion is inferred from two premises.

Categorical syllogism
A deductive argument consisting of three categorical propositions that contain exactly three terms, each of which occurs in exactly two of the propositions.

Standard form
The form in which a syllogism is said to be when its premises and conclusion are all standard-form categorical propositions (**A**, **E**, **I**, or **O**) and are arranged in standard order (major premise, then minor premise, then conclusion).

chapter 6

a specified *standard order*. The importance of this standard form will become evident when we turn to the task of testing the validity of syllogisms.

To explain the order of the premises that is required to put any syllogism into standard form, we need the *logical names* of the *premises* of the syllogism, and the names of the *terms* of the syllogism, and we must understand why those names—very useful and very important—are assigned to them. This is the next essential step in our analysis of categorical syllogisms. In this chapter, for the sake of brevity, we will refer to categorical syllogisms simply as "syllogisms," even though there are other kinds of syllogisms that will be discussed in later chapters.

A. Terms of the Syllogism: Major, Minor, and Middle

The three categorical propositions in our example argument above contain exactly three terms: *heroes, soldiers*, and *cowards*. To identify the terms by name, we look to the conclusion of the syllogism, which of course contains exactly two terms. The conclusion in our sample is an **O** proposition, "Some soldiers are not heroes." The term that occurs as the *predicate* of the conclusion ("heroes," in this case) is called the **major term** of the syllogism. The term that occurs as the *subject* of the conclusion ("soldiers," in this case) is called the **minor term** of the syllogism. The third term of the syllogism ("cowards," in this case), which never occurs in the conclusion but always appears in both premises, is called the **middle term**.

The premises of a syllogism also have names. Each premise is named after the term that appears both in it and in the conclusion. The major term and the minor term must each occur in a different premise. The premise containing the major term is called the **major premise**. In the example, "heroes" is the major term, so the premise containing "heroes"—"No heroes are cowards"—is the major premise. It is the major premise not because it appears first, but only because it is the premise that contains the major term; it would be the major premise no matter in what order the premises were written.

The premise containing the minor term is called the **minor premise**. In the example, "soldiers" is the minor term, so the premise containing "soldiers"— "Some soldiers are cowards"—is the minor premise. It is the minor premise not because of its position, but because it is the premise that contains the minor term.

Major term
The term that occurs as the predicate term of the conclusion in a standard-form categorical syllogism.

Minor term
The term that occurs as the subject term of the conclusion in a standard-form categorical syllogism.

Middle term
In a standard-form categorical syllogism (which must contain exactly three terms), the term that appears in both premises but does not appear in the conclusion.

Major premise
In a standard-form categorical syllogism, the premise that contains the major term.

Minor premise
In a standard-form categorical syllogism, the premise that contains the minor term.

> **overview**
>
> ### The Parts of a Standard-Form Categorical Syllogism
>
> | | |
> |---|---|
> | Major Term | The predicate term of the conclusion. |
> | Minor Term | The subject term of the conclusion. |
> | Middle Term | The term that appears in both premises but not in the conclusion. |
> | Major Premise | The premise containing the major term. |
> | Minor Premise | The premise containing the minor term. |

A syllogism is in standard form, we said, when its premises are arranged in a specified standard order. Now we can state that order: *In a standard-form syllogism, the major premise is always stated first, the minor premise second, and the conclusion last.* The reason for the importance of this order will soon become clear.

B. The Mood of the Syllogism

Every syllogism has a mood. The **mood** of a syllogism is determined by the types (**A**, **E**, **I**, or **O**) of standard-form categorical propositions it contains. The mood of the syllogism is therefore represented by three letters, and those three letters are always given in standard-form order. That is, the first letter names the type of the syllogism's major premise; the second letter names the type of the syllogism's minor premise; the third letter names the type of the syllogism's conclusion. In our example syllogism, the major premise ("No heroes are cowards") is an **E** proposition; the minor premise ("Some soldiers are cowards") is an **I** proposition; the conclusion ("Some soldiers are not heroes") is an **O** proposition. Therefore the mood of this syllogism is **EIO**.

C. The Figure of the Syllogism

The mood of a standard-form syllogism is not enough, by itself, to characterize its logical form. This can be shown by comparing two syllogisms, A and B, with the same mood, which are logically very different.

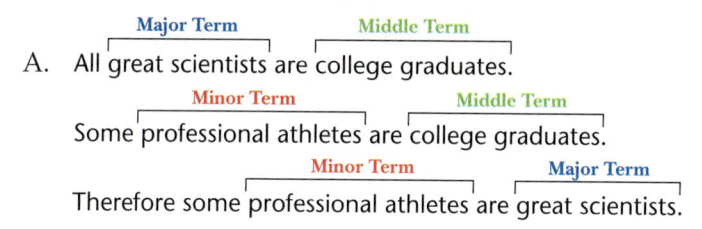

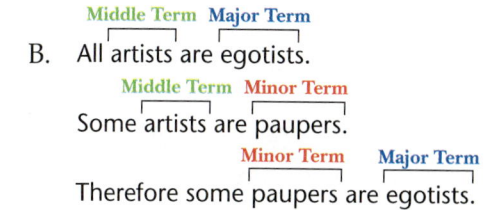

A. All [great scientists]{Major Term} are [college graduates]{Middle Term}.
Some [professional athletes]{Minor Term} are [college graduates]{Middle Term}.
Therefore some [professional athletes]{Minor Term} are [great scientists]{Major Term}.

B. All [artists]{Middle Term} are [egotists]{Major Term}.
Some [artists]{Middle Term} are [paupers]{Minor Term}.
Therefore some [paupers]{Minor Term} are [egotists]{Major Term}.

Both of these are of mood **AII**, but one of them is valid and the other is not. The difference in their forms can be shown most clearly if we display their logical "skeletons" by abbreviating the minor terms as *S* (subject of the conclusion), the major terms as *P* (predicate of the conclusion), and the middle terms as *M*. Using the three-dot symbol "∴" for "therefore," we get these skeletons:

A. All *P* is *M*.
   Some *S* is *M*.
   ∴ Some *S* is *P*.

B. All *M* is *P*.
   Some *M* is *S*.
   ∴ Some *S* is *P*.

Mood
A characterization of categorical syllogisms, determined by the forms of the standard-form categorical propositions it contains. Since there are just four forms of propositions, **A**, **E**, **I**, and **O**, and each syllogism contains exactly three such propositions, there are exactly 64 moods, each mood identified by the three letters of its constituent propositions, **AAA**, **AAI**, **AAE**, and so on, to **OOO**.

These are very different. In the one labeled A, the middle term, *M*, is the predicate term of both premises; but in the one labeled B, the middle term, *M*, is the subject term of both premises. Syllogism B will be seen to be a valid argument; syllogism A, on the other hand, is invalid.

These examples show that although the form of a syllogism is partially described by its mood (**AII** in both of these cases), syllogisms that have the same mood may differ importantly in their forms, depending on the relative positions of their middle terms. To describe the form of a syllogism completely we must state its *mood* (the three letters of its three propositions) *and* its *figure*—where by **figure** we mean the position of the middle term in its premises.

Syllogisms can have four—and only four—possible different figures:

1. The middle term may be the subject term of the major premise and the predicate term of the minor premise; or
2. The middle term may be the predicate term of both premises; or
3. The middle term may be the subject term of both premises; or
4. The middle term may be the predicate term of the major premise and the subject term of the minor premise.

These different possible positions of the middle term constitute the first, second, third, and fourth figures, respectively. Every syllogism must have one or another of these four figures. The characters of these figures may be visualized more readily when the figures are schematized as in the following array, in which reference to mood is suppressed and the quantifiers and copulas are not shown—but the relative positions of the terms of the syllogism are brought out:

M — P	P — M	M — P	P — M
S — M	S — M	M — S	M — S
∴ S — P	∴ S — P	∴ S — P	∴ S — P
First Figure	Second Figure	Third Figure	Fourth Figure

Any standard-form syllogism is completely described when we specify its mood and its figure. The syllogism we have been using as an example is in the second figure; "cowards," the middle term, is the predicate term of both premises. Its mood, as we pointed out, is **EIO**. So it is completely described as being a syllogism of the form **EIO–2**. It is a valid syllogism, as we noted; every valid syllogistic form, as we shall see, has it own name. The name of this form, **EIO–2**, is *Festino*. We say of this syllogism that it is "in *Festino*."

Here is another example:

No *M* is *P*.
All *S* is *M*.
∴ No *S* is *P*.

Figure
The position of the middle term in the premises of a standard-form categorical syllogism.

This syllogism is in the first figure (its middle term is the subject of the major premise and the predicate of the minor premise); its mood is **EAE**. So we may characterize it completely as **EAE–1**, a form whose unique name is *Celarent*. Any syllogism of this form is "in *Celarent*," just as any syllogism of the earlier form is "in *Festino*." Because *Celarent* (**EAE–1**) and *Festino* (**EIO–2**) are known to be *valid* forms, we may conclude that whenever we encounter an argument in one of these forms, it too is valid.

With these analytical tools we can identify every possible categorical syllogism by mood and figure. If we were to list all the possible moods, beginning with **AAA, AAE, AAI, AAO, AEA, AEE**, . . ., and so on, continuing until every possibility had been named, we would eventually (upon reaching **OOO**) have enumerated sixty-four possible moods. Each mood can occur in each of the four figures; $4 \times 64 = 256$. It is certain, therefore, that there are exactly 256 distinct forms that standard-form syllogisms may assume.

Of these 256 possible forms, as we shall see, only a few are valid forms. Each of those valid forms has a unique name, as will be explained.

EXERCISES

Rewrite each of the following syllogisms in standard form, and name its mood and figure. (*Procedure:* first, identify the conclusion; second, note its predicate term, which is the major term of the syllogism; third, identify the major premise, which is the premise containing the major term; fourth, verify that the other premise is the minor premise by checking to see that it contains the minor term, which is the subject term of the conclusion; fifth, rewrite the argument in standard form—major premise first, minor premise second, conclusion last; sixth, name the mood and figure of the syllogism.)

EXAMPLE

1. No nuclear-powered submarines are commercial vessels, so no warships are commercial vessels, because all nuclear-powered submarines are warships.

SOLUTION

Step 1. The conclusion is "No warships are commercial vessels."
Step 2. "Commercial vessels" is the predicate term of this conclusion and is therefore the major term of the syllogism.
Step 3. The major premise, the premise that contains this term, is "No nuclear-powered submarines are commercial vessels."
Step 4. The remaining premise, "All nuclear-powered submarines are warships," is indeed the minor premise, because it does contain the subject term of the conclusion, "warships."

Step 5. In standard form this syllogism is written thus:

> No nuclear-powered submarines are commercial vessels.
> All nuclear-powered submarines are warships.
> Therefore no warships are commercial vessels.

Step 6. The three propositions in this syllogism are, in order, **E**, **A**, and **E**. The middle term, "nuclear-powered submarines," is the subject term of both premises, so the syllogism is in the *third* figure. The mood and figure of the syllogism therefore are **EAE–3**.

2. Some evergreens are objects of worship, because all fir trees are evergreens, and some objects of worship are fir trees.

3. All artificial satellites are important scientific achievements; therefore some important scientific achievements are not U.S. inventions, inasmuch as some artificial satellites are not U.S. inventions.

4. No television stars are certified public accountants, but all certified public accountants are people of good business sense; it follows that no television stars are people of good business sense.

*5. Some conservatives are not advocates of high tariff rates, because all advocates of high tariff rates are Republicans, and some Republicans are not conservatives.

6. All CD players are delicate mechanisms, but no delicate mechanisms are suitable toys for children; consequently, no CD players are suitable toys for children.

7. All juvenile delinquents are maladjusted individuals, and some juvenile delinquents are products of broken homes; hence some maladjusted individuals are products of broken homes.

8. No stubborn individuals who never admit a mistake are good teachers, so, because some well-informed people are stubborn individuals who never admit a mistake, some good teachers are not well-informed people.

9. All proteins are organic compounds, hence all enzymes are proteins, as all enzymes are organic compounds.

*10. No sports cars are vehicles intended to be driven at moderate speeds, but all automobiles designed for family use are vehicles intended to be driven at moderate speeds, from which it follows that no sports cars are automobiles designed for family use.

6.2 The Formal Nature of Syllogistic Argument

In all deductive logic we aim to discriminate valid arguments from invalid ones; in classical logic this becomes the task of discriminating valid syllogisms from invalid ones. It is reasonable to assume that the constituent propositions of a syllogism are all contingent—that is, that no one of those propositions is necessarily true, or necessarily false. Under this assumption, the validity or invalidity of any syllogism depends entirely on its *form*. Validity and invalidity are completely independent of the specific content of the argument or its subject matter. Thus any syllogism of the form **AAA–1**

> All *M* is *P*.
> All *S* is *M*.
> ∴ All *S* is *P*.

is valid, regardless of its subject matter. The name of this syllogism's form is *Barbara*; no matter what terms are substituted for the letters *S*, *P*, and *M*, the resulting argument, "in *Barbara*," will always be valid. If we substitute "Athenians" and "humans" for *S* and *P*, and "Greeks" for *M*, we obtain this valid argument:

> All Greeks are humans.
> All Athenians are Greeks.
> ∴ All Athenians are humans.

If we substitute the terms "soaps," "water-soluble substances," and "sodium salts" for the letters *S*, *P*, and *M* in the same form, we obtain

> All sodium salts are water-soluble substances.
> All soaps are sodium salts.
> Therefore all soaps are water-soluble substances.

which also is valid.

A valid syllogism is valid in virtue of its *form* alone, and so we call it formally valid. We assume throughout that its constituent propositions are themselves contingent, that is, neither logically true (e.g., "All easy chairs are chairs") nor logically false (e.g., "Some easy chairs are not chairs"). The reason for the assumption is this: If it contained either a logically false premise or a logically true conclusion, then the argument would be valid regardless of its syllogistic form—valid in that it would be logically impossible for its premises to be true and its conclusion false. (We also assume that the only logical relations among the terms of the syllogism are those asserted or entailed by its premises. The point of these restrictions is to limit our considerations in this chapter and the next to syllogistic arguments alone and to exclude other kinds of arguments whose validity turns on more complex logical considerations that are not appropriate to introduce at this place.)

If any syllogism is valid in virtue of its form alone, *any other syllogism having that same form will also be valid*; and if a syllogism is invalid, *any other syllogism having that same form will also be invalid*. The common recognition of this fact is attested to by the frequent use of logical analogies in argumentation. Suppose that we are presented with the argument

> All liberals are proponents of national health insurance.
> Some members of the administration are proponents of national health insurance.
> Therefore some members of the administration are liberals.

and felt (justifiably) that, regardless of the truth or falsehood of its constituent propositions, the argument is invalid. The best way to expose its fallacious character is to construct another argument that has exactly the same form but whose invalidity is immediately apparent. We might seek to expose the given argument by replying: You might as well argue that

> All rabbits are very fast runners.
> <u>Some horses are very fast runners.</u>
> Therefore some horses are rabbits.

We might continue: You cannot seriously defend this argument, because here there is no question about the facts. The premises are known to be true and the conclusion is known to be false. Your argument is of the same pattern as this analogous one about horses and rabbits. This one is invalid—so *your* argument is invalid.

This is an excellent method of arguing; the logical analogy is one of the most powerful weapons that can be used in debate.

Underlying the method of logical analogy is the fact that the validity or invalidity of such arguments as the categorical syllogism is a purely formal matter. Any fallacious argument can be proved to be invalid by finding a second argument that has exactly the same form and is known to be invalid by the fact that its premises are known to be true while its conclusion is known to be false. (It should be remembered that an invalid argument may very well have a true conclusion—that an argument is invalid simply means that its conclusion is not logically implied or necessitated by its premises.)

This method of testing the validity of arguments has serious limitations, however. Sometimes a logical analogy is difficult to "think up" on the spur of the moment. There are far too many invalid forms of syllogistic argument (well over two hundred!) for us to prepare and remember refuting analogies of each of them in advance. Moreover, although being able to think of a logical analogy with true premises and a false conclusion proves its form to be invalid, *not* being able to think of one does not prove the form valid, for it may merely reflect the limitations of our thinking. There may be an invalidating analogy even though we are not able to think of it. A more effective method of establishing the formal validity or invalidity of syllogisms is required. The explanation of effective methods of testing syllogisms is the object of the remaining sections of this chapter.

EXERCISES

Refute, by the method of constructing logical analogies, any of the following arguments that are invalid:

EXAMPLE

1. All business executives are active opponents of increased corporation taxes, for all active opponents of increased corporation taxes are members of the chamber of commerce, and all members of the chamber of commerce are business executives.

SOLUTION

One possible refuting analogy is this: All bipeds are astronauts, for all astronauts are humans and all humans are bipeds.

2. No medicines that can be purchased without a doctor's prescription are habit-forming drugs, so some narcotics are not habit-forming drugs, because some narcotics are medicines that can be purchased without a doctor's prescription.
3. No Republicans are Democrats, so some Democrats are wealthy stockbrokers, because some wealthy stockbrokers are not Republicans.
4. No college graduates are persons having an IQ of less than 70, but all persons who have an IQ of less than 70 are morons, so no college graduates are morons.
*5. All fireproof buildings are structures that can be insured at special rates, so some structures that can be insured at special rates are not wooden houses, because no wooden houses are fireproof buildings.
6. All blue-chip securities are safe investments, so some stocks that pay a generous dividend are safe investments, because some blue-chip securities are stocks that pay a generous dividend.
7. Some pediatricians are not specialists in surgery, so some general practitioners are not pediatricians, because some general practitioners are not specialists in surgery.
8. No intellectuals are successful politicians, because no shy and retiring people are successful politicians, and some intellectuals are shy and retiring people.
9. All trade union executives are labor leaders, so some labor leaders are conservatives in politics, because some conservatives in politics are trade union executives.
*10. All new automobiles are economical means of transportation, and all new automobiles are status symbols; therefore some economical means of transportation are status symbols.

6.3 Venn Diagram Technique for Testing Syllogisms

In Chapter 5 we explained the use of two-circle Venn diagrams to represent standard-form categorical propositions. In order to test a categorical syllogism using Venn diagrams, one must first represent both of its premises in one diagram. That requires drawing *three* overlapping circles, for the two premises of a standard-form syllogism contain three different terms—minor term, major term, and middle term—which we abbreviate as S, P, and M, respectively. We first draw two circles, just as we did to diagram a single proposition, and then we draw a third circle beneath, overlapping both of the first two. We label the three circles S, P, and M, in that order. Just as one circle labeled S diagrammed both the class S and the class $\bar{S}$, and as two overlapping circles labeled S and P diagrammed four classes (SP, S$\bar{P}$, $\bar{S}$P, and $\bar{S}\bar{P}$), three overlapping circles, labeled S, P, and M, diagram eight classes: S$\bar{P}\bar{M}$, SP$\bar{M}$, $\bar{S}$P$\bar{M}$, S$\bar{P}$M, SPM, $\bar{S}$PM, $\bar{S}\bar{P}$M, and $\bar{S}\bar{P}\bar{M}$. These are represented by the eight parts into which the three circles divide the plane, as shown in Figure 6-1.

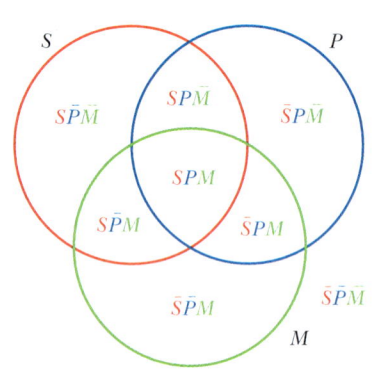

Figure 6-1

Figure 6-1 can be interpreted, for example, in terms of the various different classes determined by the class of all Swedes (S), the class of all peasants (P), and the class of all musicians (M). SPM is the product of these three classes, which is the class of all Swedish peasant musicians. SP$\bar{M}$ is the product of the first two and the complement of the third, which is the class of all Swedish peasants who are not musicians. S$\bar{P}$M is the product of the first and third and the complement of the second: the class of all Swedish musicians who are not peasants. S$\bar{P}\bar{M}$ is the product of the first and the complements of the others: the class of all Swedes who are neither peasants nor musicians. Next, $\bar{S}$PM is the product of the second and third classes with the complements of the first: the class of all peasant musicians who are not Swedes. $\bar{S}$P$\bar{M}$ is the product of the second class with the complements of the other two: the class of all peasants who are neither Swedes nor musicians. $\bar{S}\bar{P}$M is the product of the third class and the complements of the first two: the class of all musicians who are neither Swedes nor peasants. Finally, $\bar{S}\bar{P}\bar{M}$ is the product of the complements of the three original classes: the class of all things that are neither Swedes nor peasants nor musicians.

If we focus our attention on just the two circles labeled P and M, it is clear that by shading out, or by inserting an x, we can diagram any standard-form categorical proposition whose two terms are P and M, regardless of which is the subject term and which is the predicate. Thus, to diagram the proposition "All M is P" (M$\bar{P}$ = 0), we shade out all of M that is not contained in (or overlapped by) P. This area, it is seen, includes both the portions labeled S$\bar{P}$M and $\bar{S}\bar{P}$M. The diagram then becomes Figure 6-2.

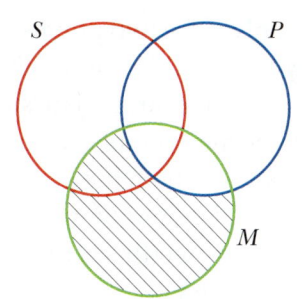

6.3 Venn Diagram Technique for Testing Syllogisms

Figure 6-2

If we focus our attention on just the two circles S and M, by shading out, or by inserting an x, we can diagram any standard-form categorical proposition whose terms are S and M, regardless of the order in which they appear in it. To diagram the proposition "All S is M" ($S\bar{M} = 0$) we shade out all of S that is not contained in (or overlapped by) M. This area, it is seen, includes both the portions labeled $S\bar{P}\bar{M}$ and $SP\bar{M}$. The diagram for this proposition will appear as Figure 6-3.

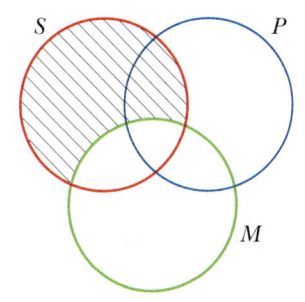

Figure 6-3

The advantage of using three overlapping circles is that it allows us to diagram two propositions together—on the condition, of course, that only three different terms occur in them. Thus diagramming both "All M is P" and "All S is M" at the same time gives us Figure 6-4.

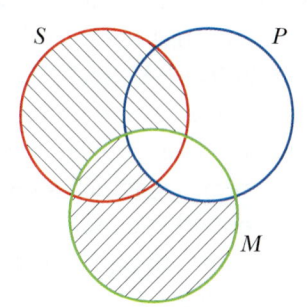

Figure 6-4

This is the diagram for both premises of the syllogism **AAA–1**:

All M is P.
All S is M.
∴ All S is P.

This syllogism is valid if and only if the two premises imply or entail the conclusion—that is, if together they say what is said by the conclusion. Consequently, diagramming the premises of a valid argument should suffice to diagram its conclusion also, with no further marking of the circles needed. To diagram the conclusion "All S is P" is to shade out both the portion labeled $S\bar{P}\bar{M}$ and the portion labeled $S\bar{P}M$. Inspecting the diagram that represents the two premises, we see that it also diagrams the conclusion (this is true even though the region $SP\bar{M}$ has been shaded, because the only region in class S that can still have members lies within class P—hence, "All S is P"). From this we can conclude that **AAA–1** is a valid syllogism.

Let us now apply the Venn diagram test to an obviously invalid syllogism, one containing three **A** propositions in the second figure:

> All dogs are mammals.
> All cats are mammals.
> Therefore all cats are dogs.

Diagramming both premises gives us Figure 6-5.

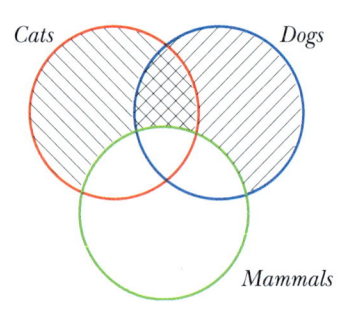

Figure 6-5

In this diagram, where the class of all cats corresponds to S, the class of all dogs corresponds to P, and the class of all mammals corresponds to M, the portions corresponding to $S\bar{P}\bar{M}$, $SP\bar{M}$, and $\bar{S}P\bar{M}$, have been shaded out. But the conclusion has not been diagrammed, because the part $S\bar{P}M$ has been left unshaded, and to diagram the conclusion both $S\bar{P}\bar{M}$ and $S\bar{P}M$ must be shaded. Thus we see that diagramming both the premises of a syllogism of form **AAA–2** does not suffice to diagram its conclusion, which proves that the conclusion says something more than is said by the premises, which shows that the premises do not imply the conclusion. An argument whose premises do not imply its conclusion is invalid, so our diagram proves that the given syllogism is invalid. (It proves more: that any syllogism of the form **AAA–2** is invalid.)

When we use a Venn diagram to test a syllogism with one universal premise and one particular premise, it is important to *diagram the universal premise first*. Thus, in testing the **AII–3** syllogism,

> All artists are egotists.
> Some artists are paupers.
> Therefore some paupers are egotists.

we should diagram the universal premise, "All artists are egotists," before inserting an *x* to diagram the particular premise, "Some artists are paupers." Properly diagrammed, the syllogism looks like Figure 6-6.

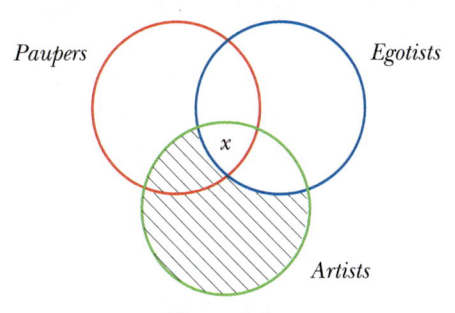

Figure 6-6

Had we tried to diagram the particular premise first, before the region $S\bar{P}M$ was shaded out along with $\bar{S}\bar{P}M$ in diagramming the universal premise, we would not have known whether to insert an *x* in SPM or in $S\bar{P}M$ or in both. Had we put it in $S\bar{P}M$ or on the line separating it from SPM, the subsequent shading of $S\bar{P}M$ would have obscured the information the diagram was intended to exhibit. Now that the information contained in the premises has been inserted into the diagram, we can examine it to see whether the conclusion already has been diagrammed. If the conclusion, "Some paupers are egotists," has been diagrammed, there will be an *x* somewhere in the overlapping part of the circles labeled "Paupers" and "Egotists." This overlapping part consists of both of the regions $SP\bar{M}$ and SPM, which together constitute SP. There is an *x* in the region SPM, so there is an *x* in the overlapping part SP. What the conclusion of the syllogism says has already been diagrammed by the diagramming of its premises; therefore the syllogism is valid.

Let us consider still another example, the discussion of which will bring out another important point about the use of Venn diagrams. Let's say we are testing the argument

> All great scientists are college graduates.
> Some professional athletes are college graduates.
> Therefore some professional athletes are great scientists.

After diagramming the universal premise first (Figure 6-7) by shading out both regions $SP\bar{M}$ and $\bar{S}P\bar{M}$,

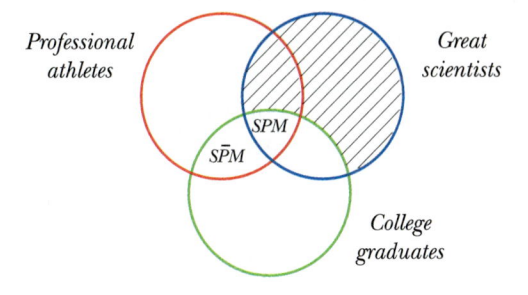

Figure 6-7

we may still be puzzled about where to put the x needed in order to diagram the particular premise. That premise is "Some professional athletes are college graduates," so an x must be inserted somewhere in the overlapping part of the two circles labeled "Professional athletes" and "College graduates." That overlapping part, however, contains two regions, SPM and $S\bar{P}M$. In which of these should we put an x? The premises do not tell us, and if we make an arbitrary decision to place it in one rather than the other, we would be inserting more information into the diagram than the premises warrant—which would spoil the diagram's use as a test for validity. Placing x's in each of them would also go beyond what the premises assert. Yet by placing an x on the line that divides the overlapping region SM into the two parts SPM and $S\bar{P}M$, we can diagram exactly what the second premise asserts without adding anything to it. Placing an x on the line between two regions indicates that there is something that belongs in one of them, but does not indicate which one. The completed diagram of both premises thus looks like Figure 6-8.

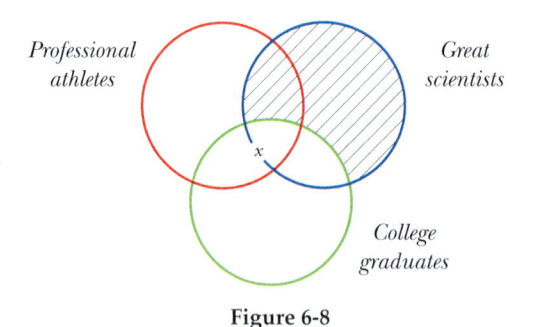

Figure 6-8

When we inspect this diagram of the premises to see whether the conclusion of the syllogism has already been diagrammed in it, we find that it has not. For the conclusion, "Some professional athletes are great scientists," to be diagrammed, an x must appear in the overlapping part of the two upper circles, either in $SP\bar{M}$ or in SPM. The first of these is shaded out and certainly contains no x. The diagram does not show an x in SPM either. True, there must be a member of *either* SPM or $S\bar{P}M$, but the diagram does not tell us that it is in the former rather than the latter and so, for all the premises tell us, the conclusion may be false. We do not know that the conclusion is false, only that it is not asserted or implied by the premises. The latter is enough, however, to let us know that the argument is invalid. The diagram suffices to show not only that the given syllogism is invalid, but that *all* syllogisms of the form **AII–2** are invalid.

The general technique of using Venn diagrams to test the validity of any standard-form syllogism may be summarized as follows. First, label the circles of a three-circle Venn diagram with the syllogism's three terms. Next, diagram both premises, diagramming the universal one first if there is one universal and one particular, and being careful, in diagramming a particular proposition, to put an *x* on a line if the premises do not determine on which side of the line it should go. Finally, inspect the diagram to see whether the diagram of the premises contains a diagram of the conclusion: If it does, the syllogism is valid; if it does not, the syllogism is invalid.

What is the theoretical rationale for using Venn diagrams to distinguish valid from invalid syllogisms? The answer to this question divides into two parts. The first part has to do with the formal nature of syllogistic argument as explained in Section 6.2. It was shown there that one legitimate test of the validity or invalidity of a syllogism is to establish the validity or invalidity of a different syllogism that has exactly the same form. This technique is basic to the use of Venn diagrams.

The explanation of *how* the diagrams serve this purpose constitutes the second part of the answer to our question. Ordinarily, a syllogism will be about classes of objects that are not all present, such as the class of all musicians, or great scientists, or sodium salts. The relations of inclusion or exclusion among such classes may be reasoned about and may be empirically discoverable in the course of scientific investigation. But they certainly are not open to direct inspection, because not all members of the classes involved are ever present at one time to be inspected. We can, however, examine situations of our own making, in which the only classes concerned contain by their very definitions only things that are present and open to direct inspection. We can argue syllogistically about such situations of our own making. Venn diagrams are devices for expressing standard-form categorical propositions, but they also are situations of our own making, patterns of graphite or ink on paper, or lines of chalk on blackboards. The propositions they express can be interpreted as referring to the diagrams themselves. An example can help to make this clear. Suppose we have a particular syllogism whose terms denote various kinds of people who are successful, interested in their work, and able to concentrate, and who may be scattered widely over all parts of the world:

> All successful people are people who are keenly interested in their work.
> No people who are keenly interested in their work are people whose attention is easily distracted when they are working.
> Therefore no people whose attention is easily distracted when they are working are successful people.

Its form is **AEE–4**, and it may be schematized as

> All *P* is *M*.
> No *M* is *S*.
> ∴ No *S* is *P*.

We may test it by constructing the Venn diagram shown in Figure 6-9, in which regions $SP\bar{M}$ and $\bar{S}P\bar{M}$ are shaded out to express the first premise, and $S\bar{P}M$ and SPM are shaded out to express the second premise.

Examining Figure 6-9, we find that *SP* (which consists of the regions *SPM* and $SP\bar{M}$) has been shaded out, so the syllogism's conclusion has already been

Visual Logic

Where do I place the *x* in a Venn diagram?

In the Venn diagram representing a categorical syllogism, the three terms of the syllogism (minor, major, and middle) are represented by three interlocking circles labeled *S*, *P*, and *M* (the choice of *S* and *P* reflects the fact that the minor and major term of a syllogism correspond to the Subject and Predicate terms of its conclusion).

Diagram of three circles, *S*, *P*, and *M*, with nothing else showing

When one of the premises of a syllogism calls for an *x* to be placed on a line in such a Venn diagram, we may ask: Which line? And why? Answer: The *x* is always placed *on the line of the circle designating the class not mentioned in that premise.*

Example: Suppose you are given as premise, "Some *S* is *M*." You may not be able to determine whether the *x* representing that "some" is a *P* or is not a *P*—so the *x* goes on the line of the *P* circle, thus:

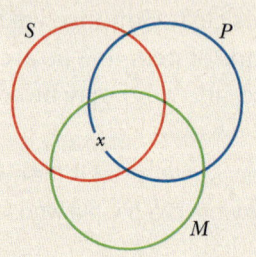

Diagram of three circles with *x* on the *P* circle

Another example: Suppose you are given as premise, "Some *M* is not *P*." You may not be able to determine whether the *M* that is not *P* is an *S* or is not an *S*—so the *x* goes on the line of the *S* circle, thus:

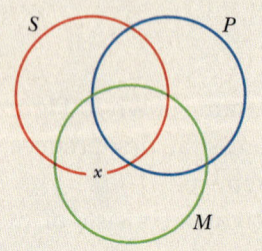

Diagram of three circles with *x* on the *S* circle

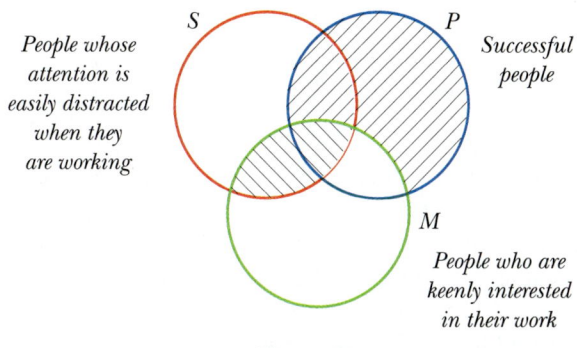

Figure 6-9

People whose attention is easily distracted when they are working — S

Successful people — P

People who are keenly interested in their work — M

diagrammed. How does this tell us that the given syllogism is valid? This syllogism concerns large classes of remote objects: There are many people whose attention is easily distracted when they are working, and they are scattered far and wide. However, we can construct a syllogism of the same form that involves objects that are immediately present and directly available for our inspection. These objects are the points within the unshaded portions of the circles labeled *S*, *P*, and *M* in our Venn diagram.

Here is the new syllogism:

> All points within the unshaded part of the circle labeled *P* are points within the unshaded part of the circle labeled *M*.
>
> No points within the unshaded part of the circle labeled *M* are points within the <u>unshaded part of the circle labeled *S*.</u>
>
> Therefore no points within the unshaded part of the circle labeled *S* are points within the unshaded part of the circle labeled *P*.

This new syllogism refers to nothing remote; it is about the parts of a situation we ourselves have created: the Venn diagram we have drawn. All the parts and all the possibilities of inclusion and exclusion among these classes are immediately present to us and directly open to inspection. We can literally *see* all the possibilities here, and know that because all the points of *P* are also points of *M*, and because *M* and *S* have no points in common, *S* and *P* cannot possibly have any points in common. Because the new syllogism refers only to classes of points in the diagram, it can be literally *seen* to be valid by looking at the things it talks about. The original syllogism about classes of people has exactly the same form as this second one, so we are assured by the formal nature of syllogistic argument that the original syllogism is also valid. The explanation is exactly the same for Venn diagram proofs of the invalidity of invalid syllogisms; there, too, we test the original syllogism indirectly by testing directly a second syllogism that has exactly the same form and referring to the diagram that exhibits that form.

Biography

John Venn

Born in Hull, Yorkshire, England, the son and grandson of Church of England evangelicals, John Venn (1834–1923) earned his degree in mathematics at Gonville and Caius College, Cambridge University, in 1857—whereupon he became a Fellow of that College, remaining closely associated with it all his life. He was ordained a priest in 1859 but, coming to question his faith, he soon returned to Cambridge to teach logic and probability theory.

It is the diagrams named after him that assure John Venn a special place in the history of logic. The idea that logical relations might be represented visually with diagrams had been pursued in some degree by the German philosopher Gottfried Leibniz in the seventeenth century, and then again by the prolific Swiss mathematician Leonhard Euler in the eighteenth century. But it was Venn who developed the system of visual representation so that it could be used readily and effectively in logic. His diagrams reinforce our understanding of the relations among propositions, and they provide a reliable and simple method for determining the validity or invalidity of syllogisms.

Consider a circle as the visual representation of a set, or a class of things. Two intersecting circles represent the relations of two sets; three intersecting circles can thus represent the three categories (or terms) of the categorical syllogism. Venn had studied the work of George Boole, from which he advanced. With the Boolean interpretation of propositions, all of the 256 possible syllogistic forms—combinations of syllogistic moods and figures—can be represented on a simple, three-circle Venn diagram. It is a device as powerful as it is elegant. The "Venn diagram technique" for establishing the validity or invalidity of syllogisms is discussed in detail in Part II of this book.

Venn had some influence also on the development of statistics, and the theory of probability. He rejected the long-standing notion that the probability of an event is to be understood as the degree of "rational belief" in its occurrence. He insisted instead upon the objectivity of what came to be called the *frequency theory* of probability, thus making the determination of the probability of an event an empirical matter, as explained in the final chapter of this book. He remained active as teacher, inventor, and historian of his College until his death in 1923, but his great book was *Symbolic Logic* (1881), in which the analysis of categorical propositions was refined, and the force of categorical arguments made visually vivid. His drive for clarity and simplicity establishes John Venn as a great and permanent friend of the student of logic. ∎

mylogiclab

EXERCISES

A. Write out each of the following syllogistic forms, using S and P as the subject and predicate terms of the conclusion, and M as the middle term. (Refer to the chart of the four syllogistic figures, if necessary, on p. 235.) Then test the validity of each syllogistic form using a Venn diagram.

EXAMPLE

1. AEE–1

SOLUTION

We are told that this syllogism is in the first figure, and therefore the middle term, M, is the subject term of the major premise and the predicate term of the minor premise. (See chart on p. 235.) The conclusion of the syllogism is an **E** proposition and therefore reads: No S is P. The first (major) premise (which contains the predicate term of the conclusion) is an **A** proposition, and therefore reads: All M is P. The second (minor) premise (which contains the subject term of the conclusion) is an **E** proposition and therefore reads: No S is M. This syllogism therefore reads as follows:

All M is P.
No S is M.
Therefore no S is P.

Tested by means of a Venn diagram, as in Figure 6-10, this syllogism is shown to be invalid.

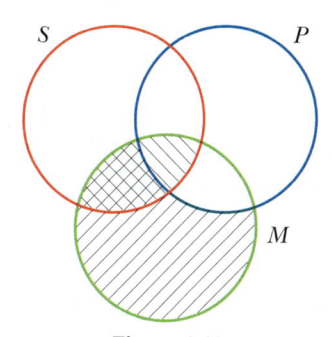

Figure 6-10

2. EIO–2	6. OAO–2	*10. IAI–4	14. OAO–4
3. OAO–3	7. AOO–1	11. AOO–3	*15. EIO–1
4. AOO–4	8. EAE–3	12. EAE–1	
*5. EIO–4	9. EIO–3	13. IAI–1	

B. Put each of the following syllogisms into standard form, name its mood and figure, and test its validity using a Venn diagram:

*1. Some reformers are fanatics, so some idealists are fanatics, because all reformers are idealists.
2. Some philosophers are mathematicians; hence some scientists are philosophers, because all scientists are mathematicians.
3. Some mammals are not horses, for no horses are centaurs, and all centaurs are mammals.
4. Some neurotics are not parasites, but all criminals are parasites; it follows that some neurotics are not criminals.

*5. All underwater craft are submarines; therefore no submarines are pleasure vessels, because no pleasure vessels are underwater craft.

   6. No criminals were pioneers, for all criminals are unsavory persons, and no pioneers were unsavory persons.

   7. No musicians are astronauts; all musicians are baseball fans; consequently, no astronauts are baseball fans.

   8. Some Christians are not Methodists, for some Christians are not Protestants, and some Protestants are not Methodists.

   9. No people whose primary interest is in winning elections are true liberals, and all active politicians are people whose primary interest is in winning elections, which entails that no true liberals are active politicians.

   *10. No weaklings are labor leaders, because no weaklings are true liberals, and all labor leaders are true liberals.

6.4 Syllogistic Rules and Syllogistic Fallacies

A syllogism may fail to establish its conclusion in many different ways. To help avoid common errors we set forth rules—six of them—to guide the reasoner; any given standard-form syllogism can be evaluated by observing whether any one of these rules has been violated. Mastering the rules by which syllogisms may be evaluated also enriches our understanding of the syllogism itself; it helps us to see how syllogisms work and to see why they fail to work if the rules are broken.

A violation of any one of these rules is a mistake, and it renders the syllogism invalid. Because it is a mistake of that special *kind*, we call it a fallacy; and because it is a mistake in the *form* of the argument, we call it a *formal fallacy* (to be contrasted with the *informal* fallacies described in Chapter 4). In reasoning with syllogisms, one must scrupulously avoid the fallacies that violations of the rules invariably yield. Each of these formal fallacies has a traditional name, explained below.

Rule 1. Avoid four terms.

> *A valid standard-form categorical syllogism must contain exactly three terms, each of which is used in the same sense throughout the argument.*

In every categorical syllogism, the conclusion asserts a relationship between two terms, the subject (minor term) and the predicate (major term). Such a conclusion can be justified only if the premises assert the relationship of each of those two terms to the same third term (middle term). If the premises fail to do this consistently, the needed connection of the two terms in the conclusion cannot be established, and the argument fails. So every valid categorical syllogism must involve three terms—no more and no less. If more than three terms are involved, the syllogism is invalid. The fallacy thus committed is called the **fallacy of four terms**.

The mistake that commonly underlies this fallacy is equivocation: using one word or phrase with two different meanings. Most often it is the middle term whose meaning is thus shifted, in one direction to connect it with the minor term, in a different direction to connect it with the major term. In doing this the two terms of the conclusion are connected with two different terms (rather than with the same middle term), and so the relationship asserted by the conclusion is not established. Because it is the middle term that is most often manipulated, this fallacy is sometimes called "the fallacy of the ambiguous middle." However, this name is not generally applicable, because one (or more) of the other terms may have its meaning shifted as well. Ambiguities may result in as many as five or six different terms being involved, but the mistake retains its traditional name: the fallacy of four terms.

When the expression *categorical syllogism* was defined at the beginning of this chapter, we noted that by its nature every syllogism must have three and only three terms. (The term *syllogism* is sometimes defined more broadly than it has been in this book. The informal fallacy of equivocation, explained and warned against in Chapter 4, may arise in many different argumentative contexts, of course.) So this rule ("Avoid four terms") may be regarded as a reminder to make sure that the argument being appraised really is a categorical syllogism.

Rule 2. Distribute the middle term in at least one premise.

*A term is "distributed" in a proposition when (as was explained in Section 5.4) the proposition refers to **all** members of the class designated by that term. If the middle term is not distributed in at least one premise, the connection required by the conclusion cannot be made.*

Historian Barbara Tuchman observed that many early critics of anarchism relied on the following "unconscious syllogism":

> All Russians were revolutionists.
> All anarchists were revolutionists.
> Therefore, all anarchists were Russians.[1]

This syllogism is plainly invalid. Its mistake is that it asserts a connection between anarchists and Russians by relying on the links between each of those classes and the class of revolutionists—but revolutionists is an *un*distributed term in both of the premises. The first premise does not refer to all revolutionists, and neither does the second. "Revolutionists" is the middle term in this argument, and if the middle term is not distributed in at least one premise of a syllogism, that syllogism cannot be valid. The fallacy this syllogism commits is called the **fallacy of the undistributed middle**.

What underlies this rule is the need to *link* the minor and the major terms. If they are to be linked by the middle term, either the subject or the predicate of the

Fallacy of four terms
The formal fallacy that is committed when a syllogism is constructed with more than three terms.

Fallacy of the undistributed middle
The formal fallacy that is committed when the middle term of a syllogism is not distributed in at least one premise.

conclusion must be related to the *whole* of the class designated by the middle term. If that is not so, it is possible that each of the terms in the conclusion may be connected to a different part of the middle term, and not necessarily connected with each other.

This is precisely what happens in the syllogism given in the preceding example. The Russians are included in a *part* of the class of revolutionists (by the first premise), and the anarchists are included in a *part* of the class of revolutionists (by the second premise)—but *different* parts of this class (the middle term of the syllogism) may be involved, and so the middle term does not successfully link the minor and major terms of the syllogism. In a valid syllogism, *the middle term must be distributed in at least one premise*.

Rule 3. Any term distributed in the conclusion must be distributed in the premises.

> *To refer to* **all** *members of a class is to say more about that class than is said when only* **some** *of its members are referred to. Therefore, when the conclusion of a syllogism distributes a term that was undistributed in the premises, it says more about that term than the premises did. But a valid argument is one whose premises logically entail its conclusion, and for that to be true the conclusion must not assert any more than is asserted in the premises. A term that is distributed in the conclusion but is not distributed in the premises is therefore a sure mark that the conclusion has gone beyond its premises and has reached too far. This is called the* **fallacy of illicit process**.

The conclusion may overreach with respect to either the minor term (its subject) or the major term (its predicate). So there are two different forms of illicit process, and different names have been given to the two formal fallacies involved. They are

> Illicit process of the major term (an **illicit major**).
> Illicit process of the minor term (an **illicit minor**).

To illustrate an illicit process of the major term, consider this syllogism:

> All dogs are mammals.
> <u>No cats are dogs.</u>
> Therefore no cats are mammals.

The reasoning is obviously bad, but where is the mistake? The mistake is in the conclusion's assertion about *all* mammals, saying that all of them fall outside the class of cats. Bear in mind that an **A** proposition distributes its subject term but does not distribute its predicate term. Hence the premises make no assertion about *all* mammals—so the conclusion illicitly goes beyond what the premises assert. Because "mammals" is the major term in this syllogism, the fallacy here is that of an illicit major.

To illustrate the illicit process of the minor term, consider this syllogism:

> All traditionally religious people are fundamentalists.
> <u>All traditionally religious people are opponents of abortion.</u>
> Therefore all opponents of abortion are fundamentalists.

Fallacy of illicit process
The formal fallacy that is committed when a term that is distributed in the conclusion is not distributed in the corresponding premise.

Again we sense quickly that something is wrong with this argument, and what is wrong is this: The conclusion makes an assertion about *all* opponents of abortion, but the premises make no such assertion; they say nothing about *all* abortion opponents. So the conclusion here goes illicitly beyond what the premises warrant. In this case "opponents of abortion" is the minor term, so the fallacy is that of an illicit minor.

Rule 4. Avoid two negative premises.

> Any negative proposition (**E** or **O**) denies class inclusion; it asserts that some or all of one class is excluded from the whole of the other class. Two premises asserting such exclusion cannot yield the linkage that the conclusion asserts, and therefore cannot yield a valid argument. The mistake is named the **fallacy of exclusive premises**.

Understanding the mistake identified here requires some reflection. Suppose we label the minor, major, and middle terms of the syllogism *S*, *P*, and *M*, respectively. What can two negative premises tell us about the relations of these three terms? They can tell us that *S* (the subject of the conclusion) is wholly or partially excluded from all or part of *M* (the middle term), and that *P* (the predicate of the conclusion) is wholly or partially excluded from all or part of *M*. However, any one of these relations may very well be established no matter how *S* and *P* are related. The negative premises cannot tell us that *S* and *P* are related by inclusion or by exclusion, partial or complete. Two negative premises (where *M* is a term in each) simply cannot justify the assertion of *any* relationship whatever between *S* and *P*. Therefore, if both premises of a syllogism are negative, the argument must be invalid.

Rule 5. If either premise is negative, the conclusion must be negative.

> If the conclusion is affirmative—that is, if it asserts that one of the two classes, S or P, is wholly or partly contained in the other—it can only be inferred from premises that assert the existence of a third class that contains the first and is itself contained in the second. However, class inclusion can be stated only by affirmative propositions. Therefore, an affirmative conclusion can follow validly only from two affirmative premises. The mistake here is called the **fallacy of drawing an affirmative conclusion from a negative premise**.

If an affirmative conclusion requires two affirmative premises, as has just been shown, we can know with certainty that if either of the premises is negative, the conclusion must also be negative, or the argument is not valid.

Unlike some of the fallacies identified here, this fallacy is not common, because any argument that draws an affirmative conclusion from negative premises will be instantly recognized as highly implausible. Even an illustration of the mistake will appear strained:

No poets are accountants.
<u>Some artists are poets.</u>
Therefore some artists are accountants.

Fallacy of exclusive premises
The formal fallacy that is committed when both premises in a syllogism are negative propositions (**E** or **O**).

Immediately it will be seen that the *ex*clusion of poets and accountants, asserted by the first premise of this syllogism, cannot justify *any* valid inference regarding the inclusion of artists and accountants.

Rule 6. From two universal premises no particular conclusion may be drawn.

*In the Boolean interpretation of categorical propositions (explained in Section 5.7), universal propositions (**A** and **E**) have no existential import, but particular propositions (**I** and **O**) do have such import. Wherever the Boolean interpretation is supposed, as in this book, a rule is needed that precludes passage from premises that have no existential import to a conclusion that does have such import.*

This final rule is not needed in the traditional or Aristotelian account of the categorical syllogism, because that traditional account paid no attention to the problem of existential import. However, when existential import is carefully considered, it will be clear that if the premises of an argument do not assert the existence of anything at all, the conclusion will be unwarranted when, from it, the existence of some thing may be inferred. The mistake is called the **existential fallacy**.

Here is an example of a syllogism that commits this fallacy:

> All household pets are domestic animals.
> No unicorns are domestic animals.
> Therefore some unicorns are not household pets.

If the conclusion of this argument were the universal proposition, "No unicorns are household pets," the syllogism would be perfectly valid for all. And because, under the traditional interpretation, existential import may be inferred from universal as well as from particular propositions, it would not be problematic (in that traditional view) to say that the conclusion in the example given here is simply a "weaker" version of the conclusion we all agree is validly drawn.

In our Boolean view, however, the conclusion of the example ("Some unicorns are not household pets"), because it is a particular proposition, is not just "weaker," it is very different. It is an **O** proposition, a particular proposition, and thus has an existential import that the **E** proposition ("No unicorns are household pets") cannot have. Reasoning that is acceptable under the traditional view is therefore unacceptable under the Boolean view because, from the Boolean perspective, that reasoning commits the existential fallacy—a mistake that cannot be made under the traditional interpretation.

Another interesting consequence of the difference between the traditional and the Boolean interpretation of categorical propositions is this: In the traditional view there is a need for a rule that states the converse of Rule 5 ("If either premise is negative, the conclusion must be negative"). The converse states sim-

Existential fallacy
The formal fallacy that is committed when, in a standard-form categorical syllogism, a particular conclusion is inferred from two universal premises.

ply that "If the conclusion of a valid syllogism is negative, at least one premise must be negative." That is indisputable, because if the conclusion is negative, it denies inclusion. But affirmative premises assert inclusion; therefore affirmative premises cannot entail a negative conclusion. This corollary is unnecessary in the Boolean interpretation because the rule precluding the existential fallacy (Rule 6) will, in the presence of the other rules, suffice to invalidate any syllogism with affirmative premises and a negative conclusion.

The six rules given here are intended to apply only to standard-form categorical syllogisms. In this realm they provide an adequate test for the validity of any argument. If a standard-form categorical syllogism violates any one of these rules, it is invalid; if it conforms to all of these rules, it is valid.

overview

Syllogistic Rules and Fallacies

Rule	Associated Fallacy
1. Avoid four terms.	Four terms
2. Distribute the middle term in at least one premise.	Undistributed middle
3. Any term distributed in the conclusion must be distributed in the premises.	Illicit process of the major term (illicit major); illicit process of the minor term (illicit minor)
4. Avoid two negative premises.	Exclusive premises
5. If either premise is negative, the conclusion must be negative.	Drawing an affirmative conclusion from a negative premise
6. No particular conclusion may be drawn from two universal premises.	Existential fallacy

Flowchart for Applying the Six Syllogistic Rules

The following chart captures the process for working through the six rules of validity for categorical syllogisms:

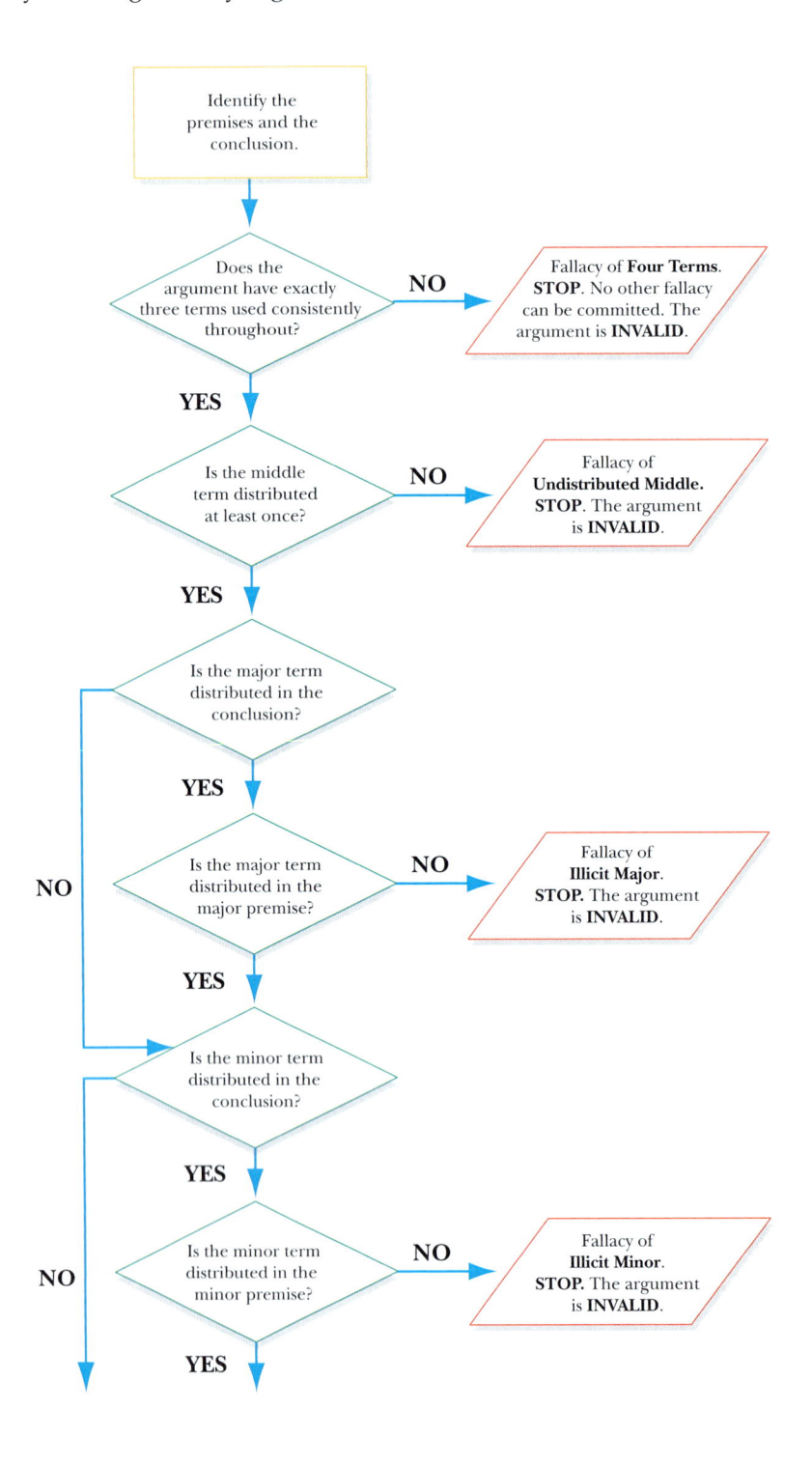

- Identify the premises and the conclusion.
- Does the argument have exactly three terms used consistently throughout?
  - **NO** → Fallacy of **Four Terms**. **STOP**. No other fallacy can be committed. The argument is **INVALID**.
  - **YES** ↓
- Is the middle term distributed at least once?
  - **NO** → Fallacy of **Undistributed Middle**. **STOP**. The argument is **INVALID**.
  - **YES** ↓
- Is the major term distributed in the conclusion?
  - **NO** → (continues to minor term check)
  - **YES** ↓
- Is the major term distributed in the major premise?
  - **NO** → Fallacy of **Illicit Major**. **STOP**. The argument is **INVALID**.
  - **YES** ↓
- Is the minor term distributed in the conclusion?
  - **NO** → (continues)
  - **YES** ↓
- Is the minor term distributed in the minor premise?
  - **NO** → Fallacy of **Illicit Minor**. **STOP**. The argument is **INVALID**.
  - **YES** ↓

6.4 Syllogistic Rules and Syllogistic Fallacies

[Flowchart:

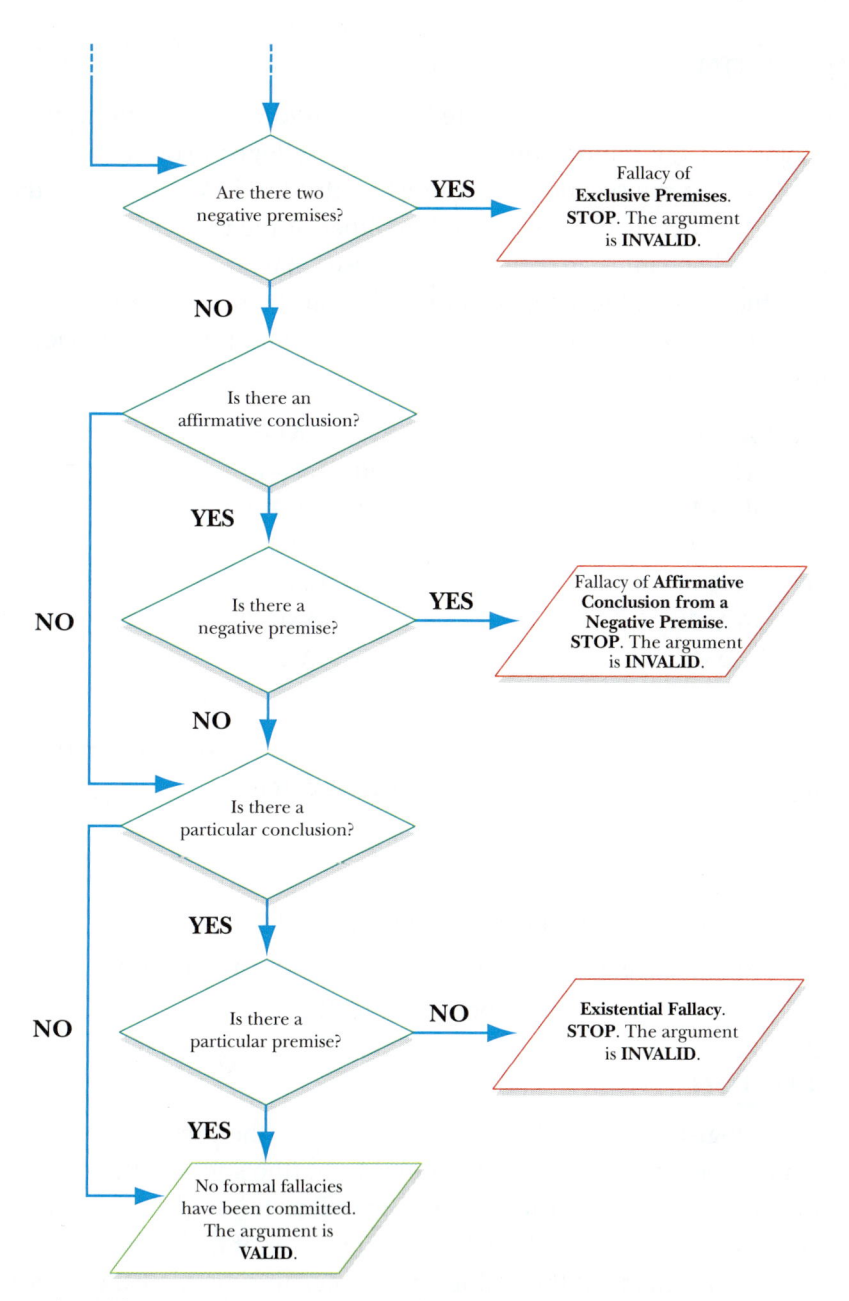

- Are there two negative premises? — YES → Fallacy of **Exclusive Premises**. STOP. The argument is **INVALID**.
- NO ↓
- Is there an affirmative conclusion?
  - YES ↓
    - Is there a negative premise? — YES → Fallacy of **Affirmative Conclusion from a Negative Premise**. STOP. The argument is **INVALID**.
    - NO ↓
  - NO → (to particular conclusion check)
- Is there a particular conclusion?
  - YES ↓
    - Is there a particular premise? — NO → **Existential Fallacy**. STOP. The argument is **INVALID**.
    - YES ↓
  - NO → (to valid)
- No formal fallacies have been committed. The argument is **VALID**.]

Adapted from Daniel E. Flage, *Essentials of Logic*, 2e (Englewood Cliffs, NJ: Prentice Hall, 1995).

EXERCISES

A. Identify the rule that is broken by invalid syllogisms of the following forms, and name the fallacy that each commits:

EXAMPLE

1. AAA–2

SOLUTION

Any syllogism in the second figure has the middle term as predicate of both the major and the minor premise. Thus any syllogism consisting of three **A** propositions, in the second figure, must read: All *P* is *M*; all *S* is *M*; therefore all *S* is *P*. Since *M* is not distributed in either of the premises in that form, it cannot validly be inferred from such premises that all *S* is *P*. Thus every syllogism of the form **AAA–2** violates the rule that the middle term must be distributed in at least one premise, thereby committing the fallacy of the undistributed middle.

2. EAA–1
3. IAO–3
4. OEO–4
*5. AAA–3
6. IAI–2
7. OAA–3
8. EAO–4
9. OAI–3
*10. IEO–1
11. EAO–3
12. AII–2
13. EEE–1
14. OAO–2
*15. IAA–3

B. Identify the rule that is broken by each invalid syllogism you can find in the following exercises, and name the fallacy that is committed:

EXAMPLE

1. All textbooks are books intended for careful study.
   <u>Some reference books are books intended for careful study.</u>
   Therefore some reference books are textbooks.

SOLUTION

In this syllogism, "textbooks" is the major term (the predicate of the conclusion) and "reference books" is the minor term (the subject of the conclusion). "Books intended for careful study" is therefore the middle term, and it appears as the predicate of both premises. In neither of the premises is this middle term distributed, so the syllogism violates the rule that the middle term must be distributed in at least one premise, thereby committing the fallacy of the undistributed middle.

2. All criminal actions are wicked deeds.
   <u>All prosecutions for murder are criminal actions.</u>
   Therefore all prosecutions for murder are wicked deeds.
3. No tragic actors are idiots.
   <u>Some comedians are not idiots.</u>
   Therefore some comedians are not tragic actors.

4. Some parrots are not pests.
   All parrots are pets.
   Therefore no pets are pests.

*5. All perpetual motion devices are 100 percent efficient machines.
   All 100 percent efficient machines are machines with frictionless bearings.
   Therefore some machines with frictionless bearings are perpetual motion devices.

6. Some good actors are not powerful athletes.
   All professional wrestlers are powerful athletes.
   Therefore all professional wrestlers are good actors.

7. Some diamonds are precious stones.
   Some carbon compounds are not diamonds.
   Therefore some carbon compounds are not precious stones.

8. Some diamonds are not precious stones.
   Some carbon compounds are diamonds.
   Therefore some carbon compounds are not precious stones.

9. All people who are most hungry are people who eat most.
   All people who eat least are people who are most hungry.
   Therefore all people who eat least are people who eat most.

*10. Some spaniels are not good hunters.
   All spaniels are gentle dogs.
   Therefore no gentle dogs are good hunters.

C. Identify the rule that is broken by any of the following syllogisms that are invalid, and name the fallacy that is committed:

EXAMPLE

1. All chocolate éclairs are fattening foods, because all chocolate éclairs are rich desserts, and some fattening foods are not rich desserts.

SOLUTION

In this syllogism the conclusion is affirmative ("all chocolate éclairs are fattening foods"), while one of the premises is negative ("some fattening foods are not rich desserts"). The syllogism therefore is invalid, violating the rule that if either premise is negative the conclusion must also be negative, thereby committing the fallacy of drawing an affirmative conclusion from a negative premise.

234   **CHAPTER 6**   Categorical Syllogisms

2. All inventors are people who see new patterns in familiar things, so all inventors are eccentrics, because all eccentrics are people who see new patterns in familiar things.

3. Some snakes are not dangerous animals, but all snakes are reptiles, therefore some dangerous animals are not reptiles.

4. Some foods that contain iron are toxic substances, for all fish containing mercury are foods that contain iron, and all fish containing mercury are toxic substances.

*5. All opponents of basic economic and political changes are outspoken critics of the liberal leaders of Congress, and all right-wing extremists are opponents of basic economic and political changes. It follows that all outspoken critics of the liberal leaders of Congress are right-wing extremists.

6. No writers of lewd and sensational articles are honest and decent citizens, but some journalists are not writers of lewd and sensational articles; consequently, some journalists are honest and decent citizens.

7. All supporters of popular government are democrats, so all supporters of popular government are opponents of the Republican Party, inasmuch as all Democrats are opponents of the Republican Party.

8. No coal-tar derivatives are nourishing foods, because all artificial dyes are coal-tar derivatives, and no artificial dyes are nourishing foods.

9. No coal-tar derivatives are nourishing foods, because no coal-tar derivatives are natural grain products, and all natural grain products are nourishing foods.

*10. All people who live in London are people who drink tea, and all people who drink tea are people who like it. We may conclude, then, that all people who live in London are people who like it.

6.5 Exposition of the Fifteen Valid Forms of the Categorical Syllogism

The *mood* of a syllogism is its character as determined by the forms (**A**, **E**, **I**, or **O**) of the three propositions it contains. There are sixty-four possible moods of the categorical syllogism—that is, sixty-four possible sets of three propositions: **AAA, AAI, AAE**, and so on, to . . . **EOO, OOO**.

The *figure* of a syllogism is its logical shape, as determined by the position of the middle term in its premises. So there are four possible figures, which can be most clearly grasped if one has in mind a chart, or an iconic representation, of the four possibilities, as exhibited in the Overview table:

overview

The Four Figures

	First Figure	Second Figure	Third Figure	Fourth Figure
Schematic Representation	M – P S – M ∴ S – P	P – M S – M ∴ S – P	M – P M – S ∴ S – P	P – M M – S ∴ S – P
Description	The middle term is the subject of the major premise and the predicate of the minor premise.	The middle term is the predicate of both the major and minor premises.	The middle term is the subject of both the major and minor premises.	The middle term is the predicate of the major premise and the subject of the minor premise.

It will be seen that:

- In the first figure the middle term is the subject of the major premise and the predicate of the minor premise;

- In the second figure the middle term is the predicate of both premises;

- In the third figure the middle term is the subject of both premises;

- In the fourth figure the middle term is the predicate of the major premise and the subject of the minor premise.

Each of the sixty-four moods can appear in each of the four figures. The mood and figure of a given syllogism, taken together, uniquely determine the logical form of that syllogism. Therefore there are (as noted earlier) exactly 256 (that is, 64 × 4) possible forms of the standard-form categorical syllogism.

The vast majority of these forms are not valid. We can eliminate every form that violates one or more of the syllogistic rules set forth in the preceding section. The forms that remain after this elimination are the only valid forms of the categorical syllogism. Of the 256 possible forms, there are exactly fifteen forms that cannot be eliminated and thus are valid. It should be borne in mind that we adopt here the Boolean interpretation of categorical propositions, according to which universal propositions (**A** and **E** propositions) do not have existential import. The classical interpretation of categorical propositions, according to which all the classes to which propositions refer do have members, makes acceptable some inferences that are found here to be invalid. Under that older interpretation, for example, it is plausible to infer the subaltern from its corresponding superaltern—to infer an **I** proposition from its corresponding **A** proposition, and an **O** proposition from its corresponding **E** proposition. This makes plausible the claim that there are other valid syllogisms (so-called "weakened syllogisms") that are not considered valid here. Compelling reasons for the rejection of that older interpretation (and hence the justification of our stricter standards for valid syllogisms) were given at some length in Section 5.7.

To advance the mastery of syllogistics, classical logicians gave a unique name to every valid syllogism, each characterized completely by mood and figure. Understanding this small set of valid forms, and knowing the name of each, is very useful when putting syllogistic reasoning to work. Each name, carefully devised, contained three vowels representing (in standard-form order: major premise, minor premise, conclusion) the mood of the syllogism named. Where there are valid syllogisms of a given mood but in different figures, a unique name was assigned to each. Thus, for example, a syllogism of the mood **EAE** in the first figure was named *Celarent*, whereas a syllogism of the mood **EAE** in the second figure, also valid, was named *Cesare*. The principles that governed the construction of those traditional names, the selection and placement of consonants as well as vowels, were quite sophisticated. Some of these conventions relate to the place of the weakened syllogisms noted just above and are therefore not acceptable in the Boolean interpretation we adopt. Some other conventions remain acceptable. For example, the letter *s* that follows the vowel *e* indicates that when that **E** proposition is converted *simpliciter*, or simply (as all **E** propositions will convert), then that syllogism reduces to, or is transformed into, another syllogism of the same mood in the first figure, which is viewed as the most basic figure. To illustrate, *Festino*, in the second figure, reduces, when its major premise is converted simply, to *Ferio*; and *Cesare*, in the second figure, reduces to *Celarent*, and so on. The possibility of these and other reductions explains why the names of groups of syllogisms begin with the same consonant. The intricate details of the classical naming system need not be fully recounted here.

These names had (and still have) a very practical purpose: If one knows that only certain combinations of mood and figure are valid, and can recognize by name those valid arguments, the merit of any syllogism in a given figure, or of a given mood, can be determined almost immediately. For example, the mood **AOO** is valid only in the second figure. That unique form (**AOO–2**) is known as *Baroko*. Here is an example of *Baroko*:

> All good mathematicians have creative intellects.
> Some scholars do not have creative intellects.
> Therefore some scholars are not good mathematicians.

With practice one comes to recognize the *cadence* of the different valid forms. One who is familiar with *Baroko* and able to discern it readily may be confident that a syllogism of this mood presented in any other figure may be rejected as invalid.

The standard form of the categorical syllogism is the key to the system. A neat and efficient method of identifying the few valid syllogisms from the many possible syllogisms is at hand, but it depends on the assumption that the propositions of the syllogism in question are in (or can be put into) standard order—major premise, minor premise, then conclusion. The unique identification of each valid syllogism relies on the specification of its mood, and its mood is determined by the letters characterizing its three constituent propositions in that standard order. If the premises of a valid syllogism were to be set forth in a different order, then that syllogism would remain valid, of course; the Venn diagram technique can prove this. However, much would be lost. Our ability to identify syllogisms uniquely, and with that identification our ability to comprehend the forms of those syllogisms fully and to test their validity crisply, all rely on their being in standard form.*

Classical logicians studied these forms closely, and they became fully familiar with their structure and their logical "feel." This elegant system, finely honed, enabled reasoners confronting syllogisms in speech or in texts to recognize immediately those that were valid, and to detect with confidence those that were not. For centuries it was common practice to defend the solidity of reasoning in progress by giving the names of the forms of the valid syllogisms being relied on. The ability to provide these identifications even in the midst of heated oral disputes was considered a mark of learning and acumen, and it gave evidence that the chain of deductive reasoning being relied on was indeed unbroken. Once the theory of the syllogism has been fully mastered, this practical skill can be developed with profit and pleasure.

Syllogistic reasoning was so very widely employed, and so highly regarded as the most essential tool of scholarly argument, that the logical treatises of its original and greatest master, Aristotle, were venerated for more than a thousand years. His analytical account of the syllogism still carries the simple name that conveys respect and awe: the *Organon*, the *Instrument*.

Valid syllogisms are powerful weapons in controversy, but the effectiveness of those weapons depends, of course, on the truth of the premises. A great theologian, defiant in battling scholars who resisted his reform of the catholic Church, wrote: "They may attack me with an army of six hundred syllogisms. . ."[2]

As students of this remarkable logical system, our proficiency in syllogistics may be only moderate—but we will nevertheless find it useful to have before us a synoptic account of all the valid syllogisms. There are fifteen valid syllogisms under the Boolean interpretation. In the older tradition, in which reasoning from universal premises to particular conclusions was believed to be correct, the number of valid syllogisms (each uniquely named) was of course more than fifteen. To illustrate, if an **I** proposition may be inferred from its corresponding **A** proposition (as we think mistaken), the valid syllogism known as *Barbara* (**AAA–1**) will

---

*The burdensome consequences of ignoring standard form have been eloquently underscored by Keith Burgess-Jackson in his essay, "Why Standard Form Matters," October 2003.

have a putatively valid "weakened" sister, *Barbari* (**AAI–1**); and if an **O** proposition may be inferred from its corresponding **E** proposition (as we think mistaken), the valid syllogism known as *Camestres* (**AEE–2**) will have a putatively valid "weakened" brother, *Camestrop* (**AEO–2**).

These fifteen valid syllogisms may be divided by figure into four groups:

> ### overview
>
> ### The Fifteen Valid Forms of the Standard-Form Categorical Syllogism
>
> In the first figure (in which the middle term is the subject of the major premise and the predicate of the minor premise):
>
> 1. **AAA–1**   *Barbara*
> 2. **EAE–1**   *Celarent*
> 3. **AII–1**   *Darii*
> 4. **EIO–1**   *Ferio*
>
> In the second figure (in which the middle term is the predicate of both premises):
>
> 5. **AEE–2**   *Camestres*
> 6. **EAE–2**   *Cesare*
> 7. **AOO–2**   *Baroko*
> 8. **EIO–2**   *Festino*
>
> In the third figure (in which the middle term is the subject of both premises):
>
> 9. **AII–3**   *Datisi*
> 10. **IAI–3**   *Disamis*
> 11. **EIO–3**   *Ferison*
> 12. **OAO–3**   *Bokardo*
>
> In the fourth figure (in which the middle term is the predicate of the major premise and the subject of the minor premise):
>
> 13. **AEE–4**   *Camenes*
> 14. **IAI–4**   *Dimaris*
> 15. **EIO–4**   *Fresison*

EXERCISES

At the conclusion of Section 6.3, in exercise group B (on pages 223–224), ten syllogisms were to be tested using Venn diagrams. Of these ten syllogisms, numbers 1, 4, 6, 9, and 10 are valid. What is the *name* of each of these five valid syllogisms?

EXAMPLE

Number 1 is **IAI–3** (*Disamis*).

Appendix: Deduction of the Fifteen Valid Forms of the Categorical Syllogism

In Section 6.5 the fifteen valid forms of the categorical syllogism were identified and precisely characterized. The unique name of each syllogism is also given there—a name assigned in view of its unique combination of mood and figure. The summary account of these fifteen syllogisms appears in the Overview immediately preceding.

It is possible to *prove* that these, and only these, are the valid forms of the categorical syllogism. This proof—*the deduction of the valid forms of the categorical syllogism*—is presented as an appendix, rather than in the body of the chapter, because mastering it is not essential for the student of logic. However, understanding it can give one a deeper appreciation of the *system* of syllogistics. For those who derive satisfaction from the intricacies of analytical syllogistics, thinking through this deduction will be a pleasing, if somewhat arduous challenge.

We emphasize that if the chief aims of study are to recognize, understand, and apply the valid forms of the syllogism, as exhibited in Section 6.5, this appendix may be bypassed.

The deduction of the fifteen valid syllogisms is not easy to follow. Those who pursue it must keep two things very clearly in mind: (1) The *rules of the syllogism*, six basic rules set forth in Section 6.4, are the essential tools of the deduction; and (2) the *four figures of the syllogism*, as depicted in the Overview in Section 6.5 (p. 235) are referred to repeatedly as the rules are invoked.

We have seen that there are 256 *possible* forms of the syllogism, sixty-four moods (or combinations of the four categorical propositions) in each of the four figures. The deduction of the fifteen valid syllogisms proceeds by *eliminating* the syllogisms that violate one of the basic rules and that thus cannot be valid.

The conclusion of every syllogism is a categorical proposition, either **A**, or **E**, or **I**, or **O**. We begin by dividing all the possible syllogistic forms into four groups, each group having a conclusion with a different form (**A**, **E**, **I**, or **O**). Every syllogism must of course fall into one of these four groups. Taking each of the four groups in turn, we ask what characteristics a valid syllogism with such a conclusion must possess. That is, we ask what forms are *excluded* by one or more of the syllogistic rules if the conclusion is an **A** proposition, and if the conclusion is an **E** proposition, and so on.

After excluding all those invalid syllogisms, only the valid syllogisms remain. To assist in visualization, we note in the margin as we proceed the moods and figures, and the names, of the fifteen valid categorical syllogisms.

Case 1. If the conclusion of the syllogism is an **A** proposition

In this case, neither premise can be an **E** or an **O** proposition, because if either premise is negative, the conclusion must be negative (Rule 5). Therefore the two premises must be **I** or **A** propositions. The minor premise cannot be an **I** proposition because the minor term (the subject of the conclusion, which is an **A**) is distributed in the conclusion, and therefore if the minor premise were an **I** proposition, a term would be distributed in the conclusion that is not distributed in the premises, violating Rule 3. The two premises, major and minor, cannot be **I** and **A**, because if they were, the middle term of the syllogism would not be distributed in either premise, violating Rule 2. So the two premises (if the conclusion is an **A**) must both be **A** as well, which means that the only possible valid mood is **AAA**. But in the second figure **AAA** again results in the middle term being distributed in neither premise; and in both the third figure and the fourth figure **AAA** results in a term being distributed in the conclusion that is not distributed in the premise in which it appears. Therefore, if the conclusion of the syllogism is an **A** proposition, the only valid form it can take is **AAA** in the first figure. This valid form, **AAA–1**, is the syllogism traditionally given the name *Barbara*.

Summary of Case 1: If the syllogism has an **A** conclusion, there is only one possibly valid form: **AAA–1**—*Barbara*.

Case 2. If the conclusion of the syllogism is an **E** proposition

Both the subject and the predicate of an **E** proposition are distributed, and therefore all three terms in the premises of a syllogism having such a conclusion must be distributed, and this is possible only if one of the premises is also an **E**. Both premises cannot be **E** propositions, because two negative premises are never allowed (Rule 4), and the other premise cannot be an **O** proposition because then both premises would also be negative. Nor can the other premise be an **I** proposition, for if it were, a term distributed in the conclusion would then not be distributed in the premises, violating Rule 3. So the other premise must be an **A**, and the two premises must be either **AE** or **EA**. The only possible moods (if the conclusion of the syllogism is an **E** proposition) are therefore **AEE** and **EAE**.

If the mood is **AEE**, it cannot be either in the first figure or in the third figure, because in either of those cases a term distributed in the conclusion would then not be distributed in the premises. Therefore, the mood **AEE** is possibly valid only in the second figure, **AEE–2** (traditionally called *Camestres*), or in the fourth figure, **AEE–4** (traditionally called *Camenes*). If the mood is **EAE**, it cannot be in the third figure or in the fourth figure, because again that would mean that a term distributed in the conclusion would not be distributed in the premises, which leaves as valid only the first figure, **EAE–1** (traditionally called *Celarent*), and the second figure, **EAE–2** (traditionally called *Cesare*.)

Summary of Case 2: If the syllogism has an **E** conclusion, there are only four possibly valid forms: **AEE–2**, **AEE–4**, **EAE–1**, and **EAE–2**—*Camestres, Camenes, Celarent*, and *Cesare*, respectively.

Case 3. If the conclusion is an **I** proposition

In this case, neither premise can be an **E** or an **O**, because if either premise is negative, the conclusion must be negative (Rule 5). The two premises cannot both be **A**, because a syllogism with a particular conclusion cannot have two universal premises (Rule 6). Neither can both premises be **I**, because the middle

---

Barbara
The traditional name for the valid syllogism with the mood and figure
AAA–1

Camestres
The traditional name for the valid syllogism with the mood and figure
AEE–2

Camenes
The traditional name for the valid syllogism with the mood and figure
AEE–4

Celarent
The traditional name for the valid syllogism with the mood and figure
EAE–1

Cesare
The traditional name for the valid syllogism with the mood and figure
EAE–2

term must be distributed in at least one premise (Rule 2). So the premises must be either **AI** or **IA**, and therefore the only possible moods with an **I** conclusion are **AII** and **IAI**.

AII is not possibly valid in the second figure or in the fourth figure because the middle term must be distributed in at least one premise. The only valid forms remaining for the mood **AII**, therefore, are **AII–1** (traditionally called *Darii*) and **AII–3** (traditionally called *Datisi*). If the mood is **IAI**, it cannot be **IAI–1** or **IAI–2**, because they also would violate the rule that requires the middle term to be distributed in at least one premise. This leaves as valid only **IAI–3** (traditionally called *Disamis*), and **IAI–4** (traditionally called *Dimaris*).

Summary of Case 3: If the syllogism has an **I** conclusion, there are only four possibly valid forms: **AII–1**, **AII–3**, **IAI–3**, and **IAI–4**—*Darii, Datisi, Disamis,* and *Dimaris*, respectively.

Case 4. If the conclusion is an **O** proposition

In this case, the major premise cannot be an **I** proposition, because any term distributed in the conclusion must be distributed in the premises. So the major premise must be either an **A** or an **E** or an **O** proposition.

Suppose the major premise is an **A**. In that case, the minor premise cannot be either an **A** or an **E**, because two universal premises are not permitted when the conclusion (an **O**) is particular. Neither can the minor premise then be an **I**, because if it were, either the middle term would not be distributed at all (a violation of Rule 2), or a term distributed in the conclusion would not be distributed in the premises. So, if the major premise is an **A**, the minor premise has to be an **O**, yielding the mood **AOO**. In the fourth figure, **AOO** cannot possibly be valid, because in that case the middle term would not be distributed, and in the first figure and the third figure **AOO** cannot possibly be valid either, because that would result in terms being distributed in the conclusion that were not distributed in the premises. For the mood **AOO**, the only possibly valid form remaining, if the major premise is an **A**, is therefore in the second figure, **AOO–2** (traditionally called *Baroko*).

Now suppose (if the conclusion is an **O**) that the major premise is an **E**. In that case, the minor premise cannot be either an **E** or an **O**, because two negative premises are not permitted. Nor can the minor premise be an **A**, because two universal premises are precluded if the conclusion is particular (Rule 6). This leaves only the mood **EIO**—and this mood is valid in all four figures, traditionally known as *Ferio* (**EIO–1**), *Festino* (**EIO–2**), *Ferison* (**EIO–3**), and *Fresison* (**EIO–4**).

Finally, suppose (if the conclusion is an **O**) that the major premise is also an **O** proposition. Then, again, the minor premise cannot be an **E** or an **O**, because two negative premises are forbidden. The minor premise cannot be an **I**, because then the middle term would not be distributed, or a term that is distributed in the conclusion would not be distributed in the premises. Therefore, if the major premise is an **O**, the minor premise must be an **A**, and the mood must be **OAO**. But **OAO–1** is eliminated, because in that case the middle term would not be distributed. **OAO–2** and **OAO–4** are also eliminated, because in both a

---

Darii
The traditional name for the valid syllogism with the mood and figure **AII–1**

Datisi
The traditional name for the valid syllogism with the mood and figure **AII–3**

Disamis
The traditional name for the valid syllogism with the mood and figure **IAI–3**

Dimaris
The traditional name for the valid syllogism with the mood and figure **IAI–4**

Baroko
The traditional name for the valid syllogism with the mood and figure **AOO–2**

Ferio
The traditional name for the valid syllogism with the mood and figure **EIO–1**

Festino
The traditional name for the valid syllogism with the mood and figure **EIO–2**

Ferison
The traditional name for the valid syllogism with the mood and figure **EIO–3**

Fresison
The traditional name for the valid syllogism with the mood and figure **EIO–4**

term distributed in the conclusion would then not be distributed in the premises. This leaves as valid only **OAO–3** (traditionally known as *Bokardo*).

Summary of Case 4: If the syllogism has an **O** conclusion, there are only six possibly valid forms: **AOO–2, EIO–1, EIO–2, EIO–3, EIO–4,** and **OAO–3**—*Baroko, Ferio, Festino, Ferison, Fresison,* and *Bokardo*.

This analysis has demonstrated, by elimination, that there are exactly fifteen valid forms of the categorical syllogism: one if the conclusion is an **A** proposition, four if the conclusion is an **E** proposition, four if the conclusion is an **I** proposition, and six if the conclusion is an **O** proposition. Of these fifteen valid forms, four are in the first figure, four are in the second figure, four are in the third figure, and three are in the fourth figure. This completes the deduction of the fifteen valid forms of the standard-form categorical syllogism.

EXERCISES

For students who enjoy the complexities of analytical syllogistics, here follow some theoretical questions whose answers can all be derived from the systematic application of the six rules of the syllogism set forth in Section 6.4. Answering these questions will be much easier if you have fully grasped the deduction of the fifteen valid syllogistic forms presented in this appendix.

EXAMPLE

1. Can any standard-form categorical syllogism be valid that contains exactly three terms, each of which is distributed in both of its occurrences?

SOLUTION

No, such a syllogism cannot be valid. If each of the three terms were distributed in both of its occurrences, all three of the syllogism's propositions would have to be **E** propositions, and the mood of the syllogism would thus be **EEE**, which violates Rule 4, which forbids two negative premises.

2. In what mood or moods, if any, can a first-figure standard-form categorical syllogism with a particular conclusion be valid?
3. In what figure or figures, if any, can the premises of a valid standard-form categorical syllogism distribute both the major and minor terms?
4. In what figure or figures, if any, can a valid standard-form categorical syllogism have two particular premises?
*5. In what figure or figures, if any, can a valid standard-form categorical syllogism have only one term distributed, and that one only once?
6. In what mood or moods, if any, can a valid standard-form categorical syllogism have just two terms distributed, each one twice?
7. In what mood or moods, if any, can a valid standard-form categorical syllogism have two affirmative premises and a negative conclusion?
8. In what figure or figures, if any, can a valid standard-form categorical syllogism have a particular premise and a universal conclusion?
9. In what mood or moods, if any, can a second-figure standard-form categorical syllogism with a universal conclusion be valid?

Bokardo
The traditional name for the valid syllogism with the mood and figure **OAO–3**

*10. In what figure or figures, if any, can a valid standard-form categorical syllogism have its middle term distributed in both premises?

11. Can a valid standard-form categorical syllogism have a term distributed in a premise that appears undistributed in the conclusion?

chapter 6 Summary

In this chapter we have examined the standard-form categorical syllogism: its elements, its forms, its validity, and the rules governing its proper use.

In Section 6.1, the major, minor, and middle terms of a syllogism were identified:

- **Major term:** the predicate of the conclusion
- **Minor term:** the subject of the conclusion
- **Middle term:** the third term appearing in both premises but not in the conclusion.

We identified major and minor premises as those containing the major and minor terms, respectively. We specified that a categorical syllogism is in standard form when its propositions appear in precisely this order: major premise first, minor premise second, and conclusion last.

We also explained in Section 6.1 how the mood and figure of a syllogism are determined.

The mood of a syllogism is determined by the three letters identifying the forms of its three propositions, **A, E, I,** or **O**. There are sixty-four possible different moods.

The figure of a syllogism is determined by the position of the middle term in its premises. The four possible figures are described and named thus:

- **First figure:** The middle term is the subject term of the major premise and the predicate term of the minor premise.
  Schematically: *M–P, S–M,* therefore *S–P.*
- **Second figure:** The middle term is the predicate term of both premises.
  Schematically: *P–M, S–M,* therefore *S–P.*
- **Third figure:** The middle term is the subject term of both premises.
  Schematically: *M–P, M–S,* therefore *S–P.*
- **Fourth figure:** The middle term is the predicate term of the major premise and the subject term of the minor premise.
  Schematically: *P–M, M–S,* therefore *S–P.*

In Section 6.2, we explained how the mood and figure of a standard-form categorical syllogism jointly determine its logical form. Because each of the sixty-four moods may appear in all four figures, there are exactly 256 standard-form categorical syllogisms, of which only a few are valid.

In Section 6.3, we explained the Venn diagram technique for testing the validity of syllogisms, using overlapping circles appropriately marked or shaded to exhibit the meaning of the premises.

In Section 6.4, we explained the six essential rules for standard-form syllogisms and named the fallacy that results when each of these rules is broken:

- **Rule 1.** A standard-form categorical syllogism must contain exactly three terms, each of which is used in the same sense throughout the argument.
  Violation: Fallacy of four terms.
- **Rule 2.** In a valid standard-form categorical syllogism, the middle term must be distributed in at least one premise.
  Violation: Fallacy of undistributed middle.
- **Rule 3.** In a valid standard-form categorical syllogism, if either term is distributed in the conclusion, then it must be distributed in the premises.
  Violation: Fallacy of the illicit major, or fallacy of the illicit minor.
- **Rule 4.** No standard-form categorical syllogism having two negative premises is valid.
  Violation: Fallacy of exclusive premises.
- **Rule 5.** If either premise of a valid standard-form categorical syllogism is negative, the conclusion must be negative.
  Violation: Fallacy of drawing an affirmative conclusion from a negative premise.
- **Rule 6.** No valid standard-form categorical syllogism with a particular conclusion can have two universal premises.
  Violation: Existential fallacy.

In Section 6.5, we presented an exposition of the fifteen valid forms of the categorical syllogism, identifying their moods and figures, and explaining their traditional Latin names:

AAA–1 (*Barbara*); **EAE–1** (*Celarent*); **AII–1** (*Darii*); **EIO–1** (*Ferio*); **AEE–2** (*Camestres*); **EAE–2** (*Cesare*); **AOO–2** (*Baroko*); **EIO–2** (*Festino*); **AII–3** (*Datisi*); **IAI–3** (*Disamis*); **EIO–3** (*Ferison*); **OAO–3** (*Bokardo*); **AEE–4** (*Camenes*); **IAI–4** (*Dimaris*); **EIO–4** (*Fresison*).

In the Appendix to Chapter 6 (which may be bypassed), we presented the deduction of the fifteen valid forms of the categorical syllogism, demonstrating, through a process of elimination, that only those fifteen forms can avoid all violations of the six basic rules of the syllogism.

END NOTES

[1] Barbara Tuchman, *The Proud Tower* (New York: Macmillan, 1966.)
[2] Erasmus, *The Praise of Folly*, 1511.

For additional exercises and tutorials about concepts covered in this chapter, log in to MyLogicLab at *www.mylogiclab.com* and select your current textbook.

Syllogisms in Ordinary Language

7.1 Syllogistic Arguments
7.2 Reducing the Number of Terms to Three
7.3 Translating Categorical Propositions into Standard Form
7.4 Uniform Translation
7.5 Enthymemes
7.6 Sorites
7.7 Disjunctive and Hypothetical Syllogisms
7.8 The Dilemma

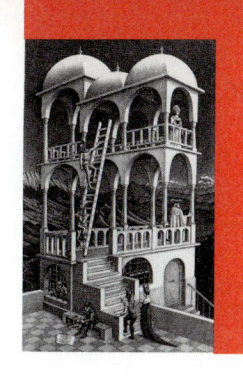

7.1 Syllogistic Arguments

In ordinary discourse the arguments we encounter rarely appear as neatly packaged, standard-form categorical syllogisms. So the syllogistic arguments that arise in everyday speech cannot always be readily tested. They can be tested, however, if we *put* them into standard form—and we can generally do that by reformulating their constituent propositions. The term **syllogistic argument** refers to any argument that either is a standard-form categorical syllogism or *can be reformulated as* a standard-form categorical syllogism without any loss or change of meaning.

We want to be able to test the validity of syllogistic arguments. If they are fallacious or misleading, that will be most easily detected, as Immanuel Kant pointed out, when they are set out in correct syllogistic form. The process of reformulation is therefore important, because the effective tests discussed in Chapter 6—Venn diagrams and the rules for categorical syllogisms—cannot be applied directly until the syllogism is in standard form. Putting it into standard form is called **reduction** (or **translation**) **to standard form**. When we reformulate (or reduce) a loosely put argument that appears in ordinary language into a classical syllogism, the resulting argument is called a *standard-form translation* of the original argument. Effecting this reformulation can present some difficulties.

We already know the tests for validity (Venn diagrams and the rules for syllogisms). What we need, to evaluate syllogistic arguments using these tests, are techniques for *translating* syllogistic arguments from their loose forms into standard form. With these techniques in hand, we can first *translate* the argument into standard form, and then *test* that argument using the Venn diagram method or the syllogistic rules.

To describe the various techniques for reduction to standard form, we begin by noting the kinds of problems that create the need for them—that is, by noting different ways in which a syllogistic argument in ordinary language may *deviate*

chapter 7

Syllogistic argument Any argument that either is a standard-form categorical syllogism or can be reformulated as a standard-form categorical syllogism without any change of meaning.

Reduction to standard form The translation of syllogistic arguments in any form into the standard form in which they can be tested for validity; also called translation to standard form.

245

from a standard-form categorical argument. Understanding those deviations, we can proceed to counteract them.

First deviation. The premises and conclusion of an argument in ordinary language may appear in an *order* that is not the order of the standard-form syllogism. This difficulty is easily remedied by reordering the premises: The major premise is put first, the minor premise second, and the conclusion third. (Recall that the major premise is the premise that contains the term that is the predicate term of the conclusion, whereas the minor premise contains the term that is the subject term of the conclusion.)

Second deviation. A standard-form categorical syllogism always has exactly three terms. The premises of an argument in ordinary language may appear to involve *more than three terms*—but that appearance may be deceptive. If the number of terms can be reduced to three without loss of meaning, the reduction to standard form may be successful.

Third deviation. The *component propositions* of a syllogistic argument in ordinary language *may not all be standard-form propositions*. This deviation is very common, but if the components can be converted into standard-form propositions without loss of meaning, the reduction to standard form may be successful.

To cope with the second and third of these deviant patterns, there are known techniques, which will now be explained.

7.2 Reducing the Number of Terms to Three

A valid syllogism must have exactly three terms. If more than three terms seem to be involved in an argument of apparently syllogistic form, it may be possible to translate the argument into a standard-form categorical syllogism that is equivalent to it but that contains only three terms and is perfectly valid. How can that be done?

One way is by *eliminating synonyms*. A synonym of one of the terms in the syllogism is not really a fourth term, but only another way of referring to one of the three classes involved. So we begin by eliminating synonyms, if any appear. For example, the following syllogistic argument appears to contain six terms:

> No wealthy persons are vagrants.
> All lawyers are rich people.
> Therefore no attorneys are tramps.

However, "wealthy" and "rich" are synonyms, as are "lawyer" and "attorney," and also "vagrant" and "tramp." If the synonyms are eliminated, the argument becomes

> No wealthy persons are vagrants.
> All lawyers are wealthy persons.
> Therefore no lawyers are vagrants.

This argument in standard form, **EAE–1** (*Celarent*), is plainly valid.

A second way to reduce the number of terms to three is by *eliminating class complements*, a concept explained in Section 5.6. We illustrate this using the following syllogistic argument, whose propositions are standard-form categorical propositions:

> All mammals are warm-blooded animals.
> No lizards are warm-blooded animals.
> Therefore all lizards are nonmammals.

On the surface, this argument appears to be invalid, because it seems to have four terms—and it also draws an affirmative conclusion from a negative premise, which breaks one of the rules of the syllogism.

This argument, however, is perfectly valid when it is translated into standard form. We can reduce the number of terms to three, because two of the terms in it ("mammals" and "nonmammals") are complements of one another. So, by obverting the conclusion (to *obvert* a proposition, we change its quality and replace the predicate term by its complement), we get "No lizards are mammals." Using this valid immediate inference, we derive the following standard-form translation of the original argument:

> All mammals are warm-blooded animals.
> No lizards are warm-blooded animals.
> Therefore no lizards are mammals.

which is logically equivalent to the original because it has identically the same premises and a logically equivalent conclusion. This standard-form translation conforms to all the syllogistic rules and thus is known to be valid. Its form is **AEE–2** (*Camestres*).

There may be more than one translation of a syllogistic argument into standard form, but if any one of those translations yields a valid syllogism, all the others must be valid as well. Thus, for example, the preceding illustrative argument can also be reduced to standard form in a different (but logically equivalent) way. This time we leave the conclusion unchanged and work with the premises. We take the contrapositive of the first premise, and we obvert the second premise. We then get:

> All non(warm-blooded animals) are nonmammals.
> All lizards are non(warm-blooded animals).
> Therefore all lizards are nonmammals.

This is also a valid translation; its form is **AAA–1** (*Barbara*), and it conforms to all the rules of the syllogism.

Any syllogistic argument that appears to contain four terms can be reduced to standard form (that is, can be translated into a logically equivalent standard-form categorical syllogism) if one of its terms is the complement of one of the other three. Likewise, reduction from an argument with five terms is possible if two of its terms are complements of other terms in the argument; and even arguments with as many as six terms may be reduced to standard form if three of those terms are complements of other terms in the argument. The key to such reductions is to use the valid immediate inferences discussed in Chapter 5: conversion, obversion, and contraposition.

More than one immediate inference may be needed to reduce the argument to standard form. Consider this example:

> No nonresidents are citizens.
> All noncitizens are nonvoters.
> Therefore all voters are residents.

The argument has six terms, but it is valid, and that can be shown by reducing it to standard form, which can be done in more than one way. Perhaps the most natural reduction is to convert and then obvert the first premise. This yields "All citizens are residents." Then take the contrapositive of the second premise, which yields "All voters are citizens." The argument is then in standard form:

> All citizens are residents.
> All voters are citizens.
> Therefore all voters are residents.

The middle term ("citizens") is the subject term of the major premise and the predicate term of the minor premise, so the syllogism is in the first figure. Its three propositions are universal affirmatives. This is a syllogism in *Barbara*, **AAA–1**, and it is plainly valid.

EXERCISES

Translate the following syllogistic arguments into standard form, and test their validity by using either Venn diagrams or the syllogistic rules set forth in Chapter 6.

EXAMPLE

1. Some preachers are persons of unfailing vigor. No preachers are nonintellectuals. Therefore some intellectuals are persons of unfailing vigor.

SOLUTION

This argument may be translated into: Some preachers are persons of unfailing vigor. (Some P is V.) All preachers are intellectuals. (By obversion: All P is I.) Therefore some intellectuals are persons of unfailing vigor. (Some I is V.) Shown on a Venn diagram, this syllogism is seen to be valid:

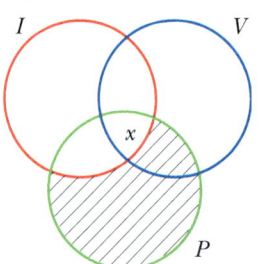

2. Some metals are rare and costly substances, but no welder's materials are nonmetals; hence some welder's materials are rare and costly substances.

 3. Some Asian nations were nonbelligerents, because all belligerents were allies either of Germany or Britain, and some Asian nations were not allies of either Germany or Britain.

 4. Some nondrinkers are athletes, because no drinkers are persons in perfect physical condition, and some people in perfect physical condition are not nonathletes.

 *5. All things inflammable are unsafe things, so all things that are safe are nonexplosives, because all explosives are flammable things.

 6. All worldly goods are changeable things, for no worldly goods are things immaterial, and no material things are unchangeable things.

 7. All those who are neither members nor guests of members are those who are excluded; therefore no nonconformists are either members or guests of members, for all those who are included are conformists.

 8. All mortals are imperfect beings, and no humans are immortals, whence it follows that all perfect beings are nonhumans.

 9. All things present are nonirritants; therefore no irritants are invisible objects, because all visible objects are absent things.

 *10. All useful things are objects no more than six feet long, because all difficult things to store are useless things, and no objects over six feet long are easy things to store.

7.3 Translating Categorical Propositions into Standard Form

It was noted in Section 7.1 that syllogistic arguments in ordinary language may deviate from standard-form categorical syllogisms not only because they may appear to contain more than three terms (as discussed in Section 7.2), but also because the component propositions of the syllogism in ordinary language may not all be standard-form propositions. **A**, **E**, **I**, and **O** propositions are clearly somewhat stilted, and many syllogistic arguments in everyday life contain nonstandard-form propositions. To reduce these arguments to standard form requires that their constituent propositions be translated into standard form.

It would be very convenient if there were some neat list of rules that we could use to effect such translations. Unfortunately, ordinary language is too rich and too multiform to permit the compilation of such a set of rules. Different sorts of transformation are called for in different settings, and to know what is called for we must, in every case, understand fully the given nonstandard-form proposition that needs to be reformulated. If we understand the proposition, we can reformulate it without losing or changing its meaning.

Although no complete set of rules can be given, we can describe a number of well-tested methods for translating nonstandard propositions of different sorts. These methods—nine of them will be presented in this section—must be regarded as guides rather than rules; they are *techniques* with which nonstandard-form propositions of certain describable kinds can be reformulated into standard-form propositions that may serve as constituents of syllogistic arguments.

I. **Singular Propositions.** Some propositions affirm or deny that a specific individual or object belongs to a given class—for example, "Socrates is a philosopher," and "This table is not an antique." These are called **singular propositions**. Such propositions do not affirm or deny the inclusion of one class in another (as standard-form propositions do), but we can nevertheless *interpret* a singular proposition as a proposition dealing with classes and their interrelations. We do this in the following way.

To every individual object there corresponds a unique **unit class** (one-membered class) whose only member is that object itself. Then, to assert that an object s belongs to a class P is logically equivalent to asserting that the unit class S containing just that object s is wholly included in the class P. And to assert that an object s does *not* belong to a class P is logically equivalent to asserting that the unit class S containing just that object s is wholly excluded from the class P.

It is customary to make this interpretation automatically, without any notational adjustment. Thus it is customary to take any affirmative singular proposition of the form "s is P" as if it were already expressed as the logically equivalent **A** proposition, "All S is P," and we similarly understand any negative singular proposition, "s is not P," as an alternative formulation of the logically equivalent **E** proposition, "No S is P"—in each case understanding S to designate the unit class whose only member is the object s. Thus no explicit translations are provided for singular propositions; traditionally they have been classified as **A** and **E** propositions as they stand. As Kant remarked, "Logicians are justified in saying that, in the employment of judgments in syllogisms, singular judgments can be treated like those that are universal."[1]

The situation, however, is not quite so simple. Bear in mind that particular propositions have existential import, but universal propositions do not. Using this Boolean interpretation (for reasons explained in Section 5.7), we find that if singular propositions are treated mechanically as **A** and **E** propositions in syllogistic arguments, and we check the validity of those arguments using Venn diagrams or the rules set forth in Chapter 6, serious difficulties arise.

In some cases, obviously valid two-premise arguments containing singular propositions translate into valid categorical syllogisms, such as when

Singular proposition
A proposition that asserts that a particular individual has (or does not have) some specified attribute.

Unit class
A class with only one member.

7.3 Translating Categorical Propositions into Standard Form

All *H* is *M*.		All *H* is *M*.
s is an *H*.	goes into the obviously valid	All *S* is *H*.
∴ *s* is an *M*.	**AAA–1** categorical syllogism in *Barbara*	∴ All *S* is *M*.

In other cases, however, obviously valid two-premise arguments containing singular propositions translate into categorical syllogisms that are invalid, such as when

s is *M*.		All *S* is *M*.
s is *H*.	goes into the invalid	All *S* is *H*.
∴ Some *H* is *M*.	**AAI–3** categorical syllogism	∴ Some *H* is *M*.

which commits the existential fallacy, violating Rule 6.

On the other hand, if we translate singular propositions into particular propositions, there is the same kind of difficulty. In some cases, obviously valid two-premise arguments containing singular propositions translate into *valid* categorical syllogisms, such as when

All *H* is *M*.		All *H* is *M*.
s is an *H*.	goes into the obviously valid	Some *S* is *H*.
∴ *s* is an *M*.	**AII–1** categorical syllogism in *Darii*	∴ Some *S* is *M*.

In other cases, however, obviously valid two-premise arguments containing singular propositions translate into categorical syllogisms that are *invalid*, such as when

s is *M*.		Some *S* is *M*.
s is *H*.	goes into the invalid	Some *S* is *H*.
∴ Some *H* is *M*.	**III–3** categorical syllogism	∴ Some *H* is *M*.

which commits the fallacy of the undistributed middle, violating Rule 2.

The difficulty arises from the fact that a singular proposition contains more information than is contained in any single one of the four standard-form categorical propositions. If "*s* is *P*" is construed as "All *S* is *P*," then what is lost is the existential import of the singular proposition, the fact that *S* is not empty. But if "*s* is *P*" is construed as "Some *S* is *P*," then what is lost is the universal aspect of the singular proposition, which distributes its subject term, the fact that *all S* is *P*.

The solution to the difficulty is to construe singular propositions as conjunctions of standard-form categorical propositions. An affirmative singular proposition is equivalent to the conjunction of the related **A** and **I** categorical propositions. Thus "*s* is *P*" is equivalent to "All *S* is *P*" *and* "Some *S* is *P*." A negative singular proposition is equivalent to the conjunction of the related **E** and **O** categorical propositions. Thus "*s* is not *P*" is equivalent to "No *S* is *P*" *and* "Some *S* is not *P*." Venn diagrams for affirmative and negative singular propositions are shown in Figure 7-1.

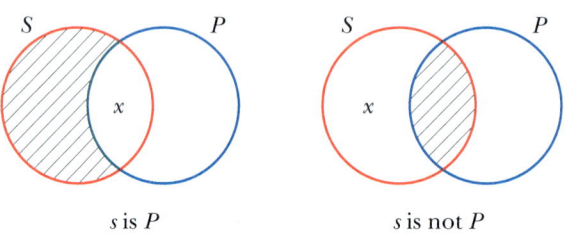

s is *P*   *s* is not *P*

Figure 7-1

In applying the syllogistic rules to evaluate a syllogistic argument containing singular propositions, we must take account of *all* the information contained in those singular propositions, both distribution and existential import.

If we keep in mind the existential import of singular propositions when we invoke the syllogistic rules or apply Venn diagrams to test the validity of syllogistic arguments, it is acceptable practice to regard singular propositions as universal (**A** or **E**) propositions.

II. **Categorical Propositions That Have Adjectives or Adjectival Phrases as Predicates, Rather than Substantives or Class Terms.** For example, "Some flowers are beautiful" and "No warships are available for active duty" are categorical propositions, yet they must be translated into standard-form categorical propositions; they deviate from standard form only in that their predicates, "beautiful" and "available for active duty," designate *attributes* rather than classes. However, every attribute *determines* a class, the class of things having that attribute, so every such proposition corresponds to a logically equivalent proposition that is in standard form. The two examples cited correspond to the **I** and **E** propositions "Some flowers are beautiful things" and "No warships are things available for active duty." When a categorical proposition is in standard form except that it has an adjectival predicate instead of a predicate term, the translation into standard form is made by replacing the adjectival predicate with a term designating the class of all objects of which the adjective may truly be predicated.

III. **Categorical Propositions Whose Main Verbs Are Other than the Standard-Form Copula "To Be."** Examples of this very common type are "All people seek recognition" and "Some people drink Greek wine." The usual method of translating such a statement into standard form is to regard all of it, except the subject term and the quantifier, as naming a class-defining characteristic. Those words can then be replaced by a term designating the class determined by that class-defining characteristic and may be linked to the subject with a standard copula. Thus the two examples given translate into the standard-form categorical propositions, "All people are seekers of recognition" and "Some people are Greek-wine drinkers."

IV. **Statements in Which the Standard-Form Ingredients Are All Present But Not Arranged in Standard-Form Order.** Two examples are "Racehorses are all thoroughbreds" and "All is well that ends well." In such cases, we first decide which is the subject term and then rearrange the words to express a standard-form categorical proposition. Such translations are usually quite straightforward. It is clear that the two preceding statements translate into the **A** propositions "All racehorses are thoroughbreds" and "All things that end well are things that are well."

V. **Categorical Propositions Whose Quantities Are Indicated by Words Other than the Standard-Form Quantifiers "All," "No," and "Some."** Statements beginning with the words "every" and "any" are easily translated. The propositions "Every dog has its day" and "Any contribution will be appreciated" reduce to "All dogs are creatures that have their days" and "All contributions are things that are appreciated." Similar to "every" and "any" are "everything" and "anything." Paralleling these, but clearly restricted to classes of persons, are "everyone," "anyone," "whoever," "whosoever," "who," "one who," and the like. These should present no difficulty.

The grammatical particles "a" and "an" may also serve to indicate quantity, but whether they are being used to mean "all" or "some" depends largely on the context. Thus "A bat is a mammal" and "An elephant is a pachyderm" are reasonably interpreted as meaning "All bats are mammals" and "All elephants are pachyderms." But "A bat flew in the window" and "An elephant escaped" quite clearly do not refer to all bats or all elephants; they are properly reduced to "Some bats are creatures that flew in the window" and "Some elephants are creatures that escaped."

The particle "the" may be used to refer either to a particular individual or to all the members of a class. There is little danger of ambiguity here, for such a statement as "The whale is a mammal" translates in almost any context into the **A** proposition "All whales are mammals," whereas the singular proposition "The first president was a military hero" is already in standard form as an **A** proposition (a singular proposition having existential import), as discussed in the first subparagraph of this section.

In some contexts the article "the" is deliberately omitted to achieve desired ambiguity. When United Nations Resolution 242 was adopted, calling for the return of "territory" captured by Israel in the Six-Day War in 1967, it was formally agreed that the English version of the Resolution would be authoritative, because the Resolution when expressed in French would require the definite article (*le territoire*), of which the English translation is "the territory," meaning all the territory captured, which is precisely what the agreed-upon English version carefully refrains from saying. The omission of the definite article in English can be logically significant.

Although affirmative statements beginning with "every" and "any" are translated into "All S is P," negative statements beginning with "not every" and "not any" are quite different. Their translations are much less obvious and require great care. Thus, for example, "Not every S is P" means that some S is not P, whereas "Not any S is P" means that no S is P.

VI. **Exclusive Propositions.** Categorical propositions involving the words "only" or "none but" are often called **exclusive propositions**, because in general they assert that the predicate applies exclusively to the subject named. Examples of such usages are "Only citizens can vote" and "None but the brave deserve the fair." The first translates into the standard-form categorical proposition, "All those who can vote are citizens," and the second into the standard-form categorical proposition, "All those who deserve the fair are those who are brave." Propositions beginning with "only" or "none but" usually translate into **A** propositions using this general rule: Reverse the subject and the predicate, and replace the "only" with "all." Thus "Only S is P" and "None but S's are P's" are usually understood to express "All P is S."

There are some contexts in which "only" and "none but" are used to convey some further meaning. "Only S is P" or "None but S's are P's" may suggest either that "All S is P" or that "Some S is P." This is not always the case, however. Where context helps to determine meaning, attention must be paid to it, of course. But in the absence of such additional information, the translations first suggested are adequate.

VII. **Categorical Propositions That Contain No Words to Indicate Quantity.** Two examples are "Dogs are carnivorous" and "Children are present." Where there is no quantifier, what the sentence is intended to express may be doubtful. We may be able to determine its meaning only by examining the context in which it occurs, and that examination usually will clear up our doubts. In the first example it is very probable that "Dogs are carnivorous" refers to *all* dogs, and is to be translated as "All dogs are carnivores."

Exclusive propositions
Propositions that assert that the predicate applies exclusively to the subject named. Example: "None but generals wear stars" asserts that the predicate, "wearing stars," applies only to generals.

In the second example, on the other hand, it is plain that only *some* children are referred to, and thus the standard-form translation of "Children are present" is "Some children are beings who are present."

VIII. **Propositions That Do Not Resemble Standard-Form Categorical Propositions But Can Be Translated into Standard Form.** Some examples are "Not all children believe in Santa Claus," "There are white elephants," "There are no pink elephants," and "Nothing is both round and square." On reflection, these propositions will be seen to be logically equivalent to, and therefore to translate into, the following standard-form propositions: "Some children are not believers in Santa Claus," "Some elephants are white things," "No elephants are pink things," and "No round objects are square objects."

IX. **Exceptive Propositions.** Some examples of **exceptive propositions** are "All except employees are eligible," "All but employees are eligible," and "Employees alone are not eligible." Translating exceptive propositions into standard form is somewhat complicated, because propositions of this kind (much like singular propositions) make *two* assertions rather than one. Each of the logically equivalent examples just given asserts not merely that *all nonemployees are eligible* but also (in the usual context) that *no employees are eligible*. Where "employees" is abbreviated to S and "eligible persons" to P, these two propositions can be written as "All non-S is P" and "No S is P." These are clearly independent and together assert that S and P are complementary classes.

Each of these exceptive propositions is *compound* and therefore cannot be translated into a single standard-form categorical proposition. Rather, each must be translated into an explicit conjunction of two standard-form categoricals. Thus the three illustrative propositions about eligibility translate identically into "All nonemployees are eligible persons, and no employees are eligible persons."

It should be noted that some arguments depend for their validity on numerical or quasi-numerical information that cannot be put into standard form. Such arguments may have constituent propositions that mention quantity more specifically than standard-form propositions do, usually by the use of quantifiers such as "one," "two," "three," "many," "a few," "most," and so on. When such specific quantitative information is critical to the validity of the argument in which it is mentioned, the argument itself is *asyllogistic* and therefore requires a more complicated analysis than that provided by the simple theory of the categorical syllogism. Yet some quasi-numerical quantifiers occur in arguments that do lend themselves to syllogistic analysis. These include "almost all," "not quite all," "all but a few," and "almost everyone." Propositions in which these phrases appear as quantifiers may be treated like the explicitly exceptive propositions just described. Thus the following exceptive propositions with quasi-numerical quantifiers are also compound: "Almost all students were at the dance,"

Exceptive proposition
A proposition that asserts that all members of some class, with the exception of the members of one of its subclasses, are members of some other class. Exceptive propositions are in reality compound, because they assert both a relation of class inclusion, and a relation of class exclusion. Example: "All persons except employees are eligible" is an exceptive proposition in which it is asserted both that "All nonemployees are eligible" and that "No employees are eligible."

"Not quite all students were at the dance," "All but a few students were at the dance," and "Only some students were at the dance." Each of these *affirms* that *some students were at the dance* and *denies* that *all students were at the dance*. The quasi-numerical information they present is irrelevant from the point of view of syllogistic inference, and all are translated as "Some students are persons who were at the dance, and some students are not persons who were at the dance."

Because exceptive propositions are not categorical propositions but conjunctions, arguments containing them are not syllogistic arguments as we are using that term. But they may nevertheless be susceptible to syllogistic analysis and appraisal. How an argument containing an exceptive proposition should be tested depends on the exceptive proposition's position in the argument. If it is a premise, then the argument may have to be given two separate tests. For example, consider the argument:

> Everyone who saw the game was at the dance.
> Not quite all the students were at the dance.
> So some students didn't see the game.

Its first premise and its conclusion are categorical propositions, which are easily translated into standard form. Its second premise, however, is an exceptive proposition, not simple but compound. To discover whether its premises imply its conclusion, first test the syllogism composed of the first premise of the given argument, the first half of its second premise, and its conclusion. In standard form, we have

> All persons who saw the game are persons who were at the dance.
> Some students are persons who were at the dance.
> Therefore some students are not persons who saw the game.

The standard-form categorical syllogism is of form **AIO–2** and commits the fallacy of the undistributed middle, violating Rule 2. However, the original argument is not yet proved to be invalid, because the syllogism just tested contains only part of the premises of the original argument. We now have to test the categorical syllogism composed of the first premise and the conclusion of the original argument together with the second half of the second premise. In standard form we then get a very different argument:

> All persons who saw the game are persons who were at the dance.
> Some students are not persons who were at the dance.
> Therefore some students are not persons who saw the game.

This is a standard-form categorical syllogism in *Baroko*, **AOO–2**, and it is easily shown to be valid. Hence the original argument is valid, because the conclusion is the same, and the premises of the original argument *include* the premises of this valid standard-form syllogism. Thus, to test the validity of an argument, one of whose premises is an exceptive proposition, may require testing two different standard-form categorical syllogisms.

If the premises of an argument are both categorical propositions, and its conclusion is exceptive, then we know it to be invalid, for although the two categorical premises may imply one or the other half of the compound conclusion, they cannot imply them both. Finally, if an argument contains exceptive propositions as both premises and conclusion, all possible syllogisms constructible out of the original argument may have to be tested to determine its validity. Enough has been explained to enable the student to cope with such situations.

It is important to acquire facility in translating the many varieties of non-standard-form propositions into standard form, because the tests of validity that we have developed—Venn diagrams and the syllogistic rules—can be applied directly only to standard-form categorical syllogisms.

EXERCISES

Translate the following into standard-form categorical propositions:

EXAMPLE

1. Roses are fragrant.

SOLUTION

Standard-form translation: All roses are fragrant things.

2. Orchids are not fragrant.
3. Many a person has lived to regret a misspent youth.
4. Not everyone worth meeting is worth having as a friend.
*5. If it's a Junko, it's the best that money can buy.
6. If it isn't a real beer, it isn't a Bud.
7. Nothing is both safe and exciting.
8. Only brave people have ever won the Congressional Medal of Honor.
9. Good counselors are not universally appreciated.
*10. He sees not his shadow who faces the sun.
11. To hear her sing is an inspiration.
12. He who takes the sword shall perish by the sword.
13. Only members can use the front door.
14. Nobody doesn't like Sara Lee.
*15. The Young Turks support no candidate of the Old Guard.
16. All styles are good, except the tiresome.
17. They also serve who only stand and wait.
18. Happy indeed is she who knows her own limitations.
19. A thing of beauty is a joy forever.

*20. He prayeth well who loveth well.

21. All that glitters is not gold.

22. None think the great unhappy but the great.

23. He jests at scars that never felt a wound.

24. Whatsoever a man soweth, that shall he also reap.

*25. A soft answer turneth away wrath.

7.4 Uniform Translation

For a syllogistic argument to be testable, it must be expressed in propositions that together contain exactly three terms. Sometimes this aim is difficult to accomplish and requires a more subtle approach than those suggested in the preceding sections. Consider the proposition, "The poor always you have with you." It clearly does not assert that *all* the poor are with you, or even that *some* (particular) poor are *always* with you. There are alternative methods of reducing this proposition to standard form, but one perfectly natural route is by way of the key word "always." This word means "at all times" and suggests the standard-form categorical proposition, "All times are times when you have the poor with you." The word "times," which appears in both the subject and the predicate terms, may be regarded as a **parameter**, an auxiliary symbol that is helpful in expressing the original assertion in standard form.

Care should be taken not to introduce and use parameters in a mechanical, unthinking fashion. One must be guided always by an understanding of the proposition to be translated. Thus the proposition, "Smith always wins at billiards," pretty clearly does not assert that Smith is incessantly, at all times, winning at billiards! It is more reasonable to interpret it as meaning that Smith wins at billiards whenever he plays. And so understood, it translates directly into "All times when Smith plays billiards are times when Smith wins at billiards."

Not all parameters need be temporal. To translate some propositions into standard form, the words "places" and "cases" can be introduced as parameters. Thus "Where there is no vision the people perish" and "Jones loses a sale whenever he is late" translate into "All places where there is no vision are places where the people perish" and "All cases in which Jones is late are cases in which Jones loses a sale."

The introduction of parameters often is requisite for the **uniform translation** of all three constituent propositions of a syllogistic argument into standard form. Because a categorical syllogism contains exactly three terms, to test a syllogistic argument we must translate its constituent propositions into standard-form categorical propositions that contain just three terms. The elimination of synonyms and the applications of conversion, obversion, and contraposition have already been discussed in Section 7.2. However, for many syllogistic arguments, the number of terms cannot be reduced to three either by eliminating synonyms or by applying conversion, obversion, or contraposition. Here uniform translation

Parameter
An auxiliary symbol or phrase that is introduced in translating statements uniformly, helping to express a syllogism with exactly three terms, so that it may be accurately tested.

Uniform translation
Techniques (often requiring the use of auxiliary symbols) making possible the reformulation of a syllogistic argument into standard form, so that it may be accurately tested.

requires the introduction of a parameter—the *same* parameter—into all three of the constituent propositions. Consider the following argument:

> Soiled paper plates are scattered only where careless people have picnicked.
> There are soiled paper plates scattered about here.
> Therefore careless people must have been picnicking here.

This argument is perfectly valid, but before it can be proved valid by our diagrams or rules, its premises and conclusion must be translated into standard-form categorical propositions involving only three terms. The second premise and the conclusion may be translated most naturally into "Some soiled paper plates are things that are scattered about here" and "Some careless people are those who have been picnicking here," but these two statements contain four different terms. To reduce the argument to standard form, we begin with the first premise, which requires a parameter for its standard-form expression, and then we use the same parameter in translating the second premise and the conclusion into standard form. The word "where" in the first premise suggests that the parameter "places" can be used. If this parameter is used to obtain uniform standard-form translations of all three propositions, the argument translates into:

> All places where soiled paper plates are scattered are places where careless people have picnicked.
> This place is a place where soiled paper plates are scattered.
> Therefore this place is a place where careless people have picnicked.

This standard-form categorical syllogism is in *Barbara* with mood and figure **AAA–1** and has already been proved valid.

The notion of standardizing expressions through the use of a parameter is not an altogether easy one to grasp, but some syllogistic arguments cannot be translated into standard-form categorical syllogisms by any other method. Another example may help to make clear the technique involved. Let us take the argument:

> The hounds bay wherever a fox has passed, so the fox must have taken another path, because the hounds are quiet.

First, we must understand what is asserted in the given argument. We may take the statement that the hounds are quiet as asserting that the hounds are not baying here and now. This step is part of the necessary process of eliminating synonyms, because the first assertion makes explicit reference to the baying of hounds. And in the same manner we may understand the conclusion that the fox must have taken another path as asserting that the fox did not pass *here*. The word "wherever" in the first assertion should suggest that the parameter "places" can be used in its translation. The standard-form translation thus arrived at is

> All places where a fox has passed are places where the hounds bay.
> This place is not a place where the hounds bay.
> Therefore this place is not a place where a fox has passed.

This standard-form categorical syllogism is in *Camestres*, with mood and figure **AEE–2**, and its validity is easy to establish.

260 CHAPTER 7 Syllogisms in Ordinary Language

EXERCISES

A. Translate the following propositions into standard form, using parameters where necessary.

EXAMPLE

1. He groans whenever he is reminded of his loss.

SOLUTION

Standard-form translation: All times when he is reminded of his loss are times when he groans.

2. She never drives her car to work.
3. He walks where he chooses.
4. He always orders the most expensive item on the menu.
*5. She does not give her opinion unless she is asked to do so.
6. She tries to sell life insurance wherever she may happen to be.
7. His face gets red when he gets angry.
8. If he is asked to say a few words, he talks for hours.
9. Error of opinion may be tolerated where reason is left free to combat it.
*10. People are never so likely to settle a question rightly as when they discuss it freely.

B. For each of the following arguments,
   a. Translate the argument into standard form.
   b. Name the mood and figure of its standard-form translation.
   c. Test its validity using a Venn diagram. If it is valid, give its traditional name.
   d. If it is invalid, name the fallacy it commits.

EXAMPLE

1. Since all knowledge comes from sensory impressions and since there's no sensory impression of substance itself, it follows logically that there is no knowledge of substance.
   —Robert M. Pirsig, *Zen and the Art of Motorcycle Maintenance*
   (New York: Bantam, 1975)

SOLUTION

a. Standard-form translation:

No things derived from sensory impressions are items of knowledge of substance itself.

<u>All items of knowledge are things derived from sensory impressions.</u>
Therefore, no items of knowledge are items of knowledge of substance itself.

b. Mood and figure: **EAE–1**

c. Valid; *Celarent*

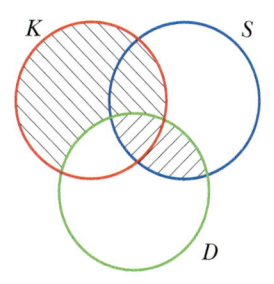

2. . . . no names come in contradictory pairs; but all predicables come in contradictory pairs; therefore no name is a predicable.
   —Peter Thomas Geach, *Reference and Generality*
   (Ithaca, NY: Cornell University Press, 1980)

3. Barcelona Traction was unable to pay interest on its debts; bankrupt companies are unable to pay interest on their debts; therefore, Barcelona Traction must be bankrupt.
   —John Brooks, "Annals of Finance," *The New Yorker*, 28 May 1979

4. Extremism in defense of liberty, or virtue, or whatever is always a vice—because extremism is but another name for fanaticism which is a vice by definition.
   —Irving Kristol, "The Environmentalist Crusade,"
   The Wall Street Journal, 16 December 1974

*5. All syllogisms having two negative premises are invalid. Some valid syllogisms are sound. Therefore some unsound arguments are syllogisms having two negative premises.

6. Not all is gold that glitters, for some base metals glitter, and gold is not a base metal.

7. Where there's smoke there's fire, so there's no fire in the basement, because there's no smoke there.

8. It seems that mercy cannot be attributed to God. For mercy is a kind of sorrow, as Damascene says. But there is no sorrow in God; and therefore there is no mercy in Him.
   —Thomas Aquinas, *Summa Theologiae*, I, question 21, art. 3

9. . . . because intense heat is nothing else but a particular kind of painful sensation; and pain cannot exist but in a perceiving being; it follows that no intense heat can really exist in an unperceiving corporeal substance.
   —George Berkeley, *Three Dialogues Between Hylas and Philonous,
   in Opposition to Sceptics and Atheists*, 1713

10. Only those who ignore the facts are likely to be mistaken. No one who is truly objective is likely to be mistaken. Hence no one who ignores the facts is truly objective.

11. All bridge players are people. All people think. Therefore all bridge players think.

—Oswald and James Jacoby, "Jacoby on Bridge,"
Syndicated Column, 5 November 1966

12. Whenever I'm in trouble, I pray. And since I'm always in trouble, there is not a day when I don't pray.

—Isaac Bashevis Singer, interview in *The New York Times*

13. The after-image is not in physical space. The brain-process is. So the after-image is not a brain-process.

—J. J. C. Smart, "Sensations and Brain Processes,"
Philosophical Review, April 1959

14. It must have rained lately, because the fish are not biting, and fish never bite after a rain.

*15. . . . it is obvious that irrationals are uninteresting to engineers, since they are concerned only with approximations, and all approximations are rational.

—G. H. Hardy, *A Mathematician's Apology*
(Cambridge: Cambridge University Press, 1940)

16. Since to fight against neighbors is an evil, and to fight against the Thebans is to fight against neighbors, it is clear that to fight against the Thebans is an evil.

—Aristotle, *Prior Analytics*

17. According to Aristotle, none of the products of Nature are due to chance. His proof is this: That which is due to chance does not reappear constantly nor frequently, but all products of Nature reappear either constantly or at least frequently.

—Moses Maimonides, *The Guide for the Perplexed*, 1180

18. Not all who have jobs are temperate in their drinking. Only debtors drink to excess. So not all the unemployed are in debt.

19. It will be a good game tomorrow, for the conference title is at stake, and no title contest is ever dull.

*20. Bill didn't go to work this morning, because he wore a sweater, and he never wears a sweater to work.

21. Cynthia must have complimented Henry, because he is cheerful whenever Cynthia compliments him, and he's cheerful now.

22. There must be a strike at the factory, for there is a picket line there, and pickets are present only at strikes.

23. Epidemiology is not merely the study of epidemics of infectious disease; it is the broad examination of the rates and patterns of disease in the

community. By almost any standard drug abuse can be regarded as a disease; accordingly it can be profitably investigated by the methods of epidemiology.
—"Science and the Citizen," *Scientific American*, February 1975

24. Since morals, therefore, have an influence on the actions and affections, it follows, that they cannot be deriv'd from reason; and that because reason alone, as we have already prov'd, can never have any such influence.
—David Hume, *A Treatise of Human Nature*, 1739

*25. All valid syllogisms distribute their middle terms in at least one premise, so this syllogism must be valid, for it distributes its middle term in at least one premise.

26. No valid syllogisms have two negative premises. No syllogisms on this page are invalid. Therefore no syllogisms on this page have two negative premises.

27. Good poll numbers raise money. Good press gets you good poll numbers. Good press gets you money.
—an advisor to Elizabeth Dole, during her campaign for the Republican presidential nomination, quoted in *The New York Times*, 15 April 2000

28. There are plants growing here, and since vegetation requires water, water must be present.

29. No one present is out of work. No members are absent. Therefore all members are employed.

*30. The competition is stiff, for there is a great deal of money involved, and there is never easy competition where much money is at stake.

31. There are handsome men, but only man is vile, so it is false that nothing is both vile and handsome.

32. What is simple cannot be separated from itself. The soul is simple; therefore, it cannot be separated from itself.
—Duns Scotus, *Oxford Commentary on the Sentences of Peter Lombard*, 1302

33. Although he complains whenever he is sick, his health is excellent, so he won't complain.

34. We... define a metaphysical sentence as a sentence which purports to express a genuine proposition, but does, in fact, express neither a tautology nor an empirical hypothesis. And as tautologies and empirical hypotheses form the entire class of significant propositions, we are justified in concluding that all metaphysical assertions are nonsensical.
—Alfred J. Ayer, *Language, Truth, and Logic*, 1936

*35. This syllogism is valid, for all invalid syllogisms commit an illicit process, and this syllogism commits no illicit process.

7.5 Enthymemes

Syllogistic arguments occur frequently, but their premises and conclusions are not always stated explicitly. Often only part of the argument is expressed, the rest being "understood." Thus one may justify the conclusion that "Jones is a citizen" by mentioning only the one premise, "Jones is a native-born American." As stated, the argument is incomplete, but the missing premise is easily supplied from knowledge of the Constitution of the United States. If the missing premise were stated, the completed argument would appear as

> All native-born Americans are citizens.
> Jones is a native-born American.
> Therefore Jones is a citizen.

Stated in full, the argument is a categorical syllogism of form **AAA–1**, *Barbara*, and is perfectly valid. An argument that is stated incompletely, part being "understood" or only "in the mind," is called an **enthymeme**. An incompletely stated argument is characterized as being *enthymematic*.

In everyday discourse, and even in science, many inferences are expressed enthymematically. The reason is easy to understand. A large body of propositions can be presumed to be common knowledge, and many speakers and writers save themselves trouble by not repeating well-known and perhaps trivially true propositions that their hearers or readers can perfectly well be expected to supply for themselves. Moreover, it is not at all unusual for an argument to be *rhetorically* more powerful and persuasive when stated enthymematically than when enunciated in complete detail. As Aristotle wrote in his *Rhetoric*, "Speeches that . . . rely on enthymemes excite the louder applause."

Because an enthymeme is incomplete, its omitted parts must be taken into account when testing its validity. Without the missing premise, the inference is invalid. However, when the unexpressed premise is easily supplied, in all fairness it ought to be included as part of the argument when one is appraising it. In such a case, one assumes that the maker of the argument did have more in mind than was stated explicitly. In most cases there is no difficulty in supplying the tacit premise that the speaker (or writer) intended but did not express. Thus, for example, as he explains the solution to the mystery in "The Adventure of Silver Blaze," Sherlock Holmes formulates an argument of which one critical premise is left unstated yet is very plainly supposed:

> A dog was kept in the stalls, and yet, though someone had been in and fetched out a horse, the dog had not barked. . . . Obviously the visitor was someone whom the dog knew well.

We all understand very well what is tacit here, that the dog would have barked had the visitor been a stranger. In fairness to the author, Arthur Conan Doyle, that premise must be seen as part of Holmes's argument.

In supplying a suppressed premise, a cardinal principle is that the proposition must be one that speakers can safely presume their hearers to accept as true. Thus it would be foolish to suggest taking the conclusion itself as a suppressed

---

Enthymeme
An argument that is stated incompletely, the unstated part of it being taken for granted. An enthymeme may be of the first, second, or third order, depending upon whether the unstated proposition is the major premise, the minor premise, or the conclusion of the argument.

premise, for if the arguer could have expected the auditors to accept that proposition as a premise, without proof, it would have been idle to present it to them as the conclusion of an argument.

Any kind of argument can be expressed enthymematically, but the kinds of enthymemes that have been most extensively studied are incompletely expressed syllogistic arguments. We confine our attention to these in the remainder of this section. Enthymemes traditionally have been divided into different *orders*, according to which part of the syllogism is left unexpressed. A **first-order enthymeme** is one in which the syllogism's major premise is not stated. The preceding example is of the first order. A **second-order enthymeme** is one in which only the major premise and the conclusion are stated, the minor premise being suppressed. An example of this type is "All students are opposed to the new regulations, so all sophomores are opposed to them." Here the minor premise is easily supplied, being the obviously true proposition, "All sophomores are students." A **third-order enthymeme** is one in which both premises are stated, but the conclusion is left unexpressed. An example of this type is the following:

> Our ideas reach no farther than our experience: we have no experience of divine attributes and operations: I need not conclude my syllogism: you can draw the inference yourself.[2]

Two steps are involved in testing an enthymeme for validity: The first is to supply the missing part of the argument, the second is to test the resulting syllogism. Formulating the unstated proposition fairly may require sensitivity to the context and an understanding of the intentions of the speaker. Consider the following argument: "No true Christian is vain, but some churchgoers are vain." It is the conclusion that remains unstated, so this is plainly a third-order syllogism. What is the intended conclusion? If the speaker intends to imply only that "Some churchgoers are not true Christians," the argument is valid (**EIO–2**, *Festino*). However, if the speaker's intention is to establish that "Some true Christians are not churchgoers," the enthymeme is invalid (**IEO–2**), because in that case the fallacy of illicit process of the major term is committed.

Usually, the context indicates unambiguously what the unstated proposition is. For example, in a U.S. Supreme Court opinion in which federal legislation regulating intrastate violence motivated by gender (the Violence Against Women Act) was held unconstitutional, the critical argument of the majority was expressed thus:

> Gender-motivated crimes of violence are not, in any sense of the phrase, economic activity.... Thus far in our nation's history our cases have upheld Commerce Clause regulation of intrastate activity only where that activity is economic in nature.[3]

The proposition that is understood but not stated in this argument is assuredly its conclusion: that gender-motivated crimes of violence may not be regulated by Congress under the long-existing precedent of Supreme Court cases.

To test this third-order enthymeme, we reformulate the argument so that its premises and (tacit) conclusion are in standard form. The major premise (the premise containing the predicate of the conclusion) is stated first; then mood and figure are identified:

First-order enthymeme
An incompletely stated syllogism in which the proposition that is taken for granted but not stated is the major premise.

Second-order enthymeme
An incompletely stated syllogism in which the proposition that is taken for granted but not stated is the minor premise.

Third-order enthymeme
An incompletely stated syllogism in which the proposition that is taken for granted but not stated is the conclusion.

Major premise: All activities that may be regulated by Congress under the precedent of Supreme Court cases are economic activities.
Minor premise: No intrastate gender-motivated crimes of violence are economic activities.
Conclusion (unstated but clearly indicated by the context): No intrastate gender-motivated crimes of violence are activities that may be regulated by Congress under the precedent of Supreme Court cases.

The mood of this syllogism is **AEE**; it is in the second figure because the middle term is the predicate of both premises. Its form is therefore *Camestres*, a valid syllogistic argument.

In some cases a third-order enthymeme may seem to be invalid without regard to context—for example, when both premises are negative, or when both premises are particular propositions, or when their common term is undistributed. In such cases, no syllogistic conclusion could follow validly, and hence such enthymemes are invalid in any context.

If it is one of the premises of the argument that is missing, it may be possible to make the argument valid only by adding a premise that is highly implausible—and pointing this out is certainly a legitimate criticism of an enthymematic argument. An even more crushing criticism, of course, would be to show that no additional premise, however implausible, can transform the enthymeme into a valid categorical syllogism.

The difference between enthymemes and normal syllogisms is essentially rhetorical, not logical. No new logical principles need be introduced in dealing with enthymemes, and they must be tested, ultimately, by the same methods that apply to standard-form categorical syllogisms.

EXERCISES

For each of the following enthymematic arguments:
  a. Formulate the plausible premise or conclusion, if any, that is missing but understood.
  b. Write the argument in standard form, including the missing premise or conclusion needed to make the completed argument valid—if possible—using parameters if necessary.
  c. Name the order of the enthymeme.
  d. If the argument is not valid even with the understood premise included, name the fallacy that it commits.

EXAMPLE

1. Transgenic animals are manmade and as such are patentable.
—Alan E. Smith, cited in *Genetic Engineering* (San Diego, CA: Greenhaven Press, 1990)

SOLUTION

a. The premise understood but not stated here is that whatever is man-made is patentable.

b. Standard-form translation:

> All manmade things are patentable things.
> <u>All transgenic animals are manmade things.</u>
> Therefore, all transgenic animals are patentable things.

c. The enthymeme is first-order, because the premise taken as understood is the major premise of the completed argument.

d. This is a valid syllogism of the form **AAA–1**, *Barbara*.

2. Abraham Beame... campaigned for mayor—as has been mentioned in recent weeks more often and with more irony than he might have wished—on the slogan "If you don't know the buck, you don't know the job—and Abe knows the buck."

—*The New Yorker*, 26 August 1974

3. Although these textbooks purport to be a universal guide to learning of great worth and importance—there is a single clue that points to another direction. In the six years I taught in city and country schools, no one ever stole a textbook.

—W. Ron Jones, *Changing Education*, Winter 1974

4. As a matter of fact, man, like woman, is flesh, therefore passive, the plaything of his hormones and of the species, the restless prey of his desires.

—Simone de Beauvoir, *The Second Sex*, 1949

*5. You never lose respect for a man who is a vicious competitor, and you never hate a man you respect.

—Pancho Gonzalez, former U.S. tennis champion

6. ... I am an Idealist, since I believe that all that exists is spiritual.

—John McTaggart Ellis McTaggart, *Philosophical Studies*, 1922

7. And why not become a perfect anthropomorphite? Why not assert the deity or deities to be corporeal, and to have eyes, a nose, mouth, ears, etc.? Epicurus maintained that no man had ever seen reason but in a human figure; therefore, the gods must have a human figure. And this argument, which is deservedly so much ridiculed by Cicero, becomes, according to you, solid and philosophical.

—David Hume, *Dialogues Concerning Natural Religion*, part V, 1779

8. Small countries tend to remember history especially well, since it often turns out badly for them.

—Marc Falcoff, "Semper Fidel," *The New Republic*, 3 July 1989

9. It must have rained lately, because the fish just aren't biting.

*10. It is not likely that the lies, misstatements, and omissions in President Carter's book are the result of ignorance. They must be the result, therefore, of malevolence.
—Facts and Logic about the Middle East, www.factsandlogic.org (2007)

11. No enthymemes are complete, so this argument is incomplete.

12. The chairman of the Student Conduct Legislative Council [at Stanford] argued that free speech rights extend only to victimized minorities, since the white majority does not need such protections.
—Nat Hentoff, "Stanford and the Speech Police," *The Washington Post*, 30 July 1990

13. Only demonstrative proof should be able to make you abandon the theory of the Creation; but such a proof does not exist in Nature.
—Moses Maimonides, *The Guide for the Perplexed*, 1180

14. It is probably true that the least destructive nuclear weapons are the most dangerous, because they make it easier for a nuclear war to begin.
—Freeman Dyson, "Reflections: Weapons and Hope," *The New Yorker*, 6 February 1984

*15. Man tends to increase at a greater rate than his means of subsistence; consequently he is occasionally subject to a severe struggle for existence.
—Charles Darwin, *The Descent of Man*, 1871

16. No internal combustion engines are free from pollution; but no internal combustion engine is completely efficient. You may draw your own conclusion.

17. A nation without a conscience is a nation without a soul. A nation without a soul is a nation that cannot live.
—Winston Churchill

18. Liberty means responsibility. That is why most men dread it.
—George Bernard Shaw, *Maxims for Revolutionists*, 1903

19. Who controls the past controls the future. Who controls the present controls the past.
—George Orwell, *1984*

*20. Productivity is desirable because it betters the condition of the vast majority of the people.
—Stephen Miller, "Adam Smith and the Commercial Republic," *The Public Interest*, Fall 1980

21. Advertisements perform a vital function in almost any society, for they help to bring buyers and sellers together.
—Burton M. Leiser, *Liberty, Justice, and Morals*, 1986

22. Logic is a matter of profound human importance precisely because it is empirically founded and experimentally applied.
—John Dewey, *Reconstruction in Philosophy*, 1920

23. *Iphigeneia at Aulis* is a tragedy because it demonstrates inexorably how human character, with its itch to be admired, combines with the malice

of heaven to produce wars which no one in his right mind would want and which turn out to be utterly disastrous for everybody.
—George E. Dimock, Jr., Introduction to *Iphigeneia at Aulis* by Euripides, 1992

24. . . . the law does not expressly permit suicide, and what it does not expressly permit it forbids.
—Aristotle, *Nicomachean Ethics*

*25. The man who says that all things come to pass by necessity cannot criticize one who denies that all things come to pass by necessity: for he admits that this too happens of necessity.
—Epicurus, Fragment XL, Vatican Collection

7.6 Sorites

Sometimes a single categorical syllogism will not suffice to account for our ability to draw a desired conclusion from a group of premises. Thus, from the premises

> All diplomats are tactful.
> Some government officials are diplomats.
> All government officials are people in public life.

one cannot draw the conclusion

> Some people in public life are tactful.

using a *single* syllogistic inference. Yet the indicated conclusion is entailed by the stated premises. To derive it requires two syllogisms rather than one. A stepwise process of argumentation must be resorted to, in which each step is a separate categorical syllogism. When stated explicitly, the required argument is

> All diplomats are tactful individuals.
> <u>Some government officials are diplomats.</u>
> Therefore some government officials are tactful individuals.
> <u>All government officials are people in public life.</u>
> Therefore some people in public life are tactful individuals.

This argument is not a syllogism but a *chain* of categorical syllogisms, connected by the conclusion of the first, which is a premise of the second. This chain has only two links, but more extended arguments may consist of a greater number. Because a chain is no stronger than its weakest link, an argument of this type is valid if, and only if, all of its constituent syllogisms are valid.

Where such an argument is expressed enthymematically, with only the premises and the final conclusion stated, it is called a **sorites** (pronounced sō-rī́-tēz—from the Greek, *soros*, meaning "heap or pile"; a sorites is a pile of syllogisms). Sorites may have three, four, or *any* number of premises. Some are very lengthy indeed. The following example is drawn from the *Monadology* of the philosopher Gottfried Leibniz:

Sorites
An argument whose conclusion is inferred from its premises by a *chain* of syllogistic inferences in which the conclusion of each inference serves as a premise for the next, and the conclusion of the last syllogism is the conclusion of the entire argument.

The human soul is a thing whose activity is thinking. A thing whose activity is thinking is one whose activity is immediately apprehended, and without any representation of parts therein. A thing whose activity is immediately apprehended without any representation of parts therein is a thing whose activity does not contain parts. A thing whose activity does not contain parts is one whose activity is not motion. A thing whose activity is not motion is not a body. What is not a body is not in space. What is not in space is insusceptible of motion. What is insusceptible of motion is indissoluble (for dissolution is a movement of parts). What is indissoluble is incorruptible. What is incorruptible is immortal. Therefore the human soul is immortal.

This sorites contains ten premises! Any sorites may be tested by making its intermediate conclusions or steps explicit, then testing separately the various categorical syllogisms thus obtained. If we ignore the possibility that an equivocation is present, then the validity of Leibniz's sorites is easily verified.

It is convenient to note here, in connection with the exercises at the end of this section, that a sorites is in standard form when all of its propositions are in standard form, when each term occurs exactly twice, and when every proposition (except the last) has a term in common with the proposition that immediately follows it. Thus one standard-form translation of Lewis Carroll's sorites

(1) Everyone who is sane can do Logic.
(2) No lunatics are fit to serve on a jury.
(3) <u>None of your sons can do Logic.</u>
Therefore none of your sons is fit to serve on a jury.

is

(2') All persons fit to serve on a jury are sane persons.
(1') All sane persons are persons who can do Logic.
(3') <u>No sons of yours are persons who can do Logic.</u>
Therefore no sons of yours are persons fit to serve on a jury.

One can test it by stating the suppressed subconclusion explicitly and then testing the resulting categorical syllogisms.

EXERCISES

A. Translate the propositions of the following sorites into standard form, and test the validity of each sorites.[4]

EXAMPLE

1. (1) Babies are illogical.
(2) Nobody is despised who can manage a crocodile.
(3) <u>Illogical persons are despised.</u>
Therefore babies cannot manage crocodiles.

SOLUTION

Standard-form translation:

(1') All babies are illogical persons.
(3') All illogical persons are despised persons.
(2') <u>No persons who can manage crocodiles are despised persons.</u>
Therefore, no babies are persons who can manage crocodiles.

This sorites consists of two syllogisms, as follows:

All *I* is *D*.	No *M* is *D*.
All *B* is *I*.	All *B* is *D*.
Therefore all *B* is *D*.	Therefore no *B* is *M*.

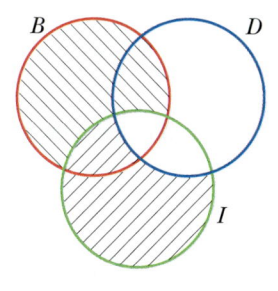

Valid, Barbara   *Valid, Cesare*

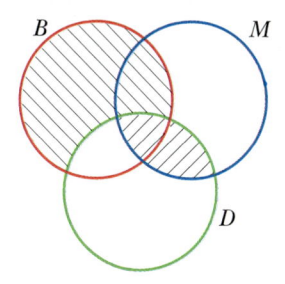

2. (1) No experienced person is incompetent.
(2) Jenkins is always blundering.
(3) <u>No competent person is always blundering.</u>
Therefore Jenkins is inexperienced.

3. (1) The only books in this library that I do not recommend for reading are unhealthy in tone.
(2) The bound books are all well written.
(3) All the romances are healthy in tone.
(4) <u>I do not recommend that you read any of the unbound books.</u>
Therefore all the romances in this library are well written.

4. (1) Only profound scholars can be dons at Oxford.
(2) No insensitive souls are great lovers of music.
(3) No one whose soul is not sensitive can be a Don Juan.
(4) <u>There are no profound scholars who are not great lovers of music.</u>
Therefore all Oxford dons are Don Juans.

*5. (1) No interesting poems are unpopular among people of real taste.
(2) No modern poetry is free from affectation.
(3) All your poems are on the subject of soap bubbles.
(4) No affected poetry is popular among people of real taste.
(5) <u>Only a modern poem would be on the subject of soap bubbles.</u>
Therefore all your poems are uninteresting.

272   CHAPTER 7   Syllogisms in Ordinary Language

    6. (1) None but writers are poets.
        (2) Only military officers are astronauts.
        (3) Whoever contributes to the new magazine is a poet.
        (4) <u>Nobody is both a military officer and a writer.</u>
        Therefore not one astronaut is a contributor to the new magazine.

B. Each of the following sets of propositions can serve as premises for a valid sorites. For each, find the conclusion and establish that the argument is valid.

  *1. (1) No one reads the *Times* unless he is well educated.
      (2) No hedgehogs can read.
      (3) Those who cannot read are not well educated.

  2. (1) All puddings are nice.
     (2) This dish is a pudding.
     (3) No nice things are wholesome.

  3. (1) The only articles of food that my doctor allows me are such as are not very rich.
     (2) Nothing that agrees with me is unsuitable for supper.
     (3) Wedding cake is always very rich.
     (4) My doctor allows me all articles of food that are suitable for supper.

  4. (1) All my daughters are slim.
     (2) No child of mine is healthy who takes no exercise.
     (3) All gluttons who are children of mine are fat.
     (4) No son of mine takes any exercise.

  *5. (1) When I work a logic example without grumbling, you may be sure it is one that I can understand.
      (2) These sorites are not arranged in regular order, like the examples I am used to.
      (3) No easy example ever makes my head ache.
      (4) I can't understand examples that are not arranged in regular order, like those I am used to.
      (5) I never grumble at an example, unless it gives me a headache.

7.7 Disjunctive and Hypothetical Syllogisms

Propositions are *categorical* when they affirm or deny the inclusion or exclusion of categories or classes. Syllogisms, arguments consisting of two premises and a conclusion, are called categorical when the propositions they contain are categorical. Up to this point our analysis has been of categorical syllogisms only. However, a syllogism may contain propositions that are not categorical. Such cases are not called categorical syllogisms but are instead named on the basis of the kind of propositions they contain. Here we look briefly at some other kinds of propositions and the syllogisms to which they give rise.

The categorical propositions with which we are familiar are *simple* in the sense that they have a single component, which affirms or denies some class relation. In contrast, some propositions are *compound*, in that they contain more than one component, each of which is itself some other proposition.

Consider first the *disjunctive* (or alternative) proposition. An example is "She was driven either by stupidity or by arrogance." Its two components are "she was driven by stupidity" and "she was driven by arrogance." The disjunctive proposition contains those two component propositions, which are called its *disjuncts*. The disjunctive proposition does not categorically affirm the truth of either one of its disjuncts, but says that at least one of them is true, allowing for the possibility that both may be true.

If we have a disjunction as one premise, and as another premise the denial or contradictory of one of its two disjuncts, then we can validly infer that the other disjunct in that disjunction is true. Any argument of this form is a valid **disjunctive syllogism**. A letter writer, critical of a woman nominated for high office by President George W. Bush, wrote:

> In trying to cover up her own illegal alien peccadillo or stonewall her way out of it, she was driven either by stupidity or arrogance. She's obviously not stupid; her plight must result, then, from her arrogance.[5]

As we use the term in this section, not every disjunctive syllogism is valid. The argument

> She was either arrogant or stupid.
> She was arrogant.
> Therefore she was not stupid.

is an example of what may be called an invalid disjunctive syllogism. We readily see that, even if the premises were true, she may have been arrogant *and* stupid. The truth of one disjunct of a disjunction does not imply the falsehood of the other disjunct, because both disjuncts of a disjunction can be true. We have a valid disjunctive syllogism, therefore, only where the categorical premise contradicts one disjunct of the disjunctive premise and the conclusion affirms the other disjunct of the disjunctive premise.

An objection might be raised at this point, based on such an argument as the following:

> Either Smith is in New York or Smith is in Paris.
> Smith is in New York.
> Therefore Smith is not in Paris.

Here the categorical premise affirms one disjunct of the stated disjunction, and the conclusion contradicts the other disjunct, yet the conclusion seems to follow validly. Closer analysis shows, however, that the stated disjunction plays no role in the argument. The conclusion follows enthymematically from the second, categorical premise, with the unexpressed additional premise

Disjunctive Syllogism
A syllogism in which one of the premises is a disjunction, the other premise is the denial or the contradictory of one of the two disjuncts in the first premise, and the conclusion is the statement that the other disjunct in that first premise is true.

being the obviously true proposition that "Smith cannot be both in New York and in Paris," which can be stated in disjunctive form as

> Either Smith is not in New York or Smith is not in Paris.

When this tacit premise is supplied and the superfluous original disjunction is discarded, the resulting argument is easily seen to be a valid disjunctive syllogism. The apparent exception is not really an exception, and the objection is groundless.

The second kind of compound proposition we consider is the *conditional* (or *hypothetical*) proposition, an example of which is "If the first native is a politician, then the first native lies." A conditional proposition contains two component propositions: The one following the "if" is the *antecedent*, and the one following the "then" is the *consequent*. A syllogism that contains conditional propositions exclusively is called a **pure hypothetical syllogism**; for example,

> If the first native is a politician, then he lies.
> If he lies, then he denies being a politician.
> Therefore if the first native is a politician, then he denies being a politician.

Pure hypothetical syllogism
A syllogism that contains only hypothetical propositions.

Mixed hypothetical syllogism
A syllogism that contains one conditional (or hypothetical) premise, and one categorical premise.

Modus ponens
A mixed hypothetical syllogism in which the first premise is a conditional proposition, the second premise affirms the antecedent of that conditional, and the conclusion affirms the consequent of that conditional.

Fallacy of affirming the consequent
A fallacy in which, from the truth of the consequent of a conditional proposition, the conclusion is reached that the antecedent of that conditional is true.

In this argument it can be observed that the first premise and the conclusion have the same antecedent, that the second premise and the conclusion have the same consequent, and that the consequent of the first premise is the same as the antecedent of the second premise. It should be clear that any pure hypothetical syllogism whose premises and conclusion have their component parts so related is a valid argument.

A syllogism that has one conditional premise and one categorical premise is called a **mixed hypothetical syllogism**. Two valid forms of the mixed hypothetical syllogism have been given special names. The first is illustrated by

> If the second native told the truth, then only one native is a politician.
> The second native told the truth.
> Therefore only one native is a politician.

Here the categorical premise affirms the antecedent of the conditional premise, and the conclusion affirms its consequent. Any argument of this form is valid and is said to be in the *affirmative mood* or **modus ponens** (from the Latin *ponere*, meaning "to affirm"). One must not confuse the valid form *modus ponens* with the clearly invalid form displayed by the following argument:

> If Bacon wrote *Hamlet*, then Bacon was a great writer.
> Bacon was a great writer.
> Therefore Bacon wrote *Hamlet*.

This argument differs from *modus ponens* in that its categorical premise affirms the consequent, rather than the antecedent, of the conditional premise. Any argument of this form is said to commit the **fallacy of affirming the consequent**.

The other valid form of mixed hypothetical syllogism is illustrated by:

> If the one-eyed prisoner saw two red hats, then he could tell the color of the hat on his own head.
> The one-eyed prisoner could not tell the color of the hat on his own head.
> Therefore the one-eyed prisoner did not see two red hats.

Here the categorical premise denies the consequent of the conditional premise, and the conclusion denies its antecedent. Any argument of this form is valid and is said to be in the form **modus tollens** (from the Latin *tollere*, meaning "to deny"). One must not confuse the valid form *modus tollens* with the clearly invalid form displayed by the following argument:

> If Carl embezzled the college funds, then Carl is guilty of a felony.
> Carl did not embezzle the college funds.
> Therefore Carl is not guilty of a felony.

This argument differs from *modus tollens* in that its categorical premise denies the antecedent, rather than the consequent, of the conditional premise. Any argument of this form is said to commit the **fallacy of denying the antecedent**.

Modus tollens
A mixed hypothetical syllogism in which the first premise is a conditional proposition, the second premise is the denial of the consequent of that conditional, and the conclusion is the denial of the antecedent of that conditional.

Fallacy of denying the antecedent A fallacy in which, from the negation of the antecedent of a conditional proposition, the conclusion is reached that the consequent of that conditional is false.

overview

Principal Kinds of Syllogisms

1. **Categorical syllogisms**, which contain only categorical propositions affirming or denying the inclusion or exclusion of categories. Example:

   All *M* is *P*.
   All *S* is *M*.
   Therefore all *S* is *P*.

2. **Disjunctive syllogisms**, which contain a compound, disjunctive (or alternative) premise asserting the truth of at least one of two alternatives, and a premise that asserts the falsity of one of those alternatives. Example:

   Either *P* is true or *Q* is true.
   P is not true.
   Therefore *Q* is true.

3. **Hypothetical syllogisms**, which contain one or more compound, hypothetical (or conditional) propositions, each affirming that if one of its components (the antecedent) is true then the other of its components (the consequent) is true. Two subtypes are distinguished:

   A. **Pure hypothetical syllogisms** contain conditional propositions only. Example:

   If *P* is true, then *Q* is true.
   If *Q* is true, then *R* is true.
   Therefore if *P* is true, then *R* is true.

> **B. Mixed hypothetical syllogisms** contain both a conditional premise and a categorical premise.
>   If the categorical premise affirms the truth of the antecedent of the conditional premise, and the consequent of that conditional premise is the conclusion of the argument, the form is valid and is called *modus ponens*. Example:
>
> If *P* is true, then *Q* is true.
> *P* is true.
> Therefore *Q* is true.
>
>   If the categorical premise affirms the falsity of the consequent of the conditional premise, and the falsity of the antecedent of that conditional premise is the conclusion of the argument, the form is valid and is called *modus tollens*. Example:
>
> If *P* is true, then *Q* is true.
> *Q* is false.
> Therefore *P* is false.

EXERCISES

Identify the form of each of the following arguments and state whether the argument is valid or invalid:

EXAMPLE

1. If a man could not have done otherwise than he in fact did, then he is not responsible for his action. But if determinism is true, it is true of every action that the agent could not have done otherwise. Therefore, if determinism is true, no one is ever responsible for what he does.
   —Winston Nesbit and Stewart Candlish, "Determinism and the Ability to Do Otherwise," *Mind*, July 1978

SOLUTION

This is a pure hypothetical syllogism. Valid.

2. Men, it is assumed, act in economic matters only in response to pecuniary compensation or to force. Force in the modern society is largely, although by no means completely, obsolete. So only pecuniary compensation remains of importance.
   —John Kenneth Galbraith, *The New Industrial State* (Boston: Houghton Mifflin, 1967)

3. If each man had a definite set of rules of conduct by which he regulated his life he would be no better than a machine. But there are no such rules, so men cannot be machines.
   —A. M. Turing, "Computing Machinery and Intelligence," *Mind*, volume 59, 1950

4. If the second native told the truth, then the first native denied being a politician. If the third native told the truth, then the first native denied being a politician. Therefore if the second native told the truth, then the third native told the truth.

*5. If the one-eyed prisoner does not know the color of the hat on his own head, then the blind prisoner cannot have on a red hat. The one-eyed prisoner does not know the color of the hat on his own head. Therefore the blind prisoner cannot have on a red hat.

6. If all three prisoners have on white hats, then the one-eyed prisoner does not know the color of the hat on his own head. The one-eyed prisoner does not know the color of the hat on his own head. Therefore all three prisoners have on white hats.

7. The stranger is either a knave or a fool. The stranger is a knave. Therefore the stranger is no fool.

8. If the first native is a politician, then the third native tells the truth. If the third native tells the truth, then the third native is not a politician. Therefore if the first native is a politician, then the third native is not a politician.

9. Mankind, he said, judging by their neglect of him, have never, as I think, at all understood the power of Love. For if they had understood him they would surely have built noble temples and altars, and offered solemn sacrifices in his honor; but this is not done.
—Plato, *Symposium*

*10. I have already said that he must have gone to King's Pyland or to Capleton. He is not at King's Pyland, therefore he is at Capleton.
—Arthur Conan Doyle, *The Adventure of Silver Blaze*

11. If then, it is agreed that things are either the result of coincidence or for an end, and that these cannot be the result of coincidence or spontaneity, it follows that they must be for an end.
—Aristotle, *Physics*

12. There is no case known (neither is it, indeed, possible) in which a thing is found to be the efficient cause of itself; for in such a case it would be prior to itself, which is impossible.
—Thomas Aquinas, *Summa Theologiae*, I, question 2, article 3

13. Either wealth is an evil or wealth is a good; but wealth is not an evil; therefore wealth is a good.
—Sextus Empiricus, *Against the Logicians*, second century CE

14. I *do* know that this pencil exists; but I could not know this, if Hume's principles were true; *therefore*, Hume's principles, one or both of them, are false.
—G. E. Moore, *Some Main Problems of Philosophy* (New York: Allen & Unwin, 1953)

*15. It is clear that we mean something, and something different in each case, by such words [as *substance, cause, change*, etc.]. If we did not we could not use them consistently, and it is obvious that on the whole we do consistently apply and withhold such names.

—C. D. Broad, *Scientific Thought*, 1923

16. If number were an idea, then arithmetic would be psychology. But arithmetic is no more psychology than, say, astronomy is. Astronomy is concerned, not with ideas of the planets, but with the planets themselves, and by the same token the objects of arithmetic are not ideas either.

—Gottlob Frege, *The Foundations of Arithmetic*, 1893

17. . . . If a mental state is to be identical with a physical state, the two must share all properties in common. But there is one property, spatial localizability, that is not so shared; that is, physical states and events are located in space, whereas mental events and states are not. Hence, mental events and states are different from physical ones.

—Jaegwon Kim, "On the Psycho-Physical Identity Theory," *American Philosophical Quarterly*, 1966

18. When we regard a man as morally responsible for an act, we regard him as a legitimate object of moral praise or blame in respect of it. But it seems plain that a man cannot be a legitimate object of moral praise or blame for an act unless in willing the act he is in some important sense a "free" agent. Evidently free will in some sense, therefore, is a precondition of moral responsibility.

—C. Arthur Campbell, *In Defence of Free Will*, 1938

19. In spite of the popularity of the finite-world picture, however, it is open to a devastating objection. In being finite the world must have a limiting boundary, such as Aristotle's outermost sphere. That is impossible, because a boundary can only separate one part of space from another. This objection was put forward by the Greeks, reappeared in the scientific skepticism of the early Renaissance and probably occurs to any schoolchild who thinks about it today. If one accepts the objection, one must conclude that the universe is infinite.

—J. J. Callahan, "The Curvature of Space in a Finite Universe," *Scientific American*, August 1976

*20. Total pacifism might be a good principle if everyone were to follow it. But not everyone does, so it isn't.

—Gilbert Harman, *The Nature Of Morality*, 1977

Dilemma
A common form of argument in ordinary discourse in which it is claimed that a choice must be made between two alternatives, both of which are (usually) bad.

7.8 The Dilemma

The **dilemma** is a common form of argument in ordinary language. It is, in essence, an argumentative device in which syllogisms on the same topic are combined, sometimes with devastating effect. Each of the constituent syllogisms may be quite ordinary, and therefore the dilemma is not of special importance from a strictly logical point of view. The premises of the syllogisms so combined are formulated disjunctively, and devised in a way designed to trap the opponent by forcing him to accept one or the other of the disjuncts. Thus the opponent is forced to accept the truth of the conclusion of one or the other of the syllogisms combined. When this is done successfully, the dilemma can prove to be a powerful instrument of persuasion.

People often say somewhat loosely that a person is "in" a dilemma (or "impaled on the horns of a dilemma") when that person must choose between two alternatives, both of which are bad or unpleasant. The dilemma is a form of argument intended to put one's opponent in just that kind of position. In debate, one uses a dilemma to offer alternative positions to one's adversary, from which a choice must be made, and then to prove that no matter which choice is made, the adversary is committed to an unacceptable conclusion.

The distinguished physicist Richard Feynman, recounting his experiences in the 1986 investigation of the catastrophic explosion of the *Challenger* space shuttle, was caustic in his criticism of mismanagement by administrators in the National Aeronautics and Space Administration (NASA). He said:

> Every time we talked to higher-level managers, they kept saying they didn't know anything about the problems below them. . . . Either the group at the top didn't know, in which case they should have known, or they did know, in which case they were lying to us.[6]

An attack of this kind is designed to push the adversaries (in this case the NASA administrators) into a corner and there annihilate them. The only explicitly stated premise of the argument is a disjunction, but one of the disjuncts must obviously be true: Either they knew or they didn't know about the problems below them. And whichever disjunct is chosen, the result for the adversary is very bad. The conclusion of a dilemma can itself be a disjunction (for example, "Either the NASA administrators did not know what they should have known, or they lied") in which case we call the dilemma a **complex dilemma**. But the conclusion may also be a categorical proposition, in which case we call it a **simple dilemma**.

A dilemma need not always have an unpleasant conclusion. An example of one with a happy conclusion is provided by the following simple dilemma:

> If the blest in heaven have no desires, they will be perfectly content; so they will be also if their desires are fully gratified; but either they will have no desires, or have them fully gratified; therefore they will be perfectly content.

The premises of a dilemma need not be stated in any special order; the disjunctive premise that offers the alternatives may either precede or follow the other. The consequences of those alternatives may be stated in a conjunctive proposition or in two separate propositions. An argument in dilemma form is often expressed enthymematically; that is, its conclusion generally is thought to be so obvious that it scarcely needs to be spelled out. This is well illustrated in a passage from a letter of President Abraham Lincoln, defending the Emancipation Proclamation that freed the slaves of the Confederacy:

Complex dilemma
An argument consisting of (a) a disjunction, (b) two conditional premises linked by a conjunction, and (c) a conclusion that is not a single categorical proposition (as in a simple dilemma) but a disjunction, a pair of (usually undesirable) alternatives.

Simple dilemma
An argument designed to push the adversary to choose between two alternatives, the (usually undesirable) conclusion in either case being a single categorical proposition.

But the proclamation, as law, either is valid, or is not valid. If it is not valid, it needs no retraction, If it is valid, it cannot be retracted, any more than the dead can be brought to life.[7]

Three ways of evading or refuting the conclusion of a dilemma have been given special names, two of them relating to the fact that a dilemma has two (or more) "horns." These three ways of defeating a dilemma are known as "going (or escaping) between the horns," "taking (or grasping) it by the horns," and "rebutting it by means of a counterdilemma." Note that these are not ways to prove the dilemma invalid; rather, they are ways in which one seeks to avoid its conclusion without challenging the formal validity of the argument.

One escapes between the horns of a dilemma by rejecting its disjunctive premise. This method is often the easiest way to evade the conclusion of a dilemma, for unless one half of the disjunction is the explicit contradictory of the other, the disjunction may very well be false. For example, one justification sometimes offered for giving grades to students is that recognizing good work will stimulate the students to study harder. Students may criticize this theory using the following dilemma:

> If students are fond of learning, they need no stimulus, and if they dislike learning, no stimulus will be of any avail. But any student either is fond of learning or dislikes it. Therefore a stimulus is either needless or of no avail.

This argument is formally valid, but one can evade its conclusion by *going between the horns*. The disjunctive premise is false, for students have all kinds of attitudes toward learning: Some may be fond of it, many dislike it, and many are indifferent. For that third group a stimulus may be both needed and of some avail. Going between the horns does not prove the conclusion to be false but shows merely that the argument does not provide adequate grounds for accepting that conclusion.

When the disjunctive premise is unassailable, as when the alternatives exhaust the possibilities, it is impossible to escape between the horns. Another method of evading the conclusion must be sought. One such method is to *grasp the dilemma by the horns*, which involves rejecting the premise that is a conjunction. To deny a conjunction, we need only deny one of its parts. When we grasp the dilemma by the horns, we attempt to show that at least one of the conditionals is false. The dilemma just above, attacking the use of grades in school, relies on the conditional "If students are fond of learning, they need no stimulus." The proponent of grading may grasp this dilemma by the horns and argue that even students who are fond of learning may sometimes need stimulus, and that the additional stimulus provided by grades promotes careful study by even the most diligent students. There may be good response to this, of course—but the original dilemma has been grasped firmly by the horns.

Rebutting a dilemma by means of a counterdilemma is the most ingenious method of all, but it is seldom cogent, for reasons that will appear presently. To rebut a given dilemma in this way, one constructs another dilemma whose conclusion is opposed to the conclusion of the original. *Any* counterdilemma may be used in rebuttal, but ideally it should be built up out of the same ingredients (categorical propositions) that the original dilemma contained.

A classical example of this elegant kind of rebuttal concerns the legendary argument of an Athenian mother attempting to persuade her son not to enter politics:

> If you say what is just, men will hate you; and if you say what is unjust, the gods will hate you; but you must either say the one or the other; therefore you will be hated.

Her son rebutted that dilemma with the following one:

> If I say what is just, the gods will love me; and if I say what is unjust, men will love me. I must say either the one or the other. Therefore I shall be loved!

In public discussion, where the dilemma is one of the strongest weapons of controversy, the use of a rebuttal of this kind, which derives an opposite conclusion from almost the same premises, is a mark of great rhetorical skill. If we examine the dilemma and rebutting counterdilemma more closely, we see that their conclusions are not as opposed as they might at first have seemed.

The conclusion of the first dilemma is that the son will be hated (by men or by the gods), whereas that of the rebutting dilemma is that the son will be loved (by the gods or by men). However, these two conclusions are perfectly compatible. The rebutting counterdilemma serves merely to establish a conclusion different from that of the original. Both conclusions may very well be true together, so no refutation has been accomplished. But in the heat of controversy analysis is unwelcome, and if such a rebuttal occurred in a public debate, the average audience might agree that the rebuttal was an effective reply to the original argument.

That this sort of rebuttal does not refute the argument but only directs attention to a different aspect of the same situation is perhaps more clearly shown in the case of the following dilemma, advanced by an "optimist":

> If I work, I earn money, and if I am idle, I enjoy myself. Either I work or I am idle. Therefore either I earn money or I enjoy myself.

A "pessimist" might offer the following counterdilemma:

> If I work, I don't enjoy myself, and if I am idle, I don't earn money. Either I work or I am idle. Therefore either I don't earn money or I don't enjoy myself.

These conclusions represent merely different ways of viewing the same facts; they do not constitute a disagreement over what the facts are.

No discussion of dilemmas would be complete unless it mentioned the celebrated lawsuit between Protagoras and Euathlus. Protagoras, a teacher who lived in Greece during the fifth century BCE, specialized in teaching the art of pleading before juries. Euathlus wanted to become a lawyer, but not being able to pay the required tuition, he made an arrangement according to which Protagoras would teach him but not receive payment until Euathlus won his first case. When Euathlus finished his course of study, he delayed going into practice. Tired of waiting for his money, Protagoras brought suit against his former pupil for the tuition money that was owed. Unmindful of the adage that the lawyer who tries his own case has a fool for a client, Euathlus decided to plead his own case in court. When the trial began, Protagoras presented his side of the case in a crushing dilemma:

> If Euathlus loses this case, then he must pay me (by the judgment of the court); if he wins this case, then he must pay me (by the terms of the contract). He must either lose or win this case. Therefore Euathlus must pay me.

The situation looked bad for Euathlus, but he had learned well the art of rhetoric. He offered the court the following counterdilemma in rebuttal:

> If I win this case, I shall not have to pay Protagoras (by the judgment of the court); if I lose this case, I shall not have to pay Protagoras (by the terms of the contract, for then I shall not yet have won my first case). I must either win or lose this case. Therefore I do not have to pay Protagoras![8]

Had you been the judge, how would you have decided?

Note that the conclusion of Euathlus's rebutting dilemma is not compatible with the conclusion of Protagoras's original dilemma. One conclusion is the explicit denial of the other. However, it is rare that a counterdilemma stands in this relation to the dilemma against which it is directed. When it does, the premises involved are themselves inconsistent, and it is this implicit contradiction that the two dilemmas make explicit.

EXERCISES

Discuss the various arguments that might be offered to refute each of the following:

EXAMPLE

1. If we interfere with the publication of false and harmful doctrines, we shall be guilty of suppressing the liberties of others, whereas if we do not interfere with the publication of such doctrines, we run the risk of losing our own liberties. We must either interfere or not interfere with the publication of false and harmful doctrines. Hence we must either be guilty of suppressing the liberties of others or else run the risk of losing our own liberties.

SOLUTION

It is impossible to go between the horns. It is possible to grasp it by either horn, arguing either (a) that liberties do not properly include the right to publish false and harmful doctrines or (b) that we run no risk of losing our own liberties if we vigorously oppose false and harmful doctrines with true and helpful ones. It could plausibly be rebutted (but not refuted) by the use of its ingredients to prove that "we must either be guiltless of suppressing the liberties of others or else run no risk of losing our own liberties."

2. If you tell me what I already understand, you do not enlarge my understanding, whereas if you tell me something that I do not understand,

then your remarks are unintelligible to me. Whatever you tell me must be either something I already understand or something that I do not understand. Hence whatever you say either does not enlarge my understanding or else is unintelligible to me.

3. If the conclusion of a deductive argument goes beyond the premises, then the argument is invalid, while if the conclusion of a deductive argument does not go beyond the premises, then the argument brings nothing new to light. The conclusion of a deductive argument must either go beyond the premises or not go beyond them. Therefore either deductive arguments are invalid or they bring nothing new to light.

4. If a deductive argument is invalid, it is without value, whereas a deductive argument that brings nothing new to light is also without value. Either deductive arguments are invalid or they bring nothing new to light. Therefore deductive arguments are without value.

*5. If the general had been loyal, he would have obeyed his orders, and if he had been intelligent, he would have understood them. The general either disobeyed his orders or else did not understand them. Therefore the general must have been either disloyal or unintelligent.

6. If he was disloyal, then his dismissal was justified, and if he was unintelligent, then his dismissal was justified. He was either disloyal or unintelligent. Therefore his dismissal was justified.

7. If the several nations keep the peace, the United Nations is unnecessary, while if the several nations go to war, the United Nations will have been unsuccessful in its purpose of preventing war. Now, either the several nations keep the peace or they go to war. Hence the United Nations is unnecessary or unsuccessful.

8. If people are good, laws are not needed to prevent wrongdoing, whereas if people are bad, laws will not succeed in preventing wrongdoing. People are either good or bad. Therefore either laws are not needed to prevent wrongdoing or laws will not succeed in preventing wrongdoing.

9. Archbishop Morton, Chancellor under Henry VII, was famous for his method of extracting "contributions" to the king's purse. A person who lived extravagantly was forced to make a large contribution, because it was obvious that he could afford it. Someone who lived modestly was forced to make a large contribution because it was clear that he must have saved a lot of money on living expenses. Whichever way he turned he was said to be "caught on Morton's fork."
—Dorothy Hayden, *Winning Declarer Play* (New York: Harper & Row, 1969)

*10. All political action aims at either preservation or change. When desiring to preserve, we wish to prevent a change to the worse; when desiring to change, we wish to bring about something better. All political action is then guided by some thought of better and worse.
—Leo Strauss, *What Is Political Philosophy?*, 1959

11. If a thing moves, it moves either in the place where it is or in that where it is not; but it moves neither in the place where it is (for it remains therein) nor in that where it is not (for it does not exist therein); therefore nothing moves.

—Sextus Empiricus, *Against the Physicists*

12. And what a life should I lead, at my age, wandering from city to city, ever changing my place of exile, and always being driven out! For I am quite sure that wherever I go, there, as here, the young men will flock to me; and if I drive them away, their elders will drive me out at their request; and if I let them come, their fathers and friends will drive me out for their sakes.

—Plato, *Apology*

13. If Socrates died, he died either when he was living or when he was dead. But he did not die while living; for assuredly he was living, and as living he had not died. Nor did he die when he was dead, for then he would be twice dead. Therefore Socrates did not die.

—Sextus Empiricus, *Against the Physicists*

14. Inevitably, the use of the placebo involved built-in contradictions. A good patient–doctor relationship is essential to the process, but what happens to that relationship when one of the partners conceals important information from the other? If the doctor tells the truth, he destroys the base on which the placebo rests. If he doesn't tell the truth, he jeopardizes a relationship built on trust.

—Norman Cousins, *Anatomy of an Illness*

***15.** The decision of the Supreme Court in *U.S. v. Nixon* (1974), handed down the first day of the Judiciary Committee's final debate, was critical. If the President defied the order, he would be impeached. If he obeyed the order, it was increasingly apparent, he would be impeached on the evidence.

—Victoria Schuck, "Watergate," *The Key Reporter*, Winter 1975–1976

16. If we are to have peace, we must not encourage the competitive spirit, whereas if we are to make progress, we must encourage the competitive spirit. We must either encourage or not encourage the competitive spirit. Therefore we shall either have no peace or make no progress.

17. The argument under the present head may be put into a very concise form, which appears altogether conclusive. Either the mode in which the federal government is to be constructed will render it sufficiently dependent on the people, or it will not. On the first supposition, it will be restrained by that dependence from forming schemes obnoxious to their constituents. On the other supposition, it will not possess the confidence of the people, and its schemes of usurpation will be easily defeated by the State governments, who will be supported by the people.

—James Madison, *The Federalist Papers*, no. 46, 1788

18. . . . a man cannot enquire either about that which he knows, or about that which he does not know; for if he knows, he has no need to enquire; and if not, he cannot; for he does not know the very subject about which he is to enquire.

—Plato, *Meno*

19. We tell clients to try to go through the entire first interview without even mentioning money. If you ask for a salary that is too high, the employer concludes that he can't afford you. If you ask for one that is too low, you're essentially saying, "I'm not competent enough to handle the job that you're offering."

—James Challenger, "What to Do—and Not to Do—When Job Hunting," *U.S. News & World Report*, 6 August 1984

*20. "Pascal's wager" is justifiably famous in the history of religion and also of betting. Pascal was arguing that agnostics—people unsure of God's existence—are best off betting that He does exist. If He does but you end up living as an unbeliever, then you could be condemned to spend eternity in the flames of Hell. If, on the other hand, He doesn't exist but you live as a believer, you suffer no corresponding penalty for being in error. Obviously, then, bettors on God start out with a big edge.

—Daniel Seligman, "Keeping Up," *Fortune*, 7 January 1985

chapter 7 Summary

In this chapter we have examined syllogistic argument as it is used in ordinary language, exhibiting the different guises in which syllogisms appear and showing how they may be best understood, used, and evaluated.

In Section 7.1, we explained the need for techniques to translate syllogistic arguments of any form into standard form, and we identified the ways in which syllogistic arguments may deviate from standard-form categorical syllogisms.

In Section 7.2, we explained how syllogisms in ordinary language appearing to have more than three terms may sometimes have the number of terms in them appropriately reduced to three—by elimination of synonyms, and by elimination of complementary classes.

In Section 7.3, we explained how the propositions of a syllogistic argument, when not in standard form, may be translated into standard form to allow the syllogism to be tested either by Venn diagrams or by use of the rules governing syllogisms. Nonstandard propositions of nine different kinds were examined, and the methods for translating each kind were explained and illustrated:

1. Singular propositions
2. Propositions having adjectives as predicates
3. Propositions having main verbs other than the copula "to be"
4. Statements having standard-form ingredients, but not in standard-form order

5. Propositions having quantifiers other than "all," "no," and "some"
6. Exclusive propositions, using "only" or "none but"
7. Propositions without words indicating quantity
8. Propositions not resembling standard-form propositions at all
9. Exceptive propositions, using "all except" or similar expressions

In Section 7.4, we explained how the uniform translation of propositions into standard form, essential for testing, may be assisted by the use of parameters.

In Sections 7.5 and 7.6, we explained enthymemes (syllogistic arguments in which one of the constituent propositions has been suppressed), and sorites (in which a chain of syllogisms may be compressed into a cluster of linked propositions).

In Section 7.7, we explained syllogisms other than categorical: disjunctive syllogisms and hypothetical syllogisms, so called because they contain disjunctive or hypothetical premises.

In Section 7.8, we discussed the rhetorical use of dilemmas, disjunctive arguments that give to the adversary a choice of alternatives neither of which is acceptable. We explained and illustrated the three possible patterns of rhetorical response: going between the horns of the dilemma, grasping the dilemma by its horns, or devising a counterdilemma.

END NOTES

[1] Immanuel Kant, *Critique of Pure Reason*, 1787, The Analytic of Concepts, chap. 1, sec. 2. More than a century later, Bertrand Russell presented a very different interpretation of singular propositions and universal propositions, and he later argued (in *My Philosophical Development*, 1959) that logic "cannot get far" until the two forms are seen to be "completely different," because the one (the singular) attributes a predicate to a named subject, while the other (the universal) expresses a relation between two predicates. Russell's interpretation had by that time become central to the theory of quantification in modern symbolic logic, discussed at length in Chapter 10; Kant's observation pertained to the use of singular propositions in traditional syllogisms, which he knew to be very powerful logical instruments.

[2] David Hume, *Dialogues Concerning Natural Religion*, part 2 (1779).

[3] *U.S. v. Morrison*, 529 U.S. 598 (2000).

[4] All the exercises in Section 7.6, except 4 and 6 under A, are taken, with little or no modification, from Lewis Carroll's *Symbolic Logic* (New York: C. N. Potter, 1977).

[5] Peter Bertocci, "Plight Must Come from Arrogance," *Ann Arbor* (MI) *News*, 19 January 2001.

[6] James Gleick, *Genius: The Life and Science of Richard Feynman* (New York: Pantheon Books, 1992).

[7] Letter to James C. Conkling, 26 August 1863.

[8] E. P. Northrop, *Riddles in Mathematics: A Book of Paradoxes* (Melbourne, FL: Krieger Publishing, 1975).

For additional exercises and tutorials about concepts covered in this chapter, log in to MyLogicLab at *www.mylogiclab.com* and select your current textbook.

Symbolic Logic

- **8.1** Modern Logic and Its Symbolic Language
- **8.2** The Symbols for Conjunction, Negation, and Disjunction
- **8.3** Conditional Statements and Material Implication
- **8.4** Argument Forms and Refutation by Logical Analogy
- **8.5** The Precise Meaning of "Invalid" and "Valid"
- **8.6** Testing Argument Validity Using Truth Tables
- **8.7** Some Common Argument Forms
- **8.8** Statement Forms and Material Equivalence
- **8.9** Logical Equivalence
- **8.10** The Three "Laws of Thought"

8.1 Modern Logic and Its Symbolic Language

To have a full understanding of deductive reasoning we need a general theory of deduction. A general theory of deduction will have two objectives: (1) to explain the relations between premises and conclusions in deductive arguments, and (2) to provide techniques for discriminating between valid and invalid deductions. Two great bodies of logical theory have sought to achieve these ends. The first, called classical (or Aristotelian) logic, was examined in Chapters 5 through 7. The second, called modern, symbolic, or mathematical logic, is the subject in this and the following two chapters.

Although these two great bodies of theory have similar aims, they proceed in very different ways. Modern logic does not build on the system of syllogisms discussed in preceding chapters. It does not begin with the analysis of categorical propositions. It does seek to discriminate valid from invalid arguments, although it does so using very different concepts and techniques. Therefore we must now begin afresh, developing a modern logical system that deals with some of the very same issues dealt with by traditional logic—and does so even more effectively.

Modern logic begins by first identifying the fundamental logical connectives on which deductive arguments depend. Using these connectives, a general account of such arguments is given, and methods for testing the validity of arguments are developed.

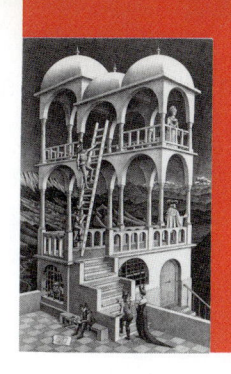

This analysis of deduction requires an artificial symbolic language. In a natural language—English or any other—there are peculiarities that make exact logical analysis difficult: Words may be vague or equivocal, the construction of arguments may be ambiguous, metaphors and idioms may confuse or mislead, emotional appeals may distract—problems discussed in Part I of this book. These difficulties can be largely overcome with an artificial language in which logical relations can be formulated with precision. The most fundamental elements of this modern symbolic language will be introduced in this chapter.

Symbols greatly facilitate our thinking about arguments. They enable us to get to the heart of an argument, exhibiting its essential nature and putting aside what is not essential. Moreover, with symbols we can perform, almost mechanically, with the eye, some logical operations which might otherwise demand great effort. It may seem paradoxical, but a symbolic language therefore helps us to accomplish some intellectual tasks without having to think too much. The Indo-Arabic numerals we use today (1, 2, 3, . . .) illustrate the advantages of an improved symbolic language. They replaced cumbersome Roman numerals (I, II, III, . . .), which are very difficult to manipulate. To multiply 113 by 9 is easy; to multiply CXIII by IX is not so easy. Even the Romans, some scholars contend, were obliged to find ways to symbolize numbers more efficiently.

Classical logicians did understand the enormous value of symbols in analysis. Aristotle used symbols as variables in his own analyses, and the refined system of Aristotelian syllogistics uses symbols in very sophisticated ways, as the preceding chapters have shown. However, much real progress has been made, mainly during the twentieth century, in devising and using logical symbols more effectively.

The modern symbolism with which deduction is analyzed differs greatly from the classical. The relations of classes of things are not central for modern logicians as they were for Aristotle and his followers. Instead, logicians look now to the internal structure of propositions and arguments, and to the logical links—very few in number—that are critical in all deductive argument. Modern symbolic logic is therefore not encumbered, as Aristotelian logic was, by the need to transform deductive arguments into syllogistic form, an often laborious task explained in the immediately preceding chapter.

The system of modern logic we now begin to explore is in some ways less elegant than analytical syllogistics, but it is more powerful. There are forms of deductive argument that syllogistics cannot adequately address. Using the approach taken by modern logic, with its more versatile symbolic language, we can pursue the aims of deductive analysis directly and we can penetrate more deeply. The logical symbols we shall now explore permit more complete and more efficient achievement of the central aim of deductive logic: discriminating between valid and invalid arguments.

8.2 The Symbols for Conjunction, Negation, and Disjunction

In this chapter we shall be concerned with relatively simple arguments such as:

> The blind prisoner has a red hat or the blind prisoner has a white hat.
> The blind prisoner does not have a red hat.
> Therefore the blind prisoner has a white hat.

and

If Mr. Robinson is the brakeman's next-door neighbor, then Mr. Robinson lives halfway between Detroit and Chicago.
Mr. Robinson does not live halfway between Detroit and Chicago.
Therefore Mr. Robinson is not the brakeman's next-door neighbor.

Every argument of this general type contains at least one compound statement. In studying such arguments we divide all statements into two general categories: simple and compound. A **simple statement** does not contain any other statement as a component. For example, "Charlie is neat" is a simple statement. A **compound statement** does contain another statement as a component. For example, "Charlie is neat and Charlie is sweet" is a compound statement, because it contains two simple statements as components. Of course, the components of a compound statement may themselves be compound. In formulating definitions and principles in logic, one must be very precise. What appears simple often proves more complicated than had been supposed. The notion of a "component of a statement" is a good illustration of this need for caution.

One might suppose that a component of a statement is simply a part of a statement that is itself a statement. But this account does not define the term with enough precision, because one statement may be a *part* of a larger statement and yet not be a *component* of it in the strict sense. For example, consider the statement: "The man who shot Lincoln was an actor." Plainly the last four words of this statement are a part of it, and could indeed be regarded as a statement; it is either true or it is false that Lincoln was an actor. But the statement that "Lincoln was an actor," although undoubtedly a part of the larger statement, is not a *component* of that larger statement.

We can explain this by noting that, for part of a statement to be a *component* of that statement, two conditions must be satisfied: (1) The part must be a statement in its own right; *and* (2) if the part is replaced in the larger statement by any other statement, the result of that replacement must be meaningful—it must make sense.

The first of these conditions is satisfied in the Lincoln example, but the second is not. Suppose the part "Lincoln was an actor" is replaced by "there are lions in Africa." The result of this replacement is nonsense: "The man who shot there are lions in Africa." The term *component* is not a difficult one to understand, but—like all logical terms—it must be defined accurately and applied carefully.

A. Conjunction

There are several types of compound statements, each requiring its own logical notation. The first type of compound statement we consider is the conjunction. We can form the **conjunction** of two statements by placing the word "and" between them; the two statements so combined are called **conjuncts**. Thus the compound statement, "Charlie is neat and Charlie is sweet," is a conjunction whose first conjunct is "Charlie is neat" and whose second conjunct is "Charlie is sweet."

Simple statement
A statement that does not contain any other statement as a component.

Compound statement
A statement that contains two or more statements as components.

Component
A part of a compound statement that is itself a statement, and is of such a nature that, if replaced in the larger statement by any other statement, the result will be meaningful.

Conjunction
A truth-functional connective meaning "and," symbolized by the dot, •. A statement of the form $p \cdot q$ is true if and only if p is true and q is true.

Conjunct
Each one of the component statements connected in a conjunctive statement

The word "and" is a short and convenient word, but it has other uses besides connecting statements. For example, the statement, "Lincoln and Grant were contemporaries," is not a conjunction, but a simple statement expressing a relationship. To have a unique symbol whose only function is to connect statements conjunctively, we introduce the **dot** " • " as our symbol for conjunction. Thus the previous conjunction can be written as "Charlie is neat • Charlie is sweet." More generally, where p and q are any two statements whatever, their conjunction is written $p \cdot q$. In some books, other symbols are used to express conjunction, such as "∧" or "&".

We know that every statement is either true or false. Therefore we say that every statement has a **truth value**, where the truth value of a true statement is *true*, and the truth value of a false statement is *false*. Using this concept, we can divide compound statements into two distinct categories, according to whether the truth value of the compound statement is determined wholly by the truth values of its components, or is determined by anything other than the truth values of its components.

We apply this distinction to conjunctions. The truth value of the conjunction of two statements is determined wholly and entirely by the truth values of its two conjuncts. If both its conjuncts are true, the conjunction is true; otherwise it is false. For this reason a conjunction is said to be a *truth-functional compound statement*, and its conjuncts are said to be **truth-functional components** of it.

Not every compound statement is truth-functional. For example, the truth value of the compound statement, "Othello believes that Desdemona loves Cassio," is not in any way determined by the truth value of its component simple statement, "Desdemona loves Cassio," because it could be true that Othello believes that Desdemona loves Cassio, regardless of whether she does or not. So the component, "Desdemona loves Cassio," is not a truth-functional component of the statement, "Othello believes that Desdemona loves Cassio," and the statement itself is not a truth-functional compound statement.

For our present purposes we define a component of a compound statement as being a **truth-functional component** if, when the component is replaced in the compound by any different statements having the same truth value as each other, the different compound statements produced by those replacements also have the same truth values as each other. Now a compound statement is defined as being a **truth-functional compound statement** if all of its components are truth-functional components of it.[1]

We shall be concerned only with those compound statements that are truth-functionally compound. In the remainder of this book, therefore, we shall use the term *simple statement* to refer to any statement that is not truth-functionally compound.

A conjunction is a truth-functional compound statement, so our dot symbol is a **truth-functional connective**. Given any two statements, p and q, there are only four possible sets of truth values they can have. These four possible cases, and the truth value of the conjunction in each, can be displayed as follows:

Dot The symbol for conjunction, •, meaning "and."

Truth value The status of any statement as true or false (**T** or **F**).

Truth-functional component Any component of a compound statement whose replacement there by any other statement having the same truth value would leave the truth value of the compound statement unchanged.

Truth-functional compound statement A compound statement whose truth value is determined wholly by the truth values of its components.

Truth-functional connective Any logical connective (e.g., conjunction, disjunction, material implication and material equivalence) between the components of a truth-functionally compound statement.

Where p is true and q is true, $p \cdot q$ is true.
Where p is true and q is false, $p \cdot q$ is false.
Where p is false and q is true, $p \cdot q$ is false.
Where p is false and q is false, $p \cdot q$ is false.

If we represent the truth values "true" and "false" by the capital letters **T** and **F**, the determination of the truth value of a conjunction by the truth values of its conjuncts can be represented more compactly and more clearly by means of a **truth table**:

p	q	$p \cdot q$
T	T	T
T	F	F
F	T	F
F	F	F

This truth table can be taken as defining the dot symbol, because it explains what truth values are assumed by $p \cdot q$ in every possible case.

We abbreviate simple statements by capital letters, generally using for this purpose a letter that will help us remember which statement it abbreviates. Thus we may abbreviate "Charlie is neat and Charlie is sweet" as $N \cdot S$. Some conjunctions, both of whose conjuncts have the same subject term—for example, "Byron was a great poet and Byron was a great adventurer"—are more briefly and perhaps more naturally stated in English by placing the "and" between the predicate terms and not repeating the subject term, as in "Byron was a great poet and a great adventurer." For our purposes, we regard the latter as formulating the same statement as the former and symbolize either one as $P \cdot A$. If both conjuncts of a conjunction have the same predicate term, as in "Lewis was a famous explorer and Clark was a famous explorer," the conjunction is usually abbreviated in English by placing the "and" between the subject terms and not repeating the predicate, as in "Lewis and Clark were famous explorers." Either formulation is symbolized as $L \cdot C$.

As shown by the truth table defining the dot symbol, a conjunction is true if and only if both of its conjuncts are true. The word "and" has another use in which it does not merely signify (truth-functional) conjunction, but has the sense of "and subsequently," meaning temporal succession. Thus the statement, "Jones entered the country at New York and went straight to Chicago," is significant and might be true, whereas "Jones went straight to Chicago and entered the country at New York" is hardly intelligible. There is quite a difference between "He took off his shoes and got into bed" and "He got into bed and took off his shoes."* Such examples show the desirability of having a special symbol with an exclusively truth-functional conjunctive use.

Truth table An array on which all possible truth values of compound statements are displayed, through the display of all possible combinations of the truth values of their simple components. A truth table may be used to define truth-functional connectives; it may also be used to test the validity of many deductive arguments.

---

*In *The Victoria Advocate*, Victoria, Texas, 27 October 1990, appeared the following report: "Ramiro Ramirez Garza, of the 2700 block of Leary Lane, was arrested by police as he was threatening to commit suicide and flee to Mexico."

Note that the English words "but," "yet," "also," "still," "although," "however," "moreover," "nevertheless," and so on, and even the comma and the semicolon, can also be used to conjoin two statements into a single compound statement, and in their conjunctive sense they can all be represented by the dot symbol.

B. Negation

The **negation** (or contradictory or denial) of a statement in English is often formed by the insertion of a "not" in the original statement. Alternatively, one can express the negation of a statement in English by prefixing to it the phrase "it is false that" or "it is not the case that." It is customary to use the symbol "$\sim$", called a **curl** or a **tilde**, to form the negation of a statement. (Again, some books use the symbol "—" for negation.) Thus, where M symbolizes the statement "All humans are mortal," the various statements "Not all humans are mortal," "Some humans are not mortal," "It is false that all humans are mortal," and "It is not the case that all humans are mortal" are all symbolized as $\sim M$. More generally, where p is any statement whatever, its negation is written $\sim p$. Some logicians treat the curl as another connective, but since it does not actually connect two or more units, it is sufficient to note that it performs an operation—reversing truth value—on a single unit, and thus may be referred to as an operator. It is a *truth-functional* operator, of course. The negation of any true statement is false, and the negation of any false statement is true. This fact can be presented very simply and clearly by means of a truth table:

p	$\sim p$
T	F
F	T

This truth table may be regarded as the definition of the negation "$\sim$" symbol.

C. Disjunction

The **disjunction** (or alternation) of two statements is formed in English by inserting the word "or" between them. The two component statements so combined are called *disjuncts* (or *alternatives*).

The English word "or" is ambiguous, having two related but distinguishable meanings. One of them is exemplified in the statement, "Premiums will be waived in the event of sickness or unemployment." The intention here is obviously that premiums are waived not only for sick persons and for unemployed persons, but also for persons who are both sick and unemployed. This sense of the word "or" is called weak or inclusive. An **inclusive disjunction** is true if one or the other or both disjuncts are true; only if both disjuncts are false is their inclusive disjunction false. The inclusive "or" has the sense of "either, possibly both." Where precision is at a premium, as in contracts and other legal documents, this sense is often made explicit by the use of the phrase "and/or."

The word "or" is also used in a strong or exclusive sense, in which the meaning is not "at least one" but "at least one and at most one." Where a restaurant

---

Negation
Denial; symbolized by the tilde or curl. $\sim p$ simply means "it is not the case that p," and may be read as "not-p."

Curl or **tilde**
The symbol for negation, $\sim$. It appears immediately before (to the left of) what is negated or denied.

Disjunction
A truth-functional connective meaning "or"; components so connected are called *disjuncts*. There are two types of disjunction: inclusive and exclusive.

Inclusive disjunction
A truth-functional connective between two components called disjuncts. A compound statement asserting inclusive disjunction is true when at least one of the disjuncts (that is, one or both) is true. Normally called simply "disjunction," it is also called "weak disjunction" and is symbolized by the wedge, $\vee$.

lists "salad or dessert" on its dinner menu, it is clearly meant that, for the stated price of the meal, the diner may have one or the other but not both. Where precision is at a premium and the exclusive sense of "or" is intended, the phrase "but not both" is often added.

We interpret the inclusive disjunction of two statements as an assertion that at least one of the statements is true, and we interpret their **exclusive disjunction** as an assertion that at least one of the statements is true but not both are true. Note that the two kinds of disjunction have a part of their meanings in common. This partial common meaning, that at least one of the disjuncts is true, is the whole meaning of the inclusive "or" and a part of the meaning of the exclusive "or."

Although disjunctions are stated ambiguously in English, they are unambiguous in Latin. Latin has two different words corresponding to the two different senses of the English word "or." The Latin word *vel* signifies weak or inclusive disjunction, and the Latin word *aut* corresponds to the word "or" in its strong or exclusive sense. It is customary to use the initial letter of the word *vel* to stand for "or" in its weak, inclusive sense. Where p and q are any two statements whatever, their weak or inclusive disjunction is written $p \vee q$. Our symbol for inclusive disjunction, called a **wedge** (or, less frequently, a *vee*) is also a truth-functional connective. A weak disjunction is false only if both of its disjuncts are false. We may regard the wedge as being defined by the following truth table:

p	q	$p \vee q$
T	T	T
T	F	T
F	T	T
F	F	F

The first specimen argument presented in this section was a *disjunctive syllogism*. (A syllogism is a deductive argument consisting of two premises and a conclusion. The term *disjunctive syllogism* is being used in a narrower sense here than it was in Chapter 7.)

> The blind prisoner has a red hat or the blind prisoner has a white hat.
> The blind prisoner does not have a red hat.
> Therefore the blind prisoner has a white hat.

Its form is characterized by saying that its first premise is a disjunction; its second premise is the negation of the first disjunct of the first premise; and its conclusion is the same as the second disjunct of the first premise. It is evident that the disjunctive syllogism, so defined, is valid on either interpretation of the word "or"—that is, regardless of whether an inclusive or exclusive disjunction is intended. The typical valid argument that has a disjunction for a premise is, like the disjunctive syllogism, valid on either interpretation of the word "or," so a simplification may be effected by translating the English word "or" into our logical symbol "∨"—*regardless of which meaning of the English word "or" is intended*. Only a close examination of the context, or an explicit questioning of the speaker or writer, can reveal which sense of "or" is intended. This problem, often impos-

Exclusive disjunction or **strong disjunction**
A logical relation meaning "or" that may connect two component statements. A compound statement asserting exclusive disjunction says that at least one of the disjuncts is true *and* that at least one of the disjuncts is false. It is contrasted with an "inclusive" (or "weak") disjunction, which says that at least one of the disjuncts is true and that they may both be true.

Wedge
The symbol for weak (inclusive) disjunction, ∨. Any statement of the form $p \vee q$ is true if p is true, or if q is true, or if both p and q are true.

sible to resolve, can be avoided if we agree to treat any occurrence of the word "or" as inclusive. On the other hand, if it is stated explicitly that the disjunction is intended to be exclusive—by means of the added phrase "but not both," for example—we have the symbolic machinery to formulate that additional sense, as will be shown directly.

Where both disjuncts have either the same subject term or the same predicate term, it is often natural to compress the formulation of their disjunction in English by placing the "or" so that there is no need to repeat the common part of the two disjuncts. Thus, "Either Smith is the owner or Smith is the manager" might equally well be stated as "Smith is either the owner or the manager," and either one is properly symbolized as $O \lor M$. And "Either Red is guilty or Butch is guilty" may be stated as "Either Red or Butch is guilty"; either one may be symbolized as $R \lor B$.

The word "unless" is often used to form the disjunction of two statements. Thus, "You will do poorly on the exam unless you study" is correctly symbolized as $P \lor S$, because that disjunction asserts that one of the disjuncts is true, and hence that if one of them is false, the other must be true. Of course, you may study and do poorly on the exam.

The word "unless" is sometimes used to convey more information; it may mean (depending on context) that one or the other proposition is true but that not both are true. That is, "unless" may be intended as an exclusive disjunction. Thus it was noted by Ted Turner that global warming will put New York under water in one hundred years and "will be the biggest catastrophe the world has ever seen—unless we have nuclear war." Here the speaker did mean that at least one of the two disjuncts is true, but of course they cannot both be true. Other uses of "unless" are ambiguous. When we say, "The picnic will be held unless it rains," we surely do mean that the picnic will be held if it does not rain. Do we mean that it will not be held if it does rain? That may be uncertain. It is wise policy to treat every disjunction as weak or inclusive unless it is certain that an exclusive disjunction is meant. "Unless" is best symbolized simply with the wedge ($\lor$).

D. Punctuation

In English, **punctuation** is absolutely required if complicated statements are to be clear. Many different punctuation marks are used, without which many sentences would be highly ambiguous. For example, quite different meanings attach to "The teacher says John is a fool" when it is given different punctuations: "The teacher," says John, "is a fool"; or "The teacher says 'John is a fool.'" Punctuation is equally necessary in mathematics. In the absence of a special convention, no number is uniquely denoted by $2 \times 3 + 5$, although when it is made clear how its constituents are to be grouped, it denotes either 11 or 16: the first when punctuated $(2 \times 3) + 5$, the second when punctuated $2 \times (3 + 5)$. To avoid ambiguity, and to make meaning clear, punctuation marks in mathematics appear in the form of parentheses, ( ), which are used to group individual symbols; brackets,

Punctuation
The parentheses, brackets, and braces used in mathematics and logic to eliminate ambiguity.

[ ], which are used to group expressions that include parentheses; and braces, { }, which are used to group expressions that include brackets.

In the language of symbolic logic those same punctuation marks—parentheses, brackets, and braces—are equally essential, because in logic compound statements are themselves often compounded together into more complicated ones. Thus $p \cdot q \vee r$ is ambiguous: it might mean the conjunction of p with the disjunction of q with r, or it might mean the disjunction whose first disjunct is the conjunction of p and q and whose second disjunct is r. We distinguish between these two different senses by punctuating the given formula as $p \cdot (q \vee r)$ or else as $(p \cdot q) \vee r$. That the different ways of punctuating the original formula do make a difference can be seen by considering the case in which p is false and q and r are both true. In this case the second punctuated formula is true (because its second disjunct is true), whereas the first one is false (because its first conjunct is false). Here the difference in punctuation makes all the difference between truth and falsehood, for different punctuations can assign different truth values to the ambiguous $p \cdot q \vee r$.

The word "either" has a variety of different meanings and uses in English. It has conjunctive force in the sentence, "There is danger on either side." More often it is used to introduce the first disjunct in a disjunction, as in "Either the blind prisoner has a red hat or the blind prisoner has a white hat." There it contributes to the rhetorical balance of the sentence, but it does not affect its meaning. Perhaps the most important use of the word "either" is to punctuate a compound statement. Thus the sentence

> The organization will meet on Thursday and Anand will be elected or the election will be postponed.

is ambiguous. This ambiguity can be resolved in one direction by placing the word "either" at its beginning, or in the other direction by inserting the word "either" before the name "Anand." Such punctuation is effected in our symbolic language by parentheses. The ambiguous formula $p \cdot q \vee r$ discussed in the preceding paragraph corresponds to the ambiguous sentence just examined. The two different punctuations of the formula correspond to the two different punctuations of the sentence effected by the two different insertions of the word "either."

The negation of a disjunction is often formed by use of the phrase "neither–nor." Thus the statement, "Either Fillmore or Harding was the greatest U.S. president," can be contradicted by the statement, "Neither Fillmore nor Harding was the greatest U.S. president." The disjunction would be symbolized as $F \vee H$, and its negation as either $\sim(F \vee H)$ or as $(\sim F) \cdot (\sim H)$. (The logical equivalence of these two symbolic formulas will be discussed in Section 8.9.) It should be clear that to deny a disjunction, which states that one or another statement is true, requires that both statements be stated to be false.

The word "both" in English has a very important role in logical punctuation, and it deserves the most careful attention. When we say "Both Jamal and Derek are not . . ." we are saying, as noted just above, that "Neither Jamal nor Derek is . . .";

we are applying the negation to each of them. But when we say "Jamal and Derek are not both . . ." we are saying something very different; we are applying the negation to the pair of them taken together, saying that "it is not the case that they are both . . . ." This difference is very substantial. Entirely different meanings arise when the word "both" is placed differently in the English sentence. Consider the great difference between the meanings of

> Jamal and Derek will not both be elected.

and

> Jamal and Derek will both not be elected.

The first denies the conjunction $J \cdot D$ and may be symbolized as $\sim(J \cdot D)$. The second says that each one of the two will not be elected, and is symbolized as $\sim(J) \cdot \sim(D)$. Merely changing the position of the two words "both" and "not" alters the logical force of what is asserted.

Of course, the word "both" does not always have this role; sometimes we use it only to add emphasis. When we say that "Both Lewis and Clark were great explorers," we use the word only to state more emphatically what is said by "Lewis and Clark were great explorers." When the task is logical analysis, the punctuational role of "both" must be very carefully determined.

In the interest of brevity—that is, to decrease the number of parentheses required—it is convenient to establish the convention that, in any formula, the negation symbol will be understood to apply to the smallest statement that the punctuation permits. Without this convention, the formula $\sim p \vee q$ is ambiguous, meaning either $(\sim p) \vee q$, or $\sim(p \vee q)$. By our convention we take it to mean the first of these alternatives, for the curl can (and therefore by our convention does) apply to the first component, p, rather than to the larger formula, $p \vee q$.

Given a set of punctuation marks for our symbolic language, it is possible to write not just conjunctions, negations, and weak disjunctions in that language, but exclusive disjunctions as well. The exclusive disjunction of p and q asserts that at least one of them is true but not both are true, which is written as $(p \vee q) \cdot \sim(p \cdot q)$. Another way of expressing the exclusive disjunction is "$\underline{\vee}$".

The truth value of any compound statement constructed from simple statements using only the curl and the truth-functional connectives—dot and wedge—is completely determined by the truth or falsehood of its component simple statements. If we know the truth values of simple statements, the truth value of any truth-functional compound of them is easily calculated. In working with such compound statements we always begin with their inmost components and work outward. For example, if A and B are true statements and X and Y are false statements, we calculate the truth value of the compound statement $\sim[\sim(A \cdot X) \cdot (Y \vee \sim B)]$ as follows: Because X is false, the conjunction $A \cdot X$ is false, and so its negation $\sim(A \cdot X)$ is true. B is true, so its negation $\sim B$ is false, and because Y is also false, the disjunction of Y with $\sim B$, $Y \vee \sim B$, is false. The bracketed formula $[\sim(A \cdot X) \cdot (Y \vee \sim B)]$ is the conjunction of a true with a false statement and is therefore false. Hence its negation, which is the entire statement, is true. Such a stepwise procedure always enables us to determine the truth value of a compound statement from the truth values of its components.

In some circumstances we may be able to determine the truth value of a truth-functional compound statement even if we cannot determine the truth or falsehood of one of its component simple statements. We may do this by first calculating the truth value of the compound statement on the assumption that a given simple component is true, and then by calculating the truth value of the compound statement on the assumption that the same simple component is false. If both calculations yield the same truth value for the compound statement in question, we have determined the truth value of the compound statement without having to determine the truth value of its unknown component, because we know that the truth value of any component cannot be other than true or false. Truth tables allow us to expand this method to cases with more than one undetermined component.

overview

Punctuation in Symbolic Notation

The statement

> I will study hard and pass the exam or fail

is ambiguous. It could mean "I will study hard and pass the exam or I will fail the exam" or "I will study hard and I will either pass the exam or fail it."

The symbolic notation

$$S \cdot P \vee F$$

is similarly ambiguous. Parentheses resolve the ambiguity. In place of "I will study hard and pass the exam or I will fail the exam," we get

$$(S \cdot P) \vee F$$

and in place of "I will study hard and I will either pass the exam or fail it," we get

$$S \cdot (P \vee F)$$

EXERCISES

A. Using the truth-table definitions of the dot, the wedge, and the curl, determine which of the following statements are true:

*1. Rome is the capital of Italy ∨ Rome is the capital of Spain.

2. ~(London is the capital of England • Stockholm is the capital of Norway).

3. ~London is the capital of England • ~Stockholm is the capital of Norway.

4. ~(Rome is the capital of Spain ∨ Paris is the capital of France).

298   **CHAPTER 8**   Symbolic Logic

*5. ~Rome is the capital of Spain ∨ ~Paris is the capital of France.

6. London is the capital of England ∨ ~London is the capital of England.

7. Stockholm is the capital of Norway • ~Stockholm is the capital of Norway.

8. (Paris is the capital of France • Rome is the capital of Spain) ∨ (Paris is the capital of France • ~Rome is the capital of Spain).

9. (London is the capital of England ∨ Stockholm is the capital of Norway) • (~Rome is the capital of Italy • ~Stockholm is the capital of Norway).

*10. Rome is the capital of Spain ∨ ~(Paris is the capital of France • Rome is the capital of Spain).

11. Rome is the capital of Italy • ~(Paris is the capital of France ∨ Rome is the capital of Spain).

12. ~(~Paris is the capital of France • ~Stockholm is the capital of Norway).

13. ~[~(~Rome is the capital of Spain ∨ ~Paris is the capital of France) ∨ ~(~Paris is the capital of France ∨ Stockholm is the capital of Norway)].

14. ~[~(~London is the capital of England • Rome is the capital of Spain) • ~(Rome is the capital of Spain • ~Rome is the capital of Spain)].

*15. ~[~(Stockholm is the capital of Norway ∨ Paris is the capital of France) ∨ ~(~London is the capital of England • Rome is the capital of Spain)].

16. Rome is the capital of Spain ∨ (~London is the capital of England ∨ London is the capital of England).

17. Paris is the capital of France • ~(Paris is the capital of France • Rome is the capital of Spain).

18. London is the capital of England • ~(Rome is the capital of Italy • Rome is the capital of Italy).

19. (Stockholm is the capital of Norway ∨ ~Paris is the capital of France) ∨ ~(~Stockholm is the capital of Norway • ~London is the capital of England).

*20. (Paris is the capital of France ∨ ~Rome is the capital of Spain) ∨ ~(~Paris is the capital of France • ~Rome is the capital of Spain).

21. ~[~(Rome is the capital of Spain • Stockholm is the capital of Norway) ∨ ~(~Paris is the capital of France ∨ ~Rome is the capital of Spain)].

22. ~[~(London is the capital of England • Paris is the capital of France) ∨ ~(~Stockholm is the capital of Norway ∨ ~Paris is the capital of France)].

23. ~[(~Paris is the capital of France ∨ Rome is the capital of Italy) • ~(~Rome is the capital of Italy ∨ Stockholm is the capital of Norway)].

24. ~[(~Rome is the capital of Spain ∨ Stockholm is the capital of Norway) • ~(~Stockholm is the capital of Norway ∨ Paris is the capital of France)].

*25. ~[(~London is the capital of England • Paris is the capital of France) ∨ ~(~Paris is the capital of France • Rome is the capital of Spain)].

B. If *A*, *B*, and *C* are true statements and *X*, *Y*, and *Z* are false statements, which of the following are true?

* 1. ~A ∨ B
2. ~B ∨ X
3. ~Y ∨ C
4. ~Z ∨ X
* 5. (A • X) ∨ (B • Y)
6. (B • C) ∨ (Y • Z)
7. ~(C • Y) ∨ (A • Z)
8. ~(A • B) ∨ (X • Y)
9. ~(X • Z) ∨ (B • C)
*10. ~(X • ~Y) ∨ (B • ~C)
11. (A ∨ X) • (Y ∨ B)
12. (B ∨ C) • (Y ∨ Z)
13. (X ∨ Y) • (X ∨ Z)
14. ~(A ∨ Y) • (B ∨ X)
*15. ~(X ∨ Z) • (~X ∨ Z)
16. ~(A ∨ C) ∨ ~(X • ~Y)
17. ~(B ∨ Z) • ~(X ∨ ~Y)
18. ~[(A ∨ ~C) ∨ (C ∨ ~A)]
19. ~[(B • C) • ~(C • B)]
*20. ~[(A • B) ∨ ~(B • A)]
21. [A ∨ (B ∨ C)] • ~[(A ∨ B) ∨ C]
22. [X ∨ (Y • Z)] ∨ ~[(X ∨ Y) • (X ∨ Z)]
23. [A • (B ∨ C)] • ~[(A • B) ∨ (A • C)]
24. ~{[(~A • B) • (~X • Z)] • ~[(A • ~B) ∨ ~(~Y • ~Z)]}
*25. ~{~[(B • ~C) ∨ (Y • ~Z)] • [(~B ∨ X) ∨ (B ∨ ~Y)]}

C. Using the letters *E*, *I*, *J*, *L*, and *S* to abbreviate the simple statements, "Egypt's food shortage worsens," "Iran raises the price of oil," "Jordan requests more U.S. aid," "Libya raises the price of oil," and "Saudi Arabia buys five hundred more warplanes," symbolize these statements.

*1. Iran raises the price of oil but Libya does not raise the price of oil.
2. Either Iran or Libya raises the price of oil.
3. Iran and Libya both raise the price of oil.
4. Iran and Libya do not both raise the price of oil.
*5. Iran and Libya both do not raise the price of oil.
6. Iran or Libya raises the price of oil but they do not both do so.
7. Saudi Arabia buys five hundred more warplanes and either Iran raises the price of oil or Jordan requests more U.S. aid.
8. Either Saudi Arabia buys five hundred more warplanes and Iran raises the price of oil or Jordan requests more U.S. aid.

Conditional statement
A hypothetical statement; a compound proposition or statement of the form "If p then q."

Antecedent
In a conditional statement ("*If ... then ...*"), the component that immediately follows the "if." Sometimes called the *implicans* or the *protasis*.

Consequent
In a conditional statement ("*If ... then ...*"), the component that immediately follows the "then." Sometimes called the *implicate*, or the *apodosis*.

9. It is not the case that Egypt's food shortage worsens, and Jordan requests more U.S. aid.

*10. It is not the case that either Egypt's food shortage worsens or Jordan requests more U.S. aid.

11. Either it is not the case that Egypt's food shortage worsens or Jordan requests more U.S. aid.

12. It is not the case that both Egypt's food shortage worsens and Jordan requests more U.S. aid.

13. Jordan requests more U.S. aid unless Saudi Arabia buys five hundred more warplanes.

14. Unless Egypt's food shortage worsens, Libya raises the price of oil.

*15. Iran won't raise the price of oil unless Libya does so.

16. Unless both Iran and Libya raise the price of oil neither of them does.

17. Libya raises the price of oil and Egypt's food shortage worsens.

18. It is not the case that neither Iran nor Libya raises the price of oil.

19. Egypt's food shortage worsens and Jordan requests more U.S. aid, unless both Iran and Libya do not raise the price of oil.

*20. Either Iran raises the price of oil and Egypt's food shortage worsens, or it is not the case both that Jordan requests more U.S. aid and that Saudi Arabia buys five hundred more warplanes.

21. Either Egypt's food shortage worsens and Saudi Arabia buys five hundred more warplanes, or either Jordan requests more U.S. aid or Libya raises the price of oil.

22. Saudi Arabia buys five hundred more warplanes, and either Jordan requests more U.S. aid or both Libya and Iran raise the price of oil.

23. Either Egypt's food shortage worsens or Jordan requests more U.S. aid, but neither Libya nor Iran raises the price of oil.

24. Egypt's food shortage worsens, but Saudi Arabia buys five hundred more warplanes and Libya raises the price of oil.

*25. Libya raises the price of oil and Egypt's food shortage worsens; however, Saudi Arabia buys five hundred more warplanes and Jordan requests more U.S. aid.

8.3 Conditional Statements and Material Implication

Where two statements are combined by placing the word "if" before the first and inserting the word "then" between them, the resulting compound statement is a **conditional** statement (also called a *hypothetical*, an *implication*, or an *implicative statement*). In a conditional statement the component statement that follows the "if" is called the **antecedent** (or the *implicans* or—rarely—the *protasis*), and the component statement that follows the "then" is the **consequent** (or the *implicate* or—rarely—the *apodosis*). For example, "If Mr. Jones is the brakeman's next-door neighbor, then Mr. Jones earns exactly three times as much as the brakeman" is a conditional statement in which "Mr. Jones is the brakeman's next-door neighbor" is the antecedent and "Mr. Jones earns exactly three times as much as the brakeman" is the consequent.

A conditional statement asserts that in any case in which its antecedent is true, its consequent is also true. It does not assert that its antecedent is true, but only that if its antecedent is true, then its consequent is also true. It does not assert that its consequent is true, but only that its consequent is true if its antecedent is true. The essential meaning of a conditional statement is the relationship asserted to hold between the antecedent and the consequent, in that order. To understand the meaning of a conditional statement, then, we must understand what the relationship of implication is.

Implication plausibly appears to have more than one meaning. We found it useful to distinguish different senses of the word "or" before introducing a special logical symbol to correspond exactly to a single one of the meanings of the English word. Had we not done so, the ambiguity of the English would have infected our logical symbolism and prevented it from achieving the clarity and precision aimed at. It will be equally useful to distinguish the different senses of "implies" or "if–then" before we introduce a special logical symbol in this connection.

Consider the following four conditional statements, each of which seems to assert a different type of implication, and to each of which corresponds a different sense of "if–then":

A. If all humans are mortal and Socrates is a human, then Socrates is mortal.

B. If Leslie is a bachelor, then Leslie is unmarried.

C. If this piece of blue litmus paper is placed in acid, then this piece of blue litmus paper will turn red.

D. If State loses the homecoming game, then I'll eat my hat.

Even a casual inspection of these four conditional statements reveals that they are of quite different types. The consequent of **A** follows logically from its antecedent, whereas the consequent of **B** follows from its antecedent by the very definition of the term *bachelor*, which means "unmarried man." The consequent of **C** does not follow from its antecedent either by logic alone or by the definition of its terms; the connection must be discovered empirically, because the implication stated here is causal. Finally, the consequent of **D** does not follow from its antecedent either by logic or by definition, nor is there any causal law involved. Statement **D** reports a decision of the speaker to behave in the specified way under the specified circumstances.

These four conditional statements are different in that each asserts a different type of implication between its antecedent and its consequent. But they are not completely different; all assert types of implication. Is there any identifiable common meaning, any partial meaning that is common to these admittedly different

Implication
The relation that holds between the antecedent and the consequent of a true conditional or hypothetical statement.

types of implication, although perhaps not the whole or complete meaning of any one of them?

The search for a common partial meaning takes on added significance when we recall our procedure in working out a symbolic representation for the English word "or." In that case, we proceeded as follows: First, we emphasized the difference between the two senses of the word, contrasting inclusive with exclusive disjunction. The inclusive disjunction of two statements was observed to mean that at least one of the statements is true, and the exclusive disjunction of two statements was observed to mean that at least one of the statements is true but not both are true. Second, we noted that these two types of disjunction had a common *partial* meaning. This partial common meaning—that at least one of the disjuncts is true—was seen to be the *whole* meaning of the weak, inclusive "or," and a *part* of the meaning of the strong, exclusive "or." We then introduced the special symbol "∨" to represent this common partial meaning (which is the entire meaning of "or" in its inclusive sense). Third, we noted that the symbol representing the common partial meaning is an adequate translation of either sense of the word "or" for the purpose of retaining the disjunctive syllogism as a valid form of argument. It was admitted that translating an exclusive "or" into the symbol "∨" ignores and loses part of the word's meaning. The part of its meaning that is preserved by this translation is all that is needed for the disjunctive syllogism to remain a valid form of argument. Because the disjunctive syllogism is typical of arguments involving disjunction, with which we are concerned here, this partial translation of the word "or," which may abstract from its "full" or "complete" meaning in some cases, is wholly adequate for our present purposes.

Now we wish to proceed in the same way, this time in connection with the English phrase "if–then." The first part is already accomplished: We have already emphasized the differences among four senses of the "if–then" phrase corresponding to four different types of implication. We are now ready for the second step, which is to discover a sense that is at least a part of the meaning of all four types of implication.

We approach this problem by asking: What circumstances suffice to establish the falsehood of a given conditional statement? Under what circumstances should we agree that the conditional statement

> If this piece of blue litmus paper is placed in that acid solution, then this piece of blue litmus paper will turn red.

is false? It is important to realize that this conditional does not assert that any blue litmus paper is actually placed in the solution, or that any litmus paper actually turns red. It asserts merely that *if* this piece of blue litmus paper is placed in the solution, *then* this piece of blue litmus paper will turn red. It is proved false if this piece of blue litmus paper is actually placed in the solution and does not turn red. The acid test, so to speak, of the falsehood of a conditional statement is available when its antecedent is true, because if its consequent is false while its antecedent is true, the conditional itself is thereby proved false.

Any conditional statement, "If p then q," is known to be false if the conjunction $p \cdot \sim q$ is known to be true—that is, if its antecedent is true and its consequent is false. For a conditional to be true, then, the indicated conjunction must be false; that is, its negation ~(p • ~q) must be true. In other words, for any conditional, "If p then q," to be true, the statement ~(p • ~q), which is the negation of the conjunction of its antecedent with the negation of its consequent, must also be true. We may then regard ~(p • ~q) as a part of the meaning of "If p then q."

Every conditional statement means to deny that its antecedent is true and its consequent false, but this need not be the whole of its meaning. A conditional such as **A** on page 301 also asserts a logical connection between its antecedent and consequent, as **B** asserts a definitional connection, **C** a causal connection, and **D** a decisional connection. No matter what type of implication is asserted by a conditional statement, part of its meaning is the negation of the conjunction of its antecedent with the negation of its consequent.

We now introduce a special symbol to represent this common partial meaning of the "if–then" phrase. We define the new symbol "⊃", called a **horseshoe** (other systems employ the symbol " → " to express this relation), by taking $p \supset q$ as an abbreviation of ~(p • ~q). The exact significance of the ⊃ symbol can be indicated by means of a truth table:

p	q	~q	p • ~q	~(p • ~q)	p ⊃ q
T	T	F	F	T	T
T	F	T	T	F	F
F	T	F	F	T	T
F	F	T	F	T	T

Here the first two columns are the guide columns; they simply lay out all possible combinations of truth and falsehood for p and q. The third column is filled in by reference to the second, the fourth by reference to the first and third, and the fifth by reference to the fourth; the sixth is identical to the fifth by definition.

The symbol ⊃ is not to be regarded as denoting *the* meaning of "if–then," or standing for the relation of implication. That would be impossible, for there is no single meaning of "if–then"; there are several meanings. There is no unique relation of implication to be thus represented; there are several different implication relations. Nor is the symbol ⊃ to be regarded as somehow standing for all the meanings of "if–then." These are all different, and any attempt to abbreviate all of them by a single logical symbol would render that symbol ambiguous—as ambiguous as the English phrase "if–then" or the English word "implication." The symbol ⊃ is completely unambiguous. What $p \supset q$ abbreviates is ~(p • ~q), whose meaning is included in the meanings of each of the various kinds of implications considered but does not constitute the entire meaning of any of them.

We can regard the symbol ⊃ as representing another kind of implication, and it will be expedient to do so, because a convenient way to read $p \supset q$ is "If p, then q." But it is not the same kind of implication as any of those mentioned earlier. It is called **material implication** by logicians. In giving it a special name, we

Horseshoe
The symbol for material implication, ⊃.

Material implication
A truth-functional relation (symbolized by the horseshoe, ⊃) that may connect two statements. The statement "p materially implies q" is true when either p is false, or q is true.

admit that it is a special notion, not to be confused with other, more usual, types of implication.

Not all conditional statements in English need assert one of the four types of implication previously considered. Material implication constitutes a fifth type that may be asserted in ordinary discourse. Consider the remark, "If Hitler was a military genius, then I'm a monkey's uncle." It is quite clear that it does not assert logical, definitional, or causal implication. It cannot represent a decisional implication, because it scarcely lies in the speaker's power to make the consequent true. No "real connection," whether logical, definitional, or causal, obtains between antecedent and consequent here. A conditional of this sort is often used as an emphatic or humorous method of denying its antecedent. The consequent of such a conditional is usually a statement that is obviously or ludicrously false. And because no true conditional can have both its antecedent true and its consequent false, to affirm such a conditional amounts to denying that its antecedent is true. The full meaning of the present conditional seems to be the denial that "Hitler was a military genius" is true when "I'm a monkey's uncle" is false. Because the latter is so obviously false, the conditional must be understood to deny the former.

The point here is that no "real connection" between antecedent and consequent is suggested by a material implication. All it asserts is that it is not the case that the antecedent is true when the consequent is false. Note that the material implication symbol is a truth-functional connective, like the symbols for conjunction and disjunction. As such, it is defined by the following truth table:

p	q	$p \supset q$
T	T	T
T	F	F
F	T	T
F	F	T

As thus defined by the truth table, the symbol $\supset$ has some features that may at first appear odd: The assertion that a false antecedent materially implies a true consequent is true; and the assertion that a false antecedent materially implies a false consequent is also true. This apparent strangeness can be dissipated in part by the following considerations. Because the number 2 is smaller than the number 4 (a fact notated symbolically as $2 < 4$), it follows that any number smaller than 2 is smaller than 4. The conditional formula

If $x < 2$, then $x < 4$.

is true for any number x whatsoever. If we focus on the numbers 1, 3, and 4, and replace the number variable x in the preceding conditional formula by each of them in turn, we can make the following observations. In

If $1 < 2$, then $1 < 4$.

both antecedent and consequent are true, and of course the conditional is true. In

If $3 < 2$, then $3 < 4$.

Visual Logic

Material Implication

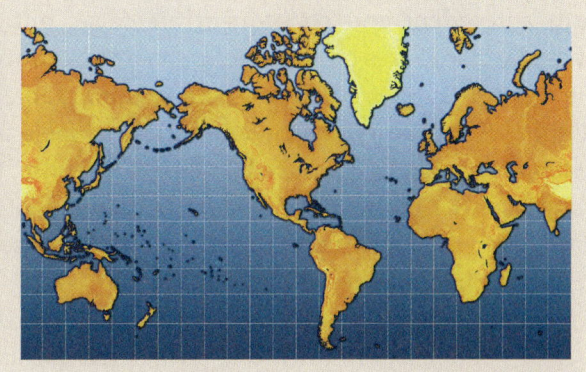

Source: Photodisc/Getty Images   *Source:* Photodisc/Getty Images

"If the moon is made of green cheese, then the Earth is flat."

This proposition, in the form $G \supset F$, is a material implication. A material implication is true when the antecedent (the "if" clause) is false. Therefore a material implication is true when the antecedent is false and the consequent is also false, as in this illustrative proposition.

Source: Photodisc/Getty Images   *Source:* Photodisc/Getty Images

"If the moon is made of green cheese, then the Earth is round."

This proposition, in the similar form $G \supset R$, is also a material implication. A material implication is true when the antecedent (the "if" clause) is false. Therefore a material implication is true when the antecedent is false and the consequent is true, as in this illustrative proposition.

A material implication is false only if the antecedent is true and the consequent is false. Therefore a material implication is true whenever the antecedent is false, whether the consequent is false or true.

the antecedent is false and the consequent is true, and of course the conditional is again true. In

$$\text{If } 4 < 2, \text{ then } 4 < 4.$$

both antecedent and consequent are false, but the conditional remains true. These last two cases correspond to the third and fourth rows of the table defining the symbol ⊃. So it is not particularly remarkable or surprising that a conditional should be true when the antecedent is false and the consequent is true, or when antecedent and consequent are both false. Of course, there is no number that is smaller than 2 but not smaller than 4; that is, there is no true conditional statement with a true antecedent and a false consequent. This is exactly what the defining truth table for ⊃ lays down.

Now we propose to translate *any* occurrence of the "if–then" phrase into our logical symbol ⊃. This proposal means that in translating conditional statements into our symbolism, we treat them all as merely material implications. Of course, most conditional statements assert more than that a merely material implication holds between their antecedents and consequents. So our proposal amounts to suggesting that we ignore, or put aside, or "abstract from," part of the meaning of a conditional statement when we translate it into our symbolic language. How can this proposal be justified?

The previous proposal to translate both inclusive and exclusive disjunctions by means of the symbol ∨ was justified on the grounds that the validity of the disjunctive syllogism was preserved even if the additional meaning that attaches to the exclusive "or" was ignored. Our present proposal to translate all conditional statements into the merely material implication symbolized by ⊃ may be justified in exactly the same way. Many arguments contain conditional statements of various kinds, but the validity of all valid arguments of the general type with which we will be concerned is preserved even if the additional meanings of their conditional statements are ignored. This remains to be proved, of course, and will occupy our attention in the next section.

Conditional statements can be formulated in a variety of ways. The statement

> If he has a good lawyer, then he will be acquitted.

can equally well be stated without the use of the word "then" as

> If he has a good lawyer, he will be acquitted.

The order of the antecedent and consequent can be reversed, provided that the "if" still directly precedes the antecedent, as

> He will be acquitted if he has a good lawyer.

It should be clear that, in any of the examples just given, the word "if" can be replaced by such phrases as "in case," "provided that," "given that," or "on condition that," without any change in meaning. Minor adjustments in the phrasings of antecedent and consequent permit such alternative phrasings of the same conditional as

> That he has a good lawyer implies that he will be acquitted.

or

> His having a good lawyer entails his acquittal.

A shift from active to passive voice may accompany a reversal of order of antecedent and consequent, yielding the logically equivalent

> His being acquitted is implied (or entailed) by his having a good lawyer.

Other variations are possible:

> There is no way he won't be acquitted if he has a good lawyer.

Any of these is symbolized as $L \supset A$.

The notions of necessary and sufficient conditions provide other formulations of conditional statements. For any specified event, many circumstances are necessary for it to occur. Thus, for a normal car to run, it is necessary that there be fuel in its tank, that its spark plugs be properly adjusted, that its oil pump be working, and so on. So if the event occurs, every one of the conditions necessary for its occurrence must have been fulfilled. Hence to say

> That there is fuel in its tank is a necessary condition for the car to run.

can equally well be stated as

> The car runs only if there is fuel in its tank.

which is another way of saying that

> If the car runs then there is fuel in its tank.

Any of these is symbolized as $R \supset F$. Usually "*q* is a *necessary condition* for *p*" is symbolized as $p \supset q$. Likewise, "*p only if q*" is also symbolized as $p \supset q$.

For a specified situation there may be many alternative circumstances, any one of which is sufficient to produce that situation. For a purse to contain more than a dollar, for example, it is sufficient for it to contain five quarters, or eleven dimes, or twenty-one nickels, and so on. If any one of these circumstances obtains, the specified situation will be realized. Hence, to say "That the purse contains five quarters is a sufficient condition for it to contain more than a dollar" is to say "If the purse contains five quarters then it contains more than a dollar." In general, "*p* is a *sufficient condition* for *q*" is symbolized as $p \supset q$.

To illustrate: Recruiters for the Wall Street investment firm Goldman Sachs (where annual bonuses are commonly in the millions) grill potential employees repeatedly. Those who survive the grilling are invited to the firm's offices for a full day of interviews, culminating in a dinner with senior Goldman Sachs executives. As reported recently, "Agile brains and near-perfect grades are necessary but not sufficient conditions for being hired. Just as important is fitting in."[2]

If p is a sufficient condition for q, we have $p \supset q$, and q must be a necessary condition for p. If p is a necessary condition for q, we have $q \supset p$, and q must be a sufficient condition for p. Hence, if p is necessary and sufficient for q, then q is sufficient and necessary for p.

Not every statement containing the word "if" is a conditional. None of the following statements is a conditional: "There is food in the refrigerator if you want some," "Your table is ready, if you please," "There is a message for you if you're interested," "The meeting will be held even if no permit is obtained." The presence or absence of particular words is never decisive. In every case, one must understand what a given sentence means, and then restate that meaning in a symbolic formula.

EXERCISES

A. If A, B, and C are true statements and X, Y, and Z are false statements, determine which of the following are true, using the truth tables for the horseshoe, the dot, the wedge, and the curl.

*1. $A \supset B$
2. $A \supset X$
3. $B \supset Y$
4. $Y \supset Z$
*5. $(A \supset B) \supset Z$
6. $(X \supset Y) \supset Z$
7. $(A \supset B) \supset C$
8. $(X \supset Y) \supset C$
9. $A \supset (B \supset Z)$
*10. $X \supset (Y \supset Z)$
11. $[(A \supset B) \supset C] \supset Z$
12. $[(A \supset X) \supset Y] \supset Z$
13. $[A \supset (X \supset Y)] \supset C$
14. $[A \supset (B \supset Y)] \supset X$
*15. $[(X \supset Z) \supset C] \supset Y$
16. $[(Y \supset B) \supset Y] \supset Y$
17. $[(A \supset Y) \supset B] \supset Z$
18. $[(A \cdot X) \supset C] \supset [(A \supset C) \supset X]$
19. $[(A \cdot X) \supset C] \supset [(A \supset X) \supset C]$
*20. $[(A \cdot X) \supset Y] \supset [(X \supset A) \supset (A \supset Y)]$
21. $[(A \cdot X) \vee (\sim A \cdot \sim X)] \supset [(A \supset X) \cdot (X \supset A)]$
22. $\{[A \supset (B \supset C)] \supset [(A \cdot B) \supset C]\} \supset [(Y \supset B) \supset (C \supset Z)]$
23. $\{[(X \supset Y) \supset Z] \supset [Z \supset (X \supset Y)]\} \supset [(X \supset Z) \supset Y]$
24. $[(A \cdot X) \supset Y] \supset [(A \supset X) \cdot (A \supset Y)]$
*25. $[A \supset (X \cdot Y)] \supset [(A \supset X) \vee (A \supset Y)]$

B. Symbolize the following, using capital letters to abbreviate the simple statements involved.

*1. If Argentina mobilizes, then if Brazil protests to the UN, then Chile will call for a meeting of all the Latin American states.

2. If Argentina mobilizes, then either Brazil will protest to the UN or Chile will call for a meeting of all the Latin American states.

3. If Argentina mobilizes, then Brazil will protest to the UN and Chile will call for a meeting of all the Latin American states.
4. If Argentina mobilizes, then Brazil will protest to the UN, and Chile will call for a meeting of all the Latin American states.
5. If Argentina mobilizes and Brazil protests to the UN, then Chile will call for a meeting of all the Latin American states.
*6. If either Argentina mobilizes or Brazil protests to the UN, then Chile will call for a meeting of all the Latin American states.
7. Either Argentina will mobilize or if Brazil protests to the UN, then Chile will call for a meeting of all the Latin American states.
8. If Argentina does not mobilize, then either Brazil will not protest to the UN or Chile will not call for a meeting of all the Latin American states.
9. If Argentina does not mobilize, then neither will Brazil protest to the UN nor will Chile call for a meeting of all the Latin American states.
*10. It is not the case that if Argentina mobilizes, then both Brazil will protest to the UN, and Chile will call for a meeting of all the Latin American states.
11. If it is not the case that Argentina mobilizes, then Brazil will not protest to the UN, and Chile will call for a meeting of all the Latin American states.
12. Brazil will protest to the UN if Argentina mobilizes.
13. Brazil will protest to the UN only if Argentina mobilizes.
14. Chile will call for a meeting of all the Latin American states only if both Argentina mobilizes and Brazil protests to the UN.
*15. Brazil will protest to the UN only if either Argentina mobilizes or Chile calls for a meeting of all the Latin American states.
16. Argentina will mobilize if either Brazil protests to the UN or Chile calls for a meeting of all the Latin American states.
17. Brazil will protest to the UN unless Chile calls for a meeting of all the Latin American states.
18. If Argentina mobilizes, then Brazil will protest to the UN unless Chile calls for a meeting of all the Latin American states.
19. Brazil will not protest to the UN unless Argentina mobilizes.
*20. Unless Chile calls for a meeting of all the Latin American states, Brazil will protest to the UN.
21. Argentina's mobilizing is a sufficient condition for Brazil to protest to the UN.
22. Argentina's mobilizing is a necessary condition for Chile to call for a meeting of all the Latin American states.
23. If Argentina mobilizes and Brazil protests to the UN, then both Chile and the Dominican Republic will call for a meeting of all the Latin American states.

24. If Argentina mobilizes and Brazil protests to the UN, then either Chile or the Dominican Republic will call for a meeting of all the Latin American states.

*25. If neither Chile nor the Dominican Republic calls for a meeting of all the Latin American states, then Brazil will not protest to the UN unless Argentina mobilizes.

8.4 Argument Forms and Refutation by Logical Analogy

The central task of deductive logic, we have said, is discriminating valid arguments from invalid ones. If the premises of a valid argument are true (we explained in the very first chapter), its conclusion *must* be true. If the conclusion of a valid argument is false, at least one of the premises must be false. In short, the premises of a valid argument give *incontrovertible proof* of the conclusion drawn.

This informal account of validity must now be made more precise. To do this we introduce the concept of an *argument form*. Consider the following two arguments, which plainly have the same logical form. Suppose we are presented with the first of these arguments:

> If Bacon wrote the plays attributed to Shakespeare, then Bacon was a great writer.
> Bacon was a great writer.
> Therefore Bacon wrote the plays attributed to Shakespeare.

We may agree with the premises but disagree with the conclusion, judging the argument to be invalid. One way of proving invalidity is by the method of logical analogy. "You might as well argue," we could retort, "that

> If Washington was assassinated, then Washington is dead.
> Washington is dead.
> Therefore Washington was assassinated.

You cannot seriously defend this argument," we would continue, "because here the premises are known to be true and the conclusion is known to be false. This argument is obviously invalid; your argument is of the same form, so yours is also invalid." This type of refutation is very effective.

This method of **refutation by logical analogy** points the way to an excellent general technique for testing arguments. To prove the invalidity of an argument, it suffices to formulate another argument that (1) has exactly the same form as the first and (2) has true premises and a false conclusion. This method is based on the fact that validity and invalidity are purely formal characteristics of arguments, which is to say that any two arguments that have exactly the same form are either both valid or both invalid, regardless of any differences in the subject matter with which they are concerned. Here we assume that the simple statements involved are neither logically true (e.g., "All chairs are chairs") nor logical-

Refutation by logical analogy
A method that shows the invalidity of an argument by presenting another argument that has the same form, but whose premises are known to be true and whose conclusion is known to be false.

ly false (e.g., "Some chairs are nonchairs"). We also assume that the only logical relations among the simple statements involved are those asserted or entailed by the premises. The point of these restrictions is to limit our considerations, in this chapter and the next, to truth-functional arguments alone, and to exclude other kinds of arguments whose validity turns on more complex logical considerations that are not appropriately introduced at this point.

A given argument exhibits its form very clearly when the simple statements that appear in it are abbreviated by capital letters. Thus we may abbreviate the statements, "Bacon wrote the plays attributed to Shakespeare," "Bacon was a great writer," "Washington was assassinated," and "Washington is dead," by the letters B, G, A, and D, respectively, and using the familiar three-dot symbol "$\therefore$" for "therefore," we may symbolize the two preceding arguments as

$$B \supset G \qquad\qquad A \supset D$$
$$G \qquad \text{and} \qquad D$$
$$\therefore B \qquad\qquad \therefore A$$

So written, their common form is easily seen.

To discuss forms of arguments rather than particular arguments having those forms, we need some method of symbolizing argument forms themselves. To achieve such a method, we introduce the notion of a **variable**. In the preceding sections we used capital letters to symbolize particular simple statements. To avoid confusion, we use small, or lowercase, letters from the middle part of the alphabet, p, q, r, s, . . ., as statement variables. A **statement variable**, as we shall use the term, is simply a letter for which, or in place of which, a statement may be substituted. Compound statements as well as simple statements may be substituted for statement variables.

We define an **argument form** as any array of symbols containing statement variables but no statements, such that when statements are substituted for the statement variables—the same statement being substituted for the same statement variable throughout—the result is an argument. For definiteness, we establish the convention that in any argument form, p shall be the first statement variable that occurs in it, and as other variables are introduced, they shall be labeled q, r, and s. Thus the expression

$$p \supset q$$
$$q$$
$$\therefore p$$

is an argument form, for when the statements B and G are substituted for the statement variables p and q, respectively, the result is the first argument in this section. If the statements A and D are substituted for the variables p and q, the result is the second argument. Any argument that results from the substitution of statements for statement variables in an argument form is called a **substitution instance** of that argument form. Any substitution instance of an argument form may be said to have that form, and any argument that has a certain form is said to be a substitution instance of that form.

Variable or **statement variable**
A place-holder; a letter (by convention, any of the lower case letters, beginning with p, q, etc.) for which a statement may be substituted.

Argument form
An array of symbols exhibiting logical structure; it contains no statements but it contains statement variables. These variables are arranged in such a way that when statements are consistently substituted for the statement variables, the result is an argument.

Substitution instance
Any argument that results from the substitution of statements for the statement variables of a given argument form.

For any argument there are usually several argument forms that have the given argument as a substitution instance. For example, the first argument of this section,

$B \supset G$
G
$\therefore B$

is a substitution instance of each of the four argument forms

$p \supset q$	$p \supset q$	$p \supset q$	p
q	r	r	q
$\therefore p$	$\therefore p$	$\therefore s$	$\therefore r$

Thus we obtain the given argument by substituting B for p and G for q in the first argument form; by substituting B for p and G for both q and r in the second; B for both p and s and G for both q and r in the third; and $B \supset G$ for p, G for q, and B for r in the fourth. Of these four argument forms, the first corresponds more closely to the structure of the given argument than do the others. It does so because the given argument results from the first argument form by substituting a different simple statement for each different statement variable in it. We call the first argument form the specific form of the given argument. Our definition of the specific form of a given argument is the following: If an argument is produced by substituting consistently a different simple statement for each different statement variable in an argument form, that argument form is the **specific form** of the given argument. For any given argument, there is a unique argument form that is the specific form of that argument.

EXERCISES

Here follow a group of arguments (Group **A**, lettered **a–o**) and a group of argument forms (Group **B**, numbered **1–24**). For each of the arguments (in Group **A**), indicate which of the argument forms (in Group **B**), if any, have the given argument as a substitution instance. In addition, for each given argument (in Group **A**), indicate which of the argument forms (in Group **B**), if any, is the specific form of that argument.

EXAMPLES

Argument **a** in Group **A**: Examining all the argument forms in Group **B**, we find that the only one of which Argument **a** is a *substitution instance* is Number **3**. Number **3** is also the *specific form* of Argument **a**.

Argument **j** in Group **A**: Examining all the argument forms in Group **B**, we find that Argument **j** is a *substitution instance* of *both* Number **6** and Number **23**. But *only* Number **23** is the *specific form* of Argument **j**.

Argument **m** in Group **A**: Examining all the argument forms in Group **B**, we find that Argument **m** is a *substitution instance* of *both* Number **3** and Number **24**. But there is *no* argument form in Group **B** that is the *specific form* of Argument **m**.

Specific form
When referring to a given argument, the argument form from which the argument results when a different simple statement is substituted *consistently* for each different statement variable in that form.

Group A—Arguments

a. $A \cdot B$
$\therefore A$

b. $C \supset D$
$\therefore C \supset (C \cdot D)$

c. E
$\therefore E \vee F$

d. $G \supset H$
$\sim H$
$\therefore \sim G$

*e. I
J
$\therefore I \cdot J$

f. $(K \supset L) \cdot (M \supset N)$
$K \vee M$
$\therefore L \vee N$

g. $O \supset P$
$\sim O$
$\therefore \sim P$

h. $Q \supset R$
$Q \supset S$
$\therefore R \vee S$

i. $T \supset U$
$U \supset V$
$\therefore V \supset T$

j. $(W \cdot X) \supset (Y \cdot Z)$
$\therefore (W \cdot X) \supset [(W \cdot X) \cdot (Y \cdot Z)]$

k. $A \supset B$
$\therefore (A \supset B) \vee C$

l. $(D \vee E) \cdot \sim F$
$\therefore D \vee E$

m. $[G \supset (G \cdot H)] \cdot [H \supset (H \cdot G)]$
$\therefore G \supset (G \cdot H)$

n. $(I \vee J) \vee (I \cdot J)$
$\sim (I \vee J)$
$\therefore \sim (I \cdot J)$

*o. $(K \supset L) \cdot (M \supset N)$
$\therefore K \supset L$

Group B—Argument Forms

*1. $p \supset q$
$\therefore \sim q \supset \sim p$

2. $p \supset q$
$\therefore \sim p \supset \sim q$

3. $p \cdot q$
$\therefore p$

4. p
$\therefore p \vee q$

*5. p
$\therefore p \supset q$

6. $p \supset q$
$\therefore p \supset (p \cdot q)$

7. $(p \vee q) \supset (p \cdot q)$
$\therefore (p \supset q) \cdot (q \supset p)$

8. $p \supset q$
$\sim p$
$\therefore \sim q$

9. $p \supset q$
$\sim q$
$\therefore \sim p$

*10. p
q
$\therefore p \cdot q$

11. $p \supset q$
$p \supset r$
$\therefore q \vee r$

12. $p \supset q$
$q \supset r$
$\therefore r \supset p$

13. $p \supset (q \supset r)$
$p \supset q$
$\therefore p \supset r$

14. $p \supset (q \cdot r)$
$(q \vee r) \supset \sim p$
$\therefore \sim p$

*15. $p \supset (q \supset r)$
$q \supset (p \supset r)$
$\therefore (p \vee q) \supset r$

16. $(p \supset q) \cdot (r \supset s)$
$p \vee r$
$\therefore q \vee s$

17. $(p \supset q) \cdot (r \supset s)$
$\sim q \vee \sim s$
$\therefore \sim p \vee \sim s$

18. $p \supset (q \supset r)$
$q \supset (r \supset s)$
$\therefore p \supset s$

19. $p \supset (q \supset r)$
$(q \supset r) \supset s$
$\therefore p \supset s$

*20. $(p \supset q) \cdot [(p \cdot q) \supset r]$
$p \supset (r \supset s)$
$\therefore p \supset s$

21. $(p \supset q) \supset (p \cdot q)$
$\sim (p \vee q)$
$\therefore \sim (p \cdot q)$

22. $(p \vee q) \supset (p \cdot q)$
$(p \cdot q)$
$\therefore p \vee q$

23. $(p \cdot q) \supset (r \cdot s)$
$\therefore (p \cdot q) \supset [(p \cdot q) \cdot (r \cdot s)]$

24. $(p \supset q) \cdot (r \supset s)$
$\therefore p \supset q$

8.5 The Precise Meaning of "Invalid" and "Valid"

We are now in a position to address with precision the central questions of deductive logic:

1. *What precisely is meant* by saying that an argument form is invalid, or valid?
2. *How do we decide* whether a deductive argument form is invalid, or valid?

The first of these questions is answered in this section, the second in the following section.

One can proceed by relying upon the technique of refutation by logical analogy (refutation by logical analogy was discussed in Section 6.2). The term **invalid** as applied to argument forms may be defined as follows: *An argument form is invalid if and only if it has at least one substitution instance with true premises and a false conclusion.* If the specific form of a given argument has any substitution instance whose premises are true and whose conclusion is false, then the given argument is invalid. This fact—that any argument whose specific form is an invalid argument form is an invalid argument—provides the basis for refutation by logical analogy. A given argument is proved invalid if a refuting analogy can be found for it.

"Thinking up" refuting analogies may not always be easy. Happily, it is not necessary, because for arguments of this type there is a simpler, purely mechanical test based on the same principle. Given any argument, we can test the specific form of that argument, because its invalidity would determine the invalidity of the argument.

The test described above can also be used to show validity. Any argument form that is not invalid must be valid. Hence *an argument form is* **valid** *if and only if it has no substitution instances with true premises and a false conclusion*. Because validity is a formal notion, an argument is valid if and only if the specific form of that argument is a *valid argument form*.

8.6 Testing Argument Validity Using Truth Tables

Knowing exactly what it means to say that an argument is valid, or invalid, we can now devise a method for testing the validity of every truth-functional argument. Our method, using a truth table, is very simple and very powerful. It is simply an application of the analysis of argument forms just given.

To test an argument form, we examine all possible substitution instances of it to see if any one of them has true premises and a false conclusion. Of course, any argument form has an infinite number of substitution instances, but we need not worry about having to examine them one at a time. We are interested only in the truth or falsehood of their premises and conclusions, so we need

---

Invalid
Not valid; characterizing a deductive argument that fails to provide conclusive grounds for the truth of its conclusion. Every deductive argument is either valid or invalid.

Valid
A deductive argument is said to be *valid* when its premises, if they were all true, would provide conclusive grounds for the truth of its conclusion. Validity is a formal characteristic; it applies only to arguments, as distinguished from truth, which applies to propositions.

consider only the truth values involved. The arguments that concern us here contain only simple statements and compound statements that are built up out of simple statements using the curl and the truth-functional connectives symbolized by the dot, wedge, and horseshoe. Hence we obtain all possible substitution instances whose premises and conclusions have different truth values by examining all possible different arrangements of truth values for the statements that can be substituted for the different statement variables in the argument form to be tested.

When an argument form contains just two different statement variables, p and q, all of its substitution instances are the result of either substituting true statements for both p and q, or a true statement for p and a false one for q, or a false one for p and a true one for q, or false statements for both p and q. These different cases are assembled most conveniently in the form of a truth table. To decide the validity of the argument form

$p \supset q$
q
$\therefore p$

we can construct the following truth table:

p	q	$p \supset q$
T	T	T
T	F	F
F	T	T
F	F	T

Each row of this table represents a whole class of substitution instances. The **T**'s and **F**'s in the two initial or guide columns represent the truth values of the statements substituted for the variables p and q in the argument form. We fill in the third column by referring to the initial or guide columns and the definition of the horseshoe symbol. The third column heading is the first "premise" of the argument form, the second column is the second "premise" and the first column is the "conclusion." In examining this truth table, we find that in the third row there are **T**'s under both premises and an **F** under the conclusion, which indicates that there is at least one substitution instance of this argument form that has true premises and a false conclusion. This row suffices to show that the argument form is invalid. Any argument of this specific form (that is, any argument the specific argument form of which is the given argument form) is said to commit the fallacy of affirming the consequent, since its second premise affirms the consequent of its conditional first premise.

Truth tables, although simple in concept, are powerful tools. In using them to establish the validity or the invalidity of an argument form, it is critically important that the table first be constructed correctly. To construct the truth table correctly, there must be a guide column for each statement variable in the argument form—p, q, r, and so on. The array must exhibit all the possible combinations of the truth and falsity of all these variables, so there must be a number of horizontal rows sufficient to do this: four rows if there are two variables, eight rows if there are three variables, and so on. There must be a vertical column for each of the premises and for the conclusion, as well as a column for each of the symbolic expressions out of which the premises and conclusion are built. The construction of a truth table in this fashion is essentially a mechanical task; it requires only careful counting and the careful placement of **T**'s and **F**'s in the appropriate columns, all governed by our understanding of the curl and the several truth-functional connectives—the dot, the wedge, the horseshoe—and the circumstances under which each truth-functional compound is true and the circumstances under which it is false.

Once the table has been constructed and the completed array is before us, it is essential to read it correctly, that is, to use it correctly to make the appraisal of the argument form in question. We must note carefully which columns are those representing the premises of the argument being tested, and which column represents the conclusion of that argument. In testing the argument just above, which we found to be invalid, we noted that it was the second and third columns of the truth table that represented the premises, while the conclusion was represented by the first (leftmost) column. Depending on which argument form we are testing, and the order in which we have placed the columns as the table was built, it is possible for the premises and the conclusion to appear in any order at the top of the table. Their position to the right or to the left is not significant; we, who use the table, must understand which column represents what, and we must understand what we are in search of. *Is there any one case*, we ask ourselves, *any single row in which all the premises are true and the conclusion false?* If there is such a row, the argument form is invalid; if there is no such row, the argument form must be valid. After the full array has been neatly and accurately set forth, great care in reading the truth table accurately is of the utmost importance.

8.7 Some Common Argument Forms

A. Common Valid Forms

Some valid argument forms are exceedingly common and may be intuitively understood. These may now be precisely identified. They should be recognized wherever they appear, and they may be called by their widely accepted names: (1) **Disjunctive Syllogism**, (2) *Modus Ponens*, (3) *Modus Tollens*, and (4) **Hypothetical Syllogism**.

Disjunctive Syllogism

One of the simplest argument forms relies on the fact that in every true disjunction, at least one of the disjuncts must be true. Therefore, if one of them is false, the other must be true. Arguments in this form are exceedingly common. When a candidate for a high appointed office was forced to withdraw her candidacy because of a tax violation involving one of her employees, a critic wrote: "In trying to cover up her own illegal alien peccadillo, or stonewall her way out of it, she was driven either by stupidity or arrogance. She's obviously not stupid; her plight must result, then, from her arrogance."[3]

We symbolize the disjunctive syllogism as

$p \vee q$
$\sim p$
$\therefore q$

and to show its validity we construct the following truth table:

p	q	p ∨ q	~p
T	T	T	F
T	F	T	F
F	T	T	T
F	F	F	T

Here, too, the initial or guide columns exhibit all possible different truth values of statements that may be substituted for the variables p and q. We fill in the third column by referring to the first two, and the fourth by reference to the first alone. Now the third row is the only one in which **T**'s appear under both premises (the third and fourth columns), and there a **T** also appears under the conclusion (the second column). The truth table thus shows that the argument form has no substitution instance having true premises and a false conclusion, and thereby proves the validity of the argument form being tested. As used in this chapter, the term **disjunctive syllogism** is the name of an elementary argument form, here proved valid. This form is always valid, of course, and therefore, in modern logic, *disjunctive syllogism* always refers to an elementary argument form that is valid. In traditional logic, however, the term *disjunctive syllogism* is used more broadly, to refer to any syllogism that contains a disjunctive premise; some such syllogisms may of course be invalid. One must be clear whether the expression is being used in the broader or the narrower sense. Here we use it in the narrower sense.

Here, as always, it is essential that the truth table be *read* accurately; the column representing the conclusion (second from the left) and the columns representing the premises (third and fourth from the left) must be carefully identified. Only by using those three columns correctly can we reliably determine the validity (or invalidity) of the argument form in question. Note that the very same truth table could be used to test the validity of a very different argument form, one whose premises are represented by the second and third columns and whose conclusion is represented by the fourth column. That argument form, as we can see from the top row of the table, is invalid. The truth-table technique provides a completely mechanical method for testing the validity of any argument of the general type considered here.

Disjunctive Syllogism
A valid argument form in which one premise is a disjunction, another premise is the denial of one of the two disjuncts, and the conclusion is the truth of the other disjunct. Symbolized as: $p \vee q$, $\sim p$, therefore q.

CHAPTER 8   Symbolic Logic

We are now in a position to justify our proposal to translate any occurrence of the "if–then" phrase into our material implication symbol, ⊃. In Section 8.3, the claim was made that all valid arguments of the general type with which we are concerned here that involve "if–then" statements remain valid when those statements are interpreted as affirming merely material implications. Truth tables can be used to substantiate this claim, and will justify our translation of "if–then" into the horseshoe symbol.

Modus Ponens

The simplest type of intuitively valid argument involving a conditional statement is illustrated by the argument:

> If the second native told the truth, then only one native is a politician.
> The second native told the truth.
> Therefore only one native is a politician.

The specific form of this argument, known as *Modus Ponens* ("the method of putting, or affirming"), is

$p \supset q$
p
$\therefore q$

and is proved valid by the following truth table:

p	q	p ⊃ q
T	T	T
T	F	F
F	T	T
F	F	T

Here the two premises are represented by the third and first columns, and the conclusion is represented by the second. Only the first row represents substitution instances in which both premises are true, and the **T** in the second column shows that in these arguments the conclusion is true also. This truth table establishes the validity of any argument of the form *modus ponens*.

Modus Tollens

If a conditional statement is true, then if the consequent is false, the antecedent must also be false. The argument form that relies on this is very commonly used to establish the falsehood of some proposition under attack. To illustrate: A distinguished rabbi, insisting that the Book of Genesis was never meant to be a scientific treatise, presented this crisp argument:

> A literal reading of Genesis would lead one to conclude that the world is less than 6,000 years old and that the Grand Canyon could have been carved by the global flood 4,500 years ago. Since this is impossible, a literal reading of Genesis must be wrong.[4]

The argument may be symbolized as

$p \supset q$
$\sim q$
$\therefore \sim p$

Modus Ponens
An elementary valid argument form according to which, if the truth of a hypothetical premise is assumed, and the truth of the antecedent of that premise is also assumed, we may conclude that the consequent of that premise is true. Symbolized as: $p \supset q$, p, therefore q.

Modus Tollens
An elementary valid argument form according to which, if the truth of a hypothetical premise is assumed, and the falsity of the consequent of that premise is also assumed, we may conclude that the antecedent of that premise is false. Symbolized as $p \supset q$, $\sim q$, therefore $\sim p$.

The validity of this argument form, called *modus tollens* ("the method of taking away, or denying"), may be shown by the following truth table:

p	q	$p \supset q$	$\sim q$	$\sim p$
T	T	T	F	F
T	F	F	T	F
F	T	T	F	T
F	F	T	T	T

Here again, there is no substitution instance, no line, on which the premises, $p \supset q$ and $\sim q$, are both true and the conclusion, $\sim p$, is false.

Hypothetical Syllogism

Another common type of intuitively valid argument contains only conditional statements. Here is an example:

> If the first native is a politician, then the first native lies.
> If the first native lies, then the first native denies being a politician.
> Therefore if the first native is a politician, then the first native denies being a politician.

The specific form of this argument is

$p \supset q$
$q \supset r$
$\therefore p \supset r$

This argument, called a **Hypothetical Syllogism** (or, as in Chapter 7, a Pure Hypothetical Syllogism), contains three distinct statement variables, so the truth table must have three initial (or guide) columns and requires eight rows to list all possible substitution instances. Besides the initial columns, three additional columns are needed: two for the premises, the third for the conclusion. The table is

p	q	r	$p \supset q$	$q \supset r$	$p \supset r$
T	T	T	T	T	T
T	T	F	T	F	F
T	F	T	F	T	T
T	F	F	F	T	F
F	T	T	T	T	T
F	T	F	T	F	T
F	F	T	T	T	T
F	F	F	T	T	T

In constructing it, we fill in the fourth column by looking back to the first and second, the fifth by reference to the second and third, and the sixth by reference to the first and third. Examining the completed table, we observe that the premises are true only in the first, fifth, seventh, and eighth rows, and that in all of these the conclusion is also true. This truth table establishes the validity of the

Hypothetical Syllogism
A syllogism that contains a hypothetical proposition as a premise. If the syllogism contains hypothetical propositions exclusively, it is called a "pure" hypothetical syllogism; if the syllogism contains one conditional and one categorical premise, it is called a "mixed" hypothetical syllogism.

argument form and proves that the hypothetical syllogism remains valid when its conditional statements are translated by means of the horseshoe symbol.

Enough examples have been provided to illustrate the proper use of the truth-table technique for testing arguments, and perhaps enough have been given to show that the validity of any valid argument involving conditional statements is preserved when its conditionals are translated into merely material implications. Any doubts that remain can be allayed by the reader's translating and testing similar examples.

The arguments that concern us here contain only simple statements and compound statements that are built up out of simple statements using the curl and the truth-functional connectives symbolized by the dot, wedge, and horseshoe. As more complicated argument forms are considered, larger truth tables are required to test them, because a separate initial or guide column is required for each different statement variable in the argument form. Only two are required for a form with just two variables, and that table will have four rows. But three initial columns are required for a form with three variables, such as the hypothetical syllogism, and such truth tables have eight rows. To test the validity of an argument form, such as that of the *Constructive Dilemma*,

$(p \supset q) \cdot (r \supset s)$
$p \vee r$
$\therefore q \vee s$

which contains four distinct statement variables, a truth table with four initial columns and sixteen rows is required. In general, to test an argument form containing n distinct statement variables we need a truth table with n initial columns and 2^n rows.

B. Common Invalid Forms

Two invalid argument forms deserve special notice because they superficially resemble valid forms and therefore often tempt careless writers or readers. The *fallacy of affirming the consequent*, discussed also in Section 7.7, is symbolized as

$p \supset q$
q
$\therefore p$

Although the shape of this form is something like that of *modus ponens*, the two argument forms are very different, and this form is not valid. It is well illustrated in a "bogus syllogism" about the dictatorial president of Iraq, the late Saddam Hussein. Here is that syllogism, as recounted by Orlando Patterson (Mr. Patterson's wording of the syllogism is very slightly different but has exactly the same logical force). Its invalidity does indeed render it bogus: "If one is a terrorist one is a tyrant who hates freedom. Saddam Hussein is a tyrant who hates freedom. Therefore Saddam Hussein is a terrorist."[5] Let us suppose that the hypothetical first premise is true and that the second premise describing Saddam Hussein is also true. But that second premise affirms (about Saddam Hussein as one tyrant) only the consequent of the preceding hypothetical. The argument plainly commits the fallacy of affirming the consequent.

Another invalid form, called the *fallacy of denying the antecedent*, has a shape somewhat like that of *modus tollens* and may be symbolized as

p ⊃ q
~p
∴ ~q

An example of this fallacy is the campaign slogan used by a candidate for mayor of New York City some years ago: "If you don't know the buck, you don't know the job—and Abe knows the buck." The unstated conclusion to which the voter was deliberately tempted was that "Abe knows the job"—a proposition that does not follow from the stated premises.

Both of these common fallacies may readily be shown to be invalid by means of truth tables. In each case there is one line of the truth table in which the premises of these fallacious arguments are all true, but the conclusion is false.

C. Substitution Instances and Specific Forms

A given argument can be a substitution instance of several different argument forms, as we noted earlier when defining *argument form*. Hence the valid disjunctive syllogism examined on page 288, which may be symbolized as

R ∨ W
~R
∴ W

is a substitution instance of the valid argument form

p ∨ q
~p
∴ q

and is *also* a substitution instance of the *in*valid argument form

p
q
∴ r

It is obvious, in this last form, that from two premises, *p* and *q*, we cannot validly infer *r*. So it is clear that a valid argument can be a substitution instance of a valid argument form and of an invalid argument form. Therefore, in determining whether any given argument is valid, *we must look to the specific form of the argument in question*. Only the specific form of the argument reveals the full logical structure of that argument, and because it does, we know that if the specific form of an argument is valid, the argument itself must be valid.

In the illustration just given, we see an argument (R ∨ W, ~R, ∴ W), and two argument forms of which that argument could be a substitution instance. The first of these argument forms (p ∨ q, ~p, ∴ q) is valid, and because that form is the specific form of the given argument, its validity establishes that the given argument is valid. The second of these argument forms is invalid, but because it is not the specific form of the given argument, it cannot be used to show that the given argument is invalid.

This point should be emphasized: An argument form that is valid can have only valid arguments as substitution instances. That is, all of the substitution instances of a valid form *must* be valid. This is proved by the truth-table proof of validity for the valid argument form, which shows that there is no possible substitution instance of a valid form that has true premises and a false conclusion.

EXERCISES

A. Use truth tables to prove the validity or invalidity of each of the argument forms in Section 8.4, Group B, page 313.

B. Use truth tables to determine the validity or invalidity of each of the following arguments:

*1. $(A \lor B) \supset (A \cdot B)$
$A \lor B$
$\therefore A \cdot B$

2. $(C \lor D) \supset (C \cdot D)$
$C \cdot D$
$\therefore C \lor D$

3. $E \supset F$
$F \supset E$
$\therefore E \lor F$

4. $(G \lor H) \supset (G \cdot H)$
$\sim(G \cdot H)$
$\therefore \sim(G \lor H)$

*5. $(I \lor J) \supset (I \cdot J)$
$\sim(I \lor J)$
$\therefore \sim(I \cdot J)$

6. $K \lor L$
K
$\therefore \sim L$

7. $M \lor (N \cdot \sim N)$
M
$\therefore \sim(N \cdot \sim N)$

8. $(O \lor P) \supset Q$
$Q \supset (O \cdot P)$
$\therefore (O \lor P) \supset (O \cdot P)$

9. $(R \lor S) \supset T$
$T \supset (R \cdot S)$
$\therefore (R \cdot S) \supset (R \lor S)$

10. $U \supset (V \lor W)$
$(V \cdot W) \supset \sim U$
$\therefore \sim U$

C. Use truth tables to determine the validity or invalidity of the following arguments:

*1. If Angola achieves stability, then both Botswana and Chad will adopt more liberal policies. But Botswana will not adopt a more liberal policy. Therefore Angola will not achieve stability.

2. If Denmark refuses to join the European Community, then, if Estonia remains in the Russian sphere of influence, then Finland will reject a free-trade policy. Estonia will remain in the Russian sphere of influence. So if Denmark refuses to join the European Community, then Finland will reject a free-trade policy.

3. If Greece strengthens its democratic institutions, then Hungary will pursue a more independent policy. If Greece strengthens its democratic institutions, then the Italian government will feel less threatened. Hence, if Hungary pursues a more independent policy, the Italian government will feel less threatened.

4. If Japan continues to increase the export of automobiles, then either Korea or Laos will suffer economic decline. Korea will not suffer economic decline. It follows that if Japan continues to increase the export of automobiles, then Laos will suffer economic decline.

* 5. If Montana suffers a severe drought, then, if Nevada has its normal light rainfall, Oregon's water supply will be greatly reduced. Nevada does have its normal light rainfall. So if Oregon's water supply is greatly reduced, then Montana suffers a severe drought.

6. If equality of opportunity is to be achieved, then those people previously disadvantaged should now be given special opportunities. If those people previously disadvantaged should now be given special opportunities, then some people receive preferential treatment. If some people receive preferential treatment, then equality of opportunity is not to be achieved. Therefore equality of opportunity is not to be achieved.

7. If terrorists' demands are met, then lawlessness will be rewarded. If terrorists' demands are not met, then innocent hostages will be murdered. So either lawlessness will be rewarded or innocent hostages will be murdered.

8. If people are entirely rational, then either all of a person's actions can be predicted in advance or the universe is essentially deterministic. Not all of a person's actions can be predicted in advance. Thus, if the universe is not essentially deterministic, then people are not entirely rational.

9. If oil consumption continues to grow, then either oil imports will increase or domestic oil reserves will be depleted. If oil imports increase and domestic oil reserves are depleted, then the nation eventually will go bankrupt. Therefore, if oil consumption continues to grow, then the nation eventually will go bankrupt.

*10. If oil consumption continues to grow, then oil imports will increase and domestic oil reserves will be depleted. If either oil imports increase or domestic oil reserves are depleted, then the nation will soon be bankrupt. Therefore, if oil consumption continues to grow, then the nation will soon be bankrupt.

8.8 Statement Forms and Material Equivalence

A. Statement Forms and Statements

We now make explicit a notion that was tacitly assumed in the preceding section, the notion of a *statement form*. There is an exact parallel between the relation of argument to argument form, on the one hand, and the relation of statement to statement form, on the other. The definition of a statement form makes this evident: A **statement form** is any sequence of symbols containing statement variables but no statements, such that when statements are substituted for the statement variables—the same statement being substituted for the same statement variable throughout—the result is a statement. Thus $p \lor q$ is a statement form, because when statements are substituted for the variables p and q, a statement results. The resulting statement is a disjunction, so $p \lor q$ is called a **disjunctive statement form**. Analogously, $p \cdot q$ and $p \supset q$ are respectively called *conjunctive* and *conditional statement forms*, and $\sim p$ is called a *negation statement form* or a *denial statement form*. Just as any argument of a certain form is

Statement form
A sequence of symbols containing no statements, but containing statement variables connected in such a way that when statements are consistently substituted for the statement variables, the result is a statement.

Disjunctive statement form A statement form symbolized as $p \lor q$; its substitution instances are disjunctive statements.

Biography

Charles Sanders Peirce

Viewed by many as the most original and creative of American logicians—Bertrand Russell called him "certainly the greatest American thinker ever"— Charles Sanders Peirce (1839–1914) made contributions of such complexity and variety to the fields of logic and mathematics that it is not easy to summarize them. For him, what we call logic was the formal branch of the theory of signs, *semiotics*—a study of which he was the founder.

The son of a professor of mathematics and astronomy at Harvard, Peirce was fascinated by logic from the time that he read Whateley's *Elements of Logic* at the age of 12. He received his BA and MA from Harvard, but was despised by one of his instructors, Charles William Eliot, who—as President of Harvard for forty years—made it virtually impossible for Peirce to obtain the academic employment that he sought.

It was in the United States Coast Survey that Peirce was chiefly employed until, at the age of 40, he was appointed lecturer in logic at the recently established Johns Hopkins University. This position he held for five years, but he lost it as a consequence of assorted marital and sexual scandals in which he became involved. Academic employment was from that time denied him everywhere. Peirce was an odd man, with an odd manner; he was not very likeable, or sociable, or cooperative; his conduct was often irresponsible. Probably he suffered from some serious psychological disabilities.

As a thinker, however, he was productive in science and mathematics and philosophy, as well as in logic. He wrote prolifically; some of his writings have not yet been published. He defended the frequency theory of probability, contending that science can achieve no more than statistical probabilities—never certainties. He worked on infinitesimals and on the theory of mathematical continua. He developed the logic of relations ("If X is taller than Y, and Y is taller than Z, then X is taller than Z"). He refined quantification theory. He created a three-value logic in which "undetermined" was the third value. He improved truth tables. He devised symbols for new logical operations. He was one of the first to see that Boolean calculations could be carried out in the material world using electrical switches. Some scholars who much later participated in the design and construction of the first electronic computers credited their insights to the suggestiveness of Peirce's writings. The American logician C. I. Lewis wrote that "the contributions of C. S. Peirce to symbolic logic are more numerous and varied than those of any other writer."

In philosophy Peirce is most famous as the founder, with John Dewey, of the American movement we call *pragmatism*, which was for him essentially a theory of truth: A proposition is true if it works satisfactorily, and the meaning of a proposition is to be found in the practical consequences of accepting it. He reported that he learned philosophy, when he was a college student, by reading every day a few pages of Immanuel Kant's *Critique of Pure Reason*, a work he studied regularly for ten years.

Peirce was a man of most peculiar habits. He spent his inheritance on land and a large house in eastern Pennsylvania that he could not afford to maintain.

Charles Sanders Peirce (1839–1914)
A great American logician, philosopher, mathematician, and scientist.
Source: Photodisc/Getty Images

> He lived beyond his means; he relied upon his friends to pay his debts and his taxes. During the last years of his life he could not afford to heat his home in winter, and lived largely on old bread donated by a local baker. In that big house in Milford, Pennsylvania, he died in 1914 at the age of 74.

said to be a substitution instance of that argument form, so any statement of a certain form is said to be a *substitution instance* of that statement form. Just as we distinguished the *specific form* of a given argument, so we distinguish the **specific form** of a given statement as that statement form from which the statement results by substituting consistently a different simple statement for each different statement variable. Thus $p \vee q$ is the *specific form* of the statement, "The blind prisoner has a red hat or the blind prisoner has a white hat."

B. Tautologous, Contradictory, and Contingent Statement Forms

The statement, "Lincoln was assassinated" (symbolized as L), and the statement, "Either Lincoln was assassinated or else he wasn't" (symbolized as $L \vee \sim L$), are both obviously true. But, we would say, they are true "in different ways" or have "different kinds" of truth. Similarly, the statement, "Washington was assassinated" (symbolized as W), and the statement "Washington was both assassinated and not assassinated" (symbolized as $W \cdot \sim W$), are both plainly false—but they also are false "in different ways" or have "different kinds" of falsehood. These differences in the "kinds" of truth or of falsehood are important and very great.

That the statement L is true, and that the statement W is false, are historical facts—facts about the way events did happen. There is no logical necessity about them. Events might have occurred differently, and therefore the truth values of such statements as L and W must be discovered by an empirical study of history. But the statement $L \vee \sim L$, although true, is not a truth of history. There is logical necessity here: Events could not have been such as to make it false, and its truth can be known independently of any particular empirical investigation. The statement $L \vee \sim L$ is a logical truth, a formal truth, true in virtue of its form alone. It is a substitution instance of a statement form all of whose substitution instances are true statements.

A statement form that has only true substitution instances is called a *tautologous statement form*, or a **tautology**. To show that the statement form $p \vee \sim p$ is a tautology, we construct the following truth table:

p	$\sim p$	$p \vee \sim p$
T	F	T
F	T	T

There is only one initial or guide column to this truth table, because the form we are considering contains only one statement variable. Consequently, there are

Specific form When referring to a given statement, the statement form from which the statement results when a different simple statement is substituted *consistently* for each different statement variable in that form.

Tautology A statement form all of whose substitution instances must be true.

only two rows, which represent all possible substitution instances. There are only **T**'s in the column under the statement form in question, and this fact shows that all of its substitution instances are true. Any statement that is a substitution instance of a tautologous statement form is true in virtue of its form, and is itself said to be tautologous, or a tautology.

A statement form that has only false substitution instances is said to be **self-contradictory**, or a **contradiction**, and is logically false. The statement form $p \cdot \sim p$ is self-contradictory, because only **F**'s occur under it in its truth table, signifying that all of its substitution instances are false. Any statement, such as $W \cdot \sim W$, which is a substitution instance of a self-contradictory statement form, is false in virtue of its form and is itself said to be self-contradictory, or a contradiction.

Statement forms that have both true and false statements among their substitution instances are called **contingent statement forms**. Any statement whose specific form is contingent is called a *contingent statement*. (It will be recalled that we are assuming here that no simple statements are either logically true or logically false. Only contingent simple statements are admitted here. See page 211.) Thus p, $\sim p$, $p \cdot q$, $p \vee q$, and $p \supset q$ are all contingent statement forms, and such statements as L, $\sim L$, $L \cdot W$, $L \vee W$, and $L \supset W$ are contingent statements, because their truth values are dependent, or contingent, on their contents rather than on their forms alone.

Not all statement forms are so obviously tautological or self-contradictory or contingent as the simple examples cited. For example, the statement form $[(p \supset q) \supset p] \supset p$ is not at all obvious, though its truth table will show it to be a tautology. It even has a special name, *Peirce's Law*.

C. Material Equivalence

Material equivalence is a truth-functional connective, just as disjunction and material implication are truth-functional connectives. The truth value of any statement formed by linking two statements with a truth-functional connective, as explained earlier, depends on (is a function of) the truth or falsity of the statements it connects. Thus, we say that the disjunction of A and B is true if either A is true or B is true or if they are both true. **Material equivalence** is the truth-functional connective that asserts that the statements it connects have the *same* truth value. Two statements that are equivalent in truth value, therefore, are materially equivalent. One straightforward definition is this: Two statements are *materially equivalent* when they are both true, or both false.

Just as the symbol for disjunction is the wedge, and the symbol for material implication is the horseshoe, there is also a special symbol for material equivalence, the three-bar sign or **tribar**, "$\equiv$". (Some systems employ the symbol "$\leftrightarrow$"). And just as we gave truth-table definitions for the wedge and the horseshoe, we can do so for the three-bar sign. Here is the truth table for material equivalence, $\equiv$:

p	q	$p \equiv q$
T	T	T
T	F	F
F	T	F
F	F	T

Contradiction
A statement form all of whose substitution instances are false.

Contingent
Being neither tautologous nor self-contradictory. A contingent statement may be true or false; a contingent statement form has some true and some false substitution instances.

Material equivalence
A truth-functional relation (symbolized by the three-bar sign, $\equiv$) that may connect two statements. Two statements are materially equivalent when they are both true, or when the are both false—that is, when they have the same truth value. Materially equivalent statements always materially imply one another.

Any two true statements materially imply one another; that is a consequence of the meaning of material implication. And any two false statements also materially imply one another. Therefore any two statements that are materially equivalent must imply one another, because they are either both true or both false.

Since any two statements, *A* and *B*, that are materially equivalent imply one another, we may infer from their material equivalence that *B* is true if *A* is true, and also that *B* is true only if *A* is true. The converse also holds: *A* is true if *B* is true, and *A* is true only if *B* is true. Because both of these relations are entailed by material equivalence, we can read the three-bar sign, ≡ , to say "if and only if."

In everyday discourse we use this logical relation only occasionally. I will go to the championship game, one may say, if and only if I can acquire a ticket. I will go if I do acquire a ticket, but I can go only if I acquire a ticket. So my going to the game, and my acquiring a ticket to the game, are materially equivalent.

Every implication is a conditional statement, as we noted earlier. Two statements, *A* and *B*, that are materially equivalent entail the truth of the conditional $A \supset B$, and also entail the truth of the conditional $B \supset A$. Because the implication goes both ways when material equivalence holds, a statement of the form $A \equiv B$ is often called a **biconditional**.

There are four truth-functional connectives on which deductive arguments commonly depend: *conjunction, disjunction, material implication,* and *material equivalence.* Our discussion of the four is now complete.

overview

The Four Truth-Functional Connectives

Truth-Functional Connective	Symbol (Name of Symbol)	Proposition Type	Names of Components of Propositions of That Type	Example
And	· (dot)	Conjunction	Conjuncts	Carol is mean **and** Bob sings the blues. $C \cdot B$
Or	∨ (wedge)	Disjunction	Disjuncts	Carol is mean **or** Tyrell is a music lover. $C \vee T$
If–then	⊃ (horseshoe)	Conditional	Antecedent, Consequent	**If** Bob sings the blues, **then** Myrna gets moody. $B \supset M$
If and only if	≡ (tribar)	Biconditional	Components	Myrna gets moody **if and only if** Bob sings the blues. $M \equiv B$

NOTE: "Not" is not a connective, but is a truth-functional operator, so it is omitted here.

D. Arguments, Conditional Statements, and Tautologies

To every argument there corresponds a conditional statement whose antecedent is the conjunction of the argument's premises and whose consequent is the argument's conclusion. Thus, an argument having the form of *modus ponens,*

$p \supset q$
p
$\therefore q$

may be expressed as a conditional statement of the form $[(p \supset q) \cdot p] \supset q$. If the argument expressed as a conditional has a valid argument form, then its conclusion must in every case follow from its premises, and therefore the conditional statement of it may be shown on a truth table to be a tautology. That is, the statement that the conjunction of the premises implies the conclusion will (if the argument is valid) have all and only true instances.

Truth tables are powerful devices for the evaluation of arguments. An argument form is valid if and only if its truth table has a **T** under the conclusion in every row in which there are **T**'s under all of its premises. This follows from the precise meaning of *validity*. Now, if the conditional statement expressing that argument form is made the heading of one column of the truth table, an **F** can occur in that column only in a row in which there are **T**'s under all the premises and an **F** under the conclusion. But there will be no such row if the argument is valid. Hence only **T**'s will occur under a conditional statement that corresponds to a valid argument, and that conditional statement *must* be a tautology. We may therefore say that an argument form is valid if, and only if, its expression in the form of a conditional statement (of which the antecedent is the conjunction of the premises of the given argument form, and the consequent is the conclusion of the given argument form) is a tautology.

For every *invalid* argument of the truth-functional variety, however, the corresponding conditional statement will not be a tautology. The statement that the conjunction of its premises implies its conclusion is (for an invalid argument) either contingent or contradictory.

EXERCISES

A. For each statement in the left-hand column, indicate which, if any, of the statement forms in the right-hand column have the given statement as a substitution instance, and indicate which, if any, is the specific form of the given statement.

1. $A \lor B$
2. $C \cdot \sim D$
3. $\sim E \supset (F \cdot G)$
4. $H \supset (I \cdot J)$
* 5. $(K \cdot L) \lor (M \cdot N)$
6. $(O \lor P) \supset (P \cdot Q)$
7. $(R \supset S) \lor (T \cdot \sim U)$
8. $V \supset (W \lor \sim W)$
9. $[(X \supset Y) \supset X] \supset X$
*10. $Z \equiv \sim\sim Z$

a. $p \cdot q$
b. $p \supset q$
c. $p \lor q$
d. $p \cdot \sim q$
e. $p \equiv q$
f. $(p \supset q) \lor (r \cdot s)$
g. $[(p \supset q) \supset r] \supset s$
h. $[(p \supset q) \supset p] \supset p$
i. $(p \cdot q) \lor (r \cdot s)$
j. $p \supset (q \lor \sim r)$

B. Use truth tables to characterize the following statement forms as tautologous, self-contradictory, or contingent.

*1. $[p \supset (p \supset q)] \supset q$
2. $p \supset [(p \supset q) \supset q]$
3. $(p \cdot q) \cdot (p \supset \sim q)$
4. $p \supset [\sim p \supset (q \vee \sim q)]$
*5. $p \supset [p \supset (q \cdot \sim q)]$
6. $(p \supset p) \supset (q \cdot \sim q)$
7. $[p \supset (q \supset r)] \supset [(p \supset q) \supset (p \supset r)]$
8. $[p \supset (q \supset p)] \supset [(q \supset q) \supset \sim (r \supset r)]$
9. $\{[(p \supset q) \cdot (r \supset s)] \cdot (p \vee r)\} \supset (q \vee s)$
*10. $\{[(p \supset q) \cdot (r \supset s)] \cdot (q \vee s)\} \supset (p \vee r)$

C. Use truth tables to decide which of the following biconditionals are tautologies.

*1. $(p \supset q) \equiv (\sim q \supset \sim p)$
2. $(p \supset q) \equiv (\sim p \supset \sim q)$
3. $[(p \supset q) \supset r] \equiv [(q \supset p) \supset r]$
4. $[p \supset (q \supset r)] \equiv [q \supset (p \supset r)]$
*5. $p \equiv [p \cdot (p \vee q)]$
6. $p \equiv [p \vee (p \cdot q)]$
7. $p \equiv [p \cdot (p \supset q)]$
8. $p \equiv [p \cdot (q \supset p)]$
9. $p \equiv [p \vee (p \supset q)]$
*10. $(p \supset q) \equiv [(p \vee q) \equiv q]$
11. $p \equiv [p \vee (q \cdot \sim q)]$
12. $p \equiv [p \cdot (q \cdot \sim q)]$
13. $p \equiv [p \cdot (q \vee \sim q)]$
14. $p \equiv [p \vee (q \vee \sim q)]$
*15. $[p \cdot (q \vee r)] \equiv [(p \cdot q) \cdot (p \cdot r)]$
16. $[p \cdot (q \vee r)] \equiv [(p \vee q) \cdot (p \vee r)]$
17. $[p \vee (q \cdot r)] \equiv [(p \cdot q) \vee (p \cdot r)]$
18. $[p \vee (q \cdot r)] \equiv [(p \vee q) \cdot (p \vee r)]$
19. $[(p \cdot q) \supset r] \equiv [p \supset (q \supset r)]$
*20. $[(p \supset q) \cdot (q \supset p)] \equiv [(p \cdot q) \vee (\sim p \cdot \sim q)]$

8.9 Logical Equivalence

At this point we introduce a new relation, important and very useful, but not a connective, and somewhat more complicated than any of the truth-functional connectives just discussed.

Statements are materially equivalent when they have the same truth value. Because two materially equivalent statements are either both true, or both false, we can readily see that they must (materially) imply one another, because a false antecedent (materially) implies any statement, and a true consequent is (materially) implied by any statement. We may therefore read the three-bar sign, $\equiv$, as "if and only if."

However, statements that are merely materially equivalent most certainly cannot be substituted for one another. Knowing that they are materially equivalent, we know only that their truth values are the same. The statements, "Jupiter is larger than the Earth" and "Tokyo is the capital of Japan," are materially equivalent because they are both true, but we obviously cannot replace one with the other. Similarly, the statements, "All spiders are poisonous" and "No spiders are poisonous," are materially equivalent simply because they are both false, but they certainly cannot replace one another!

There are many circumstances, however, in which we must express the relationship that does permit mutual replacement. Two statements can be equivalent in a sense much stronger than that of material equivalence. They may be equivalent in the sense that any proposition that incorporates one of them could just as well incorporate the other. If there is no possible case in which one of these statements is true while the other is false, those statements are *logically equivalent.*

Of course, any two statements that are logically equivalent are materially equivalent as well, for they obviously have the same truth value. Indeed, if two statements are logically equivalent, they are materially equivalent under all circumstances—and this explains the short but powerful definition of **logical equivalence**: *Two statements are logically equivalent if the statement of their material equivalence is a tautology.* That is, the statement that they have the same truth value is itself necessarily true. This is why, to express this very strong logical relationship, we use the three-bar symbol with a small T immediately above it, $\stackrel{T}{\equiv}$, indicating that the logical relationship is of such a nature that the material equivalence of the two statements is a tautology. Because material equivalence is a biconditional (the two statements implying one another), we may think of this symbol of logical equivalence, $\stackrel{T}{\equiv}$, as expressing a tautological biconditional.

Some simple logical equivalences that are very commonly used will make this relation, and its great power, very clear. It is a commonplace that p and $\sim\sim p$ mean the same thing; "he is aware of that difficulty" and "he is not unaware of that difficulty" are two statements with the same content. In substance, either of these expressions may be replaced by the other because they both say the same thing. This principle of **double negation**, whose truth is obvious to all, may be exhibited in a truth table, where the material equivalence of two statement forms is shown to be a tautology:

p	$\sim p$	$\sim\sim p$	$p \stackrel{T}{\equiv} \sim\sim p$
T	F	T	T
F	T	F	T

This truth table proves that p and $\sim\sim p$ are *logically equivalent*. This very useful logical equivalence, double negation, is symbolized as

$$p \stackrel{T}{\equiv} \sim\sim p$$

The difference between *material equivalence* on the one hand and *logical equivalence* on the other hand is very great and very important. The former is a truth-functional connective, $\equiv$, which may be true or false depending only on the truth or falsity of the elements it connects. But the latter, logical equivalence, $\stackrel{T}{\equiv}$, is not a mere connective, and it expresses a relation between two statements that is not truth-functional. Two statements are logically equivalent only when it is

Logical equivalence
When referring to truth-functional compound propositions, the relationship that holds between two propositions when the statement of their material equivalence is a tautology. A very strong relation; statements that are logically equivalent must have the same meaning, and may therefore replace one another wherever they occur.

Double negation
An expression of the logical equivalence of any symbol and the negation of the negation of that symbol. Symbolized as $p \stackrel{T}{\equiv} \sim\sim p$.

absolutely impossible for them to have different truth values. However, if they *always* have the same truth value, logically equivalent statements may be substituted for one another in any truth-functional context without changing the truth value of that context. By contrast, two statements are materially equivalent if they merely *happen* to have the same truth value, even if there are no factual connections between them. Statements that are merely materially equivalent certainly may not be substituted for one another!

There are two well-known logical equivalences (that is, logically true biconditionals) of great importance because they express the interrelations among conjunction and disjunction, and their negations. Let us examine these two logical equivalences more closely.

First, what will serve to deny that a disjunction is true? Any disjunction $p \vee q$ asserts no more than that at least one of its two disjuncts is true. One cannot contradict it by asserting that at least one is false; one must (to deny it) assert that both disjuncts are false. Therefore, asserting the *negation of the disjunction* $(p \vee q)$ is logically equivalent to asserting the *conjunction of the negations of p and of q*. To show this in a truth table, we may formulate the biconditional, $\sim(p \vee q) \equiv (\sim p \cdot \sim q)$, place it at the top of its own column, and examine its truth value under all circumstances, that is, in each row.

p	q	p ∨ q	~(p ∨ q)	~p	~q	~p • ~q	~(p ∨ q) $\stackrel{T}{\equiv}$ (~p • ~q)
T	T	T	F	F	F	F	T
T	F	T	F	F	T	F	T
F	T	T	F	T	F	F	T
F	F	F	T	T	T	T	T

Of course we see that, whatever the truth values of p and of q, this biconditional must always be true. It is a tautology. Because the statement of that material equivalence is a tautology, we conclude that its two component statements are logically equivalent. We have proved that

$$\sim(p \vee q) \stackrel{T}{\equiv} (\sim p \cdot \sim q)$$

Similarly, asserting the conjunction of p and q asserts that both are true, so to contradict this assertion we need merely assert that at least one is false. Thus, asserting the negation of the conjunction $(p \cdot q)$ is logically equivalent to asserting the disjunction of the negations of p and of q. In symbols, the biconditional, $\sim(p \cdot q) \equiv (\sim p \vee \sim q)$ may be shown, in a truth table, to be a tautology. Such a table proves that

$$\sim(p \cdot q) \stackrel{T}{\equiv} (\sim p \vee \sim q)$$

These two tautologous biconditionals, or logical equivalences, are known as De Morgan's theorems, because they were formally stated by the mathematician and logician Augustus De Morgan (1806–1871). **De Morgan's theorems** can be formulated in English thus:

a. The negation of the disjunction of two statements is logically equivalent to the conjunction of the negations of the two statements;

De Morgan's theorems
Two expressions of logical equivalence. The first states that the negation of a disjunction is logically equivalent to the conjunction of the negations of its disjuncts: $\sim(p \vee q) \stackrel{T}{\equiv} (\sim p \cdot \sim q)$. The second states that the negation of a conjunction is logically equivalent to the disjunction of the negations of its conjuncts: $\sim(p \cdot q) \stackrel{T}{\equiv} (\sim p \vee \sim q)$.

and

b. The negation of the conjunction of two statements is logically equivalent to the disjunction of the negations of the two statements.

These theorems of De Morgan are exceedingly useful.

Another important logical equivalence is very helpful when we seek to manipulate truth-functional connectives. Material implication, $\supset$, was defined (in Section 8.3) as an abbreviated way of saying $\sim(p \cdot \sim q)$. That is, "*p* materially implies *q*" simply means, by definition, that it is not the case that *p* is true while *q* is false. In this definition we see that the *definiens*, $\sim(p \cdot \sim q)$, is the denial of a conjunction. And by De Morgan's theorem we know that any such denial is logically equivalent to the disjunction of the denials of the conjuncts; that is, we know that $\sim(p \cdot \sim q)$ is logically equivalent to $(\sim p \vee \sim\sim q)$; and this expression in turn, applying the principle of double negation, is logically equivalent to $\sim p \vee q$. Logically equivalent expressions mean the same thing, and therefore the original *definiens* of the horseshoe, $\sim(p \cdot \sim q)$, may be replaced with no change of meaning by the simpler expression $\sim p \vee q$. This gives us a very useful definition of material implication: $p \supset q$ is logically equivalent to $\sim p \vee q$. In symbols we write:

$$(p \supset q) \stackrel{T}{\equiv} (\sim p \vee q)$$

This definition of material implication is widely relied on in the formulation of logical statements and the analysis of arguments. Manipulation is often essential, and manipulation is more efficient when the statements we are working with have the same central connective. With the simple definition of the horseshoe we have just established, $(p \supset q) \stackrel{T}{\equiv} (\sim p \vee q)$, statements in which the horseshoe is the connective can be conveniently replaced by statements in which the wedge is the connective; and likewise, statements in disjunctive form may be readily replaced by statements in implicative form. When we seek to present a formal proof of the validity of deductive arguments, replacements of this kind are very useful indeed.

Before going on to the methods of testing for validity and invalidity in the next section, it is worthwhile to pause for a more thorough consideration of the meaning of material implication. Implication is central in argument but, as we noted earlier, the word "implies" is highly ambiguous. *Material* implication, on which we rely in this analysis, is only one sense of that word, although it is a very important sense, of course. The definition of material implication explained just above makes it clear that when we say, in this important sense, that "*p* implies *q*," we are saying no more than that "either *q* is true or *p* is false."

Asserting the "if–then" relation in this sense has consequences that may seem paradoxical. For in this sense we can say, *correctly*, "If a statement is true, then it is implied by any statement whatever." Because it is true that the earth is round, it follows that "The moon is made of green cheese implies that the earth is round." This appears to be very curious, especially because it also follows that "The moon is not made of green cheese implies that the earth is round." Our precise understanding of material implication also entitles us to say, *correctly*, "If a statement is false, then it implies any statement whatever." Because it is false that the moon is made of green cheese, it follows that "The moon is made of green cheese implies that the earth is round," and this is the more curious when we realize that it also follows that "The moon is made of green cheese implies that the earth is *not* round."

Why do these true statements seem so curious? It is because we recognize that the shape of the earth and the cheesiness of the moon are utterly irrelevant to each other. As we normally use the word "implies," a statement cannot imply some other statement, false or true, to which it is utterly irrelevant. That is the case when "implies" is used in most of its everyday senses. And yet those "paradoxical" statements in the preceding paragraph are indeed true, and not really problematic at all, because they use the word "implies" in the logical sense of "material implication." The precise meaning of material implication we have made very clear; we understand that to say p materially implies q is only to say that either p is false or q is true.

What needs to be borne in mind is this: *Meaning*—subject matter—is strictly irrelevant to material implication. *Material implication is a truth function.* Only the truth and falsity of the antecedent and the consequent, not their content, are relevant here. There is nothing paradoxical in stating that any disjunction is true that contains one true disjunct. Well, when we say that "The moon is made of green cheese (materially) implies that the earth is round," we know that to be logically equivalent to saying "Either the moon is not made of green cheese or the earth is round"—a disjunction that is most certainly true. And any disjunction we may confront in which "The moon is not made of green cheese" is the first disjunct will certainly be true, no matter what the second disjunct asserts. So, yes, "The moon is made of green cheese (materially) implies that the earth is square" because that is logically equivalent to "The moon is not made of green cheese or the earth is square." A false statement materially implies any statement whatever. A true statement is materially implied by any statement whatever.

Every occurrence of "if–then" should be treated, we have said, as a material implication, and represented with the horseshoe, $\supset$. The justification of this practice, its logical expediency, is the fact that doing so preserves the validity of all valid arguments of the type with which we are concerned in this part of our logical studies. Other symbolizations have been proposed, adequate to other types of implication, but they belong to more advanced parts of logic, beyond the scope of this book.

8.10 The Three "Laws of Thought"

Some early thinkers, after having defined logic as "the science of the laws of thought," went on to assert that there are exactly three *basic* laws of thought, laws so fundamental that obedience to them is both the necessary and the sufficient condition of correct thinking. These three have traditionally been called:

- The **principle of identity**. This principle asserts that *if any statement is true, then it is true*. Using our notation we may rephrase it by saying that the principle of identity asserts that every statement of the form $p \supset p$ must be true, that every such statement is a tautology.

Principle of identity
The principle that asserts that if any statement is true then it is true.

- The **principle of noncontradiction**. This principle asserts that *no statement can be both true and false*. Using our notation we may rephrase it by saying that the principle of noncontradiction asserts that every statement of the form $p \cdot \sim p$ must be false, that every such statement is self-contradictory.
- The **principle of excluded middle**. This principle asserts that *every statement is either true or false*. Using our notation we may rephrase it by saying that the principle of excluded middle asserts that every statement of the form $p \vee \sim p$ must be true, that every such statement is a tautology.

It is obvious that these three principles are indeed true—logically true—but the claim that they deserve privileged status as the most fundamental laws of thought is doubtful. The first (identity) and the third (excluded middle) are tautologies, but there are many other tautologous forms whose truth is equally certain. The second (noncontradiction) is by no means the only self-contradictory form of statement.

We do use these principles in completing truth tables. In the initial columns of each row of a table we place either a **T** or an **F**, being guided by the principle of excluded middle. Nowhere do we put both **T** and **F**, being guided by the principle of noncontradiction. Once having put a **T** under a symbol in a given row, being guided by the principle of identity, when we encounter that symbol in other columns of that row, we regard it as still being assigned a **T**. So we could regard the three laws of thought as principles governing the construction of truth tables.

Nevertheless, in regarding the entire system of deductive logic, these three principles are no more important or fruitful than many others. Indeed, there are tautologies that are more fruitful than they for purposes of deduction, and in that sense more important than these three, such as De Morgan's theorems, which are more applicable in a system of natural deduction than these more abstract principles. Nonetheless, these principles are useful in guiding informal argumentation, in which axiomatic deductive systems seldom obtain. A more extended treatment of this point lies beyond the scope of this book.[6]

Some thinkers, believing themselves to have devised a new and different logic, have claimed that these three principles are in fact not true, and that obedience to them has been needlessly confining. But these criticisms have been based on misunderstandings.

The principle of identity has been attacked on the ground that things change, and are always changing. Thus, for example, statements that were true of the United States when it consisted of the thirteen original states are no longer true of the United States today, which has fifty states. But this does not undermine the principle of identity. The sentence, "There are only thirteen states in the United States," is incomplete, an elliptical formulation of the statement that "There were only thirteen states in the United States *in 1790*"—and that statement is as true today as it was in 1790. When we confine our attention to complete, nonelliptical formulations of propositions, we see that their truth (or falsity) does not change over time. The principle of identity is true, and it does not interfere with our recognition of continuing change.

Principle of noncontradiction
The principle that asserts that no statement can be both true and false.

Principle of excluded middle
The principle that asserts that any statement is either true or false.

The principle of noncontradiction has been attacked by Hegelians and Marxists on the grounds that genuine contradiction is everywhere pervasive, that the world is replete with the inevitable conflict of contradictory forces. That there are conflicting forces in the real word is true, of course—but to call these conflicting forces "contradictory" is a loose and misleading use of that term. Labor unions and the private owners of industrial plants may indeed find themselves in conflict—but neither the owner nor the union is the "negation" or the "denial" or the "contradictory" of the other. The principle of noncontradiction, understood in the straightforward sense in which it is intended by logicians, is unobjectionable and perfectly true.

The principle of excluded middle has been the object of much criticism, because it leads to a "two-valued orientation," which implies that things in the world must be either "white or black," and which thereby hinders the realization of compromise and less-than-absolute gradations. This objection also arises from misunderstanding. Of course the statement "This is black" cannot be jointly true with the statement "This is white"—where "this" refers to exactly the same thing. However, although these two statements cannot both be true, they can both be false. "This" may be neither black nor white; the two statements are *contraries*, not contradictories. The contradictory of the statement "This is white" is the statement "It is not the case that this is white" and (if "white" is used in precisely the same sense in both of these statements) one of them must be true and the other false. The principle of excluded middle is inescapable.

All three of these "laws of thought" are unobjectionable—so long as they are applied to statements containing unambiguous, nonelliptical, and precise terms. Plato appealed explicitly to the principle of noncontradiction in Book IV of his *Republic* (at numbers 436 and 439); Aristotle discussed all three of these principles in Books IV and XI of his *Metaphysics*. Of the principle of noncontradiction, Aristotle wrote: "That the same attribute cannot at the same time belong and not belong to the same subject and in the same respect" is a principle "which everyone must have who understands anything that is," and which "everyone must already have when he comes to a special study." It is, he concluded, "the most certain of all principles." The "laws of thought" may not deserve the honorific status assigned to them by some philosophers, but they are indubitably true.

chapter 8 Summary

This chapter has presented the fundamental concepts of modern symbolic logic.

In Section 8.1, we explained the general approach of modern symbolic logic and its need for an artificial symbolic language.

In Section 8.2, we introduced and defined the symbols for negation (the curl: ~); and for the truth-functional connectives of conjunction (the dot: •) and disjunction (the wedge: ∨). We also explained logical punctuation.

In Section 8.3, we discussed the different senses of implication and defined the truth-functional connective material implication (the horseshoe: ⊃).

In Section 8.4, we explained the formal structure of arguments, defined argument forms, and explained other concepts essential in analyzing deductive arguments.

In Section 8.5, we gave a precise account of valid and invalid argument forms.

In Section 8.6, we explained the truth-table method of testing the validity of argument forms.

In Section 8.7, we identified and described a few very common argument forms, some valid and some invalid.

In Section 8.8, we explained the formal structure of statements and defined essential terms for dealing with statement forms. We introduced tautologous, contradictory, and contingent statement forms, and defined a fourth truth-functional connective, material equivalence (three bars: ≡).

In Section 8.9, we introduced and defined a powerful new relation, logical equivalence, using the symbol $\stackrel{T}{\equiv}$. We explained why statements that are logically equivalent may be substituted for one another, while statements that are merely materially equivalent cannot replace one another. We introduced several logical equivalences of special importance: De Morgan's theorems, the principle of double negation, and the definition of material implication.

In Section 8.10, we discussed certain logical equivalences that have been thought by many to be fundamental in all reasoning: the principle of identity, the principle of noncontradiction, and the principle of excluded middle.

END NOTES

[1] Somewhat more complicated definitions have been proposed by David H. Sanford in "What Is a Truth Functional Component?" *Logique et Analyse* 14 (1970): 483–486.

[2] "The Firm," *The New Yorker*, 8 March 1999.

[3] Peter J. Bertocci, "Chavez' Plight Must Come from Arrogance," *The New York Times*, 19 January 2001.

[4] Rabbi Ammiel Hirsch, "Grand Canyon," *The New York Times*, 10 October 2005.

[5] Orlando Patterson, "The Speech Misheard Round the World," *The New York Times*, 22 January 2005.

[6] For further discussion of these matters, the interested reader can consult I. M. Copi and J. A. Gould, editors, *Readings on Logic*, 2nd edition (New York: Macmillan, 1972), part 2; and I. M. Copi and J. A. Gould, editors, *Contemporary Philosophical Logic* (New York: St. Martin's Press, 1978), part 8.

PEARSON mylogiclab

For additional exercises and tutorials about concepts covered in this chapter, log in to MyLogicLab at *www.mylogiclab.com* and select your current textbook.

Methods of Deduction

- **9.1** Formal Proof of Validity
- **9.2** The Elementary Valid Argument Forms
- **9.3** Formal Proofs of Validity Exhibited
- **9.4** Constructing Formal Proofs of Validity
- **9.5** Constructing More Extended Formal Proofs
- **9.6** Expanding the Rules of Inference: Replacement Rules
- **9.7** The System of Natural Deduction
- **9.8** Constructing Formal Proofs Using the Nineteen Rules of Inference
- **9.9** Proof of Invalidity
- **9.10** Inconsistency
- **9.11** Indirect Proof of Validity
- **9.12** Shorter Truth-Table Technique

9.1 Formal Proof of Validity

In theory, truth tables are adequate to test the validity of any argument of the general type we have considered. In practice, however, they become unwieldy as the number of component statements increases. A more efficient method of establishing the validity of an extended argument is to deduce its conclusion from its premises by a sequence of elementary arguments, each of which is known to be valid. This technique accords fairly well with ordinary methods of argumentation.

Consider, for example, the following argument:

If Anderson was nominated, then she went to Boston.
If she went to Boston, then she campaigned there.
If she campaigned there, she met Douglas.
Anderson did not meet Douglas.
Either Anderson was nominated or someone more eligible was selected.
Therefore someone more eligible was selected.

The validity of this argument may be intuitively obvious, but let us consider the matter of proof. The discussion will be facilitated by translating the argument into symbolism as

$A \supset B$
$B \supset C$
$C \supset D$
$\sim D$
$A \lor E$
$\therefore E$

To establish the validity of this argument by means of a truth table requires a table with thirty-two rows, because five different simple statements are involved. Instead, we can prove the argument valid by deducing its conclusion using a

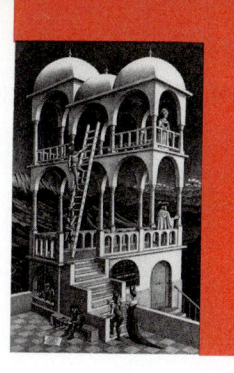

338   CHAPTER 9 Methods of Deduction

sequence of just four elementary valid arguments. From the first two premises, $A \supset B$ and $B \supset C$, we validly infer that $A \supset C$ using a Hypothetical Syllogism. From $A \supset C$ and the third premise, $C \supset D$, we validly infer that $A \supset D$ as another Hypothetical Syllogism. From $A \supset D$ and the fourth premise, $\sim D$, we validly infer that $\sim A$ by *Modus Tollens*. From $\sim A$ and the fifth premise, $A \vee E$, as a Disjunctive Syllogism we validly infer E, the conclusion of the original argument. That the conclusion can be deduced from the five premises of the original argument by four elementary valid arguments proves the original argument to be valid. Here the elementary valid argument forms Hypothetical Syllogism (H.S.), *Modus Tollens* (M.T.), and Disjunctive Syllogism (D.S.) are used as **rules of inference** whose application allows conclusions to be validly inferred or deduced from premises.

This method of deriving the conclusion of a deductive argument—using rules of inference successively to prove the validity of the argument—is as reliable as the truth-table method discussed in Chapter 8, if the rules are used with meticulous care. However, it improves on the truth-table method in two ways: It is vastly more efficient, as has just been shown; and it enables us to follow the flow of the reasoning process from the premises to the conclusion and is therefore much more intuitive and more illuminating. The method is often called **natural deduction**. Using natural deduction, we can provide a *formal proof* of the validity of any argument that is valid.

A *formal proof of validity* is given by writing the premises and the statements that we deduce from them in a single column, and setting off in another column, to the right of each such statement, its "justification," or the reason we give for including it in the proof. It is convenient to list all the premises first and to write the conclusion either on a separate line, or slightly to one side and separated by a diagonal line from the premises. If all the statements in the column are numbered, the "justification" for each statement consists of the numbers of the preceding statements from which it is inferred, together with the abbreviation for the rule of inference by which it follows from them. The formal proof of the example argument is written as

> 1. $A \supset B$
> 2. $B \supset C$
> 3. $C \supset D$
> 4. $\sim D$
> 5. $A \vee E$
> $\therefore E$
> 6. $A \supset C$ 1, 2, H.S.
> 7. $A \supset D$ 6, 3, H.S.
> 8. $\sim A$ 7, 4, M.T.
> 9. E 5, 8, D.S.

Rules of inference
The rules that permit valid inferences from statements assumed as premises. Twenty-three rules of inference are set forth in this book: nine elementary valid argument forms, ten logical equivalences whose members may replace one another, and four rules governing instantiation and generalization in quantified logic.

Natural deduction
A method of proving the validity of a deductive argument by using the rules of inference.

Formal proof of validity
A sequence of statements each of which is either a premise of a given argument, or follows from the preceding statements of the sequence by one of the rules of inference, or by logical equivalence, where the last statement in the sequence is the conclusion of the argument whose validity is proved.

Elementary valid argument
Any one of a set of specified deductive arguments that serve as rules of inference and that may therefore be used in constructing a formal proof of validity.

We define a **formal proof of validity** of a given argument as a *sequence of statements, each of which either is a premise of that argument or follows from preceding statements of the sequence by an elementary valid argument or by a logical equivalence, such that the last statement in the sequence is the conclusion of the argument whose validity is being proved.*

We define an **elementary valid argument** as *any argument that is a substitution instance of an elementary valid argument form*. Note that *any* substitution instance of an elementary valid argument form is an elementary valid argument. Thus the argument

$(A \cdot B) \supset [C \equiv (D \vee E)]$
$A \cdot B$
$\therefore C \equiv (D \vee E)$

is an elementary valid argument because it is a substitution instance of the elementary valid argument form *Modus Ponens* (M.P.). It results from

$p \supset q$
p
$\therefore q$

by substituting $A \cdot B$ for p and $C \equiv (D \vee E)$ for q, and it is therefore of that form even though *Modus Ponens* is not the *specific form* of the given argument.

Modus Ponens is a very elementary valid argument form indeed, but what *other* valid argument forms are considered to be rules of inference? We begin with a list of just nine rules of inference that can be used in constructing formal proofs of validity. With their aid, formal proofs of validity can be constructed for a wide range of more complicated arguments. The names provided are for the most part standard, and the use of their abbreviations permits formal proofs to be set down with a minimum of writing.

Biography

Kurt Gödel

The Institute for Advanced Study was founded in Princeton, NJ, in 1930. Two of its first members were Albert Einstein and John von Neumann; a third early member was the powerful Austrian logician, Kurt Gödel (1906–1978), who by 1931, at the age of 25, had published his two "Incompleteness Theorems." He became the *enfant terrible* in the world of formal logic.

To understand Gödel's impact in the logical world one must bear in mind the great project that for decades had been the program of modern logicians: to prove that all mathematics is founded upon logic and can be derived from a few basic logical axioms. Russell and Whitehead had sought to culminate this undertaking with *Principia Mathematica* (1910–13). Success would require that the logical system devised be both *consistent* and *complete*. But Gödel demonstrated, in a paper entitled "On Formally Undecidable Propositions of *Principia Mathematica* and Related Systems," that any axiomatic system powerful enough to describe the arithmetic of the natural numbers, if it is internally consistent, *cannot* be complete. Moreover (as von Neumann had also seen), the consistency of the axioms cannot be established within the system itself. If, in any such axiomatized system, there must always be at least one true but unprovable statement, the search for some set of logical axioms that would be sufficient to ground all of mathematics was doomed. Proving this, Kurt Gödel became one of the most respected logicians of the twentieth century.

Gödel was born in the city of Brno, in what was then Austria-Hungary. As a boy his insatiable curiosity led to his nickname, *Herr Warum*—"Mr. Why." Excelling both in languages and in mathematics in his early schooling, he moved to Vienna at the age of 18, where he associated closely with a number of Jewish philosophers and mathematicians. In 1933, after the publication of his incompleteness papers, Gödel visited the U.S. for the first time; he lectured at the Institute for Advanced Study and befriended Albert Einstein there. That same year Hitler came to power in Germany. When Austria was absorbed by Nazi Germany in 1938, Gödel's circumstances in Vienna were fraught with danger; he planned a circuitous escape to America by way of the Trans-Siberian railway and Japan. When the Second World War began, in September of 1939, Gödel fled. Before the end of that year he was safe in Princeton, where he continued his work as a distinguished member of the faculty at the Institute for Advanced Study. Albert Einstein, who had become his regular companion, confided to friends that he continued to visit the Institute chiefly "to have the privilege of walking home with Gödel."

Kurt Gödel was one of the many superb scholars—physicists, philosophers, logicians, mathematicians, literary figures, and thinkers of every sort—who enriched American intellectual life as a consequence of Hitler's obsessive determination to kill or expel all European Jews. The horrors of Nazi oppression proved a warped blessing to the United States. Gödel proudly became an American citizen in 1947. He made a close study of the Constitution of the United States, and here he remained until his death, in Princeton, in 1978. ∎

9.2 The Elementary Valid Argument Forms

Our object is to build a set of logical rules—rules of inference—with which we can prove the validity of deductive arguments if they are valid. We began with a few elementary valid argument forms that have already been introduced—*Modus Ponens*, for example, and Disjunctive Syllogism. These are indeed simple and common, but we need a set of rules that is more powerful. The rules of inference may be thought of as a logical toolbox, from which the tools may be taken, as needed, to prove validity. What else is needed for our toolbox? How shall we expand the list of rules of inference?

The needed rules of inference consist of two sets, each set containing rules of a different kind. The first is a set of elementary valid argument forms. The second set consists of a small group of elementary logical equivalences. In this section we discuss only the elementary valid argument forms.

To this point we have become acquainted with four elementary valid argument forms:

1. *Modus Ponens* (M.P.) $p \supset q$
p
$\therefore q$

2. *Modus Tollens* (M.T.) $p \supset q$
$\sim q$
$\therefore \sim p$

3. *Hypothetical Syllogism* (H.S.) $p \supset q$
$q \supset r$
$\therefore p \supset r$

4. *Disjunctive Syllogism* (D.S.) $p \vee q$
$\sim p$
$\therefore q$

For an effective logical toolbox we need to add five more. Let us examine these additional argument forms—each of which is valid and can be readily proved valid using a truth table.

> **5.** Rule 5 is called **Constructive Dilemma** (C.D.). It is symbolized as
>
> $(p \supset q) \cdot (r \supset s)$
> $p \lor r$
> $\therefore q \lor s$

A dilemma is an argument in which one of two alternatives must be chosen. In this argument form the alternatives are the antecedents of the two conditional propositions $p \supset q$ and $r \supset s$. We know from *Modus Ponens* that if we are given $p \supset q$ and p, we may infer q; and if we are given $r \supset s$ and r, we may infer s. Therefore if we are given both $p \supset q$, and $r \supset s$, and either p or r (that is, either of the antecedents), we may infer validly either q or s (that is, one or the other of the consequents). Constructive Dilemma is, in effect, a combination of two arguments in *Modus Ponens* form, and it is most certainly valid, as a truth table can make evident. We add Constructive Dilemma (C.D.) to our tool box.

> **6. Absorption** (Abs.)
>
> $p \supset q$
> $\therefore p \supset (p \cdot q)$

Any proposition p always implies itself, of course. Therefore, if we know that $p \supset q$, we may validly infer that p implies both itself and q. That is all that Absorption says. Why (one may ask) do we need so elementary a rule? The need for it will become clearer as we go on; in short, we need it because it will be very convenient, even essential at times, to carry the p across the horseshoe. In effect, Absorption makes the principle of identity, one of the basic logical principles discussed in Section 8.10, always available for our use. We add Absorption (Abs.) to our logical toolbox.

The next two elementary valid argument forms are intuitively very easy to grasp if we understand the logical connectives explained earlier.

> **7. Simplification** (Simp.)
>
> $p \cdot q$
> $\therefore p$

Simplification says only that if two propositions, p and q, are true when they are conjoined ($p \cdot q$), we may validly infer that one of them, p, is true by itself. We simplify the expression before us; we "pull" p from the conjunction and stand it on its own. Because we are given that $p \cdot q$, we know that both p and q must be true; we may therefore know with certainty that p is true.

What about q? Isn't q true for exactly the same reason? Yes, it is. Then why does the elementary argument form, Simplification, conclude only that p is true? The reason is that we want to keep our toolbox uncluttered. The rules of inference

Constructive dilemma (C.D.)
A rule of inference; one of nine elementary valid argument forms. Constructive dilemma permits the inference that if $(p \supset q) \cdot (r \supset s)$ is true, and $p \lor r$ is also true, then $q \lor s$ must be true.

must always be applied *exactly* as they appear. We surely need a rule that will enable us to take conjunctions apart, but we do not need two such rules; one will suffice. When we may need to "pull" some *q* from a conjunction we will be able to put it where *p* is now, and then use only the one rule, Simplification, which we add to our toolbox.

> **8. Conjunction** (Conj.)
>
> p
> q
> $\therefore p \cdot q$

Conjunction says only that if two propositions, *p* and *q*, are known to be true, we can put them together into one conjunctive expression, $p \cdot q$. We may conjoin them. If they are true separately, they must also be true when they are conjoined. In this case the order presents no problem, because we may always treat the one we seek to put on the left as *p*, and the other as *q*. That *joint* truth is what a conjunction asserts. We add Conjunction (Conj.) to our logical toolbox.

The last of the nine elementary valid argument forms is also a straightforward consequence of the meaning of the logical connectives—in this case, disjunction.

> **9. Addition** (Add.)
>
> p
> $\therefore p \vee q$

Any disjunction must be true if either of its disjuncts is true. That is, $p \vee q$ is true if *p* is true, or if *q* is true, or if they are both true. That is what disjunction means. It obviously follows from this that if we know that some proposition, *p*, is true, we also know that either it is true or some other—any other!—proposition is true. So we can construct a disjunction, $p \vee q$, using the one proposition known to be true as *p*, and adding to it (in the logical, disjunctive sense) any proposition we care to. We call this logical addition. The additional proposition, *q*, is not conjoined to *p*; it is used with *p* to build a disjunction that we may know with certainty to be true because one of the disjuncts, *p*, is known to be true. The disjunction we thus build will be true *no matter what that added proposition asserts*—no matter how absurd or wildly false it may be! We know that Michigan is north of Florida. Therefore we know that either Michigan is north of Florida *or* the moon is made of green cheese! Indeed, we know that either Michigan is north of Florida *or* 2 + 2 = 5. The truth or falsity of the added proposition does not affect the truth of the disjunction we build, because that disjunction is made certainly true by the truth of the disjunct with which we began. Therefore, if we are given *p* as true, we may validly infer *for any q whatever* that $p \vee q$. This principle, Addition (Add.), we add to our logical toolbox.

Our set of nine elementary valid argument forms is now complete.

All nine of these argument forms are very plainly valid. Any one of them whose validity we may doubt can be readily proved to be valid using a truth table. Each of them is simple and intuitively clear; as a set we will find them powerful as we go on to construct formal proofs for the validity of more extended arguments.

overview

Rules of Inference: Elementary Valid Argument Forms

Name	Abbreviation	Form
1. *Modus Ponens*	M.P.	$p \supset q$ p $\therefore q$
2. *Modus Tollens*	M.T.	$p \supset q$ $\sim q$ $\therefore \sim p$
3. Hypothetical Syllogism	H.S.	$p \supset q$ $q \supset r$ $\therefore p \supset r$
4. Disjunctive Syllogism	D.S.	$p \vee q$ $\sim p$ $\therefore q$
5. Constructive Dilemma	C.D.	$(p \supset q) \cdot (r \supset s)$ $p \vee r$ $\therefore q \vee s$
6. Absorption	Abs.	$p \supset q$ $\therefore p \supset (p \cdot q)$
7. Simplification	Simp.	$p \cdot q$ $\therefore p$
8. Conjunction	Conj.	p q $\therefore p \cdot q$
9. Addition	Add.	p $\therefore p \vee q$

Two features of these elementary argument forms must be emphasized. First, *they must be applied with exactitude*. An argument that one proves valid using *Modus Ponens* must have that exact form: $p \supset q$, p, therefore q. Each statement variable must be replaced by some statement (simple or compound) consistently and accurately. Thus, for example, if we are given $(C \vee D) \supset (J \vee K)$ and $(C \vee D)$, we may infer $(J \vee K)$ by *Modus Ponens*. But we may not infer $(K \vee J)$ by *Modus Ponens*, even though it may be true. The elementary argument form must be fitted *precisely* to the argument with which we are working. No shortcut—no fudging of any kind—is permitted, because we seek to know with certainty that the outcome of our reasoning is valid, and that can be known only if we can demonstrate that *every link in the chain of our reasoning is absolutely solid*.

Second, these elementary valid argument forms must be applied to *the entire lines* of the larger argument with which we are working. Thus, for example, if we are given [(X • Y) ⊃ Z] • T, we cannot validly infer X by Simplification. X is one of the conjuncts of a conjunction, but that conjunction is part of a more complex expression. X may not be true even if that more complex expression is true. We may only infer that if X and Y are both true, then Z is true. Simplification applies only to the entire line, which must be a conjunction; its conclusion is the left side (and only the left side) of that conjunction. So, from this same line, [(X • Y) ⊃ Z] • T, we may validly infer (X • Y) ⊃ Z by Simplification. But we may not infer T by Simplification, even though it may be true.

Formal proofs in deductive logic have crushing power, but they possess that power only because, when they are correct, there can be not the slightest doubt of the validity of each inference drawn. The tiniest gap destroys the power of the whole.

The nine elementary valid argument forms we have given should be committed to memory. They must be always readily in mind as we go on to construct formal proofs. Only if we comprehend these elementary argument forms fully, and can apply them immediately and accurately, may we expect to succeed in devising formal proofs of the validity of more extended arguments.

EXERCISES

Here follows a set of twenty elementary valid arguments. They are valid because each of them is exactly in the form of one of the nine elementary valid argument forms. For each of them, state the rule of inference by which its conclusion follows from its premise or premises.

EXAMPLE

1. (A • B) ⊃ C
 ∴ (A • B) ⊃ [(A • B) • C]

SOLUTION

Absorption. If (A • B) replaces *p*, and C replaces *q*, this argument is seen to be exactly in the form *p* ⊃ *q*, therefore *p* ⊃ (*p* • *q*).

*1. (A • B) ⊃ C
 ∴ (A • B) ⊃ [(A • B) • C]

2. (D ∨ E) • (F ∨ G)
 ∴ D ∨ E

3. H ⊃ I
 ∴ (H ⊃ I) ∨ (H ⊃ ~I)

4. ~(J • K) • (L ⊃ ~M)
 ∴ ~(J • K)

*5. [N ⊃ (O • P)] • [Q ⊃ (O • R)]
 N ∨ Q
 ∴ (O • P) ∨ (O • R)

6. (X ∨ Y) ⊃ ~(Z • ~A)
 ~~(Z • ~A)
 ∴ ~(X ∨ Y)

7. $(S \equiv T) \vee [(U \cdot V) \vee (U \cdot W)]$
   $\sim(S \equiv T)$
   $\therefore (U \cdot V) \vee (U \cdot W)$

8. $\sim(B \cdot C) \supset (D \vee E)$
   $\sim(B \cdot C)$
   $\therefore D \vee E$

9. $(F \equiv G) \supset \sim(G \cdot \sim F)$
   $\sim(G \cdot \sim F) \supset (G \supset F)$
   $\therefore (F \equiv G) \supset (G \supset F)$

*10. $(I \equiv H) \supset \sim(H \cdot \sim I)$
   $\sim(H \cdot \sim I) \supset (H \supset I)$
   $\therefore (I \equiv H) \supset (H \supset I)$

11. $(A \supset B) \supset (C \vee D)$
    $A \supset B$
    $\therefore C \vee D$

12. $[E \supset (F \equiv \sim G)] \vee (C \vee D)$
    $\sim[E \supset (F \equiv \sim G)]$
    $\therefore C \vee D$

13. $(C \vee D) \supset [(J \vee K) \supset (J \cdot K)]$
    $\sim[(J \vee K) \supset (J \cdot K)]$
    $\therefore \sim(C \vee D)$

14. $\sim[L \supset (M \supset N)] \supset \sim(C \vee D)$
    $\sim[L \supset (M \supset N)]$
    $\therefore \sim(C \vee D)$

*15. $(J \supset K) \cdot (K \supset L)$
    $L \supset M$
    $\therefore [(J \supset K) \cdot (K \supset L)] \cdot (L \supset M)$

16. $Q \supset (O \vee R)$
    $N \supset (O \vee P)$
    $\therefore [Q \supset (O \vee R)] \cdot [N \supset (O \vee P)]$

17. $(S \supset T) \supset (U \supset V)$
    $\therefore (S \supset T) \supset [(S \supset T) \cdot (U \supset V)]$

18. $(W \cdot \sim X) \equiv (Y \supset Z)$
    $\therefore [(W \cdot \sim X) \equiv (Y \supset Z)] \vee (X \equiv \sim Z)$

19. $[(H \cdot \sim I) \supset C] \cdot [(I \cdot \sim H) \supset D]$
    $(H \cdot \sim I) \vee (I \cdot \sim H)$
    $\therefore C \vee D$

*20. $(C \vee D) \supset [(O \supset P) \supset Q]$
    $[(O \supset P) \supset Q] \supset \sim(C \vee D)$
    $\therefore (C \vee D) \supset \sim(C \vee D)$

9.3 Formal Proofs of Validity Exhibited

We have defined a formal proof of validity for a given argument as a sequence of statements, each of which either is a premise of that argument or follows from preceding statements of the sequence by an elementary valid argument or by a logical equivalence, such that the last statement in the sequence is the conclusion of the argument whose validity is being proved. Our task will be to build such sequences, to prove the validity of arguments with which we are confronted.

Doing this can be a challenge. Before attempting to construct such sequences, it will be helpful to become familiar with the look and character of formal proofs. In this section we examine a number of complete formal proofs, to see how they work and to get a "feel" for constructing them.

Our first step is not to devise such proofs, but to understand and appreciate them. A sequence of statements is put before us in each case. Every statement in that sequence will either be a premise or follow from preceding statements in the sequence using one of the elementary valid argument forms—just as in the illustration that was presented in Section 9.1. When we confront such a proof, but the rule of inference that justifies each step in the proof is not given, we know (having been told that these are completed proofs) that every line in the proof that is not itself a premise can be deduced from the preceding lines. To understand those deductions, the nine elementary valid argument forms must be kept in mind.

Let us look at some proofs that exhibit this admirable solidity. Our first example is Exercise 1 in the set of exercises on pages 347–348.

EXAMPLE 1

1. $A \cdot B$
2. $(A \lor C) \supset D$
   $\therefore A \cdot D$
3. A
4. $A \lor C$
5. D
6. $A \cdot D$

The first two lines of this proof are seen to be premises, because they appear before the "therefore" symbol ($\therefore$); what appears immediately to the right of that symbol is the conclusion of this argument, $A \cdot D$. The very last line of the sequence is (as it must be if the formal proof is correct) that same conclusion, $A \cdot D$. What about the steps between the premises and the conclusion? Line 3, A, we can deduce from line 1, $A \cdot B$, by Simplification. So to the right of line 3, we put, the line number from which it comes and the rule by which it is inferred from that line, "1, Simp." Line 4 is $A \lor C$. How can that be inferred from the lines above it? We can*not* infer it from line 2 by Simplification. But we can infer it from line 3, A, by Addition. Addition tells us that if p is true, then $p \lor q$ is true, whatever q may be. Using that logical pattern precisely, we may infer from A that $A \lor C$ is true. To the right of line 4 we therefore put "3, Add." Line 5 is D. D appears in line 2 as the consequent of a conditional statement, $(A \lor C) \supset D$. We proved on line 4 that $A \lor C$ is true; now, using *Modus Ponens*, we combine this with the conditional on line 2 to prove D. To the right of line 5 we therefore write "2, 4, M.P." A has been proved true (on line 3) and D has been proved true (on line 5). We may therefore validly conjoin them, which is what line 6 asserts: $A \cdot D$. To the right of line 6 we therefore write "3, 5, Conj." This line, $A \cdot D$, is the conclusion of the argument, and it is therefore the last statement in the sequence of statements that constitutes this proof. The proof, which had been presented to us complete, has thus been "fleshed out" by specifying the justification of each step within it.

In this example, and the exercises that follow, every line of each proof can be justified by using one of the elementary valid argument forms in our logical toolbox. No other inferences of any kind are permitted, however plausible they may seem. When we had occasion to refer to an argument form that has two premises (e.g., M.P. or D.S.), we indicated first, in the justification, the numbers of the lines used, *in the order in which they appear in the elementary valid form*. Thus, line 5 in Example 1 is justified by 2, 4, M.P.

To become proficient in the construction of formal proofs, we must become fully familiar with the shape and rhythm of the nine elementary argument forms—the first nine of the rules of inference that we will be using extensively.

EXERCISES

Each of the following exercises presents a flawless formal proof of validity for the indicated argument. For each, state the justification for each numbered line that is not a premise.

1. 1. $A \cdot B$
    2. $(A \lor C) \supset D$
       $\therefore A \cdot D$
    3. A
    4. $A \lor C$
    5. D
    6. $A \cdot D$

2. 1. $(E \lor F) \cdot (G \lor H)$
    2. $(E \supset G) \cdot (F \supset H)$
    3. $\sim G$
       $\therefore H$
    4. $E \lor F$
    5. $G \lor H$
    6. H

3. 1. $I \supset J$
    2. $J \supset K$
    3. $L \supset M$
    4. $I \lor L$
       $\therefore K \lor M$
    5. $I \supset K$
    6. $(I \supset K) \cdot (L \supset M)$
    7. $K \lor M$

4. 1. $N \supset O$
    2. $(N \cdot O) \supset P$
    3. $\sim (N \cdot P)$
       $\therefore \sim N$
    4. $N \supset (N \cdot O)$
    5. $N \supset P$
    6. $N \supset (N \cdot P)$
    7. $\sim N$

*5. 1. $Q \supset R$
    2. $\sim S \supset (T \supset U)$
    3. $S \lor (Q \lor T)$
    4. $\sim S$
       $\therefore R \lor U$
    5. $T \supset U$
    6. $(Q \supset R) \cdot (T \supset U)$
    7. $Q \lor T$
    8. $R \lor U$

6. 1. $W \supset X$
    2. $(W \supset Y) \supset (Z \lor X)$
    3. $(W \cdot X) \supset Y$
    4. $\sim Z$
       $\therefore X$
    5. $W \supset (W \cdot X)$
    6. $W \supset Y$
    7. $Z \lor X$
    8. X

7. 1. $(A \lor B) \supset C$
    2. $(C \lor B) \supset [A \supset (D \equiv E)]$
    3. $A \cdot D$
       $\therefore D \equiv E$
    4. A
    5. $A \lor B$
    6. C
    7. $C \lor B$
    8. $A \supset (D \equiv E)$
    9. $D \equiv E$

8. 1. $F \supset \sim G$
    2. $\sim F \supset (H \supset \sim G)$
    3. $(\sim I \lor \sim H) \supset \sim \sim G$
    4. $\sim I$
       $\therefore \sim H$
    5. $\sim I \lor \sim H$
    6. $\sim \sim G$
    7. $\sim F$
    8. $H \supset \sim G$
    9. $\sim H$

9. 1. $I \supset J$
    2. $I \lor (\sim\sim K \cdot \sim\sim J)$
    3. $L \supset \sim K$
    4. $\sim(I \cdot J)$
       $\therefore \sim L \lor \sim J$
    5. $I \supset (I \cdot J)$
    6. $\sim I$
    7. $\sim\sim K \cdot \sim\sim J$
    8. $\sim\sim K$
    9. $\sim L$
    10. $\sim L \lor \sim J$

*10. 1. $(L \supset M) \supset (N \equiv O)$
    2. $(P \supset \sim Q) \supset (M \equiv \sim Q)$
    3. $\{[(P \supset \sim Q) \lor (R \equiv S)] \cdot (N \lor O)\} \supset [(R \equiv S) \supset (L \supset M)]$
    4. $(P \supset \sim Q) \lor (R \equiv S)$
    5. $N \lor O$
       $\therefore (M \equiv \sim Q) \lor (N \equiv O)$
    6. $[(P \supset \sim Q) \lor (R \equiv S)] \cdot (N \lor O)$
    7. $(R \equiv S) \supset (L \supset M)$
    8. $(R \equiv S) \supset (N \equiv O)$
    9. $[(P \supset \sim Q) \supset (M \equiv \sim Q)] \cdot [(R \equiv S) \supset (N \equiv O)]$
    10. $(M \equiv \sim Q) \lor (N \equiv O)$

9.4 Constructing Formal Proofs of Validity

Now we turn to one of the central tasks of deductive logic: proving formally that valid arguments really are valid. In the preceding sections we examined formal proofs that needed only to be supplemented by the justifications of the steps taken. From this point, however, we will confront arguments whose formal proofs must be constructed. This is an easy task for many arguments, a more challenging task for some. Whether the proof needed is short and simple, or long and complex, the rules of inference are in every case our instruments. Success requires *mastery* of these rules. Having the list of rules before one will probably not be sufficient. One must be able to call on the rules "from within" as the proofs are being devised. The ability to do this will grow rapidly with practice, and yields many satisfactions.

Let us begin by constructing proofs for simple arguments. The only rules needed (or available for our use) are the nine elementary valid argument forms with which we have been working. This limitation we will later overcome, but even with only these nine rules in our logical toolbox, very many arguments can be formally proved valid. We begin with arguments that require, in addition to the premises, no more than two additional statements.

We will look first at two examples, the first two in the set of exercises on pages 349–350.

In the first example consider the argument:

1. A
2. B
   $\therefore (A \lor C) \cdot B$

The conclusion of this argument $(A \lor C) \cdot B$ is a conjunction; we see immediately that the second conjunct, B, is readily at hand as a premise in line 2. All that is now needed is the statement of the disjunction, $(A \lor C)$, which may then be conjoined with B to complete the proof. $(A \lor C)$ is easily obtained from the premise

A, in line 1; we simply add *C* using the rule Addition, which tells us that to any given *p* with a truth value of *true* we may add (disjunctively) any *q* whatever. In this example we have been told that *A* is true, so we may infer by this rule that $A \vee C$ must be true. The third line of this proof is "3. $A \vee C$, 1, Add." In line 4 we can conjoin this disjunction (line 3) with the premise *B* (line 2): "4. $(A \vee C) \cdot B$, 3, 2, Conj." This final line of the sequence is the conclusion of the argument being proved. The formal proof is complete.

Here is a second example of an argument whose formal proof requires only two additional lines in the sequence:

1. $D \supset E$
2. $D \cdot F$
   $\therefore E$

The conclusion of this argument, *E*, is the consequent of the conditional statement $D \supset E$, which is given as the first premise. We know that we will be able to infer the truth of *E* by *Modus Ponens* if we can establish the truth of *D*. We can establish the truth of *D*, of course, by Simplification from the second premise, $D \cdot F$. So the complete formal proof consists of the following four lines:

1. $D \supset E$
2. $D \cdot F$ / $\therefore E$
3. D 2, Simp.
4. E 1, 3, M.P.

In each of these examples, and in all the exercises immediately following, a formal proof for each argument may be constructed by adding just two additional statements. This will be an easy task *if the nine elementary valid argument forms are clearly in mind*. Remember that the final line in the sequence of each proof is always the conclusion of the argument being proved.

EXERCISES

1. *A*
   B
   $\therefore (A \vee C) \cdot B$

2. $D \supset E$
   $D \cdot F$
   $\therefore E$

3. *G*
   H
   $\therefore (G \cdot H) \vee I$

4. $J \supset K$
   J
   $\therefore K \vee L$

*5. $M \vee N$
    $\sim M \cdot \sim O$
    $\therefore N$

6. $P \cdot Q$
   R
   $\therefore P \cdot R$

7. $S \supset T$
   $\sim T \cdot \sim U$
   $\therefore \sim S$

8. $V \vee W$
   $\sim V$
   $\therefore W \vee X$

9. $Y \supset Z$
   Y
   $\therefore Y \cdot Z$

*10. $A \supset B$
   $(A \cdot B) \supset C$
   $\therefore A \supset C$

11. $D \supset E$
   $(E \supset F) \cdot (F \supset D)$
   $\therefore D \supset F$

12. $(G \supset H) \cdot (I \supset J)$
   G
   $\therefore H \vee J$

13. $\sim(K \cdot L)$
   $K \supset L$
   $\therefore \sim K$

14. $(M \supset N) \cdot (M \supset O)$
   $N \supset O$
   $\therefore M \supset O$

*15. $(P \supset Q) \cdot (R \supset S)$
   $(P \vee R) \cdot (Q \vee R)$
   $\therefore Q \vee S$

16. $(T \supset U) \cdot (T \supset V)$
   T
   $\therefore U \vee V$

17. $(W \vee X) \supset Y$
   W
   $\therefore Y$

18. $(Z \cdot A) \supset (B \cdot C)$
   $Z \supset A$
   $\therefore Z \supset (B \cdot C)$

19. $D \supset E$
   $[D \supset (D \cdot E)] \supset (F \supset \sim G)$
   $\therefore F \supset \sim G$

*20. $(\sim H \vee I) \vee J$
   $\sim(\sim H \vee I)$
   $\therefore J \vee \sim H$

21. $(K \supset L) \supset M$
   $\sim M \cdot \sim(L \supset K)$
   $\therefore \sim(K \supset L)$

22. $(N \supset O) \supset (P \supset Q)$
   $[P \supset (N \supset O)] \cdot [N \supset (P \supset Q)]$
   $\therefore P \supset (P \supset Q)$

23. $R \supset S$
   $S \supset (S \cdot R)$
   $\therefore [R \supset (R \cdot S)] \cdot [S \supset (S \cdot R)]$

24. $[T \supset (U \vee V)] \cdot [U \supset (T \vee V)]$
   $(T \vee U) \cdot (U \vee V)$
   $\therefore (U \vee V) \vee (T \vee V)$

*25. $(W \cdot X) \supset (Y \cdot Z)$
   $\sim[(W \cdot X) \cdot (Y \cdot Z)]$
   $\therefore \sim(W \cdot X)$

26. $A \supset B$
   $A \vee C$
   $C \supset D$
   $\therefore B \vee D$

27. $(E \cdot F) \vee (G \supset H)$
   $I \supset G$
   $\sim(E \cdot F)$
   $\therefore I \supset H$

28. $J \vee \sim K$
   $K \vee (L \supset J)$
   $\sim J$
   $\therefore L \supset J$

29. $(M \supset N) \cdot (O \supset P)$
   $N \supset P$
   $(N \supset P) \supset (M \vee O)$
   $\therefore N \vee P$

*30. $Q \supset (R \vee S)$
   $(T \cdot U) \supset R$
   $(R \vee S) \supset (T \cdot U)$
   $\therefore Q \supset R$

9.5 Constructing More Extended Formal Proofs

Arguments whose formal proof requires only two additional statements are quite simple. We now advance to construct formal proofs of the validity of more complex arguments. However, the process will be the same: The target for the final statement of the sequence will always be the conclusion of the argument, and the rules of inference will always be our only logical tools.

9.5 Constructing More Extended Formal Proofs

Let us look closely at an example—the first exercise of Set A below, an argument whose proof requires three additional statements:

1. $A \lor (B \supset A)$
2. $\sim A \cdot C$
   $\therefore \sim B$

In devising the proof of this argument (as in most cases), we need some plan of action, some strategy with which we can progress, using our rules, toward the conclusion sought. Here that conclusion is $\sim B$. We ask ourselves: Where in the premises does B appear? Only as the antecedent of the hypothetical $(B \supset A)$, which is a component of the first premise. How might $\sim B$ be derived? Using *Modus Tollens*, we can infer it from $B \supset A$ if we can establish that hypothetical separately and morever establish $\sim A$. Both of those needed steps can be readily accomplished. $\sim A$ is inferred from line 2 by Simplification:

3. $\sim A$ 2, Simp.

We can then apply $\sim A$ to line 1, using Disjunctive Syllogism to infer $(B \supset A)$:

4. $(B \supset A)$ 1, 3, D.S.

The proof may then be completed using *Modus Tollens* on lines 4 and 3:

5. $\sim B$ 4, 3, M.T.

The strategy used in this argument is readily devised. In the case of some proofs, devising the needed strategy will not be so simple, but it is usually helpful to ask: What statement(s) will enable one to infer the conclusion? What statement(s) will enable one to infer *that*? One continues to move backward from the conclusion toward the premises given.

EXERCISES

A. For each of the following arguments, it is possible to provide a formal proof of validity by adding just three statements to the premises. Writing these out, carefully and accurately, will strengthen your command of the rules of inference, a needed preparation for the construction of proofs that are more extended and more complex.

1. $A \lor (B \supset A)$
   $\sim A \cdot C$
   $\therefore \sim B$

2. $(D \lor E) \supset (F \cdot G)$
   D
   $\therefore F$

3. $(H \supset I) \cdot (H \supset J)$
   $H \cdot (I \lor J)$
   $\therefore I \lor J$

4. $(K \cdot L) \supset M$
   $K \supset L$
   $\therefore K \supset [(K \cdot L) \cdot M]$

*5. $N \supset [(N \cdot O) \supset P]$
   $N \cdot O$
   $\therefore P$

6. $Q \supset R$
   $R \supset S$
   $\sim S$
   $\therefore Q \cdot \sim R$

7. $T \supset U$
   $V \lor \sim U$
   $\sim V \cdot \sim W$
   $\therefore \sim T$

8. $\sim X \supset Y$
   $Z \supset X$
   $\sim X$
   $\therefore Y \cdot \sim Z$

9. $(A \lor B) \supset \sim C$
   $C \lor D$
   A
   $\therefore D$

*10. $E \lor \sim F$
   $F \lor (E \lor G)$
   $\sim E$
   $\therefore G$

11. $(H \supset I) \cdot (J \supset K)$
   $K \lor H$
   $\sim K$
   $\therefore I$

12. $L \lor (M \supset N)$
   $\sim L \supset (N \supset O)$
   $\sim L$
   $\therefore M \supset O$

13. $(P \supset Q) \cdot (Q \supset P)$
   $R \supset S$
   $P \lor R$
   $\therefore Q \lor S$

14. $(T \supset U) \cdot (V \supset W)$
   $(U \supset X) \cdot (W \supset Y)$
   T
   $\therefore X \lor Y$

*15. $(Z \cdot A) \supset B$
   $B \supset A$
   $(B \cdot A) \supset (A \cdot B)$
   $\therefore (Z \cdot A) \supset (A \cdot B)$

Formal proofs most often require more than two or three lines to be added to the premises. Some are very lengthy. Whatever their length, however, the same process and the same strategic techniques are called for in devising the needed proofs. In this section we rely entirely on the nine elementary valid argument forms that serve as our rules of inference.

As we begin to construct longer and more complicated proofs, let us look closely at an example of such proofs—the first exercise of Set B on page 353. It is not difficult, but it is more extended than those we have worked with so far.

1. $A \supset B$
2. $A \lor (C \cdot D)$
3. $\sim B \cdot \sim E$
   $\therefore C$

The strategy needed for the proof of this argument is not hard to see: To obtain C we must break apart the premise in line 2; to do that we will need $\sim A$; to establish $\sim A$ we will need to apply *Modus Tollens* to line 1 using $\sim B$. Therefore we continue the sequence with the fourth line of the proof by applying Simplification to line 3:

1. $A \supset B$
2. $A \lor (C \cdot D)$
3. $\sim B \cdot \sim E$ / $\therefore C$
4. $\sim B$ 3, Simp.

Using line 4 we can obtain $\sim A$ from line 1:

5. $\sim A$ 1, 4, M.T.

With ~A established we can break line 2 apart, as we had planned, using D.S.:

    6. $C \cdot D$ 2, 5, D.S.

The conclusion may be pulled readily from the sixth line by Simplification.

    7. C 6, Simp.

Seven lines (including the premises) are required for this formal proof. Some proofs require very many more lines than this, but the object and the method remain always the same.

It sometimes happens, as one is devising a formal proof, that a statement is correctly inferred and added to the numbered sequence but turns out not to be needed; a solid proof may be given without using that statement. In such a case it is usually best to rewrite the proof, eliminating the unneeded statement. However, if the unneeded statement is retained, and the proof remains accurately constructed using other statements correctly inferred, the inclusion of the unneeded statement (although perhaps inelegant) does not render the proof incorrect. Logicians tend to prefer shorter proofs, proofs that move to the conclusion as directly as the rules of inference permit. But if, as one is constructing a more complicated proof, it becomes apparent that some much earlier statement(s) has been needlessly inferred, it may be more efficient to allow such statement(s) to remain in place, using (as one goes forward) the more extended numbering that that inclusion makes necessary. *Logical solidity* is the critical objective. A solid formal proof, one in which *each step is correctly derived and the conclusion is correctly linked to the premises by an unbroken chain of arguments using the rules of inference correctly*, remains a proof—even if it is not as crisp and elegant as some other proof that could be devised.

EXERCISES

B. For each of the following arguments, a formal proof of validity can be constructed without great difficulty, although some of the proofs may require a sequence of eight or nine lines (including premises) for their completion.

1. $A \supset B$
    $A \vee (C \cdot D)$
    $\sim B \cdot \sim E$
    $\therefore C$

2. $(F \supset G) \cdot (H \supset I)$
    $J \supset K$
    $(F \vee J) \cdot (H \vee L)$
    $\therefore G \vee K$

3. $(\sim M \cdot \sim N) \supset (O \supset N)$
    $N \supset M$
    $\sim M$
    $\therefore \sim O$

4. $(K \vee L) \supset (M \vee N)$
    $(M \vee N) \supset (O \cdot P)$
    K
    $\therefore O$

*5. $(Q \supset R) \cdot (S \supset T)$
    $(U \supset V) \cdot (W \supset X)$
    $Q \vee U$
    $\therefore R \vee V$

6. $W \supset X$
    $(W \cdot X) \supset Y$
    $(W \cdot Y) \supset Z$
    $\therefore W \supset Z$

354 CHAPTER 9 Methods of Deduction

7. $A \supset B$
$C \supset D$
$A \lor C$
$\therefore (A \cdot B) \lor (C \cdot D)$

8. $(E \lor F) \supset (G \cdot H)$
$(G \lor H) \supset I$
E
$\therefore I$

9. $J \supset K$
$K \lor L$
$(L \cdot {\sim}J) \supset (M \cdot {\sim}J)$
${\sim}K$
$\therefore M$

*10. $(N \lor O) \supset P$
$(P \lor Q) \supset R$
$Q \lor N$
${\sim}Q$
$\therefore R$

In the study of logic, our aim is to evaluate arguments in a natural language, such as English. When an argument in everyday discourse confronts us, we can prove it to be valid (if it really is valid) by first translating the statements (from English, or from any other natural language) into our symbolic language, and then constructing a formal proof of that symbolic translation. The symbolic version of the argument may reveal that the argument is, in fact, more simple (or possibly more complex) than one had supposed on first hearing or reading it. Consider the following example (the first in the set of exercises that immediately follow):

1. If either Gertrude or Herbert wins, then both Jens and Kenneth lose. Gertrude wins. Therefore Jens loses. (*G*—Gertrude wins; *H*—Herbert wins; *J*—Jens loses; *K*—Kenneth loses.)

Abbreviations for each statement are provided in this context because, without them, those involved in the discussion of these arguments would be likely to employ various abbreviations, making communication difficult. Using the abbreviations suggested greatly facilitates discussion.

Translated from the English into symbolic notation, this first argument appears as

1. $(G \lor H) \supset (J \cdot K)$
2. G $/\therefore J$

The rest of the formal proof of this argument is short and straightforward:

3. $G \lor H$ 2, Add.
4. $J \cdot K$ 1, 3, M.P.
5. J 4, Simp.

EXERCISES

C. Each of the following arguments in English may be similarly translated, and for each, a formal proof of validity (using only the nine elementary valid argument forms as rules of inference) may be constructed. These proofs vary in length, some requiring a sequence of thirteen statements (including the premises) to complete the formal proofs. The suggested abbreviations should be used for the sake of clarity. Bear in mind that, as one proceeds to produce a formal proof of an argument presented in a natural language, it is of the utmost importance that *the translation into symbolic notation of the statements appearing discursively in the argument be perfectly accurate*; if it is not, one will be working with an argument that is *different* from the original one, and in that case any proof devised will be useless, being not applicable to the original argument.

1. If either Gertrude or Herbert wins, then both Jens and Kenneth lose. Gertrude wins. Therefore Jens loses. (*G*—Gertrude wins; *H*—Herbert wins; *J*—Jens loses; *K*—Kenneth loses.)

2. If Adriana joins, then the club's social prestige will rise; and if Boris joins, then the club's financial position will be more secure. Either Adriana or Boris will join. If the club's social prestige rises, then Boris will join; and if the club's financial position becomes more secure, then Wilson will join. Therefore either Boris or Wilson will join. (*A*—Adriana joins; *S*—The club's social prestige rises; *B*—Boris joins; *F*—The club's financial position is more secure; *W*—Wilson joins.)

3. If Brown received the message, then she took the plane; and if she took the plane, then she will not be late for the meeting. If the message was incorrectly addressed, then Brown will be late for the meeting. Either Brown received the message or the message was incorrectly addressed. Therefore either Brown took the plane or she will be late for the meeting. (*R*—Brown received the message; *P*—Brown took the plane; *L*—Brown will be late for the meeting; *T*—The message was incorrectly addressed.)

4. If Nihar buys the lot, then an office building will be constructed; whereas if Payton buys the lot, then it will be quickly sold again. If Rivers buys the lot, then a store will be constructed; and if a store is constructed, then Thompson will offer to lease it. Either Nihar or Rivers will buy the lot. Therefore either an office building or a store will be constructed. (*N*—Nihar buys the lot; *O*—An office building will be constructed; *P*—Payton buys the lot; *Q*—The lot will be quickly sold again; *R*—Rivers buys the lot; *S*—A store will be constructed; *T*—Thompson will offer to lease the store.)

*5. If rain continues, then the river rises. If rain continues and the river rises, then the bridge will wash out. If the continuation of rain would cause the bridge to wash out, then a single road is not sufficient for the town. Either a single road is sufficient for the town or the traffic engineers have made a mistake. Therefore the traffic engineers have made a mistake. (*C*—Rain continues; *R*—The river rises; *B*—The bridge washes out; *S*—A single road is sufficient for the town; *M*—The traffic engineers have made a mistake.)

6. If Jonas goes to the meeting, then a complete report will be made; but if Jonas does not go to the meeting, then a special election will be required. If a complete report is made, then an investigation will be launched. If Jonas's going to the meeting implies that a complete report will be made, and the making of a complete report implies that an investigation will be launched, then either Jonas goes to the meeting and an investigation is launched or Jonas does not go to the meeting and no investigation is launched. If Jonas goes to the meeting and an investigation is launched, then some members will have to stand trial. But if Jonas does not go to the meeting and no investigation is launched, then the organization will

disintegrate very rapidly. Therefore either some members will have to stand trial or the organization will disintegrate very rapidly. (*J*—Jonas goes to the meeting; *R*—A complete report is made; *E*—A special election is required; *I*—An investigation is launched; *T*—Some members have to stand trial; *D*—The organization disintegrates very rapidly.)

7. If Ann is present, then Bill is present. If Ann and Bill are both present, then either Charles or Doris will be elected. If either Charles or Doris is elected, then Elmer does not really dominate the club. If Ann's presence implies that Elmer does not really dominate the club, then Florence will be the new president. So Florence will be the new president. (*A*—Ann is present; *B*—Bill is present; *C*—Charles will be elected; *D*—Doris will be elected; *E*—Elmer really dominates the club; *F*—Florence will be the new president.)

8. If Mr. Jones is the manager's next-door neighbor, then Mr. Jones's annual earnings are exactly divisible by 3. If Mr. Jones's annual earnings are exactly divisible by 3, then $40,000 is exactly divisible by 3. But $40,000 is not exactly divisible by 3. If Mr. Robinson is the manager's next-door neighbor, then Mr. Robinson lives halfway between Detroit and Chicago. If Mr. Robinson lives in Detroit, then he does not live halfway between Detroit and Chicago. Mr. Robinson lives in Detroit. If Mr. Jones is not the manager's next-door neighbor, then either Mr. Robinson or Mr. Smith is the manager's next-door neighbor. Therefore Mr. Smith is the manager's next-door neighbor. (*J*—Mr. Jones is the manager's next-door neighbor; *E*—Mr. Jones's annual earnings are exactly divisible by 3; *T*—$40,000 is exactly divisible by 3; *R*—Mr. Robinson is the manager's next-door neighbor; *H*—Mr. Robinson lives halfway between Detroit and Chicago; *D*—Mr. Robinson lives in Detroit; *S*—Mr. Smith is the manager's next-door neighbor.)

9. If Mr. Smith is the manager's next-door neighbor, then Mr. Smith lives halfway between Detroit and Chicago. If Mr. Smith lives halfway between Detroit and Chicago, then he does not live in Chicago. Mr. Smith is the manager's next-door neighbor. If Mr. Robinson lives in Detroit, then he does not live in Chicago. Mr. Robinson lives in Detroit. Mr. Smith lives in Chicago or else either Mr. Robinson or Mr. Jones lives in Chicago. If Mr. Jones lives in Chicago, then the manager is Jones. Therefore the manager is Jones. (*S*—Mr. Smith is the manager's next-door neighbor; *W*—Mr. Smith lives halfway between Detroit and Chicago; *L*—Mr. Smith lives in Chicago; *D*—Mr. Robinson lives in Detroit; *I*—Mr. Robinson lives in Chicago; *C*—Mr. Jones lives in Chicago; *B*—The manager is Jones.)

10. If Smith once beat the editor at billiards, then Smith is not the editor. Smith once beat the editor at billiards. If the manager is Jones, then Jones is not the editor. The manager is Jones. If Smith is not the editor and Jones is not the editor, then Robinson is the editor. If the manager is Jones and Robinson is the editor, then Smith is the publisher. Therefore Smith is the publisher. (*O*—Smith once beat the editor at billiards; *M*—Smith is the editor; *B*—The manager is Jones; *N*—Jones is the editor; *F*—Robinson is the editor; *G*—Smith is the publisher.)

9.6 Expanding the Rules of Inference: Replacement Rules

The nine elementary valid argument forms with which we have been working are powerful tools of inference, but they are not powerful enough. There are very many valid truth-functional arguments whose validity cannot be proved using only the nine rules thus far developed. We need to expand the set of rules, to increase the power of our logical toolbox.

To illustrate the problem, consider the following simple argument, which is plainly valid:

> If you travel directly from Chicago to Los Angeles, you must cross the Mississippi River. If you travel only along the Atlantic seaboard, you will not cross the Mississippi River. Therefore if you travel directly from Chicago to Los Angeles, you will not travel only along the Atlantic seaboard.

Translated into symbolic notation, this argument appears as

$D \supset C$
$A \supset \sim C$
$/ \therefore D \supset \sim A$

This conclusion certainly does follow from the given premises. But, try as we may, there is no way to prove that it is valid using only the elementary valid argument forms. Our logical toolbox is not fully adequate.

What is missing? Chiefly, what is missing is the ability to replace one statement by another that is logically equivalent to it. We need to be able to put, in place of any given statement, any other statement whose meaning is exactly the same as that of the statement being replaced. We need rules that identify legitimate replacements precisely.

Such rules are available to us. Recall that the only compound statements that concern us here (as we noted in Section 8.2) are *truth-functional* compound statements, and in a truth-functional compound statement, if we replace any component by another statement having the same truth value, the truth value of the compound statement remains unchanged. Therefore we may accept as an additional principle of inference what may be called the general **rule of replacement**—a rule that permits us to infer from any statement the result of replacing any component of that statement by any other statement that is logically equivalent to the component replaced.

The correctness of such replacements is intuitively obvious. To illustrate, the principle of *Double Negation* (D.N.) asserts that p is logically equivalent to $\sim\sim p$. Using the rule of replacement we may say, correctly, that from the statement $A \supset \sim\sim B$, any one of the following statements may be validly inferred:

$A \supset B$,
$\sim\sim A \supset \sim\sim B$,
$\sim\sim(A \supset \sim\sim B)$, and even
$A \supset \sim\sim\sim\sim B$.

When we put any one of these in place of $A \supset \sim\sim B$, we do no more than exchange one statement for another that is its logical equivalent.

This rule of replacement is a powerful enrichment of our rules of inference. In its general form, however, its application is problematic because its content is not

Rule of replacement
A rule that permits us to infer from any statement the result of replacing any component of that statement by any other statement that is logically equivalent to the component replaced.

definite; we are not always sure what statements are indeed logically equivalent to some other statements, and thus (if we have the rule only in its general form) we may be unsure whether that rule applies in a given case. To overcome this problem in a way that makes the rule of replacement applicable with indubitable accuracy, we make the rule *definite* by listing *ten specific logical equivalences* to which the rule of replacement may certainly be applied. Each of these equivalences—they are all logically true biconditionals—will serve as a separate rule of inference. We list the ten logical equivalences here, as ten rules, and we number them consecutively to follow the first nine rules of inference already set forth in the preceding sections of this chapter.

Biography

Augustus De Morgan

Augustus De Morgan (1806–1871), an influential deductive logician, and John Stuart Mill, an influential inductive logician, were strict contemporaries—born in the same year and dying two years apart. Oddly, both their fathers were employed by the East India Company. De Morgan was born in India; when the family returned to England he proved to be a very precocious student in languages and in mathematics. At Cambridge University De Morgan received the Bachelor of Arts degree.

In those days one could be a candidate for the Master of Arts degree at Cambridge, and for a Fellowship there, only if one signed an oath accepting the doctrines of the Church of England. Such theological tests for academic degrees were abolished in Cambridge and Oxford late in the nineteenth century, but that was too late for De Morgan, whose religious views and moral integrity obliged him to refuse to sign such an oath. He therefore had to leave Cambridge. He continued his work in London; at the age of 22 he was appointed professor of mathematics at University College, London, an institution founded with a commitment to religious neutrality. He opened his professorship there with a famous lecture on the nature of the study of mathematics, its difficulties and its potential. De Morgan's papers in logic and mathematics, written during his long academic career in London, often became the focus of discussion among the scholars of the London Mathematical Society, of which he was the first president.

In his book *Formal Logic* (1847), De Morgan noted a great deficiency of classical logic. The Aristotelians say that from two particular propositions, "Some Ps are As" and "Some Ps are Bs," nothing can be validly deduced about the relations of As and Bs. They say (as we saw in our explication of Aristotelian syllogisms in chapter 6 of this book) that in a valid syllogism the middle term must be *distributed*—that it must be taken universally in at least one of the premises to effect the link that makes the deduction possible. De Morgan pointed out that this is not correct *if* we know that "*Most* Ps are As" and "*Most* Ps are Bs." With some *quantitative* premises we can deduce a connection between As and Bs. Suppose, for example, a ship had been sunk on which there were 1,000 passengers, of whom 700 drowned. If we know that 500 passengers were in their cabins at the time of the tragedy, it follows of necessity that at least 200 passengers were drowned in their cabins. This he called the *numerically definite syllogism*.

De Morgan also advanced the field called the *logic of relatives*. Identity and difference are relations to which logicians have given great attention, but there are other relations, such as equality, affinity, and especially *equivalence*, that also deserve the logician's attention, as De Morgan showed.

Two logical equivalences, widely useful and intuitively clear, received from De Morgan their time-honored formulation and carry his name: *De Morgan's Theorems* (explained in Chapter 8 and in this chapter), which remain a permanent and prominent instrument in deductive reasoning. ∎

overview

The Rules of Replacement: Logically Equivalent Expressions

Any of the following logically equivalent expressions may replace each other wherever they occur.

Name	Abbreviation	Form
10. De Morgan's theorems	De M.	$\sim(p \cdot q) \stackrel{T}{\equiv} (\sim p \vee \sim q)$ $\sim(p \vee q) \stackrel{T}{\equiv} (\sim p \cdot \sim q)$
11. Commutation	Com.	$(p \vee q) \stackrel{T}{\equiv} (q \vee p)$ $(p \cdot q) \stackrel{T}{\equiv} (q \cdot p)$
12. Association	Assoc.	$[p \vee (q \vee r)] \stackrel{T}{\equiv} [(p \vee q) \vee r]$ $[p \cdot (q \cdot r)] \stackrel{T}{\equiv} [(p \cdot q) \cdot r]$
13. Distribution	Dist.	$[p \cdot (q \vee r)] \stackrel{T}{\equiv} [(p \cdot q) \vee (p \cdot r)]$ $[p \vee (q \cdot r)] \stackrel{T}{\equiv} [(p \vee q) \cdot (p \vee r)]$
14. Double Negation	D.N.	$p \stackrel{T}{\equiv} \sim\sim p$
15. Transposition	Trans.	$(p \supset q) \stackrel{T}{\equiv} (\sim q \supset \sim p)$
16. Material Implication	Impl.	$(p \supset q) \stackrel{T}{\equiv} (\sim p \vee q)$
17. Material Equivalence	Equiv.	$(p \equiv q) \stackrel{T}{\equiv} [(p \supset q) \cdot (q \supset p)]$ $(p \equiv q) \stackrel{T}{\equiv} [(p \cdot q) \vee (\sim p \cdot \sim q)]$
18. Exportation	Exp.	$[(p \cdot q) \supset r] \stackrel{T}{\equiv} [p \supset (q \supset r)]$
19. Tautology	Taut.	$p \stackrel{T}{\equiv} (p \vee p)$ $p \stackrel{T}{\equiv} (p \cdot p)$

Let us now examine each of these ten logical equivalences. We will use them frequently and will rely on them in constructing formal proofs of validity, and therefore we must grasp their force as deeply, and control them as fully, as we do the nine elementary valid argument forms. We take these ten in order, giving for each the name, the abbreviation commonly used for it, and its exact logical form(s).

> **10. De Morgan's Theorems**   (De M.) $\sim(p \cdot q) \stackrel{T}{\equiv} (\sim p \lor \sim q)$
>
> $\sim(p \lor q) \stackrel{T}{\equiv} (\sim p \cdot \sim q)$

This logical equivalence was explained in detail in Section 8.9. De Morgan's theorems have two variants. One variant asserts that when we deny that two propositions are *both* true, that is logically equivalent to asserting that either one of them is false, or the other one is false, or they are both false. (The negation of a conjunction is logically equivalent to the disjunction of the negations of the conjuncts.) The second variant of De Morgan's theorems asserts that when we deny that *either* of two propositions is true, that is logically equivalent to asserting that both of them are false. (The negation of a disjunction is logically equivalent to the conjunction of the negations of the disjuncts.)

These two biconditionals are tautologies, of course. That is, the expression of the material equivalence of the two sides of each is *always* true, and thus can have no false substitution instance. All ten of the logical equivalences now being recognized as rules of inference are tautological biconditionals in exactly this sense.

> **11. Commutation**   (Com.) $(p \lor q) \stackrel{T}{\equiv} (q \lor p)$
>
> $(p \cdot q) \stackrel{T}{\equiv} (q \cdot p)$

These two equivalences simply assert that the *order* of statement of the elements of a conjunction, or of a disjunction, does not matter. We are always permitted to turn them around, to *commute* them, because, whichever order happens to appear, the meanings remain exactly the same.

Recall that Rule 7, Simplification, permitted us to pull p from the conjunction $p \cdot q$, but not q. Now, with Commutation, we can always replace $p \cdot q$ with $q \cdot p$—so that, with Simplification and Commutation both at hand, we can readily establish the truth of each of the conjuncts in any conjunction we know to be true.

> **12. Association**   (Assoc.) $[p \lor (q \lor r)] \stackrel{T}{\equiv} [(p \lor q) \lor r]$
>
> $[p \cdot (q \cdot r)] \stackrel{T}{\equiv} [(p \cdot q) \cdot r]$

These two equivalences do no more than allow us to group statements differently. If we know three different statements to be true, to assert that p is true along with q

and r clumped, is logically equivalent to asserting that p and q clumped is true along with r. Equivalence also holds if the three are grouped as disjuncts: p or the disjunction of $q \vee r$, is a grouping logically equivalent to the disjunction $p \vee q$, or r.

13. Distribution (Dist.) $\quad [p \cdot (q \vee r)] \stackrel{T}{\equiv} [(p \cdot q) \vee (p \cdot r)]$
$\quad\quad\quad\quad\quad\quad\quad\quad\quad\quad\quad\quad [p \vee (q \cdot r)] \stackrel{T}{\equiv} [(p \vee q) \cdot (p \vee r)]$

Of all the rules permitting replacement, this one may be the least obvious—but it too is a tautology, of course. It also has two variants. The first variant asserts merely that the *conjunction* of one statement with the disjunction of two other statements is logically equivalent to a disjunction whose first disjunct is the conjunction of the first statement with the second and whose second disjunct is the conjunction of the first statement with the third. The second variant asserts merely that the *disjunction* of one statement with the conjunction of two others is logically equivalent to the conjunction of the disjunction of the first and the second and the disjunction of the first and the third. The rule is named *Distribution* because it *distributes* the first element of the three, exhibiting its logical connections with each of the other two statements separately.

14. Double Negation (D.N.) $\quad p \stackrel{T}{\equiv} \sim\sim p$

Intuitively clear to everyone, this rule simply asserts that any statement is logically equivalent to the negation of the negation of that statement.

15. Transposition (Trans.) $\quad (p \supset q) \stackrel{T}{\equiv} (\sim q \supset \sim p)$

This logical equivalence permits us to turn any conditional statement around. We know that if any conditional statement is true, then if its consequent is false its antecedent must also be false. Therefore any conditional statement is logically equivalent to the conditional statement asserting that the negation of its consequent implies the negation of its antecedent. Clearly, Transposition, in the form of a logical equivalence, expresses the logical force of the elementary argument form *modus tollens*.

16. Material Implication (Impl.) $\quad (p \supset q) \stackrel{T}{\equiv} (\sim p \vee q)$

This logical equivalence does no more than formulate the definition of material implication explained in Section 8.3 as a replacement that can serve as a rule of inference. There we saw that $p \supset q$ simply means that either the antecedent, p, is false or the consequent, q, is true.

As we go on to construct formal proofs, this definition of material implication will become very important, because it is often easier to manipulate or combine two statements if they have the same basic form—that is, if they are both in disjunctive form, or if they are both in implicative form. If one is in disjunctive form and the other is in implicative form, we can, using this rule, transform one of them into the form of the other. This will be very convenient.

> **17. Material Equivalence** (Equiv.)　　$(p \equiv q) \stackrel{T}{\equiv} [(p \cdot q) \vee (\sim p \cdot \sim q)]$
>
> $(p \equiv q) \stackrel{T}{\equiv} [(p \supset q) \cdot (q \supset p)]$

The two variants of this rule simply assert the two essential meanings of material equivalence, explained in detail in Section 8.8. There we explained that two statements are materially equivalent if they both have the same truth value; therefore (first variant) the assertion of their material equivalence (with the tribar, $\equiv$) is logically equivalent to asserting that they are both true, or that they are both false. We also explained at that point that if two statements are both true, they must materially imply one another, and likewise if they are both false, they must materially imply one another; therefore (second variant) the statement that they are materially equivalent is *logically* equivalent to the statement that they imply one another.

> **18. Exportation** (Exp.)　　$[(p \cdot q) \supset r] \stackrel{T}{\equiv} [p \supset (q \supset r)]$

This replacement rule states a logical biconditional that is intuitively clear upon reflection: If one asserts that two propositions conjoined are known to imply a third, that is logically equivalent to asserting that if one of those two propositions is known to be true, then the truth of the other must imply the truth of the third. Like all the others, this logical equivalence may be readily confirmed using a truth table.

> **19. Tautology** (Taut.)　　$p \stackrel{T}{\equiv} (p \vee p)$
>
> $p \stackrel{T}{\equiv} (p \cdot p)$

The two variants of this last rule are patently obvious but very useful. They say simply that any statement is logically equivalent to the disjunction of itself with itself, and that any statement is logically equivalent to the conjunction of itself with itself. It sometimes happens that, as the outcome of a series of inferences, we learn that either the proposition we seek to establish is true or that it is true. From this disjunction we may readily infer (using this rule) that the proposition in question is true. The same applies to the conjunction of a statement with itself.

It should be noted that the word "tautology" is used in three different senses. It can mean (1) a *statement form* all of whose substitution instances are true; in this sense the statement form $(p \supset q) \supset [p \supset (p \supset q)]$ is a tautology. It can mean (2) a *statement*—for example, $(A \supset B) \supset [A \supset (A \supset B)]$ whose specific form is a tautology in sense (1). It can also mean (3) *the particular logical equivalence* we have just introduced, number 19 in our list of rules of inference.

9.6 Expanding the Rules of Inference: Replacement Rules

As we look back on these ten rules, we should be clear about what it is they make possible. They are not rules of "substitution" as that term is correctly used; we *substitute* statements for statement variables, as when we say that $A \supset B$ is a substitution instance of the expression $p \supset q$. In such operations we may substitute *any* statement for any statement variable so long as it is substituted for *every other* occurrence of that statement variable. But when these listed rules of replacement are applied, we exchange, or *replace*, a component of one statement *only by a statement that we know (by one of these ten rules) to be logically equivalent to that component*. For example, by transposition we may replace $A \supset B$ by $\sim B \supset \sim A$. These rules permit us to replace one occurrence of that component without having to replace any other occurrence of it.

EXERCISES

The following set of arguments involves, in each case, one step only, in which one of the ten logical equivalences set forth in this section has been employed. Here are two examples, the first two in the exercise set immediately following.

EXAMPLE 1

$(A \supset B) \cdot (C \supset D)$
$\therefore (A \supset B) \cdot (\sim D \supset \sim C)$

SOLUTION

The conclusion of this simple argument is exactly like its premise, except for the fact that the second conjunct in the premise, $(C \supset D)$, has been replaced by the logically equivalent expression $(\sim D \supset \sim C)$. That replacement is plainly justified by the rule we call Transposition (Trans.):

$(p \supset q) \stackrel{T}{\equiv} (\sim q \supset \sim p)$

EXAMPLE 2

$(E \supset F) \cdot (G \supset \sim H)$
$(\sim E \vee F) \cdot (G \supset \sim H)$

SOLUTION

In this case the conclusion differs from the premise only in the fact that the conditional statement $(E \supset F)$ has been replaced, as first conjunct, by the disjunctive statement $(\sim E \vee F)$. The rule permitting such a replacement, Material Implication (Impl.), has the form

$$(p \supset q) \stackrel{T}{\equiv} (\sim p \vee q)$$

For each of the following one-step arguments, state the one rule of inference by which its conclusion follows from its premise.

1. $(A \supset B) \cdot (C \supset D)$
   $\therefore (A \supset B) \cdot (\sim D \supset \sim C)$

2. $(E \supset F) \cdot (G \supset \sim H)$
   $\therefore (\sim E \vee F) \cdot (G \supset \sim H)$

3. $[I \supset (J \supset K)] \cdot (J \supset \sim I)$
   $\therefore [(I \cdot J) \supset K] \cdot (J \supset \sim I)$

4. $[L \supset (M \vee N)] \vee [L \supset (M \vee N)]$
   $\therefore L \supset (M \vee N)$

*5. $O \supset [(P \supset Q) \cdot (Q \supset P)]$
   $\therefore O \supset (P \equiv Q)$

6. $\sim(R \vee S) \supset (\sim R \vee \sim S)$
   $\therefore (\sim R \cdot \sim S) \supset (\sim R \vee \sim S)$

7. $(T \vee \sim U) \cdot [(W \cdot \sim V) \supset \sim T]$
   $\therefore (T \vee \sim U) \cdot [W \supset (\sim V \supset \sim T)]$

8. $(X \vee Y) \cdot (\sim X \vee \sim Y)$
   $\therefore [(X \vee Y) \cdot \sim X] \vee [(X \vee Y) \cdot \sim Y]$

9. $Z \supset (A \supset B)$
   $\therefore Z \supset (\sim\sim A \supset B)$

*10. $[C \cdot (D \cdot \sim E)] \cdot [(C \cdot D) \cdot \sim E]$
   $\therefore [(C \cdot D) \cdot \sim E] \cdot [(C \cdot D) \cdot \sim E]$

11. $(\sim F \vee G) \cdot (F \supset G)$
   $\therefore (F \supset G) \cdot (F \supset G)$

12. $(H \supset \sim I) \supset (\sim I \supset \sim J)$
   $\therefore (H \supset \sim I) \supset (J \supset I)$

13. $(\sim K \supset L) \supset (\sim M \vee \sim N)$
   $\therefore (\sim K \supset L) \supset \sim(M \cdot N)$

14. $[(\sim O \vee P) \vee \sim Q] \cdot [\sim O \vee (P \vee \sim Q)]$
   $\therefore [\sim O \vee (P \vee \sim Q)] \cdot [\sim O \vee (P \vee \sim Q)]$

15. $[(R \vee \sim S) \cdot \sim T] \vee [(R \vee \sim S) \cdot U]$
   $\therefore (R \vee \sim S) \cdot (\sim T \vee U)$

16. $[V \supset \sim(W \vee X)] \supset (Y \vee Z)$
   $\therefore \{[V \supset \sim(W \vee X)] \cdot [V \supset \sim(W \vee X)]\} \supset (Y \vee Z)$

17. $[(\sim A \cdot B) \cdot (C \vee D)] \vee [\sim(\sim A \cdot B) \cdot \sim(C \vee D)]$
   $\therefore (\sim A \cdot B) \equiv (C \vee D)$

18. $[\sim E \vee (\sim\sim F \supset G)] \cdot [\sim E \vee (F \supset G)]$
   $\therefore [\sim E \vee (F \supset G)] \cdot [\sim E \vee (F \supset G)]$

19. $[H \cdot (I \vee J)] \vee [H \cdot (K \supset \sim L)]$
   $\therefore H \cdot [(I \vee J) \vee (K \supset \sim L)]$

*20. $(\sim M \vee \sim N) \supset (O \supset \sim\sim P)$
   $\therefore \sim(M \cdot N) \supset (O \supset \sim\sim P)$

9.7 The System of Natural Deduction

The nineteen rules of inference that have been set forth (nine elementary argument forms and ten logical equivalences) are all the rules that are needed in truth-functional logic. Together they constitute a system of natural deduction that is compact and readily mastered, but nonetheless *complete*.* This means that, using this set of rules, one can construct a formal proof of validity for *any* valid truth-functional argument.[1]

---

*This kind of completeness of a set of rules can be proved. One method of proving such completeness may be found in I. M. Copi, *Symbolic Logic*, 5th edition (New York: Macmillan, 1979), Chapter 8.

9.7 The System of Natural Deduction

Two seeming flaws of this list of nineteen rules deserve attention. First, the set is somewhat *redundant*, in the sense that these nineteen do not constitute the minimum that would suffice for the construction of formal proofs of validity for extended arguments. To illustrate this we might note that *Modus Tollens* could be dropped from the list without any real weakening of our proof apparatus, because any line that depends on that rule can be justified by appealing instead to other rules in our list. Suppose, for example, we know that $A \supset D$ is true, and that $\sim D$ is true, and suppose we want to deduce that $\sim A$ is true. *Modus Tollens* allows us to do that directly. If *Modus Tollens* were not included in the list of rules, we would still have no trouble deducing $\sim A$ from $A \supset D$ and $\sim D$; we would simply need to insert the intermediate line, $\sim D \supset \sim A$, which follows from $A \supset D$ by Transposition (Trans.), then obtain $\sim A$ from $\sim D \supset \sim A$ by *Modus Ponens* (M.P.). We keep *Modus Tollens* in the list because it is such a commonly used and intuitively obvious rule of inference. Others among the nineteen rules are redundant in this sense.

Second, the list of nineteen rules may also be said to be *deficient* in one sense. Because the set of rules is short, there are some arguments that, although they are simple and intuitively valid, require several steps to prove. To illustrate this point, consider the argument

$A \lor B$
$\sim B$
  $/ \therefore A$

which is obviously valid. Its form, equally valid, is

$p \lor q$
$\sim q$
  $/ \therefore p$

However, this elementary argument form has not been included as a rule of inference. No single rule of inference will serve in this case, so we must construct the proof using two rules of inference, commuting the first premise, and then applying Disjunctive Syllogism, thus:

1. $A \lor B$
2. $\sim B$ $/ \therefore A$
3. $B \lor A$ 1, Com.
4. A 3, 2, D.S.

One may complain that the system is in this way clumsy, at times obliging a slow and tortuous path to a proof that ought to be easy and direct. There is good reason for this clumsiness. We certainly want a set of rules that is complete, as this set is; but we also want a set of rules that is short and easily mastered. We could add rules to our set—additional equivalences, or additional valid argument forms—but with each such addition our logical toolbox would become more congested and more difficult to command. We could delete some rules (e.g., *Modus Tollens*, as noted above), but with each such deletion the set, although smaller, would become even more clumsy, requiring extended proofs for very simple arguments. Long experience has taught that this set of nineteen rules serves as an ideal compromise: a list of rules of inference that is short enough to master fully, yet long enough to do all that one may need to do with reasonable efficiency.

There is an important difference between the first nine and the last ten rules of inference. *The first nine rules can be applied only to whole lines of a proof*. Thus, in a formal proof of validity, the statement A can be inferred from the statement $A \cdot B$ by Simplification only if $A \cdot B$ constitutes a whole line. It is obvious that A cannot be inferred validly either from $(A \cdot B) \supset C$ or from $C \supset (A \cdot B)$, because the latter two statements can be true while A is false. And the statement $A \supset C$ does not follow from the statement $(A \cdot B) \supset C$ by Simplification or by any other rule of inference. It does not follow at all, for if A is true and B and C are both false, $(A \cdot B) \supset C$ is true but $A \supset C$ is false. Again, although $A \vee B$ follows from A by Addition, we cannot infer $(A \vee B) \supset C$ from $A \supset C$ by Addition or by any other rule of inference. For if A and C are both false and B is true, $A \supset C$ is true but $(A \vee B) \supset C$ is false. On the other hand, *any of the last ten rules can be applied either to whole lines or to parts of lines*. Not only can the statement $A \supset (B \supset C)$ be inferred from the whole line $(A \cdot B) \supset C$ by Exportation, but from the line $[(A \cdot B) \supset C] \vee D$ we can infer $[A \supset (B \supset C)] \vee D$ by Exportation. By replacement, logically equivalent expressions can replace each other wherever they occur, even where they do not constitute whole lines of a proof. But the first nine rules of inference can be used only with whole lines of a proof serving as premises.

The notion of *formal proof* is an *effective* notion, which means that it can be decided quite mechanically, in a finite number of steps, whether or not a given sequence of statements constitutes a formal proof (with reference to a given list of rules of inference). No thinking is required, either in the sense of thinking about what the statements in the sequence "mean" or in the sense of using logical intuition to check any step's validity. Only two things are required. The first is the ability to see that a statement occurring in one place is precisely the same as a statement occurring in another, for we must be able to check that some statements in the proof are premises of the argument being proved valid and that the last statement in the proof is the conclusion of that argument. The second thing that is required is the ability to see whether a given statement has a certain pattern—that is, to see if it is a substitution instance of a given statement form.

Thus, any question about whether the numbered sequence of statements on page 365 is a formal proof of validity can easily be settled in a completely mechanical fashion. That lines 1 and 2 are the premises and line 4 is the conclusion of the given argument is obvious on inspection. That 3 follows from preceding lines by one of the given rules of inference can be decided in a finite number of steps—even where the notation "1, Com." is not written at the side. The explanatory notation in the second column is a help and should always be included, but it is not, strictly speaking, a necessary part of the proof itself. At every line, there are only finitely many preceding lines and only finitely many rules of inference or reference forms to be consulted. Although it is time-consuming, it can be verified by inspection and comparison of shapes that 3 does not follow from 1 and 2 by

Modus Ponens, or by *Modus Tollens*, or by a Hypothetical Syllogism . . . and so on, until in following this procedure we come to the question of whether 3 follows from 1 by the principle of Commutation, and there we see, simply by looking at the forms, that it does. In the same way, the legitimacy of *any* statement in *any* formal proof can be tested in a finite number of steps, none of which involves anything more than comparing forms or shapes.

To preserve this effectiveness, we require that only one step be taken at a time. One might be tempted to shorten a proof by combining steps, but the space and time saved are negligible. More important is the effectiveness we achieve by taking each step by means of one single rule of inference.

Although a formal proof of validity is effective in the sense that it can be mechanically decided, of any given sequence, whether it is a proof, *constructing* a formal proof is not an effective procedure. In this respect, formal proofs differ from truth tables. The making of truth tables is completely mechanical: given any argument of the sort with which we are now concerned, we can always construct a truth table to test its validity by following the simple rules of procedure set forth in Chapter 8. But we have no effective or mechanical rules for the construction of formal proofs. Here we must think, or "figure out," where to begin and how to proceed. Nevertheless, proving an argument valid by constructing a formal proof of its validity is much easier than the purely mechanical construction of a truth table with perhaps hundreds or even thousands of rows.

Although we have no purely mechanical rules for constructing formal proofs, some rough-and-ready rules of thumb or hints on procedure may be suggested. The first is simply to begin deducing conclusions from the given premises by the given rules of inference. As more and more of these subconclusions become available as premises for further deductions, the greater is the likelihood of being able to see how to deduce the conclusion of the argument to be proved valid. Another hint is to try to eliminate statements that occur in the premises but not in the conclusion. Such elimination can proceed, of course, only in accordance with the rules of inference, but the rules contain many techniques for eliminating statements. Simplification is such a rule, whereby the right-hand conjunct can be dropped from a whole line that is a conjunction. Commutation is a rule that permits switching the left-hand conjunct of a conjunction over to the right-hand side, from which it can be dropped by Simplification. The "middle" term q can be eliminated by a Hypothetical Syllogism given two statements of the patterns $p \supset q$ and $q \supset r$. Distribution is a useful rule for transforming a disjunction of the pattern $p \vee (q \cdot r)$ into the conjunction $(p \vee q) \cdot (p \vee r)$, whose right-hand conjunct can then be eliminated by Simplification. Another rule of thumb is to introduce by means of Addition a statement that occurs in the conclusion but not in any premise. Yet another method, often very productive, is to work backward from the conclusion by looking for some statement or statements from which it can be deduced, and then trying to deduce those intermediate statements from the premises. There is, however, no substitute for practice as a method of acquiring facility in the construction of formal proofs.

9.8 Constructing Formal Proofs Using the Nineteen Rules of Inference

Having now a set of nineteen rules at our disposal, rather than just nine, the task of constructing formal proofs becomes somewhat more complicated. The objective remains the same, of course, but the process of devising the proof involves inspection of a larger intellectual toolbox. The unbroken logical chain that we devise, leading ultimately to the conclusion, may now include steps justified by either an elementary valid argument form or a logical equivalence. Any given proof is likely to employ rules of both kinds. The balance or order of their use is determined only by the logical need encountered as we implement the strategy that leads to the consummation of the proof.

Following is a set of flawless formal proofs, each of which relies on rules of both kinds. To become accustomed to the use of the full set of rules, we examine each of these proofs to determine what rule has been used to justify each step in that proof, noting that justification to the right of each line. We begin with two examples.

EXAMPLE 1

1. $A \supset B$
2. $C \supset \sim B$
   $\therefore A \supset \sim C$
3. $\sim\sim B \supset \sim C$
4. $B \supset \sim C$
5. $A \supset \sim C$

overview

The Rules of Inference

Nineteen rules of inference are specified for use in constructing formal proofs of validity. They are as follows:

Elementary Valid Argument Forms

1. **Modus Ponens** (M.P.)

   $p \supset q, p, \therefore q$

2. **Modus Tollens** (M.T.)

   $p \supset q, \sim q, \therefore \sim p$

Logically Equivalent Expressions

10. **De Morgan's theorems** (De M.)

    $\sim(p \cdot q) \stackrel{T}{\equiv} (\sim p \vee \sim q)$
    $\sim(p \vee q) \stackrel{T}{\equiv} (\sim p \cdot \sim q)$

11. **Commutation** (Com.)

    $(p \vee q) \stackrel{T}{\equiv} (q \vee p)$
    $(p \cdot q) \stackrel{T}{\equiv} (q \cdot p)$

3. Hypothetical Syllogism (H.S.)

$p \supset q, q \supset r, \therefore p \supset r$

4. Disjunctive Syllogism (D.S.)

$p \vee q, \sim p, \therefore q$

5. Constructive Dilemma (C.D.)

$(p \supset q) \cdot (r \supset s), p \vee r, \therefore q \vee s$

6. Absorption (Abs.)

$p \supset q, \therefore p \supset (p \cdot q)$

7. Simplification (Simp.)

$p \cdot q, \therefore p$

8. Conjunction (Conj.)

$p, q, \therefore p \cdot q$

9. Addition (Add.)

$p, \therefore p \vee q$

12. Association (Assoc.)

$[p \vee (q \vee r)] \stackrel{T}{\equiv} [(p \vee q) \vee r]$

$[p \cdot (q \cdot r)] \stackrel{T}{\equiv} [(p \cdot q) \cdot r]$

13. Distribution (Dist.)

$[p \cdot (q \vee r)] \stackrel{T}{\equiv} [(p \cdot q) \vee (p \cdot r)]$

$[p \vee (q \cdot r)] \stackrel{T}{\equiv} [(p \vee q) \cdot (p \vee r)]$

14. Double Negation (D.N.)

$p \stackrel{T}{\equiv} \sim\sim p$

15. Transposition (Trans.)

$(p \supset q) \stackrel{T}{\equiv} (\sim q \supset \sim p)$

16. Material Implication (Impl.)

$(p \supset q) \stackrel{T}{\equiv} (\sim p \vee q)$

17. Material Equivalence (Equiv.)

$(p \equiv q) \stackrel{T}{\equiv} [(p \supset q) \cdot (q \supset p)]$

$(p \equiv q) \stackrel{T}{\equiv} [(p \cdot q) \vee (\sim p \cdot \sim q)]$

18. Exportation (Exp.)

$[(p \cdot q) \supset r] \stackrel{T}{\equiv} [p \supset (q \supset r)]$

19. Tautology (Taut.)

$p \stackrel{T}{\equiv} (p \vee p)$

$p \stackrel{T}{\equiv} (p \cdot p)$

SOLUTION

Line 3 is simply line 2 transposed; we write beside line 3: 2, Trans.

Line 4 is simply line 3 with $\sim\sim B$ replaced by B, so we write beside line 4: 3, D.N.

Line 5 applies the Hypothetical Syllogism argument form to lines 1 and 4. We write beside line 5: 1, 4, H.S.

EXAMPLE 2

1. $(D \cdot E) \supset F$
2. $(D \supset F) \supset G \quad \therefore E \supset G$
3. $(E \cdot D) \supset F$
4. $E \supset (D \supset F)$
5. $E \supset G$

SOLUTION

Line 3 merely commutes $(D \cdot E)$ from line 1; we write: 1, Com.
Line 4 applies Exportation to line 3; we write: 3, Exp.
Line 5 applies Hypothetical Syllogism to lines 4 and 2; we write: 4, 2, H.S.

EXERCISES

A. For each numbered line that is not a premise in each of the formal proofs that follow, state the rule of inference that justifies it.

1.
   1. $A \supset B$
   2. $C \supset \sim B$
      $\therefore A \supset \sim C$
   3. $\sim\sim B \supset C$
   4. $B \supset \sim C$
   5. $A \supset \sim C$

2.
   1. $(D \cdot E) \supset F$
   2. $(D \supset F) \supset G$
      $\therefore E \supset G$
   3. $(E \cdot D) \supset F$
   4. $E \supset (D \supset F)$
   5. $E \supset G$

3.
   1. $(H \lor I) \supset [J \cdot (K \cdot L)]$
   2. I
      $\therefore J \cdot K$
   3. $I \lor H$
   4. $H \lor I$
   5. $J \cdot (K \cdot L)$
   6. $(J \cdot K) \cdot L$
   7. $J \cdot K$

4.
   1. $(M \lor N) \supset (O \cdot P)$
   2. $\sim O$
      $\therefore \sim M$
   3. $\sim O \lor \sim P$
   4. $\sim(O \cdot P)$
   5. $\sim(M \lor N)$
   6. $\sim M \cdot \sim N$
   7. $\sim M$

*5.
   1. $(Q \lor \sim R) \lor S$
   2. $\sim Q \lor (R \cdot \sim Q)$
      $\therefore R \supset S$
   3. $(\sim Q \lor R) \cdot (\sim Q \lor \sim Q)$
   4. $(\sim Q \lor \sim Q) \cdot (\sim Q \lor R)$
   5. $\sim Q \lor \sim Q$
   6. $\sim Q$
   7. $Q \lor (\sim R \lor S)$
   8. $\sim R \lor S$
   9. $R \supset S$

6.
   1. $T \cdot (U \lor V)$
   2. $T \supset [U \supset (W \cdot X)]$
   3. $(T \cdot V) \supset \sim(W \lor X)$
      $\therefore W \equiv X$
   4. $(T \cdot U) \supset (W \cdot X)$
   5. $(T \cdot V) \supset (\sim W \cdot \sim X)$
   6. $[(T \cdot U) \supset (W \cdot X)] \cdot [(T \cdot V) \supset (\sim W \cdot \sim X)]$
   7. $(T \cdot U) \lor (T \cdot V)$
   8. $(W \cdot X) \lor (\sim W \cdot \sim X)$
   9. $W \equiv X$

7. 1. $Y \supset Z$
   2. $Z \supset [Y \supset (R \lor S)]$
   3. $R \equiv S$
   4. $\sim(R \cdot S)$
      $\therefore \sim Y$
   5. $(R \cdot S) \lor (\sim R \cdot \sim S)$
   6. $\sim R \cdot \sim S$
   7. $\sim(R \lor S)$
   8. $Y \supset [Y \supset (R \lor S)]$
   9. $(Y \cdot Y) \supset (R \lor S)$
   10. $Y \supset (R \lor S)$
   11. $\sim Y$

8. 1. $A \supset B$
   2. $B \supset C$
   3. $C \supset A$
   4. $A \supset \sim C$
      $\therefore \sim A \cdot \sim C$
   5. $A \supset C$
   6. $(A \supset C) \cdot (C \supset A)$
   7. $A \equiv C$
   8. $(A \cdot C) \lor (\sim A \cdot \sim C)$
   9. $\sim A \lor \sim C$
   10. $\sim(A \cdot C)$
   11. $\sim A \cdot \sim C$

9. 1. $(D \cdot E) \supset \sim F$
   2. $F \lor (G \cdot H)$
   3. $D \equiv E$
      $\therefore D \supset G$
   4. $(D \supset E) \cdot (E \supset D)$
   5. $D \supset E$
   6. $D \supset (D \cdot E)$
   7. $D \supset \sim F$
   8. $(F \lor G) \cdot (F \lor H)$
   9. $F \lor G$
   10. $\sim\sim F \lor G$
   11. $\sim F \supset G$
   12. $D \supset G$

*10. 1. $(I \lor \sim\sim J) \cdot K$
   2. $[\sim L \supset \sim(K \cdot J)] \cdot [K \supset (I \supset \sim M)]$
      $\therefore \sim(M \cdot \sim L)$
   3. $[(K \cdot J) \supset L] \cdot [K \supset (I \supset \sim M)]$
   4. $[(K \cdot J) \supset L] \cdot [(K \cdot I) \supset \sim M]$
   5. $(I \lor J) \cdot K$
   6. $K \cdot (I \lor J)$
   7. $(K \cdot I) \lor (K \cdot J)$
   8. $(K \cdot J) \lor (K \cdot I)$
   9. $L \lor \sim M$
   10. $\sim M \lor L$
   11. $\sim M \lor \sim\sim L$
   12. $\sim(M \cdot \sim L)$

We now advance to the construction of formal proofs using the full set of rules of inference. We begin with simple arguments whose proofs require only two statements added to the premises. Each of those statements, of course, may be justified by either an elementary valid argument form or by one of the rules of replacement. We begin with two examples, the first two exercises of Set B, immediately following.

EXAMPLE 1

1. $A \supset \sim A$
   $\therefore \sim A$

SOLUTION

The first step in this proof, obviously, must manipulate the single premise. What can we do with it that will be helpful? If we apply Material Implication (Impl.), we will obtain a statement, $\sim A \lor \sim A$, to which we can apply the valid argument form Tautology (Taut.), and that will yield the conclusion we seek. So the proof is

1. $A \supset \sim A$
   $\therefore \sim A$
2. $\sim A \lor \sim A$ 1, Impl.
3. $\sim A$ 2, Taut.

EXAMPLE 2

1. $B \cdot (C \cdot D)$
   $\therefore C \cdot (D \cdot B)$

SOLUTION

In this proof we need only rearrange the statements, whose conjunction is given as true. In the first step we can commute the main conjunction of the first premise, which will yield $(C \cdot D) \cdot B$. Then we need only regroup the three statements by Association. So the proof is

1. $B \cdot (C \cdot D)$
   $\therefore C \cdot (D \cdot B)$
2. $(C \cdot D) \cdot B$ 1, Com.
3. $C \cdot (D \cdot B)$ 2, Assoc.

In this proof, as in all formal proofs, the last line of the sequence we construct *is* the conclusion we are aiming to deduce.

EXERCISES

B. For each of the following arguments, adding just two statements to the premises will produce a formal proof of its validity. Construct a formal proof for each of these arguments.

In these formal proofs, and in all the proofs to follow in later sections, note to the right of each line the rule of inference that justifies that line of the proof. It is most convenient if the justification specifies first the number of the line (or lines) being used, and then the name (abbreviated) of the rule of inference that has been applied to those numbered lines.

1. $A \supset \sim A$
   $\therefore \sim A$

2. $B \cdot (C \cdot D)$
   $\therefore C \cdot (D \cdot B)$

3. E
   $\therefore (E \vee F) \cdot (E \vee G)$

4. $H \vee (I \cdot J)$
   $\therefore H \vee I$

*5. $\sim K \vee (L \supset M)$
   $\therefore (K \cdot L) \supset M$

6. $(N \cdot O) \supset P$
   $\therefore (N \cdot O) \supset [N \cdot (O \cdot P)]$

7. $Q \supset [R \supset (S \supset T)]$
   $Q \supset (Q \cdot R)$
   $\therefore Q \supset (S \supset T)$

8. $U \supset \sim V$
   V
   $\therefore \sim U$

9. $W \supset X$
   $\sim Y \supset \sim X$
   $\therefore W \supset Y$

*10. $Z \supset A$
   $\sim A \vee B$
   $\therefore Z \supset B$

11. $C \supset \sim D$
   $\sim E \supset D$
   $\therefore C \supset \sim\sim E$

12. $F \equiv G$
   $\sim(F \cdot G)$
   $\therefore \sim F \cdot \sim G$

13. $H \supset (I \cdot J)$
   $I \supset (J \supset K)$
   $\therefore H \supset K$

14. $(L \supset M) \cdot (N \supset M)$
   $L \vee N$
   $\therefore M$

*15. $(O \vee P) \supset (Q \vee R)$
   $P \vee O$
   $\therefore Q \vee R$

16. $(S \cdot T) \vee (U \cdot V)$
   $\sim S \vee \sim T$
   $\therefore U \cdot V$

17. $(W \cdot X) \supset Y$
   $(X \supset Y) \supset Z$
   $\therefore W \supset Z$

18. $(A \vee B) \supset (C \vee D)$
   $\sim C \cdot \sim D$
   $\therefore \sim(A \vee B)$

19. $(E \cdot F) \supset (G \cdot H)$
   $F \cdot E$
   $\therefore G \cdot H$

*20. $I \supset [J \vee (K \vee L)]$
   $\sim[(J \vee K) \vee L]$
   $\therefore \sim I$

21. $(M \supset N) \cdot (\sim O \vee P)$
   $M \vee O$
   $\therefore N \vee P$

22. $(\sim Q \supset \sim R) \cdot (\sim S \supset \sim T)$
   $\sim\sim(\sim Q \vee \sim S)$
   $\therefore \sim R \vee \sim T$

23. $\sim[(U \supset V) \cdot (V \supset U)]$
   $(W \equiv X) \supset (U \equiv V)$
   $\therefore \sim(W \equiv X)$

24. $(Y \supset Z) \cdot (Z \supset Y)$
   $\therefore (Y \cdot Z) \vee (\sim Y \cdot \sim Z)$

*25. $A \vee B$
   $C \vee D$
   $\therefore [(A \vee B) \cdot C] \vee [(A \vee B) \cdot D]$

26. $[(E \vee F) \cdot (G \vee H)] \supset (F \cdot I)$
   $(G \vee H) \cdot (E \vee F)$
   $\therefore F \cdot I$

27. $(J \cdot K) \supset [(L \cdot M) \vee (N \cdot O)]$
   $\sim(L \cdot M) \cdot \sim(N \cdot O)$
   $\therefore \sim(J \cdot K)$

28. $(P \supset Q) \supset [(R \vee S) \cdot (T \equiv U)]$
   $(R \vee S) \supset [(T \equiv U) \supset Q]$
   $\therefore (P \supset Q) \supset Q$

374   CHAPTER 9 Methods of Deduction

29. $[V \cdot (W \lor X)] \supset (Y \supset Z)$
    $\sim(Y \supset Z) \lor (\sim W \equiv A)$
    $\therefore [V \cdot (W \lor X)] \supset (\sim W \equiv A)$

*30. $\sim[(B \supset \sim C) \cdot (\sim C \supset B)]$
    $(D \cdot E) \supset (B \equiv \sim C)$
    $\therefore \sim(D \cdot E)$

As we advance to arguments whose formal proofs require three lines added to the premises, it becomes important to devise a strategy for determining the needed sequence. Most such arguments remain fairly simple, but the path to the proof may sometimes be less than obvious. Again we begin with two examples, the first two exercises of Set C, which follows the examples.

EXAMPLE 1

1. $\sim A \supset A$
   $\therefore A$

SOLUTION

We have only one premise with which to work. It is often fruitful to convert conditional statements into disjunctive statements. Doing that with line 1 (using Impl.) will yield $\sim\sim A$ as the first of the disjuncts; that component may be readily replaced with A; then, applying the argument form Tautology will give us what we aim for. The proof is

> 1. $\sim A \supset A$
> $\therefore A$
> 2. $\sim\sim A \lor A$ 1, Impl.
> 3. $A \lor A$ 2, D.N.
> 4. A 3, Taut.

EXAMPLE 2

1. $\sim B \lor (C \cdot D)$
   $\therefore B \supset C$

SOLUTION

The single premise in this argument contains the statement D. We need a proof whose conclusion is $B \supset C$, and therefore we must somehow eliminate that D. How can we do that? We can break apart the statement $(C \cdot D)$ by distributing the statement $\sim B$. Distribution asserts, in one of its variants, that $[p \lor (q \cdot r)] \stackrel{\mathrm{T}}{=} [(p \lor q) \cdot (p \lor r)]$. Applied to line 1, that replacement will yield $(\sim B \lor C) \cdot (\sim B \lor D)$. These two expressions are just conjoined, so by simplification we may extract $(\sim B \lor C)$. This statement may be replaced, using Impl., by $B \supset C$, which is the conclusion sought. The proof is

9.8 Constructing Formal Proofs Using the Nineteen Rules of Inference

> 1. ~B ∨ (C • D)
> ∴ B ⊃ C
> 2. (~B ∨ C) • (~B ∨ D)   1, Dist.
> 3. ~B ∨ C   2, Simp.
> 4. B ⊃ C   3, Impl.

EXERCISES

C. For each of the following arguments, a formal proof may be constructed by adding just three statements to the premises. Construct a formal proof of validity for each of them.

1. ~A ⊃ A
   ∴ A

2. ~B ∨ (C • D)
   ∴ B ⊃ C

3. E ∨ (F • G)
   ∴ E ∨ G

4. H • (I • J)
   ∴ J • (I • H)

*5. [(K ∨ L) ∨ M] ∨ N
   ∴ (N ∨ K) ∨ (L ∨ M)

6. O ⊃ P
   P ⊃ ~P
   ∴ ~O

7. Q ⊃ (R ⊃ S)
   Q ⊃ R
   ∴ Q ⊃ S

8. T ⊃ U
   ~(U ∨ V)
   ∴ ~T

9. W • (X ∨ Y)
   ~W ∨ ~X
   ∴ W • Y

*10. (Z ∨ A) ∨ B
   ~A
   ∴ Z ∨ B

11. (C ∨ D) ⊃ (E • F)
    D ∨ C
    ∴ E

12. G ⊃ H
    H ⊃ G
    ∴ (G • H) ∨ (~G • ~H)

13. (I ⊃ J) • (K ⊃ L)
    I ∨ (K • M)
    ∴ J ∨ L

14. (N • O) ⊃ P
    (~P ⊃ ~O) ⊃ Q
    ∴ N ⊃ Q

*15. [R ⊃ (S ⊃ T)] • [(R • T) ⊃ U]
    R • (S ∨ T)
    ∴ T ∨ U

Formal proofs of validity sometimes require many steps or lines in the needed sequence. We will find that certain *patterns* of inference are encountered repeatedly in longer proofs. It is wise to become familiar with these recurring patterns.

This may be nicely illustrated using the first two exercises of Set D, which follows immediately below. First, suppose that a given statement, A, is known to be false. The next stage of the proof may require that we prove that some different statement, say B, is implied by the truth of the statement that we know is

376   CHAPTER 9 Methods of Deduction

false. This can be easily proved, and the pattern is not uncommon. Put formally, how may we infer $A \supset B$ from $\sim A$? Let us examine the argument.

EXAMPLE 1

1. $\sim A$
   $\therefore A \supset B$

SOLUTION

If $\sim A$ is known to be true, as here, then A must be false. A false statement materially implies any other statement. So $A \supset B$ must be true, whatever B may assert, if we know that $\sim A$ is true. In this case, $\sim A$ is given as premise; we only need to add the desired B and then apply Implication. The proof of the argument (or the proof segment, when it is a part of some longer proof) is

> 1. $\sim A$
> $\therefore A \supset B$
> 2. $\sim A \lor B$ 1, Add.
> 3. $A \supset B$ 2, Impl.

EXAMPLE 2

1. C
   $\therefore D \supset C$

This pattern arises very frequently. The truth of some statement C is known. In this case it is given as a premise; in some longer proof we might have established its truth at some other point in the sequence. We know that a true statement is materially implied by any statement whatever. Therefore any statement we choose, D, must imply C. Put formally, how may we infer $D \supset C$ from C?

SOLUTION

D does not appear in the premise but it does appear in the conclusion, so we must somehow get D into the sequence of steps. We could simply add D, but that won't succeed—because after commuting that disjunction, and replacing it, using Impl., with a conditional, we wind up with $\sim D \supset C$, which is certainly not the conclusion we were after. We want $D \supset C$. To obtain this needed result we must, in the very first step, add $\sim D$ rather than D. This we certainly may do, because Addition permits us to add disjunctively any statement whatever to a statement we know to be true. Then, applying Com. and Impl. will give us what we seek. The formal proof of the argument in this case (or the proof segment, when it occurs as part of a longer proof) is

1. C
   ∴ D ⊃ C

2. C ∨ ~D	1, Add.
3. ~D ∨ C	2, Com.
4. D ⊃ C	3, Impl.

EXERCISES

D. Each of the exercises immediately below exhibits a commonly recurring pattern. Constructing the formal proof in each case will take some ingenuity, and (in a few cases) the proof will require eight or nine lines. However, most of these proofs will present little difficulty, and devising the strategies needed to produce them is excellent practice. Construct a formal proof for each of the following arguments.

1. ~A
   ∴ A ⊃ B

2. C
   ∴ D ⊃ C

3. E ⊃ (F ⊃ G)
   ∴ F ⊃ (E ⊃ G)

4. H ⊃ (I • J)
   ∴ H ⊃ I

*5. K ⊃ L
   ∴ K ⊃ (L ∨ M)

6. N ⊃ O
   ∴ (N • P) ⊃ O

7. (Q ∨ R) ⊃ S
   ∴ Q ⊃ S

8. T ⊃ U
   T ⊃ V
   ∴ T ⊃ (U • V)

9. W ⊃ X
   Y ⊃ X
   ∴ (W ∨ Y) ⊃ X

*10. Z ⊃ A
   Z ∨ A
   ∴ A

When, after substantial practice, one has become well familiar with the nineteen rules of inference, and is comfortable in applying them, it is time to tackle formal proofs that are longer and more convoluted. The three sets of exercises that follow will present some challenges, but devising these formal proofs will be a source of genuine satisfaction. The great mathematician, G. H. Hardy, long ago observed that there is a natural and widespread thirst for intellectual "kick"—and that "nothing else has quite the kick" that solving logical problems has.

Arguments in a natural language, as in the last two sets, need no further explanation. After translating them into symbolic notation, using the suggested abbreviations, the procedure for constructing the proofs is no different from that used when we begin with an argument formulated in symbols. Before adventuring further in the realm of logical proofs, it will be helpful to examine, from Exercise Set E, two examples of the kinds of formal proofs we will be dealing with from this point forward.

The arguments presented in all these sets of exercises are valid. Therefore, because the system of nineteen rules we have devised is known to be complete,

CHAPTER 9 Methods of Deduction

we may be certain that a formal proof for each one of those arguments can be constructed. Nevertheless, the path from the premises to the conclusion may be far from obvious. In each case, some plan of action must be devised as one goes forward.

We illustrate the need for a plan of attack, and the way in which such a plan may be devised, by examining very closely two of the exercises—the first and the last—in Set **E**, which follows on page 380.

EXAMPLE 1

1. $A \supset {\sim}B$
   ${\sim}(C \cdot {\sim}A)$
   $\therefore C \supset {\sim}B$

SOLUTION

In this argument the conclusion unites a statement that appears in the second premise, C, with a statement that appears in the first premise, ${\sim}B$. How shall we effect that unification? The first premise is a conditional whose consequent, ${\sim}B$, is also the consequent of the conclusion. The second premise contains the negation of the antecedent of the first premise, ${\sim}A$. If we can manipulate the second premise to emerge with $C \supset A$, we can achieve the needed unification with H.S. We can do that. If we apply De M. to the second premise we will get a disjunction that, when replaced by a conditional using Impl., will be one short step away from the conditional needed. The formal proof is

1. $A \supset {\sim}B$		
2. ${\sim}(C \cdot {\sim}A)$		
$\therefore C \supset {\sim}B$		
3. ${\sim}C \vee {\sim}{\sim}A$	2, De M.	
4. $C \supset {\sim}{\sim}A$	3, Impl.	
5. $C \supset A$	4, D.N.	
6. $C \supset {\sim}B$	5, 1, H.S.	

Note that in this proof, as in many, a somewhat different sequence can be devised that leads to the same successful result. Line 3 is a needed first step. But we could have kept the disjunction on line 4, at that point only replacing ${\sim}{\sim}A$ by A:

4. ${\sim}C \vee A$	3, D.N.	Replacement of this by a conditional is then needed.
5. $C \supset A$	4, Impl.	H.S. then again concludes the proof:
6. $C \supset {\sim}B$	5, 1, H.S.	

The difference between the two sequences, in this case, is chiefly one of order. Sometimes there are alternative proofs using quite different strategies altogether.

Let us examine, as our final explication of the detail of formal proofs, one of the longer arguments in Set **E**, exercise 20, in which devising the strategy needed is more challenging.

EXAMPLE 2

1. $(R \vee S) \supset (T \cdot U)$
2. $\sim R \supset (V \supset \sim V)$
3. $\sim T$
   $\therefore \sim V$

The conclusion we seek, $\sim V$, appears only in the second of the three premises, and even there it is buried in a longer compound statement. How may we prove it? We notice that the consequent of the second premise ($V \supset \sim V$) is a conditional that, if replaced by a disjunction, yields $\sim V \vee \sim V$, which in turn yields $\sim V$ independently, by Taut. Might we obtain ($V \supset \sim V$) by M.P.? For that we need $\sim R$. R appears in the first premise, as part of a disjunction; if we can obtain the negation of that disjunction, we may derive $\sim R$. To obtain the negation of that disjunction we need the negation of the consequent of the first premise, so M.T. may be applied. It can be seen that the negation of that consequent ($T \cdot U$) should be available, because the third premise asserts $\sim T$, and if $\sim T$ is true, then ($T \cdot U$) surely is false. How may we show this? We look at the negation that we seek: $\sim(T \cdot U)$. This is logically equivalent to $\sim T \vee \sim U$. We can establish $\sim T \vee \sim U$ simply by adding $\sim U$ to $\sim T$. All the elements of the plan are before us; we need only put them into a logical sequence that is watertight. This is not at all difficult once the strategy has been devised. We begin by building the negation of the consequent of the first premise, then derive the negation of the antecedent of that premise, then obtain $\sim R$. With $\sim R$ we establish ($V \supset \sim V$) by M.P., and the conclusion we want to prove is at hand. The actual lines of the formal proof are

1. $(R \vee S) \supset (T \cdot U)$
2. $\sim R \supset (V \supset \sim V)$
3. $\sim T$
   $\therefore \sim V$
4. $\sim T \vee \sim U$ 3, Add.
5. $\sim(T \cdot U)$ 4, De M.
6. $\sim(R \vee S)$ 1, 5, M.T.
7. $\sim R \cdot \sim S$ 6, De M.
8. $\sim R$ 7, Simp.
9. $V \supset \sim V$ 2, 8, M.P.
10. $\sim V \vee \sim V$ 9, Impl.
11. $\sim V$ 10, Taut. Q.E.D.

At the conclusion of a proof it is traditional practice to place the letters Q.E.D.—a minor exhibition of pride in the form of an acronym for the Latin expression, *Quod erat demonstrandum*—"What was to be demonstrated."

EXERCISES

E. Construct a formal proof of validity for each of the following arguments:

1. $A \supset \sim B$
   $\sim(C \cdot \sim A)$
   $\therefore C \supset \sim B$

2. $(D \cdot \sim E) \supset F$
   $\sim(E \vee F)$
   $\therefore \sim D$

3. $(G \supset \sim H) \supset I$
   $\sim(G \cdot H)$
   $\therefore I \vee \sim H$

4. $(J \vee K) \supset \sim L$
   L
   $\therefore \sim J$

*5. $[(M \cdot N) \cdot O] \supset P$
   $Q \supset [(O \cdot M) \cdot N]$
   $\therefore \sim Q \vee P$

6. $R \vee (S \cdot \sim T)$
   $(R \vee S) \supset (U \vee \sim T)$
   $\therefore T \supset U$

7. $(\sim V \supset W) \cdot (X \supset W)$
   $\sim(\sim X \cdot V)$
   $\therefore W$

8. $[(Y \cdot Z) \supset A] \cdot [(Y \cdot B) \supset C]$
   $(B \vee Z) \cdot Y$
   $\therefore A \vee C$

9. $\sim D \supset (\sim E \supset \sim F)$
   $\sim(F \cdot \sim D) \supset \sim G$
   $\therefore G \supset E$

*10. $[H \vee (I \vee J)] \supset (K \supset J)$
   $L \supset [I \vee (J \vee H)]$
   $\therefore (L \cdot K) \supset J$

11. $M \supset N$
    $M \supset (N \supset O)$
    $\therefore M \supset O$

12. $(P \supset Q) \cdot (P \vee R)$
    $(R \supset S) \cdot (R \vee P)$
    $\therefore Q \vee S$

13. $T \supset (U \cdot V)$
    $(U \vee V) \supset W$
    $\therefore T \supset W$

14. $(X \vee Y) \supset (X \cdot Y)$
    $\sim(X \vee Y)$
    $\therefore \sim(X \cdot Y)$

*15. $(Z \supset Z) \supset (A \supset A)$
    $(A \supset A) \supset (Z \supset Z)$
    $\therefore A \supset A$

16. $\sim B \vee [(C \supset D) \cdot (E \supset D)]$
    $B \cdot (C \vee E)$
    $\therefore D$

17. $\sim F \vee \sim[\sim(G \cdot H) \cdot (G \vee H)]$
    $(G \supset H) \supset [(H \supset G) \supset I]$
    $\therefore F \supset (F \cdot I)$

18. $J \vee (\sim J \cdot K)$
    $J \supset L$
    $\therefore (L \cdot J) \equiv J$

19. $(M \supset N) \cdot (O \supset P)$
    $\sim N \vee \sim P$
    $\sim(M \cdot O) \supset Q$
    $\therefore Q$

*20. $(R \vee S) \supset (T \cdot U)$
    $\sim R \supset (V \supset \sim V)$
    $\sim T$
    $\therefore \sim V$

F. Construct a formal proof of validity for each of the following arguments, in each case using the suggested notation:

*1. Either the manager didn't notice the change or else he approves of it. He noticed it all right. So he must approve of it. (N, A)

2. The oxygen in the tube either combined with the filament to form an oxide or else it vanished completely. The oxygen in the tube could not have vanished completely. Therefore the oxygen in the tube combined with the filament to form an oxide. (C, V)

3. If a political leader who sees her former opinions to be wrong does not alter her course, she is guilty of deceit; and if she does alter her course,

9.8 Constructing Formal Proofs Using the Nineteen Rules of Inference   **381**

she is open to a charge of inconsistency. She either alters her course or she doesn't. Therefore either she is guilty of deceit or else she is open to a charge of inconsistency. (*A, D, I*)

4. It is not the case that she either forgot or wasn't able to finish. Therefore she was able to finish. (*F, A*)

*5. If the litmus paper turns red, then the solution is acid. Hence if the litmus paper turns red, then either the solution is acid or something is wrong somewhere. (*R, A, W*)

6. She can have many friends only if she respects them as individuals. If she respects them as individuals, then she cannot expect them all to behave alike. She does have many friends. Therefore she does not expect them all to behave alike. (*F, R, E*)

7. If the victim had money in his pockets, then robbery wasn't the motive for the crime. But robbery or vengeance was the motive for the crime. The victim had money in his pockets. Therefore vengeance must have been the motive for the crime. (*M, R, V*)

8. Napoleon is to be condemned if he usurped power that was not rightfully his own. Either Napoleon was a legitimate monarch or else he usurped power that was not rightfully his own. Napoleon was not a legitimate monarch. So Napoleon is to be condemned. (*C, U, L*)

9. If we extend further credit on the Wilkins account, they will have a moral obligation to accept our bid on their next project. We can figure a more generous margin of profit in preparing our estimates if they have a moral obligation to accept our bid on their next project. Figuring a more generous margin of profit in preparing our estimates will cause our general financial condition to improve considerably. Hence a considerable improvement in our general financial condition will follow from our extension of further credit on the Wilkins account. (*C, M, P, I*)

*10. If the laws are good and their enforcement is strict, then crime will diminish. If strict enforcement of laws will make crime diminish, then our problem is a practical one. The laws are good. Therefore our problem is a practical one. (*G, S, D, P*)

11. Had Roman citizenship guaranteed civil liberties, then Roman citizens would have enjoyed religious freedom. Had Roman citizens enjoyed religious freedom, there would have been no persecution of the early Christians. But the early Christians were persecuted. Hence Roman citizenship could not have guaranteed civil liberties. (*G, F, P*)

12. If the first disjunct of a disjunction is true, the disjunction as a whole is true. Therefore if both the first and second disjuncts of the disjunction are true, then the disjunction as a whole is true. (*F, W, S*)

13. If the new courthouse is to be conveniently located, it will have to be situated in the heart of the city; and if it is to be adequate to its function, it will have to be built large enough to house all the city offices. If the new courthouse is situated in the heart of the city and is built large enough to house all the city offices, then its cost will run to over $10 million. Its cost cannot

382 CHAPTER 9 Methods of Deduction

exceed $10 million. Therefore either the new courthouse will have an inconvenient location or it will be inadequate to its function. (C, H, A, L, O)

14. Jalana will come if she gets the message, provided that she is still interested. Although she didn't come, she is still interested. Therefore she didn't get the message. (C, M, I)

*15. If the Mosaic account of the cosmogony (the account of the creation in Genesis) is strictly correct, the sun was not created until the fourth day. And if the sun was not created until the fourth day, it could not have been the cause of the alternation of day and night for the first three days. But either the word "day" is used in Scripture in a different sense from that in which it is commonly accepted now or else the sun must have been the cause of the alternation of day and night for the first three days. Hence it follows that either the Mosaic account of the cosmogony is not strictly correct or else the word "day" is used in Scripture in a different sense from that in which it is commonly accepted now. (M, C, A, D)

16. If the teller or the cashier had pushed the alarm button, the vault would have locked automatically, and the police would have arrived within three minutes. Had the police arrived within three minutes, the robbers' car would have been overtaken. But the robbers' car was not overtaken. Therefore the teller did not push the alarm button. (T, C, V, P, O)

17. If people are always guided by their sense of duty, they must forgo the enjoyment of many pleasures; and if they are always guided by their desire for pleasure, they must often neglect their duty. People are either always guided by their sense of duty or always guided by their desire for pleasure. If people are always guided by their sense of duty, they do not often neglect their duty; and if they are always guided by their desire for pleasure, they do not forgo the enjoyment of many pleasures. Therefore people must forgo the enjoyment of many pleasures if and only if they do not often neglect their duty. (D, F, P, N)

18. Although world population is increasing, agricultural production is declining and manufacturing output remains constant. If agricultural production declines and world population increases, then either new food sources will become available or else there will be a radical redistribution of food resources in the world unless human nutritional requirements diminish. No new food sources will become available, yet neither will family planning be encouraged nor will human nutritional requirements diminish. Therefore there will be a radical redistribution of food resources in the world. (W, A, M, N, R, H, P)

19. Either the robber came in the door, or else the crime was an inside one and one of the servants is implicated. The robber could come in the door only if the latch had been raised from the inside; but one of the servants is surely implicated if the latch was raised from the inside. Therefore one of the servants is implicated. (D, I, S, L)

*20. If I pay my tuition, I won't have any money left. I'll buy a computer only if I have money. I won't learn to program computers unless I buy a computer. But if I don't pay tuition, I can't enroll in classes; and if I don't enroll in classes I certainly won't buy a computer. I must either pay my tuition or not pay my tuition. So I surely will not learn to program computers! (*P, M, C, L, E*)

G. The five arguments that follow are also valid, and a proof of the validity of each of them is called for. However, these proofs will be somewhat more difficult to construct than those in earlier exercises, and students who find themselves stymied from time to time ought not become discouraged. What may appear difficult on first appraisal may come to seem much less difficult with continuing efforts. Familiarity with the nineteen rules of inference, and repeated practice in applying those rules, are the keys to the construction of these proofs.

1. If you study the humanities, then you will develop an understanding of people, and if you study the sciences, then you will develop an understanding of the world about you. So if you study either the humanities or the sciences, then you will develop an understanding either of people or of the world about you. (*H, P, S, W*)

2. If you study the humanities, then you will develop an understanding of people, and if you study the sciences then you will develop an understanding of the world about you. So if you study both the humanities and the sciences, you will develop an understanding both of people and of the world about you. (*H, P, S, W*)

3. If you have free will, then your actions are not determined by any antecedent events. If you have free will, then if your actions are not determined by any antecedent events, then your actions cannot be predicted. If your actions are not determined by any antecedent events, then if your actions cannot be predicted then the consequences of your actions cannot be predicted. Therefore if you have free will, then the consequences of your actions cannot be predicted. (*F, A, P, C*)

4. Socrates was a great philosopher. Therefore either Socrates was happily married or else he wasn't. (*G, H*)

5. If either Socrates was happily married or else he wasn't, then Socrates was a great philosopher. Therefore Socrates was a great philosopher. (*H, G*)

9.9 Proof of Invalidity

For an invalid argument there is, of course, no formal proof of validity. However, if we fail to discover a formal proof of validity for a given argument, this failure does not prove that the argument is invalid and that no such proof can be constructed. It may mean only that we have not tried hard enough. Our inability to find a proof of validity may be caused by the fact that the argument is not valid, but it may be caused instead by our own lack of ingenuity—as a consequence of the noneffective character of the process of proof construction. Not being able to construct a formal proof of its validity does not prove an argument to be invalid. What *does* constitute a proof that a given argument is invalid?

The method about to be described is closely related to the truth-table method, although it is a great deal shorter. It will be helpful to recall how an invalid argument form is proved invalid by a truth table. If a single case (row) can be found in which truth values are assigned to the statement variables in such a way that the premises are made true and the conclusion false, then the argument form is invalid. If we can somehow make an assignment of truth values to the simple component statements of an argument that will make its premises true and its conclusion false, then making that assignment will suffice to prove the argument invalid. To make such an assignment is, in effect, what the truth table does. If we can make such an assignment of truth values without actually constructing the whole truth table, much work will be eliminated.

Consider this argument:

> If the governor favors public housing, then she is in favor of restricting the scope of private enterprise.
> If the governor were a socialist, then she would be in favor of restricting the scope of private enterprise.
> Therefore if the governor favors public housing, then she is a socialist.

It is symbolized as

$F \supset R$
$S \supset R$
$\therefore F \supset S$

and we can prove it invalid without having to construct a complete truth table. First we ask, "What assignment of truth values is required to make the conclusion false?" It is clear that a conditional is false only if its antecedent is true and its consequent false. Hence assigning the truth value *true* to F and *false* to S will make the conclusion $F \supset S$ false. Now if the truth value *true* is assigned to R, both premises are made true, because a conditional is true whenever its consequent is true. We can say, then, that if the truth value *true* is assigned to F and to R, and the truth value *false* is assigned to S, the argument will have true premises and a false conclusion and is thus proved to be invalid.

This method of proving invalidity is an alternative to the truth-table method of proof. The two methods are closely related, however, and the essential connection between them should be noticed. In effect, what we did when we made the indicated assignment of truth values was to construct one row of the given argument's truth table. The relationship can perhaps be seen more clearly when the truth-value assignments are written out horizontally:

F	R	S	$F \supset R$	$S \supset R$	$F \supset S$
True	True	False	True	True	False

In this configuration they constitute one row (the second) of the truth table for the given argument. *An argument is proved invalid by displaying at least one row of its truth table in which all its premises are true but its conclusion is false.* Consequently, we need not examine all rows of its truth table to discover an argument's invalidity: discovering a single row in which its premises are all true and its conclusion false will suffice. This method of proving invalidity is a method of constructing such a row without having to construct the entire truth table.*

The present method is shorter than writing out an entire truth table, and the amount of time and work saved is exponentially greater for arguments involving a greater number of component simple statements. For arguments with a considerable number of premises, or with premises of considerable complexity, the needed assignment of truth values may not be so easy to make. There is no mechanical method of proceeding, but some hints may prove helpful.

It is most efficient to proceed by assigning those values seen immediately to be essential if invalidity is to be proved. Thus, any premise that simply asserts the truth of some statement S suggests the immediate assignment of **T** to S (or **F** if the falsehood of S was asserted as premise), because we know that all the premises must be made true. The same principle applies to the statements in the conclusion, except that the assignments of truth values there must make the conclusion false. Thus a conclusion of the form $A \supset B$ suggests the immediate assignment of **T** to A and **F** to B, and a conclusion in the form $A \lor B$ suggests the immediate assignment of **F** to A and **F** to B, because only those assignments could result in a proof of invalidity.

Whether one ought to begin by seeking to make the premises true or by seeking to make the conclusion false depends on the structure of the propositions; usually it is best to begin wherever assignments can be made with greatest confidence. Of course, there will be many circumstances in which the first assignments have to be arbitrary and tentative. A certain amount of trial and error is likely to be needed. Even so, this method of proving invalidity is almost always shorter and easier than writing out a complete truth table.

EXERCISES

Prove the invalidity of each of the following by the method of assigning truth values.

*1. $A \supset B$
  $C \supset D$
  $A \lor D$
  $\therefore B \lor C$

2. $\sim(E \cdot F)$
  $(\sim E \cdot \sim F) \supset (G \cdot H)$
  $H \supset G$
  $\therefore G$

3. $I \lor \sim J$
  $\sim(\sim K \cdot L)$
  $\sim(\sim I \cdot \sim L)$
  $\therefore \sim J \supset K$

4. $M \supset (N \lor O)$
  $N \supset (P \lor Q)$
  $Q \supset R$
  $\sim(R \lor P)$
  $\therefore \sim M$

---

*The whole truth table (were we to construct it) would of course test the validity of the *specific form* of the argument in question. If it can be shown that the specific form of an argument is invalid, we may infer that the argument having that specific form is an invalid argument. The method described here differs only in that truth values here are assigned directly to premises and conclusion; nonetheless, the relation between this method and the truth-table method applied in Chapter 8 is very close.

386   CHAPTER 9   Methods of Deduction

*5. $S \supset (T \supset U)$
$V \supset (W \supset X)$
$T \supset (V \cdot W)$
$\sim(T \cdot X)$
$\therefore S \equiv U$

7. $D \supset (E \vee F)$
$G \supset (H \vee I)$
$\sim E \supset (I \vee J)$
$(I \supset G) \cdot (\sim H \supset \sim G)$
$\sim J$
$\therefore D \supset (G \vee I)$

9. $(S \supset T) \cdot (T \supset S)$
$(U \cdot T) \vee (\sim T \cdot \sim U)$
$(U \vee V) \vee (S \vee T)$
$\sim U \supset (W \cdot X)$
$(V \supset \sim S) \cdot (\sim V \supset \sim Y)$
$X \supset (\sim Y \supset \sim X)$
$(U \vee S) \cdot (V \vee Z)$
$\therefore X \cdot Z$

6. $A \equiv (B \vee C)$
$B \equiv (C \vee A)$
$C \equiv (A \vee B)$
$\sim A$
$\therefore B \vee C$

8. $K \supset (L \cdot M)$
$(L \supset N) \vee \sim K$
$O \supset (P \vee \sim N)$
$(\sim P \vee Q) \cdot \sim Q$
$(R \vee \sim P) \vee \sim M$
$\therefore K \supset R$

*10. $A \supset (B \supset \sim C)$
$(D \supset B) \cdot (E \supset A)$
$F \vee C$
$G \supset \sim H$
$(I \supset G) \cdot (H \supset J)$
$I \equiv \sim D$
$(B \supset H) \cdot (\sim H \supset D)$
$\therefore E \equiv F$

9.10 Inconsistency

An argument is proved invalid if truth values can be assigned to make all of its premises true and its conclusion false. If a deductive argument is not invalid, it must be valid. So, if truth values *cannot* be assigned to make the premises true and the conclusion false, then the argument must be valid. This follows from the definition of validity, but it has this curious consequence: Any argument whose premises are inconsistent must be valid.

In the following argument, for example, the premises appear to be totally irrelevant to the conclusion:

> If the airplane had engine trouble, it would have landed at Bend.
> If the airplane did not have engine trouble, it would have landed at Creswell.
> The airplane did not land at either Bend or Creswell.
> Therefore the airplane must have landed in Denver.

Here is its symbolic translation:

$A \supset B$
$\sim A \supset C$
$\sim(B \vee C)$
$\therefore D$

Any attempt to assign truth values to its component simple statements in such a way as to make the conclusion false and the premises all true is doomed to failure. Even if we ignore the conclusion and attend only to the premises, we find that there is no way to assign truth values to their components such that the premises will all be true. No truth-value assignment can make them all true because they are inconsistent with one another. Their conjunction is self-contradictory, being a substitution instance of a self-contradictory statement form. If we were to construct a truth table for this argument, we would find that in every row at least one of the premises is false. Because there is no row in which the premises are all true, there is no row in which the premises are all true and the conclusion false. Hence the truth table for this argument would establish that it is in fact valid. Of course, we can also provide a formal proof of its validity:

1. $A \supset B$
2. $\sim A \supset C$
3. $\sim(B \vee C)/ \therefore D$
4. $\sim B \cdot \sim C$ 3, De M.
5. $\sim B$ 4, Simp.
6. $\sim A$ 1, 5, M.T.
7. C 2, 6, M.P.
8. $\sim C \cdot \sim B$ 4, Com.
9. $\sim C$ 8, Simp.
10. $C \vee D$ 7, Add.
11. D 10, 9, D.S.

In this proof, lines 1 through 9 are devoted to making explicit the inconsistency that is implicitly contained in the premises. That inconsistency emerges clearly in line 7 (which asserts C) and line 9 (which asserts $\sim C$). Once this explicit contradiction has been expressed, the conclusion follows swiftly using Add. and D.S.

Thus we see that if a set of premises is inconsistent, those premises will validly yield any conclusion, no matter how irrelevant. The essence of the matter is more simply shown with the following outrageous argument, whose openly inconsistent premises allow us to infer—validly!—an irrelevant and absurd conclusion.

Today is Sunday.
Today is not Sunday.
Therefore the moon is made of green cheese.

In symbols we have

1. S
2. $\sim S/ \therefore M$

The rest of the formal proof of its validity is almost immediately obvious:

3. $S \vee M$ 1, Add.
4. M 3, 2, D.S.

Of course, an argument that is valid because its premises are inconsistent cannot possibly be *sound*—for if the premises are inconsistent with each other, they cannot possibly be all true. By such an argument, therefore, it is not possible to establish any conclusion to be true, because we know that at least one of the premises must be false.

How can such meager premises make any argument in which they occur valid? The premises of a valid argument imply its conclusion not merely in the sense of "material" implication, but *logically*, or strictly. In a valid argument it is logically impossible for the premises to be true when the conclusion is false—and this is the situation that obtains when it is logically impossible for the premises to be true, putting the conclusion aside. What we have shown is this: *Any argument with inconsistent premises is valid, regardless of what its conclusion may be*. Its validity may be established by a truth table, or as we saw above, by a formal proof in which the contradiction is first formally expressed (for example, S and ~S), the desired conclusion is then added to one side of the contradiction (for example, S ∨ M), and that desired conclusion (for example, M) is then inferred by Disjunctive Syllogism using the other side of the contradiction (for example, ~S).

This discussion helps to explain why consistency is prized so highly. One reason is that two inconsistent statements cannot both be true. In a courtroom, therefore, cross-examination often aims to bring a hostile witness to contradict himself. If a witness makes inconsistent assertions, not all that he says can be true, and his credibility is seriously undermined. When it has been once established that a witness has lied under oath (or is perhaps thoroughly confused), no testimony of that witness can be fully trusted. Lawyers say: *Falsus in unum, falsus in omnibus*—"untrustworthy in one thing, untrustworthy in all."

Another, deeper reason why inconsistency is so repugnant is that—as we have seen—any and every conclusion follows logically from inconsistent statements taken as premises. Inconsistent statements are not "meaningless"; their trouble is just the opposite—they mean too much. They mean everything, in the sense of *implying* everything, and if *everything* is asserted, half of what is asserted is surely *false*, because every statement has a denial.

We are thus provided with an answer to the old riddle: What happens when an irresistible force meets an immovable object? The situation described by the riddle involves a contradiction. An irresistible force can meet an immovable object only if both exist; there must be an irresistible force and there must also be an immovable object. But if there is an irresistible force, there can be no immovable object. Let us make the contradiction explicit: There is an immovable object, and there is no immovable object. From these inconsistent premises, any conclusion may validly be inferred. So the correct answer to the question, "What happens when an irresistible force meets an immovable object?" is "Everything!"

Inconsistency, devastating when found among the premises of an argument, can be highly amusing. Everett Dirksen, leader of the Republican Party in the U.S. Senate for a decade in the twentieth century, enjoyed describing himself as "a man of fixed and unbending principles, the first of which is to be flexible at all times."[2] When an internal contradiction, not recognized by a speaker, yields unseen absurdity, we call the statement an "Irish bull." Writes the schoolboy, for example, "The climate of the Australian interior is so bad that the inhabitants don't live there anymore." Yogi Berra, famous for his Irish bulls, observed that a certain restaurant, once very popular, had become "so crowded that nobody goes there anymore." He also said, "When you see a fork in the road, take it."

As a matter of logic, in an internally inconsistent set, not all the propositions can be true. But human beings are not always logical and do utter, and sometimes may even believe, two propositions that contradict one another. This may seem difficult to do, but we are told by Lewis Carroll, a very reliable authority in such matters, that the White Queen in *Alice in Wonderland* made a regular practice of believing six impossible things before breakfast!

EXERCISES

For each of the following, either construct a formal proof of validity or prove invalidity by the method of assigning truth values to the simple statements involved.

*1. $(A \supset B) \cdot (C \supset D)$
   $\therefore (A \cdot C) \supset (B \vee D)$

2. $(E \supset F) \cdot (G \supset H)$
   $\therefore (E \vee G) \supset (F \cdot H)$

3. $I \supset (J \vee K)$
   $(J \cdot K) \supset L$
   $\therefore I \supset L$

4. $M \supset (N \cdot O)$
   $(N \vee O) \supset P$
   $\therefore M \supset P$

*5. $[(X \cdot Y) \cdot Z] \supset A$
   $(Z \supset A) \supset (B \supset C)$
   B
   $\therefore X \supset C$

6. $[(D \vee E) \cdot F] \supset G$
   $(F \supset G) \supset (H \supset I)$
   H
   $\therefore D \supset I$

7. $(J \cdot K) \supset (L \supset M)$
   $N \supset \sim M$
   $\sim(K \supset \sim N)$
   $\sim(J \supset \sim L)$
   $\therefore \sim J$

8. $(O \cdot P) \supset (Q \supset R)$
   $S \supset \sim R$
   $\sim(P \supset \sim S)$
   $\sim(O \supset Q)$
   $\therefore \sim O$

9. $T \supset (U \cdot V)$
   $U \supset (W \cdot X)$
   $(T \supset W) \supset (Y \equiv Z)$
   $(T \supset U) \supset \sim Y$
   $\sim Y \supset (\sim Z \supset X)$
   $\therefore X$

*10. $A \supset (B \cdot C)$
   $B \supset (D \cdot E)$
   $(A \supset D) \supset (F \equiv G)$
   $A \supset (B \supset \sim F)$
   $\sim F \supset (\sim G \supset E)$
   $\therefore E$

B. For each of the following, either construct a formal proof of validity or prove invalidity by the method of assigning truth values to the simple statements involved. In each case, use the notation in parentheses.

*1. If the linguistics investigators are correct, then if more than one dialect was present in ancient Greece, then different tribes came down at different times from the north. If different tribes came down at different times from the north, they must have come from the Danube River valley. But archaeological excavations would have revealed traces of different tribes there if different tribes had come down at different times from the north, and archaeological excavations have revealed no such traces there. Hence if more than one dialect was present in ancient Greece, then the linguistics investigators are not correct. (C, M, D, V, A)

2. If there are the ordinary symptoms of a cold and the patient has a high temperature, then if there are tiny spots on his skin, he has measles. Of course the patient cannot have measles if his record shows that he has

had them before. The patient does have a high temperature and his record shows that he has had measles before. Besides the ordinary symptoms of a cold, there are tiny spots on his skin. I conclude that the patient has a viral infection. (O, T, S, M, R, V)

3. If God were willing to prevent evil, but unable to do so, he would be impotent; if he were able to prevent evil, but unwilling to do so, he would be malevolent. Evil can exist only if God is either unwilling or unable to prevent it. There is evil. If God exists, he is neither impotent nor malevolent. Therefore God does not exist. (W, A, I, M, E, G)

4. If I buy a new car this spring or have my old car fixed, then I'll get up to Canada this summer and stop off in Duluth. I'll visit my parents if I stop off in Duluth. If I visit my parents, they'll insist on my spending the summer with them. If they insist on my spending the summer with them, I'll be there till autumn. But if I stay there till autumn, then I won't get to Canada after all! So I won't have my old car fixed. (N, F, C, D, V, I, A)

*5. If Salome is intelligent and studies hard, then she will get good grades and pass her courses. If Salome studies hard but lacks intelligence, then her efforts will be appreciated; and if her efforts are appreciated, then she will pass her courses. If Salome is intelligent, then she studies hard. Therefore Salome will pass her courses. (I, S, G, P, A)

6. If there is a single norm for greatness of poetry, then Milton and Edgar Guest cannot both be great poets. If either Pope or Dryden is regarded as a great poet, then Wordsworth is certainly no great poet; but if Wordsworth is no great poet, then neither is Keats nor Shelley. But after all, even though Edgar Guest is not, Dryden and Keats are both great poets. Hence there is no single norm for greatness of poetry. (N, M, G, P, D, W, K, S)

7. If the butler were present, he would have been seen; and if he had been seen, he would have been questioned. If he had been questioned, he would have replied; and if he had replied, he would have been heard. But the butler was not heard. If the butler was neither seen nor heard, then he must have been on duty; and if he was on duty, he must have been present. Therefore the butler was questioned. (P, S, Q, R, H, D)

8. If the butler told the truth, then the window was closed when he entered the room; and if the gardener told the truth, then the automatic sprinkler system was not operating on the evening of the murder. If the butler and the gardener are both lying, then a conspiracy must exist to protect someone in the house and there would have been a little pool of water on the floor just inside the window. We know that the window could not have been closed when the butler entered the room. There was a little pool of water on the floor just inside the window. So if there is a conspiracy to protect someone in the house, then the gardener did not tell the truth. (B, W, G, S, C, P)

9. Their chief would leave the country if she feared capture, and she would not leave the country unless she feared capture. If she feared capture and left the country, then the enemy's espionage network would be demoral-

ized and powerless to harm us. If she did not fear capture and remained in the country, it would mean that she was ignorant of our own agents' work. If she is really ignorant of our agents' work, then our agents can consolidate their positions within the enemy's organization; and if our agents can consolidate their positions there, they will render the enemy's espionage network powerless to harm us. Therefore the enemy's espionage network will be powerless to harm us. (L, F, D, P, I, C)

*10. If the investigators of extrasensory perception are regarded as honest, then considerable evidence for extrasensory perception must be admitted; and the doctrine of clairvoyance must be considered seriously if extrasensory perception is tentatively accepted as a fact. If considerable evidence for extrasensory perception is admitted, then it must be tentatively accepted as a fact and an effort must be made to explain it. The doctrine of clairvoyance must be considered seriously if we are prepared to take seriously that class of phenomena called occult; and if we are prepared to take seriously that class of phenomena called occult, a new respect must be paid to mediums. If we pursue the matter further, then if a new respect must be paid to mediums, we must take seriously their claims to communicate with the dead. We do pursue the matter further, but still we are practically committed to believing in ghosts if we take seriously the mediums' claims to communicate with the dead. Hence if the investigators of extrasensory perception are regarded as honest, we are practically committed to believing in ghosts. (H, A, C, F, E, O, M, P, D, G)

11. If we buy a lot, then we will build a house. If we buy a lot, then if we build a house we will buy furniture. If we build a house, then if we buy furniture we will buy dishes. Therefore if we buy a lot, we will buy dishes. (L, H, F, D)

12. If your prices are low, then your sales will be high, and if you sell quality merchandise, then your customers will be satisfied. So if your prices are low and you sell quality merchandise, then your sales will be high and your customers satisfied. (L, H, Q, S)

13. If your prices are low, then your sales will be high, and if you sell quality merchandise, then your customers will be satisfied. So if either your prices are low or you sell quality merchandise, then either your sales will be high or your customers will be satisfied. (L, H, Q, S)

14. If Jordan joins the alliance, then either Algeria or Syria boycotts it. If Kuwait joins the alliance, then either Syria or Iraq boycotts it. Syria does not boycott it. Therefore if neither Algeria nor Iraq boycotts it, then neither Jordan nor Kuwait joins the alliance. (J, A, S, K, I)

*15. If either Jordan or Algeria joins the alliance, then if either Syria or Kuwait boycotts it, then although Iraq does not boycott it, Yemen boycotts it. If either Iraq or Morocco does not boycott it, then Egypt will join the alliance. Therefore if Jordan joins the alliance, then if Syria boycotts it, then Egypt will join the alliance. (J, A, S, K, I, Y, M, E)

C. If any truth-functional argument is valid, we have the tools to prove it valid; and if it is invalid, we have the tools to prove it invalid. Prove each of the following arguments valid or invalid. These proofs will be more difficult to construct than in preceding exercises, but they will offer greater satisfaction.

1. If the president cuts Social Security benefit payments, he will lose the support of the senior citizens; and if he cuts defense spending, he will lose the support of the conservatives. If the president loses the support of either the senior citizens or the conservatives, then his influence in the Senate will diminish. But his influence in the Senate will not diminish. Therefore the president will not cut either Social Security benefits or defense spending. (B, S, D, C, I)

2. If inflation continues, then interest rates will remain high. If inflation continues, then if interest rates remain high then business activity will decrease. If interest rates remain high, then if business activity decreases then unemployment will rise. So if unemployment rises, then inflation will continue. (I, H, D, U)

3. If taxes are reduced, then inflation will rise, but if the budget is balanced, then unemployment will increase. If the president keeps his campaign promises, then either taxes are reduced or the budget is balanced. Therefore if the president keeps his campaign promises, then either inflation will rise or unemployment will increase. (T, I, B, U, K)

4. Weather predicting is an exact science. Therefore either it will rain tomorrow or it won't. (W, R)

5. If either it will rain tomorrow or it won't rain tomorrow, then weather predicting is an exact science. Therefore weather predicting is an exact science. (R, W)

9.11 Indirect Proof of Validity

Contradictory statements cannot both be true. Therefore, a statement added to the premises that makes it possible to deduce a contradiction must entail a falsehood. This gives rise to another method of proving validity. Suppose we *assume* (for the purposes of the proof only) the *denial* of what is to be proved. Suppose, using that assumption, we can derive a contradiction. That contradiction will show that when we denied what was to be proved we were brought to absurdity. We will have established the desired conclusion *indirectly*, with a proof by *reductio ad absurdum*.

An *indirect proof of validity* is written out by stating as an additional assumed premise the negation of the conclusion. If we can derive an explicit contradiction from the set of premises thus augmented, the argument with which we began must be valid. The method is illustrated with the following argument:

1. $A \supset (B \cdot C)$
2. $(B \lor D) \supset E$
3. $D \lor A$
   $\therefore E$

In the very next line we make explicit our assumption (for the purpose of the indirect proof) of the denial of the conclusion.

    4. ~E   I.P. (Indirect Proof)

With the now enlarged set of premises we can, using the established rules of inference, bring out an explicit contradiction, thus:

    5. ~(B ∨ D)   2, 4, M.T.
    6. ~B • ~D   5, De M.
    7. ~D • ~B   6, Com.
    8. ~D   7, Simp.
    9. A   3, 8, D.S.
    10. B • C   1, 9, M.P.
    11. B   10, Simp.
    12. ~B   6, Simp.
    13. B • ~B   11, 12, Conj.

The last line of the proof is an explicit contradiction, which is a demonstration of the absurdity to which we were led by assuming ~E in line 4. This contradiction, formally and explicitly expressed in the last line, exhibits the absurdity and completes the proof.

    This method of indirect proof strengthens our machinery for testing arguments by making it possible, in some circumstances, to prove validity more quickly than would be possible without it. We can illustrate this by first constructing a direct formal proof of the validity of an argument, and then demonstrating the validity of that same argument using an indirect proof. In the following example, the proof without the *reductio ad absurdum* is on the left and requires fifteen steps; the proof using the *reductio ad absurdum* is on the right and requires only eight steps. An exclamation point (!) is used to indicate that a given step is derived after the assumption advancing the indirect proof had been made.

1. (H ⊃ I) • (J ⊃ K)
2. (I ∨ K) ⊃ L
3. ~L
∴ ~(H ∨ J)

4. ~(I ∨ K)	2, 3, M.T.	!4 ~~(H ∨ J)	I.P. (Indirect Proof)
5. ~I • ~K	4, De M.	!5 H ∨ J	4, D.N.
6. ~I	5, Simp.	!6 I ∨ K	1, 5, C.D.
7. H ⊃ I	1, Simp.	!7 L	2, 6, M.P.
8. ~H	7, 6, M.T.	!8 L • ~L	7, 3, Conj.
9. (J ⊃ K) • (H ⊃ I)	1, Com.		
10. J ⊃ K	9, Simp.		
11. ~K • ~I	5, Com.		
12. ~K	11, Simp.		
13. ~J	10, 12, M.T.		
14. ~H • ~J	8, 13, Conj.		
15. ~(H ∨ J)	14, De M.		

EXERCISES

A. For each of the following arguments, construct an indirect proof of validity.

1. 1. A ∨ (B • C)
   2. A ⊃ C
   ∴ C

2. 1. (G ∨ H) ⊃ ~G
   ∴ ~G

3. 1. (D ∨ E) ⊃ (F ⊃ G)
   2. (~G ∨ H) ⊃ (D • F)
   ∴ G

4. 1. (M ∨ N) ⊃ (O • P)
   2. (O ∨ Q) ⊃ (~R • S)
   3. (R ∨ T) ⊃ (M • U)
   ∴ ~R

*5. 1. D ⊃ (Z ⊃ Y)
   2. Z ⊃ (Y ⊃ ~Z)
   ∴ ~D ∨ ~Z

6. 1. (O ∨ P) ⊃ (D • E)
   2. (E ∨ L) ⊃ (Q ∨ ~D)
   3. (Q ∨ Z) ⊃ ~(O • E)
   ∴ ~O

7. 1. (F ∨ G) ⊃ (D • E)
   2. (E ∨ H) ⊃ Q
   3. (F ∨ H)
   ∴ Q

8. 1. B ⊃ [(O ∨ ~O) ⊃ (T ∨ U)]
   2. U ⊃ ~(G ∨ ~G)
   ∴ B ⊃ T

B. For each of the following two arguments, construct an indirect proof of validity.

1. If a sharp fall in the prime rate of interest produces a rally in the stock market, then inflation is sure to come soon. But if a drop in the money supply produces a sharp fall in the prime rate of interest, then early inflation is equally certain. So inflation will soon be upon us. (F, R, I, D)

2. If precipitation levels remain unchanged and global warming intensifies, ocean levels will rise and some ocean ports will be inundated. But ocean ports will not be inundated if global warming intensifies. Therefore either precipitation levels will not remain unchanged or global warming will not intensify. (L, G, O, P)

C. For the following argument, construct both (a) a direct formal proof of validity and (b) an indirect proof of validity. Compare the lengths of the two proofs.

1. (V ⊃ ~W) • (X ⊃ Y)
2. (~W ⊃ Z) • (Y ⊃ ~A)
3. (Z ⊃ ~B) • (~A ⊃ C)
4. V • X
∴ ~B • C

9.12 Shorter Truth-Table Technique

There is still another method of testing the validity of arguments. We have seen how an argument may be proved *invalid* by assigning truth values to its component simple statements in such a way as to make all its premises true and its conclusion false. It is of course impossible to make such assignments if the argument is valid. So we can prove the *validity* of an argument by showing that *no such set of truth values can be assigned*. We do this by showing that its premises can be made true, and its conclusion false, only by assigning truth values *inconsistently*—that is, only with an assignment of values such that some component statement is assigned both a T and an F. In other words, if the truth value T is assigned to each premise of a valid argument, and the truth value F is assigned to its conclusion, this will necessitate assigning both T and F to some component statement—which is, of course, a contradiction. Here again we use the general method of *reductio ad absurdum*.

For example, we can very quickly prove the validity of the argument

$(A \lor B) \supset (C \cdot D)$
$(D \lor E) \supset G$
$\therefore A \supset G$

by first assigning T to each premise and F to the conclusion. But assigning F to the conclusion requires that T be assigned to *A* and F be assigned to *G*. Because T is assigned to *A*, the antecedent of the first premise is true, and because the premise as a whole has been assigned T, its consequent must be true also—so T must be assigned to both *C* and *D*. Because T is assigned to *D*, the antecedent of the second premise is true, and because the premise as a whole has been assigned T, its consequent must also be true, so T must be assigned to *G*. But we have already been forced to assign F to *G*, in order to make the conclusion false. Hence the argument would be invalid only if the statement *G* were both false and true, which is obviously impossible. Proving the validity of an argument with this "shorter truth-table technique" is one version of the use of *reductio ad absurdum*, reducing to the absurd—but instead of using the rules of inference, it uses truth-value assignments.

This *reductio ad absurdum* method of assigning truth values is often the quickest method of testing arguments, but it is more readily applied in some arguments than in others, depending on the kinds of propositions involved. Its easiest application is when F is assigned to a disjunction (in which case both of the disjuncts must be assigned F) or T to a conjunction (in which case both of the conjuncts must be assigned T). When assignments to simple statements are thus forced, the absurdity (if there is one) is quickly exposed. But where the method calls for T to be assigned to a disjunction, we cannot be sure which disjunct is true; and where F must be assigned to a conjunction, we cannot be sure which conjunct is false; in such cases we must make various "trial assignments," which slows the process and diminishes the advantage of this method. However, it remains the case that the *reductio ad absurdum* method of proof is often the most efficient means in testing the validity of a deductive argument.

EXERCISES

A. Use the *reductio ad absurdum* method of assigning truth values (the shorter truth-table technique) to determine the validity or invalidity of the arguments in Exercise Set B, on page 322.

B. Do the same for the arguments in Exercise Set C, on pages 322–323.

chapter 9 Summary

In this chapter we explained various methods with which the validity or invalidity of deductive arguments may be proved.

In Section 9.1 we introduced and explained the notion of a formal proof of validity, and we listed the first nine rules of inference with which formal proofs may be constructed.

In Section 9.2 we examined in detail the elementary valid argument forms that constitute the first nine rules of inference, and illustrated their use in simple arguments.

In Section 9.3 we illustrated the ways in which the elementary valid argument forms can be used to build formal proofs of validty.

In Section 9.4 we began the process of constructing formal proofs of validity, using only the first nine rules of inference.

In Section 9.5 we illustrated the ways in which the first nine rules of inference can be used to construct more extended formal proofs of validity.

In Section 9.6 we introduced the general rule of replacement, and expanded the rules of inference by adding ten logical equivalences, each of which permits the replacement of one logical expression by another having exactly the same meaning.

In Section 9.7 we discussed the features of the system of natural deduction that contains nineteen rules of inference.

In Section 9.8 we began the enterprise of building formal proofs of validity using all nineteen rules of inference: nine elementary valid argument forms, and ten logical equivalences permitting replacement.

In Section 9.9 we explained the method of proving invalidity when deductive arguments are not valid.

In Section 9.10 we discussed inconsistency, explaining why any argument with inconsistent premises cannot be sound, but will be valid.

In Section 9.11 we explained and illustrated indirect proof of validity.

In Section 9.12 we explained and illustrated the shorter truth-table technique for proving validity.

END NOTES

[1] See also John A. Winnie, "The Completeness of Copi's System of Natural Deduction," *Notre Dame Journal of Formal Logic* 11 (July 1970), pages 379–382.

[2] Recounted by George Will, in *Newsweek,* 27 October 2003.

Quantification Theory

- **10.1** The Need for Quantification
- **10.2** Singular Propositions
- **10.3** Universal and Existential Quantifiers
- **10.4** Traditional Subject–Predicate Propositions
- **10.5** Proving Validity
- **10.6** Proving Invalidity
- **10.7** Asyllogistic Inference

10.1 The Need for Quantification

Many valid deductive arguments cannot be tested using the logical techniques of the preceding two chapters. Therefore we must now enhance our analytical tools. We do this with *quantification*, a twentieth-century development chiefly credited to Gottlob Frege (1848–1945), a great German logician and the founder of modern logic. His discovery of quantification has been called the deepest single technical advance ever made in logic.

Biography

Gottlob Frege

One of the founders of modern symbolic logic, and also of analytic philosophy, Gottlob Frege (1848–1925) began as a mathematician. He came to believe that mathematics grows out of logic, and sought to devise the symbolic language with which this could be shown.

Frege was born in Wismar, a Hanseatic seaport in Germany east of Hamburg, the son of creative schoolmasters. His interest in the logic of language was first stimulated by his father's textbook designed to teach German teenagers the deep structure of their language. Frege studied mathematics and physics at the University of Jena, becoming a close friend of his teachers there. The great center for the study of mathematics in those days was the University of Göttingen, where Frege continued his studies, obtaining his doctorate in geometry in 1873.

It was logic, however, that became his consuming interest. His great work is called *Concept Script [Begriffsschrift]: A Formal Language for Pure Thought, Modeled on that of Arithmetic* (1879). The problem he confronted can be seen in this way: Logicians had long dealt with the basic connectives by which propositions are tied together—*and, or, if...then,*

explained in Chapters 8 and 9 in this book—but they had not yet devised the language with which to express fully, and to manipulate, expressions involving the concepts "some" and "all." Propositions having forms like "Some women overcome every hurdle," and "Some hurdles are overcome by every woman," as they might occur in argument, did not yield to the logical language then at hand. A new way to express these concepts accurately, a new *Begriffsschrift*, had to be invented.

Frege did this. Within this new formal language his development of *quantification*, explained and applied in this chapter, became a turning point in modern logic. Virtually all twentieth-century logicians were influenced by Frege's work. His larger objective was to show how logic provides the most fundamental principles of all inference, an enterprise later advanced by Bertrand Russell, with whom he corresponded. In Frege's *Foundations of Arithmetic* (1884) he sought to explain these connections in non-symbolic terms; later, in his *Basic Laws of Arithmetic* (1893 and 1903), he advanced the great project by building upon symbolic axioms derived from his earlier *Concept Script*.

Is the quality of a logician's work to be judged in the light of his character and political views? Bitter and introverted, Frege hated Catholics, hated the French, hated socialists, and above all hated the Jews, whose total expulsion from Germany he actually helped to plan. In his diary he made it clear that Adolf Hitler was his hero. Frege died in 1925, a loyal Nazi. ■

PEARSON
mylogiclab

To understand how quantification increases the power of logical analysis, we must recognize the limitations of the methods we have developed so far. The preceding chapters have shown that we can test deductive arguments effectively—but only arguments of one certain type, those whose validity depends entirely on the ways in which simple statements are truth-functionally combined into compound statements. Applying elementary argument forms and the rule of replacement, we draw inferences that permit us to discriminate valid from invalid arguments of that type. This we have done extensively.

When we confront arguments built of propositions that are *not* compound, however, those techniques are not adequate; they cannot *reach* the critical elements in the reasoning process. Consider, for example, the ancient argument

> All humans are mortal.
> Socrates is human.
> Therefore Socrates is mortal.[*]

---

[*]It was to arguments of this type that the classical or Aristotelian logic was primarily devoted, as described in Chapters 5 and 6. Those traditional methods, however, do not possess the generality or power of the newer symbolic logic and cannot be extended to cover all deductive arguments of the kinds we are likely to confront.

This argument is obviously valid. However, using the methods so far introduced we can only symbolize it as

A
H
∴ M

and on this analysis it appears not to be valid. What is wrong here? The difficulty arises from the fact that the validity of this argument, which is intuitively clear, depends on the *inner logical structure* of its premises, and that inner structure cannot be revealed by the system we have developed thus far for symbolizing statements. The symbolization immediately above, plainly too blunt, is the best we can do without quantifiers. That is because the propositions in this valid argument are not compound, and the techniques presented thus far, which are designed to deal with compound statements, cannot deal adequately with noncompound statements. A method is needed with which noncompound statements can be described and symbolized in such a way that their inner logical structure will be revealed. The theory of **quantification** provides that method.

Quantification enables us to interpret noncompound premises as compound statements, without loss of meaning. With that interpretation we can then use all the elementary argument forms and the rule of replacement (as we have done with compound statements), drawing inferences and proving validity or invalidity—after which the compound conclusion reached may be transformed (again using quantification) back into the noncompound form with which we began. This technique adds very greatly to the power of our analytical machinery.

The methods of deduction developed earlier remain fundamental; quantifiers do not alter the rules of inference in any way. What has gone before may be called the *logic of propositions*. We now proceed, using some additional symbolization, to apply these rules of inference more widely, in what is called the *logic of predicates*. The inner structure of propositions—the relations of subjects and predicates—is brought to the surface and made accessible by *quantifiers*. Introducing this symbolization is the next essential step.

10.2 Singular Propositions

We begin with the simplest kind of noncompound statement, illustrated by the second premise of the illustrative argument above, "Socrates is human." Statements of this kind have traditionally been called *singular propositions*. An **affirmative singular proposition** asserts that a particular individual has some specified attribute. "Socrates" is the subject term in the present example (as ordinary grammar and traditional logic both agree), and "human" is the predicate term. The subject term denotes a particular individual; the predicate term designates some attribute that individual is said to have.

The same subject term, obviously, can occur in different singular propositions. One may assert that "Socrates is mortal," or "Socrates is fat," or "Socrates is wise," or "Socrates is beautiful." Of these assertions, some are true (the first and the third), and some are false (the second and fourth). Similarly, the very

Quantification
A method for describing and symbolizing noncompound statements by reference to their inner logical structure; the modern theory used in the analysis of what were traditionally called **A**, **E**, **I**, and **O** propositions.

Affirmative singular proposition
A proposition in which it is asserted that a particular individual has some specified attribute.

same predicate term can occur in different singular propositions. The term *human* is a predicate that appears in each of the following: "Aristotle is human," "Brazil is human," "Chicago is human," and "O'Keeffe is human"—of which the first and fourth are true, while the second and third are false.

An "individual" in this symbolism can refer not only to persons, but to any individual *thing*, such as a country, a book, a city, or anything of which an *attribute* (such as "human" or "heavy") can be predicated. Attributes do not have to be adjectives (such as "mortal" or "wise") as in our examples thus far, but can also be nouns (such as "a human"). In grammar the distinction between adjective and noun is important, of course, but in this context it is not significant. We do not need to distinguish between "Socrates is mortal" and "Socrates is a mortal." Predicates can also be verbs, as in "Aristotle writes," which can be expressed alternatively as "Aristotle is a writer." The critical first step is to distinguish between the subject and the predicate terms, between the individuals and the attributes they may be said to have. We next introduce two different kinds of symbols for referring to *individuals* and to *attributes*.

To denote individuals we use (following a very widely adopted convention) small, or lowercase, letters, from a through w. These symbols are **individual constants**. In any particular context in which they may occur, each will designate one particular individual throughout the whole of that context. It is usually convenient to denote an individual by the first letter of its (or his or her) name. We may use the letter s to denote Socrates, a to denote Aristotle, b to denote Brazil, c to denote Chicago, and so forth.

Capital letters are used to symbolize attributes that individuals may have, and again it is convenient to use the first letter of the attribute referred to: H for human, M for mortal, F for fat, W for wise, and so forth.

Singular propositions can now be symbolized. By writing an attribute symbol immediately to the left of an individual symbol, we symbolize the singular proposition affirming that the individual named has the attribute specified. Thus the singular proposition, "Socrates is human," will be symbolized simply as Hs. Of course, Ha symbolizes "Aristotle is human," Hb symbolizes "Brazil is human," Hc symbolizes "Chicago is human," and so forth.

It is important to note the pattern that is common to these terms. Each begins with the same attribute symbol, H, and is followed by a symbol for some individual, s or a or b or c, and so forth. We could write the pattern as "H—", where the dash to the right of the predicate symbol is a place marker for some individual symbol. This pattern we symbolize as Hx. We use Hx [sometimes written as $H(x)$] to symbolize the common pattern of all singular propositions that attribute "being human" to some individual. The letter x is called an **individual variable**—it is simply a place marker, indicating where the various individual letters a through w (the individual constants) may be written. When one of those constants does appear in place of x, we have a singular proposition. The letter x is available to serve as the variable because, by convention, a through w are the only letters we allow to serve as individual constants.

Let us examine the symbol Hx more closely. It is called a *propositional function*. We define a **propositional function** as an expression that (1) contains an individ-

Individual constant
A symbol (by convention, normally a lower case letter, a through w) used in logical notation to denote an individual.

Individual variable
A symbol (by convention, normally the lower case x or y) that serves as a placeholder for an individual constant.

Propositional function
In quantification theory, an expression that contains an individual variable and becomes a statement when that variable is replaced with an individual constant. A propositional function can also become a statement by the process of generalization.

ual variable and (2) becomes a statement when an individual constant is substituted for the individual variable.* So a propositional function is not itself a proposition, although it can become one by substitution. Individual constants may be thought of as the proper names of individuals. Any singular proposition is a substitution instance of a propositional function; it is the result of substituting some individual constant for the individual variable in that propositional function.

A propositional function normally has some true substitution instances and some false substitution instances. If *H* symbolizes human, *s* symbolizes Socrates, and *c* symbolizes Chicago, then *Hs* is true and *Hc* is false. With the substitution made, what confronts us is a proposition; before the substitution is made, we have only the propositional *function*. There are an unlimited number of such propositional functions, of course: *Hx*, and *Mx*, and *Bx*, and *Fx*, and *Wx*, and so on. We call these propositional functions *simple predicates*, to distinguish them from more complex propositional functions to be introduced in following sections. A **simple predicate** is a propositional function that has some true and some false substitution instances, each of which is an affirmative singular proposition.

10.3 Universal and Existential Quantifiers

A singular proposition affirms that some individual thing has a given predicate, so it is the substitution instance of some propositional function. If the predicate is *M* for mortal, or *B* for beautiful, we have the simple predicates *Mx* or *Bx*, which assert humanity or beauty of nothing in particular. If we substitute Socrates for the variable *x*, we get singular propositions, "Socrates is mortal," or "Socrates is beautiful." But we might wish to assert that the attribute in question is possessed by more than a single individual. We might wish to say that "Everything is mortal," or that "Something is beautiful." These expressions contain predicate terms, but they are not singular propositions because they do not refer specifically to any particular individuals. These are *general* propositions.

Let us look closely at the first of these general propositions, "Everything is mortal." It may be expressed in various ways that are logically equivalent. We could express it by saying "All things are mortal." We could express it by saying:

> Given any individual thing whatever, it is mortal.

In this latter formulation the word "it" is a pronoun that refers back to the word "thing" that precedes it. We can use the letter *x*, our individual variable, in place of both the pronoun and its antecedent. So we can rewrite the first general proposition as

> Given any *x*, *x* is mortal.

Alternatively, using the notation for predicates we introduced in the preceding section, we may write

> Given any *x*, *Mx*.

Simple predicate
In quantification theory, a propositional function having some true and some false substitution instances, each of which is an affirmative singular proposition.

---
*Some writers regard "propositional functions" as the meanings of such expressions, but here we define them to be the expressions themselves.

Universal quantifier In quantification theory, a symbol, (x), used before a propositional function to assert that the predicate following the symbol is true of everything. Thus "(x) Fx" means "Given any x, F is true of it."

We know that *Mx* is a propositional function, not a proposition. But here, in this last formulation, we have an expression that *contains Mx*, and that clearly *is* a proposition. The phrase "Given any *x*" is customarily symbolized by "(*x*)", which is called the **universal quantifier**. That first general proposition may now be completely symbolized as

(*x*) *Mx*

which says "Everything is mortal."

This analysis shows that we can convert a propositional function into a proposition not only by *substitution*, but also by *generalization*, or *quantification*.

Biography

Alfred North Whitehead

Alfred North Whitehead was born in 1861 in Ramsgate, Kent, England, the son of an Anglican minister. After graduating from a prestigious "public" school, he studied mathematics at Trinity College, Cambridge, eventually becoming a Fellow of that College. The famous economist, John Maynard Keynes, was one of his students in mathematics there. Another of his students, Bertrand Russell, was later to become his collaborator and co-author.

Together, Whitehead and Russell produced, after the labor of a decade, the highly influential treatise, *Principia Mathematica*, in three volumes: 1910, 1912, and 1913. In this work the derivation of mathematics from basic logical principles, earlier attempted by Gottlob Frege in Germany, was at last carried out. That joint product is one of the most significant achievements of twentieth-century logic.

The friendship of Whitehead and Russell was split sharply by political differences. Whitehead had three sons, one of whom was killed in the First World War. Russell was a pacifist who actively objected to British participation in the war. The two authors were divided deeply and permanently. When a new edition of *Principia Mathematica* was published in 1927, Whitehead refused to contribute to it.

After a full career teaching mathematics and logic at Cambridge University, and then in London, Whitehead turned to the development of metaphysical and historical themes. In 1924 he was invited to Harvard to teach philosophy; he did so, and spent the remainder of his life in America. *Science and the Modern World* (1925) was his penetrating account of the role of science and mathematics in the rise of Western civilization. *Process and Reality* (1929) presented his metaphysical views, in which, in the tradition of the ancient philosopher Heraclitus, "all things flow," and nothing is stable. Truths, he therefore thought, could be no more than half-truths. But it is as creative logician, and as Russell's collaborator in the writing of *Principia Mathematica*, that he is best known. Whitehead died in Cambridge, Massachusetts, in 1947. ∎

PEARSON
mylogiclab

Consider now the second general proposition we had entertained: "Something is beautiful." This may also be expressed as

> There is at least one thing that is beautiful.

In this latter formulation, the word "that" is a relative pronoun referring to the word "thing." Using our individual variable x once again in place of both the pronoun "that" and its antecedent "thing," we may rewrite the second general proposition as

> There is at least one x such that x is beautiful.

Or, using the notation for predicates, we may write

> There is at least one x such that Bx.

Once again we see that, although Bx is a propositional function and not a proposition, we have here an expression that contains Bx that *is* a proposition. The phrase "there is at least one x such that" is customarily symbolized as "$(\exists x)$"; the $\exists$ is called the **existential quantifier**. Thus the second general proposition may be completely symbolized as

$(\exists x)Bx$

which says "Something is beautiful."

Thus we see that propositions may be formed from propositional functions either by **instantiation**, that is, by substituting an individual constant for its individual variable, or by **generalization**, that is, by placing a universal or existential quantifier before the propositional function.

Now consider: The *universal* quantification of a propositional function, $(x)Mx$, is true if and only if *all* the substitution instances of the function are true; that is what universality means here. It is also clear that the *existential* quantification of a propositional function, $(\exists x)Mx$, is true if and only if its propositional function has *at least one* true substitution instance. To understand quantified propositions and how they relate to one another, we will show how the traditional square of opposition can be represented in terms of quantified propositions. To do this, for the rest of this section we will assume (what no one will deny) that there exists at least one individual. Under this very weak assumption, every propositional function must have at least one substitution instance, an instance that may or may not be true. But it is certain that, under this assumption, if the *universal* quantification of a propositional function is true, then the *existential* quantification of it must also be true. That is, if every x is M, then, if there exists at least one thing, that thing is M.

Up to this point, only affirmative singular propositions have been given as substitution instances of propositional functions. Mx (x is mortal) is a propositional function. Ms is an instance of it, an affirmative singular proposition that says "Socrates is mortal." But not all propositions are affirmative. One may deny that Socrates is mortal, saying $\sim Ms$, "Socrates is not mortal." If Ms is a substitution instance of Mx, then $\sim Ms$ may be regarded as a substitution instance of the

Existential quantifier
In quantification theory, a symbol, $\exists$, used before a propositional function to assert that the function has one or more true substitution instances. Thus "$(\exists x)Fx$" means "there exists an x such that F is true of it."

Instantiation
In quantification theory, the process of substituting an individual constant for an individual variable, thereby converting a propositional function into a proposition.

Generalization
In quantification theory, the process of forming a proposition from a propositional function by placing a universal quantifier or an existential quantifier before it.

propositional function ~Mx. Thus we may enlarge our conception of propositional functions, beyond the simple predicates introduced in the preceding section, to permit them to contain the negation symbol, "~".

With the negation symbol at our disposal, we may now enrich our understanding of quantification as follows. We begin with the general proposition

> Nothing is perfect.

which we can paraphrase as

> Everything is imperfect.

which in turn may be written as

> Given any individual thing whatever, it is not perfect.

which can be rewritten as

> Given any x, x is not perfect.

If P symbolizes the attribute of being perfect, we can use the notation just developed (the quantifier and the negation sign) to express this proposition ("Nothing is perfect.") as (x)~Px.

Now we are in a position to list and illustrate a series of important connections between universal and existential quantification.

First, the (universal) general proposition "Everything is mortal" is *denied* by the (existential) general proposition "Something is not mortal." Using symbols, we may say that $(x)Mx$ is denied by $(\exists x)$~Mx. Because each of these is the denial of the other, we may certainly say (prefacing the one with a negation symbol) that the biconditional

$$\sim(x)Mx \stackrel{T}{\equiv} (\exists x)\sim Mx$$

is necessarily, logically true.

Second, "Everything is mortal" expresses exactly what is expressed by "There is nothing that is not mortal"—which may be formulated as another biconditional, also logically true:

$$(x)Mx \stackrel{T}{\equiv} \sim(\exists x)\sim Mx$$

Third, it is clear that the (universal) general proposition, "Nothing is mortal," is *denied* by the (existential) general proposition, "Something is mortal." In symbols we say that (x)~Mx is denied by $(\exists x)Mx$. And because each of these is the denial of the other, we may certainly say (again prefacing the one with a negation symbol) that the biconditional

$$\sim(x)\sim Mx \stackrel{T}{\equiv} (\exists x)Mx$$

is necessarily, logically true.

Fourth, "Everything is not mortal" expresses exactly what is expressed by "There is nothing that is mortal"—which may be formulated as a logically true biconditional:

$$(x)\sim Mx \stackrel{T}{\equiv} \sim(\exists x)Mx$$

These four logically true biconditionals set forth the interrelations of universal and existential quantifiers. We may replace any proposition in which the quantifier is prefaced by a negation sign (using these logically true biconditionals) with another logically equivalent proposition in which the quantifier is not prefaced by a negation sign. We list these four biconditionals again, now replacing the illustrative predicate M (for mortal) with the symbol Φ (the Greek letter *phi*), which will stand for *any* simple predicate whatsoever.

$$[(\exists x)\sim\Phi x] \stackrel{T}{\equiv} [\sim(x)\Phi x]$$
$$[(x)\Phi x] \stackrel{T}{\equiv} [\sim(\exists x)\sim\Phi x]$$
$$[(\exists x)\Phi x] \stackrel{T}{\equiv} [\sim(x)\sim\Phi x]$$
$$[(x)\sim\Phi x] \stackrel{T}{\equiv} [\sim(\exists x)\Phi x]$$

Graphically, the general connections between universal and existential quantification can be described in terms of the square array shown in Figure 10-1.

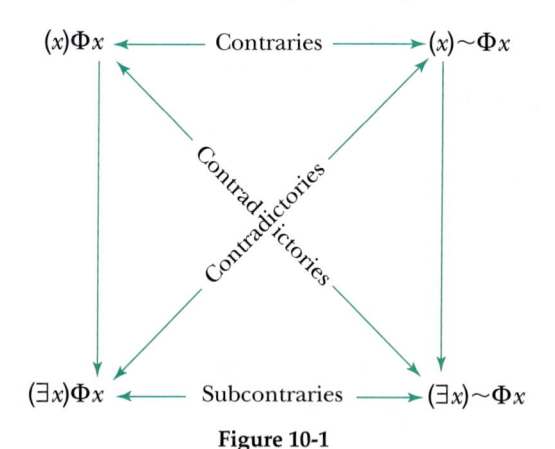

```
    (x)Φx  ←———— Contraries ————→  (x)~Φx
        ↑  ↖                   ↗  ↑
        │     Contradictories    │
        │         ╳              │
        │     Contradictories    │
        ↓  ↙                   ↘  ↓
    (∃x)Φx ←——— Subcontraries ———→ (∃x)~Φx
```

Figure 10-1

Continuing to assume the existence of at least one individual, we can say, referring to this square, that:

1. The two top propositions are *contraries;* that is, they may both be false but they cannot both be true.
2. The two bottom propositions are *subcontraries;* that is, they may both be true but they cannot both be false.
3. Propositions that are at opposite ends of the diagonals are *contradictories,* of which one must be true and the other must be false.
4. On each side of the square, the truth of the lower proposition is implied by the truth of the proposition directly above it.

10.4 Traditional Subject–Predicate Propositions

Using the existential and universal quantifiers, and with an understanding of the square of opposition in Figure 10-1, we are now in a position to analyze (and to use accurately in reasoning) the four types of general propositions that have been traditionally emphasized in the study of logic. The standard illustrations of these four types are the following:

> All humans are mortal. (universal affirmative: **A**)
> No humans are mortal. (universal negative: **E**)
> Some humans are mortal. (particular affirmative: **I**)
> Some humans are not mortal. (particular negative: **O**)

Each of these types is commonly referred to by its letter: the two affirmative propositions, **A** and **I** (from the Latin *affirmo*, "I affirm"); and the two negative propositions, **E** and **O** (from the Latin *nego*, "I deny").*

In symbolizing these propositions by means of quantifiers, we are led to a further enlargement of our conception of a propositional function. Turning first to the **A** proposition, "All humans are mortal," we proceed by means of successive paraphrasings, beginning with

> Given any individual thing whatever, if it is human then it is mortal.

The two instances of the pronoun "it" clearly refer to their common antecedent, the word "thing." As in the early part of the preceding section, because those three words have the same (indefinite) reference, they can be replaced by the letter x and the proposition rewritten as

> Given any x, if x is human then x is mortal.

Now using our previously introduced notation for "if–then," we can rewrite the preceding as

> Given any x, x is human $\supset$ x is mortal.

Finally, using our now-familiar notation for propositional functions and quantifiers, the original **A** proposition is expressed as

$$(x)(Hx \supset Mx)$$

In our symbolic translation, the **A** proposition appears as the universal quantification of a new kind of propositional function. The expression $Hx \supset Mx$ is a propositional function that has as its substitution instances neither affirmative nor negative singular propositions, but conditional statements whose antecedents and consequents are singular propositions that have the same subject term. Among the substitution instances of the propositional function $Hx \supset Mx$ are the conditional statements $Ha \supset Ma$, $Hb \supset Mb$, $Hc \supset Mc$, $Hd \supset Md$, and so on.

There are also propositional functions whose substitution instances are conjunctions of singular propositions that have the same subject terms. Thus the conjunctions $Ha \cdot Ma$, $Hb \cdot Mb$, $Hc \cdot Mc$, $Hd \cdot Md$, and so on, are substitution instances

---

*An account of the traditional analysis of these four types of propositions was presented in Chapter 5.

of the propositional function $Hx \cdot Mx$. There are also propositional functions such as $Wx \lor Bx$, whose substitution instances are disjunctions such as $Wa \lor Ba$ and $Wb \lor Bb$. In fact, any truth-functionally compound statement whose simple component statements are singular propositions that all have the same subject term may be regarded as a substitution instance of a propositional function containing some or all of the various truth-functional connectives and operators (dot, wedge, horseshoe, three-bar equivalence, and curl), in addition to the simple predicates $Ax, Bx, Cx, Dx, \ldots$. In our translation of the **A** proposition as $(x)(Hx \supset Mx)$, the parentheses serve as punctuation marks. They indicate that the universal quantifier (x) "applies to" or "has within its scope" the entire (complex) propositional function $Hx \supset Mx$.

Before going on to discuss the other traditional forms of categorical propositions, it should be observed that our symbolic formula $(x)(Hx \supset Mx)$ translates not only the standard-form proposition, "All H's are M's," but any other English sentence that has the same meaning. When, for example, a character in Henrik Ibsen's play, *Love's Comedy*, says, "A friend married is a friend lost," that is just another way of saying, "All friends who marry are friends who are lost." There are many ways, in English, of saying the same thing.

Here is a list, not exhaustive, of different ways in which we commonly express universal affirmative propositions in English:

H's are M's.
An H is an M.
Every H is M.
Each H is M.
Any H is M.
No H's are not M.
Everything that is H is M.
Anything that is H is M.
If anything is H, it is M.
If something is H, it is M.
Whatever is H is M.
H's are all M's.
Nothing is an H unless it is an M.
Nothing is an H but not an M.

To evaluate an argument we must understand the language in which the propositions of that argument are expressed. Some English idioms are a little misleading, using a temporal term when no reference to time is intended. Thus the proposition, "H's are always M's," is ordinarily understood to mean simply that *all* H's are M's. Again, the same meaning may be expressed using abstract nouns: "Humanity implies (or entails) mortality" is correctly symbolized as an **A** proposition. That the language of symbolic logic has a single expression for the common meaning of a considerable number of English sentences may be regarded as an advantage of symbolic logic over English for cognitive or informative purposes—although admittedly a disadvantage from the point of view of rhetorical power or poetic expressiveness.

Quantification of the A Proposition

The **A** proposition, "All humans are mortal," asserts that if anything is a human, then it is mortal. In other words, for any given thing x, *if x is a human, then x is mortal*. Substituting the horseshoe symbol for "if–then," we get

> Given any x, x is a human $\supset$ x is mortal.

In the notation for propositional functions and quantifiers this becomes

$$(x) [Hx \supset Mx]$$

Quantification of the E Proposition

The **E** proposition, "No humans are mortals," asserts that if anything is human, then it is not mortal. In other words, for any given thing x, *if x is a human, then x is not mortal*. Substituting the horseshoe symbol for "if–then," we get:

> Given any x, x is a human $\supset$ x is not mortal.

In the notation for propositional functions and quantifiers, this becomes

$$(x) [Hx \supset {\sim} Mx]$$

This symbolic translation expresses not only the traditional **E** form in English, but also such diverse ways of saying the same thing as "There are no H's that are M," "Nothing is both an H and an M," and "H's are never M."

Quantification of the I Proposition

The **I** proposition, "Some humans are mortal," asserts that there is at least one thing that is a human *and* is mortal. In other words, there is at least one x such that x is a human *and* x is mortal. Substituting the dot symbol for conjunction, we get

> There is at least one x such that x is a human $\cdot$ x is mortal.

In the notation for propositional functions and quantifiers, this becomes

$$(\exists x) [Hx \cdot Mx]$$

Quantification of the O Proposition

The **O** proposition, "Some humans are not mortal," asserts that there is at least one thing that is a human and is not mortal. In other words, there is at least one x such that x is human *and* x is not mortal. Substituting the dot symbol for conjunction we get

> There is at least one x such that x is a human $\cdot$ x is not mortal.

In the notation for propositional functions and quantifiers, this becomes

$$(\exists x) [Hx \cdot {\sim} Mx]$$

10.4 Traditional Subject–Predicate Propositions

Where the Greek letters *phi* (Φ) and *psi* (Ψ) are used to represent any predicates whatever, the four general subject–predicate propositions of traditional logic may be represented in a square array as shown in Figure 10-2.

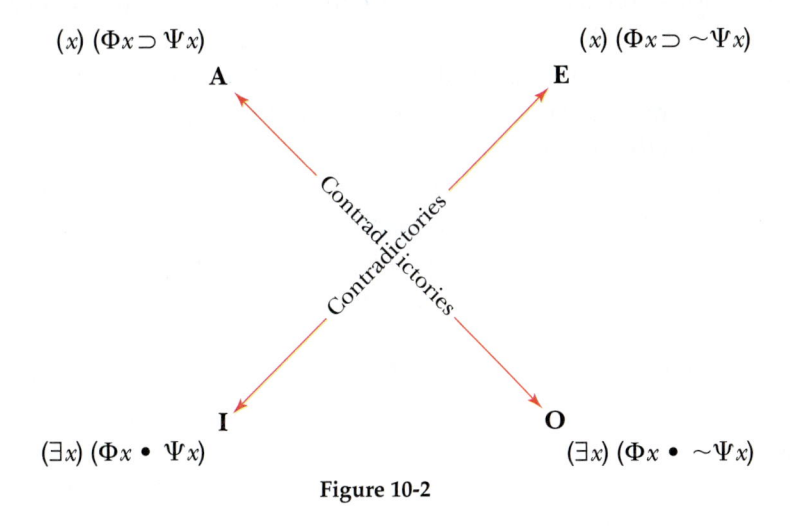

$$(x)(\Phi x \supset \Psi x) \quad \mathbf{A} \qquad \mathbf{E} \quad (x)(\Phi x \supset \sim\Psi x)$$

Contradictories

$$(\exists x)(\Phi x \cdot \Psi x) \quad \mathbf{I} \qquad \mathbf{O} \quad (\exists x)(\Phi x \cdot \sim\Psi x)$$

Figure 10-2

The relations displayed in Figure 10-2 match those displayed in Figure 5-2. For example, we have seen that **A** and **O** are contradictories, each being the denial of the other; and we have seen that **E** and **I** are also contradictories.

Thus far we have worked under the weak assumption that there exists at least one individual. Under this assumption, we expect an **I** proposition to follow from its corresponding **A** proposition, and an **O** from its corresponding **E**. But in fact our new formulation of universal categorical propositions as conditionals neatly incorporates the Boolean interpretation so that, say, an **A** proposition may very well be true while its corresponding **I** proposition is false. This we will now explain.

Where Φx is a propositional function that has no true substitution instances, then no matter what kinds of substitution instances the propositional function Ψx may have, the universal quantification of the (complex) propositional function $\Phi x \supset \Psi x$ will be true. For example, consider the propositional function, "x is a centaur," which we abbreviate as Cx. Because there are no centaurs, every substitution instance of Cx is false, that is, $Ca, Cb, Cc, \ldots$ are all false. Hence every substitution instance of the complex propositional function $Cx \supset Bx$ will be a conditional statement whose antecedent is false. The substitution instances $Ca \supset Ba, Cb \supset Bb, Cc \supset Bc, \ldots$ are therefore all true, because any conditional statement asserting a material implication must be true if its antecedent is false. Because all its substitution instances are true, the universal quantification of the propositional function $Cx \supset Bx$, which is the **A** proposition $(x)(Cx \supset Bx)$, is true. But the corresponding **I** proposition $(\exists x)(Cx \cdot Bx)$ is false, because the propositional function $Cx \cdot Bx$ has no true substitution instances. That $Cx \cdot Bx$ has no true substitution instances follows from the fact that Cx has no true substitution instances. The various substitution instances of $Cx \cdot Bx$ are $Ca \cdot Ba, Cb \cdot Bb, Cc \cdot Bc, \ldots$ each of which is a conjunction whose first conjunct is false, because $Ca, Cb, Cc, \ldots$ are all false. Because all its substitution instances are false, the existential quantification of the propositional function $Cx \cdot Bx$, which is the **I** proposition $(\exists x)(Cx \cdot Bx)$, is false. Hence an **A** proposition may be true while its corresponding **I** proposition is false.

This analysis shows also why an **E** proposition may be true while its corresponding **O** proposition is false. If we replace the propositional function Bx by the propositional function $\sim Bx$ in the preceding discussion, then $(x)(Cx \supset \sim Bx)$ may be true while $(\exists x)(Cx \cdot \sim Bx)$ will be false because, of course, there are no centaurs.

The key to the matter is this: **A** propositions and **E** propositions do not assert or suppose that anything exists; they assert only that (*if* one thing *then* another) is the case. But **I** propositions and **O** propositions do suppose that some things exist; they assert that (this *and* the other) is the case. The existential quantifier in **I** and **O** propositions makes a critical difference. It would plainly be a mistake to infer the existence of anything from a proposition that does not assert or suppose the existence of anything.

If we make the general assumption that there exists at least one individual, then $(x)(Cx \supset Bx)$ does imply $(\exists x)(Cx \supset Bx)$. But the latter is not an **I** proposition. The **I** proposition, "Some centaurs are beautiful," is symbolized as $(\exists x)(Cx \cdot Bx)$, which says that there is at least one centaur that is beautiful. But what is symbolized as $(\exists x)(Cx \supset Bx)$ can be rendered in English as "There is at least one thing such that, if it is a centaur, then it is beautiful." It does not say that there is a centaur, but only that there is an individual that either is not a centaur or is beautiful. This proposition would be false in only two possible cases: first, if there were no individuals at all; and second, if all individuals were centaurs and none of them were beautiful. We rule out the first case by making the explicit (and obviously true) assumption that there is at least one individual in the universe; and the second case is so extremely implausible that any proposition of the form $(\exists x)(\Phi x \supset \Psi x)$ is bound to be quite trivial, in contrast to the significant **I** form $(\exists x)(\Phi x \cdot \Psi x)$. The foregoing should make clear that, although in English the **A** and **I** propositions "All humans are mortal" and "Some humans are mortal" differ only in their initial words, "all" and "some," their difference in meaning is not confined to the matter of universal versus existential quantification, but goes deeper than that. The propositional functions quantified to yield **A** and **I** propositions are not just differently quantified; they are different propositional functions, one containing " $\supset$ ", the other " $\cdot$ ". In other words, **A** and **I** propositions are not as much alike as they appear in English. Their differences are brought out very clearly in the notation of propositional functions and *quantifiers*.

For purposes of logical manipulation we can work best with formulas in which the negation sign, if one appears at all, applies only to simple predicates. So we will want to replace formulas in ways that have this result. This we can do quite readily. We know from the rule of replacement established in Chapter 9 that we are always entitled to replace an expression by another that is logically equivalent to it; and we have at our disposal four logical equivalences (listed in Section 10.3) in which each of the propositions in which the quantifier is negated is shown equivalent to another proposition in which the negation sign applies directly to the predicates. Using the rules of inference with which we have long been familiar, we can shift negation signs so that, in the end, they no longer apply to compound expressions but apply only to simple predicates. Thus, for example, the formula

$$\sim(\exists x)(Fx \cdot \sim Gx)$$

can be successively rewritten. First, when we apply the fourth logical equivalence given on page 405, it is transformed into

$$(x)\sim(Fx \cdot \sim Gx)$$

Then when we apply De Morgan's theorem, it becomes

$$(x)(\sim Fx \vee \sim\sim Gx)$$

Next, the principle of Double Negation gives us

$$(x)(\sim Fx \vee Gx)$$

And finally, when we invoke the definition of Material Implication, the original formula is rewritten as the **A** proposition

$$(x)(Fx \supset Gx)$$

We call a formula in which negation signs apply only to simple predicates a **normal-form formula**.

Before turning to the topic of inferences involving noncompound statements, the reader should acquire some practice in translating noncompound statements from English into logical symbolism. The English language has so many irregular and idiomatic constructions that there can be no simple rules for translating an English sentence into logical notation. What is required in each case is that the meaning of the sentence be understood and then restated in terms of propositional functions and quantifiers.

EXERCISES

A. Translate each of the following into the logical notation of propositional functions and quantifiers, in each case using the abbreviations suggested and making each formula begin with a quantifier, not with a negation symbol.

EXAMPLE

1. No beast is without some touch of pity. (*Bx*: *x* is a beast; *Px*: *x* has some touch of pity.)

SOLUTION

$$(x)(Bx \supset Px)$$

2. Sparrows are not mammals. (*Sx*: *x* is a sparrow; *Mx*: *x* is a mammal.)
3. Reporters are present. (*Rx*: *x* is a reporter; *Px*: *x* is present.)
4. Nurses are always considerate. (*Nx*: *x* is a nurse; *Cx*: *x* is considerate.)
*5. Diplomats are not always rich. (*Dx*: *x* is a diplomat; *Rx*: *x* is rich.)
6. "To swim is to be a penguin." (*Sx*: *x* swims; *Px*: *x* is a penguin.)
   —Christina Slagar, curator, Monterey Bay Aquarium, 17 January 2003

Normal-form formula
A formula in which negation signs apply to simple predicates only.

7. No boy scout ever cheats. (*Bx*: *x* is a boy scout; *Cx*: *x* cheats.)

8. Only licensed physicians can charge for medical treatment. (*Lx*: *x* is a licensed physician; *Cx*: *x* can charge for medical treatment.)

9. Snake bites are sometimes fatal. (*Sx*: *x* is a snake bite; *Fx*: *x* is fatal.)

*10. The common cold is never fatal. (*Cx*: *x* is a common cold; *Fx*: *x* is fatal.)

11. A child pointed his finger at the emperor. (*Cx*: *x* is a child; *Px*: *x* pointed his finger at the emperor.)

12. Not all children pointed their fingers at the emperor. (*Cx*: *x* is a child; *Px*: *x* pointed his finger at the emperor.)

13. All that glitters is not gold. (*Gx*: *x* glitters; *Ax*: *x* is gold.)

14. None but the brave deserve the fair. (*Bx*: *x* is brave; *Dx*: *x* deserves the fair.)

*15. Only citizens of the United States can vote in U.S. elections. (*Cx*: *x* is a citizen of the United States; *Vx*: *x* can vote in U.S. elections.)

16. Citizens of the United States can vote only in U.S. elections. (*Ex*: *x* is an election in which citizens of the United States can vote; *Ux*: *x* is a U.S. election.)

17. Not every applicant was hired. (*Ax*: *x* is an applicant; *Hx*: *x* was hired.)

18. Not any applicant was hired. (*Ax*: *x* is an applicant; *Hx*: *x* was hired.)

19. Nothing of importance was said. (*Lx*: *x* is of importance; *Sx*: *x* was said.)

*20. They have the right to criticize who have a heart to help. (*Cx*: *x* has the right to criticize; *Hx*: *x* has a heart to help.)

B. Translate each of the following into the logical notation of propositional functions and quantifiers, in each case making the formula begin with a quantifier, *not* with a negation symbol.

1. Nothing is attained in war except by calculation.
   —Napoleon Bonaparte

2. No one doesn't believe in laws of nature.
   —Donna Haraway, *The Chronicle of Higher Education,* 28 June 1996

3. He only earns his freedom and existence who daily conquers them anew.
   —Johann Wolfgang von Goethe, *Faust,* Part II

4. No man is thoroughly miserable unless he be condemned to live in Ireland.
   —Jonathan Swift

*5. Not everything good is safe, and not everything dangerous is bad.
   —David Brooks, in *The Weekly Standard,* 18 August 1997

6. There isn't any business we can't improve.
   —Advertising slogan, Ernst and Young, Accountants

7. A problem well stated is a problem half solved.
   —Charles Kettering, former research director for General Motors

8. There's not a single witch or wizard who went bad who wasn't in Slytherin.
   —J. K. Rowling, in *Harry Potter and the Sorcerer's Stone*

9. Everybody doesn't like something, but nobody doesn't like Willie Nelson.
   —Steve Dollar, Cox News Service

*10. No man but a blockhead ever wrote except for money.
   —Samuel Johnson

C. For each of the following, find a normal-form formula that is logically equivalent to the given one:

*1. $\sim(x)(Ax \supset Bx)$

2. $\sim(x)(Cx \supset \sim Dx)$

3. $\sim(\exists x)(Ex \cdot Fx)$

4. $\sim(\exists x)(Gx \cdot \sim Hx)$

*5. $\sim(x)(\sim Ix \vee Jx)$

6. $\sim(x)(\sim Kx \vee \sim Lx)$

7. $\sim(\exists x)[\sim(Mx \vee Nx)]$

8. $\sim(\exists x)[\sim(Ox \vee \sim Px)]$

9. $\sim(\exists x)[\sim(\sim Qx \vee Rx)]$

*10. $\sim(x)[\sim(Sx \cdot \sim Tx)]$

11. $\sim(x)[\sim(\sim Ux \cdot \sim Vx)]$

12. $\sim(\exists x)[\sim(\sim Wx \vee Xx)]$

Biography

John von Neumann

Logic is absolutely central in the design of computers. John von Neumann (1903–1957), a Hungarian-American mathematician and logician, helped to bring logic into all our lives through his work on the intellectual architecture of computers.

Von Neumann's intellect was utterly remarkable; he inspired awe among his colleagues, who regarded him as among the greatest mathematicians of modern history. As a very young boy in Hungary, under the direction of private tutors, he had mastered arithmetic, algebra, analytic geometry and trigonometry. He taught himself calculus. He exhibited prodigious skills in learning languages, including classical Greek and Latin, and in memorizing vast bodies of material. The speed and depth of his mental calculations, even as a youth, were staggering.

At the age of 22 he received his Ph.D. in mathematics in Hungary, and that same year his diploma in chemical engineering in Switzerland. He lectured at the University of Berlin, and then in Hanover. When his father died, in 1929, the family emigrated to the United States, where von Neumann anglicized his first name (from Janos to John). He was invited to Princeton University and became one of the first four professors selected, in 1933, for the Institute for Advanced Study there (two of the others being Albert Einstein and Kurt Gödel). Von Neumann remained a mathematics professor at the Institute until his early death from cancer in 1957. He lived an active social life in Princeton. He loved good clothes and fine cars; he loved eating, and drinking, and telling jokes. He threw great parties and enjoyed the good life. He was warmly liked and enormously admired by his friends and colleagues.

Von Neumann's contributions in logic began with his work on the axiomatization of set theory. The advance of the theory of sets was dealt a blow when Gödel proved that axiomatic systems are necessarily incomplete, in the sense that they cannot prove every truth that is expressible in their own language. Von Neumann wrote to Gödel, calling to his attention that it can also be proved that it is impossible for the usual axiomatic systems to demonstrate their own consistency; this became what is now called Gödel's second incompleteness theorem.

As a theoretical mathematician von Neumann contributed significantly to the development of the atomic bomb during the Second World War. During that war the first general-purpose electronic computer, ENIAC (Electronic Numerical Integrator and Computer) was designed at the University of Pennsylvania. Logicians (including the distinguished Michigan logician, Arthur Burks) were its creators. As the war ended, a second and more advanced computer project was undertaken there: EDVAC (Electronic Discrete Variable Automatic Computer). John von Neumann was called in to assist in its development. He did so, summarizing and improving all logical computer design, and writing, in 1945, the *First Draft of a Report on the EDVAC*. That computer (physically huge, although exceedingly weak in comparison to electronic computers now commonly at hand) was actually built. It was completed at the U.S Army's Ballistic Research Laboratory in Maryland in 1949, and ran successfully, day and night, for about ten years, from 1951 to 1961. John von Neumann, as logician, had played a key role in the birth of the computer age. His very last work, written while he was in the hospital in 1956, and published posthumously, was entitled: *The Computer and the Brain*. ∎

10.5 Proving Validity

To construct formal proofs of validity for arguments whose validity turns on the inner structures of noncompound statements that occur in them, we must expand our list of rules of inference. Only four additional rules are required, and they will be introduced in connection with arguments for which they are needed.

Consider the first argument we discussed in this chapter: "All humans are mortal. Socrates is human. Therefore Socrates is mortal." It is symbolized as

$(x)(Hx \supset Mx)$
Hs
$\therefore Ms$

The first premise affirms the truth of the universal quantification of the propositional function $Hx \supset Mx$. Because the universal quantification of a propositional function is true if and only if all of its substitution instances are true, from the first premise we can infer any desired substitution instance of the propositional function $Hx \supset Mx$. In particular, we can infer the substitution instance $Hs \supset Ms$.

Biography

Bertrand Russell

Bertrand Arthur William, Lord Russell, The Right Honourable The Earl Russell (1872–1970), was one of the most remarkable thinkers of recent centuries. His grandfather had been England's Prime Minister, befriended by Queen Victoria; his parents, religious skeptics who endorsed free love, died by the time he was four. Placed then in the custody of his grandparents, he encountered early in life the most prominent thinkers and writers of those days.

Independently wealthy, Russell studied mathematics at Trinity College, Cambridge, becoming eventually a Fellow of that College. His sexual and familial adventures were many and daring; he later ran a progressive school in which nudity for all was the rule. He married four times: in 1894, then in 1921, then again in 1936, and finally—at last happily at the age of 80—in 1952. The radical sexual freedom that he professed he practiced unrelentingly and without shame.

In the very early years of the twentieth century, inspired by Frege, Russell developed his own logicism—the view that mathematics grows out of logic —which he first formulated in *The Principles of Mathematics* in 1903. Collaborating closely with the mathematician, Alfred North Whitehead, he pursued tenaciously the project of proving that this derivation could indeed be carried out, overcoming the problems that Frege's work had failed to solve. Ten years of arduous labor resulted in their publication of one of the towering works of modern logic: *Principia Mathematica* (3 volumes: 1910, 1912, 1913).

Russell was a pacifist; he was dismissed from Trinity College and went to prison for his activism against British participation in the First World War. But he campaigned actively against Hitler in the Second World War, and came to believe that although war is indeed always a very great evil, there are times when it is the lesser of the evils we confront. He campaigned against the brutality of the Stalinist regime in the Soviet Union in the years following WWII, and then again, engaging in deliberate, orderly civil disobedience, against America's involvement in the Vietnam War.

He taught at the City College of New York, and at the University of Chicago, and at the University of California at Los Angeles—but even in America he was often hounded because of his radical opinions on matters of sex, very publicly expressed. He is famous for the clarity and beauty of his prose, nowhere more evident than in his *History of Western Philosophy* (1945), which became a world-wide best seller. In 1950 he was awarded the Nobel Prize in Literature. Over the years Russell had many very distinguished students whom he admired and who carried on his work. Among them were the philosopher, Ludwig Wittgenstein, at Trinity College, Cambridge, and later the logician, Irving Copi, at the University of Chicago.

Colorful, prolific, creative, passionate and courageous, Bertrand Russell was not only one of the great modern logicians; he was one of the most extraordinary intellectual figures of his time. ■

From that and the second premise *Hs*, the conclusion *Ms* follows directly by *Modus Ponens*.

If we add to our list of rules of inference the principle that *any substitution instance of a propositional function can validly be inferred from its universal quantification*, then we can give a formal proof of the validity of the given argument by reference to the expanded list of elementary valid argument forms. This new rule of inference is the principle of **Universal Instantiation**[*] and is abbreviated as U.I. Using the Greek letter *nu* (ν) to represent any individual symbol whatever, we state the new rule as

> U.I.: $(x)(\Phi x)$
> $\therefore \Phi \nu$ (where ν is any individual symbol)

Universal Instantiation (U.I.) In quantification theory, a rule of inference that permits the valid inference of any substitution instance of a propositional function from the universal quantification of the propositional function.

---

[*]This rule, and the three following, are variants of rules for natural deduction that were devised independently by Gerhard Gentzen and Stanislaw Jaskowski in 1934.

A formal proof of validity may now be written as

> 1. $(x)(Hx \supset Mx)$
> 2. Hs
> $\therefore Ms$
> 3. $Hs \supset Ms$ 1, U.I.
> 4. Ms 3, 2, M.P.

The addition of U.I. strengthens our proof apparatus considerably, but more is required. The need for additional rules governing quantification arises in connection with arguments such as "All humans are mortal. All Greeks are human. Therefore all Greeks are mortal." The symbolic translation of this argument is

$(x)(Hx \supset Mx)$
$(x)(Gx \supset Hx)$
$\therefore (x)(Gx \supset Mx)$

Here both the premises and the conclusion are general propositions rather than singular ones, universal quantifications of propositional functions rather than substitution instances of them. From the two premises, by U.I., we may validly infer the following pairs of conditional statements:

$$\left\{\begin{array}{l} Ga \supset Ha \\ Ha \supset Ma \end{array}\right\} \left\{\begin{array}{l} Gb \supset Hb \\ Hb \supset Mb \end{array}\right\} \left\{\begin{array}{l} Gc \supset Hc \\ Hc \supset Mc \end{array}\right\} \left\{\begin{array}{l} Gd \supset Hd \\ Hd \supset Md \end{array}\right\} \ldots$$

and by successive uses of the principle of the Hypothetical Syllogism we may validly infer the conclusions:

$$Ga \supset Ma, Gb \supset Mb, Gc \supset Mc, Gd \supset Md, \ldots$$

If $a, b, c, d, \ldots$ were all the individuals that exist, it would follow that from the truth of the premises one could validly infer the truth of all substitution instances of the propositional function $Gx \supset Mx$. The universal quantification of a propositional function is true if and only if all its substitution instances are true, so we can go on to infer the truth of $(x)(Gx \supset Mx)$, which is the conclusion of the given argument.

The preceding paragraph may be thought of as containing an *informal* proof of the validity of the given argument, in which the principle of the hypothetical syllogism and two principles governing quantification are appealed to. But it describes indefinitely long sequences of statements: the lists of all substitution instances of the two propositional functions quantified universally in the premises, and the list of all substitution instances of the propositional function whose universal quantification is the conclusion. A *formal* proof cannot contain such indefinitely, perhaps even infinitely, long sequences of statements, so some method must be sought for expressing those indefinitely long sequences in some finite, definite fashion.

A method for doing this is suggested by a common technique of elementary mathematics. A geometer, seeking to prove that *all* triangles possess a certain attribute, may begin with the words "Let *ABC* be any arbitrarily selected triangle." Then the geometer begins to reason about the triangle *ABC* and establishes that it has the attribute in question. From this she concludes that *all* triangles have that attribute. Now what justifies her final conclusion? Granted of the particular triangle *ABC* that *it* has the attribute, why does it follow that *all* triangles do? The answer to this question is easily given. If no assumption other than its triangularity is made about the triangle *ABC*, then the symbol "*ABC*" can be taken as denoting any triangle one pleases. Then the geometer's argument establishes that *any* triangle has the attribute in question, and if *any* triangle has it, then *all* triangles do. We now introduce a notation analogous to the geometer's in talking about "any arbitrarily selected triangle *ABC*." This will avoid the pretense of listing an indefinite or infinite number of substitution instances of a propositional function, for instead we shall talk about *any* substitution instance of the propositional function.

We shall use the (hitherto unused) lowercase letter *y* to denote any arbitrarily selected individual. We shall use it in a way similar to that in which the geometer used the letters *ABC*. Because the truth of *any* substitution instance of a propositional function follows from its universal quantification, we can infer the substitution instance that results from replacing *x* by *y*, where *y* denotes "any arbitrarily selected" individual. Thus we may begin our formal proof of the validity of the given argument as follows:

1. $(x)(Hx \supset Mx)$
2. $(x)(Gx \supset Hx)$
    $\therefore (x)(Gx \supset Mx)$
3. $Hy \supset My$ 1, U.I.
4. $Gy \supset Hy$ 2, U.I.
5. $Gy \supset My$ 4, 3, H.S.

From the premises we have deduced the statement $Gy \supset My$, which in effect, because *y* denotes "any arbitrarily selected individual," asserts the truth of *any* substitution instance of the propositional function $Gx \supset Mx$. Because *any* substitution instance is true, all substitution instances must be true, and hence the universal quantification of that propositional function is true also. We may add this principle to our list of rules of inference, stating it as follows: *From the substitution instance of a propositional function with respect to the name of any arbitrarily selected individual, one can validly infer the universal quantification of that propositional function.* This new principle permits us to *generalize*, that is, to go from a special substitution instance to a generalized or universally quantified expression, so we refer to it as the principle of **Universal Generalization** and abbreviate it as U.G. It is stated as

U.G.: Φy (where y denotes "any arbitrarily selected individual")
        $\therefore (x)(\Phi x)$

The sixth and final line of the formal proof already begun may now be written (and justified) as

       6. $(x)(Gx \supset Mx)$ 5, U.G.

Let us review the preceding discussion. In the geometer's proof, the only assumption made about *ABC* is that it is a triangle; hence what is proved true of *ABC* is proved true of *any* triangle. In our proof, the only assumption made about *y* is that it is an individual; hence what is proved true of *y* is proved true of *any* individual. The symbol *y* is an individual symbol, but it is a very special one. Typically it is introduced into a proof by using U.I., and only the presence of *y* permits the use of U.G.

Here is another valid argument, the demonstration of whose validity requires the use of U.G. as well as U.I.: "No humans are perfect. All Greeks are humans. Therefore no Greeks are perfect."* The formal proof of its validity is:

     1. $(x)(Hx \supset \sim Px)$
     2. $(x)(Gx \supset Hx)$
       $\therefore (x)(Gx \supset \sim Px)$
     3. $Hy \supset \sim Py$ 1, U.I.
     4. $Gy \supset Hy$ 2, U.I.
     5. $Gy \supset \sim Py$ 4, 3, H.S.
     6. $(x)(Gx \supset \sim Px)$ 5, U.G.

There may seem to be some artificiality about the preceding. It may be argued that distinguishing carefully between $(x)(\Phi x)$ and Φy, so that they are not treated as identical but must be inferred from each other by U.I. and U.G., is to insist on a distinction without a difference. But there certainly is a *formal* difference

Universal Generalization (U.G.) In quantification theory, a rule of inference that permits the valid inference of a generalized, or universally quantified, expression from an expression that is given as true of any arbitrarily selected individual.

---

*This is an appropriate point to observe that, for arguments of some kinds, the traditional syllogistic analysis can establish validity as efficiently as modern quantified logic. A classical logician would quickly identify this syllogism as having the mood **EAE** in the first figure—necessarily of the form *Celarent*, and therefore immediately seen to be valid. See Section 6.5 for a summary exposition of the valid standard-form categorical syllogisms.

420   CHAPTER 10 Quantification Theory

between them. The statement $(x)(Hx \supset Mx)$ is a noncompound statement, whereas $Hy \supset My$ is compound, being a conditional. From the two noncompound statements $(x)(Gx \supset Hx)$ and $(x)(Hx \supset Mx)$, no relevant inference can be drawn by means of the original list of nineteen rules of inference. But from the compound statements $Gy \supset Hy$ and $Hy \supset My$, the indicated conclusion $Gy \supset My$ follows by a Hypothetical Syllogism. The principle of U.I. is used to get from noncompound statements, to which our earlier rules of inference do not usefully apply, to compound statements, to which they *can* be applied to derive the desired conclusion. The quantification principles thus augment our logical apparatus to make it capable of validating arguments essentially involving noncompound (generalized) propositions as well as the other (simpler) kind of argument discussed in earlier chapters. On the other hand, in spite of this formal difference, there must be a logical equivalence between $(x)(\Phi x)$ and Φy, or the rules U.I. and U.G. would not be valid. Both the difference and the logical equivalence are important for our purpose of validating arguments by reference to a list of rules of inference. The addition of U.I. and U.G. to our list strengthens it considerably.

The list must be expanded further when we turn to arguments that involve existential propositions. A convenient example with which to begin is "All criminals are vicious. Some humans are criminals. Therefore some humans are vicious." It is symbolized as

$(x)(Cx \supset Vx)$
$(\exists x)(Hx \cdot Cx)$
$\therefore (\exists x)(Hx \cdot Vx)$

The existential quantification of a propositional function is true if and only if the function has at least one true substitution instance. Hence whatever attribute may be designated by Φ, $(\exists x)(\Phi x)$ says that there is at least one individual that has the attribute Φ. If an individual constant (other than the special symbol y) is used nowhere earlier in the context, we may use it to denote either the individual that has the attribute Φ, or some one of the individuals that have Φ if there are several. Knowing that there is such an individual, say, a, we know that Φa is a true substitution instance of the propositional function Φx. Hence we add to our list of rules of inference this principle: *From the existential quantification of a propositional function, we may infer the truth of its substitution instance with respect to any individual constant (other than y) that occurs nowhere earlier in that context.* The new rule of inference is the principle of **Existential Instantiation** and is abbreviated as "E.I." It is stated as:

Existential Instantiation (E.I.)
In quantification theory, a rule of inference that says that we may (with some restrictions) validly infer from the existential quantification of a propositional function the truth of its substitution instance with respect to any individual constant that does not occur earlier in that context.

E.I.: $(\exists x)(\Phi x)$ [where ν is any individual constant (other than y)
        $\therefore \Phi\nu$ having no previous occurrence in the context]

Granted the additional rule of inference E.I., we may begin a demonstration of the validity of the stated argument:

1. $(x)(Cx \supset Vx)$	
2. $(\exists x)(Hx \cdot Cx)$	
$\therefore (\exists x)(Hx \cdot Vx)$	
3. $Ha \cdot Ca$	2, E.I.
4. $Ca \supset Va$	1, U.I.
5. $Ca \cdot Ha$	3, Com.
6. Ca	5, Simp.
7. Va	4, 6, M.P.
8. Ha	3, Simp.
9. $Ha \cdot Va$	8, 7, Conj.

Thus far we have deduced $Ha \cdot Va$, which is a substitution instance of the propositional function whose existential quantification is asserted by the conclusion. Because the existential quantification of a propositional function is true if and only if it has at least one true substitution instance, we add to our list of rules of inference the principle that *from any true substitution instance of a propositional function we may validly infer the existential quantification of that propositional function*. This fourth and final rule of inference is the principle of **Existential Generalization**, abbreviated as E.G. and stated as

E.G.: $\Phi \nu$ (where ν is any individual symbol)
$\therefore (\exists x)(\Phi x)$

The tenth and final line of the demonstration already begun may now be written (and justified) as

10. $(\exists x)(Hx \cdot Vx)$ 9, E.G.

The need for the indicated restriction on the use of E.I. can be seen by considering the obviously invalid argument, "Some alligators are kept in captivity. Some birds are kept in captivity. Therefore some alligators are birds." If we failed to heed the restriction on E.I. that a substitution instance of a propositional function inferred by E.I. from the existential quantification of that propositional function can contain only an individual symbol (other than *y*) *that has no previous occurrence in the context*, then we might proceed to construct a "proof" of validity for this invalid argument. Such an erroneous "proof" might proceed as follows:

1. $(\exists x)(Ax \cdot Cx)$.	
2. $(\exists x)(Bx \cdot Cx)$	
$\therefore (\exists x)(Ax \cdot Bx)$	
3. $Aa \cdot Ca$	1, E.I.
4. $Ba \cdot Ca$	2, E.I. *(wrong!)*
5. Aa	3, Simp.
6. Ba	4, Simp.
7. $Aa \cdot Ba$	5, 6, Conj.
8. $(\exists x)(Ax \cdot Bx)$	7, E.G.

Existential Generalization (E.G.)
In quantification theory, a rule of inference that says that from any true substitution instance of a propositional function we may validly infer the existential quantification of the function.

The error in this "proof" occurs at line 4. From the second premise $(\exists x)(Bx \cdot Cx)$, we know that there is at least one thing that is both a bird and kept in captivity. *If* we were free to assign it the name *a* in line 4, we could, of course, assert $Ba \cdot Ca$. But we are not free to make any such assignment of *a*, for it has already been pre-empted in line 3 to serve as the name for an alligator that is kept in captivity. To avoid errors of this sort, we must obey the indicated restriction whenever we use E.I. The preceding discussion should make clear that in any demonstration requiring the use of both E.I. and U.I., E.I. should always be used first.

For more complicated modes of argumentation, especially those that involve relations, certain additional restrictions must be placed on our four quantification rules. But for arguments of the present sort, traditionally called *categorical syllogisms*, the present restrictions are sufficient to prevent mistakes.

overview

Rules of Inference: Quantification

Name	Abbreviation	Form	Effect
Universal Instantiation	U.I.	$(x)(\Phi x)$ $\therefore \Phi\nu$ (where ν is any individual symbol)	Any substitution instance of a propositional function can be validly inferred from its universal quantification.
Universal Generalization	U.G.	Φy $\therefore (x)(\Phi x)$ (where y denotes "any arbitrarily selected individual")	From the substitution instance of a propositional function with respect to the name of any arbitrarily selected individual, one may validly infer the universal quantification of that propositional function.
Existential Instantiation	E.I.	$(\exists x)(\Phi x)$ $\therefore \Phi\nu$ (where ν is any individual constant, other than y, having no previous occurrence in the context)	From the existential quantification of a propositional function, we may infer the truth of its substitution instance with respect to any individual constant (other than y) that occurs nowhere earlier in the context.
Existential Generalization	E.G.	$\Phi\nu$ $\therefore (\exists x)(\Phi x)$ (where ν is any individual symbol)	From any true substitution instance of a propositional function, we may validly infer the existential quantification of that propositional function.

EXERCISES

A. Construct a formal proof of validity for each of the following arguments:

EXAMPLE

1. $(x)(Ax \supset \sim Bx)$
   $(\exists x)(Cx \cdot Ax)$
   $\therefore (\exists x)(Cx \cdot \sim Bx)$

SOLUTION

The conclusion of this argument is an existentially quantified statement. Plainly, the last step will therefore be the application of E.G. To obtain the line needed, we will first have to instantiate the premises, applying E.I. to the second premise and U.I. to the first premise. The restriction on the use of E.I. makes it essential that we apply E.I. *before* we apply U.I., so that we may use the same individual constant, say *a*, for both. The proof looks like this:

1. $(x)(Ax \supset \sim Bx)$
2. $(\exists x)(Cx \cdot Ax)$
   $\therefore (\exists x)(Cx \cdot \sim Bx)$
3. $Ca \cdot Aa$ 2, E.I.
4. $Aa \supset \sim Ba$ 1, U.I.
5. $Aa \cdot Ca$ 3, Com.
6. Aa 5, Simp.
7. $\sim Ba$ 4, 6, M.P.
8. Ca 3, Simp.
9. $Ca \cdot \sim Ba$ 8, 7, Conj.
10. $(\exists x)(Cx \cdot \sim Bx)$ 9, E.G.

1. $(x)(Ax \supset \sim Bx)$

2. $(x)(Dx \supset \sim Ex)$
   $(x)(Fx \supset Ex)$
   $\therefore (x)(Fx \supset \sim Dx)$

3. $(x)(Gx \supset Hx)$
   $(x)(Ix \supset \sim Hx)$
   $\therefore (x)(Ix \supset \sim Gx)$

4. $(\exists x)(Jx \cdot Kx)$
   $(x)(Jx \supset Lx)$
   $\therefore (\exists x)(Lx \cdot Kx)$

*5. $(x)(Mx \supset Nx)$
   $(\exists x)(Mx \cdot Ox)$
   $\therefore (\exists x)(Ox \cdot Nx)$

6. $(\exists x)(Px \cdot \sim Qx)$
   $(x)(Px \supset Rx)$
   $\therefore (\exists x)(Rx \cdot \sim Qx)$

7. $(x)(Sx \supset \sim Tx)$
   $(\exists x)(Sx \cdot Ux)$
   $\therefore (\exists x)(Ux \cdot \sim Tx)$

8. $(x)(Vx \supset Wx)$
   $(x)(Wx \supset \sim Xx)$
   $\therefore (x)(Xx \supset \sim Vx)$

9. $(\exists x)(Yx \cdot Zx)$
   $(x)(Zx \supset Ax)$
   $\therefore (\exists x)(Ax \cdot Yx)$

*10. $(x)(Bx \supset \sim Cx)$
   $(\exists x)(Cx \cdot Dx)$
   $\therefore (\exists x)(Dx \cdot \sim Bx)$

11. $(x)(Fx \supset Gx)$
   $(\exists x)(Fx \cdot \sim Gx)$
   $\therefore (\exists x)(Gx \cdot \sim Fx)$

B. Construct a formal proof of validity for each of the following arguments, in each case using the suggested notations:

*1. No athletes are bookworms. Carol is a bookworm. Therefore Carol is not an athlete. (*Ax, Bx, c*)

2. All dancers are exuberant. Some fencers are not exuberant. Therefore some fencers are not dancers. (*Dx, Ex, Fx*)

3. No gamblers are happy. Some idealists are happy. Therefore some idealists are not gamblers. (*Gx, Hx, Ix*)

4. All jesters are knaves. No knaves are lucky. Therefore no jesters are lucky. (*Jx, Kx, Lx*)

*5. All mountaineers are neighborly. Some outlaws are mountaineers. Therefore some outlaws are neighborly. (*Mx, Nx, Ox*)

6. Only pacifists are Quakers. There are religious Quakers. Therefore pacifists are sometimes religious. (*Px, Qx, Rx*)

7. To be a swindler is to be a thief. None but the underprivileged are thieves. Therefore swindlers are always underprivileged. (*Sx, Tx, Ux*)

8. No violinists are not wealthy. There are no wealthy xylophonists. Therefore violinists are never xylophonists. (*Vx, Wx, Xx*)

9. None but the brave deserve the fair. Only soldiers are brave. Therefore the fair are deserved only by soldiers. (*Dx: x* deserves the fair; *Bx: x* is brave; *Sx: x* is a soldier)

*10. Everyone that asketh receiveth. Simon receiveth not. Therefore Simon asketh not. (*Ax, Rx, s*)

10.6 Proving Invalidity

To prove the invalidity of an argument involving quantifiers, we can use the method of refutation by logical analogy. For example, the argument, "All conservatives are opponents of the administration; some delegates are opponents of the administration; therefore some delegates are conservatives," is proved invalid by the analogy, "All cats are animals; some dogs are animals; therefore some dogs are cats," which is obviously invalid, because its premises are known to be true and its conclusion is known to be false. Such analogies, however, are not always easy to devise. Some more effective method of proving invalidity is desirable.

In Chapter 9 we developed a method of proving invalidity for arguments involving truth-functional compound statements. That method consisted of making truth-value assignments to the component simple statements in arguments, in such a way as to make the premises true and the conclusions false. That method can be adapted for arguments involving quantifiers. The adaptation involves our general assumption that there is at least one individual. For an argument involving quantifiers to be valid, it must be impossible for its premises to be true and its conclusion false as long as there is at least one individual.

The general assumption that there is at least one individual is satisfied if there is exactly one individual, or if there are exactly two individuals, or exactly three individuals, or . . . .

If any one of these assumptions about the exact number of individuals is made, there is an equivalence between general propositions and truth-functional compounds of singular propositions. If there is exactly one individual, say a, then

$$(x)(\Phi x) \stackrel{T}{\equiv} \Phi a \stackrel{T}{\equiv} (\exists x)(\Phi x)$$

If there are exactly two individuals, say a and b, then

$$(x)(\Phi x) \stackrel{T}{\equiv} [\Phi a \cdot \Phi b] \quad \text{and} \quad (\exists x)(\Phi x) \stackrel{T}{\equiv} [\Phi a \vee \Phi b]$$

If there are exactly three individuals, say a, b, and c, then

$$(x)(\Phi x) \stackrel{T}{\equiv} [\Phi a \cdot \Phi b \cdot \Phi c] \quad \text{and} \quad (\exists x)(\Phi x) \stackrel{T}{\equiv} [\Phi a \vee \Phi b \vee \Phi c]$$

In general, if there are exactly n individuals, say $a, b, c, \ldots, n$, then

$$(x)(\Phi x) \stackrel{T}{\equiv} [\Phi a \cdot \Phi b \cdot \Phi c \cdot \ldots \cdot \Phi n] \quad \text{and} \quad (\exists x)(\Phi x) \stackrel{T}{\equiv} [\Phi a \vee \Phi b \vee \Phi c \vee \ldots \vee \Phi n]$$

These biconditionals are true as a consequence of our definitions of the universal and existential quantifiers. No use is made here of the four quantification rules explained in Section 10.5.

An argument involving quantifiers is valid if, *and only if*, it is valid no matter how many individuals there are, provided there is at least one. So an argument involving quantifiers is proved invalid if there is a possible universe or *model* containing at least one individual such that the argument's premises are true and its conclusion false *of that model*. Consider the argument, "All mercenaries are undependable. No guerrillas are mercenaries. Therefore no guerrillas are undependable." It may be symbolized as

$(x)(Mx \supset Ux)$
$(x)(Gx \supset \sim Mx)$
$\therefore (x)(Gx \supset \sim Ux)$

If there is exactly one individual, say a, this argument is logically equivalent to

$Ma \supset Ua$
$Ga \supset \sim Ma$
$\therefore Ga \supset \sim Ua$

The latter can be proved invalid by assigning the truth value *true* to Ga and Ua and *false* to Ma. (This assignment of truth values is a shorthand way of describing the *model* in question as containing only the one individual, a, which is a guerrilla and undependable but is not a mercenary.) Hence the original argument is not valid for a model containing exactly one individual, and it is therefore *invalid*. Similarly, we can prove the invalidity of the first argument mentioned in this section (on p. 425) by describing a model containing exactly one individual, *a*, so that *Aa* and *Da* are assigned the value *true* and *Ca* is assigned the value *false*.*

Some arguments, for example,

(∃x)Fx
∴ (x)Fx

may be valid for any model in which there is exactly one individual, but invalid for a model containing two or more individuals. Such arguments must also count as invalid, because a valid argument must be valid regardless of how many individuals there are, so long as there is at least one. Another example of this kind of argument is "All collies are affectionate. Some collies are watchdogs. Therefore all watchdogs are affectionate." Its symbolic translation is

(x)(Cx ⊃ Ax)
(∃x)(Cx • Wx)
∴ (x)(Wx ⊃ Ax)

For a model containing exactly one individual, *a*, it is logically equivalent to

Ca ⊃ Aa
Ca • Wa
∴ Wa ⊃ Aa

which is valid. But for a model containing two individuals, *a* and *b*, it is logically equivalent to

(Ca ⊃ Aa) • (Cb ⊃ Ab)
(Ca • Wa) ∨ (Cb • Wb)
∴ (Wa ⊃ Aa) • (Wb ⊃ Ab)

which is proved invalid by assigning *true* to *Ca*, *Aa*, *Wa*, *Wb*, and *false* to *Cb* and *Ab*. Hence the original argument is not valid for a model containing exactly two individuals, and it is therefore *invalid*. For any invalid argument of this general type, it is possible to describe a model containing some definite number of individuals for which its logically equivalent truth-functional argument can be proved invalid by the method of assigning truth values.

It should be emphasized again: In moving from a given argument involving general propositions to a truth-functional argument (one that is logically equivalent to the given argument for a specified model), no use is made of our four quantification rules. Instead, each statement of the truth-functional argument is logically equivalent to the corresponding general proposition of the given argument, and that logical equivalence is shown by the biconditionals formulated earlier in this section on page 425, whose logical truth for the model in question follows from the very definitions of the universal and existential quantifiers.

---

*Here we assume that the simple predicates *Ax*, *Bx*, *Cx*, *Dx*, . . ., occurring in our propositions are neither necessary, that is, logically true of all individuals (for example, *x* is identical with itself), nor impossible, that is, logically false of all individuals (for example, *x* is different from itself). We also assume that the only logical relations among the simple predicates involved are those asserted or logically implied by the premises. The point of these restrictions is to permit us to assign truth values arbitrarily to the substitution instances of these simple predicates without any inconsistency—for of course, a correct description of any model must be consistent.

The procedure for proving the invalidity of an argument containing general propositions is the following. First, consider a one-element model containing only the individual *a*. Then, write out the logically equivalent truth-functional argument for that model, which is obtained by moving from each general proposition (quantified propositional function) of the original argument to the substitution instance of that propositional function with respect to *a*. If the truth-functional argument can be proved invalid by assigning truth values to its component simple statements, that suffices to prove the original argument invalid. If that cannot be done, next consider a two-element model containing the individuals *a* and *b*. In order to obtain the logically equivalent truth-functional argument for this larger model, one can simply join each original substitution instance with respect to *a* to a new substitution instance of the same propositional function with respect to *b*. This "joining" must be in accord with the logical equivalences stated on page 425; that is, where the original argument contains a *universally* quantified propositional function, $(x)(\Phi x)$, the new substitution instance Φb is combined with the first substitution instance Φa by *conjunction* ("•"); but where the original argument contains an *existentially* quantified propositional function, $(\exists x)(\Phi x)$, the new substitution instance Φb is combined with the first substitution instance Φa by *disjunction* ("∨"). The preceding example illustrates this procedure. If the new truth-functional argument can be proved invalid by assigning truth values to its component simple statements, that suffices to prove the original argument invalid. If that cannot be done, next consider a three-element model containing the individuals *a, b,* and *c*. And so on. None of the exercises in this book requires a model containing more than three elements.

EXERCISES

In the following exercises, no model containing more than two elements is required.

A. Prove the invalidity of the following:

EXAMPLE

1. $(\exists x)(Ax \cdot Bx)$
   $(\exists x)(Cx \cdot Bx)$
   $\therefore (x)(Cx \supset \sim Ax)$

SOLUTION

We first construct a model (or possible universe, represented below by a rectangular box) containing exactly one individual, *a*. We then exhibit the logically equivalent propositions in that model. Thus,

$(\exists x)(Ax \cdot Bx)$ logically $Aa \cdot Ba$
$(\exists x)(Cx \cdot Bx)$ equivalent $Ca \cdot Ba$
$\therefore (x)(Cx \supset \sim Ax)$ in $\boxed{a}$ to $\therefore Ca \supset \sim Aa$

We may prove the argument invalid in this model by assigning truth values as follows:

Aa	Ba	Ca
T	T	T

Because the argument has been proved invalid in this model, the argument has been proved invalid.

1. (∃x)(Ax • Bx)
   (∃x)(Cx • Bx)
   ∴ (x)(Cx ⊃ ~Ax)

2. (x)(Dx ⊃ ~Ex)
   (x)(Ex ⊃ Fx)
   ∴ (x)(Fx ⊃ ~Dx)

3. (x)(Gx ⊃ Hx)
   (x)(Gx ⊃ Ix)
   ∴ (x)(Ix ⊃ Hx)

4. (∃x)(Jx • Kx)
   (∃x)(Kx • Lx)
   ∴ (∃x)(Lx • Jx)

*5. (∃x)(Mx • Nx)
   (∃x)(Mx • Ox)
   ∴ (x)(Ox ⊃ Nx)

6. (x)(Px ⊃ ~Qx)
   (x)(Px ⊃ ~Rx)
   ∴ (x)(Rx ⊃ ~Qx)

7. (x)(Sx ⊃ ~Tx)
   (x)(Tx ⊃ Ux)
   ∴ (∃x)(Ux • ~Sx)

8. (∃x)(Vx • ~Wx)
   (∃x)(Wx • ~Xx)
   ∴ (∃x)(Xx • ~Vx)

9. (∃x)(Yx • Zx)
   (∃x)(Ax • Zx)
   ∴ (∃x)(Ax • ~Yx)

*10. (∃x)(Bx • ~Cx)
    (x)(Dx ⊃ ~Cx)
    ∴ (x)(Dx ⊃ Bx)

B. Prove the invalidity of the following, in each case using the suggested notation:

*1. All anarchists are bearded. All communists are bearded. Therefore all anarchists are communists. (Ax, Bx, Cx)

2. No diplomats are extremists. Some fanatics are extremists. Therefore some diplomats are not fanatics. (Dx, Ex, Fx)

3. All generals are handsome. Some intellectuals are handsome. Therefore some generals are intellectuals. (Gx, Hx, Ix)

4. Some journalists are not kibitzers. Some kibitzers are not lucky. Therefore some journalists are not lucky. (Jx, Kx, Lx)

*5. Some malcontents are noisy. Some officials are not noisy. Therefore no officials are malcontents. (Mx, Nx, Ox)

6. Some physicians are quacks. Some quacks are not responsible. Therefore some physicians are not responsible. (Px, Qx, Rx)

7. Some politicians are leaders. Some leaders are not orators. Therefore some orators are not politicians. (Px, Lx, Ox)

8. None but the brave deserve the fair. Every soldier is brave. Therefore none but soldiers deserve the fair. (Dx: x deserves the fair; Bx: x is brave; Sx: x is a soldier)

9. If anything is metallic, then it is breakable. There are breakable ornaments. Therefore there are metallic ornaments. (Mx, Bx, Ox)

*10. Only students are members. Only members are welcome. Therefore all students are welcome. *(Sx, Mx, Wx)*

10.7 Asyllogistic Inference

All the arguments considered in the preceding two sections were of the form traditionally called *categorical syllogisms*. These consist of two premises and a conclusion, each of which is analyzable either as a singular proposition or as one of the **A**, **E**, **I**, or **O** varieties. We turn now to the problem of evaluating somewhat more complicated arguments. These require no greater logical apparatus than has already been developed, yet they are **asyllogistic arguments**; that is, they cannot be reduced to standard-form categorical syllogisms, and therefore evaluating them requires a more powerful logic than was traditionally used in testing categorical syllogisms.

In this section we are still concerned with general propositions, formed by quantifying propositional functions that contain only a single individual variable. In the categorical syllogism, the only kinds of propositional functions quantified were of the forms $\Phi x \supset \Psi x$, $\Phi x \supset {\sim}\Psi x$, $\Phi x \cdot \Psi x$, and $\Phi x \cdot {\sim}\Psi x$. Now we shall be quantifying propositional functions with more complicated internal structures. An example will help make this clear. Consider the argument

> Hotels are both expensive and depressing.
> Some hotels are shabby.
> Therefore some expensive things are shabby.

This argument, for all its obvious validity, is not amenable to the traditional sort of analysis. True enough, it could be expressed in terms of **A** and **I** propositions by using the symbols Hx, Bx, Sx, and Ex to abbreviate the propositional functions "*x* is a hotel," "*x* is both expensive and depressing," "*x* is shabby," and "*x* is expensive," respectively.* Using these abbreviations, we might propose to symbolize the given argument as

$(x)(Hx \supset Bx)$
$(\exists x)(Hx \cdot Sx)$
$\therefore (\exists x)(Ex \cdot Sx)$

Forcing the argument into the straitjacket of the traditional **A** and **I** forms in this way obscures its validity. The argument just given in symbols is invalid, although the original argument is perfectly valid. A notation restricted to categorical propositions here obscures the logical connection between Bx and Ex. A more adequate analysis is obtained by using Hx, Sx, and Ex, as explained, plus Dx as an abbreviation for "*x* is depressing." By using these symbols, the original argument can be translated as

1. $(x)[Hx \supset (Ex \cdot Dx)]$
2. $(\exists x)(Hx \cdot Sx)$
$\therefore (\exists x)(Ex \cdot Sx)$

Asyllogistic argument
An argument in which one or more of the component propositions is of a form more complicated than the form of the **A**, **E**, **I**, and **O** propositions of the categorical syllogism, and whose analysis therefore requires logical tools more powerful than those provided by Aristotelian logic.

---

*This would, however, violate the restriction stated in the footnote on page 426.

Thus symbolized, a demonstration of its validity is easily constructed. One such demonstration proceeds as follows:

3. $Hw \cdot Sw$	2, E.I.
4. $Hw \supset (Ew \cdot Dw)$	1, U.I.
5. Hw	3, Simp.
6. $Ew \cdot Dw$	4, 5, M.P.
7. Ew	6, Simp.
8. $Sw \cdot Hw$	3, Com.
9. Sw	8, Simp.
10. $Ew \cdot Sw$	7, 9, Conj.
11. $(\exists x)(Ex \cdot Sx)$	10, E.G.

In symbolizing general propositions that result from quantifying more complicated propositional functions, care must be taken not to be misled by the deceptiveness of ordinary English. One cannot translate from English into our logical notation by following any formal or mechanical rules. In every case, *one must understand the meaning of the English sentence, and then symbolize that meaning in terms of propositional functions and quantifiers.*

Three locutions of ordinary English that are sometimes troublesome are the following. First, note that a statement such as "All athletes are either very strong or very quick" is *not* a disjunction, although it contains the connective "or." It definitely does *not* have the same meaning as "Either all athletes are very strong or all athletes are very quick." The former is properly symbolized—using obvious abbreviations—as

$$(x)[Ax \supset (Sx \lor Qx)]$$

whereas the latter is symbolized as

$$(x)(Ax \supset Sx) \lor (x)(Ax \supset Qx)$$

Second, note that a statement such as "Oysters and clams are delicious," while it *can* be stated as the conjunction of two general propositions—"Oysters are delicious and clams are delicious"—also can be stated as a single noncompound general proposition, in which case the word "and" is properly symbolized by the "$\lor$" rather than by the "$\cdot$". The stated proposition is symbolized as

$$(x)[(Ox \lor Cx) \supset Dx]$$

not as

$$(x)[(Ox \cdot Cx) \supset Dx]$$

For to say that oysters and clams are delicious is to say that anything is delicious that is *either* an oyster *or* a clam, *not* to say that anything is delicious that is *both* an oyster *and* a clam.

Third, what are called *exceptive* propositions require very careful attention. Such propositions—for example, "All except previous winners are eligible"—may be treated as the conjunction of two general propositions. Using the example just given, we might reasonably understand the proposition to assert both that previous winners are not eligible, *and* that those who are not previous winners are eligible. It is symbolized as:

$$(x)(Px \supset {\sim}Ex) \cdot (x)({\sim}Px \supset Ex)$$

The same exceptive proposition may also be translated as a noncompound general proposition that is the universal quantification of a propositional function containing the symbol for material equivalence "≡" (a biconditional), and symbolized thus:

$$(x)(Ex \equiv \sim Px)$$

which can also be rendered in English as "Anyone is eligible if and only if that person is not a previous winner." Exceptive propositions are most conveniently regarded as quantified biconditionals.

Whether a proposition is in fact exceptive is sometimes difficult to determine. A recent controversy requiring resolution by a federal court panel illustrates this contextual difficulty. The Census Act, a law that establishes the rules for the conduct of the national census every ten years, contains the following passage:

> Sec. 195. Except for the determination of population for purposes of apportionment of Representatives in Congress among the several States, the Secretary [of Commerce] shall, if he considers it feasible, authorize the use of the statistical method known as "sampling" in carrying out the provisions of this title.

For the 2000 census, which did determine population for the purposes of apportionment, the Census Bureau sought to use the sampling technique, and was sued by the House of Representatives, which claimed that the passage quoted here prohibits sampling in such a census. The Bureau defended its plan, contending that the passage authorizes the use of sampling in some contexts, but in apportionment contexts leaves the matter undetermined. Which interpretation of that exceptive provision in the statute is correct?

The court found the House position correct, writing:

> Consider the directive "except for my grandmother's wedding dress, you shall take the contents of my closet to the cleaners." It is . . . likely that the granddaughter would be upset if the recipient of her directive were to take the wedding dress to the cleaners and subsequently argue that she had left this decision to his discretion. The reason for this result . . . is because of our background knowledge concerning wedding dresses: We know they are extraordinarily fragile and of deep sentimental value to family members. We therefore would not expect that a decision to take [that] dress to the cleaners would be purely discretionary.
>
> The apportionment of Congressional representatives among the states is the wedding dress in the closet. . . . The apportionment function is the "sole constitutional function of the decennial enumeration." The manner in which it is conducted may impact not only the distribution of representatives among the states, but also the balance of political power within the House. . . . This court finds that the Census Act prohibits the use of statistical sampling to determine the population for the purpose of apportionment of representatives among the states. . . .*

The exceptive proposition in this statute is thus to be understood as asserting the conjunction of two propositions: (1) that the use of sampling is not permitted in the context of apportionment, and (2) that in all other contexts sampling

---
*Decided by a specially appointed Voting Rights Act panel of three judges on 24 August 1998.

is discretionary. A controversial sentence in exceptive form must be interpreted in its context.

In Section 10.5, our list of rules of inference was expanded by four, and we showed that the expanded list was sufficient to demonstrate the validity of categorical syllogisms when they are valid. We have just seen that the same expanded list suffices to establish the validity of asyllogistic arguments of the type described. Now we may observe that, just as the expanded list was sufficient to establish *validity* in asyllogistic arguments, so also the method of proving syllogisms invalid (explained in Section 10.6) by describing possible nonempty universes, or models, is sufficient to prove the *invalidity* of asyllogistic arguments of the present type as well. The following asyllogistic argument,

> Managers and superintendents are either competent workers or relatives of the owner.
> Anyone who dares to complain must be either a superintendent or a relative of the owner.
> Managers and foremen alone are competent workers.
> Someone did dare to complain.
> Therefore some superintendent is a relative of the owner.

may be symbolized as

$(x)[(Mx \lor Sx) \supset (Cx \lor Rx)]$
$(x)[Dx \supset (Sx \lor Rx)]$
$(x)[(Mx \lor Fx) \equiv Cx]$
$(\exists x)Dx$
$\therefore (\exists x)(Sx \cdot Rx)$

and we can prove it invalid by describing a possible universe or model containing the single individual *a* and assigning the truth value *true* to Ca, Da, Fa, Ra, and the truth value *false* to Sa.

EXERCISES

A. Translate the following statements into logical symbolism, in each case using the abbreviations suggested:

EXAMPLE

1. Apples and oranges are delicious and nutritious. *(Ax, Ox, Dx, Nx)*

SOLUTION

The meaning of this proposition clearly is that if anything is *either* an apple or an orange it is *both* delicious and nutritious. Hence it is symbolized as

$(x)[(Ax \lor Ox) \supset (Dx \cdot Nx)]$

2. Some foods are edible only if they are cooked. *(Fx, Ex, Cx)*
3. No car is safe unless it has good brakes. *(Cx, Sx, Bx)*

4. Any tall man is attractive if he is dark and handsome. *(Tx, Mx, Ax, Dx, Hx)*

*5. A gladiator wins if and only if he is lucky. *(Gx, Wx, Lx)*

6. A boxer who wins if and only if he is lucky is not skillful. *(Bx, Wx, Lx, Sx)*

7. Not all people who are wealthy are both educated and cultured. *(Px, Wx, Ex, Cx)*

8. Not all tools that are cheap are either soft or breakable. *(Tx, Cx, Sx, Bx)*

9. Any person is a coward who deserts. *(Px, Cx, Dx)*

*10. To achieve success, one must work hard if one goes into business, or study continuously if one enters a profession. *(Ax: x achieves success; Wx: x works hard; Bx: x goes into business; Sx: x studies continuously; Px: x enters a profession)*

11. An old European joke goes like this: In America, everything is permitted that is not forbidden. In Germany, everything is forbidden that is not permitted. In France, everything is permitted even if it's forbidden. In Russia, everything is forbidden even if it's permitted. *(Ax: x is in America; Gx: x is in Germany; Fx: x is in France; Rx: x is in Russia; Px: x is permitted; Nx: x is forbidden)*

B. For each of the following, either construct a formal proof of validity or prove it invalid. If it is to be proved invalid, a model containing as many as three elements may be required.

*1. $(x)[(Ax \lor Bx) \supset (Cx \cdot Dx)]$
$\therefore (x)(Bx \supset Cx)$

2. $(\exists x)\{(Ex \cdot Fx) \cdot [(Ex \lor Fx) \supset (Gx \cdot Hx)]\}$
$\therefore (x)(Ex \supset Hx)$

3. $(x)\{[Ix \supset (Jx \cdot \sim Kx)] \cdot [Jx \supset (Ix \supset Kx)]\}$
$(\exists x)[(Ix \cdot Jx) \cdot \sim Lx]$
$\therefore (\exists x)(Kx \cdot Lx)$

4. $(x)[(Mx \cdot Nx) \supset (Ox \lor Px)]$
$(x)[(Ox \cdot Px) \supset (Qx \lor Rx)]$
$\therefore (x)[(Mx \lor Ox) \supset Rx]$

*5. $(\exists x)(Sx \cdot Tx)$
$(\exists x)(Ux \cdot \sim Sx)$
$(\exists x)(Vx \cdot \sim Tx)$
$\therefore (\exists x)(Ux \cdot Vx)$

6. $(x)[Wx \supset (Xx \supset Yx)]$
$(\exists x)[Xx \cdot (Zx \cdot \sim Ax)]$
$(x)[Wx \supset Yx) \supset (Bx \supset Ax)]$
$\therefore (\exists x)(Zx \cdot \sim Bx)$

7. ($\exists x$)[$Cx \cdot \sim(Dx \supset Ex)$]
   (x)[($Cx \cdot Dx$) $\supset Fx$]
   ($\exists x$)[$Ex \cdot \sim(Dx \supset Cx)$]
   (x)($Gx \supset Cx$)
   $\therefore$ ($\exists x$)($Gx \cdot \sim Fx$)

8. (x)($Hx \supset Ix$)
   (x)[($Hx \cdot Ix$) $\supset Jx$]
   (x)[$\sim Kx \supset (Hx \lor Ix)$]
   (x)[($Jx \lor \sim Jx$) $\supset (Ix \supset Hx)$]
   $\therefore$ (x)($Jx \lor Kx$)

9. (x){($Lx \lor Mx$) $\supset$ {[($Nx \cdot Ox$) $\lor Px$] $\supset Qx$}}
   ($\exists x$)($Mx \cdot \sim Lx$)
   (x){[($Ox \supset Qx$) $\cdot \sim Rx$] $\supset Mx$}
   ($\exists x$)($Lx \cdot \sim Mx$)
   $\therefore$ ($\exists x$)($Nx \supset Rx$)

*10. (x)[($Sx \lor Tx$) $\supset \sim(Ux \lor Vx)$]
   ($\exists x$)($Sx \cdot \sim Wx$)
   ($\exists x$)($Tx \cdot \sim Xx$)
   (x)($\sim Wx \supset Xx$)
   $\therefore$ ($\exists x$)($Ux \cdot \sim Vx$)

C. For each of the following, either construct a formal proof of its validity or prove it invalid, in each case using the suggested notation:

   *1. Acids and bases are chemicals. Vinegar is an acid. Therefore vinegar is a chemical. (Ax, Bx, Cx, Vx)

   2. Teachers are either enthusiastic or unsuccessful. Teachers are not all unsuccessful. Therefore there are enthusiastic teachers. (Tx, Ex, Ux)

   3. Argon compounds and sodium compounds are either oily or volatile. Not all sodium compounds are oily. Therefore some argon compounds are volatile. (Ax, Sx, Ox, Vx)

   4. No employee who is either slovenly or discourteous can be promoted. Therefore no discourteous employee can be promoted. (Ex, Sx, Dx, Px)

   *5. No employer who is either inconsiderate or tyrannical can be successful. Some employers are inconsiderate. There are tyrannical employers. Therefore no employer can be successful. (Ex, Ix, Tx, Sx)

   6. There is nothing made of gold that is not expensive. No weapons are made of silver. Not all weapons are expensive. Therefore not everything is made of gold or silver. (Gx, Ex, Wx, Sx)

   7. There is nothing made of tin that is not cheap. No rings are made of lead. Not everything is either tin or lead. Therefore not all rings are cheap. (Tx, Cx, Rx, Lx)

   8. Some prize fighters are aggressive but not intelligent. All prize fighters wear gloves. Prize fighters are not all aggressive. Any slugger is aggressive. Therefore not every slugger wears gloves. (Px, Ax, Ix, Gx, Sx)

   9. Some photographers are skillful but not imaginative. Only artists are photographers. Photographers are not all skillful. Any journeyman is skillful. Therefore not every artist is a journeyman. (Px, Sx, Ix, Ax, Jx)

   *10. A book is interesting only if it is well written. A book is well written only if it is interesting. Therefore any book is both interesting and well written if it is either interesting or well written. (Bx, Ix, Wx)

D. Do the same (as in Set C) for each of the following:
   *1. All citizens who are not traitors are present. All officials are citizens. Some officials are not present. Therefore there are traitors. (Cx, Tx, Px, Ox)
   2. Doctors and lawyers are professional people. Professional people and executives are respected. Therefore doctors are respected. (Dx, Lx, Px, Ex, Rx)
   3. Only lawyers and politicians are members. Some members are not college graduates. Therefore some lawyers are not college graduates. (Lx, Px, Mx, Cx)
   4. All cut-rate items are either shopworn or out of date. Nothing shopworn is worth buying. Some cut-rate items are worth buying. Therefore some cut-rate items are out of date. (Cx, Sx, Ox, Wx)
   *5. Some diamonds are used for adornment. Only things worn as jewels or applied as cosmetics are used for adornment. Diamonds are never applied as cosmetics. Nothing worn as a jewel is properly used if it has an industrial application. Some diamonds have industrial applications. Therefore some diamonds are not properly used. (Dx, Ax, Jx, Cx, Px, Ix)
   6. No candidate who is either endorsed by labor or opposed by the *Tribune* can carry the farm vote. No one can be elected who does not carry the farm vote. Therefore no candidate endorsed by labor can be elected. (Cx, Lx, Ox, Fx, Ex)
   7. No metal is friable that has been properly tempered. No brass is properly tempered unless it is given an oil immersion. Some of the ashtrays on the shelf are brass. Everything on the shelf is friable. Brass is a metal. Therefore some of the ashtrays were not given an oil immersion. (Mx: x is metal; Fx: x is friable; Tx: x is properly tempered; Bx: x is brass; Ox: x is given an oil immersion; Ax: x is an ashtray; Sx: x is on the shelf)
   8. Anyone on the committee who knew the nominee would vote for the nominee if free to do so. Everyone on the committee was free to vote for the nominee except those who were either instructed not to by the party caucus or had pledged support to someone else. Everyone on the committee knew the nominee. No one who knew the nominee had pledged support to anyone else. Not everyone on the committee voted for the nominee. Therefore the party caucus had instructed some members of the committee not to vote for the nominee. (Cx: x is on the committee; Kx: x knows the nominee; Vx: x votes for the nominee; Fx: x is free to vote for the nominee; Ix: x is instructed by the party caucus not to vote for the nominee; Px: x had pledged support to someone else)
   9. All logicians are deep thinkers and effective writers. To write effectively, one must be economical if one's audience is general, and comprehensive if one's audience is technical. No deep thinker has a technical audience if he has the ability to reach a general audience. Some logicians are comprehensive rather than economical. Therefore not all logicians have the ability to reach a general audience. (Lx: x is a logician; Dx: x is a deep

thinker; *Wx*: *x* is an effective writer; *Ex*: *x* is economical; *Gx*: *x*'s audience is general; *Cx*: *x* is comprehensive; *Tx*: *x*'s audience is technical; *Ax*: *x* has the ability to reach a general audience)

*10. Some criminal robbed the Russell mansion. Whoever robbed the Russell mansion either had an accomplice among the servants or had to break in. To break in, one would either have to smash the door or pick the lock. Only an expert locksmith could have picked the lock. Had anyone smashed the door, he would have been heard. Nobody was heard. If the criminal who robbed the Russell mansion managed to fool the guard, he must have been a convincing actor. No one could rob the Russell mansion unless he fooled the guard. No criminal could be both an expert locksmith and a convincing actor. Therefore some criminal had an accomplice among the servants. (*Cx*: *x* is a criminal; *Rx*: *x* robbed the Russell mansion; *Sx*: *x* had an accomplice among the servants; *Bx*: *x* broke in; *Dx*: *x* smashed the door; *Px*: *x* picked the lock; *Lx*: *x* is an expert locksmith; *Hx*: *x* was heard; *Fx*: *x* fooled the guard; *Ax*: *x* is a convincing actor)

11. If anything is expensive it is both valuable and rare. Whatever is valuable is both desirable and expensive. Therefore if anything is either valuable or expensive then it must be both valuable and expensive. (*Ex*: *x* is expensive; *Vx*: *x* is valuable; *Rx*: *x* is rare; *Dx*: *x* is desirable)

12. Figs and grapes are healthful. Nothing healthful is either illaudable or jejune. Some grapes are jejune and knurly. Some figs are not knurly. Therefore some figs are illaudable. (*Fx*: *x* is a fig; *Gx*: *x* is a grape; *Hx*: *x* is healthful; *Ix*: *x* is illaudable; *Jx*: *x* is jejune; *Kx*: *x* is knurly)

13. Figs and grapes are healthful. Nothing healthful is both illaudable and jejune. Some grapes are jejune and knurly. Some figs are not knurly. Therefore some figs are not illaudable. (*Fx*: *x* is a fig; *Gx*: *x* is a grape; *Hx*: *x* is healthful; *Ix*: *x* is illaudable; *Jx*: *x* is jejune; *Kx*: *x* is knurly)

14. Gold is valuable. Rings are ornaments. Therefore gold rings are valuable ornaments. (*Gx*: *x* is gold; *Vx*: *x* is valuable; *Rx*: *x* is a ring; *Ox*: *x* is an ornament)

*15. Oranges are sweet. Lemons are tart. Therefore oranges and lemons are sweet or tart. (*Ox*: *x* is an orange; *Sx*: *x* is sweet; *Lx*: *x* is a lemon; *Tx*: *x* is tart)

16. Socrates is mortal. Therefore everything is either mortal or not mortal. (*s*: Socrates; *Mx*: *x* is mortal)

chapter 10 Summary

In Section 10.1, we explained that the analytical techniques of the previous chapters are not adequate to deal with arguments whose validity depends on the inner logical structure of noncompound propositions. We described quantification as a theory that, with some additional symbolization, enables us to exhibit this inner structure and thereby greatly enhances our analytical powers.

In Section 10.2, we explained singular propositions and introduced the symbols for an individual variable x, for individual constants (lowercase letters a through w), and for attributes (capital letters). We introduced the concept of a propositional function, an expression that contains an individual variable and becomes a statement when an individual constant is substituted for the individual variable. A proposition may thus be obtained from a propositional function by the process of instantiation.

In Section 10.3, we explained how propositions also can be obtained from propositional functions by means of generalization, that is, using quantifiers such as "everything," "nothing," and "some." We introduced the universal quantifier (x), meaning "given any x," and the existential quantifier $(\exists x)$, meaning "there is at least one x such that." On a square of opposition, we showed the relations between universal and existential quantification.

In Section 10.4, we showed how each of the four main types of general propositions,

- **A:** universal affirmative propositions
- **E:** universal negative propositions
- **I:** particular affirmative propositions
- **O:** particular negative propositions

is correctly symbolized by propositional functions and quantifiers. We also explained the modern interpretation of the relations of **A, E, I,** and **O** propositions.

In Section 10.5, we expanded the list of rules of inference, adding four additional rules:

- Universal Instantiation, U.I.
- Universal Generalization, U.G.
- Existential Instantiation, E.I.
- Existential Generalization, E.G.

and showed how, by using these and the other nineteen rules set forth earlier, we can construct a formal proof of validity of deductive arguments that depend on the inner structure of noncompound propositions.

In Section 10.6, we explained how the method of refutation by logical analogy can be used to prove the invalidity of arguments involving quantifiers by creating a model, or possible universe, containing exactly one, or exactly two, or exactly three (etc.) individuals and the restatement of the constituent propositions of an argument in that possible universe. An argument involving quantifiers is proved invalid if we can exhibit a possible universe containing at least one individual, such that the argument's premises are true and its conclusion is false in that universe.

In Section 10.7, we explained how we can symbolize and evaluate asyllogistic arguments, those containing propositions not reducible to **A, E, I,** and **O** propositions, or singular propositions. We noted the complexity of exceptive propositions and other propositions whose logical meaning must first be understood and then rendered accurately with propositional functions and quantifiers.

LOGIC IN THE REAL WORLD

JENNA'S WORLD

Jenna is a college sophomore, and she is not happy. She signed up for *Logic* thinking that the course catalog said *Logging*.

I want to be a lumberjack.

I have an axe.

1. Which of the following propositions is indicated by the following Venn diagram? (More than one may be correct.)

   Things that Jenna hates — kickball, culottes, cheese, The Volturi, ex-boyfriends

   Spiders — Gross, this thing is full of spiders!

   Logs

   A. No spiders are things that Jenna hates.
   B. Some spiders are things that Jenna hates.
   C. All spiders are things that Jenna hates.
   D. If Jenna does not hate something, then that thing is not a spider.
   E. If Jenna hates something, then it is a spider.

How I Remember A, E, I, and O Propositions *by Jenna*

A – ALL of some kind of thing is some other kind of thing. (This one actually starts with A!)

E – EXACTLY ZERO of some kind of thing is some other kind of thing!

I – INTERESTINGLY, *some* of some kind of thing is some other kind of thing, but I'm not telling you which ones!

O – OMG! You know how we've been thinking that some kind of thing was some other kind of thing? Turns out *some* of it is *NOT*!

2. Identify each as an **A**, **E**, **I**, or **O** proposition, and write out in words the statement represented in each case:

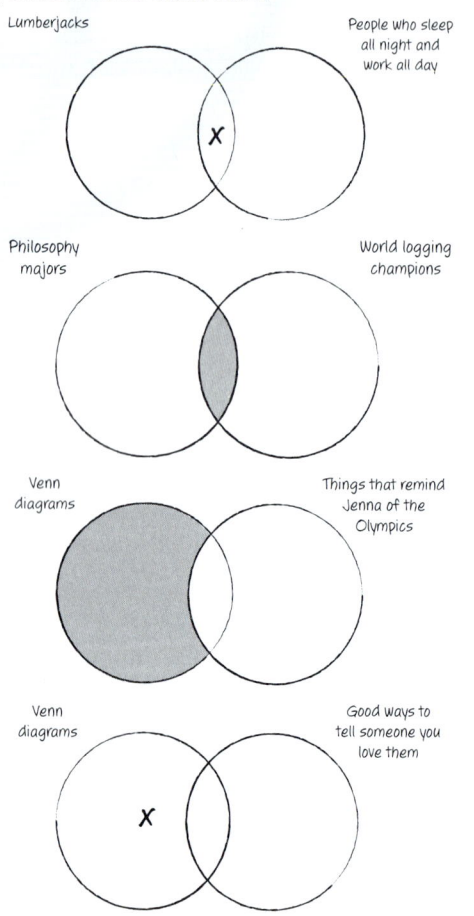

   Lumberjacks — People who sleep all night and work all day (X in intersection)

   Philosophy majors — World logging champions (shaded intersection)

   Venn diagrams — Things that remind Jenna of the Olympics (left circle shaded)

   Venn diagrams — Good ways to tell someone you love them (X in left circle only)

3. Put this syllogism into standard form, being sure to translate each proposition into a categorical statement:

   Obviously, some girls wish that their colleges offered courses on logging. Of all of the people who want to be lumberjacks, some are girls. Some people who want to be lumberjacks wish their colleges offered courses on logging.

4. Is the syllogism presented in Exercise 3 valid? If so, what is its name?

   A. Barbara
   B. Cesare
   C. Ferison
   D. The syllogism is not valid.

5. Translate this sorites into standard form, being sure to translate each proposition into a categorical statement:

   All the dining hall employees totally smell like cheese. Everyone who has found Jenna's framed portrait of Joe Biden a little off-putting is someone who has been allowed into her dorm room. No dining hall employees have found Jenna's framed portrait of Joe Biden a little off-putting, since no one who smells like cheese is allowed in Jenna's dorm room.

6. Write the sorites above as two syllogisms (symbolizing each term with a capital letter), and give the name of each syllogism (see page 238 for a reminder of the fifteen valid syllogisms).

Symbolize each of the following using capital letters to abbreviate the simple statements involved and using the horseshoe, the dot, the wedge, and the curl.

7. Jenna will fail logic unless she masters enthymemes or if she spends too much time thinking about Robert Pattinson.

8. If a spider crawls in Jenna's bed while she is sleeping, she will either sleep through it or else wake up, scream, and then call her parents in the middle of the night.

9. If it is the case that if Jenna fails logic, her parents will take away her credit card, then Jenna will either study harder or hire a tutor, but if it is not the case that if Jenna fails logic her parents will take away her credit card, then Jenna will skip class and stare at her belly button instead.

10. If $\sim A \supset B$, is it true that $\sim\sim\sim B \supset \sim\sim A$?

440 CHAPTER 10 Quantification Theory

TXT MSGS FROM JENNA TO HER BF

1:05 if u rlly luv me, u will plan 2 help me w/these proofs

1:11 either u dont plan 2 help me w/these proofs or u have been kidnapped!

2:44 if u rlly luv me, u have been kidnapped!!! OMG!

7:50 I L U.

7:52 I dont know what that means.

11. Jenna's first three text messages form a valid argument; adding just one statement to the premises will produce a formal proof of its validity. If L = you really love me, H = you will plan to help me with these proofs, and K = you have been kidnapped, write the statements in logic notation, add the missing components to the proof, and name the rules of inference that have been applied. Disregard "OMG!"

I will make an asyllogistic inference! It will have spiders.

x is a spider — $(x)(Sx \supset Bx)$ — x both has 8 legs and is unwelcome in a lunch box

$(\exists x)(Sx \cdot Rx)$ — x is radioactive

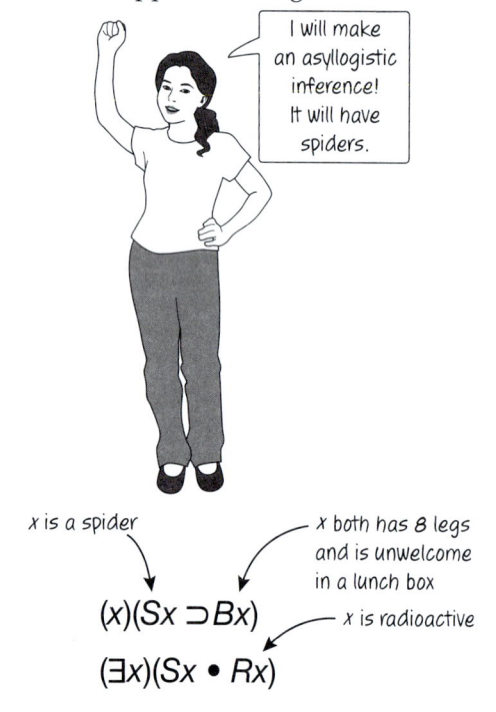

12. An asyllogistic inference can be made from the two premises shown above. Which of the following would be an appropriate conclusion, if Ex means "x has eight legs" and Ux means "x is unwelcome in a lunch box"? (More than one answer may be correct.)

A. $(\exists x)(Rx \cdot Ux)$ B. $(\exists x)(Rx \cdot Ex)$
C. $(x)(Rx \supset Ux)$ D. $(x)(\sim Sx \supset \sim Ux)$

Solutions

1. C and D only. From the Venn diagram, we can see that the area for spiders that is not in the overlap is shaded; therefore, there are NO spiders that are not things that Jenna hates. That is, Jenna hates all spiders. From that statement, the contrapositive follows: If all spiders are things that Jenna hates, then if something is a thing Jenna does not hate, that thing is not a spider.

2. *I*–Some lumberjacks are people who sleep all night and work all day.
   E–No philosophy majors are world logging champions.
   A–All Venn diagrams are things that remind Jenna of the Olympics.
   O–Some Venn diagrams are not good ways to tell someone you love them.

3. As a syllogism composed of categorical propositions, Jenna's argument is:
   Some people who want to be lumberjacks are students who wish that their colleges offered courses on logging.
   Some girls are people who want to be lumberjacks.
   Therefore, some girls are students who wish that their colleges offered courses on logging.

4. The argument is not valid. You can determine this by noting that it is not one of the fifteen valid forms of the standard-form categorical syllogism (page 238), or you may simply use common sense: Imagine that there are 1,000 people who want to be lumberjacks. It is possible, from the argument, that 10 of them are girls, or all of them are girls (we just know that there are some). We also know that *some* of the people who want to be lumberjacks (maybe 10 of them, maybe 1,000 of them) wish that their colleges offered courses on logging. So, while it is certainly possible that some girls are students who wish that their colleges offered courses

on logging, it is also perfectly possible that, out of 1,000 people who want to be lumberjacks, 10 of them are girls, and a *different* 10 of them are students who wish that their colleges offered courses on logging. There is no guarantee of an overlap.

5. Here is the sorites in standard form (the numbers show the position of each statement in the original sorites):

   (1) All people who work in the dining hall are people who smell like cheese.
   (4) No one who smells like cheese is allowed in Jenna's dorm room.
   (2) All people who have found Jenna's framed portrait of Joe Biden a little off-putting are people who have been allowed into her dorm room.
   (3) Therefore, no dining hall employees are people who have found Jenna's framed portrait of Joe Biden a little off-putting.

6. All D is S.
   No S is A.
   Therefore, no D is A.
   Camenes

   No D is A.
   All O is A.
   Therefore, no D is O.
   Cesare

7. If M = Jenna masters enthymemes, R = Jenna spends too much time thinking about Robert Pattinson, and F = Jenna will fail logic:

   $(\sim M \vee R) \supset F$

8. If C = A spider crawls in Jenna's bed while she is sleeping, L = Jenna will sleep through it, W = Jenna will wake up, S = Jenna will scream, and P = Jenna will call her parents in the middle of the night:

   $C \supset [L \vee (W \cdot S \cdot P)]$

9. If F = Jenna fails logic, P = Jenna's parents will take away her credit card, S = Jenna will study harder, H = Jenna will hire a tutor, K = Jenna will skip class, and B = Jenna will stare at her belly button:

   $(F \supset P) \supset (S \vee H) \cdot \sim(F \supset P) \supset (K \cdot B)$

10. Yes. If $\sim A \supset B$, then it is true, from contraposition, that $\sim B \supset A$. Three negatives is the same as one negative (that is, since a double negative is a positive, three negatives is really one positive and one negative, and hence, a negative) and two negatives is the same as zero negatives, so $\sim\sim\sim B \supset \sim\sim A$ is equivalent to $\sim B \supset A$.

11. Jenna's original argument may be written:

    $L \supset H$
    $\sim H \vee K$
    $\therefore L \supset K$

    As a correct proof:

    1. $L \supset H$
    2. $\sim H \vee K$
       $\therefore L \supset K$
    3. $H \supset K$ 2, Implication
    4. $L \supset K$ 1, 3, Hypothetical Syllogism

12. A and B only. Consider conclusion A, placed correctly below Jenna's premises:

$(x)(Sx \supset Bx)$	All spiders both have eight legs and are unwelcome in a lunch box.
$(\exists x)(Sx \cdot Rx)$	Some spiders are radioactive.
$\therefore (\exists x)(Rx \cdot Ux)$	Therefore, some things that are radioactive are unwelcome in a lunch box.

    This is plainly true, and since Bx included both Ex and Ux, it is clear that Ex can serve just as well as Ux in the conclusion, so B is also true:

$(x)(Sx \supset Bx)$	All spiders both have eight legs and are unwelcome in a lunch box.
$(\exists x)(Sx \cdot Rx)$	Some spiders are radioactive.
$\therefore (\exists x)(Rx \cdot Ex)$	Therefore, some things that are radioactive have eight legs.

    However, choices C and D state, unjustifiably, that ALL x that are radioactive are unwelcome in a lunch box, and that anything that is not a spider is not unwelcome in a lunch box—that is, that anything other than a spider would be totally fine in a lunch box. These conclusions are not implied by the argument.

Some point of view is assumed in all our perceptions. In looking at a picture the viewer craves some unambiguous perspective. The coherent representation of things (artists learned long ago) calls for a single vanishing point with which the horizontal lines of the picture are laid out. *Relativity*, by M. C. Escher, exhibits the perplexing result when that coherence is missing. There is no single point of view in this picture; the ceiling for one is the floor for another. Which way is up? Three perspectives throw all into confusion.

Knowledge also supposes some steady context, some accepted theory or shared perspective within which explanations can cohere. But the quest for new knowledge brings new perspectives that, because they are sometimes inconsistent with what had been long supposed, confuse and perplex us. New theories replace the old; what we had thought were facts become suppositions, possibly false. Inductive inquiry is not as dizzy as the world in Escher's *Relativity*, but his picture is a provocative reminder of the uncertainty of what we think we know.

M. C. Escher's *Relativity* © 2004 The M.C. Escher Company. Baarn, Holland. All rights reserved. *www.mcescher.com*

part III

Induction

SECTION A	**ANALOGY AND CAUSATION**
chapter 11	Analogical Reasoning
chapter 12	Causal Reasoning
SECTION B	**SCIENCE AND PROBABILITY**
chapter 13	Science and Hypothesis
chapter 14	Probability

> The contrary of every matter of fact is still possible, because it can never imply a contradiction, and is conceived by the mind with the same facility and distinctness, as if ever so conformable to reality. That the sun will not rise tomorrow is no less intelligible a proposition, and implies no more contradiction than the affirmation, that it will rise. . . . It may, therefore, be a subject worthy of curiosity, to enquire what is the nature of that evidence which assures us of any real existence and matter of fact, beyond the present testimony of our senses, or the records of our memory.
>
> —*David Hume*

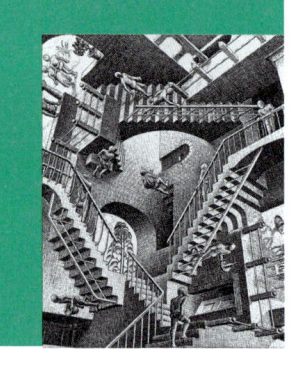

Analogical Reasoning

11.1 Induction and Deduction Revisited

11.2 Argument by Analogy

11.3 Appraising Analogical Arguments

11.4 Refutation by Logical Analogy

11.1 Induction and Deduction Revisited

Arguments are built on premises that are believed, or assumed, to be true. Some premises we establish by deductive arguments that have preceded, but very many of the premises on which we must rely cannot be established by deduction. Our reasoning process usually begins with the accepted truth of some "matters of fact," in David Hume's phrase. To establish matters of fact we must rely on reasoning that is *inductive*.

Induction thus provides the starting points—the foundation—for the reasoning that concerns us most. We reason to establish truths in our everyday lives, to learn facts about our society, to understand the natural world. Deduction is certainly powerful in enabling us to move from known (or assumed) propositions to other propositions that those premises entail, but in the search for truths with which our reasoning must begin, it is insufficient.

The inductive arguments with which we establish matters of fact differ fundamentally from the deductive arguments that were the concern in Part II of this book. One essential contrast between the two families of argument (noted much earlier in our discussion of basic logical concepts, Section 1.5) lies in the relation of the premises to the conclusion in the arguments of the two great families. In *deductive arguments*, the claim is made that conclusions follow with certainty from their premises. That claim is appropriate because any deductive argument, if it is good, brings to light in its conclusion what was already buried in its premises. The relation between premises and conclusion, in deduction, is one of *logical necessity*. In every deductive argument, if it is valid and if its premises are true, its conclusion *must* be true.

In *inductive arguments*—the concern of this chapter and those that follow—the relation between premises and conclusion is not one of logical necessity. The claim of certainty is not made. The terms *valid* and *invalid* simply do not apply. This does not mean that inductive arguments are always weak; sometimes they are very strong indeed, and fully deserve our confidence. Scientists now assert without reservation (for example) that smoking is a cause of cancer. This is true, but it is a truth that cannot be known with the demonstrative certainty of a valid syllogism. If *p or q* is true, and *not p* is true, we may conclude that *q* must be the case, beyond all doubt. It is a truth we establish as an inescapable consequence of the relations of the concepts involved. Empirical truths—about the consequences of smoking, or the causes of cancer, and all others of that sort—cannot satisfy the standard of deductive certainty. By that standard, as one distinguished medical investigator observes, "No one will ever be able to prove that smoking causes cancer, or that anything causes anything."[1]

In the realm of induction, as we seek new knowledge of facts about the world, nothing is beyond all doubt. We must rely on arguments that support their conclusions only as *probable*, or probably true. Some such arguments are of only moderate worth; others are very powerful, as we shall see. The strengths and weaknesses of inductive arguments, and the techniques for the evaluation of such arguments, are the focus of Part III of this book.

Arguments grounded on *analogies*, aiming to establish particular conclusions, are examined first, in this chapter. Arguments that go beyond particulars, aiming to establish generally applicable *causal laws*, are examined in the following chapter. The uses of hypotheses and their confirmation in developing *scientific theories* follows in Chapter 13; and we conclude, in Chapter 14, with an analysis of the concept of *probability* itself, the conceptual instrument with which inductive conclusions are commonly expressed.

11.2 Argument by Analogy

The most common type of inductive argument relies on *analogy*. If I report that I got very good service from a computer of a certain make and model, you may infer that a new computer of the same make and model will serve you well. That conclusion has some degree of probability, but the argument is far from compelling. When a new book is called to my attention and I infer that I will enjoy reading it because I have read and enjoyed other books by the same author, I may have my confidence in that author strengthened when I read the book—or I may be disappointed. Analogy is the common ground of our everyday inferences from past experience to what the future will hold.

Here follow two more analogical arguments, carefully formulated. The first concludes, on the basis of what we commonly think to be prudent and fair, that it would be prudent and fair to adopt now a major change in public policy:

> Some people look on preemployment testing of teachers as unfair—a kind of double jeopardy. "Teachers are already college graduates," they say. "Why should they be tested?" That's easy. Lawyers are college graduates and graduates of professional school, too, but they have to take a bar exam. And a number of other professions ask prospective members to prove that they know their stuff by taking and passing examinations: accountants, actuaries, doctors, architects. There is no reason why teachers shouldn't be required to do this too.[2]

The second illustration is an argument—entirely plausible when first presented two centuries ago—whose conclusion is very probably false:

> We may observe a very great similitude between this earth which we inhabit, and the other planets, Saturn, Jupiter, Mars, Venus, and Mercury. They all revolve around the sun, as the earth does, although at different distances and in different periods. They borrow all their light from the sun, as the earth does. Several of them are known to revolve around their axis like the earth, and by that means, must have a like succession

of day and night. Some of them have moons, that serve to give them light in the absence of the sun, as our moon does to us. They are all, in their motions, subject to the same law of gravitation, as the earth is. From all this similitude, it is not unreasonable to think that those planets may, like our earth, be the habitation of various orders of living creatures. There is some probability in this conclusion from analogy.[3]

Neither these arguments, nor those everyday inferences we draw about computers and books and the like, are demonstratively valid. Their conclusions are not claimed to follow from their premises with logical necessity, and they obviously do not follow with certainty. What is appropriate for judging the employability of lawyers and doctors may not be appropriate for judging the employability of teachers. The earth is very likely to be the only inhabited planet in our solar system. Your new computer may prove unsuitable for the work you do, and I may find my favorite author's latest book intolerably dull. In all such arguments it is plainly possible—logically possible—that although the premises are true, the conclusions are false. Arguments by analogy are not to be classified as either valid or invalid; probability is all that is claimed for them.

In addition to their use in arguments, analogies are very often used nonargumentatively, for the purpose of lively description. The literary uses of analogy in metaphor and simile are tremendously helpful to the writer who strives to create a vivid picture in the reader's mind. In the continuing controversy in the United States over immigration, for example, one writer expressed his views with a forceful analogy:

> I'm a third-generation American. I don't know all the legal details about how my grandparents got here. But I do know that they worked very hard, paid their taxes, and raised a son who served his country. Americans being against immigration is like a house being against its bricks.[4]

Analogy is also used in explanation, when something that may not be familiar to the reader is made somewhat more intelligible by being compared to something else, presumably more familiar, to which it has certain similarities. When Eric Lander, the director of the Genome Center at the Massachusetts Institute of Technology, sought to explain the huge eventual impact of the Human Genome Project, analogy was one of the devices he used to enhance the understanding of those unfamiliar with genetic research:

> The genome project is wholly analogous to the creation of the periodic table in chemistry. Just as Mendeleev's arrangement of the chemical elements in the periodic table made coherent a previously unrelated mass of data, so the tens of thousands of genes in present-day organisms will all turn out to be made from combinations of a much smaller number of simpler genetic modules or elements, the primordial genes, so to speak.[5]

Analogies—But No Arguments

Nonargumentative analogies are commonly encountered in the writing of high-school students—and some of these are quite funny. We pause for a chuckle:

1. She grew on him like she was a colony of *E. coli* and he was room-temperature Canadian beef.

> 2. McMurphy fell 12 stories, hitting the pavement like a Hefty bag filled with vegetable soup.
> 3. Her hair glistened in the rain, like a nose hair after a sneeze.
> 4. Even in his last years, Grandpappy had a mind like a steel trap, only one that had been left out so long it had rusted shut.
> 5. He was deeply in love. When she spoke he thought he heard bells, like a garbage truck backing up.
> 6. She had a deep, throaty, genuine laugh, like that sound a dog makes just before it throws up.
> 7. His thoughts tumbled in his head, making and breaking alliances like underpants in a dryer without Cling Free.
> 8. The hailstones leaped from the pavement, just like maggots when you fry them in hot grease.
> 9. The ballerina rose gracefully *en pointe* and extended one slender leg behind her, like a dog at a fire hydrant.
> 10. She walked into my office like a centipede with 98 missing legs.

The use of analogies in description and explanation is not the same as their use in argument, though in some cases it may not be easy to decide which use is intended. *But* whether used argumentatively or otherwise, analogy is not difficult to define. To draw an **analogy** between two or more entities is to indicate one or more respects in which they are similar.

This definition explains what an analogy is, but there is still the problem of characterizing an *argument* by analogy. Let us analyze the structure of a particular analogical argument, using a very simple example. Consider the argument that a new car whose purchase I am now contemplating will be very satisfactory because my old car, of the same make and model, has long given very satisfactory service. The two entities that are said to be similar are two cars. Three points of analogy are involved, three respects in which the two entities are said to resemble each other: first, in being cars; second, in being of the same make and model; and third, in serving me well.

The three points of analogy do not play identical roles in the argument, however. The first two occur in the premises, whereas the third occurs both in the premises and in the conclusion. The given argument may be described as having premises that assert, first, that two things are similar in two respects, and second, that one of those things has a further characteristic, from which the conclusion is drawn that the other thing also has that further characteristic.

Analogical argument is one of the most fundamental tools of appellate courts. The inference in a particular case before a court may be shown to be very much like some other inference drawn previously, and if it was clearly correct in that earlier case, it is held to be correct in this one too. In 2004, the U.S. Supreme Court decided unanimously a case requiring the interpretation of the Sixth

Analogy
A parallel drawn between two (or more) entities by indicating one or more respects in which they are similar.

Amendment of the U.S. Constitution, which gives to every criminal defendant the right "to be confronted with the witnesses against him." Does this forbid the use, at a defendant's trial, of testimony from a witness who is not available for cross-examination, even if the trial judge believes that testimony to be reliable? Yes, said Justice Antonin Scalia, delivering the opinion of the Court, it does. The right to cross-examine adverse witnesses was firmly established in the English common law at the time our Constitution was adopted. Justice Scalia's subsequent analogy epitomizes the argument of the Court:

> Admitting statements deemed reliable by a judge is fundamentally at odds with the right of confrontation. Dispensing with confrontation because testimony is obviously reliable is akin to dispensing with a jury trial because a defendant is obviously guilty. This is not what the Sixth Amendment prescribes."[6]

Analogical argument is also common in political controversy. Sometimes the analogy is effective, sometimes it is far-fetched. The threat of global warming, and the need of our country to respond concretely to that threat, was argued heatedly before the Congress of the United States, in 2007, by former presidential candidate Al Gore, who described the danger as a "planetary emergency." Against those who thought him to be exaggerating the dangers, he then argued:

> The planet has a fever. If your baby has a fever you go to the doctor. If the doctor says you need to intervene here, you don't say "I read a science-fiction novel that says it's not a problem." You take action.[7]

Not every analogical argument need concern exactly two things or exactly three different characteristics, of course. Thus the argument presented earlier, suggesting that other planets in our solar system may well be inhabited, draws analogies among six things (the then-known planets) in some eight respects. Apart from these numerical differences, however, all analogical arguments have the same general structure or pattern. Every **analogical argument** *proceeds from the similarity of two or more things in one or more respects to the similarity of those things in some further respect*. Schematically, where *a*, *b*, *c*, and *d* are any entities and *P*, *Q*, and *R* are any attributes or "respects," an analogical argument may be represented as having the form

a, *b*, *c*, *d* all have the attributes *P* and *Q*.

a, *b*, *c* all have the attribute *R*.

Therefore *d* probably has the attribute *R*.

In identifying, and especially in appraising, analogical arguments, it may be found helpful to recast them into this form.

Analogical argument
A kind of inductive argument in which, from the fact that two entities are alike in some respect(s), it is concluded that they are also alike in some other respect(s).

EXERCISES

All of the following passages contain analogies. Distinguish those passages that contain analogical arguments from those that make nonargumentative uses of analogy.

EXAMPLE

1. A Man ought no more to value himself for being wiser than a Woman, if he owes his Advantage to a better Education, than he ought to boast of his Courage for beating a Man when his hands were bound.

   —Mary Astell, *An Essay in Defence of the Female Sex*, 1721

SOLUTION

This is an analogical argument. The analogy drawn here is between beating a man when his hands are bound and being wiser than a woman as a consequence of a better education, one party having an enormous advantage in both cases. In the first case, it is plain that one with such an advantage ought not to boast of his courage; in the second case (this argument concludes), it is equally inappropriate for one with such an advantage to boast of his relative wisdom.

2. "I'm not anti-Semitic, I'm just anti-Zionist" is the equivalent of "I'm not anti-American, I just think the United States shouldn't exist."

   —Benjamin Netanyahu, *A Place Among the Nations*, (New York: Bantam Books, 1993)

3. Instead of investing in the future, we throw money away on absurd luxuries, finance corrupt and hostile oil-rich countries, pollute our atmosphere and increase our trade deficit. Sort of like driving a Hummer to the shopping mall.

   —Eric Buckvar, "A Wasteful Society," *The New York Times*, 23 March 2007

4. The British are less rigid about punctuation and related matters, such as footnote and bibliographic form, than Americans are. An Englishwoman lecturing Americans on semicolons is a little like an American lecturing the French on sauces.

   —Louis Menand, "Bad Comma," *The New Yorker*, 28 June 2004

5. Studies show that girls get better grades in high school and college than boys—yet only about 35 percent of National Merit Scholarship winners are girls. The Executive Director of FairTest contends that the "inequity is due solely to gender bias in the test used to select eligible students." But the spokeswoman for the National Merit Scholarship Corporation, Elaine Detweiler, replies "We don't really know why girls do worse on the exams. To blame the test for the difference between how boys and girls perform is like blaming a yardstick that boys are taller than girls."

   —"Merit Test Defended," *The Los Angeles Times*, 26 May 1993

6. The famous chemist and biologist Justus von Liebig dismissed the germ theory with a shrug of the shoulders, regarding Pasteur's view that microbes could cause fermentation as ridiculous and naive as the opinion of a child "who would explain the rapidity of the Rhine current by attributing it to the violent movement of the many millwheels at Maintz."

   —René Dubos, *Pasteur and Modern Science* (New York: Da Capo Press, 1988)

7. Talking about Christianity without saying anything about sin is like discussing gardening without saying anything about weeds.
   —The Rev. Lord Soper, quoted in *The New York Times*, 24 December 1998

8. Men and women may have different reproductive strategies, but neither can be considered inferior or superior to the other, any more than a bird's wings can be considered superior or inferior to a fish's fins.
   —David M. Buss, "Where Is Fancy Bred? In the Genes or in the Head?" *The New York Times*, 1 June 1999

9. "This is a matter of national spirit," said Marjorie Wilson, coordinator of the Kangaroo Protection Cooperative, an Australian wildlife group. "We believe here that we have enough meat in this country to satisfy people without them having to eat their national symbol. You Americans don't cook your bald eagles, do you?"
   —"Battling over a National Symbol," *The New York Times*, 10 July 1995

10. One sure thing is that melting sea ice cannot be implicated in the coastal flooding that many global warming models have projected. Just as melting ice cubes do not cause a glass of water to overflow, melting sea ice does not increase oceanic volume. Any future rise in sea level would result from glaciers melting on land.
    —Walter Gibbs, "Research Predicts Summer Doom for Northern Icecap," *The New York Times*, 11 July 2000

11. Thomas Henry Huxley, Charles Darwin's nineteenth-century disciple, presented this analogy: "Consciousness would appear to be related to the mechanism of the body simply as a collateral product of its working and to be completely without any power of modifying that working, as the steam whistle which accompanies the work of a locomotive is without influence upon its machinery."

12. The Elgin Marbles—17 figures and 56 panels that once decorated the Parthenon, on the Acropolis in Athens—were taken from the Parthenon in 1801 by Thomas Bruce, the seventh Earl of Elgin, and brought to the British Museum, in London. The Greeks say that he stole them; the British say that they were properly acquired, by purchase. Some Britons urged that the Marbles be returned to Greece in time for the Olympic Games to be held in Athens in 2004. Said one of the leaders of the Labor Party: "The Parthenon without the Elgin Marbles is like a smile missing a tooth."

13. The Feminists decided to examine the institution of marriage as it is set up by law in order to find out whether or not it did operate in women's favor. It became increasingly clear to us that the institution of marriage "protects" women in the same way that the institution of slavery was said to "protect" blacks—that is, that the word "protection" in this case is simply a euphemism for oppression.
    —Sheila Cronan, "Marriage," in Anne Koedt, Ellen Levine, and Anita Rapone, eds., *Radical Feminism* (New York: Quadrangle Books, 1976)

14. Wittgenstein used to compare thinking with swimming: just as in swimming our bodies have a natural tendency to float on the surface so that it requires great physical exertion to plunge to the bottom, so in thinking it requires great mental exertion to force our minds away from the superficial, down into the depth of a philosophical problem.
    —George Pitcher, *The Philosophy of Wittgenstein* (Englewood Cliffs, NJ: 1964)

15. A person without a goal is like a computer without a program. And that's an ugly piece of furniture.
    —Steve Danish, "Getting a Life," *The New York Times*, March 1998

16. The quest for usable energy from fusion involves the use of interlocked magnetic fields to contain very hot (180 million degrees Fahrenheit) and highly compressed (to a density 20 times that of lead) electrically charged plasma (a kind of gas) within a vacuum chamber. The plasma must never touch the solid walls of its container, for if it does it instantly loses its heat and can never be coaxed into undergoing fusion. One scientific report put the problem this way:

    > Everything depends on keeping the plasma's magnetic bottle tightly stoppered . . . [but] confining a dollop of super-hot compressed plasma has proved to be harder than compressing and shaping a blob of jelly using only rubber bands. Each clever idea of the plasma physicists for solving this problem has been matched by a new challenge.

    —Malcolm W. Browne, "Reviving the Quest to Tame the Energy of the Stars," *The New York Times*, 8 June 1999

17. It is important that we make clear at this point what definition is and what can be attained by means of it. It seems frequently to be credited with a creative power; but all it accomplishes is that something is marked out in sharp relief and designated by a name. Just as the geographer does not create a sea when he draws boundary lines and says: the part of the ocean's surface bounded by these lines I am going to call the Yellow Sea, so too the mathematician cannot really create anything by his defining.
    —Gottlob Frege, *The Basic Laws of Arithmetic*, 1893

18. Children in school are like children at the doctor's. He can talk himself blue in the face about how much good his medicine is going to do them; all they think of is how much it will hurt or how bad it will taste. Given their own way, they would have none of it.

    So the valiant and resolute band of travelers I thought I was leading toward a much hoped-for destination turned out instead to be more like convicts in a chain gang, forced under threat of punishment to move along a rough path leading nobody knew where and down which they could see hardly more than a few steps ahead. School feels like this to children: it is a place where they make you go and where they tell you to do things and where they try to make your life unpleasant if you don't do them or don't do them right.
    —John Holt, *How Children Fail* (New York: Delta/Lawrence, 1964)

19. I simply can't imagine the world will ever be normal again for us. I do talk about "after the war," but it's as if I were talking about a castle in the air, something that can never come true.

    I see the eight of us in the Annex as if we were a patch of blue sky surrounded by menacing black clouds. The perfectly round spot on which we're standing is still safe, but the clouds are moving in on us, and the ring between us and the approaching danger is being pulled tighter and tighter. We're surrounded by darkness and danger, and in our desperate search for a way out we keep bumping into each other. We look at the fighting down below and the peace and beauty up above. In the meantime, we've been cut off by the dark mass of clouds, so that we can go neither up nor down. It looms before us like an impenetrable wall, trying to crush us, but not yet able to. I can only cry out and implore, "Oh, ring, ring, open wide and let us out!"

    —Anne Frank, from *The Diary of a Young Girl*, 8 November 1943

20. Unfortunately, the diary [of H. L. Mencken] reveals a man who was shockingly anti-Semitic and racist, to the point where his stature as a giant of American letters may be in danger. . . . I would draw a comparison with Richard Wagner, a virulent anti-Semite. One can still listen to Wagner's operas and appreciate their artistic beauty. The work is separated from the man. Or is it?

    —Gwinn Owens, "Mencken—Getting a Bum Rap?" *The New York Times*, 13 December 1989

11.3 Appraising Analogical Arguments

Some analogical arguments are much more cogent than others. Although no argument by analogy can be deductively valid, some such arguments yield conclusions that are very probably true, whereas others are very weak indeed. Analogical arguments are evaluated as better or worse depending on the degree of probability with which, relying on the premises they put forward, their conclusions may be affirmed.

Two commonplace examples will help to exhibit the features of analogical arguments that make them better or worse. Suppose you choose to purchase a given pair of shoes because other pairs like it have given you satisfaction in the past; and suppose you select a dog of a given breed because other dogs of that same breed have exhibited the characteristics that you prize. In both cases, analogical arguments have been relied on. To appraise the strength of these sample arguments, and indeed of all analogical arguments, six criteria may be distinguished.

1. **Number of entities.** If my past experience with shoes of a certain kind is limited to only one pair that I wore and liked, I will be disappointed although not surprised by an apparently similar pair that I find flawed

in unexpected ways. But if I have repeatedly purchased shoes just like those, I may reasonably suppose that the next pair will be as good as the ones worn earlier. Several experiences of the same kind with an item of just that sort will support the conclusion—that the purchase will be satisfying—much more than will a single instance. Each instance may be thought of as an additional entity, and the number of entities is the first criterion in evaluating an analogical argument.

As a general rule, *the larger the number of entities*—that is, cases in our experience—*the stronger the argument*. However, there is no simple ratio between that number and the probability of the conclusion. Six happy experiences with golden retrievers, intelligent and sweet-tempered dogs, will lead one to conclude that the next golden retriever will also be intelligent and sweet-tempered. However, the conclusion of an analogical argument that has six instances in its premises will not be exactly three times as probable as a similar argument that has two such instances in its premises. Increasing the number of entities is important, but so are other factors.

2. **Variety of the instances in the premises.** If my previous purchases of those good shoes had been from both a department store and a specialty store, and had been made both in New York and in California, by both mail order and direct sale, I may be confident that it is the shoes themselves and not their seller that accounts for my satisfaction. If my previous golden retrievers were both males and females, acquired both as puppies from breeders and as adults from the humane society, I may be more confident that it is their breed—not their sex or age or source—that accounts for my earlier satisfaction.

We understand this criterion intuitively: *The more dissimilar the instances mentioned only in the premises of an analogical argument, the stronger is the argument.*

3. **Number of similar respects.** Among the instances in the premises there may have been various similarities: perhaps the shoes were of the same style, had the same price, were made of the same sort of leather; perhaps the dogs were of the same breed, came from the same breeder at the same age, and so on. All the respects in which the instances in the premises are like one another, and like the instance in the conclusion, increase the probability that the instance in the conclusion will have that further attribute at which the argument is aimed—giving great satisfaction in the case of the new shoes, being of a sweet disposition in the case of a new dog.

This criterion also is rooted in common sense: *The greater the number of respects in which the entity in the conclusion is similar to the entities in the premises, the more probable is that conclusion.* Again, of course, there is no simple way to decide when the number of similar respects identified is sufficient.

4. **Relevance.** As important as the number of respects shared is the kind of respects in which the instances in the premises are like the instance in the conclusion. If the new pair of shoes, like the previous pairs, is purchased on a Tuesday, that is a likeness that will have no bearing on the satisfaction they give; but if the new pair, like all the previous pairs, had the same manufacturer, that will count heavily. *Respects add to the force of the argument when they are relevant* (as style of shoe, and price, and material surely are)—*and a single highly relevant factor contributes more to the argument than a host of irrelevant similarities.*

   There will sometimes be disagreement about which attributes really are relevant in establishing the likelihood of our conclusion, but the *meaning* of relevance itself is not in dispute. One attribute is relevant to another when it is connected to that other, when there is some kind of *causal relation* between them. That is why identifying causal connections of one kind or another is critical in analogical arguments, and why establishing such connections is often crucial in determining the admissibility of evidence, as relevant or irrelevant, in a court of law.

   Analogical arguments can be probable whether they go from cause to effect or from effect to cause. They can even be probable when the attribute in the premise is neither the cause nor the effect of the conclusion's attribute, provided both are the effect of the same cause. A doctor, noting the presence of a certain symptom in her patient, may predict another symptom accurately not because either symptom is the cause of the other, but because they are jointly caused by the same disorder. The color of a manufactured product is most often irrelevant to function, but it may serve as a relevant respect in an argument when that color is very unusual, and shared by the entities in the premises and the conclusion. The color itself may contribute nothing to the function of the product, but it may serve in argument if it is known to be an attribute of the manufacturing process of a unique producer.

   The causal connections that are the key to the evaluation of analogical arguments can be discovered only empirically, by observation and experiment. The general theory of empirical investigation is the central concern of inductive logic, and will be discussed at length in the chapters that follow.

5. **Disanalogies.** A **disanalogy** is a point of difference, a respect in which the case we are reasoning about in our conclusion is distinguishable from the cases on which the argument is based. Returning to the example of the shoes, if the pair we plan to buy looks like those we had owned earlier, but is in fact much cheaper and made by a different company, those disanalogies will give us reason to doubt the satisfaction they will provide.

   What was said earlier about relevance is also important here. Disanalogies undermine analogical arguments when the points of difference

Disanalogy
In an analogical argument, a point of difference between the cases cited in the premises and the case mentioned in the conclusion.

identified are relevant—causally connected to the outcome we are seeking. Investors often purchase shares of a stock mutual fund on the basis of its successful "track record," reasoning that because earlier purchases resulted in capital appreciation, a future purchase will do so as well. However, if we learn that the person who managed the fund during the period of its profitability has just been replaced, we confront a disanalogy that substantially reduces the strength of that analogical argument.

Disanalogies weaken analogical arguments. They are therefore commonly employed in *attacking* an analogical argument. As critics, we may try to show that the case in the conclusion is different in important ways from the earlier cases, and that what was true of them is not likely to be true of the present case. In the law, where the uses of analogy are pervasive, some earlier case or cases are often offered to a court as a precedent for deciding the case at hand. The argument is analogical. Opposing counsel will seek to *distinguish* the case at hand from the earlier cases; that is, counsel will seek to show that because there is some critical difference between the facts in the case at hand and the facts in those earlier cases, the latter do not serve as good precedents in the present matter. If the differences are great—if the disanalogy is indeed critical—that may demolish the analogical argument that had been put forward.

Because disanalogies are the primary weapon against an analogical argument, whatever can ward off any potential disanalogies will strengthen the argument. This explains why variety among the instances in the premises adds force to an argument. The more the instances in the premises vary from one to another, the less likely it is that a critic will be able to point to some disanalogy between all of them and the conclusion that will weaken the argument. To illustrate, suppose that Kim Kumar comes to a university as a first-year student; ten others from her secondary school have successfully completed studies at the same university. We may argue analogically that in view of her secondary school preparation, she is likely to succeed as well. If all those other students from her school were similar to one another in some respect that bears on college study but differ from Kim in that respect, that disanalogy will undermine the argument for Kim's success. However, if we learn that the ten successful predecessors varied among themselves in many ways—in economic background, in family relations, in religious affiliation, and so on—those differences among them ward off such potential disanalogies. The argument for Kim's success is fortified—as we saw earlier—if the other students from her school serving as instances in the premises of the argument do not resemble each other closely, but exhibit substantial variety.

A confusion must be avoided: The principle that disanalogies weaken analogical arguments is to be contrasted with the principle that differences among the premises strengthen such arguments. In the former, the differences are between the instances in the premises and the instance in the conclusion; in the latter, differences are among the instances in the premises only. A disanalogy is a difference between the cases with which we have experience and the case about which a conclusion is being drawn. That conclusion (we may say in presenting the disanalogy as refutation) is not warranted, because circumstances in the critical case are not similar to circumstances in earlier cases. We say that the analogy is "strained" or that it "does not hold." But when we point to dissimilarities among the premises we are strengthening the argument by saying, in effect, that the analogy has wide force, that it holds in cases like these and in other cases, and that therefore the respects in which the instances in the premises vary are not relevant to the matter with which the conclusion is concerned.

In summary, disanalogies undermine an analogical argument; dissimilarities among the premises reinforce it. Both considerations are tied to the question of relevance: Disanalogies tend to show that there are relevant respects in which the case in the conclusion differs from those in the premises; dissimilarities among the premises tend to show that other factors, which might have been thought causally relevant to the attribute of interest, are not really relevant at all.

Note that the very first criterion identified, pertaining to the *number* of entities among which the analogy is said to hold, is also linked to relevance. The greater the number of instances appealed to, the greater is the number of dissimilarities likely to obtain among them. Increasing the number of entities is therefore desirable, but as the number of entities increases, the impact of each additional case is reduced. The dissimilarity it may provide is more likely to have been provided by earlier instances, in which case it will add little or nothing to the protection of the conclusion from damaging disanalogies.

6. **Claim that the conclusion makes.** Every argument makes the claim that its premises give reasons to accept its conclusion. It is easy to see that the more one claims, the greater the burden of sustaining that claim, and that is obviously true for every analogical argument. The *modesty of the conclusion relative to the premises* is critical in determining the merit of the inference.

    If my friend gets 30 miles to the gallon from his new car, I may infer that, were I to acquire a car of the same make and model, I would get at least 20 miles to the gallon; that conclusion is modest and therefore very probable. Were my conclusion much bolder—say, that I would get at least 29 miles to the gallon—it would be less well supported by the evidence I have. *The more modest the claim, the less burden is placed on the premises and the stronger the argument; the bolder the claim, the greater is the burden on the premises and the weaker the argument.*

An analogical argument is strengthened by reducing the claim made on the basis of the premises affirmed, or by retaining the claim unchanged while supporting it with additional or more powerful premises. Likewise, an analogical argument is weakened if its conclusion is made bolder while its premises remain unchanged, or if the claim remains unchanged while the evidence in its support is found to exhibit greater frailty.

EXERCISES

A. For each of the following arguments by analogy, six additional premises are suggested. For each of these alternative premises, decide whether its addition would make the conclusion of the resulting argument more or less probable. Identify the criterion of appraisal that justifies this judgment, and explain how that criterion applies to the given case.

EXAMPLE

1. An investor has purchased one hundred shares of oil stock every December for the past five years. In every case the value of the stock has appreciated by about 15 percent a year, and it has paid regular dividends of about 8 percent a year on the price at which she bought it. This December she decides to buy another hundred shares of oil stock, reasoning that she will probably receive modest earnings while watching the value of her new purchase increase over the years.

   a. Suppose that she had always purchased stock in eastern oil companies before, and plans to purchase stock in an eastern oil company this year, too.
   b. Suppose that she had purchased oil stocks every December for the past fifteen years, instead of for only five years.
   c. Suppose that the oil stocks previously purchased had gone up by 30 percent a year, instead of only by 15 percent.
   d. Suppose that her previous purchases of oil stock had been in foreign companies as well as in eastern, southern, and western U.S. oil companies.
   e. Suppose she learns that OPEC has decided to meet every month instead of every six months.
   f. Suppose she discovers that tobacco stocks have just raised their dividend payments.

SOLUTION

   a. More probable. *Number of similar respects*. The change provides an additional respect in which the instance in the conclusion is the same as those in the premises.

b. *More probable. Number of entities.* With this change the number of entities in the premisses is substantially increased.

c. *More probable. Claim made by the conclusion.* With this change in the premises, the conclusion, although unchanged, is now, relatively speaking, substantially more modest.

d. *More probable. Variety among the premises.* With this change, the dissimilarity among the instances in the premises is clearly established.

e. *Less probable. Disanalogy.* With this change in the premises, a significant difference between the instance in the conclusion and the instances in the premises is introduced.

f. *Neither. Relevance.* It is unlikely that the dividends paid by tobacco companies would have any impact on the profitability of oil companies or the price of their shares.

2. A faithful alumnus, heartened by State's winning its last four football games, decides to bet his money that State will win its next game, too.
   a. Suppose that since the last game, State's outstanding quarterback was injured in practice and hospitalized for the remainder of the season.
   b. Suppose that two of the last four games were played away, and that two of them were home games.
   c. Suppose that, just before the game, it is announced that a member of State's Chemistry Department has been awarded a Nobel Prize.
   d. Suppose that State had won its last *six* games rather than only four of them.
   e. Suppose that it has rained hard during each of the four preceding games, and that rain is forecast for next Saturday's game too.
   f. Suppose that each of the last four games was won by a margin of at least four touchdowns.

3. Although she was bored by the last few foreign films she saw, Charlene agrees to go to see another one this evening, fully expecting to be bored again.
   a. Suppose that Charlene also was bored by the last few American movies she saw.
   b. Suppose that the star of this evening's film has recently been accused of bigamy.
   c. Suppose that the last few foreign films that Charlene saw were Italian, and that tonight's film is Italian as well.
   d. Suppose that Charlene was so bored by the other foreign films that she actually fell asleep during the performance.
   e. Suppose that the last few foreign films she saw included an Italian, a French, an English, and a Swedish film.
   f. Suppose that tonight's film is a mystery, whereas all of those she saw before were comedies.

4. Bill has taken three history courses and found them very stimulating and valuable, so he signs up for another one, confidently expecting that it too will be worthwhile.
    a. Suppose that his previous history courses were in ancient history, modern European history, and U.S. history.
    b. Suppose that his previous history courses had all been taught by the same professor scheduled to teach the present one.
    c. Suppose that his previous history courses had all been taught by Professor Smith, and the present one is taught by Professor Jones.
    d. Suppose that Bill had found his three previous history courses to be the most exciting intellectual experiences of his life.
    e. Suppose that his previous history courses had all met at 9 A.M., and that the present one is also scheduled to meet at 9 A.M.
    f. Suppose that, in addition to the three history courses he took previously, Bill had also taken and enjoyed courses in anthropology, economics, political science, and sociology.

5. Dr. Brown has stayed at the Queen's Hotel every fall for the past six years on her annual visit to New York, and she has been quite satisfied with her accommodations there. On her visit to New York this fall, Dr. Brown goes again to the Queen's Hotel, confidently expecting to enjoy her stay there again.
    a. Suppose that when she stayed at the Queen's Hotel before, she had occupied a single room twice, shared a double room twice, and twice occupied a suite.
    b. Suppose that last spring a new manager had been put in charge of the Queen's Hotel.
    c. Suppose that she had occupied a suite on all of her previous trips and is assigned a suite this time as well.
    d. Suppose that on her previous trips she had come to New York by train, but this time she flew.
    e. Suppose that, when she stayed at the Queen's Hotel before, her quarters had been the most luxurious she had ever known.
    f. Suppose that she had stayed at the Queen's Hotel three times a year for the past six years.

B. Analyze the structure of the analogical arguments in the following passages, and evaluate them in terms of the six criteria that have been explained:

1. If you cut up a large diamond into little bits, it will entirely lose the value it had as a whole; as an army divided up into small bodies of soldiers loses all its strength. So a great intellect sinks to the level of an ordinary one, as soon as it is interrupted and disturbed, its attention distracted and drawn off from the matter in hand: for its superiority depends upon its power of concentration—of bringing all its strength to bear upon one theme, in the same way as a concave mirror collects into one point all the rays of light that strike upon it.

—Arthur Schopenhauer, "On Noise," 1851

2. It would be the height of hypocrisy if Pete Rose, one of baseball's star players, were allowed back into baseball and elected to the Hall of Fame after finally admitting that he placed bets on his team and other teams and lied about it. In coming to a decision about Rose, the Baseball Commissioner should remember that Olympic athletes who have been caught using performance-enhancing drugs are stripped permanently of their titles and medals.

—Frank Ulrich, *The New York Times*, 8 January 2004

3. Look round the world: contemplate the whole and every part of it: you will find it to be nothing but one great machine, subdivided into an infinite number of lesser machines, which again admit of subdivisions, to a degree beyond what human senses and faculties can trace and explain. All these various machines, and even their most minute parts, are adjusted to each other with an accuracy which ravishes into admiration all men who have ever contemplated them. The curious adapting of means to ends, throughout all nature, resembles exactly, though it much exceeds, the production of human contrivance, of human design, thought, wisdom, and intelligence. Since therefore the effects resemble each other, we are led to infer, by all the rules of analogy, that the causes also resemble; and that the Author of Nature is somewhat similar to the mind of men; though possessed of much larger faculties, proportioned to the grandeur of the work, which he has executed. By this argument *a posteriori*, and by this argument alone, do we prove at once the existence of a Deity, and his similarity to human mind and intelligence.

—David Hume, *Dialogues Concerning Natural Religion*, 1779

4. The philosopher Metrodorus of Chios, who lived in the fourth century BCE, was greatly interested in the heavenly bodies. He wrote: "To consider the Earth as the only populated world in infinite space is as absurd as to assert that in an entire field of millet, only one grain will grow."

5. To the casual observer porpoises and sharks are kinds of fish. They are streamlined, good swimmers, and live in the sea. To the zoologist who examines these animals more closely, the shark has gills, cold blood, and scales; the porpoise has lungs, warm blood, and hair. The porpoise is fundamentally more like man than like the shark and belongs, with man, to the mammals—a group that nurses its young with milk. Having decided that the porpoise is a mammal, the zoologist can, without further examination, predict that the animal will have a four-chambered heart, bones of a particular type, and a certain general pattern of nerves and blood vessels. Without using a microscope the zoologist can say with reasonable confidence that the red blood cells in the blood of the porpoise will lack nuclei. This ability to generalize about animal structure depends upon a system for organizing the vast amount of knowledge about animals.

—Ralph Buchsbaum, *Animals without Backbones*
(Chicago: University of Chicago Press, 1961)

6. The body is the substance of the soul; the soul is the functioning of the body. . . . The relationship of the soul to its substance is like that of sharpness to a knife, while the relationship of the body to its functioning is like that of a knife to sharpness. What is called sharpness is not the same as the knife, and what is called the knife is not the same as sharpness. Nevertheless, there can be no knife if the sharpness is discarded, nor sharpness if the knife is discarded. I have never heard of sharpness surviving if the knife is destroyed, so how can it be admitted that the soul can remain if the body is annihilated?

—Fan Chen, *Essay on the Extinction of the Soul*, in Fung Yu-Lan, *A History of Chinese Philosophy*, 1934

7. If a single cell, under appropriate conditions, becomes a person in the space of a few years, there can surely be no difficulty in understanding how, under appropriate conditions, a cell may, in the course of untold millions of years, give origin to the human race.

—Herbert Spencer, *Principles of Biology*, 1864

8. An electron is no more (and no less) hypothetical than a star. Nowadays we count electrons one by one in a Geiger counter, as we count the stars one by one on a photographic plate. In what sense can an electron be called more unobservable than a star? I am not sure whether I ought to say that I have seen an electron; but I have just the same doubt whether I have seen a star. If I have seen one, I have seen the other. I have seen a small disc of light surrounded by diffraction rings which has not the least resemblance to what a star is supposed to be; but the name "star" is given to the object in the physical world which some hundreds of years ago started a chain of causation which has resulted in this particular light-pattern. Similarly in a Wilson expansion chamber I have seen a trail not in the least resembling what an electron is supposed to be; but the name "electron" is given to the object in the physical world which has caused this trail to appear. How can it possibly be maintained that a hypothesis is introduced in one case and not in the other?

—Arthur Eddington, *New Pathways in Science*, 1939

9. Just as the bottom of a bucket containing water is pressed more heavily by the weight of the water when it is full than when it is half empty, and the more heavily the deeper the water is, similarly the high places of the earth, such as the summits of mountains, are less heavily pressed than the lowlands are by the weight of the mass of the air. This is because there is more air above the lowlands than above the mountain tops; for all the air along a mountain side presses upon the lowlands but not upon the summit, being above the one but below the other.

—Blaise Pascal, *Treatise on the Weight of the Mass of the Air*, 1653

10. Suppose that someone tells me that he has had a tooth extracted without an anaesthetic, and I express my sympathy, and suppose that I am then asked, "How do you know that it hurt him?" I might reasonably reply, "Well, I know that it would hurt me. I have been to the dentist

and know how painful it is to have a tooth stopped [filled] without an anaesthetic, let alone taken out. And he has the same sort of nervous system as I have. I infer, therefore, that in these conditions he felt considerable pain, just as I should myself."

—Alfred J. Ayer, "One's Knowledge of Other Minds," *Theoria*, 1953

11. Now if we survey the universe, so far as it falls under our knowledge, it bears a great resemblance to an animal or organized body and seems actuated with a like principle of life and motion. A continual circulation of matter in it produces no disorder: a continual waste in every part is incessantly repaired; the closest sympathy is perceived throughout the entire system: and each part or member, in performing its proper offices, operates both to its own preservation and to that of the whole. The world, therefore, I infer, is an animal, and the Deity is the soul of the world, actuating it, and actuated by it.

—David Hume, *Dialogues Concerning Natural Religion*, 1779

12. One cannot require that everything shall be defined, any more than one can require that a chemist shall decompose every substance. What is simple cannot be decomposed, and what is logically simple cannot have a proper definition.

—Gottlob Frege, "On Concept and Object," 1892

13. Most endangered or threatened species in the United States find suitable habitat on private land, and the destruction of habitat is widely recognized as the leading cause of extinctions. For these reasons, protecting wildlife without regulating the use of private land has been compared by biologists to playing the piano with just the black keys.

—John H. Cushman, Jr., "Environmentalists Gain a Victory," *The New York Times*, 30 June 1995

14. Opposing legislation that would restrict handgun ownership in the United Kingdom, the husband of Queen Elizabeth II reasoned as follows:

Look, if a cricketer, for instance, suddenly decided to go into a school and batter a lot of people to death with a cricket bat, which he could do very easily, are you going to ban cricket bats?

—Prince Philip, the Duke of Edinburgh, in an interview on the BBC, 19 December 1996

15. . . . The simplest form of the theological argument from design [was] once well known under the name "Paley's watch." Paley's form of it was just this: "If we found by chance a watch or other piece of intricate mechanism we should infer that it had been made by someone. But all around us we do find intricate pieces of natural mechanism, and the processes of the universe are seen to move together in complex relations; we should therefore infer that these too have a Maker."

B. A. D. Williams, "Metaphysical Arguments," in D. F. Pears, ed., *The Nature of Metaphysics* (New York: Macmillan, 1957)

11.4 Refutation by Logical Analogy

"You should say what you mean," [said the March Hare, reproving Alice sharply.]

"I do," Alice hastily replied; "at least—at least I mean what I say—that's the same thing, you know."

"Not the same thing a bit!" said the Hatter. "Why, you might just as well say that 'I see what I eat' is the same thing as 'I eat what I see'!"

"You might just as well say," added the March Hare, "that 'I like what I get' is the same thing as 'I get what I like'!"

"You might just as well say," added the Dormouse, which seemed to be talking in its sleep, "that 'I breathe when I sleep' is the same thing as 'I sleep when I breathe'!"

"It is the same thing with you," said the Hatter, and here the conversation dropped.
—Lewis Carroll, *Alice's Adventures in Wonderland*

The Hare, the Hatter, and the Dormouse all seek to refute Alice's claim—that meaning what you say is the same as saying what you mean—by using a *logical analogy*. The form of an argument, as distinct from its particular content, is the most important aspect of that argument from a logical point of view. Therefore, we often seek to demonstrate the weakness of a given argument by stating another argument, known to be erroneous, that has the same logical form.

In the realm of deduction, a refuting analogy for a given argument is an argument that has the same form as the given argument but whose premises are known to be true and whose conclusion is known to be false. The refuting analogy is therefore known to be invalid, and the argument under attack, because it has the same form, is thus shown to be invalid as well. This is the same principle that underlies the testing of categorical syllogisms explained in Section 6.2, and it also underlies the repeated emphasis on the centrality of logical form, as explained in Section 8.4.

In the realm of inductive argument, our present concern, the technique of *refutation by logical analogy*, can also be used to great effect. Scientific, political, or economic arguments, not purporting to be deductive, may be countered by presenting other arguments that have very similar designs and whose conclusions are known to be false or are generally believed to be improbable. Inductive arguments differ fundamentally from deductive arguments in the character of the support claimed to be given to the conclusion by the premises. All arguments, however, inductive as well as deductive, may be said to have some underlying form or pattern. If, when confronted by an inductive argument we wish to attack, we can present another inductive argument that has essentially the same form but is clearly flawed and whose conclusion is very doubtful, we throw similar doubt on the conclusion of the argument being examined.

Consider the following illustration. In two highly controversial cases before the U.S. Supreme Court in 2007,[8] the central issue was the constitutionality of the consideration of race by school boards in the assignment of students to public schools. In an editorial, the *New York Times* supported the race-conscious systems as fair, and called the objections to it "an assault on local school control." A prominent critic of the race-based systems wrote a critical response to that editorial, within which appeared the following passage:

> You argue that the race-based system "is applied to students of all races" and "does not advantage or disadvantage any particular racial group." But, of course, the same argument might have been made in defense of miscegenation statutes, which forbade blacks from marrying whites as well as whites from marrying blacks.[9]

The technique of refutation by logical analogy is here very keenly exemplified; the focus is on the *form* of the two arguments. The argument under attack has the same form as that of another argument whose unsatisfactoriness is now universally understood. We surely would not say that miscegenation statutes are acceptable because they apply equally to all races. Some policies involving the use of race by the state are not acceptable (the critic argues) even when it is true that no particular racial group is disadvantaged by that use. By highlighting such unacceptability in some well-known settings (regulations governing marriage), he strikes a sharp blow against the argument in this setting that relies on the claim that no particular racial group is disadvantaged by the race-based policy under attack.

The presentation of a refutation by logical analogy is often signaled by the appearance of some revealing phrase: "You might just as well say," or some other words having that sense. In the example just given, the telltale phrase is "the same argument might have been made. . . ." In another context, the argument that because Islamic culture had been brought to the country of Chad from without, it is no more than an Islamic overlay, is attacked with the refuting analogy of a scholar who introduces the refutation with a slightly different set of words: "One could as sensibly say that France has only a Christian overlay."[10]

When the point of the refuting analogy is manifest, no introductory phrases may be needed. The former governor of Mississippi, Kirk Fordice, argued that "It is a simple fact that the United States is a Christian nation" because "Christianity is the predominant religion in America." Journalist Michael Kinsley, with whom Fordice was debating on television, responded with these telling analogies: "Women are a majority in this country. Does that make us a female country? Or does it make us a white country because most people in this country are white?"[11]

A careless effort to refute an argument with an analogy can backfire when the allegedly refuting argument differs importantly from the target argument in ways that tend to reinforce the one that is under attack. This is illustrated by a recent exchange on the highly controversial topic of global warming. Newspaper columnist John Tierney raised some serious questions about the wisdom of immediate large-scale efforts to combat an apparent but uncertain climate trend.[12] A critic, Ray Sten, responded in this way:

> John Tierney suggests that we not worry much about climate change because its consequences are uncertain and far in the future, and in the meantime somebody may discover a technological quick fix. That's like telling a smoker not to worry because it's not certain whether he'll develop cancer, and besides, a cure may have been found by then. Call me a worry wart, but I'd quit smoking.[13]

The immediate and large-scale steps whose wisdom Tierney questions are thus likened to quitting smoking. There is, however, an important contrast between those two. Quitting has no economic costs (and even some economic benefits), while industrial changes designed to cut greenhouse gases by reducing the use of fossil fuels will probably be very costly. In presenting an analogy intended to refute Tierney, Mr. Sten (whose position on global warming may well be correct) undermines his cause by indirectly calling attention to the costs of the change he seeks to advance.

Here is a letter to the editor from Jeff Weaver, published in *The Ann Arbor (MI.) News* in July 2005:

> I find it amusing that anyone would be offended by the name or appearance of a team mascot. But apparently there are people who are devastated that there are schools with team names such as the Hurons, Chippewas, Braves, Chiefs, Seminoles, etc.
>
> I sympathize with their plight. I would also suggest that we change the name of the Pioneers of Ann Arbor Pioneer High School. My forefathers were pioneers and I'm sure they would be devastated that a school adopted their name as a team mascot. That name and mascot are a direct slap against my people.
>
> While we are at it, we had better change the names of the Cowboys, Fighting Irish, Celtics, Hoosiers, Sooners, Boilermakers, Packers, Aggies, Oilers, Mountaineers, Friars, Patriots, Volunteers and Tar Heels, to name a few, because I'm sure those names are equally demeaning and degrading to those groups as well. . . .

And here is a rather amusing one from Justice Antonin Scalia:

> Justice Scalia argues that we should never, ever, use the word "choate." Scalia wrote: "There is no such word as choate. Choate is to inchoate as sult is to insult."[14]

To conclude, here is a letter signed by a scientist from Woods Hole Oceanographic Institution, responding to the claim that there is "plentiful" water on the moon.

> No one except William S. Marshall has claimed "large quantities of water" or "plentiful lunar water." The chief scientist of the recent lunar mission described the target crater as "probably a little wetter than the Atacama desert in Chile." It's as if Martians had targeted the right South African mine, observed a diamond, and then proclaimed that "diamonds are plentiful on Earth."[15]

Refutation by analogy, when well designed, can be exceedingly effective. If the argument presented as a refuting analogy is plainly rotten, and it does indeed have the same form as that of the argument under attack, that target argument must be seriously wounded.

EXERCISES

Each of the following is intended to be a refutation by logical analogy. Identify the argument being refuted in each and the refuting analogy, and decide whether they do indeed have the same argument form.

1. Steve Brill, founder of Court TV, has no doubt that cameras belong in the courtroom, and answers some critics in the following way: "Some lawyers and judges say that TV coverage makes the system look bad. They confuse the messenger with the message. If press coverage of something makes it look bad, that is a reason to have the press coverage.

That criticism is like saying that because journalists were allowed to be with the troops in Vietnam, the Vietnam War was ruined."

—Steve Brill, "Trial: A Starting Place for Reform,"
The Ann Arbor (Mich.) *News*, 12 June 1995

2. The whole history of bolshevism, both before and after the October revolution, is full of instances of maneuvering, temporizing and compromising with other parties, bourgeois parties included! To carry on a war for the overthrow of the international bourgeoisie, a war which is a hundred times more difficult, prolonged and complicated than the most stubborn of ordinary wars between states, and to refuse beforehand to maneuver, to utilize the conflict of interests (even though temporary) among one's enemies, to refuse to temporize and compromise with possible (even though transitory, unstable, vacillating and conditional) allies—is this not ridiculous in the extreme? Is it not as though, when making a difficult ascent of an unexplored and hitherto inaccessible mountain, we were to refuse beforehand ever to move in zigzags, ever to retrace our steps, ever to abandon the course once selected to try others?

—V. I. Lenin, *"Left Wing" Communism: An Infantile Disorder*, 1920

3. The distinguished naturalist E. O. Wilson argues that humans are no more than a biological species of a certain physical composition, and that the human mind can have no characteristics attributable to nonphysical causes. This claim can no longer be disputed. "Virtually all contemporary scientists and philosophers expert on the subject agree [he writes] that the mind, which comprises consciousness and rational process, is the brain at work. . . . The brain and its satellite glands have now been probed to the point where no particular site remains that can reasonably be supposed to harbor a nonphysical mind.[16] Stephen Barr presented the following counterargument in the form of a logical analogy: "This [Wilson's argument quoted above] is on a par with Nikita Khrushchev's announcement [aiming to support atheism] that Yuri Gagarin, the first human visitor to space, had failed to locate God. Does Wilson suppose that if there were an immaterial component to the mind it would show up in a brain scan?"[17]

4. The argument against new highways is given forceful statement by three distinguished urban planners: the authors write: "The only long term solutions to traffic are public transit and coordinated land use." New highways, they argue, bring "induced traffic." So building more highways will only cause more traffic congestion, not less.[18]

    A highly critical reviewer responds to this argument as follows: "This is nonsense. . . . Long lines at a grocery store would not prompt anyone to say, "Well, we can't build any more grocery stores. That would only bring out more customers." Building more highways wouldn't lure cars. The cars come anyway."[19]

5. America's supply of timber has been increasing for decades, and the nation's forests have three times more wood today than in 1920. "We're not running out of wood, so why do we worry so much about recycling

paper?" asks Jerry Taylor, the director of natural research studies at the Cato Institute. "Paper is an agricultural product, made from trees grown specifically for paper production. Acting to conserve trees by recycling paper is like acting to conserve cornstalks by cutting back on corn consumption."

—John Tierney, "Recycling Is Garbage," *The New York Times Magazine*, 30 June 1996

6. In 1996, heated controversy arose between the states of New Jersey and New York over formal possession of Ellis Island, located at the mouth of the Hudson River near the New Jersey shore, a tiny speck of land on which so many tens of thousands of immigrants to the United States first touched American soil. An essay defending New York's claim to the historic island appeared in the *New York Times* on 23 July 1996. The following letter appeared in the same newspaper four days later:

> Clyde Haberman is right that almost every immigrant who passed through Ellis Island was bound for New York, not New Jersey. But this fact does not determine where the island is. A significant number of passengers arriving at Newark International Airport are also on their way to New York, but it would be hard to argue that New York thus has a claim on the airport. Cincinnati International Airport is in Covington, Kentucky, and presumably, few travelers are on their way to sparsely populated northern Kentucky. Would Mr. Haberman suggest that the airport belongs to Ohio?

7. Edward Rothstein suggests that poverty and injustice cannot be considered among the root causes of Islamic terrorism because Osama bin Laden is a multimillionaire. By that logic, slavery could not have caused the Civil War because Abraham Lincoln was not a slave.

—Corey Robin, "The Root Causes of Terror," *The New York Times*, 17 November 2001

8. Each of the multitude of universes may have different laws of nature. Or different values of quantities that determine how they behave, such as the speed of light. Some may be suitable for life, and some may not. All those suitable for life may have life develop. Sometimes life will evolve only into dinosaurs rather than something more intelligent. We cannot attach any meaning to the fact that a life form which could ask anthropic questions [questions about the properties that are essential for intelligent life] did develop in at least one universe. It is very much like a lottery. If you win the lottery, you may feel very grateful, but someone had to win, and no one selected who that was, except randomly. Just because a universe has a unique set of laws and parameters should not lead one to wonder whether that set was designed.

—Gordon Kane, "Anthropic Questions," *Phi Kappa Phi Forum*, Fall 2002

9. Artificial human minds will never be made (we are told) because "artificial intelligence investigation is based on advanced solid-state physics, whereas the humble human brain is a viable semiliquid system!" That is

no more reassuring than the suggestion that automobiles could never replace horses because they are made of metal, while the humble horse is a viable organic system with legs of flesh and bone.

—Michael D. Rohr, *The New York Times*, 27 March 1998

10. Modern political rhetoric [Ronald Dworkin argues] "is now extremely repetitive," and a good bit of it could be dispensed with—by law. "Every European democracy does this," the world's most highly regarded legal philosopher points out, "and Europeans are amazed that we do not."

Europeans are also amazed that we bathe as frequently as we do. What the hell kind of argument is that?

—David Tell, "Silencing Free Speech in the Name of Reform," *The Weekly Standard*, 25 November 1996

chapter 11 Summary

In this chapter we began the analysis of induction. Section 11.1 reviewed the fundamental distinction between deductive arguments, which claim certainty for their conclusions, and inductive arguments, which make no such claim. The terms *validity* and *invalidity* do not apply to inductive arguments, whose conclusions can only have some degree of probability of being true.

In Section 11.2 we explained argument by analogy. An analogy is a likeness or comparison; we draw an analogy when we indicate one or more respects in which two or more entities are similar. An argument by analogy is an argument whose premises assert the similarity of two or more entities in one or more respects, and whose conclusion is that those entities are similar in some further respect. Its conclusion, like that of every inductive argument, can be no more than probable.

Courtesy of King Features Syndicate.

In Section 11.3 we explained six criteria used in determining whether the premises of an analogical argument render its conclusion more or less probable. These criteria are:

1. The *number of entities* among which the analogy is said to hold
2. The *variety, or degree of dissimilarity*, among those entities or instances mentioned only in the premises
3. The *number of respects* in which the entities involved are said to be analogous
4. The *relevance* of the respects mentioned in the premises to the further respect mentioned in the conclusion
5. The *number and importance of disanalogies* between the instances mentioned only in the premises and the instance mentioned in the conclusion
6. The *modesty (or boldness)* of the conclusion relative to the premises

In Section 11.4 we explained refutation by logical analogy. To show that a given argument (whether inductive or deductive) is mistaken, one effective method is to present another argument, which is plainly mistaken, and whose form is the same as that of the argument under attack.

END NOTES

[1] Bert Vogelstein, "So, Smoking Causes Cancer: This Is News?" *The New York Times,* 27 October 1996.

[2] Albert Shanker, "Testing Teachers," *The New York Times*, 8 January 1995.

[3] Thomas Reid, *Essays on the Intellectual Powers of Man*, Essay 1, 1785.

[4] Andrew Massimino, "Building a Country," *The New York Times*, 5 June 2006.

[5] Eric Lander, quoted in an interview in the *The New York Times*, 10 September 1996.

[6] *Crawford v. Washington*, 541 U.S. 36 (2004).

[7] F. Barringer and A. Revkin, "A Few Spitballs Greet Professor Gore," *The Ann Arbor* (Mich.) *News*, 22 March 2007.

[8] *Parents Involved in Community Schools v. Seattle School District No. 1* (No. 05-908); and *Crystal D. Meredith, Custodial Parent v. Jefferson County (KY) Board of Education* (No. 05-915).

[9] Roger Clegg, "An Issue for the Court: Diversity in Our Schools," *The New York Times*, 11 December 2006.

[10] Bassam Abed, in a letter to *The New York Times*, 26 June 1988.

[11] "Evangelical Update," *The New York Times*, 21 November 1992.

[12] John Tierney, "Findings," *The New York Times*, 13 February 2007.

[13] Ray Sten, "Debating Climate Change," *The New York Times*, 20 February 2007.

[14] Antonin Scalia, *The New York Times Magazine*, 3 January 2010.

[15] John M. Hayes, *The New York Times*, 27 November 2009.

[16] E.O. Wilson, *Consilience* (New York: Alfred A. Knopf, 1998), p. 99.

[17] S. N. Barr, "Mindless Science," *The Weekly Standard*, 6 April 1998.

[18] A. Duany, E. Plater-Zyberk, and J. Speck, *Suburban Nation: The Rise of Sprawl and the Decline of the American Dream* (New York: North Point, 2000).

[19] F. Barnes, "Suburban Beauty: Why Sprawl Works," *The Weekly Standard*, 22 May 2000.

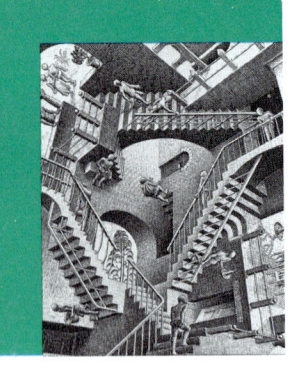

Causal Reasoning

12.1 Cause and Effect

12.2 Causal Laws and the Uniformity of Nature

12.3 Induction by Simple Enumeration

12.4 Methods of Causal Analysis

12.5 Limitations of Inductive Techniques

12.1 Cause and Effect

Induction goes far beyond analogical arguments. When we know, or think we know, that one thing is the *cause* of another, or the *effect* of another, we can reason from cause to effect, or from effect to cause. If the supposed relations between cause and effect have been correctly established, the reasoning based on those relations is very powerful.

Causal reasoning is also of the very greatest practical importance. Our ability to control our environment, to live successfully and to achieve our purposes, depends critically on our knowledge of causal connections. To cure some disease, for example, physicians must know its cause—and of course they must learn the effects (including the side effects) of the drugs they administer.

In every sphere in which we take action and seek to achieve some result, the relation of cause and effect is fundamental. David Hume, one of the keenest of all thinkers in this arena, wrote:

> All reasonings concerning matter of fact seem to be founded on the relation of Cause and Effect. By means of that relation alone we can go beyond the evidence of our memory and senses. If you were to ask a man, why he believes any matter of fact, which is absent; for instance, that his friend is in the country, or in France; he would give you a reason; and this reason would be some other fact; as a letter received from him, or the knowledge of his former resolutions and promises. A man finding a watch or any other machine in a desert island, would conclude that there had once been men in that island. All our reasonings concerning fact are of the same nature. . . . If we would satisfy ourselves, therefore, concerning the nature of that evidence, which assures us of matters of fact, we must enquire how we arrive at the knowledge of cause and effect.[1]

The methods by which we arrive at such knowledge are the central concern of this chapter. This matter is complicated, however, by the fact that there are several different meanings of the word "cause." Therefore we begin by distinguishing these meanings from one another.

Things do not just happen. Events take place *under certain conditions*, and it is an axiom in the study of nature that to understand the world in which we live we must seek to learn the conditions under which events do or do not happen. It is customary to distinguish between the *necessary* and the *sufficient* conditions for the occurrence of an event.

Causal reasoning
Inductive reasoning in which some effect is inferred from what is assumed to be its cause, or some cause is inferred from what is assumed to be its effect.

chapter 12

470

A **necessary condition** for the occurrence of a specified event is a circumstance in whose *absence* the event *cannot* occur. For example, the presence of oxygen is a necessary condition for combustion to occur. If combustion occurs, then oxygen must have been present, because in the absence of oxygen there can be no combustion.

A **sufficient condition** for the occurrence of an event is a circumstance in whose *presence* the event *must* occur. The presence of oxygen is a necessary condition for combustion, as we noted, but it is not a sufficient condition for combustion to occur—because it is obvious that oxygen can be present without combustion occurring. For almost any substance, however, there is some range of temperature such that being in that range of temperature in the presence of oxygen is a sufficient condition for the combustion of that substance. So it is clear that for the occurrence of an event there may be several necessary conditions—and all of those necessary conditions must be included in the sufficient condition of that event.

Now, the word "cause" is used (with respect to some event) sometimes to mean "the necessary condition of that event," and sometimes to mean "the sufficient condition of that event." It is most often used in the sense of necessary condition when the problem at hand is the *elimination* of some undesirable phenomenon. To eliminate it, one need only find some condition that is necessary to the existence of that phenomenon, and then eliminate that condition. What virus or bacterium is the cause of a certain illness? The physician cures the illness by administering a drug that will destroy those germs. The germs are said to be the *cause* of the disease in that they are a *necessary condition* for it—because in their absence the disease cannot occur.

However, the word "cause" is also commonly used to mean *sufficient condition*—especially when we are interested in the *production* of something desired, rather than the elimination of something undesirable. The metallurgist aims to discover what will produce greater strength in metal alloys, and when it is found that a certain process of mixed heating and cooling has that desired result, we say that such a process is the *cause* of the increased strength of the alloy. It is correct to use the word "cause" in the one sense (necessary condition), or in the other (sufficient condition), but one should be clear about which of those meanings is intended.

Closely related to *sufficient condition* is another sense of the word "cause"—when a given phenomenon *tends* to have a causative role in the production of certain outcomes. For example, it is indeed correct to say that "smoking causes lung cancer," even though smoking cigarettes may long continue without having cancer as its result. Smoking is certainly not a necessary condition of lung cancer, because many such cancers arise in the total absence of smoking. But smoking cigarettes, in conjunction with very common biological circumstances, so frequently plays a role in the development of lung cancer that we think it correct to report that smoking is a "cause" of cancer.

Necessary condition
A circumstance (or set of circumstances) in whose absence a given event cannot occur.

Sufficient condition
A circumstance (or set of circumstances) whose presence ensures the occurrence of a given event.

This points to yet another common use of the word "cause"—cause as the one factor that was critical in the occurrence of some phenomenon. An insurance company sends investigators to determine the cause of a mysterious fire. The investigators are likely to lose their jobs if they report that it was the presence of oxygen in the atmosphere that was the fire's cause—and yet of course it was (in the sense of necessary condition), for had there been no oxygen present there would have been no fire. Nor is the sufficient condition of the fire of interest to the company, for if the investigators reported that, although they had proof the fire was deliberately ignited by the policyholder, they had not yet been able to learn all the necessary conditions of the fire and therefore had not yet determined its full cause, they would certainly lose their jobs! What the company was seeking to discover was the incident or action that, in the presence of those conditions that usually prevail, *made the difference* between the occurrence and nonoccurrence of the fire.

In the real world, a huge man, forcibly resisting arrest, died shortly after having been beaten into submission by police officers in Cincinnati, Ohio, in November 2003. The county coroner investigating the death held it to be a "homicide," carefully noting that hostile or malign intent is not implied by that word. "Absent the struggle," the coroner said, "Mr. Jones would not have died at that precise moment in time, and the struggle therefore is the primary cause of his death." This sense of cause as "critical factor" is common and useful.[2]

There are subdivisions of this third sense of *cause*. When there is a causal sequence—a chain of events in which A causes B, B causes C, C causes D, and D causes E—we may regard the outcome, E, as the effect of any one of those preceding events. The death described above (symbolized by E) was caused by the struggle, the struggle (D) was caused by the resistance, the resistance (C) was caused by the arrest, the arrest (B) was caused by some violation of law (A), and so on. We distinguish between the **remote cause** and the **proximate cause** of E. The proximate cause is the event closest to it in the chain of events. The death, E, is the result of the proximate cause of the struggle, D; the other causes are remote: A more remote than B, B more remote than C, and so on.

Persons who leave school before the age of 16 are five times more likely than university graduates to die from a heart attack; and the death rate within one year of a heart attack is 3.5 percent for college graduates but 20 percent for those with fewer than eight years of formal schooling.[3] But a college education is not the proximate cause of good health, nor is ignorance the proximate cause of disease. A poor education is a link in the causal chain, often resulting in a less adequate understanding of the disease process and thus a failure to make the lifestyle changes needed to promote better medical outcomes. So it is commonly and correctly observed that poverty, affecting education almost universally, is one of the "root causes" of poor health—not its proximate cause, of course, but a remote cause that needs uprooting.

The several different senses of the word "cause" need to be distinguished. We can legitimately infer cause from effect only when by *cause* is meant *necessary condition*. We can infer effect from cause only when by *cause* is meant *sufficient condition*. When inferences are drawn both from cause to effect and from effect to cause, the word "cause" must be used in the sense of **necessary and sufficient condition**—the

---

Remote cause
In any chain of causes and effects, an event distant from the effect for which explanation is sought. Contrasted with "proximate" cause.

Proximate cause
In any chain of causes and effects, the event nearest to the event whose explanation is sought. Contrasted with "remote" causes, which are more distant in the causal chain.

Necessary and sufficient condition
The conjunction of necessary conditions for the occurrence of a given event, this conjunction being all that is needed to ensure the occurrence of the event. It is the sense in which the word *cause* is used when inferences are drawn both from cause to effect and from effect to cause.

cause regarded as the sufficient condition of the event and that sufficient condition regarded as the conjunction of all its necessary conditions. No single definition of *cause* conforms to all the different (and reasonable) uses of that word.

12.2 Causal Laws and the Uniformity of Nature

Every use of the word "cause," whether in everyday life or in science, involves or presupposes the doctrine that cause and effect are uniformly connected. We will allow that some particular circumstance was the cause of some particular effect only if we agree that any other circumstance of that type will (if the attendant circumstances are sufficiently similar) cause another effect of the same kind as the first. In other words, similar causes produce similar effects. As we use the word "cause," part of its meaning is that every occurrence of a cause producing some effect is an *instance* or *example* of the general causal law that such circumstances are always accompanied by such phenomena. If it can be shown that in another situation, after an occurrence of that supposed cause, the supposed effect did not occur, we will relinquish the belief that the one is the cause of the other.

Because every assertion that a particular circumstance was the cause of a particular phenomenon implies the existence of some causal law, every assertion of causal connection contains a critical element of *generality*. A **causal law**, as we use the term, asserts that a circumstance of such-and-such kind is invariably attended by a phenomenon of a specified kind, no matter where or when it occurs.

How can we come to know such general truths? The causal relation is not purely logical or deductive; as David Hume emphasized, it cannot be discovered by any *a priori* reasoning.* Causal laws can be discovered only empirically, *a posteriori*, by an appeal to experience. But our experiences are always of *particular* circumstances, *particular* phenomena, and *particular* sequences of them. We may observe several instances of a circumstance (call it C), and every instance that we observe may be accompanied by an instance of a certain kind of phenomenon (call it P). However, we will have experienced only some of the instances of C in the world, and our observations can therefore show us only that some cases of C are attended by P. Yet our aim is to establish a general causal relation. How are we to get from the particulars we experience to the general proposition that *all* cases of C are attended by P—which is involved in saying that C causes P?

Causal laws
Descriptive laws asserting a necessary connection between events of two kinds, of which one is the cause and the other the effect.

---

*Hume wrote: "But to convince us that all the laws of nature, and all the operations of bodies without exception, are known only by experience, the following reflections may, perhaps, suffice. Were any object presented to us, and were we required to pronounce concerning the effect, which will result from it, without consulting past observation, after what manner, I beseech you, must the mind proceed in this operation? It must invent or imagine some event, which it ascribes to the object as its effect; and it is plain that this invention must be entirely arbitrary. The mind can never possibly find the effect in the supposed cause, by the most accurate scrutiny and examination. For the effect is totally different from the cause, and consequently can never be discovered in it. . . . A stone or piece of metal raised into the air, and left without any support, immediately falls; but to consider the matter *a priori*, is there anything we can discover in this situation which can beget the idea of a downward, rather than an upward, or any other motion, in the stone or metal? . . . In vain, therefore, should we pretend to determine any single event, or infer any cause or effect, without the assistance of observation and experience." (*An Enquiry Concerning Human Understanding*, 1748, sec. IV).

12.3 Induction by Simple Enumeration

When we assert that all cases of C are attended by P—that is, when we affirm a general causal relation—we have gone beyond analogy. The process of arriving at universal propositions from the particular facts of experience is called **inductive generalization**. Suppose we dip blue litmus paper into acid and it turns red. Suppose we do this three times, or ten times, always with the same result. What conclusion do we draw? By *analogy* we may draw a *particular* conclusion about what will happen to the color of the next piece of litmus paper we dip in acid—the fourth or the eleventh. We may draw a general conclusion about what will happen to *every* piece of blue litmus paper when it is dipped in acid. If we do the latter, it is with an *inductive generalization* that our argument concludes.

When the premises of an argument report a number of instances in which two attributes (or circumstances, or phenomena) occur together, we may infer by analogy that some particular instance of one attribute will also exhibit the other attribute. By inductive generalization we might infer that every instance of the one attribute will also be an instance of the other. Inductive generalization of the form

> Instance 1 of phenomenon E is accompanied by circumstance C.
> Instance 2 of phenomenon E is accompanied by circumstance C.
> <u>Instance 3 of phenomenon E is accompanied by circumstance C.</u>
> Therefore every instance of phenomenon E is accompanied by circumstance C.

is an **induction by simple enumeration**. An induction by simple enumeration is very similar to an argument by analogy, differing only in having a more general conclusion.

Simple enumeration is often used in establishing causal connections. Where a number of instances of a phenomenon are invariably accompanied by a certain type of circumstance, it is only natural to infer the existence of a causal relationship between them. Since the circumstance of dipping blue litmus paper in acid is accompanied in all observed instances by the phenomenon of the paper turning red, we infer by simple enumeration that dipping blue litmus paper in acid is the cause of its turning red. The analogical character of such an argument is quite apparent.

Because of the great similarity between argument by simple enumeration and argument by analogy, similar criteria for appraisal apply to both. Some arguments by simple enumeration may establish their conclusions with a higher degree of probability than others. The greater the number of instances appealed to, the greater is the probability of the conclusion. The various instances or cases of phenomenon E accompanied by circumstance C are often called *confirming instances* of the causal law asserting that C causes E. The greater the number of confirming instances, the greater is the probability of the causal law—other things

---

Inductive generalization
The process of arriving at universal propositions from particular facts of experience, relying upon the principle of induction.

Induction by simple enumeration
A type of inductive generalization in which the premises are instances where phenomena of two kinds repeatedly accompany one another in certain circumstances, from which it is concluded that phenomena of those two kinds always accompany one another in such circumstances.

being equal. Thus the first criterion for analogical arguments also applies directly to arguments by simple enumeration.

In a historical report, simple enumeration can provide persuasive grounds for inferring a causal relationship. To illustrate, legislative acts called "bills of attainder," designed to savage some individual or group temporarily out of favor, are known to endanger their advocates when the pendulum of political power swings. The accuser today becomes the victim tomorrow. Condemning such a bill of attainder (aimed at Thomas Osborne, Earl of Danbury) in the British House of Lords, the Earl of Carnarvon drove the point home in 1678 with the following enumeration:

> My Lords, I understand . . . not a little of our English history, from which I have learnt the mischiefs of prosecutions such as these, and the ill fate of the prosecutors. I shall go no further back than the latter end of Queen Elizabeth's reign, at which time the Earl of Essex was run down by Sir Walter Raleigh, and your Lordships well know what became of Sir Walter Raleigh. My Lord Bacon, he ran down Sir Walter Raleigh, and your Lordships know what became of my Lord Bacon. The Duke of Buckingham, he ran down my Lord Bacon, and your Lordships know what happened to the Duke of Buckingham. Sir Thomas Wentworth, afterwards Earl of Strafford, ran down the Duke of Buckingham, and you all know what became of him. Sir Harry Vane, he ran down the Earl of Strafford, and your Lordships know what became of Sir Harry Vane. Chancellor Hyde, he ran down Sir Harry Vane, and your Lordships know what became of the Chancellor. Sir Thomas Osborne, now Earl of Danby, ran down Chancellor Hyde.
>
> What will now become of the Earl of Danby, your Lordships best can tell. But let me see that man that dare run the Earl of Danby down, and we shall soon see what will become of him.[4]

Rhetorically effective though this recounting of instances may be, it does not provide a trustworthy argument. The conclusion—that there is a causal connection between malicious accusation and subsequent destruction—appeals to six confirming instances, but by the very nature of those instances we cannot distinguish which are confirming instances of a genuine causal law and which are mere historical accidents.

The heart of the difficulty is this: The method of simple enumeration takes no account—*can* take no account—of exceptions to the causal law being suggested. Any alleged causal law may be overthrown by a single negative case, for any one disconfirming instance shows that what had been proposed as a "law" was not truly general. Exceptions *disprove* the rule—for an exception (or "negative instance") is either one in which the alleged cause is found and is not followed by the alleged effect (in this historical case, a bill of attainder whose author did not suffer a like fate), or one in which the effect is encountered while the alleged cause is absent—where (using our earlier schema) C is present without E, or E is present without C. In an argument by simple enumeration there is no place for either of these; the only legitimate premises in such an argument are reports of instances in which *both* the alleged cause and the alleged effect are present.

Four hundred years ago Sir Francis Bacon, in *The Advancement of Learning* (1605), clearly identified the shortcomings of induction by simple enumeration. He wrote: "The induction that proceeds by simple enumeration is childish; its conclusions are precarious, and exposed to peril from a contradictory instance; and it generally reaches decision on too small a number of facts, and on those only that are on hand."

It is thus a grave weakness of simple enumeration arguments that, if we confine ourselves to them exclusively, we will not look for, and are therefore unlikely even to notice, the negative or disconfirming instances that might otherwise be found. For this reason, despite their fruitfulness and value in *suggesting* causal laws, inductions by simple enumeration are not at all suitable for *testing* causal laws. Yet such testing is essential; to accomplish it we must rely upon other types of inductive arguments—and to these we turn now.

12.4 Methods of Causal Analysis

The classic formulation of the methods central to all induction were given in the nineteenth century by John Stuart Mill (in *A System of Logic*, 1843). His systematic account of these methods has led logicians to refer to them as Mill's methods of inductive inference. The techniques themselves—five are commonly distinguished—were certainly not invented by him, nor should they be thought of as merely a product of nineteenth-century thought. On the contrary, these are universal tools of scientific investigation. The names Mill gave to them are still in use, as are Mill's precise formulations of what he called the "canons of induction." These techniques of investigation are permanently useful. Present-day accounts of discoveries in the biological, social, and physical sciences commonly report the methodology used as one or another variant (or combination) of these five techniques of inductive inference called **Mill's methods**. They are:

1. The method of agreement
2. The method of difference
3. The joint method of agreement and difference
4. The method of residues
5. The method of concomitant variation

We will examine each of these in turn, presenting Mill's classic statement of each (with one exception), followed by explication and illustration. These are the techniques on which science does and will rely in the search for causal laws.

Mill's methods
The five patterns of inductive inference, analyzed and formulated by John Stuart Mill, with which hypotheses are confirmed or disconfirmed.

A. The Method of Agreement

John Stuart Mill wrote:

> If two or more instances of the phenomenon under investigation have only one circumstance in common, the circumstance in which alone all the instances agree, is the cause (or effect) of the given phenomenon.

This method goes beyond simple enumeration in that it seeks not only to discover the repeated conjunction of cause with effect, but also to identify the *only* circumstance, the *one* circumstance, that is invariably associated with the effect, or phenomenon, in which we are interested. This is an essential, and exceedingly common, tool of scientific inquiry. In searching for the cause of some deadly epidemic, for example, or in searching for the cause of some geological phenomenon, the epidemiologist or geologist will seek out the special circumstances that in every instance attend that result. In what way, they ask, do apparently differing sets of circumstances *agree*, where that result is produced?

Imagine, among the residents of some residence hall, a rash of stomach upsets, whose cause we must learn. The first line of inquiry naturally will be: What food or foods were eaten by *all* those who fell ill? Foods that were eaten by some but not all of those afflicted are not likely to be the cause of the outbreak; we want to know what circumstance can be found to be *common* to every case of the illness. Of course, what turns out to be common may not be a food; it may be the use of some infected utensil, or proximity to some noxious effluent, or other circumstance. Only when some circumstance is found in which *all* the cases of the illness agree are we on the way to the solution of the problem.

Schematically, the **method of agreement** may be represented as follows, where capital letters represent circumstances and lowercase letters denote phenomena:

A B C D occur together with *w x y z.*
A E F G occur together with *w t u v.*
Therefore *A* is the cause (or the effect) of *w.*

This method is particularly useful in identifying a *kind* of phenomenon, or a *range* of circumstances, whose investigation holds scientific promise. In molecular genetics, for example, the search for the causes of some inherited disease can be greatly narrowed down using the method of agreement. Is there a common factor among families in which some specific disorder is prevalent? By examining the genetic makeup of such families, then closing in on those genetic factors that are found in such families but are not found commonly in others, the chromosome (and sometimes the site on that chromosome) where the inherited defect lies may be identified. This has proved a very effective method in tracing the cause of some diseases.

Similarly, the fluoridation of water in developed areas around the globe was the consequence of the discovery, more than half a century ago, that in cities where the rates of dental decay were unusually low, the one circumstance in common was an unusually high level of fluorine in the water supply. To confirm the causal connection, two cities of comparable size along the Hudson River—Newburgh and Kingston, New York—were closely studied in the 1940s; Newburgh's water was treated with fluoride, Kingston's had no

Method of agreement
A pattern of inductive inference in which it is concluded that, if two or more instances of a given phenomenon have only one circumstance in common, that one common circumstance is the cause (or effect) of the phenomenon.

fluoride. The statistics proved remarkable: Children in Newburgh showed a *70 percent reduction in cavities* by the time they reached 14 years of age—and yet there were no differences between the two cities in rates of cancer, birth defects, or heart disease. The full explanation of this prevention of cavities could not be given at that time, but enough was known to justify the fluoridation of municipal water systems.

The method of agreement is widely powerful. A very promising development in the effort to help smokers break their addiction to nicotine was the discovery, reported in *Science* in 2007, that in a small number of persons who had suffered an injury to a particular region of their brains called the *insula*, the desire to smoke was immediately lost! Something in the insula appears to be a critical element in addiction. When statistical analyses of the data were completed, said a lead investigator from the University of Southern California, "it turned out that the likelihood of quitting smoking with ease after insula damage was 136 times higher than for damage anywhere else in the brain." A neuroscientist from the National Institute of Drug Abuse was enthusiastic: "To have any kind of variable produce this rate of quitting smoking is remarkable, to have it associated with a particular brain region is fantastic."[5] Thus, for addiction researchers, who are eager to apply the method of agreement to nicotine addiction, a major question has now become: "Can we learn to de-activate the insula?"[6]

In short, whenever we find *a single circumstance common to all instances* of a given phenomenon, we may rightly conclude that we have located at least the region of its cause.

The method of agreement has serious limitations, however. Looking chiefly to confirming instances, the method by itself is often insufficient to identify the cause being sought. The data available are seldom so conveniently arranged as to permit the identification of one circumstance common to all cases. When inquiry reveals more than one circumstance common to all cases, this technique alone cannot evaluate those alternative possibilities.

Although the presence of agreement between circumstance and phenomenon is often inconclusive, the *absence* of agreement may help us to determine what is *not* the cause of a phenomenon of interest. The method of agreement is in essence eliminative; it points to the fact that circumstances arising in some of the cases, but not all of the cases, of the phenomenon in which we are interested, are not likely to be its cause. Those who argue against an alleged causal relation, therefore, are likely to call attention to the absence of uniform agreement, inferring that the alleged cause can be neither the sufficient condition nor the necessary condition of that phenomenon.

After we have learned all that the method of agreement can teach, other inductive methods capable of greater refinement in the search for causes are sure to be required.

EXERCISES

Analyze each of the following scientific reports, explaining how the pattern of the method of agreement is manifested by each. Discuss, in each case, the limitations of the method of agreement as applied to that quest for a causal connection.

1. Contaminated scallions, chopped up raw in salsa that was served free to every table at a Chi-Chi's restaurant in western Pennsylvania, almost certainly caused the large outbreak of hepatitis A in the region, the Centers for Disease Control and Prevention said yesterday. Bunches of scallions (green onions) were stored together in large buckets for five days or more with the ice they had been shipped in from Mexico. As a result, even if only some bunches were tainted with the hepatitis virus when they were delivered, it would have quickly spread to all the other scallions—the ice water in the bucket becoming "hepatitis soup." The scallions were later rinsed, chopped, refrigerated for two more days, and then added to the salsa which was made in 40-quart batches and kept refrigerated for up to three days. The outbreak, which has killed three people and made 575 other Chi-Chi's patrons sick, is the nation's biggest outbreak of hepatitis A from one source. Hepatitis A is spread by fecal matter from infected people, particularly those who fail to wash their hands after using the restroom. The virus does not multiply outside the body, but it can survive in food.

   Hepatitis A is a common childhood disease in Mexico, and children commonly work on the scallion farms there; sewage-contaminated water could also have been the culprit, whether used to irrigate the scallions, or wash them, or make the ice used in shipping. How the scallions became contaminated is not known.

   —"Government Makes It Official: Blame Scallions for Outbreak,"
   The New York Times, 22 November 2003

2. Researchers at the University of California at Irvine have theorized that listening to Mozart's piano music significantly improves performance on intelligence tests. Dr. Frances H. Rauscher and her colleagues reported:

   > We performed an experiment in which students were each given three sets of standard IQ spatial reasoning tasks; each task was preceded by 10 minutes of
   >
   > 1. listening to Mozart's Sonata for Two Pianos in D major, K. 488; or
   > 2. listening to a relaxation tape; or
   > 3. silence.

Performance was improved for those tasks immediately following the first condition compared to the second two.

Test scores rose an average of 8 or 9 points following the Mozart sonata. Some of the students had reported that they liked Mozart, and some that they did not, but there were no measurable differences attributable to varying tastes. "We are testing a neurobiological model of brain function with these experiments," Dr. Rauscher said, "and we hypothesize that these patterns may be common in certain activities—chess, mathematics, and certain kinds of music. . . . Listening to such music may stimulate neural pathways important to cognition."

—Frances H. Rauscher, Gordon L. Shaw, Katherine N. Ky, "Music and Spatial Task Performance," *Nature*, 14 October 1993

3. Medical researchers have concluded not only that the timing of sexual intercourse in relation to ovulation strongly influences the chance of conception, but that conception occurs *only* when intercourse takes place during a specifiable period in the menstrual cycle. The researchers summarized their findings thus:

> We recruited 221 healthy women who were planning to become pregnant. At the same time the women stopped using birth control methods, they began collecting daily urine specimens and keeping daily records of whether they had sexual intercourse. We measured estrogen and progesterone metabolites in urine to estimate the day of ovulation.
>
> In a total of 625 menstrual cycles for which the dates of ovulation could be estimated, 192 pregnancies were initiated. . . . Two-thirds ($n = 129$) ended in live births. Conception occurred only when intercourse took place during a six-day period that ended on the estimated day of ovulation. The probability of conception ranged from 0.10 when intercourse occurred five days before ovulation to 0.33 when it occurred on the day of ovulation itself.
>
> Conclusion: Among healthy women trying to conceive, nearly all pregnancies can be attributed to intercourse during a six-day period ending on the day of ovulation.

—Allen J. Wilcox, Clarice R. Weinberg, Donna D. Baird, "Timing of Sexual Intercourse in Relation to Ovulation," *The New England Journal of Medicine*, 7 December 1995

4. A large extended family in the town of Cartago, Costa Rica, has long suffered an unusual affliction—an incurable form of genetically caused deafness. Children born into the family have a 50 percent chance of developing the disease, and learn their fate at about the age of ten, when those who have inherited a genetic mutation find that they are beginning to lose their hearing. Scientists from the University of Washington have recently traced the cause of the family's affliction to a previously unknown gene, named the diaphanous gene, that helps operate the delicate hair cells in the inner ear that respond to sound vibrations.

This gene has a single mutation appearing in the Costa Rican family, whose founder arrived in Cartago from Spain in 1713, and who suffered from this form of deafness—as have half his descendants in the eight generations since. Many in the family remain in Cartago because the family's hereditary deafness is well known and accepted there. With only a single family to be studied, and thus very few genetic differences to work with, pinpointing the gene took six years. The critical mutation involved just one of the 3,800 chemical letters that constitute the gene's DNA.

—Reported in *Science*, 14 November 1997

5. Researchers from the National Cancer Institute announced that they have found a number of genetic markers shared by gay brothers, indicating that homosexuality has genetic roots. The investigators, reporting in *Science*, 16 July 1993, have found that out of 40 pairs of gay brothers examined in their study, 33 pairs shared certain DNA sequences on their X chromosome, the chromosome men inherit only from their mothers. The implicit reasoning of this report is that, if brothers who have specific DNA sequences in common are both gay, these sequences can be considered genetic markers for homosexuality.

6. The relation between male circumcision and HIV infection has been a concern of the British medical journal, *The Lancet*, for many years. Before the turn of this century investigators studying that relation wrote, in *The Lancet*, that studies going back as far as 1989 showed a very greatly increased risk of HIV-1 infection for men who are not circumcised. The epidemiological and biological evidence that links the two, they later wrote, "has become compelling." Very recent studies in Kenya and Uganda have produced evidence that is even more compelling. In 2006, trials in those countries conducted by the U.S. National Institutes of Health were *stopped* because the results were so clear! It appeared that circumcision reduces a man's risk of contracting AIDS from heterosexual sex by about half, and therefore U.S. officials concluded that it would have been unethical to continue without offering circumcision to all 8,000 men in the trials. The final figures, reevaluated and published in *The Lancet* on 23 February 2007, are even more striking. They suggest that *circumcision reduces a man's risk of contracting AIDS by as much as 65 percent*. Dr. Anthony Fauci, of the National Institute of Allergy and Infectious Diseases, was emphatic: "Look. This is a one-time, permanent intervention that's safe when done under appropriate medical conditions. If we had an AIDS vaccine that was performing as well as this, it would be the talk of the town."

B. The Method of Difference

John Stuart Mill wrote:

> If an instance in which the phenomenon under investigation occurs and an instance in which it does not occur, have every circumstance in common save one, that one occurring only in the former, the circumstance in which alone the two instances differ, is the effect, or the cause, or an indispensable part of the cause, of the phenomenon.

This pattern focuses not on what is common among those cases in which the effect is produced, but on what is *different* between those cases in which the effect is produced and those in which it is not. If we had learned, when investigating that rash of stomach upsets described earlier, that all those who had become ill had eaten the canned pears for dessert, but that the pears had been eaten by none of those who did not become ill, we would be fairly confident that the cause of the illness had been identified.

The difference between the **method of difference** and the method of agreement is highlighted in a recent report about the role of the hormone testosterone in the aggressive conduct of males.

> Among many species, testes are mothballed most of the year, kicking into action and pouring out testosterone only during a very circumscribed mating season—precisely the time when male-male aggression soars. Impressive though they seem, these data are only correlative—[reporting only] testosterone found on the scene repeatedly when aggression has occurred.
>
> The proof comes with the knife, the performance of what is euphemistically known as a subtraction experiment. Remove the source of the testosterone in species after species, and levels of aggression plummet. Reinstate normal testosterone levels afterward with injections of synthetic testosterone and aggression returns.
>
> The subtraction and replacement paradigm gives damning proof that this hormone is involved in aggression.[7]

Testosterone makes the critical difference, clearly, but the author of this report is careful not to assert that testosterone is *the cause* of male aggression. More accurately, the report states that testosterone is surely *involved* in aggression. As Mill would put it, the hormone is *an indispensable part of the cause* of male aggression. Wherever we can identify a single factor that makes the critical difference when all else remains normal—the factor that eliminates the phenomenon in question when we remove it, or the factor that produces the phenomenon in question when we introduce it—we will pretty surely have identified the cause, or an indispensable part of the cause, of the phenomenon we are investigating.

Schematically, where again capital letters denote circumstances and lowercase letters denote phenomena, the **method of difference** may be represented as follows:

> *A B C D* occur together with *w x y z*.
> <u>*B C D* occur together with *x y z*.</u>
> Therefore *A* is the cause, or the effect, or an indispensable part of the cause of *w*.

Method of difference
A pattern of inductive inference in which, when cases in which a given phenomenon occurs differ in only one circumstance from cases in which the phenomenon does not occur, that circumstance is inferred to be causally connected to the phenomenon.

The method of difference is of central importance in scientific investigations of almost every kind. One vivid illustration of its use is the ongoing investigation by medical researchers into the effects of particular proteins suspected of being implicated in the development of certain diseases. Whether the substance under investigation really is the cause (or an indispensable part of the cause) can only be determined when we create an experimental environment in which that substance has been eliminated. Investigators sometimes are able to do just that—not in humans, of course, but in mice which are subject to the same disease and from which the gene that is known to produce the suspect protein is deleted. Animals so treated are then inbred, creating populations of what are called "knockout mice," precious in the world of contemporary medical research, in which the process relevant to the disease in question can be studied in an animal exactly like other animals subject to that disease, *except for the critical difference created by the knockout*, the absence of the substance hypothesized as cause. Such studies have resulted in some remarkable medical advances.

To illustrate: Using knockout mice, scientists have been able to identify the gene that causes inflammation—swelling, redness, and pain. The gene *MIP-1 alpha*, present in mice and in humans, was suspected of producing the protein that begins the process of inflammation. Pathologists at the University of North Carolina at Chapel Hill bred mice that *lacked* the gene *MIP-1 alpha*, and then infected those mice, as well as a control group of normal mice, with viruses known to cause influenza and other diseases. The normal mice did develop extreme inflammation as expected, but mice lacking the *MIP-1 alpha* gene had only slight inflammation. This is one big step toward the development of drugs that will allow humans to fight viral infections without painful and damaging inflammation.[8]

A famous and very dramatic illustration of the method of difference is provided by the following account of experiments confirming the true cause of yellow fever, long one of the great plagues of humankind. The experiments described here were conducted by U.S. Army doctors Walter Reed, James Carroll, and Jesse W. Lazear. Shortly before those experiments Dr. Carroll had contracted yellow fever by deliberately allowing himself to be bitten by an infected mosquito in a different experiment. Shortly after these experiments Dr. Lazear died of yellow fever; the camp in which the experiments took place was named for him.

> Experiments were devised to show that yellow fever was transmitted by the mosquito alone, all other reasonable opportunities for being infected being excluded. A small building was erected, all windows and doors and every other possible opening being absolutely mosquito-proof. A wire mosquito screen divided the room into two spaces. In one of these spaces fifteen mosquitoes, which had fed on yellow fever patients, were liberated. A nonimmune volunteer entered the room with the mosquitoes and was bitten by seven mosquitoes. Four days later, he suffered an attack of yellow fever. Two other nonimmune men slept for thirteen nights in the mosquito-free room without disturbances of any sort.

To show that the disease was transmitted by the mosquito and not through the excreta of yellow fever patients or anything which had come in contact with them, another house was constructed and made mosquito-proof. For 20 days, this house was occupied by three nonimmunes, after the clothing, bedding and eating utensils and other vessels soiled with the discharge, blood and vomitus of yellow fever patients had been placed in it. The bed clothing which they used had been brought from the beds of the patients who had died of yellow fever, without being subjected to washing or any other treatment to remove anything with which it might have been soiled. The experiment was twice repeated by other nonimmune volunteers. During the entire period all the men who occupied the house were strictly quarantined and protected from mosquitoes. None of those exposed to these experiments contracted yellow fever. That they were not immune was subsequently shown, since four of them became infected either by mosquito bites or the injection of blood from yellow fever patients.[9]

That portion of the experiment described in the first paragraph above very deliberately created a single important difference between the subjects in the two carefully enclosed spaces: the presence of mosquitoes that had fed on yellow fever patients in the one space, the absence of such mosquitoes in the other. That portion of the experiment described in the second paragraph above deliberately created a second use of the method of difference, in which the only significant difference between two groups of subjects, both of whom had submitted to very close contact with items that had been used by victims of yellow fever, was the exposure of some of them to infected mosquito bites or infected blood. Absent that circumstance, no infection arose.

Science seeks causal laws. In the never-ending efforts to confirm or to disconfirm hypothesized causal connections, the method of difference is pervasive and powerful.

EXERCISES

Analyze each of the following reports, explaining the ways in which the method of difference has been applied in the investigations recounted. Discuss the strengths and weaknesses of the method of difference as it is used in each case.

1. How critical is sleep to memory? Researchers at two universities, separately, conducted experiments in 2003 designed to determine how sleep affects our ability to remember. College-age people were trained to perform certain tasks and then tested to see how much they recalled in confronting such tasks after either a night's sleep or several hours awake. "We all have the experience of going to sleep with a question and waking up with the solution," observed one of the investigators, Prof. Danial Margoliash, of the University of Chicago. But does the sleep really help?

    It does, markedly. Not just as a matter of re-charge, but, the investigators found, because sleep *rescues* memories by storing and consolidating them deep in the brain's circuitry. At the University of

Chicago, subjects trained to understand murky speech on a voice synthesizer could regularly understand more words after a night of sleep than matched counterparts who were tested just hours after the training with no intervening sleep. And at the Harvard Medical School, one hundred subjects were trained to perform certain finger-tapping sequences that they were later asked, at various intervals, to repeat. The process of memory consolidation required one or two nights of sleep—after which the performance of the subjects improved substantially.

—Reported in *Nature*, 9 October 2003

2. The heavy use of salt is widely suspected by experts to be the cause of an epidemic of high blood pressure and many deaths from heart disease around the world. But how to prove that salt is the culprit? There are "natural experiments" in which isolated jungle or farming communities are introduced to modern civilization, move to cities, adopt high-salt diets, and commonly develop high blood pressure. But such evidence is inconclusive because many important factors change together; new stresses and many dietary changes accompany the increase in salt. How can the causal effects of salt by itself be tested?

Dr. Derek Denton, of the University of Melbourne, selected a group of normal chimpanzees, a species biologically very close to humans, in which to conduct the needed trials. A group of chimpanzees in Gabon, with normal blood pressure, were first studied in their natural state. The group was then divided in half, with one half receiving gradually increasing amounts of salt in their diet for twenty months. Normal blood pressure in a chimpanzee is 110/70. In Dr. Denton's experiment, the animals' blood pressure commonly rose as high as 150/90, and in some individuals much higher. But among animals in the control group, who received no additional salt, blood pressure did not rise. Six months after the extra salt was withdrawn from their diet, all the chimpanzees in the experimental group had the same low blood pressure they had enjoyed before the experiment. Because there was no other change in the lifestyle of those animals, the investigators concluded that changes in salt consumption caused the changes in blood pressure.

—D. Denton et al., "The Effect of Increased Salt Intake on Blood Pressure of Chimpanzees," *Nature*, October 1995

3. Does Louisiana hot sauce, the principal ingredient of the spicy New Orleans cocktail sauce commonly served with raw shellfish, kill certain bacteria found in raw oysters and clams? The answer appears to be yes. Bacteria of an infectious and sometimes fatal kind—*Vibrio vulnificus*—are found in 5 to 10 percent of raw shellfish on the market. Dr. Charles V. Sanders and his research team, from Louisiana State University Medical Center in New Orleans, added Louisiana hot sauce to cultures of *Vibrio* growing in test tubes; the sauce, even when greatly diluted, killed

V. vulnificus in five minutes or less. "I couldn't believe what happened," Dr. Sanders said. He admits that he still eats raw oysters, "but only with plenty of hot sauce."

—Reported to the Interscience Conference on Antimicrobial Agents, New Orleans, October 1993

4. In Lithuania, rear-end auto collisions happen as they do in the rest of the world; bumpers crumple, tempers flare. But drivers there do not seem to suffer the complaints so common in the United States, the headaches and lingering neck pains known as "whiplash syndrome." Dr. Harald Schrader and colleagues from University Hospital in Trondheim, Norway, without disclosing the purpose of their study, gave health questionnaires to 202 Lithuanian drivers whose cars had been struck from behind one to three years earlier in accidents of varying severity. The drivers' reports of their symptoms were compared to the reports of a control group (of the same size, same ages, and same home towns) of drivers who had not been in an accident. Thirty-five percent of the accident victims reported neck pain, but so did 33 percent of the controls; 53 percent of those who had been in an accident had headaches, but so did 50 percent of those in the control group. The researchers concluded: "No one in the study group had disabling or persistent symptoms as a result of the car accident."

What, then, can account for the explosion of whiplash cases elsewhere in the world? Drivers in the Lithuanian study did not carry personal injury insurance at the time of the study, and people there very infrequently sue one another. Most medical bills are paid by the government, and at the time of the study there were no claims to be filed, no money to be won, and nothing to be gained from a diagnosis of chronic whiplash. Chronic whiplash syndrome, the Norwegian researchers concluded, "has little validity."

—Harald Schrader et al., "Natural Evolution of Late Whiplash Syndrome Outside the Medicolegal Context," *The Lancet*, 4 May 1996

5. To determine the role of specific genes, mice are bred in which certain genes have been deleted. Such mice are called "knockout mice." When normal mice are placed in a lighted room, with dark corners, they go immediately to the dark. In one recent experiment the mice, upon entering the dark, encounter a mild electric shock, and very quickly learn to stay away from those dark regions. Mice who lack a gene called *Ras-GRF* learn to be wary just as quickly as do normal mice. But, unlike normal mice, the knockout mice throw caution to the winds the next day, and chance the dark corners again and yet again. It appears that the *Ras-GRF* gene—probably very much like the analogous gene in humans—plays a critical role in the ability of the mice to remember fear. This gene is almost certainly crucial for the survival of mammals.

—Reported in *Nature*, December 1997

6. Here is some reassuring news for those whose career plans are slightly behind schedule: It turns out that peaking too early may kill you. That's the finding of Stuart J. H. McCann, a professor of psychology at the University College of Cape Breton in Nova Scotia.

   McCann's research concerns what he calls the "precocity-longevity hypothesis." McCann analyzed the lives of 1,672 U.S. governors who served between 1789 and 1978 and found that those who were elected at relatively tender ages generally died earlier than their less precocious counterparts. Even when he controlled for the year that the governors were born, or how long they served, and what state they governed, the pattern held. No matter how he sliced the data, or ran the regressions, or accounted for various statistical biases, the story remained the same: governors elected to office at younger ages tended to have shorter lives.

   And what holds for state executives seems also to hold for other young achievers. McCann also analyzed smaller but more diverse sets of accomplished people—including American and French presidents, Canadian and British prime ministers, Nobel Laureates, signers of the Declaration of Independence, Academy Award winners, and seven centuries of Roman Catholic pontiffs. Again he found that "those who climb to the loftiest peaks in the shortest time also die younger. For the eminent, and perhaps for all, an early rise may lead to an early fall."

   —*Personality and Social Psychology Bulletin*, February 2003

7. Cholera, caused by a water-borne bacterium ingested by drinking contaminated water, is a dreadful disease; pandemics of cholera in the 19th century killed tens of thousands. The accepted view, that it was caused by breathing a filthy miasma, was doubted by John Snow, a founding member of the London Epidemiological Society. When a terrible cholera epidemic struck London in 1848–49, Snow hypothesized that bad water, from urban wells and from the Thames River, was the villain. Some water companies drew their water from within the tidal section of the Thames, where the city's sewage was also dumped, thus providing their customers with excrement-contaminated drinking water. It stank, so some of the intake pipes were shifted to points above the tideway. In 1854 cholera returned with even greater horror. Snow identified two water companies, one of which had moved its intake to a point above the tidal region of the river, the other still supplying a fecal cocktail; his data from these two districts showed a strong connection between cholera mortality and water source. Snow also identified a particular well, on Broad Street, and plotted cholera mortality house by house in the area of that well—the number of dead increasing sharply with proximity to the Broad Street pump—while a few streets away on Warwick Street there were no cholera deaths at all. Just across from the Broad Street pump was the Poland Street Workhouse, whose wretched inmates remained healthy—the workhouse had its own well. The Lion Brewery, close to the pump on Broad Street, also had its own well; its workers did not contract

cholera—they drank mainly malt liquor. The outbreak ended when Snow persuaded the authorities to *remove the handle from the Broad Street pump*. There is today a replica of the handleless pump outside a nearby pub named in honor of John Snow.

Steven Shapin, "Sick City," *The New Yorker*, 6 November 2006

C. The Joint Method of Agreement and Difference

Although Mill believed that the **joint method of agreement and difference** was an additional and separate technique, it is best understood as the combined use of the method of agreement and the method of difference in the same investigation. It can be represented schematically (capital letters again denoting circumstances, lowercase letters denoting phenomena) as follows:

$$A\ B\ C - x\ y\ z.\qquad A\ B\ C - x\ y\ z.$$
$$A\ D\ E - x\ t\ w.\qquad B\ C - y\ z.$$

Therefore A is the effect, or the cause, or an indispensable part of the cause, of x.

Because each of the two methods (agreement schematized above on the left, difference schematized on the right) affords some probability to the conclusion, their joint use affords a higher probability to that conclusion. In many scientific investigations this combination serves as an extremely powerful pattern of inductive inference.

A notable advance in medicine provides an illustration of the power of the joint method. Hepatitis A is a liver infection that afflicted tens of thousands of Americans; it spread widely among children, chiefly through contaminated food or water, and was sometimes deadly. How might it be prevented? The ideal solution, of course, would be an effective vaccine. However, an enormous difficulty faced those who would test any vaccine for hepatitis A: It was very hard to predict where outbreaks of the infection would occur, and therefore it was usually not possible to select experimental subjects in ways that would yield reliable results. This difficulty was finally overcome in the following way.

A potential vaccine was tested in a community of Hasidic Jews, Kiryas Joel, in Orange County, New York, a community that was highly unusual in that it was plagued by yearly epidemics of this infection. Almost no one escaped hepatitis A in Kiryas Joel, and nearly 70 percent of the community members had been infected by the time they were nineteen years old. Dr. Alan Werzberger, of the Kiryas Joel Institute of Medicine, and his colleagues recruited 1,037 children in that community, ages two to sixteen, who had not been exposed to the hepatitis A virus, as determined by a lack of antibodies to the virus in their blood. Half of them (519) received a single dose of the new vaccine, and among those vaccinated children not a single case of hepatitis A was reported. Of the 518 children who received dummy injections, 25 became infected with hepatitis A soon after. The vaccine for hepatitis A had been found.[10]

Liver specialists in Boston and Washington greeted this study with admiration, calling it "a great breakthrough" and a "major medical advance." What

---

Joint method of agreement and difference
A pattern of inductive inference in which the method of agreement and the method of difference are used in combination to give the conclusion a higher degree of probability.

is the pattern of inference on which this achievement relied? Both the method of agreement and the method of difference were employed, as is common in medical investigations. Among all those young residents of the community who became immune to hepatitis A, there was only one relevent circumstance *in common:* All the immunes had received the new vaccine. By itself, this strongly tended to show that the vaccine did cause that immunity. The method of difference supported this conclusion overwhelmingly: The circumstances of those who did become immune and those who did not were essentially alike *in every respect except one*, the administration of the vaccine to the immune residents.

The testing of new drugs or procedures is often conducted in what are called "double-arm" trials, one group receiving the new treatment while the other group does not, after which (in suitable cases) there may be a carefully executed crossover, in a second phase, in which those who originally did not receive the treatment do so, and those who originally did receive the treatment do not. The application of the joint method of agreement and difference underlies such investigations, which are common and exceedingly productive.

EXERCISES

Analyze each of the following reports, explaining the way in which the method of agreement and the method of difference have been jointly applied, and identifying the special force, if any, of their combination.

1. Pain can be agonizing, but it serves a useful function: It teaches people and animals to avoid dangers, and forces them to attend to wounds. Strangely, there are a very few people who never feel pain; they remain unaware of having suffered significant injuries.

    One family in northern Pakistan has several such members. One, a ten-year old boy, became famous for giving street performances in which he put knives through his arms and walked on hot coals. Tissue damage would result, but no discomfort. Geneticist C. G. Woods of Cambridge University searched for the cause of this remarkable inability to feel pain. Eventually he zeroed in on mutations in a gene, *SCN9A*, that codes for the channel through which sodium enters pain-sensing cells, critical to the pain signal. Testing with electric current, he could open and close sodium channels on some cells—but he could not open the sodium channels on those mutant cells. Said Woods, "This shows that rare diseases can still be of great importance because of the insights they give into biological processes."

    A Yale University neurologist, Stephen Waxman, observes that if researchers could craft a drug that can make these channels inactive, as they are in the Pakistani family members, millions of people worldwide who suffer from chronic pain would be wonderfully served.

    —Reported in *Nature*, 14 December 2006

2. A deadly heart ailment affecting about 1 million African American men—familial amyloid cardiomyopathy—and another that afflicts older men of all ethnicities, are known to be caused by an abnormally folded protein that builds up in the organism. Transthyretin protein, made in the liver, has four subunits. A mutation in the gene that makes two of those subunits results in the instability of the protein, its misfolding, and eventually in death. That this is indeed the cause of the ailments was shown by the fact that a liver transplant, providing a healthy version of the critical gene, can result in cure—but often that correction comes too late to stymie the misfolding that did the damage.

   A strange twist of nature, reported in *Science* in January 2003 by Dr. Jeffrey Kelly, of the Scripps Research Institute in San Diego, provided the clues to a therapy that can thwart the misfolding process. Because diseases of this kind are quite common in Portugal, families there are screened to see who has the mutated gene and is therefore at risk. One very large family was identified whose members had the mutated gene and yet never did contract the disease. It turned out that in this family, a second gene that made the other two subunits of the protein had undergone its own mutation, suppressing or reversing the disease process. Members of that family carried a cure to an inherited disease in their own genes.

   Dr. Kelly found that as a result of this further mutation the disease was prevented by the erection of a kind of barrier between the normal and the abnormal protein states. Then, by screening libraries of small molecules, he located several that, already approved by the Food and Drug Administration for other purposes, could mimic the effect of the second mutation, successfuly reversing the misfolded protein in animals.

3. Sixteen-year-old David Merrill, of Suffolk, Virginia, hypothesized that the loud sounds of hard-rock music have a bad effect on its devoted fans. He tested the theory on mice. Seventy-two mice were divided into three groups of 24, the first to be exposed to hard-rock music, the second to music by Mozart, and the third to no music at all. After allowing the mice to become accustomed to their environments, but before exposing them to the music, Merrill tested all of them in a maze, which took the mice an average of 10 minutes to complete. Then the groups were exposed to the music for 10 hours a day.

   With repeated testing the control-group mice *reduced* their time in the maze by an average of 5 minutes. Those exposed to Mozart reduced their time by 8.5 minutes. The hard-rock mice *increased* their time in the maze by 20 minutes.

   Merrill also reported that when, in an earlier attempt, he had allowed all the mice to live together, the project had to be cut short because, unlike the Mozart-listening mice, the hard-rock-listening mice killed other mice.

   —Reported in *Insight*, 8 September 1997

4. Scientists have long known that severely restricting the number of calories that mice and other organisms consume lengthens their life span. Animals on low-calorie diets typically have abnormally cool body temperatures. Does low temperature, in itself, result in longer life? The answer is yes.

   Bruno Conti, of the Scripps Research Institute in La Jolla, California, genetically engineered mice to have a faulty sense of body temperature. The alteration reduced the animals' temperatures by 0.03 to 0.05°C below normal; they were given as much food as they wanted, maintaining their normal weight. The low-temperature mice lived about 15 percent longer than normal mice did.

   —Reported in *Science*, 3 November 2006

5. At a social gathering of eighty-five faculty members, graduate students, and staff workers in the Department of Food Science at the University of Illinois in Urbana-Champaign, the partygoers served themselves ice cream. They did not know they were also the subjects of an experiment. Half the participants were given 17-ounce bowls, and half 34-ounce bowls. In addition, half were given 2-ounce spoons to scoop out their ice cream, and half were given 3-ounce serving spoons.

   With larger spoons, people served themselves 14.5 percent more, and with a larger bowl they heaped on 31 percent more. With both large spoon and large bowl these nutrition experts helped themselves to 56.8 percent more ice cream than those who used the smaller utensils. And all but three ate every bit of the ice cream they took. Smaller platters and smaller utensils may be the key to a successful diet.

   —Reported by Brian Wansink in *The American Journal of Preventive Medicine*, September 2006

D. The Method of Residues

John Stuart Mill wrote:

> Subduct from any phenomenon such part as is known by previous inductions to be the effect of certain antecedents, and the residue of the phenomenon is the effect of the remaining antecedents.

The first three methods seem to suppose that we can eliminate or produce the cause (or effect) of some phenomenon in its entirety, as indeed we sometimes can. In many contexts, however, we can only deduce the causal effect of some phenomenon by observing the *change* that it makes in a set of circumstances whose cause is already understood in part.

This method, focusing on *residues*, is well illustrated by the very simple device used to weigh truck cargos. The weight of the truck when empty is known. To determine the weight of the cargo, the entire truck is weighed with its cargo—and the weight of the cargo is then known to be the weight of the whole minus the weight of the truck. The known "antecedent," in Mill's phrase, is the recorded weight of the empty truck that must be subtracted from the reading on the scale;

492   CHAPTER 12 Causal Reasoning

the cause of the difference between that reading and the known antecedent is obviously attributable to the remaining "antecedents"—that is, to the cargo itself.

Schematically, the **method of residues** can be represented as follows:

A B C — x y z.
B is known to be the cause of y.
C is known to be the cause of z.
Therefore A is the cause of x.

A splendid illustration of the effectiveness of the method of residues is provided by one of the great chapters in the history of astronomy, the discovery of the planet Neptune:

> In 1821, Bouvard of Paris published tables of the motions of a number of planets, including Uranus. In preparing the latter he had found great difficulty in making an orbit calculated on the basis of positions obtained in the years after 1800 agree with one calculated from observations taken in the years immediately following discovery. He finally disregarded the older observations entirely and based his tables on the newer observations. In a few years, however, the positions calculated from the tables disagreed with the observed positions of the planet and by 1844 the discrepancy amounted to 2 minutes of arc. Since all the other known planets agreed in their motions with those calculated for them, the discrepancy in the case of Uranus aroused much discussion.
>
> In 1845, Leverrier, then a young man, attacked the problem. He checked Bouvard's calculations and found them essentially correct. Thereupon he felt that the only satisfactory explanation of the trouble lay in the presence of a planet somewhere beyond Uranus which was disturbing its motion. By the middle of 1846 he had finished his calculations. In September he wrote to Galle at Berlin and requested the latter to look for a new planet in a certain region of the sky for which some new star charts had just been prepared in Germany but of which Leverrier apparently had not as yet obtained copies. On the twenty-third of September Galle started the search and in less than an hour he found an object which was not on the chart. By the next night it had moved appreciably and the new planet, subsequently named Neptune, was discovered within 1° of the predicted place. This discovery ranks among the greatest achievements of mathematical astronomy.[11]

Method of residues
A pattern of inductive inference in which, when some portions of a given phenomenon are known to be the effects of certain identified antecedents, we conclude that the remaining portion of the phenomenon is the effect of the remaining antecedents.

The phenomenon under investigation here is the movement of Uranus. A great part of that phenomenon, the orbit of Uranus around the sun, was well understood at the time. Observations of Uranus approximated this calculated orbit but exhibited a puzzling residue, some perturbation of what had been calculated, for which further explanation was needed. An additional "antecedent"—that is, an additional existing factor that would account for the perturbation—was hypothesized to be another (undiscovered) planet whose gravity would, together with what was already known about the orbit of Uranus, explain that residue. Once hypothesized, that new planet, Neptune, was very quickly found.

The method of residues differs from the other methods in that it can be used with the examination of only one case, whereas the others require the examination of at least two cases. The method of residues, unlike the others, appears to depend on antecedently established causal laws, while the other methods (as Mill formulated them) do not. The method of residues is nevertheless an inductive, not a deductive, method (as some have suggested), because it yields conclusions that are only probable and cannot be *validly deduced* from their premises. An additional premise or two might transform an inference by the method of residues into a valid deductive argument, but that can be said for other inductive methods as well.

EXERCISES

Analyze each of the following arguments in terms of "antecedents" and "phenomena" to show how they follow the pattern of the method of residues:

1. For nineteen years space scientists, astronomers, and physicists have been puzzled by what appears to be a mysterious force pulling spacecraft in the direction of the sun. It was first noticed when the trajectories of two outward bound and very distant spacecraft (*Pioneer 10* and *11*, launched in 1972 and 1973) were carefully analyzed. The trajectories of two later probes (*Galileo*, launched toward Jupiter in 1989, and *Ulysses*, launched into polar orbit around the sun) have exhibited the same peculiarities: They give evidence of a weak force that perturbs their directions and velocities. This force was discovered by adding up the effects of all other known forces acting on the spacecraft and finding that something unexplained was left over.

    This force is apparently slowing the outward progress of the spacecraft speeding away from or around the sun—but in contrast to the force of gravity, the strength of this mystery force does not decline proportionally to the inverse square of a spacecraft's distance from the sun, but instead at a linear rate, which makes it very unlikely that the mystery force is a gravitational effect of the sun.

    Calculations were made using two independent methods, and data of different types, taking into account possible errors in the software and the hardware used in the measurements. A host of other possible errors were investigated and accounted for—and after ruling all of these out, a team of physicists from the Los Alamos National Laboratory announced that the mystery remained. This means that some hitherto unknown phenomenon may be at work—what physicists excitedly call "new physics."

    —Reported in *Physical Review Letters*, September 1998

2. In H. Davies' experiments on the decomposition of water by galvanism, it was found that besides the two components of water, oxygen and hydrogen, an acid and an alkali were developed at opposite poles of the machine. Since the theory of the analysis of water did not give reason to expect these products, their presence constituted a problem. Some

chemists thought that electricity had the power of producing these substances of itself. Davies conjectured that there might be some hidden cause for this part of the effect—the glass might suffer decomposition, or some foreign matter might be in the water. He then proceeded to investigate whether or not the diminution or total elimination of possible causes would change or eliminate the effect in question. Substituting gold vessels for glass ones, he found no change in the effect and concluded that glass was not the cause. Using distilled water, he found a decrease in the quantity of acid and alkali involved, yet enough remained to show that the cause was still in operation. He inferred that impurity of the water was not the sole cause, but was a concurrent cause. He then suspected that perspiration from the hands might be the cause, as it would contain salt which would decompose into acid and alkali under electricity. By avoiding such contact, he reduced the quantity of the effect still further, till only slight traces remained. These might be due to some impurity of the atmosphere decomposed by the electricity. An experiment determined this to be the case. The machine was put under an exhaust receiver and when it was thus secured from atmospheric influences, no acid or alkali was produced.

—G. Gore, *The Art of Scientific Discovery*, 1878

3. Satellite observations collected between 1992 and 2001 suggest that the upper surface of the Larsen C ice shelf, in Antarctica, dropped as much as 27 cm per year during that period. About a quarter of that shrinkage, or 7 cm, may have resulted from snow packing down into denser material called *firn*. Uncertainties about such factors as the height of the ocean tides, and the salinity of water beneath the ice shelf would account for no more than a small fraction of the remaining loss of height above water.

Therefore, concluded Andrew Shepherd, a glaciologist at the University of Cambridge in England, as much as 20 cm per year of the upper surface's drop must stem from melting. Nine-tenths of any mass of floating ice lies below the water's surface, suggesting that the Larsen C ice shelf is thinning by as much as 2 m each year.

The likely cause of this thinning is relatively warm water beneath the shelf. Even a very small temperature increase in the water below an ice shelf can make a big difference in the melting rate of the overlying ice. Larsen C is stable, and isn't shedding more icebergs than normal, Shepherd reported, but at its current rate of thinning, Larsen C could reach 200 m in thickness (the thickness at which other ice shelves have disintegrated) and therefore be susceptible to disintegration in 70 years—but if the waters in the region continue to warm, the demise of Larsen C could occur even sooner.

—Reported in *Science News*, 1 November 2003

4. Analyzing more than forty years of weather data, climatologists at the National Oceanic and Atmospheric Administration in Boulder, Colorado, recently found that the daily temperature range—the difference between the daytime maximum and the nighttime minimum temperatures—at 660 weather stations in the continental United States fluctuates in a very puzzling manner: the variation of the temperature range over the course of a week, in some regions, does not line up with any natural cycles that can be detected.

   The average temperature range for the weekends (Saturday, Sunday, and Monday) varied from the average temperature range for weekdays (Tuesday, Wednesday, Thursday, and Friday)! Fluctuations in the daily range can be caused by natural factors; storm systems moving across an area, for example, can cause such fluctuation—but there are no natural factors known to fall consistently on certain days of the week.

   The precise cause of this extraordinary pattern is not clear. However, contend the researchers (Piers M. de F. Forster and Susan Solomon), the only possible explanation for this weekend/weekday disparity is human activity and the atmospheric pollutants such activity creates.
   —Reported in *Proceedings of the National Academy of Sciences*, 30 September 2003

5. It is no longer open to discussion that air has weight. It is common knowledge that a balloon is heavier when inflated than when empty, which is proof enough. For if the air were light, the more the balloon was inflated, the lighter the whole would be, since there would be more air in it. But since, on the contrary, when more air is put in, the whole becomes heavier, it follows that each part has a weight of its own, and consequently that the air has weight.
   —Blaise Pascal, *Treatise on the Weight of the Mass of the Air*, 1653

E. The Method of Concomitant Variation

The four methods discussed so far are all *eliminative* in nature. By eliminating some possible cause or causes of a given phenomenon, they support some other causal account hypothesized. The method of agreement eliminates as possible causes those circumstances in whose absence the phenomenon can nevertheless occur; the method of difference permits the elimination of some possible causes by removing an antecedent factor shown to be critical; the joint method is eliminative in both of these ways; and the method of residues seeks to eliminate as possible causes those circumstances whose effects have already been established by previous inductions.

   However, in many situations, no one of these methods is applicable, because they involve circumstances that cannot possibly be eliminated. This is often the case in economics, in physics, in medicine, and wherever the general increase or decrease of one factor results in a concomitant increase or decrease of another— the complete elimination of either factor not being feasible.

John Stuart Mill wrote:

> Whatever phenomenon varies in any manner whenever another phenomenon varies in some particular manner is either a cause or an effect of that phenomenon or is connected with it through some fact of causation.

Concomitant variation is critical to the study of the causal impact of certain foods, for example. We cannot eliminate disease, no matter the diet; we can rarely eliminate foods of certain kinds from the diets of large populations. But we can note the effect of increasing or decreasing the intake of certain foods on the frequency of certain diseases in specified populations. One investigation of this kind examined the frequency of heart attacks compared to the frequency with which fish had been eaten by those in the study. The inductive conclusion was striking: Eating one fish meal a week reduced the risk of heart attack by 50 percent; eating just two fish meals a month reduced the risk of heart attack by 30 percent. Within some limits there appears to be a marked concomitant variation between cardiac arrests and the use of fish in the diet.[12]

Using plus and minus signs to indicate the greater or lesser degree to which a varying phenomenon is present in a given situation, the **method of concomitant variation** can be schematized as:

$$A\ B\ C - x\ y\ z.$$
$$\underline{A+\ B\ C - x+\ y\ z.}$$
Therefore A and x are causally connected.

This method is very widely used. A farmer establishes that there is a causal connection between the application of fertilizer to the soil and the size of the crop by applying different amounts to different parts of a field, then noting the concomitant variation between the amounts of the additive and the yield. A merchant seeks to verify the efficacy of advertising of different kinds by running varied advertisements at varying intervals, then noting the concomitant increase or decrease of business during some of those periods.

Concomitant variation is exemplified in the search for the causes of divorce and of other important decisions among families. Of course the cause of any particular divorce will lie in the special circumstances of that marriage and that family, but there are conditions that tend generally to contribute to the breakup of families, and concomitant variation is useful in learning what these are. Analysis of data from the U.S. Census Bureau reveals that, in every decade since the 1940s, and in every region of the country, couples who were the parents of only girls divorced more often than couples who were the parents of only boys. It happened among whites and among blacks, among those with only high school diplomas and among those with college degrees. Parents with an only child who is a girl are 6 percent more likely to split up than parents of a single boy. The gap rises to 8 percent for parents of two girls versus parents of

Method of concomitant variation
A pattern of inductive inference in which it is concluded that, when one phenomenon varies consistently with some other phenomenon in some manner, there is some causal relation between the two phenomena.

two boys, 10 percent for families with three girls, and 13 percent if there are four girls. Thousands upon thousands of U.S. divorces appear to stem partly from the number of girls in the family.

The age-old favoring of boys, overt and common in China, India, and other developing countries, is more subtle in the United States, but it remains a widespread factor in the dynamics of U.S. family life. Parents invest more in their sons, spending, when their families include a boy, an average of an additional $600 a year on housing. Fathers increase their workweeks after the birth of the first family child of either sex—but increase it by more than two hours if the child is a boy, less than one hour if it is a girl. These patterns of concomitant variation make it plain that parents have a preference for boys—a preference that will have increasingly important consequences when the technology for the selection of the sex of a baby, already known and reliable, becomes more widely available.[13]

When the increase of one phenomenon parallels the increase of another, we say that the phenomena vary *directly* with each other. However, the method permits the use of variation "in any manner," and we may also infer a causal connection when the phenomena vary *inversely*—the increase of one leading to the decrease of another. Thus economists often say that, other things remaining roughly stable, in an unregulated market an increase of the supply of some good (say, crude oil) will result in a concomitant decrease in its price. That relation does appear to be genuinely concomitant: When international tension threatens to reduce the available supply of crude oil, we note that the price of the oil almost invariably rises.

Some concomitant variations are entirely coincidental, of course. Care must be taken not to infer a causal connection from patterns of occurrence that are wholly fortuitous. But some variations that appear to be coincidental, or are otherwise puzzling, may have an obscure causal explanation. It has been shown that there is a high correlation between the number of storks found nesting in English villages and the number of babies born in each of those villages—the more storks, the more babies. Surely it is not possible that . . . No, it's not. Villages with high birth rates have more newly married couples, and therefore have more newly constructed houses. Storks, it turns out, prefer to nest beside chimneys that have not been used previously by other storks.[14] Tracing the causal chains of phenomena that vary concomitantly, we may find links in common, which is what Mill meant when he said that the phenomena may be "connected . . . through some fact of causation."

Because the method of concomitant variation permits us to adduce, as evidence, changes in the *degree* to which circumstances and phenomena are present, it greatly strengthens our set of inductive techniques. It is a *quantitative* method of inductive inference, those earlier discussed being essentially qualitative. The use of concomitant variation therefore presupposes the existence of some method of measuring or estimating, even if only roughly, the degrees to which phenomena vary.

EXERCISES

Analyze each of the following arguments in terms of the variation of the "phenomena" to show how they follow the pattern of the method of concomitant variation:

1. The notion that poverty and mental illness are intertwined is not new—but finding evidence that one begets the other has often proved difficult. New research, which coincided with the opening of a new gambling casino on an Indian reservation, appears to strengthen that link, strongly suggesting that lifting children out of poverty (as casino income did in many cases) tends to diminish some (but not all) psychiatric symptoms.

    A study published in the *Journal of the American Medical Association* in October 2003 tracked 1,420 children, ages 9 to 13, in rural North Carolina, very many of whom lived on a Cherokee Indian reservation. During the study a casino that had been opened on the reservation began distributing some of its profits to tribal families, the payments reaching about $6,000 per year by 2001. The researchers found that the rate of psychiatric symptoms among the children who had risen from poverty dropped steadily; those children were less inclined to temper tantrums, stealing, bullying, and vandalism—common symptoms of oppositional defiant disorders.

    Children whose families rose above the poverty threshold showed a 40 percent decrease in behavioral symptoms. The rate of such behaviors, after four years, dropped to the same levels found among children whose families had never been poor. But the casino payments had no effect on children whose families were nevertheless unable to rise from poverty, or on those children whose families had not been poor to begin with.

    The economic change had a significant effect on only a fraction of the children followed. This, it was hypothesized, was a consequence of the fact that, although all the families that received the payment received the same amount of money, the payments resulted in lifting only 14 percent of those families above the poverty line which, in 2002, was $14,348 for a family of three. The study suggests, said Dr. Arline Geronimus, of the University of Michigan, that poverty puts stress on families, which can increase the likelihood that children will develop behavioral problems.

2. In Finland, heart attacks occur more frequently in the eastern part of the country than in the western and southern parts. Researchers seeking to explain these differences concluded that they "cannot be explained by individual lifestyle or by genetic factors." How, then, can they be explained? A study led by Dr. Anne Kousa, of the Geological Survey of Finland, examined heart attacks that occurred in 18,946 men,

ages 35 to 74, in three different years. The researchers then correlated the incidence of heart attack in these populations with the level of water hardness—as measured by the presence of minerals in the water—in their communities. The study found that the degree of water hardness correlated directly with a lowered risk of heart attack. Drinking water rich in minerals appears to play a role in reducing heart disease.

— *Journal of Epidemiology and Community Health*, January 2004

3. When it comes to love, sex, and friendship, do birds of a feather flock together? Or is it more important that opposites attract? Dr. Claus Wedekind, of Bern University in Switzerland, hypothesized that body odor might signal that its owner had desirable immune genes—called MHC genes—that would help offspring to fight off diseases. He devised an experiment to see if human body odor correlated with MHC genes and if people could tell.

He and his team collected DNA samples for 49 female and 44 male university students. He asked the men to wear cotton T-shirts on two successive nights, to keep the shirt in a plastic bag, to use perfume-free detergents and soaps, and to avoid smelly rooms, smell-producing foods, and activities like smoking and sex that create odors. Meanwhile, the women were given a nasal spray to protect their nasal membranes from infection, and each received a copy of the Patrick Susskind novel *Perfume* to make them more conscious of odors.

When the T-shirts were collected, the women were asked to give ratings, for intensity, pleasantness, and sexiness, to three T-shirts from men whose MHC genes were similar to those of the women, and three from men whose MHC genes were dissimilar, not knowing which were which.

Women who were dissimilar to a particular male's MHC perceived his odor as more pleasant than did women whose MHC was similar to that of the test man. Odors of men with dissimilar MHC reminded the women of their own mates or former mates twice as often as did the odors of men with similar MHC.

However, if a woman was taking oral contraceptives, which partly mimic pregnancy, this predilection was reversed, and they gave higher ratings to men with similar MHC. "The Pill effect really surprised me," said Dr. Wedekind.

—*Proceedings* of the Royal Society of London, 1995

4. Stanley Coren sought to plumb the connections between sleeplessness and accidents. To do that he focused on the yearly shift to daylight time in eastern North America, when (because clocks are moved forward one hour) most people lose an hour of sleep. He compared the number of accidents then with the number on normal days, and found that on the day after the time change, in Canada, there was an 8 percent increase in accidents. Then, examining the day after the return to standard time, when people gain an hour of sleep, he found a corresponding decrease

in accidents. "What we're looking at," says the Director of the Human Chronobiology Laboratory at the University of Pittsburgh, commenting upon Soren's results, "is national jet lag."

—S. Coren, *Sleep Thieves* (New York: The Free Press, 1996)

5. Prof. Kathleen Vohs reports that two groups of college students were asked to read out loud from "a boring book on the biographies of scientists." One of the groups was obliged to wear fake expressions of delight and interest, while the other group was allowed to read the same texts naturally. Each group was afterward given a sum of money to spend on an assortment of goods, or to save. Those who had been faking delight spent 62 percent more than those who had not. Similarly, a group of students writing down their thoughts without restraint spent very much less than a similar group obliged while writing to avoid all thoughts about white bears. The more self-restraint that a person expends to control one impulse, it appears, the less self-restraint is available to control others.

—Reported in *The Journal of Consumer Research*, March 2007

6. Potassium in the urine is known to reflect potassium intake from the diet. At the Prosserman Center for Health Research in Toronto, Dr. Andrew Mente and colleagues analyzed urinary potassium as a useful clinical marker of a healthy diet. They collected urine samples from hundreds of patients and separately calculated the quality of their diets. The results were striking: as urinary potassium increased there was a steady and significant increase in diet quality score, as well as a steady decrease in body mass, blood pressure and heart rate. "This urinary marker," said Dr. Mente, "is a simple, objective, universally available measure of diet quality."

—*Urology/Nephrology News*, 20 November 2006

7. Whenever the U.S. says things that make a military conflict with Iran seem more likely, the price of oil rises, strengthening Iran's regime rather than weakening it. The more we talk about curbing Iranian power, the more difficult it gets. . . . So cooling down the martial rhetoric, even if we plan to take military action eventually, would likely bring oil prices down, making Iran weaker. . . . Lower oil prices won't, by themselves, topple the mullahs in Iran. But it's significant that, historically, when oil prices have been low, Iranian reformers have been ascendant and radicals relatively subdued, and vice versa when prices have been high. Talking tough may look like a good way of demonstrating U.S. resolve, but when tough talk makes our opponent richer and stronger we may accomplish more by saying less.

—James Surowieki, "Troubled Waters over Oil," *The New Yorker*, 19 February 2007

overview

Five Methods of Inductive Inference

1. **The method of agreement.** The one factor or circumstance that is *common* to all the cases of the phenomenon under investigation is likely to be the cause (or effect) of that phenomenon.
2. **The method of difference.** The one factor or circumstance whose absence or presence *distinguishes* all cases in which the phenomenon under investigation occurs from those cases in which it does not occur, is likely to be the cause, or part of the cause, of that phenomenon.
3. **The joint method of agreement and difference.** Although perhaps not a separate method, *the combination*, in the same investigation, *of the method of agreement and the method of difference* gives substantial probability to the inductive conclusion.
4. **The method of residues.** When some portion of the phenomenon under examination is known to be the consequence of well-understood antecedent circumstances, we may infer that *the remainder of that phenomenon is the effect of the remaining antecedents*.
5. **The method of concomitant variation.** *When the variations in one phenomenon are highly correlated with the variations in another phenomenon*, one of the two is likely to be the cause of the other, or they may be related as the products of some third factor causing both.

These are the inductive methods frequently called *Mill's methods*, most commonly used by scientists in their investigation of causal laws.

12.5 Limitations of Inductive Techniques

What do the methods explained in the preceding sections actually do for us? John Stuart Mill believed that they were instruments with which we may *discover* causal connections; also that they were canons with which causal connections may be *proved*. On both counts he overestimated their power. Inductive techniques are indeed of very great importance, but their role in science is more limited than Mill supposed.

One substantial difficulty arises from the fact that, in formulating these methods, Mill made the assumption that one can identify cases "having *only* one circumstance in common" or other cases "having *every* circumstance in common save one." But these expressions must not be taken literally; any two objects will have many circumstances in common, however different they may appear; and no two things can ever differ in only one respect—one will be farther to the north, one will be closer to the sun, and so on. Nor could we even examine all possible circumstances to determine if they differ in only one way. What the scientist has in mind as he or she applies these techniques are not all circumstances, but the sets of *relevant* circumstances—whether there is only one relevant circumstance in common, or all relevant circumstances save one in common. That is, we apply the methods to the circumstances that have some bearing on the causal connection in question.

Biography

John Stuart Mill

John Stuart Mill (1806–1873), one of the most prominent philosophers and logicians of the nineteenth century, was the beneficiary of an extraordinary education, which he recounts in detail in his famous *Autobiography* (1873). He did not attend school, but studied intensively with his father, James Mill, a very learned man and a philosopher himself. The young John Stuart was taught to read before he was two; he studied Greek at the age of three and Latin at the age of eight. Before he was fourteen he had read most of the Greek and Latin classics, had pursued extensive studies in logic and mathematics, and had been writing essays in economic theory under the direction of his father, whose guidance was demanding but also loving. James was one of a small group of brilliant liberal intellectuals, radical reformers who were at once moral thinkers and economists, led by the great English legislative reformer, Jeremy Bentham. While still a boy Mill interacted with these powerful intellects in his own living room; eventually he himself became a leader of nineteenth-century English liberalism.

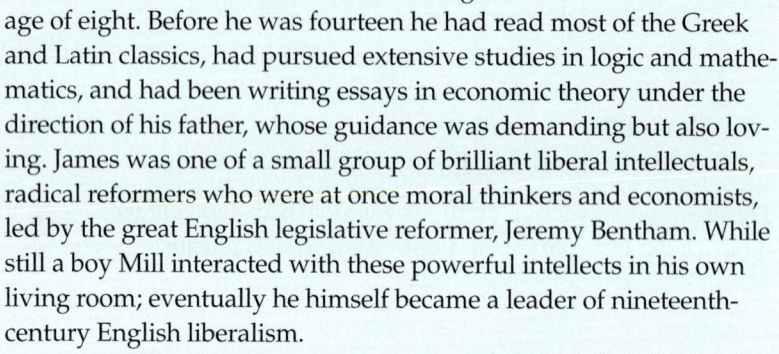

At the age of seventeen Mill followed his father as an employee of the East India Company, which was the *de facto* government of much of India during those years. Mill rose through the ranks to become, when he was 50, Chief Examiner for that company. After his retirement he was elected to Parliament, where he served with great distinction.

Mill's philosophical views remain highly relevant in the twenty-first century. His early feminism (*The Subjection of Women*, 1869) was courageous; his advocacy of self-government (*Considerations on Representative Government*, 1861) was profound; his moral views brought utilitarian thinking to the highest point it had reached in his time (*Utilitarianism*, 1861); his defense of free speech and expression (*On Liberty*, 1959) has made him, to this day, an intellectual hero around the globe.

His philosophical reputation was first established in his 30s, when he published his *System of Logic* (1843). In that work he explained why formal deductive logic, or syllogistics, cannot really add to our knowledge, although it certainly can help us to reason consistently on the basis of what we do already know. On the other hand, *inductive logic*—the logic of science—can provide, he argued, rules and guidelines for the discovery of new truths. Refining the old Baconian rules of inference, Mill formulated and explained the principles of inductive logic as no one had before him. Since the premises of any syllogism must be established or assumed before ratiocination can begin, Mill contended that induction must precede deduction and is necessarily more fundamental. In the realm of inductive logic his work was incisive and original.

John Stuart Mill was, overall, an effective and honest reformer, universally respected. He was a loving husband, an admired scholar, and a legislator of unquestioned integrity. His philosophical reasoning on a wide range of issues was subtle, his sentiments always humane. He was the most influential English-speaking philosopher and logician of the nineteenth century. ■

Which are those circumstances? We cannot learn which factors are relevant using the methods alone. In order to use the methods we must come to the context in which they are to be applied with some analysis of causal factors already in mind. The caricature of the "scientific drinker" illustrates this difficulty: He drinks Scotch and soda one night, bourbon and soda the next night, and on the following nights brandy and soda, then rum and soda, then gin and soda. What is the *cause* of his intoxication? Repeatedly inebriated, he swears never to touch soda again!

This scientific drinker did apply the method of agreement in accordance with the rules—but his doing so was to no avail because the factors that really are relevant in those antecedent circumstances had not been identified and therefore could not be manipulated. Had *alcohol* been specified as one of the factors common to all the cases, it would have been possible to eliminate soda very quickly, of course, using the method of difference.

The heroic investigation of the causes of yellow fever, discussed earlier in connection with the method of difference, confirmed the conclusion that the fever is spread by the bite of an infected mosquito. We know that *now*, just as we know now that it is alcohol and not soda that causes drunkenness. But the yellow fever experiments required insight and imagination as well as courage; the notion that the fever was spread by mosquitoes was originally thought to be silly, or absurd, or was not thought of at all. Circumstances in the real world do not come wearing tags marked "relevant" or "irrelevant." The testing of mosquito bites as cause required some earlier sorting of possibly relevant factors, to which the inductive methods might then be applied. With that prior analysis in hand, the methods can prove exceedingly helpful—but the methods by themselves, without some hypotheses in the background, are not *sufficient* instruments for scientific discovery.

Nor can the methods by themselves constitute rules for *proof*. Their application proceeds always on the basis of some antecedent hypotheses about causal factors, as noted just above, and because all circumstances cannot have been considered, attention will be confined to those believed to be the possible causes in question. However, this judgment regarding which circumstances are to be investigated may prove to have been in error. Medical scientists, for a very long time, did not consider dirty hands even as *possible* agents of infection, and so could not identify such dirtiness as the cause of disease. The failure of physicians to wash their hands (because they did not understand how infectious diseases were spread) resulted in untold misery and uncounted deaths over centuries, especially from puerperal (or childbed) fever, which was carried on the hands of doctors from mother to mother, until the proof of that disastrous causal connection was given by the Hungarian physician Ignac Semmelweis, in the middle of the nineteenth century.[15] Investigation is stymied when the investigators fail to break down the circumstances before them into the appropriate elements, elements that cannot be known in advance. Because the analyses presupposed by the application of the methods may be incorrect, or inadequate, the inferences based on those analyses may also prove to be mistaken. This dependence of induction on the merit of the underlying hypotheses shows that inductive techniques cannot by themselves provide the proof of causation that Mill had hoped for.

Yet another problem should be borne in mind: The application of inductive methods always depends on *observed* correlations, and even when the observations have been made accurately, they may be incomplete and therefore deceptive. The greater the number of observations, the greater is the likelihood that the correlation we observe is the manifestation of a genuine causal law—but no matter how great that number is, we cannot infer with certainty a causal connection among instances that have not yet been observed.

These limitations illuminate once again the great gulf between deduction and induction. A valid deductive inference constitutes a proof, or demonstration, but every inductive inference is, at best, highly probable and never demonstrative. Therefore Mill's claim that his canons are "methods of proof" must be rejected, along with the claim that they are "*the* methods of discovery."

Nevertheless, the techniques explained in this chapter are central in much of science and are very powerful. Because it is impossible for investigators to take all circumstances into account, the application of the methods must always suppose one or more causal *hypotheses* about the circumstances under investigation. Being unsure which factor(s) are the cause(s) of the phenomenon under investigation, we often formulate alternative hypotheses and subject each to testing. What the five methods of induction, being mainly eliminative in nature, enable us to determine is this: *If* some specified analysis of the antecedent circumstances is correct, one of these factors cannot be (or must be) the cause (or part of the cause) of the phenomenon in question. This may be deduced, and the deduction may be valid, but the soundness of that argument will always depend on the correctness of the antecedent analysis that had been supposed.

The methods of induction are splendid, but they can yield reliable results only when the hypothesis that they seek to confirm (or falsify) does identify correctly the circumstances that are causally relevant. The methods permit the *deduction* of those results only when that hypothesis has been assumed as a *premise* in the argument. The nature of the power these methods give us may now be seen. They are not paths for discovery; they are not rules for proof. *They are instruments for testing hypotheses*. The statements of these inductive techniques, taken together, describe the general method of controlled experiment, which is a common and indispensable tool in all of modern science.

So important is the role of hypotheses in systematic empirical investigations that the enterprise of devising and testing hypotheses may be regarded as *the* method of science—to which we turn in the next chapter.

EXERCISES

Analyze each of the following investigations, or arguments, and indicate which of the methods of causal reasoning—Mill's methods—are being used in each of them:

1. Teens who lose their virginity earlier than their peers are more likely to shoplift, destroy property, or sell drugs than their virgin counterparts, according to a recent national study of 7,000 teenagers. Those who had sex early were 20 percent more likely to engage in delinquent acts one year later compared to those whose first sexual experience occurred at the average age for their school. Those who waited longer than average to have sex had delinquency rates 50 percent lower a year later compared to average teens. Waiting appears to have a protective effect. "We're not finding that sex itself leads to delinquency; sex itself is not always a problem behavior," writes co-author and Ohio State sociologist Stacy Armour. However, "the timing of sexual initiation does matter. Kids go off on a different trajectory if they're having sex early."
—Reported in *The Journal of Youth and Adolescence*, February 2007

2. Strong evidence has been presented that a diet low in folic acid [a trace vitamin in the B complex] during pregnancy increases the chances of giving birth to a premature baby of lower than normal birth weight. Dr. Theresa Scholl [of the University of Medicine and Dentistry of New

Jersey] studied the outcomes of pregnancy for 832 women from the inner city of Camden, N.J., to determine the influence of dietary and supplementary consumption of folic acid. "We found that the women who consumed less than 240 micrograms per day of folic acid had about a two- to threefold greater risk of preterm delivery and low birth weight," she said. She reported that even small increases in the women's serum folic acid concentrations by the 28th week decreased the odds of preterm delivery as well as the chance of having a baby of low birth weight. Of the 219 women in the low-folic-acid category (receiving less than 240 micrograms a day), 44 had preterm, low-birth-weight infants. "The risks declined in direct relationship to increased serum levels of folic acid, showing that low intake is a risk factor throughout pregnancy," Dr. Scholl concluded.

—T. O. Scholl, et al., "Dietary and Serum Folate: Their Influence on the Outcome of Pregnancy," *American Journal of Clinical Nutrition*, April 1996

3. The sequence of DNA units in the genome of humans and in that of chimpanzees is 98.8 percent identical; humans and chimps shared a joint ancestor as recently as five million years ago. Relatively few genes, therefore, must define the essence of humanity, and biologists have long supposed that if they could identify genes that have *changed* in the evolutionary advance leading from that joint ancestor, they would better understand the genetic basis of how people differ from chimpanzees, and hence what makes humans human.

   This project received a significant boost in 2001 when a large London family with barely intelligible speech was found to have mutations in a gene called FOXP2. Chimpanzees also have an FOXP2 gene, but theirs is significantly different from ours. The human version shows signs of accelerated evolutionary change in the last 100,000 years, which suggests that the gene acquired a new function that helped to make human speech possible.

   —Reported by Dr. Michelle Cargill of Celera Diagnostics, Alameda, CA, and Dr. Andrew Clark, of Cornell, in *Science*, 11 December 2003

4. A simple, inexpensive and surprisingly powerful combination of treatments that all but wiped out malaria in a group of HIV-positive children in a recent study in Uganda was described at a very recent medical conference in Los Angeles. The combination—taking one inexpensive antibiotic pill each day and sleeping under an insecticide-treated mosquito net—*reduced the incidence of malaria by 97 percent* compared with a control group. The study, conducted by Dr. Anne Gasasira of Makerere University in Kampala, Uganda, found that among 561 healthy children who were not HIV-infected and who did not take the antibiotic or sleep under bed nets, there were 356 episodes of malaria. This compared with 4 episodes among 300 children who were known to be HIV-infected and received both treatments. "The findings were shockingly dramatic,"

said Dr. Elaine Abrams, a professor of pediatrics and epidemiology at Columbia University.

—Reported at the *14th Conference on Retroviruses and Opportunistic Infections*, Los Angeles, 28 February 2007

5. Some theories arise from anecdotal evidence that is difficult to confirm. In *The Left-Hander Syndrome* (New York: Bantam Books, 1992), Stanley Coren sought to evaluate the common belief that left-handed persons die sooner than right-handers. But death certificates and other public records very rarely mention the hand preferred by the deceased. What could serve as a reliable data source with which that hypothesis could be tested? Coren searched baseball records, noting which hand baseball pitchers threw with, and then recording their ages at death. Right-handed pitchers, he found, lived on average nine months longer than lefties. Then, in a follow-up study, he and a colleague telephoned the relatives of people named on death certificates in two California counties, to ask which hand the deceased favored. Right-handed people (that study found) lived an average of nine years longer than lefties.

6. It has long been recognized that taller adults hold jobs of higher status and, on average, earn more than other workers. A large number of hypotheses have been put forward to explain the association between height and earnings. In developed countries, researchers have emphasized factors such as self esteem, social dominance, and discrimination. In this paper, we offer a simpler explanation: On average, taller people earn more because they are smarter. As early as age 3—before schooling has had a chance to play a role—and throughout childhood, taller children perform significantly better on cognitive tests. The correlation between height in childhood and adulthood is approximately 0.7 for both men and women, so that tall children are much more likely to become tall adults. As adults, taller individuals are more likely to select into higher-paying occupations that require more advanced verbal and numerical skills and greater intelligence, for which they earn handsome returns. Using four data sets from the US and the UK, we find that the height premium in adult earnings can be explained by childhood scores on cognitive tests. Furthermore, we show that taller adults select into occupations that have higher cognitive skill requirements and lower physical skill demands.

—Anne Case and Christina Paxson, "Stature and Status: Height, Ability, and Labor Market Outcomes," *National Bureau of Economic Research*, Working Paper No. 12466, August 2006

7. Does the position of the arm, when blood pressure is being checked, make any difference? Researchers at the University of California at San Diego, using automated cuffs, took six readings from one hundred emergency room patients whose problems did not involve their circulatory systems. Their blood pressure was measured standing, sitting, and

lying down; in each position it was measured with the arm straight out from the body and with the arm held at the side. They found that the position of the arm had a bigger effect on the readings than the position of the body. When the arm was parallel to the body, readings were higher by as much as 14 millimeters of mercury. Dr. David A. Guss, one of the authors of the study, said that no single position was more accurate, "the most important thing is to use a consistent position from measurement to measurement."

—From the *Annals of Internal Medicine*, reported in *The New York Times*, 6 January 2004

8. Near the end of the Middle Ages, a few theologians (the "scientists" of that time) persuaded a king of France to give them permission for an experiment that had been forbidden by the Roman Catholic Church. They were allowed to weigh the soul of a criminal by measuring him both before and after his hanging. As usually happens with academics, they came up with a definite result: the soul weighed about an ounce and a half.

—John Lukacs, "Atom Smasher Is Super Nonsense," *The New York Times*, 17 June 1993

9. Undoubtedly the outstanding point of departure of industrial social psychology was the series of studies performed in the Hawthorne plant of the Western Electric Company, starting in 1927. These were conducted by three Harvard professors, Elton Mayo, F. J. Roethlisberger, and T. N. Whitehead, and by W. J. Dickson of Western Electric. The original aim of the studies was to obtain concrete data on the effects of illumination, temperature, rest periods, hours of work, wage rate, etc., upon production. A group of six girls, average workers, were chosen for the experiment; their task was the assembly of telephone relays. Almost from the beginning, unexpected results appeared: The production rate kept going up whether rest periods and hours were increased or decreased! In each experimental period, whatever its conditions, output was higher than in the preceding one. The answer seemed to lie in a number of subtle social factors.

. . . As Homans summarizes it, the increase in the girls' output rate "could not be related to any change in their conditions of work, whether experimentally induced or not. It could, however, be related to what can only be spoken of as the development of an organized social group in a peculiar and effective relation with its supervisors."

—S. Stansfeld Sargent and Robert C. Williamson, *Social Psychology*, 1966

10. Does noise have an adverse effect on those subjected involuntarily to it? When the airport at Munich, Germany, moved, researchers from the University of Hamburg, the University of Gavle in Sweden, and Cornell University took that rare opportunity to conduct a prospective study on the effects of noise, measuring the performance of students near the old airport and near the new one, before and after the move. The reading skills of students in both groups were tested, along with short-term and

long-term memory, as reported in the journal *Psychological Science*, in October 2002. After the move, improvements in memory and reading were found among students near the old airport, while among students living near the new airport, reading skills and memory performance declined.

High levels of noise do interfere with learning and development, those researchers concluded—but the brighter side of their findings was this: Most of the learning damage done by noise appeared to reverse itself when the noise was removed.

11. The mood changes that many people experience during the shorter days of winter have a physiological basis in the brain, according to a study reported in the British medical journal, *The Lancet*, in January 2003. One hundred healthy volunteers, ages 18 to 79, allowed researchers to draw blood samples, at different times of the year, from their jugular veins, to get blood as close to the brain as possible. The researchers then correlated levels of brain chemicals, especially serotonin, with the weather data—temperature, air pressure, rainfall, and sunlight—at the times of blood collection. Only sunlight had causal impact; serotonin levels were found to be lowest in the three months of winter, but varied depending on the brightness of the day. "Our findings [the researchers wrote] are further evidence for the notion that changes in release of serotonin by the brain underlie mood seasonality and seasonal affective disorder."

12. Prof. Norbert Schwartz, of the University of Michigan, conducted the following experiment. He tested the attitudes of people who had just used a University of Michigan copying machine in which, for some subjects, he had planted a dime (which they found), while for others there was no windfall dime. After using the copier, subjects were asked how happy they were about life. Those who had found a dime were consistently more upbeat about "their lives as a whole," and about the economy and many other matters. "We found," said Prof. Schwartz, "that a dime can make you happy for about twenty minutes. Then the mood wears off."

—N. Schwartz, *Well Being: Foundations of Hedonic Psychology*
(New York: Russell Sage Foundation, 1999)

13. The largest and longest-running study of American child care has found that keeping a preschooler in a day care center for a year or more increased the likelihood that the child would become disruptive in class—and that this effect persisted through the sixth grade. Every year spent in such centers for at least 10 hours per week was associated with a 1 percent higher score on a standardized assessment of problem behaviors completed by teachers. Parents' guidance, and their genes, had the strongest influence on how children behaved—but this finding about the impact of day care centers held up regardless of the child's sex, or family income, and regardless of the quality of the day care center.

—National Institute of Child Health and Human Development,
"Early Child Care and Youth Development," 26 March 2007

14. Speed kills. A report from the Insurance Institute for Highway Safety, issued in November of 2003, concluded that increased speed limits on Interstate highways led to nearly 1,900 additional deaths in 22 states from 1996 to 1999. The report is based, oddly, on a study by the Transport Safety Authority of New Zealand, working in the United States, which showed that, when the Federal cap on speed limits was placed at 65 mph, the number of deaths on U.S. highways decreased. But almost immediately after the repeal of that Federal cap on speed limits the number of deaths in the states that did not retain the 65 mph limit increased markedly, while the number of deaths in those states that retained the 65 mph limit did not increase. Drivers in states with higher speed limits, the study showed, drive faster, and where the driving is faster the number of traffic fatalities goes up.

—"Study Links Higher Speed Limits to Deaths,"
The New York Times, 24 November 2003

15. A 16-year study followed 8,867 non-smoking male professionals with normal body weight who participated in vigorous daily exercise and ate a healthy diet. Those who drank one-half to two normal servings of wine, beer, or hard liquor a day had a 41 to 62 percent reduction of heart attack risk compared with those who drank no alcohol at all. It seems clear that in moderate quantities alcoholic drinks reduce the likelihood of heart attack. This effect is found not only in those with heart disease. The lead author of the study writes: "Even in the lowest risk people, we still find a lower risk associated with moderate drinking."

—Kenneth Mukamal, "Alcohol Consumption and Risk for Coronary Heart Disease in Men with Healthy Lifestyles," *Archives of Internal Medicine*, 23 October 2006

16. For heart patients, "noetic" intervention—such as prayer and MIT (therapy relying on music, imagery, and touch)—is defined as "an intangible healing influence brought about without the use of a drug, device, or surgical procedure." 748 patients with coronary heart disease who were to undergo percutaneous coronary intervention (a type of stenting procedure), or elective cardiac catheterization, were enrolled at one of nine study sites between 1999 and 2002. To test the efficacy of noetic intervention, patients were randomized into four groups: one group (189 patients) received both offsite intercessory prayer and MIT therapy; a second group (182 patients) received intercessory prayer only; a third group (185 patients) received MIT therapy only; the fourth group (192 patients) received neither the intercessory prayer nor the MIT therapy. The interventional heart procedures were conducted according to each institution's standard practices, with a six-month period of follow-up. The prayer portion was double blinded, meaning that the patients and their care team did not know which patients were receiving intercessory prayer. The prayer groups for the study were located throughout the world and included Buddhist, Muslim, Jewish and many Christian denominations. 89 percent of the patients in this study also knew of someone praying for them outside of the study protocol.

As reported by the Duke University Medical Center, the researchers found no significant difference among the four treatment groups. Distant prayer and the bedside use of music, imagery and touch did not have a significant effect upon the primary clinical outcome of these patients undergoing medical interventions.
—"First Multicenter Trial of Intercessory Prayer," *The Lancet*, 16 July 2005

17. The impulse to share does not come naturally to one who is thinking about money. Psychologists found that subconscious reminders of money prompted people to become more independent in their work, and less likely to seek help from others or to provide it. In one experiment 52 undergraduates unscrambled sets of jumbled phrases; one group untangled phrases that were often about money, like "high salary paying," while another solved word puzzles that did not refer to money. Researchers then had the students work on a difficult abstract puzzle and offered to give help if they wanted it. Those who had been thinking about money worked on the problem by themselves an average of more than 70 percent longer than the others. Students "primed" to have money on their minds, while clearly self-reliant, were less likely than peers who had not been so primed to lend assistance, twice as slow to help another confused student, and about twice as stingy when asked to donate money to help needy students.
—Kathleen Vohs, Nicole Mead, and Miranda Goode, "The Psychological Consequences of Money," *Science*, 17 November 2006

chapter 12 Summary

In this chapter we have examined the concept of cause, the nature of causal connections, and the methods used to establish causal laws.

In Section 12.1, we examined various meanings of "cause."

In Section 12.2, we explained the supposition of the uniformity of nature, and the generality of causal laws.

In Section 12.3, we discussed induction by simple enumeration.

In Section 12.4, we recounted and illustrated the principal techniques of inductive inference, called Mill's methods, explaining their essentially eliminative nature. These five methods are:

1. The method of agreement
2. The method of difference
3. The joint method of agreement and difference
4. The method of residues
5. The method of concomitant variation

In Section 12.5, we explained the limitations and the strengths of these inductive techniques, concluding that, although they cannot do all that John Stuart Mill had claimed for them, they are profoundly important as the intellectual instruments with which scientific hypotheses are confirmed or disconfirmed.

END NOTES

[1]David Hume, *An Enquiry Concerning Human Understanding* (1748), sec. IV.

[2]Dao, "Coroner in Cincinnati Rules Man's Struggle Led to Death," *The New York Times*, 4 December 2003.

[3]Reported by Connor O'Shea of the Duke University Medical Center, at the meetings of the European Society of Cardiology in August 2000.

[4]See Zachariah Chafee, Jr., *Three Human Rights in the Constitution of 1787* (1952).

[5]Dr. Steven Grant, quoted in *Science News*, 27 January 2007.

[6]Here quoting Dr. Nora Volkow, director of the National Institute of Drug Abuse, quoted by Benedict Carey, "In Clue to Addictive Behavior, A Brain Injury Halts Smoking," *The New York Times*, 26 January 2007.

[7]Robert Sapolsky, "Testosterone Rules," *Discover*, March 1997.

[8]D. N. Cook et al., "Requirement of *MIP-1 alpha* for an Inflammatory Response to Viral Infection," *Science*, 15 September 1995.

[9]Paul Henle and William K. Frankena, *Exercises in Elementary Logic* (1940).

[10]A. Werzberger et al., "A Controlled Trial of a Formalin-Inactivated Hepatitis A Vaccine in Healthy Children," *The New England Journal of Medicine*, 13 August 1992.

[11]Edward Arthur Fath, *The Elements of Astronomy* (New York: McGraw-Hill, 1926), p. 170.

[12]D. S. Siscovick et al., "Dietary Intake and Cell Membrane Levels of Long-Chain *n*-3 Polyunsaturated Fatty Acids and the Risk of Primary Cardiac Arrest," *Journal of the American Medical Association*, 1 November 1995.

[13]The source of these data is the U.S. Census Bureau; the analysts are Gordon B. Dahl of the University of Rochester and Enrico Moretti of the University of California at Los Angeles, reporting online in *Slate*, in October 2003.

[14]J. L. Casti, *Searching for Certainty* (New York: William Morrow, 1991).

[15]See Sherwin B. Nuland, *The Doctors' Plague* (New York: W. W. Norton, 2003).

For additional exercises and tutorials about concepts covered in this chapter, log in to MyLogicLab at *www.mylogiclab.com* and select your current textbook.

Science and Hypothesis

13.1 Scientific Explanation

13.2 Scientific Inquiry: Hypothesis and Confirmation

13.3 Evaluating Scientific Explanations

13.4 Classification as Hypothesis

13.1 Scientific Explanation

To learn the truth about the world, the world must be studied scientifically. However, individual truths do not take us very far; a mere collection of facts no more constitutes a science than a collection of stones constitutes a house. The aim of science is to discover general truths (chiefly in the form of causal connections like those discussed in the preceding chapter) with which the facts we encounter can be *explained*.

What is an explanation? Every explanation gives an account, a set of statements from which the thing to be explained can be logically inferred. The best account will be the one that most reduces the problematic aspects of what was to be explained. Such an account will comprise a coherent set of general truths, or a theory. To explain some serious disease, for example, we need a coherent account of what causes that disease and how it can be treated. Is the presence or absence of some particular substance the key to the disorder? The theory explaining diabetes, for example, is a coherent account of the use of sugars by the human body and the central role, in that use, of a protein hormone called *insulin*, produced by certain special cells within the body. According to this theory, it is a deficiency of insulin (or the inability of the body to use the insulin it produces) that explains the resulting disorder in the absorption of sugars from the blood. An account of this kind (here greatly oversimplified, of course) gives a **scientific explanation** of this serious disease. Patients suffer from diabetes *because* of their insulin deficiency.

When we say "*Q* because *P*," that may express either an explanation or an argument. It expresses an argument when we are inferring the conclusion, *Q*, from the premises, *P*. It expresses an explanation when, facing the *fact* of *Q*, our reasoning moves back from that fact to discover the circumstances that led to it. Diabetes—excess sugar in the blood—is a cruel fact in the lives of many patients. We *explain* their diabetes by calling attention to the insulin deficiency that has that result. The account of the interrelated set of circumstances in which the insulin deficiency, *P*, accounts for the sugar excess, *Q*, is thus an *explanation* of that disease.

A good explanation must offer truths that are *relevant* to the fact explained. If I seek to explain my being late to work on some occasion by calling attention to the rising birth rate in Brazil, the fact thus introduced may be correct, but it is not relevant, and it therefore cannot be a satisfactory explanation of my ab-

chapter 13

Scientific explanation
A theoretical account of some fact or event, predicated upon empirical evidence and subject to revision in the light of new information.

513

sence, the event in question. In this trivial example, an explanation is sought for a single event. In science we seek explanations that are not only true and relevant, but also general. The explanations we aim for will provide an understanding of all the events of some given *kind*—say, all the occurrences of diabetes, for example.

The more facts for which a scientific theory accounts, the more powerful it is. Some theories are magnificent in their range and power. Here, for example, is a short statement of Isaac Newton's law of universal gravitation:

> Every particle of matter in the universe attracts every other particle with a force that is directly proportional to the product of the masses of the particles and inversely proportional to the square of the distance between them.

An explanation may be relevant and general, and yet not scientific. The regular motions of the planets were long thought to be accounted for by the "intelligence" that was held to reside in each planet. In some cultures, disease is "explained" as the work of an evil spirit that has invaded the body. These are certainly unscientific accounts, although the explanations they offer are general and are relevant to the facts of interest. What, then, distinguishes genuinely scientific from unscientific explanations?

There are two chief differences. The first is *attitude*. An **unscientific explanation** is presented dogmatically; the account that it gives is regarded as being unquestionably true and not improvable. The opinions of Aristotle were accepted for centuries as the ultimate authority on matters of fact. Aristotle himself appears to have been open-minded, but his views were adopted by some medieval scholars in a rigid and unscientific spirit. One of the scholars to whom Galileo offered his telescope to view the newly discovered moons of Jupiter declined to look, expressing his certainty that no real moons could possibly be seen because no mention of them could be found in Aristotle's treatise on astronomy! In contrast, the attitude of a serious scientist is undogmatic; explanations are put forward provisionally; hypotheses may be thought highly probable, but they are regarded as subject to alteration in the light of the evidence.

The vocabulary of science is sometimes misleading on this point. When what is first suggested as a "hypothesis" is well confirmed, its status may be elevated to that of a "theory"; after universal acceptance, it may be further elevated to that of a "law." However, the use of these terms is not consistent. Newton's discovery is still called the "law of gravitation," while Einstein's contribution, which improved and superseded it, is referred to as the "theory of relativity." Whatever the terms used, the attitude of genuine scientists is not dogmatic. The general propositions of science are all in essence hypotheses, never absolutely certain.

In everyday speech the word *theory* is often used to refer to a hunch, or a mere opinion. Scientists use the word differently. In physics and chemistry we refer—not dogmatically, but nevertheless with great confidence—to "quantum theory" and to the "molecular theory of matter"; in biology we rightly rely upon the "cellular theory" and the "germ theory of disease." These are sets of very well-established truths, not ungrounded speculations.

Unscientific explanation An explanation that is asserted dogmatically and regarded as unquestionable.

Evolution—the "theory of evolution"—is also an established fact; doubts about evolution expressed because it is "only a theory" are the result of this semantic misunderstanding.

The second difference concerns the *basis* for accepting the account in question. In science a hypothesis is worthy of acceptance only to the extent that there is good evidence for it. An unscientific belief may be held independently of what we should regard as evidence in its favor; the explanation is taken as simply true—perhaps because "everyone knows" that it is so, or perhaps because it is thought to have been revealed from on high. There is no reliable test of such claims, whereas in genuine science the claims for truth can be tested, and those tests lie in our experience. Thus we say that genuine science is *empirical*.

To say that a hypothesis is *testable* is at least to say that some prediction made on the basis of that hypothesis may confirm or disconfirm it. Science demands evidence. But, of course, the evidence accumulated that could confirm the hypothesis in question can never be complete, as we have earlier emphasized; *all* the evidence is never in hand. Therefore, even when that supporting evidence is very strong, some doubt must remain, and certainty is unattainable. On the negative side, however, if the evidence shows indisputably that the predictions made on the basis of that hypothesis are false, our confidence that the hypothesis must be rejected may be total. Although we cannot complete the verification of a hypothesis, we can, with closure, establish that it has been falsified. For reasons of this kind, some philosophers have held that to say of a scientific hypothesis that it is testable is also to say that it is, at least in principle, falsifiable.

The test of truth may be direct or indirect. To determine whether it is raining outside, I need only glance out the window. In general, however, the propositions offered as explanatory hypotheses are not directly testable. If my lateness at work had been explained by my claim about some traffic accident, my employer, if suspicious, might test that explanation indirectly by seeking the police accident report. An indirect test deduces, from the proposition to be tested (for example, that I was involved in an accident), some other proposition (for example, that an accident report had been submitted) capable of being tested directly. If that deduced proposition is false, the explanation that implied it is very likely to be false. If the deduced proposition is true, that provides some evidence (but not conclusive evidence) that the explanation is true, having been indirectly confirmed.

Indirect testing is never certain. It always relies on some additional premises, such as the premise that accidents of the sort I described to my employer are invariably reported to the police. But the accident report that should have been submitted in my case may not have been, so its absence does not *prove* my explanation false. Even the truth of some added premises does not render my explanation *certain*—although the successful testing of the conclusion deduced (the reality of the accident report, in this example) does corroborate the premises from which it was deduced.

Even an unscientific explanation has *some* evidence in its favor, namely, the very fact it is held to explain. The unscientific theory that the planets are inhabited by "intelligences" that cause them to move in their observed orbits can claim, as evidence, the fact that the planets do move in those orbits. However, the great difference between that hypothesis and the reliable astronomical explanation of planetary movement lies in this: For the unscientific hypothesis there is no other directly testable proposition that can be deduced from it. Any scientific explanation of a given phenomenon, on the other hand, will have directly testable propositions deducible from it *other than the proposition stating the fact to be explained*. This is what we mean when we say that an explanation is *empirically verifiable*, and such verifiability is the most essential mark of a scientific explanation.*

13.2 Scientific Inquiry: Hypothesis and Confirmation

We seek scientific explanations that are correct, and whose correctness may be empirically verified. How can we obtain these explanations? No formulas for doing science can be given, but there are stages, or distinct phases, in most scientific investigations. By identifying and describing seven such stages we may come to understand more fully how good science advances.

A. Identifying the Problem

Scientific investigation begins with a problem of some kind. By *problem* is meant only some fact, or group of facts, for which no acceptable explanation is at that time available. The sociologist confronts a puzzling trend in work or play; what accounts for it? The medical investigator confronts a puzzling disease; what causes it? The economist observes different patterns of spending or saving; what explains the variations? Some problems are quite sharply identified, as when a detective confronts a specific crime and asks: Who is the perpetrator? Some problems may arise from a gap in current understanding. Eratosthenes, librarian at Alexandria in the third century BCE, believed correctly that the Earth was a sphere, but its size was unknown. His problem was to determine the circumference of the sphere we call Earth. Reflective thinking—whether in sociology or medicine or law enforcement or physics, or any other realm—is *problem-solving* activity, as John Dewey and other modern philosophers have repeatedly emphasized. The recognition of some problem is the trigger for the science that ensues.

B. Devising Preliminary Hypotheses

Preliminary speculation is the second step—some very tentative explanation of the problem identified. Long before a full solution is in sight, some theorizing is needed to indicate the kind of evidence needed, and perhaps to indicate where such evi-

---

*This general conception of "scientific explanation" rightly applies outside the realm of what is normally thought of as the sciences, such as physics or psychology. Thus, the explanation of an event such as my lateness to work as a consequence of a traffic accident, in being indirectly testable in various ways, is in this wide sense "scientific."

dence might best be sought. The detective examines the scene of the crime, interviews suspects, seeks clues. The physician examines the patient, records data, notes irregularities. Bare facts are accumulated; they become usable clues or revealing symptoms only when they are fitted into some coherent pattern, even if that pattern is speculative and incomplete. To illustrate: Thomas Malthus had shown, in "An Essay on the Principle of Population" (1798), that the tendency of population to grow faster than the food supply keeps most people at the edge of starvation. Charles Darwin, reading this while speculating about the origin of species many years later, hit upon an exceedingly fruitful notion. He wrote:

> It at once struck me that under these circumstances favourable variations would tend to be preserved, and unfavourable ones to be destroyed. . . . Here then I had at last got a theory by which to work" (*Autobiography*, 1881).

There are too many possibly relevant facts—too much data in the world—for the scientist to collect them all. The most thorough investigator must select some facts for further study and put other facts aside as not relevant. If the Earth is a sphere, the rays of the sun will fall (at any given time) upon different points of that sphere at different angles. Might geometry help us to calculate the size of the Earth? The outline of a theory is essential because, without that, the investigator cannot decide which facts to select and pursue from the totality of facts. However incomplete or tentative, a preliminary hypothesis of some kind is needed before serious inquiry can get under way.

C. Collecting Additional Facts

The preliminary hypothesis serves to guide the search for relevant facts. As a preliminary matter, the patient is thought to have some infection, and that hypothesis puts the physician on the trail of certain kinds of data that are normally associated with infection: temperature irregularities, patterns of inflammation, and the like. The preliminary supposition that the crime was committed by a member of the household will cause the detective to inquire into the conduct of persons residing there, and so on. If the angle at which the sun's rays strike the Earth must differ at different points on the Earth's surface, one must seek, in order to apply geometric principles, at least one point at which the sun is known to be *directly* overhead at a given time. Where might that be?

The second and third steps are not fully separable, of course; in real life they are interconnected and mutually suggestive. New facts found may cause an adjustment of the preliminary hypothesis; that adjustment may lead to facts earlier not noted. The process of gathering evidence by using the preliminary hypothesis merges with the process of refining that hypothesis, leading to new findings, and so on and on.

D. Formulating the Explanatory Hypothesis

Eventually, the investigator—the scientist, or detective, or ordinary person—may come to believe that all the facts needed for solving the original problem are in hand. The task then becomes that of assembling the pieces of the puzzle in a way that makes sense of the whole. If that synthesis is successful, a hypothesis will emerge that accounts for all the data—the original set of facts that gave rise to the problem, as well as the additional facts to which earlier hypotheses had pointed. A surge in unemployment is explained by some larger theory of the labor market. The patient is found to be suffering from an identifiable infectious agent known to cause the symptoms noted in this patient's condition. An identifiable member of the household is charged by the state as the perpetrator of the crime and the case against him is formulated.

There is no mechanical way to find some overarching theory. The actual discovery, or invention, of a successful explanatory hypothesis is a process of creation, in which imagination as well as knowledge is involved. That is why those who make important scientific discoveries are so widely honored and so much admired.

What *is* the circumference of the globe? Eratosthenes learned that in the Egyptian town of Syene (now called Aswan), the sun's rays shine directly down a deep well at a given time on a particular day each year. At that same time he could measure the sun's shadows (and therefore the angle of its rays) in Alexandria; he found that the rays there deviated from the vertical by 7°. That is about one-fiftieth of the 360° of the sphere's circumference. The distance between Syene and Alexandria was known. The circumference of the entire sphere of Earth must therefore be about fifty times that distance. Eratosthenes' subsequent calculation of the Earth's circumference ("250,000 *stadia*") is believed (we are unsure of the length of a *stadium*) to have an error of less than 5 percent. He had no way to confirm that calculation, but it was impressive science for his time. Truly great scientists—such as Einstein or Newton—are understandably viewed as creative geniuses.

E. Deducing Further Consequences

A good explanatory hypothesis will be fruitful; that is, it will explain not only the facts that provoked the inquiry but many other facts as well. It is likely to suggest some facts that had not even been thought of earlier. Verification of these additional facts may strongly confirm (but, of course, cannot prove with certainty) the hypothesis that led to them.

To illustrate, the cosmological theory known as the *Big Bang* hypothesizes that the present universe began with one singular explosive event. The initial fireball would have been smooth and homogenous, lacking structure. But the universe today exhibits a great deal of structure; its visible matter is clumped into galaxies, clusters of galaxies, and so on. If the Big Bang theory is correct, the seeds of the present structure of the universe must in principle be identifiable. We need to be able to look back in time—and by observing the most distant objects in an expanding universe, astronomers actually can, in effect, look back in time, since the light being received must have left its sources billions of years ago. If, in these observations, early structures were not detectable by the most sensitive instruments, the Big Bang theory would be seriously undermined. But if such structure were detectable, the Big Bang theory would be significantly confirmed.

F. Testing the Consequences

Critical for the evaluation of every explanatory hypothesis is the accuracy of its predictions. Can the facts to which the theory points be ascertained? Often they can. If there was structure in the universe early in its expansion, as the Big Bang theory predicts, there would have to be irregularities, unevenness, that may be found in background radiation and traced to that early time. Happily, it is possible to measure that background radiation and thus to determine now, indirectly, that there were structural irregularities very shortly after the supposed Big Bang. To detect those predicted radiation irregularities, a special satellite was designed—the Cosmic Background Explorer (COBE). Using this satellite, the predicted irregularities have indeed been detected, giving very important confirmatory evidence for the truth of the Big Bang hypothesis.

Consider prediction in another context. In biology we may come to formulate the hypothesis that a particular protein is produced in mammals as a reaction to a particular enzyme, and that that enzyme is produced under the direction of a specifically identified gene. From this hypothesis we may deduce the further consequence that when that gene is absent there will be an absence, or a deficiency, of the protein in question.

To test that hypothesis we construct an experiment in which the effect of the identified gene may be measured. This can sometimes be done by breeding mice in which the critical gene has been deleted—"knockout mice." If, in such mice, the enzyme in question and the protein associated with it are indeed also absent, our hypothesis will have been strongly confirmed.[*] Much information that proves very valuable in medicine is acquired in just this way. We devise the experiment to determine whether what we thought to be true (if such-and-such were the case) really is true. To do that we must often *construct* the very special circumstances in which such-and-such has been made the case. "An experiment," as the great physicist Max Planck said, "is a question that science poses to Nature; a measurement is the recording of Nature's answer."

It is not always feasible to construct the circumstances needed to perform a test. We must then seek the circumstances needed for testing in some natural setting. That was the case in the effort to test the general theory of relativity.[†] Einstein's theory proposed that gravitation is not a force (as Newton had thought) but a curved field in the space–time continuum, created by the presence of mass. This might be proved (or disproved), Einstein suggested, by measuring the de-

---

[*]Testing of this kind relies on what we called the method of difference in Chapter 12. The many methods discussed there (Mill's methods) are intellectual tools used to confirm (or disconfirm) hypotheses.
[†]"The Foundation of the General Theory of Relativity" was published in 1916, in *Annalen der Physik*.

flection of starlight as it traveled close by the mass of the sun; the starlight needed would be visible only during a total eclipse of the sun. The testing of this prediction had to await the solar eclipse of 1919, when the sun would be silhouetted against the Hyades star cluster, for which the positions were known exactly. During that eclipse, physicist Sir Arthur Eddington stationed himself on an island off the western coast of Africa; another group of British scientists went to Brazil. The two teams measured accurately the apparent position of several of the stars in the cluster; their measurements plainly showed that light from these stars was indeed bent as it grazed the sun, and that it was bent by the exact amount of Einstein's prediction. The general theory of relativity had been very solidly confirmed.

This theory showed that space, time, and gravity are so entwined that to speak sensibly about one there must be reference to the others. Einstein struggled to go further, to develop a theory in which all of nature's forces are merged into one single, overarching theory. In this effort he did not succeed, and neither has anyone else so far.

A new approach—to construct a complete and unified theory of natural forces called *string theory*—now has many adherents. It offers a theoretical account that may unify gravity, quantum mechanics, and nature's other forces, and solve some earlier mathematical problems as well. String theory, which is based on a new conception of matter's fundamental constituents, is free of mathematical contradictions but it has not yet been confirmed.

What predictions does string theory make that might confirm it by experimental test? It may become possible to confirm the theory's predictions regarding new kinds of particles; it may become possible to test the prediction that highly energetic particle collisions will produce microscopic black holes. As this is being written, a gigantic particle accelerator, the Large Hadron Collider, is performing its first set of high-energy collisions. In a few years we may have empirical evidence to confirm the explanations given by string theory, or to disconfirm them.[*]

Evolutionary theory, as presented by Darwin in *The Origin of Species* (1859) and by very many of his successors, is now almost universally accepted as a correct explanation of the development of species of animals and plants. Predictions that can test this theory prospectively (rather than retrospectively) are difficult to devise because the natural selection hypothesized seems to require the passage of many generations. Very recently, a Harvard professor of evolutionary biology, Jonathan Losos, devised an experiment that makes speedy testing feasible. On some tiny cays in the Bahamas, where the brown lizard *Anolis sagrei* lives free of predators and reproduces rapidly, he introduced a predator whose activity would quickly result in the development (in those islands as compared to other, similar islands left unperturbed) of a lizard population with longer legs, much

---

[*]Some scientists contend that string theory makes no predictions whose testing will truly confirm or disconfirm it. See L. Smolin, *The Trouble with Physics* (Boston: Houghton Mifflin, 2006); and P. Woit, *Not Even Wrong: The Failure of String Theory* (New York: Basic Books, 2006). It may be that the theory makes predictions that are testable in principle, but not testable in practice given current technological limitations. This heated controversy is likely to continue.

better suited to running away. Selective forces operated as expected; long legs came to predominate. However, when continually preyed upon, the *Anolis* lizard climbs into trees and bushes, where short legs are much more advantageous. The further prediction was that natural selection would produce a reversal and that short legs would eventually predominate—and six months later that prediction also was confirmed. Evolution has been first manipulated and then deliberately reversed. Said Prof. Losos:

> Evolutionary biology is often caricatured as incompatible with controlled experimentation. Recent work has shown, however, that evolutionary biology can be studied on short time scales and that predictions about it can be tested experimentally. We predicted, and then demonstrated, a reversal in the direction of natural selection acting on limb length in a population of lizards. We did a controlled, replicable experiment in nature. It illustrates that evolutionary biology at its heart is no different from any other science.[1]

G. Applying the Theory

When a phenomenon is encountered, one goal is to explain it; however, people also strive to *control* those phenomena to their advantage. Not only do the theories of Newton and Einstein and their successors play a central role in our understanding of celestial phenomena, but they are also critical in our actual exploration of the solar system, and outer space beyond. Nuclear fusion is now well understood as a process; we seek to apply this understanding in producing energy on a scale we can control. Disease and disorder are understood as never before, incorporating the well-tested explanations of genetic theory and our grasp of the human genome; now we seek to put this understanding to use in clinical medicine by eliminating genetic disorders, and even by regenerating organic tissue. In this twenty-first century it is probably biological science, more than any other field, whose explanations will enhance the quality and the length of ordinary human lives.

Good practice in every sphere must be guided by good theory. Good theory must pass the test of empirical verification. Theory and practice are not two realms; they are equally critical aspects of every genuinely scientific undertaking. More than two centuries ago, Immanuel Kant wrote an incisive little book explaining why it makes no good sense to say, "that may be right in theory but it won't work in practice."[2] What is right in theory *does* work in practice, and for everything that does work in practice we may reasonably hope to discover the explanatory theory that underlies its success.

13.3 Evaluating Scientific Explanations

The same phenomenon may receive different explanations, all scientific in the sense we have described, and yet some of them may not be true. Conflicting explanations of some physical or economic phenomenon may be offered. In a criminal investigation we may hypothesize that the perpetrator was X, or was Y.

More than one hypothesis may account for the facts neatly, but not all can be true. How shall we choose among alternative scientific explanations?

Let us assume that all the alternatives are relevant and testable. How ought we determine which of the available hypotheses is the best? There are standards—going beyond relevance and testability—to which acceptable hypotheses may be expected to conform. Three criteria are most commonly used in judging the merit of competing hypotheses.

1. *Compatibility* **with previously well-established hypotheses.** Science aims at achieving a *system* of explanatory hypotheses. A satisfactory system must be internally consistent, of course. A satisfactory explanatory system cannot contain contradictory elements; if it did, the full set of propositions could not possibly be true. We progress by gradually expanding hypotheses to comprehend more and more facts, but each new hypothesis brought into the set must be compatible with those already confirmed.

    Sometimes the expansion involves only one new hypothesis, as when the aberrations in the orbit of Uranus were explained by the hypothesis that there was some other planet, uncharted at that time, whose mass was creating the aberrations. That supposition was perfectly consistent with the main body of astronomical theory at the time. A search for the mysterious object resulted in the discovery of the planet Neptune in 1846. The theory that led to that discovery *fit* very nicely with all the other theories concerning planetary movements generally accepted at that time.

    Although theoretical knowledge grows gradually, it does not always grow by adding just one new hypothesis after another in orderly fashion. Clumps of theory may be introduced; new hypotheses that are flatly inconsistent with older theories sometimes replace their predecessors outright, rather than being fitted in with them. Einstein's theory of relativity was of that sort: It shattered many of the preconceptions of the older, Newtonian theory of gravitation. In another branch of physics, it was discovered that radium atoms undergo spontaneous disintegration, and this well-confirmed fact was simply inconsistent with an older principle that matter could neither be created nor destroyed. To maintain a consistent set of hypotheses, the older principle had to be relinquished.

    The consistency of the set of scientific theories in a given field is thus achieved in different ways. However, apart from those cases in which some revolutionary theory upsets long-established principles, the first criterion for an acceptable new hypothesis is that it retain the existing consistency, be compatible with what is already known, or be reasonably believed.

    When old and new collide, the established scientific theories will not be abandoned quickly in favor of some that are shinier or more trendy. The older body of theory will be adjusted to accommodate the new if that is possible. Large-scale change will be resisted. Einstein himself always insisted that his own work was a modification of Newton's, not a rejection of it. The principle of the conservation of matter was modified by being absorbed into the more comprehensive principle of the conservation of mass–energy. An established theory has the support that it does because it explains a considerable mass of data, so it cannot be dethroned by some new hypothesis unless the new hypothesis accounts for the same facts as well as (or better than) the older one, and accounts for other known facts also.

Science advances as its theories give more comprehensive explanations, more adequate accounts of the world we encounter. When inconsistencies arise, the greater age of one hypothesis does not automatically prove it correct. If the older view has been extensively confirmed, presumption will support it. When the newer, competing view has also received extensive confirmation, mere age and priority cease to be relevant. We must then decide between the competitors on the basis of something we learn about the observable facts. The ultimate court of appeal is always experience.

2. *Predictive power.* As we have seen, every scientific hypothesis must be testable, and testability requires that some observable fact or facts be deducible from it. Alternative hypotheses will differ in the nature and extent of their predictions, and we seek the theoretical explanation that has the greater predictive power.

To illustrate: The behavior of bodies near the surface of the Earth was explained by Galileo Galilei (1564–1642) with his laws of falling bodies. The behavior of bodies far off in the solar system was explained at about that same time by the German astronomer Johannes Kepler (1571–1630), who formulated the laws of planetary motion. Using the data that had been collected by Denmark's Tycho Brahe, Kepler could account for the motions of the planets on the basis of the elliptical orbits they travel around the sun. Galileo gave a theoretically powerful account of the various phenomena of terrestrial mechanics. Kepler gave a theoretically powerful account of celestial mechanics. But the two accounts were isolated from one another. Their unification was needed; it came with Isaac Newton's theory of universal gravitation, and his three laws of motion. All the phenomena explained by Galileo and by Kepler, and many more facts besides, were explained by Newton's account of universal gravitation.

A fact that can be deduced from a given hypothesis is said to be explained by it, and may also be said to be *predicted* by it. Newton's theories had enormous predictive power. The greater the predictive power of any hypothesis, the better it contributes to our understanding of the phenomena with which it is concerned.

Earlier we described the great predictive power of Einstein's general theory of relativity, which accounts for the admiration given to it and to its creator. We also pointed out that his enterprise—the development of an overarching theory of natural forces—is held by some to approach success now in the form of what is called string theory; some predictions of great interest are claimed to be deducible from this theory. If those predictions are one day confirmed, the predictive power of string theory will elevate it to a position of the very first importance in physics and cosmology.

However, the criterion of predictive power also has a negative side. If the hypothesis predicts what does not take place, or is in some other way shown to be inconsistent with well-attested observations, that hypothesis has been *falsified* and must be rejected. A meaningful scientific hypothesis must be at least falsifiable—that is, we must know what would or might show it to be false. If there is no set of observable outcomes that will lead us to conclude that the hypothesis is false, we may seriously doubt if the hypothesis has any predictive power whatever.

Suppose we confront two different hypotheses, both of which fully explain some set of facts, both of which are testable, and both of which are compatible with the body of already established scientific theory. In such a case, it may be possible to devise a *crucial experiment* to decide between the conflicting theories. If the first hypothesis entails that, under a given set of circumstances, a specified result will occur, and the second entails that it will not, we may decide between the competitors by observing the presence or absence of that predicted result. Its appearance falsifies the second hypothesis; its nonappearance falsifies the first.

The experiment described earlier, in which the general theory of relativity was tested by making exact measurements of the starlight that passed closely by the mass of sun, was crucial in just this way. The theory of Newton and the theory of Einstein cannot both be correct. If the bending of the light is as Einstein's theory predicted, the Newtonian view is disconfirmed; if the bending of light is not observed, the general theory of relativity is disconfirmed. With good cameras, very careful observers, and a solar eclipse in which the three bodies (sun, moon, and Earth) were correctly lined up, the crucial experiment might be made. Those ideal circumstances arose on 29 May 1919. Photographs proved that Einstein was right; we do live in a curved, four-dimensional space–time continuum. Einstein became a worldwide sensation overnight.

3. **Simplicity.** Two rival hypotheses may fit equally well with established theory, and they may also have predictive power that is roughly equal. In such circumstances we are likely to favor the simpler of the two. The conflict between the Ptolemaic (Earth-centered) and the Copernican (sun-centered) theories of celestial motion was like that. Both fit well with earlier theory, and they predicted celestial movements about equally

well. Both hypotheses relied on a clumsy (and, as we now know, mistaken) device, hypothesized epicycles (smaller circles of movement on the larger orbits), in order to explain some well-established astronomical observations. But the Copernican system relied on many fewer such epicycles and was therefore much simpler. This greater simplicity contributed substantially to its acceptance by later astronomers.

Simplicity seems to be a "natural" criterion to invoke. In ordinary life also, we are inclined to accept the simplest theory that fits all the facts. Two theories about a crime may be presented at a trial; the verdict is likely to be given—perhaps ought to be given—in favor of the hypothesis that seems simpler, more natural.

"Simplicity," however, is a tricky notion. That one of the competing theories will involve a smaller number of some troubling entity (such as the epicycles in the case of Copernican astronomy) is a rare situation. Each of two theories may be simpler than the other in different ways. One may rely on a smaller number of entities, while the other may rely on simpler mathematical equations. Even "naturalness" may prove to be deceptive. Many find it more "natural" to believe that the Earth, which does not seem to be moving, really is not moving, and that the Sun, which appears to move around us, is doing just that. The lesson here is that simplicity is a criterion that is difficult to formulate and not always easy to apply.

Progress in science is never easy and rarely straightforward. No one supposes that simply by applying the seven steps of the hypothetico-deductive method (recounted in Section 13.2) to some problem he will find its solution. Correct explanatory hypotheses are often obscure and may require very elaborate theoretical machinery. Devising a final, presumably correct theory may be exceedingly difficult. Far from being mechanical, the process commonly requires, in addition to laborious observation and measurement, insight and creative imagination.

When some hypothesis already in hand is widely believed to explain the phenomena in question, a replacement for it encounters very high hurdles. The new hypothesis is likely to encounter ridicule and disdain. The new hypothesis is very probably inconsistent with the previously accepted theory, and the established view always has the upper hand. A crucial experiment, of the sort described earlier in the case of the general theory of relativity, is possible only in rare circumstances.

Contemporary physics faces a major conflict of just this kind. Between its two most powerful general theories there is an apparent conflict that cannot presently be resolved. The general theory of relativity is well confirmed. From its laws (describing gravity and how it shapes space and time), it is an apparently inevitable consequence that some collapsing, massive stars will form "black holes" from which escape would require a speed faster than light, which is impossible. The laws of quantum mechanics are also well confirmed, and they entail that information cannot ever be permanently lost, even if drawn into a black hole. Therefore, either there is some property of space and time, not now understood, that can account for the retention of that information, or there is some lawlessness in physics that can account for the permanent loss of that information. One of the two theories must need at least an amendment, but we do not yet know which one, and we do not have the means to construct an experiment that would enable us to decide between them.*

Confronted by such conflicts we will seek to apply the criteria of good scientific explanations we set forth earlier: Which of the competing theories is *simpler*? Which of the two has greater *compatibility* with previously established hypotheses? Finally, above all, which has the greater explanatory or *predictive power*? So long as definitive answers to these questions are lacking, the intellectual controversy is likely to continue unresolved.

It does happen in the history of scientific progress that such conflicts are sometimes resolved. There is no better way to exhibit the methods of science, and to exemplify the application of the criteria described here, than by recounting the observational confirmation by Galileo of the heliocentric account of the solar system—and the resulting replacement of the geocentric account that had been accepted as true for more than a thousand years.

By the early 1600s, the movement of the planets against the backdrop of the fixed stars had been so carefully studied that their apparent movements were quite accurately predictable. The Moon, also much studied, was believed by theologians to be a perfect sphere. The heavenly bodies, deemed flawless in shape and movement, were widely believed to travel in perfect circles around the Earth, which was the center of the world God had created. By 1609, Galileo had devised a telescope with 20-power magnification, its chief uses being thought at first to be maritime, or as a spyglass that could provide military advantage. With this instrument he observed the heavens, almost by accident, in January 1610. On the 7th of that month he began a long letter, reporting in detail his observations of the moon and other bodies. He wrote:

> I have observed with one of my telescopes. . . the face of the Moon, which I have been able to see very near. . . . [W]hat is there can be discerned with great distinctness, and in fact it is seen that the Moon is most evidently not at all of an even, smooth and regular surface, as a great many people believe of it and of the other heavenly bodies, but on the contrary it is rough and unequal. In short, it is shown to be such that sane reasoning cannot conclude otherwise than that it is full of prominences and cavities similar, but much larger, to the mountains and valleys spread over the Earth's surface.[3]

---

*A hypothetical experiment has been proposed: Throw a volume of the *Encyclopaedia Britannica* into a black hole. Will the information it contains be forever lost? Is such a total loss impossible? A wager, lighthearted but serious, between two distinguished Caltech physicists has been placed on the outcome. Prof. Kip Thorne bets on relativity, whose equations describe space and time and predict that from the singularity of a black hole there could *never* be any recovery. Prof. John Preskill bets on quantum mechanics, whose equations precisely describe the lives of minuscule elementary particles and predict that the information can never be *totally* lost. The stakes of the wager are a set of encyclopedias. Payoff is unlikely to come soon. Says their equally distinguished colleague, Prof. Stephen Hawking of Cambridge University, who originally was in on the bet, "In my opinion it could go either way." Hawking, but not Thorne, conceded the bet in 2004. [*Science News*, 25 September 2004]

To save the hypothesis that the Moon was indeed a perfect sphere, and thus to retain the coherence of the theological account of the heavenly bodies of which that perfection was one element, some of Galileo's critics later proposed the hypothesis—outrageously *ad hoc*—that the apparent cavities and irregularities on the surface of the Moon were, in fact, filled in by a celestial substance that was flawless and crystalline, and thus invisible through Galileo's telescope!

More than the Moon was examined by Galileo. His letter continued:

> And besides the observations of the Moon. . . many fixed stars are seen with the telescope that are not [otherwise] discerned; and only this evening I have seen Jupiter accompanied by three fixed stars, totally invisible [to the naked eye] by their smallness, and the configuration was in this form:[4]

At that point Galileo inserted a sketch that appears here as Figure 13-1, showing the three stars in a straight line, two to the east and one to the west of Jupiter; he reported that they did not extend more than one degree of longitude, but since at that time he supposed them to be fixed stars, their distances from Jupiter and from one another were indicated only very roughly.

On the following day, 8 January 1610, "led by I know not what," Galileo happened to observe Jupiter once again; the earlier positions of those "fixed stars" had fortunately been written down. His letter remained unsent; at the bottom of the sheet he wrote the following note:

> On the 8th thus: [He inserts a sketch showing Jupiter and three stars now closer to one another and nearly equidistant from one another, and *all three to the west of Jupiter*!]

This created a serious theoretical problem for Galileo, because at this time the assumption that the newly discovered stars were fixed had not been seriously doubted. Therefore their appearance on the other side of Jupiter had to be accounted for by Jupiter's movements. On the 8th he added the note:

> It [Jupiter's movement] was therefore direct and not retrograde.

If, on the 8th, Jupiter was to the east of all three stars, and the day before Jupiter had been to the west of two of them, Jupiter must have moved, and moved in a way that was *contrary* to reliable astronomical calculations! One can imagine Galileo's agitation as he waited for the observations of the following night; could his direct observations and his calculations remain so sharply inconsistent? On the 9th it was too cloudy to observe, but he was able to resume his observations the following night and to record the new pattern. On 11 January a similar pattern was observed, but on this night Galileo later wrote:

> The star nearer Jupiter was half the size of the other, and very close to the other, whereas the other evenings all three of the said stars appeared of equal size and equally far apart.

On the 12th, Jupiter apparently had moved back to the west, and two of the new "stars" were again observed to the east of the planet! Clearly, something had to give. From the accepted theories and beliefs a prediction confidently could be drawn, a deduction concerning the movements of Jupiter, which—if those three

528 CHAPTER 13 Science and Hypothesis

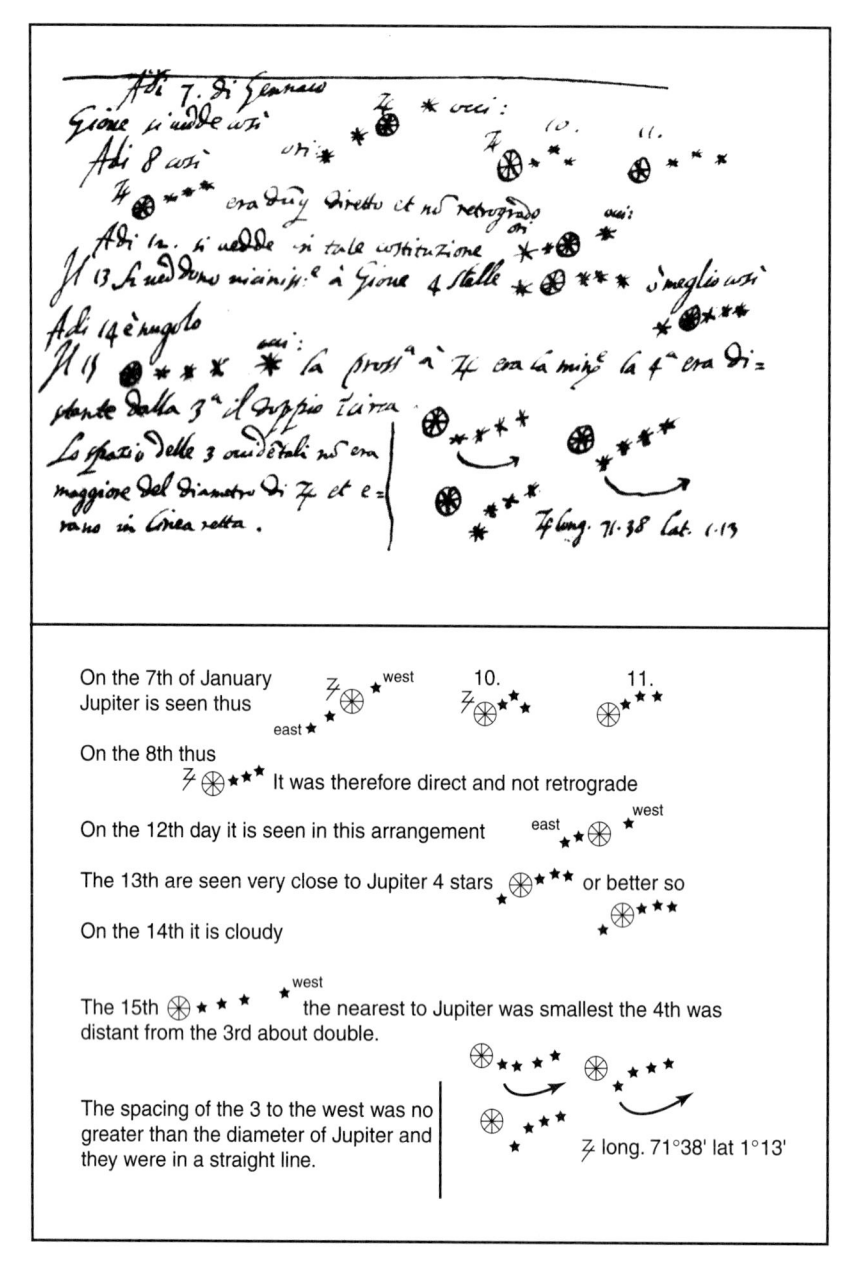

> On the 7th of January Jupiter is seen thus
>
> On the 8th thus It was therefore direct and not retrograde
>
> On the 12th day it is seen in this arrangement
>
> The 13th are seen very close to Jupiter 4 stars or better so
>
> On the 14th it is cloudy
>
> The 15th   the nearest to Jupiter was smallest the 4th was distant from the 3rd about double.
>
> The spacing of the 3 to the west was no greater than the diameter of Jupiter and they were in a straight line.
>
> ♃ long. 71°38' lat 1°13'

Figure 13-1 A photograph of the letter begun by Galileo on 7 January 1610, on which are recorded his first monumental observations of the four major satellites of Jupiter, thus confirming the Copernican account of the movement of the celestial bodies. The letter itself was to be sent to the Doge in Venice, and included a telescope with which Galileo intended to present him. On a draft of that letter which he happened to have in hand, Galileo made the critical notes of his observations, which appear on the bottom half of the sheet. The translation of the bottom half into English appears below. Courtesy of the Special Collections Library, University of Michigan.

new stars were fixed, and Galileo's observations were accurate—did not take place. One could save the belief that those new stars were fixed by somehow revamping the entire set of astronomical calculations, but these were not in serious doubt; or, one could challenge the accuracy of Galileo's observations—which is what some of his critics later sought to do, calling his telescope an instrument of the devil. Galileo himself had no doubt about what he had seen, and he grasped quickly which element in the set of accepted hypotheses had to be relinquished, to the great distress of his dogmatic opponents. His note on the observation of the 11th continued:

> . . . from which it appears that around Jupiter there are three moving stars invisible to everyone to this time.

And these three moving stars, he later wrote,

> . . . revolved round Jupiter in the same manner as Venus and Mercury revolved round the sun.

The observations of the following nights confirmed this revolutionary conclusion, which, together with his earlier observations of the moon, cast serious doubt on the account of celestial bodies that had been widely and dogmatically affirmed for many centuries.

On 13 January 1610, Galileo observed a fourth "star," and the four major satellites of Jupiter had been discovered. These observations provided very strong confirmation of the Copernican hypothesis—an account of the celestial bodies that was difficult to reconcile with the established theological doctrine of Galileo's time. Many moons of Jupiter have been discovered since, but these four moons—Ganymede, Io, Europa, and Callisto—are appropriately called "the Galilean satellites." On a clear night, when Jupiter is visible in the sky, the revolutions of the Galilean satellites around the planet may be readily confirmed with no more than an ordinary pair of binoculars.

The ultimate success of the Copernican account of the solar system was due not merely to its greater simplicity, but to its correctness, made manifest in the much larger body of facts it was able to account for, and the remarkable predictions deducible from the theory that were very soon confirmed beyond reasonable doubt.

13.4 Classification as Hypothesis

It is a mistake to suppose that hypotheses are important only in the advanced sciences, such as physics and chemistry, but play no role in the so-called descriptive sciences, such as botany or history. In fact, description itself is based on, or embodies, hypotheses. Hypotheses are as critical to the various systems of classification in biology as they are to interpretation in history, and as they are to all knowledge in the social sciences.

In the science of history the importance of hypotheses is easily shown. Many historians seek explanations of past events that can account for them and that can be confirmed by other recorded events. For some it is some larger purpose or pattern, religious or naturalistic, that explains the entire course of recorded history. For others, who reject such cosmic designs, the study of the past nevertheless reveals some historical laws that explain some past sequences and can then be used to predict some future events. Both of these two groups conceive of history as a theoretical science, not one that is merely descriptive; for both the role of hypothesis is central to the historian's enterprise.

A third group sets a more modest goal. For them the task of historians is simply to chronicle the past, to present an accurate description of past events in chronological order. Their concern is with the facts themselves, rather than with theories about the facts, so it might seem that they have no need of hypotheses.

However, past events are not so easily chronicled as this view would have us believe. The past itself simply is not available for this kind of bare description. What is available are records of the past and traces of the past. We have government archives, epic poems, the writings of earlier historians, the artifacts unearthed by archeological excavations, and so on. It is from a great variety of facts like these that historians must infer the nature of the past events they aim to describe. They cannot do this without some hypotheses. Not all hypotheses are general; some are particular, and with particular hypotheses historians seek to convert the data at hand into evidence for their account of the events in question.

Historians are detectives on a grand scale. Their methods are the same, and their difficulties too. The evidence is scanty, and much of it has been destroyed by intervening wars or natural disasters. False or misleading clues throw detectives off the scent, and similarly, many existing "records" are falsifications of the past, perhaps unintentional, such as the writings of earlier, uncritical historians. The methods of science must be used by good detectives and good historians both, and even those historians who seek to limit themselves to the bare description of past events must work from some hypotheses. They are theorists in spite of themselves.

Biologists are in a more favorable position. The facts with which they deal are present and available for inspection. To describe the flora and fauna of a region, biologists are not obliged to draw elaborate inferences, as historians are, because they can perceive the data directly. Their descriptions are not casual or random, but highly systematic. They *classify* plants and animals, and do not merely describe them. But classification and description are, at bottom, the same process. To describe an animal as carnivorous is to classify it as a carnivore; to classify it as a reptile is to describe it as reptilian. To describe any object as having a certain attribute is to classify it as a member of the class of objects having that attribute.

Scientific **classification** involves not merely a single division of objects into groups, but further subdivision of each group into subgroups and subclasses, and so on. Classification is also the tool of our inquiry when we play "Twenty Questions"—but it is a nearly universal tool, because it answers an almost universal need. Primitive people needed to sort the poisonous from the edible, the dangerous from the harmless, and so forth. We all draw distinctions, and we do so more meticulously with respect to the matters that chiefly concern us. The farmer's vegetables he will classify with greatest care, while treating all the flowers, in which he has no interest, as weeds. The florist will give delicate care to the classification of flowers, but may treat all the farmer's crops merely as "produce."

Classification
The organization and division of large collections of things into an ordered system of groups and subgroups, often used in the construction of scientific hypotheses.

Two basic motives lead us to classify things. One is practical, the other theoretical. In any library, with many thousands of volumes, books could not be found if they were not shelved according to some system of classification. The larger the number of objects with which we deal, the greater is the need to classify them. In museums, libraries, large department stores, this practical need is plain.

The theoretical object of classification is less obvious. Alternative schemes of classification are neither true nor false. Objects may be described in different ways, from different points of view. The system of classification adopted will depend on the purpose or interest of the classifier. A librarian will classify books according to their subject matter; a bookbinder according to the material of their leaves and bindings; a bibliophile by date of publication and perhaps by rarity; a shipper by weight and size—and there will be other schemes of classification as well.

What is the special interest of scientists, leading them to prefer one scheme of classification over another? The scientist seeks knowledge, not merely of this or that particular fact, but of the general laws to which the facts conform, and of their causal interrelations. One scheme of classification is better than another, from the scientific point of view, to the extent that it is more fruitful in suggesting scientific laws, and more helpful in the formulation of explanatory hypotheses.

The theoretical, or scientific, motivation for classifying objects is the desire to increase our knowledge of them, to achieve insight into their attributes, their similarities and differences, and their interrelations. Classification with a narrowly practical purpose—dangerous and harmless, or flying and swimming—will not much advance that understanding. The rattlesnake and the wild boar will go into one class, the grass snake and the domestic pig into the other; the bats and the birds will go into one class, the whales and the fishes into another. However, snakes and boars are profoundly different, whereas whales and bats are profoundly like one another. Being warm-blooded or not, bearing young alive or laying eggs, are much more important characteristics than dangerousness on which to base a system of classification.

A characteristic is important when it indicates the presence of other characteristics. When an attribute is causally connected with many other attributes, it can serve in the framing of a greater number of causal laws and of more general explanatory hypotheses. That classification scheme is best which is based on the most important characteristics of the objects to be classified. We cannot know in advance which these are, because we cannot know in advance the causal connections we aim to learn. So scientists classify *hypothetically*. Different classification schemes are tried, with the understanding that later they may be improved on or rejected. Later investigations may reveal other characteristics that are involved in a greater number of causal laws and explanatory hypotheses, and we will then revise the classification scheme so as to base our categories on it.

It is true that classification tends to be more important in the early or less developed stages of a science, but it need not diminish in importance as that science develops. Taxonomy is a legitimate, important, and still growing branch of biology, in which earlier systems of classification have been abandoned in favor of others that prove more productive. Some classificatory tools—such as the periodic table of the elements—remain valuable to the chemist.

Hypotheses in history are illuminated by these biological considerations. Historians, too, focus on what they find to be most important in increasing our understanding of past events. Life is too short to permit the description of past events in *complete* detail, so every description by a historian must be selective, recording only some features. How may that selection be made? Of course historians want to focus on what is important, ignoring the insignificant. Historians, like biologists and other scientists, regard those aspects of events as important that enter most widely into the formulation of causal laws and explanatory hypotheses—always subject to correction in the light of further research, of course. Early historians emphasized the political and military aspects of events, ignoring other attributes we now think to be important. The turn to economic and social attributes brought enormous changes in the work and the products of historians; today we go beyond economic and social issues to attend to cultural and other characteristics that are now thought to be causally related to a maximum number of others. So the decision to focus on one rather than another set of attributes embodies some hypothesis about which characteristics really are important. Some such hypotheses are required before historians can even begin to do any systematic describing of the past. It is this *hypothetical* character of classification and description that leads us to regard hypothesis as the all-pervasive method of scientific inquiry.

EXERCISES

In each of the following passages,

a. What data are to be explained?

b. What hypotheses are proposed to explain them?

c. Evaluate the hypotheses in terms of the criteria presented in Section 13.3, pages 522–525.

1. In an unusual logjam of contradictory claims, a revolutionary new model of the universe, as a soccer ball, arrived on astronomers' desks in October of 2003—at least slightly deflated.

    Based on an analysis of maps of the Big Bang, Dr. Jeffrey Weeks and colleagues, from Canton, NY, suggest that space is a kind of 12-sided hall of mirrors, in which the illusion of infinity is created by looking out and seeing multiple copies of the same stars.

    If his model is correct, Dr. Weeks said, it would rule out one variant of the Big Bang theory that asserts that our own observable universe is just a bubble among others in a realm of vastly larger extent. "It means we can just about see the whole universe now," Dr. Weeks said.

    Other astronomers, led by Dr. David Spergel of Princeton, said that their analysis of the same data had probably already ruled out the soccer-ball universe. The two groups of scientists, who have been in intense communication in recent days, disagree about whether the soccer ball universe has been refuted. But they all agree that what is amazing about this debate is that the controversy will actually be settled soon, underscoring the power of modern data to resolve issues that were once considered almost metaphysical.

In the scientific journal *Nature* Dr. Weeks wrote: "Since antiquity our ancestors have wondered whether our universe is finite or infinite. Now, after more than two millennia of speculation, observational data might finally settle this ancient question."

Dr. Weeks and his colleagues propose that the universe is 12-sided, a dodecahedron. The waves appearing in a radio map of the universe when it was very young indicate, he argues, that if you go far enough in one direction you would find yourself back where you started, like a cursor disappearing off the left side of a computer screen and reappearing on the right. Thus when cosmic radiation intersects the edges of the universe it would make identical circles on opposite sides of the sky—six pairs of circles, 35 degrees in diameter, in the case of Dr. Weeks's dodecahedron.

Dr. Max Tegmark, a cosmologist at the University of Pennsylvania, observed: "What's nice is that this is so testable. It's the truth or it's dead. The data are actually already out there; it's just a question of sifting through them. We ought to have seen those circles." So far the circles have not showed up. "Is space infinite or is it not?" Dr. Tegmark asked. "This is what got Giordano Bruno burned at the stake!"

—Reported in *Nature*, 9 October 2003

2. Population clusters—groups of persons who are found to buy the same things, get their entertainment from the same sources, exhibit similar voting patterns, and generally behave in quite similar ways—are of growing interest. Michael J. Weiss has distinguished some 62 of these clusters, which he calls "distinctive lifestyle types." He also names them and highlights some of their peculiarities.

In the *Towns and Gowns* cluster, for example, tequila is far more popular than elsewhere, and twice as many people watch the soap opera "Another World" there than do people elsewhere. In the *Military Quarters* cluster people are four times as likely to watch the TV show "Hard Copy" as the average American. Among the young, middle-class Americans in suburbia, furniture refinishing, downhill skiing, and cats are abnormally popular, while chess and tractor pulls are abnormally unpopular.

Lifestyle clusters are found useful by businesses seeking customers, by candidates seeking votes, by nonprofit organizations seeking new contributors, and so on. What may appear trivial can be very revealing. In Washington, DC, Weiss observes, "there is a fault line between the fans of Brie cheese, who tend to hold down executive jobs and write the laws, and those of Kraft Velveeta, who maintain the service economy." He asks: "What prompts some of us to eat Brie and others to devour Velveeta cheese?"

—Michael J. Weiss, *The Clustered World* (Boston: Little, Brown, 2000)

3. Monkeypox, a viral disease related to smallpox but less infectious and less deadly, was detected for the first time in the Americas in 2003. At least 20 cases have been reported, in three Midwestern states, Wisconsin, Illinois and Indiana, according to the Centers for Disease Control and Prevention.

   The patients ranged in age from 4 to 48, and became ill between 15 May and 3 June, 2003. All had direct or close contact with ill prairie dogs, which have become common household pets, and which might have caught monkeypox from another species, possibly Gambian giant pouched rats, which are imported as pets from West or Central Africa, where the disease had long occurred. Monkeypox in Africa is carried mainly by squirrels but is named after monkeys because it often kills them.

   Several patients in the American outbreak work for veterinarians or pet stores that sold prairie dogs and Gambian rats. By quickly identifying the animals that can be infected with monkeypox, health officials hope to eliminate them before the disease becomes endemic in the Americas.

   —Reported in the *The New York Times*, 9 June 2003

4. A small study of heart-disease patients testing a hypothesis so improbable that its principal investigator says he gave it a one-in-10,000 chance of succeeding has found that just a few treatments with an experimental drug, developed by Esperion Therapeutics of Ann Arbor, Michigan, reversed what may be the equivalent of years' worth of plaque in coronary arteries.

   Forty-seven heart attack patients were randomly assigned to be infused with either a concentration of a substance that mimics high density lipoprotein (or HDL, the substance that removes cholesterol from arteries) or to be infused with an inactive saline solution, which served as a control.

   After 5 weekly infusions those who got the experimental drug had a 4.2-percent decrease in the volume of plaque in their coronary arteries, while those who had saline infusions had, if anything, a slight increase in their plaque.

   "Until now," said Dr. Steven Nissen, a cardiologist at the Cleveland Clinic who directed the study, "the paradigm has been to prevent disease by lowering bad cholesterol (LDL). If you get the bad cholesterol low enough, the plaques don't build up in the artery walls. This experiment says you can also remove the disease in the wall of the artery."

   —Reported in the *Journal of the American Medical Association*, 5 November 2003

5. Boy babies tend to be about 100 grams heavier on average than girl babies, but it has never been explained, until recently, why that is so. Investigators were unsure whether the increased weight was to be explained by the fact that mothers of boys took in more energy, or because (when the fetus was male) those mothers used the energy taken in more efficiently.

   Dr. Rulla M. Tamimi, of the Harvard School of Public Health, sought to resolve this uncertainty by measuring the intake of calories. During the second trimester of their pregnancy, 244 women in Boston were asked to record their dietary intake in full detail. The data collected were later correlated with the resultant births. Women carrying boys, Dr. Tamimi found, took in (as carbohydrates, fats, or proteins) about 10 percent more calories than women carrying girls. It is intake, and not efficiency of use, that makes the difference.

   But what accounts for that difference of intake? Dr. Tamimi speculated that it may be triggered by some signal from the testosterone given off by the male fetuses.

   —Reported in the *British Medical Journal*, June 2003

6. Humans, apes, and dolphins are highly social animals with large brains; they have been shown to be aware of themselves by recognizing themselves in a mirror. Most animals pay very little attention to their reflections in a mirror. Elephants are like humans in being large-brained and empathic, but they don't share a relatively recent common ancestor with humans, like apes do. Might they also recognize an image of themselves?

   Yes, they do. Elephants at the Bronx Zoo, in New York City, inspected themselves with their trunks while staring at their reflections in a huge mirror. One of the elephants (but only one) completed the highest level of self-recognition, called the "mark test." Researchers placed a white X above one eye of each elephant. After approaching the mirror, this elephant touched the mark with her trunk 12 times in 90 seconds—confirmation that she believed that what she saw in the mirror was indeed herself.

   —Reported by Diana Reiss, of the Wildlife Conservation Society and Columbia University, in *Proceedings of the National Academy of Sciences*, November 7, 2006

7. The Nobel Prize for chemistry for 2003 was shared by Dr. Peter Agre, who encountered a new protein by serendipity. He had been studying a particular protein found in blood when he found another protein contaminating his sample. Trying to develop an antibody that would hook on to the protein he was studying, Dr. Agre found that the antibody hooked on to the contaminating protein instead—which turned out to be one of the most abundant proteins found in blood samples, although no one had identified it before.

   But what did it do? He looked for similar proteins and found some—whose functions also were not known—in the roots of plants. The situation grew "curiouser and curiouser," Dr. Agre said. Finally he tried testing whether the new protein could be a water channel. That such channels might exist had been suggested long ago—but diffusion had then seemed to explain water movement, and specific channels had never been discovered.

To test the water channel hypothesis, Dr. Agre added the gene that produced the mystery protein to the eggs of frogs. The modified eggs, placed in fresh water, quickly swelled and burst, strongly confirming that theory. "The eggs exploded like popcorn," Dr. Agre said. The newly discovered proteins, called "aquaporins," have a channel just a little wider than a water molecule, and have recently been found also in human kidneys, where water is extracted from urine and recycled.

"This really fell into our laps," Dr. Agre said when his Nobel Prize was announced. "Being lucky is an important ingredient in scientific success."

8. Early in the eighteenth century Edmund Halley asked: "Why is the sky dark at night?" This apparently naive question is not easy to answer, because if the universe had the simplest imaginable structure on the largest possible scale, the background radiation of the sky would be intense. Imagine a static infinite universe—that is, a universe of infinite size in which the stars and galaxies are stationary with respect to one another. A line of sight in any direction will ultimately cross the surface of a star, and the sky should appear to be made up of overlapping stellar disks. The apparent brightness of a star's surface is independent of its distance, so that everywhere the sky should be as bright as the surface of an average star. Since the sun is an average star, the entire sky, day and night, should be about as bright as the surface of the sun. The fact that it is not was later characterized as Olbers' paradox (after the eighteenth-century German astronomer Heinrich Olbers). The paradox applies not only to starlight but also to all other regions of the electromagnetic spectrum. It indicates that there is something fundamentally wrong with the model of a static infinite universe, but it does not specify what.

—Adrian Webster, "The Cosmic Radiation Background,"
Scientific American, August 1974

9. Swedish researchers, collaborating with colleagues in South Africa, found that dung beetles active during the day detect polarity patterns in sunlight and rely on those patterns to find their way out of great masses of elephant dung. Dr. Marie Dacke, of the University of Lund, noticed subsequently that on moonlit nights one beetle species worked (rolling dung) particularly late. Could they have been relying upon the polarization of moonlight?

Researchers set up polarizing filters to shift the moonbeams—and sure enough, the African beetle, *Scarabaeus zambesianus*, changed direction to compensate. When the polarization of the moonlight under the filter was rotated by 90 degrees, they found that beetles under that filter deviated from their course by almost exactly 90 degrees. "This is the first proof," writes Dr. Dacke in her report in *Nature* of 3 July 2003, "that any animal can use polarized moonlight for orientation."

10. For centuries (since the 1500s in Scandinavia) people have puzzled over lemmings, northern rodents whose populations surge and crash so quickly and so regularly that they inspired an enduring myth: that lemmings commit mass suicide when their numbers grow too large, pitching themselves off cliffs to their deaths in a foamy sea.

    Scientists debunked that notion decades ago, but have never been certain what causes the rapid boom-and-bust population cycles—a mystery in ecology that has been hotly debated. "There have been several dozen hypotheses," said Dr. Oliver Gilg, an ecologist at the University of Helsinki in Finland, "and scientists were sticking so closely to their hypotheses that they were almost killing each other." But Dr. Gilg, the author of a recent study published in the journal *Science*, provides a single hypothesis that his team of researchers claims provides the entire explanation.

    The rapid population cycles have nothing to do with self-annihilation, they contend, but everything to do with hungry predators. After 15 years of research they have discovered that the actions of four predator species—snowy owls, arctic foxes, seabirds called long-tailed skuas, and the weasel-like stoats—account for the four-year cycles during which lemming populations rapidly explode and then nearly disappear. After creating a model based only on those four predators, they found that the model predicted precisely the numerical fluctuation of lemming populations in nature.

    —Reported in *Science*, 31 October 2003

chapter 13 Summary

In this chapter we explored the principles that underlie the methods of science.

In Section 13.1, we distinguished scientific from unscientific explanations, the former being always hypothetical and empirically verifiable, the latter dogmatic in spirit and not testable by propositions that can be deduced from them.

In Section 13.2, we examined the method of science, relying on the confirmation of hypotheses. We identified the seven stages that may be distinguished in any scientific inquiry:

1. The identification of some problem
2. The construction of some preliminary hypothesis
3. The collection of additional data in the light of that preliminary hypothesis
4. The formulation of a fully explanatory hypothesis supported by the data collected

5. The deduction of further consequences from the explanatory hypothesis
6. The testing of the consequences deduced
7. The application of the theory developed

In Section 13.3, we explored the evaluation of alternative scientific hypotheses. We identified criteria with which we might choose between competing hypotheses:

1. The compatibility of a theory with the body of theory previously established
2. The degree of predictive or explanatory power that a new theory manifests
3. The relative simplicity of competing theories

We illustrated these criteria with events in the history of science—most notably, the replacement of the geocentric (or Ptolemaic) theory of the solar system with the heliocentric (or Copernican) theory, confirmed by the remarkable observations of Galileo Galilei.

In Section 13.4, we discussed classification, an intellectual instrument that is greatly valued in the social and biological sciences as well as in the physical sciences, noting that every classificatory scheme suggests general truths and invites the formation of explanatory hypotheses.

END NOTES

[1]D. Biello, "Island Lizards Morph in Evolutionary Experiment," *Scientific American*, 17 November 2006.

[2]*On the Old Saying: "That Might Be Right in Theory"* [*Uber den Gemeinspruch: Das mag in der Theorie richtig sein*], 1793, translated by E. B. Ashton (Philadelphia: University of Pennsylvania Press, 1974).

[3]This letter, dated 7 January 1610, apparently was written over a period of many days. It, and other notes taken by Galileo during these momentous days, are discussed in detail in Jean Meeus, "Galileo's First Records of Jupiter's Satellites," *Sky and Telescope,* February 1964; in Stillman Drake, "Galileo's First Telescopic Observations," *Journal of the History of Astronomy,* 1976, p. 153; and in Dale P. Cruikshank and David Morrison, "The Galilean Satellites of Jupiter," *Scientific American*, May 1976. A photocopy of the original sketch Galileo made to record his observations, his notes appearing on it in Italian, is reproduced in Figure 13-1, through the courtesy of the library of the University of Michigan, Ann Arbor, in whose rare-book room that precious manuscript is held.

[4]That Galileo began this letter on 7 January 1610 is clear; the exact days of that month on which he continued it, with sketches and notes, are a matter about which scholars disagree.

For additional exercises and tutorials about concepts covered in this chapter, log in to MyLogicLab at *www.mylogiclab.com* and select your current textbook.

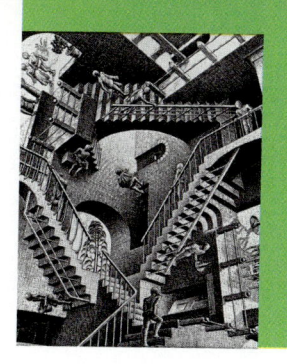

Probability

14.1 Alternative Conceptions of Probability
14.2 The Probability Calculus
14.3 Probability in Everyday Life

14.1 Alternative Conceptions of Probability

Probability is the central evaluative concept in all inductive logic. The theory of probability, as the American philosopher Charles Sanders Peirce put it, "is simply the science of logic quantitatively treated." The mathematical applications of this theory go far beyond the concerns of this book, but it is fitting to conclude our treatment of inductive logic with an analysis of the concept of probability and a brief account of its practical applications.

Scientific theories, and the causal laws that they encompass, can be no more than probable. Inductive arguments, even at their very best, fall short of the certainty that attaches to valid deductive arguments. We assign to theories, or to hypotheses of any sort, a *degree* of probability expressed discursively. As one example, we may assert, on the evidence we now have, that it is "highly probable" that Einstein's theory of relativity is correct. As another example, although we cannot be certain that there is no life on other planets in our solar system, we can say that the probability of any theory that entails such life, in the light of what we know about these planets, is very low. We do not normally assign a numerical value to the probability of theories in this sense.

However, we can and do assign numbers to the probability of events in many contexts. The number we assign to the probability of an event is called the **numerical coefficient of probability**, and that number may be very useful. How can such numbers be reliably assigned? To answer this question we must distinguish two additional senses in which the concept of "probability" is used:

1. The *a priori* conception of probability
2. The relative frequency conception of probability

We use the first of these when we toss a coin and suppose that the probability that it will show heads is $1/2$. We use the second of these when we say that the probability that an American woman of age 25 will live at least one additional year is .971. Games of chance—dice and cards—gave rise to the investigation of probability in the first sense,[*] and the uses of mortality statistics gave rise to the investigation of probability in the second sense,[†] in both cases during the seventeenth

Numerical coefficient of probability
A number that describes the likelihood, or probability, of the occurrence of an event. Its possible values range from 0 (impossibility) to 1 (certainty).

---
[*]Pierre de Fermat (1608–1665) and Blaise Pascal, both distinguished mathematicians, reflected upon probabilities when corresponding about the proper division of the stakes when a game of chance had been interrupted.

[†]Captain John Graunt published (in 1662) calculations concerning what could be inferred from death records that had been kept in London from 1592.

century. The calculations in the two cases were of different kinds, leading eventually to the two different interpretations of the coefficient of probability. Both are important.

The *a priori* **theory of probability** asks, in effect, what a rational person ought to believe about some event under consideration, and assigns a number between 0 and 1 to represent the degree of belief that is rational. If we are completely convinced that the event will take place, we assign the number 1. If we believe that the event cannot possibly happen, our belief that it will happen is assigned the number 0. When we are unsure, the number assigned will be between 0 and 1. Probability is predicated of an event according to the degree to which one rationally believes that that event will occur. Probability is predicated of a proposition according to the degree to which a completely rational person would believe it.

How (in this theory) do we determine rationally, when we are unsure, what number between 0 and 1 ought to be assigned? We are unsure, in the classical view, because our knowledge is partial; if we knew everything about a coin being flipped, we could confidently predict its trajectory and its final resting position. However, there is an enormous amount about that coin and its flip that we do not and cannot know. What we mainly know is this: The coin has two sides, and we have no reason to believe it more likely that it will come to rest on one side than on the other. So we consider all the possible outcomes that are (so far as we know) equally probable; in the case of a flipped coin there are two—heads and tails. Of the two, heads is only one. The probability of heads is therefore one over two, $1/2$, and this number, .5, is said to be the probability of the event in question.

Similarly, when a deck of randomly shuffled cards is about to be dealt, they will come off the deck in exactly the sequence they are in, determined by the outcome of the preceding shuffle, which we do not know. We know only that there are 13 cards of each suit (out of a total of 52 in the deck) and therefore the probability that the first card dealt will be a spade is 13/52, or exactly $1/4$.

This is called the *a priori theory of probability* because we make the numerical assignment, $1/4$, before we run any trials with that deck of cards. If the deck is regular and the shuffle was fair, we think it is not necessary to take a sample, but only to consider the antecedent conditions: 13 spades, 52 cards, and an honest deal. Any one card (as far as we know) has as much chance as any other of being dealt first.

To compute the probability of an event's occurring in given circumstances, we divide the number of ways it can occur by the total number of possible outcomes of those circumstances, provided that there is no reason to believe that any one of those possible outcomes is more likely than any other. The probability of an event, in the *a priori* theory of probability, is thus expressed by a fraction, whose denominator is the number of equipossible outcomes and whose numerator is the number of outcomes that will successfully yield the event in question. Such numerical assignments ("successes over possibilities") are rational, convenient, and very useful.

***A priori* theory of probability**
A theory in which the probability ascribed to a simple event is a fraction between 0 and 1, of which the denominator is the number of equipossible outcomes, and the numerator is the number of outcomes in which the event in question occurs. Thus on the *a priori* theory, the probability of drawing a spade at random from a deck of playing cards is 13/52.

There is an alternative view of probability. In this view the probability assigned to an event must depend on the *relative frequency* with which the event takes place. Earlier we suggested that the probability of a 25-year-old American woman living at least one additional year is .971. This can be learned only by examining the entire class of 25-year-old American women, and determining how many of them do indeed live, or have lived, at least one additional year. Only *after* we learn the mortality rates for that class of women can we make the numerical assignment.

We distinguish, in this theory, the *reference class* (25-year-old American women, in the example given) and the *attribute* of interest (living at least one additional year, in this example). The probability assigned is the measure of the relative frequency with which the members of the class exhibit the attribute in question. In this theory also, probability is expressed as a fraction (and often expressed in decimal form), but the denominator is in this case the number of members in the reference class and the numerator is the number of class members that have the attribute of interest. If the number of male automobile drivers in California between the ages of 16 and 24 is y, and the number of such drivers who are involved in an automobile accident in the course of a year is x, the probability of an accident among such drivers in any given year we assign as x/y. The reference class here is the set of drivers described in certain ways, and the attribute is the fact of involvement in an automobile accident within some specified period. "Rational belief" is not at issue here. In the **relative frequency theory of probability**, probability is defined as *the relative frequency with which members of a class exhibit a specified attribute.*

Note that in both theories the probabilities assigned are relative to the evidence available. For the relative frequency theory this is obvious: The probability of a given attribute must vary with the reference class chosen for the computation. If the male automobile drivers in the reference class are between the ages of 36 and 44, the relative frequency of accidents will be lower; drivers in that range have fewer accidents, and hence the computed probability of an accident will be lower. If the reference class consisted of females rather than males, that would again change the coefficient of probability. Probability is relative to the evidence.

This is also true in the *a priori* theory of probability. An event can be assigned a probability only on the basis of the evidence available to the person making the assignment. After all, a person's "rational belief" may change with changes in the knowledge that person possesses. For example, suppose that two people are watching a deck of cards being shuffled, and because of the dealer's slip, one of them happens to see that the top card is black, but cannot see the card's suit. The second observer sees nothing but the shuffle. If asked to estimate the probability of the first card's being a spade, the first observer will assign the probability $1/2$, because he knows that there are 26 black cards, of which half are spades. The second observer will assign the probability $1/4$, because he knows only that there are 13 spades in the deck of 52 cards. Different probabilities are assigned by the two observers to the same event. Neither has made a mistake; both have assigned

Relative frequency theory of probability
The view of probability in which the probability of a simple event is determined as a fraction whose denominator is the total number of members of a class, and whose numerator is the number of members of that class that are found to exhibit a particular attribute that is equivalent to the event in question.

the correct probability relative to the evidence available to each—even if the card turns out to be a club. No event has any probability in and of itself, in this view, and therefore, with different sets of evidence, the probabilities may well vary.

These two accounts of probability—the relative frequency account and the *a priori* account—are in fundamental agreement in holding that probability is relative to the evidence. They are also in agreement in holding that a numerical assignment of probability can usually be made for a given event. It is possible to reinterpret the number assigned on the *a priori* theory as being a "shortcut" estimate of relative frequency. Thus the probability that a flipped coin, if it is fair, will show heads when it comes to rest may be calculated as a relative frequency; it will be the relative frequency with which the coin does show heads when it is randomly flipped a thousand, or ten thousand times. As the number of random flips increases (supposing the coin truly balanced), the fraction representing the relative frequency of heads will continue to approach .5 more closely. We may call .5 the *limit of the relative frequency of that event*. In the light of such possible reinterpretation of numerical assignments, some theorists hold that the relative frequency theory is the more fundamental of the two. It is also true, however, that in a great many contexts the *a priori* theory is the simpler and more convenient theory to employ; we will rely chiefly on the latter as we go forward.

14.2 The Probability Calculus

The probability of single events, as we have seen, can often be determined. Knowing (or assuming) these, we can go on to calculate the probability of some complex event—an event that may be regarded as a whole of which its component single events are parts. To illustrate, the probability of drawing a spade from a shuffled deck of cards is $1/4$, as we have seen, relying on the *a priori* theory of probability. What, then, is the probability of drawing *two spades in succession* from a deck of playing cards? Drawing the first spade is the first component; drawing the second spade is the second; drawing two spades in succession is the complex event whose probability we may want to calculate. When it is known how the component events are related to each other, the probabilities of the complex event can be calculated from the probabilities of its components.

The **calculus of probability** is the branch of mathematics that permits such calculation. Here we explore only its elementary outline. Knowing the likelihood of certain outcomes in our everyday lives can be important; application of the probability calculus, therefore, can be extremely helpful. Mastery of its basic theorems is one of the most useful products of the study of logic.

The probability calculus can be most easily explained in terms of games of chance—dice, cards, and the like—because the artificially restricted universe created by the rules of such games makes possible the straightforward application of probability theorems. In this exposition, the *a priori* theory of probability is used, but all of these results can, with a minimum of reinterpretation, be expressed and justified in terms of the relative frequency theory as well.

Calculus of probability
A branch of mathematics that can be used to compute the probabilities of complex events from the probabilities of their component events.

Two elementary theorems will be discussed.

A. With the first we can calculate the probability of a complex event consisting of the *joint occurrences* of its components: the probability of two events both happening, or of all the events of a specified set happening.

B. With the second we can calculate the probability of a complex event consisting of *alternative occurrences:* the probability that at least one (that is, one or more) of a given set of alternative events will occur. We take these in turn.

A. Probability of Joint Occurrences

Suppose we wish to learn the probability of getting two heads in two flips of a coin. Call these two components *a* and *b*; there is a very simple theorem that enables us to compute the probability of *both a and b*. It is called the **product theorem**, and it involves merely multiplying the two fractions representing the probabilities of the component events. There are four distinct possible outcomes when two coins are tossed. These may be shown most clearly in a table:

First Coin	Second Coin
H	H
H	T
T	H
T	T

There is no reason to expect any one of these four cases more than another, so we regard them as equipossible. The case (two heads) about which we are asking occurs in only one of the four equipossible events, so the probability of getting two heads in two flips of a coin is $1/4$. We can *calculate* this directly: The joint occurrence of two heads is equal to the probability of getting a head on the first flip ($1/2$) multiplied by the probability of getting a head on the second flip ($1/2$), or $1/2 \times 1/2 = 1/4$. However, this simple multiplication succeeds only when the two events are **independent events**—that is, when the occurrence of the one does not affect the probability of the occurrence of the other.

The product theorem for independent events asserts that the probability of the joint occurrence of two independent events is equal to the product of their separate probabilities. It is written as

$$P(a \text{ and } b) = P(a) \times P(b)$$

where $P(a)$ and $P(b)$ are the separate probabilities of the two events, and $P(a \text{ and } b)$ designates the probability of their joint occurrence.

Applied to another case, what is the probability of getting 12 when rolling two dice? Two dice will show twelve points only if each of them shows six points. Each die has six sides, any one of which is as likely to be face up after a roll as any other. When *a* is the event of the first die showing 6, $P(a) = 1/6$. And

Product theorem
In the calculus of probability, a theorem asserting that the probability of the joint occurrence of multiple independent events is equal to the product of their separate probabilities.

Independent events
In probability theory, events so related that the occurrence or nonoccurrence of one has no effect upon the occurrence or nonoccurrence of the other.

when b is the event of the second die showing 6, $P(b) = 1/6$. The complex event of the two dice showing 12 is constituted by the joint occurrence of a and b. By the product theorem, $P(a \text{ and } b) = 1/6 \times 1/6 = 1/36$, which is the probability of getting a 12 on one roll of two dice. The same result is shown if we lay out, in a table, all the separate equipossible outcomes of the roll of two dice. There are 36 possible outcomes, and only one of them is favorable to getting 12.

We do not need to restrict ourselves to two components. The product theorem may be *generalized* to cover the joint occurrence of any number of independent events. If we draw a card from a shuffled deck, replace it and draw again, replace it again and draw a third time, the likelihood of getting a spade in each drawing is not affected by success or failure in the other drawings. (We assume that the replacement of a card is followed immediately by a reshuffling of the deck.) The probability of getting a spade in any one drawing is $13/52$, or $1/4$. The probability of getting three spades in three drawings, if the card is replaced after each drawing, is $1/4 \times 1/4 \times 1/4 = 1/64$. The general product theorem thus allows us to compute the probability of the joint occurrence of any number of independent events.

But what happens if the events are not independent? What happens if success in one case has an effect on the probability of success in another case? The examples thus far need take no account of any relationship among the component events, and yet component events may be related in ways that require more careful calculation. Consider a revised version of the example just given. Suppose we seek the probability of drawing three successive spades from a shuffled deck, *but the cards withdrawn are not replaced*. If each card drawn is not returned to the deck before the next drawing, the outcomes of the earlier drawings *do* have an effect on the outcomes of the later drawings.

If the first card drawn is a spade, then for the second draw there are only 12 spades left among a total of 51 cards, whereas if the first card is *not* a spade, then there are 13 spades left among 51 cards. Where a is the event of drawing a spade from the deck and not replacing it, and b is the event of drawing another spade from among the remaining cards, the probability of b, that is, $P(b \text{ if } a)$, is $12/51$, or $4/17$. If both a and b occur, the third draw will be made from a deck of 50 cards containing only 11 spades. If c is this last event, then $P(c \text{ if both } a \text{ and } b)$ is $11/50$. Thus, the probability that all three are spades, if three cards are drawn from a deck and not replaced, is, according to the product theorem, $13/52 \times 12/51 \times 11/50$, or $11/850$. This is less than the probability of getting three spades in three draws when the cards drawn are replaced before drawing again, which was to be expected, because replacing a spade increases the probability of getting a spade on the next draw.

The general product theorem can be applied to real-world problems of consequence, as in the following true account. A California teenager, afflicted with chronic leukemia that would soon kill her if untreated, could be saved only if a donor with matching bone marrow were found. When all efforts to locate such a donor failed, her parents decided to try to have another child, hoping that a successful bone-marrow transplant might then be possible. But the girl's father first had to have his vasectomy reversed, for which there was only a 50 percent (.5) chance of success. Even if that were successful, the mother, 45 years old at the time, would have only a .73 chance of becoming pregnant, and if she did become pregnant, there was only a one-in-four chance (.25) that the baby's marrow would match that of the afflicted daughter. Even if there were such a match, there would still be only a .7 chance that the leukemia patient would live through the needed chemotherapy and bone-marrow transplant.

The probability of a successful outcome was seen at the outset to be low, but not hopelessly low. The vasectomy was successfully reversed, and the mother did become pregnant—after which prospects improved. It turned out that the baby did possess matching bone marrow. Then, in 1992, the arduous bone-marrow transplant procedure was begun. It proved to be a complete success.[*] What was the probability of this happy outcome at the time of the parents' original decision to pursue it?

EXERCISES

EXAMPLE

1. What is the probability of getting three aces in three successive draws from a deck of cards:

    a. If each card drawn is replaced before the next drawing is made?
    b. If the cards drawn are not replaced?

SOLUTION

a. If each card drawn is *replaced* before the next drawing is made, the component events have absolutely no effect on one another and are therefore *independent*. In this case, $P(a \text{ and } b \text{ and } c) = P(a) \times P(b) \times P(c)$. There are 52 cards in the deck, of which four are aces. So the probability of drawing the first ace, $P(a)$, is $4/52$, or $1/13$. The probability of drawing the second ace, $P(b)$, is likewise $1/13$, as is the probability of drawing the third ace, $P(c)$. So the probability of the joint occurrence of a and b and c is $1/13 \times 1/13 \times 1/13$, or $1/2{,}197$.

b. If the cards drawn are *not replaced*, the component events are *dependent*, not independent. The formula is
$P(a \text{ and } b \text{ and } c) = P(a) \times P(b \text{ if } a) \times P(c \text{ if } a \text{ and } b)$.
In this case, the probability of drawing the first ace, $P(a)$, remains $4/52$, or $1/13$. But the probability of drawing a second ace if the first card drawn was an ace, $P(b \text{ if } a)$, is $3/51$, or $1/17$. And the probability of drawing a third ace if the first two cards drawn were aces, $P(c \text{ if } a \text{ and } b)$, is $2/50$, or $1/25$. The probability of the joint occurrence of these three dependent events is therefore $1/13 \times 1/17 \times 1/25$, or $1/5{,}525$.

---

[*]Anissa Ayala, the patient, was married a year after the successful transplant; the sister who saved her life, Marissa Ayala, was a flower girl at her wedding. Details of this case were reported in *Life* magazine, December 1993.

The probability of getting three successive aces in the second case is much lower than in the first, as one might expect, because without replacement the chances of getting an ace in each successive drawing are reduced by success in the preceding drawing.

2. What is the probability of getting tails every time in three tosses of a coin?

3. An urn contains 27 white balls and 40 black balls. What is the probability of getting four black balls in four successive drawings:

    a. If each ball drawn is replaced before making the next drawing?
    b. If the balls are not replaced?

4. What is the probability of rolling three dice so the total number of points that appear on their top faces is 3, three times in a row?

*5. Four men whose houses are built around a square spend an evening celebrating in the center of the square. At the end of the celebration each staggers off to one of the houses, no two going to the same house. What is the probability that each one reached his own house?

6. A dentist has her office in a building with five entrances, all equally accessible. Three patients arrive at her office at the same time. What is the probability that they all entered the building by the same door?

7. On 25 October 2003, at the Santa Anita Racetrack in Arcadia, California, Mr. Graham Stone, from Rapid City, South Dakota, won a single bet in which he had picked the winner of *six successive races*! Mr. Stone had never visited a racetrack; racing fans across the nation were stunned. The winning horses, and the odds of each horse winning, as determined just before the race in which it ran, were as follows:

Winning Horse	Odds
1. Six Perfections	5–1
2. Cajun Beat	22–1
3. Islington	3–1
4. Action This Day	26–1
5. High Chaparral	5–1
6. Pleasantly Perfect	14–1

Mr. Stone's wager cost $8; his payoff was $2,687,661.60.

The odds against such good fortune (or handicapping skill?), we might say in casual conversation, are "a million to one." Mr. Stone's payoff was at a rate far below that. Did he deserve a million-to-one payoff? How would you justify your answer?

8. In each of two closets there are three cartons. Five of the cartons contain canned vegetables. The other carton contains canned fruits: ten cans of pears, eight cans of peaches, and six cans of fruit cocktail.

Each can of fruit cocktail contains 300 chunks of fruit of approximately equal size, of which three are cherries. If a child goes into one of the closets, unpacks one of the cartons, opens a can and eats two pieces of its contents, what is the probability that two cherries will be eaten?

9. A player at draw poker holds the seven of spades and the eight, nine, ten, and ace of diamonds. Aware that all the other players are drawing three cards, he figures that any hand he could win with a flush he could also win with a straight. For which should he draw? (A *straight* consists of any five cards in numerical sequence; a *flush* consists of any five cards all of the same suit.)

*10. Four students decide they need an extra day to cram for a Monday exam. They leave town for the weekend, returning Tuesday. Producing dated receipts for hotel and other expenses, they explain that their car suffered a flat tire, and that they did not have a spare.

The professor agrees to give them a make-up exam in the form of a single written question. The students take their seats in separate corners of the exam room, silently crowing over their deceptive triumph—until the professor writes the question on the blackboard: "Which tire?"

Assuming that the students had not agreed in advance on the identification of the tire in their story, what is the probability that all four students will identify the same tire?

B. Probability of Alternative Occurrences

Sometimes we ask: What is the probability of the occurrence of *at least one* of some set of events—their alternative occurrence? This we can calculate if we know or can estimate the probability of each of the component events. The theorem we use is called the **addition theorem**.

For example, one might ask: What is the probability of drawing, from a shuffled deck of cards, *either* a spade *or* a club? Of course the probability of getting either of these outcomes will be greater than the probability of getting one of them, and certainly greater than the probability of getting the two of them jointly. In many cases, like this one, the probability of their alternative occurrence is simply the *sum* of the probabilities of the components. The probability of drawing a spade is $1/4$; the probability of drawing a club is $1/4$; the probability of drawing either a spade or a club is $1/4 + 1/4 = 1/2$. When the question concerns joint occurrence, we multiply; when the question concerns alternative occurrence, we add.

In the example just above, the two component events are **mutually exclusive**; if one of them happens, the other cannot. Drawing a spade necessarily entails the fact that a club was not drawn, and vice versa. So the addition theorem, when events are mutually exclusive, is straightforward and simple:

$$P(a \text{ or } b) = P(a) + P(b)$$

Addition theorem
In the calculus of probability, a theorem used to determine the probability of a complex event consisting of one or more alternative occurrences of simple events whose probabilities are known. The theorem applies only to mutually exclusive alternatives.

Mutually exclusive events
Events of such a nature that, if one occurs, the other(s) cannot occur at the same time. Thus, in a coin flip, the outcomes "heads" and "tails" are mutually exclusive events.

This may be generalized to any number of alternatives, *a* or *b* or *c* or .... If all the alternatives are mutually exclusive, the probability of one or another of them taking place is the sum of the probabilities of all of them.

Sometimes we may need to apply both the addition theorem and the product theorem. To illustrate, in the game of poker, a flush (five cards of the same suit) is a very strong hand. What is the probability of such a draw? We calculate first the probability of getting five cards in one given suit—say, spades. That is a joint occurrence, five component events that are certainly not independent, because each spade dealt reduces the probability of getting the next spade. Using the product theorem for dependent probabilities, we get

$$13/52 \times 12/51 \times 11/50 \times 10/49 \times 9/48 = 33/66,640$$

The same probability applies to a flush in hearts, or diamonds, or clubs. These four different flushes are mutually exclusive alternatives, so the probability of being dealt *any* flush is the sum of them: 33/66,640 + 33/66,640 + 33/66,640 + 33/66,640 = 33/16,660, a little less than .002. No wonder a flush is usually a winning hand.

Alternative events are often *not mutually exclusive*, and when they are not, the calculation becomes more complicated. Consider first an easy case: What is the probability of getting *at least* one head in two flips of a coin? The two components (getting a head on the first flip, or getting one on the second flip) are certainly not mutually exclusive; both could happen. If we simply add their probabilities, we get $1/2 + 1/2 = 1$, or certainty—and we know that the outcome we are interested in is not certain! This shows that the addition theorem is not directly applicable when the component events are not mutually exclusive. But we can use it *indirectly*, in either of two ways.

First, we can break down the set of favorable cases into mutually exclusive events and then simply add those probabilities. In the coin example, there are three favorable events: head–tail, tail–head, and head–head. The probability of each (calculated using the product theorem) is $1/4$. The probability of getting one of those three mutually exclusive events (using the addition theorem) is the sum of the three: $3/4$, or .75.

There is another way to reach the same result. We know that no outcome can be both favorable and unfavorable. Therefore the probability of the alternative complex we are asking about will be equal to the probability that not one of the component alternatives occurs, *subtracted from 1*. In the coin example, the only unfavorable outcome is tail–tail. The probability of tail–tail is $1/4$; hence the probability of a head on *at least one* flip is $1 - 1/4 = 3/4$, or .75, again. Using the notation $\bar{a}$ to designate an event that is *unfavorable* to *a*, we can formulate the theorem for alternative events, where the component events are not mutually exclusive, in this way:

$$P(a) = 1 - P(\bar{a})$$

The probability of an event's occurrence is equal to 1, minus the probability that that event will not occur.[*]

Sometimes the first method is simpler, sometimes the second. The two methods may be compared using the following illustration: Suppose we have two urns, the first containing two white balls and four black balls, the second containing three white balls and nine black balls. If one ball is drawn at random from each urn, what is the probability of drawing *at least one white ball*? Using the first method we divide the favorable cases into three mutually exclusive alternatives and then add the probabilities: (1) a white ball from the first urn and a black ball from the second $(2/6 \times 9/12 = 1/4)$; (2) a black ball from the first urn and a white ball from the second $(4/6 \times 3/12 = 1/6)$; and (3) a white ball from both urns $(2/6 \times 3/12 = 1/12)$. These being mutually exclusive we can simply add $1/4 + 1/6 + 1/12 = 1/2$. That sum is the probability of drawing at least one white ball. Using the second method we determine the probability of *failing*, which is the probability of drawing a black ball from both urns $(4/6 \times 9/12)$ and subtract that from 1. Thus we get $1 - 1/2 = 1/2$. The two methods yield the same result, of course.

Application of the probability calculus sometimes leads to a result that, although correct, differs from what we might anticipate after a casual consideration of the facts given. Such a result is called *counterintuitive*. When a problem's solution is counterintuitive, one may be led to judge probability mistakenly, and such "natural" mistakes encourage, at carnivals and elsewhere, the following wager: Three dice are to be thrown; the operator of the gambling booth offers to bet you even money (risk one dollar, and get that dollar back plus one more if you win) that no one of the three dice will show a one. There are six faces on each of the dice, each with a different number; you get three chances for an ace; superficially, this looks like a fair game.

In fact it is *not* a fair game, and hefty profits are reaped by swindlers who capitalize on that counterintuitive reality. The game would be fair only if the appearance of any given number on one of the three dice precluded its appearance on either of the other two dice. That is plainly not true. The unwary player is misled by mistakenly (and subconsciously) supposing mutual exclusivity. Of course, the numbers are not mutually exclusive; some throws will result in the same number appearing on two or three of the dice. The attempt to identify and count all possible outcomes, and then to count the outcomes in which at least one ace appears, quickly becomes frustrating. Because the appearance of any given number does not exclude the appearance of that same number on the remaining dice, the game truly is a swindle—and this becomes evident when the chances of winning are calculated by first determining the probability of *losing* and subtracting that from 1. The probability of any single *non*-ace (a 2, or 3, or 4, or 5, or 6) showing up is $5/6$. The probability of losing is that of getting three non-aces, which (because the dice are independent of one another) is $5/6 \times 5/6 \times 5/6$, which

---

*The reasoning that underlies this formulation of the theorem for alternative occurrences is as follows: The probability coefficient assigned to an event that is certain to occur is 1. For every event it is certain that either it occurs or it does not; either a or $\bar{a}$ must be true. Therefore, $P(a \text{ or } \bar{a}) = 1$. Obviously, a and $\bar{a}$ are mutually exclusive, so the probability of one or the other is equal to the sum of their probabilities; that is, $P(a \text{ or } \bar{a}) = P(a) + P(\bar{a})$. So $P(a) + P(\bar{a}) = 1$. By moving $P(\bar{a})$ to the other side of the equation and changing its sign, we get $P(a) = 1 - P(\bar{a})$.

equals $^{125}/_{216}$, or .579! The probability of the player throwing at least one ace, therefore, is $1 - {}^{125}/_{216} = {}^{91}/_{216}$, which is .421. This is a gambling game to pass up.

Let us now attempt to work out a moderately complicated problem in probability. The game of craps is played with two dice. The *shooter*, who rolls the dice, wins if a 7 or an 11 turns up on the first roll, but loses if a 2, or 3, or 12 turns up on the first roll. If one of the remaining numbers, 4, 5, 6, 8, 9, or 10, turns up on the first roll, the shooter continues to roll the dice until either that number turns up again, in which case the shooter wins, or a 7 appears, in which case the shooter loses. Craps is widely believed to be a "fair" game—that is, a game in which the shooter has an even chance of winning. Is this true? Let us calculate the probability that the shooter will win at craps.

To do this, we must first obtain the probabilities that the various numbers will occur. There are 36 different equipossible ways for two dice to fall. Only one of these ways will show a 2, so the probability here is $^1/_{36}$. Only one of these ways will show a 12, so here the probability is also $^1/_{36}$. There are two ways to throw a 3: 1–2 and 2–1, so the probability of a 3 is $^2/_{36}$. Similarly, the probability of getting an 11 is $^2/_{36}$. There are three ways to throw a 4: 1–3, 2–2, and 3–1, so the probability of a 4 is $^3/_{36}$. Similarly, the probability of getting a 10 is $^3/_{36}$. There are four ways to roll a 5 (1–4, 2–3, 3–2, and 4–1), so its probability is $^4/_{36}$, and this is also the probability of getting a 9. A 6 can be obtained in any one of five ways (1–5, 2–4, 3–3, 4–2, and 5–1), so the probability of getting a 6 is $^5/_{36}$, and the same probability exists for an 8. There are six different combinations that yield 7 (1–6, 2–5, 3–4, 4–3, 5–2, 6–1), so the probability of rolling a 7 is $^6/_{36}$.

The probability that the shooter will win on the first roll is the sum of the probability that a 7 will turn up and the probability that an 11 will turn up, which is $^6/_{36} + {}^2/_{36} = {}^8/_{36}$, or $^2/_9$. The probability of losing on the first roll is the sum of the probabilities of getting a 2, a 3, and a 12, which is $^1/_{36} + {}^2/_{36} + {}^1/_{36} = {}^4/_{36}$, or $^1/_9$. The shooter is twice as likely to win on the first roll as to lose on the first roll; however, the shooter is most likely not to do either on the first roll, but to get a 4, 5, 6, 8, 9, or 10. If one of these six numbers is thrown, the shooter is obliged to continue rolling the dice until that number is rolled again, in which case the shooter wins, or until a 7 comes up, which is a losing case. Those cases in which neither the number first thrown nor a 7 occurs can be ignored, for they are not decisive. Suppose the shooter gets a 4 on the first roll. The next *decisive* roll will show either a 4 or a 7. In a decisive roll, the equipossible cases are the three combinations that make up a 4 (1–3, 2–2, 3–1) and the six combinations that make up a 7. The probability of throwing a second 4 in the next decisive roll is therefore $^3/_9$. The probability of getting a 4 on the first roll was $^3/_{36}$, so the probability of winning by throwing a 4 on the first roll and then getting another 4 before a 7 occurs is $^3/_{36} \times {}^3/_9 = {}^1/_{36}$. Similarly, the probability of the shooter winning by throwing a 10 on the first roll and then getting another 10 before a 7 occurs is also $^3/_{36} \times {}^3/_9 = {}^1/_{36}$.

By the same line of reasoning, we can find the probability of the shooter winning by throwing a 5 on the first roll and then getting another 5 before throwing a 7. In this case, there are 10 equipossible cases for the decisive roll: the four ways to make a 5 (1–4, 2–3, 3–2, 4–1) and the six ways to make a 7. The probability of winning with a 5 is therefore $^4/_{36} \times {}^4/_{10} = {}^2/_{45}$. The probability of winning with a 9 is also $^2/_{45}$. The number 6 is still more likely to occur on the first roll, its probability being $^5/_{36}$. and it is more likely than the others mentioned to occur a second time before a 7 appears, the probability here being $^5/_{11}$. So the probability of winning with a 6 is $^6/_{36} \times {}^5/_{11} + = {}^{25}/_{396}$. And again, likewise, the probability of winning with an 8 is $^{25}/_{396}$.

There are eight different ways for the shooter to win: if a 7 or 11 is thrown on the first roll, or if one of the six numbers 4, 5, 6, 8, 9, or 10 is thrown on the first roll *and* again before a 7. These ways are all exclusive; so the total probability of the shooter's winning is the sum of the probabilities of the alternative ways in which winning is possible, and this is $^6/_{36} + {}^2/_{36} + {}^1/_{36} + {}^2/_{45} + {}^{25}/_{396} + {}^{25}/_{396} + {}^2/_{45} + {}^1/_{36} = {}^{244}/_{495}$. Expressed as a decimal fraction this is .493. This shows that in a craps game the shooter has *less* than an even chance of winning—only slightly less, to be sure, but still less than .5.

> **overview**
>
> ### The Product Theorem
>
> To calculate the probability of the *joint occurrence* of two or more events:
>
> A. If the events (say, *a* and *b*) are *independent*, the probability of their joint occurance is the simple product of their probabilities:
>
> $$P(a \text{ and } b) = P(a) \times P(b)$$
>
> B. If the events (say, *a* and *b* and *c*, etc.) are *not independent*, the probability of their joint occurance is the probability of the first event times the probability of the second event if the first occured, times the probability of the third event if the first and the second occured, etc:
>
> $$P(a \text{ and } b \text{ and } c) = P(a) \times P(b \text{ if } a) \times P(c \text{ if both } a \text{ and } b)$$
>
> ### The Addition Theorem
>
> To calculate the probability of the *alternative occurence* of two or more events:
>
> A. If the events (say, *a* and *b*) are *mutually exclusive*, the probability of at least one of them occuring is the simple addition of their probabilities:
>
> $$P(a \text{ or } b) = P(a) + P(b)$$
>
> B. If the events (say, *a* or *b* or *c*, etc.) are *not mutually exclusive*, the probability of at least one of them occuring may be determined by either
>   1. Analyzing the favorable cases into mutually exclusive events and summing the probabilities of those successful events; or
>   2. Determining the probability that no one of the alternative events will occur, and then subtracting that probability from 1.

552 CHAPTER 14 Probability

EXERCISES

B. *1. Calculate the shooter's chances of winning in a craps game by the second method; that is, compute the chances of his losing, and subtract that result from 1.

2. In drawing three cards in succession from a standard deck, what is the probability of getting at least one spade (a) if each card is replaced before making the next drawing? (b) if the cards drawn are not replaced?

3. What is the probability of getting heads at least once in three tosses of a coin?

4. If three balls are selected at random from an urn containing 5 red, 10 white, and 15 blue balls, what is the probability that they will all be the same color (a) if each ball is replaced before the next one is withdrawn? (b) if the balls selected are not replaced?

*5. If someone offers to bet you even money that you will not throw either an ace or a six on either of two successive throws of a die, should you accept the wager?

6. In a group of 30 students randomly gathered in a classroom, what is the probability that no two of those students will have the same birthday; that is, what is the probability that there will be no duplication of the same date of birth, ignoring the year and attending only to the month and the day of the month? How many students would need to be in the group in order for the probability of such a duplication to be approximately .5?

7. If the probability that a man of 25 will survive his 50th birthday is .742, and the probability that a woman of 22 will survive her 47th birthday is .801, and such a man and woman marry, what is the probability (a) that at least one of them lives at least another 25 years? (b) that only one of them lives at least another 25 years?

8. One partly filled case contains two bottles of orange juice, four bottles of cola, and four bottles of beer; another partly filled case contains three bottles of orange juice, seven colas, and two beers. A case is opened at random and a bottle selected at random from it. What is the probability that it contains a nonalcoholic drink? Had all the bottles been in one case, what is the probability that a bottle selected at random from it would contain a nonalcoholic drink?

9. A player in a game of draw poker is dealt three jacks and two small odd cards. He discards the latter and draws two cards. What is the probability that he improves his hand on the draw? (One way to improve it is to draw another jack to make four-of-a-kind; the other way to improve it is to draw any pair to make a full house.

CHALLENGE TO THE READER

The following problem has been a source of some controversy among probability theorists. Is the correct solution counterintuitive?

10. Remove all cards except aces and kings from a deck, so that only eight cards remain, of which four are aces and four are kings. From this abbreviated deck, deal two cards to a friend. If she looks at her cards and announces (truthfully) that her hand contains an ace, what is the probability that both her cards are aces? If she announces instead that one of her cards is the ace of spades, what is the probability then that both her cards are aces? Are these two probabilities the same?*

14.3 Probability in Everyday Life

In placing bets or making investments, it is important to consider not only the probability of winning or receiving a return, but also *how much* can be won on the bet or returned on the investment. These two considerations, *safety* and *productivity*, often clash; greater potential returns usually entail greater risks. The safest investment may not be the best one to make, nor may the investment that promises the greatest return *if* it succeeds. The need to reconcile safety and maximum return confronts us not only in gambling and investing, but also in choosing among alternatives in education, employment, and other spheres of life. We would like to know whether the investment—of money or of time and energy—is "worth it"—that is, whether that wager on the future is wise, all things considered. The future cannot be known, but the probabilities may be estimated. When one is attempting to compare investments, or bets, or "chancy" decisions of any kind, the concept of *expectation value* is a powerful tool to use.

Expectation value can best be explained in the context of wagers whose outcomes have known probabilities. Any bet—say, an even-money bet of $1 that heads will appear on the toss of a coin—should be thought of as a purchase; the money is spent when the bet has been made. The dollar wagered is the price of the purchase; it buys some *expectation*. If heads appears, the bettor receives a return of two dollars (one his own, the other his winnings); if tails appears, the bettor receives a $0 return. There are only two possible outcomes of this wager, a head or a tail; the probability of each is known to be $1/2$; and there is a specified return ($2 or $0) associated with each outcome. We multiply the return yielded on each possible outcome by the probability of that outcome being realized; the sum of all such products is the **expectation value** of the bet or investment. The expectation value of a one-dollar bet that heads will turn up when a fair coin is

Expectation value
In probability theory, the value of a wager or an investment; determined by multiplying each of the mutually exclusive possible returns from that wager by the probability of the return, and summing those products.

---
*For some discussion of this problem, see L. E. Rose, "Countering a Counter-Intuitive Probability," *Philosophy of Science* 39 (1972): 523–524; A. I. Dale, "On a Problem in Conditional Probability," *Philosophy of Science* 41 (1974): 204–206; R. Faber, "Re-Encountering a Counter-Intuitive Probability," *Philosophy of Science* 43 (1976): 283–285; and S. Goldberg, "Copi's Conditional Probability Problem," *Philosophy of Science* 43 (1976): 286–289.

tossed is thus equal to $(1/2 \times \$2) + (1/2 \times \$0)$, which is $1. In this case, as we know, the "odds" are even—which means that the expectation value of the purchase was equal to the purchase price.

This is not always the case. We seek investments in which the expectation value purchased will prove greater than the cost of our investment. We want the odds to be in our favor. Yet often we are tempted by wagers for which the expectation value is less, sometimes much less, than the price of the gamble.

The disparity between the price and the expectation value of a bet can be readily seen in a raffle, in which the purchase of a ticket offers a small chance at a large return. How much the raffle ticket is really worth depends on how small the chance is *and* how large the return is. Suppose that the return, if we win it, is an automobile worth $20,000, and the price of the raffle ticket is $1. If 20,000 raffle tickets are sold, of which we buy one, the probability of our winning is $1/20{,}000$. The chances of winning are thus very small, but the return if we win is very large. In this hypothetical case, the expectation value of the raffle ticket is $(1/20{,}000 \times \$20{,}000) + (19{,}999/20{,}000 \times \$0)$, or precisely $1, the purchase price of the ticket. The usual purpose of a raffle, however, is to raise money for some worthy cause, and that can happen only if more money is collected from ticket sales than is paid out in prizes. Therefore many more than 20,000 tickets—perhaps 40,000 or 80,000 or 100,000—will be sold. Suppose that 40,000 tickets are sold. The expectation value of our $1 ticket then will be $(1/40{,}000 \times \$20{,}000) + (39{,}999/40{,}000 \times \$0)$, or 50 cents. If 80,000 tickets are sold, the expectation value of the $1 ticket will be reduced to 25 cents, and so on. We may be confident that the expectation value of any raffle ticket we are asked to buy will be substantially less than the amount we are asked to pay for it.

Lotteries are very popular because of the very large prizes that may be won. States and countries conduct lotteries because every ticket purchased buys an expectation value equal to only a fraction of the ticket's price; those who run the lottery retain the difference, reaping huge profits.

The Michigan lottery, played by more than two-thirds of the citizens of that state, is typical. Different bets are offered. In one game, called the "Daily 3," the player may choose (in a "straight bet") any three-digit number from 000 to 999. After all bets are placed, a number is drawn at random and announced by the state; a player who has purchased a $1 straight-bet ticket on that winning number wins a prize of $500. The probability that the correct three digits in the correct order have been selected is 1 in 1,000; the expectation value of a $1 "Daily 3" straight-bet ticket is therefore $(1/1{,}000 \times \$500) + (999/1{,}000 \times \$0)$, or 50 cents.[*]

Lotteries and raffles are examples of great disparity between the price and the expectation value of the gambler's purchase. Sometimes the disparity is small, but the number of purchasers nevertheless ensures the profitability of the sale, as in gambling casinos, where every normal bet is one in which the purchase price is

---

[*]However imprudent a wager on the "Daily 3" may be, it is a very popular lottery—so popular that it is now run twice a day, midday and evening. One may infer that either those who purchase such lottery tickets have not thought through the expected value of their wagers, or that such wagering offers them satisfactions independent of the money value of their bets.

greater than the expectation value bought. In the preceding section we determined, using the product theorem and the addition theorem of the calculus of probability, that the dice game called craps is one in which the shooter's chance of winning is .493—just a little less than even. But that game is widely and mistakenly believed to offer the shooting player an even chance. Betting on the shooter in craps, at even money, is therefore a leading attraction in gambling casinos. Every such bet of $1 is a purchase of expectation value equal to (.493 × $2) + (.507 × $0), which is 98.6 cents. The difference of approximately a penny and a half may seem trivial, but because casinos receive that advantage (and other even greater advantages on other wagers) in thousands of bets made each day on the dice tables, they are very profitable enterprises. In the gambling fraternity, those who regularly bet on the shooter to win at craps are called paradoxically "right bettors," and among professional gamblers it is commonly said that "all right bettors die broke."

The concept of *expectation value* is of practical use in helping to decide how to save (or invest) money most wisely. Banks pay differing rates of interest on accounts of different kinds. Let us assume that the alternative bank accounts among which we choose are all government-insured, and that therefore there is no chance of a loss of the principal. At the end of a full year, the expectation value of each $1,000 savings investment, at 5 percent simple interest, is ($1,000 [the principal that we know will be returned]) + (.05 × $1,000), or $1,050 in all. To complete the calculation, this return must be multiplied by the probability of our getting it—but here we assume, because the account is insured, that our getting it is certain, so we merely multiply by 1, or $^{100}/_{100}$. If the rate of interest is 6 percent, the insured return will be $1,060, and so on. The expectation value purchased in such savings accounts is indeed greater than the deposit, the purchase price, but to get that interest income we must give up the use of our money for some period of time. The bank pays us for its use during that time because, of course, it plans to invest that money at yet higher rates of return.

Safety and productivity are considerations that are always in tension. If we are prepared to sacrifice a very small degree of safety for our savings, we may achieve a modest increase in the rate of return. For example, with that $1,000 we may purchase a corporate bond, perhaps paying 8 or even 10 percent interest, in effect lending our money to the company issuing the bond. The yield on our corporate bond may be double that of a bank savings account, but we will be running the risk—small but real—that the corporation issuing the bond will be unable to make payment when the loan we made to them falls due. In calculating the expectation value of such a bond, say at 10 percent, the amount to be returned to the investor of $1,000 is determined in precisely the same way in which we calculated the yield on a savings account. First we calculate the return, if we get it: ($1,000 [the principal]) + (10% × $1,000 [the interest]), or $1,100 total return. But in this case the probability of our getting that return is not $^{100}/_{100}$; it may be very high, but it is not 1. The fraction by which that $1,100 return therefore must be multiplied is the probability, as best we can estimate it, that the corporation will be financially sound when its bond is due for payment. If we think this probability is very high—say, .99—we may conclude that the purchase of the corporate bond at 10 percent offers an expectation value ($1,089) greater than that of the insured bank account at 5 percent ($1,050), and is therefore a wiser investment. Here is the comparison in detail:

Insured bank account at 5 percent simple interest for 1 year:
Return = (principal + interest) = ($1,000 + $50) = $1,050
Probability of return (assumed) = 1.0
Expectation value of investment in this bank account:
($1050 × 1 = $1,050) + ($0 × 0 = $0) *or* $1,050 *total*

Corporate bond at 10 percent interest, at the end of 1 year:
Return if we get it = (principal + interest) = ($1,000 + $100) = $1,100
Probability of return (estimated) = .99
Expectation value of investment in this corporate bond:
($1,100 × .99 = $1,089) + ($0 × .01 = $0) *or* $1,089 *total*

However, if we conclude that the company to which we would be lending the money is not absolutely reliable, our estimated probability of ultimate return will drop, say, to .95, and the expectation value will also drop:

Corporate bond at 10 percent interest, at the end of 1 year:
Return if we get it = (principal + interest) = ($1,000 + $100) = $1,100
Probability of return (new estimate) = .95
Expectation value of investment in this corporate bond:
($1,100 × .95 = $1,045) + ($0 × .05 = $0) *or* $1,045 *total*

If this last estimate reflects our evaluation of the company selling the bond, then we will judge the bank account, paying a lower rate of interest with much greater safety, the wiser investment.

Interest rates on bonds or on bank accounts fluctuate, of course, depending on the current rate of inflation and other factors, but the interest paid on a commercial bond is always higher than that paid on an insured bank account *because* the risk of the bond is greater; that is, the probability of its anticipated return is lower. The greater the known risk, the higher the interest rate must go to attract investors. Expectation value, in financial markets as everywhere, must take into consideration both probability (risk) and outcome (return).

When the soundness of a company enters our calculation of the expectation value of an investment in it, we must make some probability assumptions. Explicitly or implicitly, we estimate the fractions that we then think best represent the likelihoods of the possible outcomes foreseen. These are the fractions by which the returns that we anticipate in the event of these outcomes must be multiplied, before we sum the products. All such predictions are necessarily speculative, and all the outcomes calculated are therefore uncertain, of course.

When we can determine the approximate value of a given return *if* we achieve it, calculations of the kind here described enable us to determine what probability those outcomes *need* to have (given present evidence) so that our investment now will prove worthwhile. Many decisions in financial matters, and many choices in ordinary life, depend (if they are to be rational) on such estimates of probability and the resultant expected value. The calculus of probability may have application whenever we must gamble on the future.

There is no gambling system that can evade the rigor of the probability calculus. It is sometimes argued, for example, that in a game in which there are even-money stakes to be awarded on the basis of approximately equiprobable alternatives (such as tossing a coin, or betting black versus red on a roulette wheel), one can be *sure to win* by making the same bet consistently—always heads, or always the same color—and doubling the amount of money wagered after each loss. Thus, if I bet $1 on heads, and tails shows, then I should bet $2 on heads the next time, and if tails shows again, my third bet, also on heads, should be $4, and so on. One cannot fail to win by following this procedure, some suppose, because extended runs (of tails, or of the color I don't bet on) are highly improbable.* Anyway, it is said, the longest run must *sometime* end, and when it does, the person who has regularly doubled the bet will always be money ahead.

Wonderful! Why need anyone work for a living, when we can all adopt this apparently foolproof system of winning at the gaming table? Let us ignore the fact that most gaming houses put an upper limit on the size of the wager they will accept, a limit that may block the application of the doubling system. What is the real fallacy contained in this doubling prescription? A long run of tails, say, is almost certain to end sooner or later, but it may end later rather than sooner. So an adverse run may last long enough to exhaust any finite amount of money the bettor has to wager. To be certain of being able to continue doubling the bet each time, no matter how long the adverse run may continue or how large the losses are that the run has imposed, the bettor would have to begin with an infinite amount of money. Of course, a player with an infinite amount of money could not possibly win—in the sense of increasing his wealth.

Finally, there is a dangerous fallacy, in wagering or in investing, that an understanding of the calculus of probability may help us to avoid. The inevitable failure of the doubling technique underscores the truth that the probability of getting a head (or a tail) on the next toss of a fair coin cannot be affected by the outcomes of preceding tosses; each toss is an independent event. Therefore it is a foolish mistake to conclude, in flipping a coin, that because heads have appeared ten times in a row, tails is "due"—or to suppose that because certain digits have appeared frequently among winning lottery numbers that those digits are "hot." One who bets, or invests, on the supposition that some future event is made more probable, or less probable, by the frequency of the occurrence of independent events that have preceded it commits a blunder so common that it has been given a mocking name: "the gambler's fallacy."

On the other hand, if some mechanical device produces certain outcomes more frequently than others in a long-repeated pattern, one might conclude that the device is not designed (or is not functioning) as one supposed to yield outcomes that are equipossible. The dice may be loaded, or a roulette wheel (if the ball very frequently stops at the same section of the wheel) may not be balanced properly. The

---

*In fact, a long random sequence of heads and tails (or reds and blacks on a roulette wheel, etc.) will include extended runs of one result (tails, or reds, etc.) with much greater frequency than is commonly supposed. A run of a dozen heads in a row—requiring a bet of $2,048 on the 12th bet if one wagers steadily on tails in a doubling series that began with $1—is very far from rare. And after a run of twelve, of course, the chance of a thirteenth tail is $1/2$!

a priori theory of probability ("successes over outcomes") is rationally applied when we are confident that the set represented by the denominator of the fraction—the set of all outcomes—is a set of genuinely equipossible events. However, accumulated evidence may eventually cause one to conclude that the members of that set are not equipossible. At that point we may be well advised to revert to the frequency theory of probability, reasoning that the likelihood of certain outcomes is the fraction that represents the limit of the frequency with which that smaller set of outcomes has appeared. Whether we ought to apply the *a priori* theory or the relative frequency theory must depend on the evidence we have gathered and on our understanding of the calculus of probability as applied to that context.

EXERCISES

*1. In the Virginia lottery in 1992, six numbers were drawn at random from 44 numbers; the winner needed to select all six, in any order. Each ticket (with one such combination) cost $1. The total number of possible six-number combinations was 7,059,052. One week in February of that year, the jackpot in the Virginia lottery had risen to $27 million. (a) What was the expectation value of each ticket in the Virginia lottery that week?

These unusual circumstances led an Australian gambling syndicate to try to buy all of the tickets in the Virginia lottery that week. They fell short, but they were able to acquire some 5 million of the available six-number combinations. (b) What was the expectation value of their $5-million purchase? (Yes, the Aussies won!)

2. At most craps tables in gambling houses, the house will give odds of 6 to 1 against rolling a 4 the "hard way," that is, with a pair of 2's as contrasted with a 3 and a 1, which is the "easy way." A bet made on a "hard way" 4 wins if a pair of 2's show before either a 7 is rolled or a 4 is made the "easy way"; otherwise it loses. What is the expectation purchased by a $1 bet on a "hard-way" 4?

3. If the odds in craps are 8-to-1 against rolling an 8 the "hard way" (that is, with two 4's), what is the expectation purchased by a $1 bet on a "hard-way" 8?

4. What expectation does a person with $15 have who bets on heads, beginning with a $1 bet, and uses the doubling technique, if the bettor resolves to play just four times and quit?

*5. Anthrax is a disease that is nearly always deadly to cows and other animals. The nineteenth-century French veterinarian Louvrier devised a treatment for anthrax that was later shown to be totally without merit. His alleged "cure" was tried on two cows, selected at random from four cows that had received a powerful dose of anthrax microbes. Of the two he treated, one died and one recovered; of the two he left untreated, one died and one recovered. The reasons for recovery were unknown. Had Louvrier tested his "cure" on the two cows that happened to live, his treatment would have received impressive but spurious confirmation.

What was the probability of Louvrier choosing, for his test, just those two cows that chanced to live?

6. On the basis of past performance, the probability that the favorite will win the Bellevue Handicap is .46, while there is a probability of only .1 that a certain dark horse will win. If the favorite pays even money, and the odds offered are 8-to-1 against the dark horse, which is the better bet?

7. If $100 invested in the preferred stock of a certain company will yield a return of $110 with a probability of .85, whereas the probability is only .67 that the same amount invested in common stock will yield a return of $140, which is the better investment?

8. The probability of being killed by a stroke of lightning, calculated using the frequency theory of probability, is approximately 1 in 3 million. That is about 58 times greater than was the probability of winning the jackpot ($390 million) in the Mega Millions lottery in the United States in March of 2007. In that lottery the chances of buying a successful ticket were 1 in 175,711,536. One of two winning tickets was in fact purchased by a Canadian truck driver. If he had known the size of the jackpot, and that it would be divided by two winners, what would the expectation value of the $1 ticket he bought have been, at the time he bought it?

9. The following notice is a real one, distributed to all the parents in the school attended by the son of one of the authors of this book:

> **Emerson School Raffle—Up, Up and Away!**
>
> We are all winners!
> 4 lucky people will walk away with lots of cash!
>
> 1st Prize $1,000
> 2nd Prize $400
> 3rd Prize $250
> 4th Prize $100
>
> Chances of winning are good—Only 4,000 tickets were printed!
> Everyone will benefit from the great new sports equipment we will be able to buy with the money raised!

In this raffle, supposing all the tickets were sold, what was the expectation value of each ticket costing $1?

*10. The probability of the shooter winning in craps is .493, slightly less than even, as we proved in the preceding section. In casinos, a bet that the shooter will win is a bet on what is called the "Pass" line. We could all become rich, it would seem, if only we bet consistently against the shooter, on the "Don't Pass" line. But of course there is no such line; one cannot simply bet against the shooter, because the house will not take so unprofitable a bet. Yet usually one can place a bet called "Don't Pass–Bar 3," which wins if the shooter loses, unless the shooter loses by rolling a 3, in which case this bet loses also. What is the expectation value of a $100 bet on the "Don't Pass–Bar 3" line?

chapter 14 Summary

In all inductive arguments the conclusion is supported by the premises with only some degree of probability, and it is usually described simply as "more" or "less" probable in the case of scientific hypotheses. We explained in this chapter how a *quantitative* measure of probability, stated as a fraction between 0 and 1, can be assigned to many inductive conclusions.

Two alternative conceptions of probability, both permitting this quantitative assignment, were presented in Section 14.1.

- The *relative frequency* theory, according to which probability is defined as the relative frequency with which members of a class exhibit a specified attribute.
- The *a priori* theory, according to which the probability of an event occurring is determined by dividing the number of ways in which the event can occur by the number of equipossible outcomes.

Both theories accommodate the development of a calculus of probability, introduced in Section 14.2, with which the probability of a complex event can be computed if the probability of its component events can be determined. Two basic theorems, the *product theorem* and the *addition theorem*, are used in this probability calculus.

If the complex event of interest is a joint occurrence, the probability of two or more components both occurring, the *product* theorem is applied. The product theorem asserts that if the component events are *independent*, the probability of their joint occurrence is equal to the product of their separate probabilities. However, if the component events are *not independent*, the general product theorem applies, in which the probability of (*a* and *b*) is equal to the probability of (*a*) multiplied by the probability of (*b* if *a*).

If the complex event of interest is an alternative occurrence (the probability of at least one of two or more events), the *addition* theorem is applied. The addition theorem asserts that if the component events are *mutually exclusive*, their probabilities are summed to determine their alternative occurrence. But if the component events are *not mutually exclusive*, the probability of their alternative occurrence may be computed either:

- by analyzing the favorable cases into mutually exclusive events and summing the probabilities of those successes; or
- by determining the probability that the alternative occurrence will not occur and subtracting that fraction from 1.

In Section 14.3 we explained how the calculus of probability can be put to use in everyday life, allowing us to calculate the relative merits of alternative investments or wagers. We must consider both the *probability* of each of the various possible outcomes of an investment and the *return* received in the event of each. For each outcome, the anticipated return is multiplied by the fraction that represents the probability of that outcome occurring; those products are then summed to calculate the expectation value of that investment. ∎

LOGIC IN THE REAL WORLD

¡LUCHA LIBRE!

Lucha libre ("free wrestling") is a form of professional wrestling popular in Mexico and many other countries. The fighters, or *luchadores*, wear colorful masks and adopt intriguing stage names, such as *Mil Máscaras*, "Man of a Thousand Masks."

In an unprecedented move, two rival *luchadores*, *El Profesor Belicoso* (B) and *El Filósofo Enojado*, (G) have decided to settle their disputes with inductive arguments rather than body presses, headbutts, and aerial power moves.

Round 1: Argumentative Versus Nonargumentative Use of Analogy

■ Your mother is like an iceberg for five reasons: She is large, she is cold, she sank the *Titanic* in 1912, she helps polar bears travel from place to place, and global warming is melting her!

■ That is a nonargumentative use of analogy!

■ ¡*Estúpido*! Then you try!

■ ¡*Sí, señor*! Four months ago, the entire Hoyle family of Chiapas was preemptively imprisoned because they had been making suspicious maps and stockpiling supplies from the local match factory. Two weeks later, police searched the Hoyle family's iPads and found plans for a huge explosion! Since everyone who is related to you has been making suspicious maps and stockpiling supplies from the local match factory, your whole family should be jailed!

■ That is a mere personal attack, not an argumentative use of analogy! ¡*El Profesor está belicoso*!

You Be the Referee: Use your knowledge of argumentative versus nonargumentative analogies to decide who is correct and wins the point!

Round 2: Refutation by Logical Analogy

■ My divorce is a travesty! The judge has ordered me to give half of my assets to my ex-wife, saying that, by being my spotter in the gym, videotaping and analyzing the playbacks on all my fights, and making me high-protein meals for the last fifteen years, she gets half the credit for my success. However, my ex-wife and I also contributed 50/50 to our son Anolfo and the judge does not propose that we split him down the middle!

■ Your refutation by logical analogy does not share the same form as the argument you are trying to refute!

■ Oh yes it does, *señor*!

■ And besides, there are very important differences between your assets and your son Anolfo. Assets can be split down the middle, whereas a child cannot. And Anolfo has his own interests that must be taken into account, whereas your assets do not.

■ I defy you!

You Be the Referee: *El Profesor* has made two charges against *El Filósofo*—that his refutation does not share the same form as the argument he is trying to refute, and that the analogy is weak because of differences between the two things being compared. Use your knowledge of refutation by logical analogy to decide who wins the point!

Round 3: Match the Induction with the Method Being Used

■ If a man lights a match, calls his mother on the phone, and adopts a shelter dog, and then receives a care package in the mail, and then scratches his elbow, calls his mother on the phone, and buys illegal fireworks, and then receives a care package in the mail, and we conclude that the man's calling his mother on the phone is the cause of receiving the care package, which of John Stuart Mill's five techniques of inductive inference have we just used?!

■ Er... that is the method of difference! Okay, my turn!

If a *luchador* eats more tomatoes, begins exercising his pinky toe flexors, and adopts Vedic meditation practices, and then discovers that his fighting and his digestive health have improved, and then he increases his Vedic meditation and his digestive health improves to such a glorious extent that his lower body feels like that of a new, solid-gold-plated man, and we infer that the meditation has caused the enhanced digestive health, which of John Stuart Mill's five techniques of inductive inference have we just used?!

■ The method of concomitant variation!
■ Oh, really? Then what kind of variation is it, direct or inverse?
■ Um, direct!

Okay, my turn! If we know that Spandex gives you a rash and then you put on a Spandex bodysuit and I bodyslam you, and you both get a rash and get a big bruise on your stupid chest, and I conclude, since I know the bodysuit caused the rash, that my bodyslamming caused the bruise, which of John Stuart Mill's five techniques of inductive inference have I just used?

■ That would never happen! But it is the method of residues!

You Be the Referee: Each *luchador* has answered two questions. Tally up the right versus wrong answers (refer to section 12.4 in your textbook for a review of Mill's techniques) to decide who wins the point!

Round 4: Scientific Method Quiz

True or False: The first step in scientific investigation is formulating the explanatory hypothesis.

■ True!
■ False! The first step is identifying the problem!

True or False: Compatibility with previously well-established hypotheses is a justifiable way of evaluating scientific explanations.

■ True!
■ Ridiculous. The point of hypotheses is to help us find out *new* things.

True or False: When two rival hypotheses fit equally well with established theory and have roughly equal predictive power, we are likely to favor the more intricate of the two.

■ True! And fancy!
■ False! We want the simpler of the two. The simplest theory that fits the facts is likely to involve the smallest number of unjustified assumptions.

You Be the Referee: Use your knowledge of the scientific method (Chapter 13) to decide who is correct and wins the point!

Round 5: The Probability Calculus

■ If there is a 1/10 chance that you will ever win a match against a *luchador* over five feet tall, what is the probability that you will win five matches against five different *luchadores* all over five feet tall?

■ Lies! But if I only had a 1/10 chance of winning, which is totally false, then the answer is 5/10, or 1/2!

Here is one for you: If there is a 1/5 chance you can eat a taco without getting salsa all over your costume, what is the probability that you can eat three tacos without getting salsa all over your costume?

■ I will avenge my honor! But the answer is 1/125.

A question: If you have a bag of eggs and 8 of them are rotten and 2 of them are good because all your chickens are sickly, and you reach into the bag to pull out one egg, and then another egg, what are your miserable chances of getting two good eggs?

■ How dare you malign my chickens! But the answer is 1/25.

One more question: If the probability of your passing an eye test is 0.65, and the probability of your showing up to a match on time is 50%, and the probability of your rash clearing up is 3/8, and if all those things are necessary— but not sufficient—conditions for you to win a match, and even after all of that, you still only have a one-in-four chance of victory, what is the probability of your winning a match? You may use a calculator, fool!

■ I do not know! I will crush you!
■ I have crushed you already—*with my mind!*

You Be the Referee: Each *luchador* has answered (or failed to answer) two questions. Calculate the answer to each question for yourself, and award the point to the competitor who performed better overall.

Who wins the five-round battle?

Solutions

Because Part III of your textbook dealt with inductive, rather than deductive, arguments, some difference of opinion is reasonable here.

Round One

El Profesor gives five reasons why *El Filósofo*'s mother is like an iceberg. While he puts great effort into constructing an analogy, he neglects to use the analogy to make an argument; that is, there is no conclusion other than the analogy itself. A fully developed analogical argument would then say *something else* about an iceberg that could then be applied to *El Filósofo*'s mother.

El Profesor attempts to write off *El Filósofo*'s argument as a mere personal attack, but the argument is actually a well-developed analogical argument: It lays out the ways in which *El Profesor*'s family is like the Hoyle family, and then it makes an argument that the way the Hoyle family was treated is how *El Profesor*'s family should be treated. Regardless of the truth of the premises, *El Filósofo* has in fact made an argumentative use of analogy.

Point to El Filósofo!

Round Two

El Profesor alleges that *El Filósofo*'s refutation by logical analogy does not share the same form as the argument he is trying to refute. This is not true. The argument he is trying to refute is as follows: Because *El Filósofo* and his ex-wife each contributed to his career, the proceeds should be split. *El Filósofo* makes an argument that shares the same form: He and his ex-wife each contributed to Anolfo, but the conclusion that Anolfo should be split is absurd.

However, *El Profesor*'s second claim certainly has merit. Anolfo is different from *El Filósofo*'s assets in many important ways.

El Filósofo has made a flawed argument. *El Profesor* has made two criticisms of it, only one of which was correct.

A referee might reasonably award the point to El Profesor or might declare a draw.

Round Three

If a man does A, B, and C, and then receives a care package in the mail; and then does X, B, and Y, and again receives a care package in the mail; and we conclude that B was the cause of receiving the care package, we have not used the method of difference, but rather the method of agreement. *El Profesor* is ahead.

If a *luchador* does A, B, and C, and then discovers that his fighting and his digestive health have improved, and then he increases C and his digestive health improves even more, and we infer that C has caused the enhanced digestive health, we have indeed used the method of concomitant variation. *El Profesor* is also correct that this is an instance of direct variation—when one thing increases, so does the other thing.

If A and C result in X and Y, and we know that A causes X, so we "subtract out" A and X to conclude that C caused Y, we have indeed used the method of residues. Score one for *El Filósofo*.

El Profesor has wiped the floor with *El Filósofo* in this round.

Round Four

1. *El Filósofo* is correct—the first step is identifying the problem (p. 516).
2. *El Profesor* is correct—while it is true that the point of the scientific method is to be open to new information and ideas, it is also true that science aims at compiling a system of hypotheses that work together (p. 522).
3. *El Filósofo* is correct—the simpler theory is preferred (p. 524). The principle of Ockham's Razor may also be invoked (Google it!)

Point to El Filósofo!

Round Five

El Filósofo asks about the chance of winning five matches if the chance of winning each one is 1/10. *El Profesor* answers 5/10, or 1/2. However, this answer is not sensible—how could the probability of winning all five matches be greater than the probability of winning one of them? *El Profesor* has made the common mistake of adding instead of multiplying. The correct answer is 1/100,000.

El Profesor asks, "If there is a 1/5 chance you can eat a taco without getting salsa all over your costume, what is the probability that you can eat three tacos without getting salsa all over your costume?" *El Filósofo* has correctly multiplied the probabilities of three independent events: $1/5 \times 1/5 \times 1/5 = 1/125$. *El Filósofo* is most definitely ahead.

El Filósofo gives a scenario with 8 rotten and 2 good eggs, and asks for the probability of selecting the two good eggs. *El Profesor* seems to have correctly determined that the probability of selecting a good egg on the first pick is 2/10, or 1/5, but then he incorrectly assumes that the probability is the same on the second pick. He then erroneously multiplies $1/5 \times 1/5$ to get 1/25. However, after the first good egg has been selected, the chance of picking *another* good egg is no longer 2 out of 10 because there are only 9 eggs left and only 1 of them is good. The chance of getting a good egg the second time around is 1/9. The correct answer is $1/5 \times 1/9 = 1/45$. *El Filósofo* is still ahead.

Finally, *El Profesor* gives a complicated and insulting scenario with a variety of probabilities. In order to win, *El Filósofo* would have to pass an eye test ($P = 0.65$), show up on time (0.5), have his rash clear up (3/8 or 0.375), and then actually beat his opponent (0.25). While this sounds complicated, all that is necessary is simply to plug $0.65 \times 0.5 \times 0.375 \times 0.25$ into a calculator. The answer is 0.03046875, or about 3%. *El Filósofo* neglected to attempt this question, so score one for *El Profesor*.

However, El Filósofo is the only one to have answered a question correctly in round 5, so he gets the point.

El Filósofo is the clear winner in three of five rounds and wins overall! He will be receiving an extremely heavy belt as a prize. El Profesor has already challenged him to a rematch on the topic of seventeenth-century French poetry.

Appendix

A number of high-level tests of cognitive aptitude rely upon some of the operations covered in this book. The GRE (Graduate Record Examination), the LSAT (Law School Admission Test), the MAT (Miller Analogies Test), and the GMAT (Graduate Management Admission Test), as well as several IQ tests, rely heavily upon the ability to process information in ways that have been detailed herein.

Guide to the Graduate Record Examination (GRE)

The **GRE** is a test of great complexity. It consists of a verbal and a quantitative section, as well as analytical writing. The verbal section includes tests of reading comprehension, sentence completion, vocabulary (usually tested by identifying synonyms and antonyms), and analogies. The quantitative section tests the ability to solve problems of a basic mathematical nature. The only section that concerns us here is the verbal section, and in that, the analogies subtest. Analogies were covered at length in Chapter 11.

In analogy tests, the basic structure is made explicit, making it easy to isolate and identify relationships. The various types of analogies given include:

1. Definitions
2. Ascertaining defining characteristics
3. Class and member
4. Synonyms
5. Antonyms
6. Part/Whole
7. Degree
8. Function
9. Tool/Function
10. Action/Effect
11. Cause/Effect
12. Worker/Tool
13. Worker/Product
14. Worker/Workplace
15. Kind
16. Size
17. Spatial sequence
18. Time sequence
19. Part of speech
20. Symbol/Concept

The structure is of the form A:B::C:D, which is read "A is to B as C is to D." To solve problems of these sorts, the relationship between the first pair of words has to be identified. Then, among the options provided, the pair of words exhibiting the analogous relationship has to be identified. All options should be read; if more than one seems plausible, look for the one that most precisely expresses the relationship sought. More than the meanings of the individual words is involved: connotations, nuances, secondary meanings, and contextual information may be significant. For example, "right" can mean direction, or a political concept, or an ethical judgment. "Embroider" could relate to fabric, or could be used metaphorically, to suggest exaggeration.

The following examples of analogies from the GRE are from the Analogies Practice Tests, available online at *http://www.testprepreview.com*.

1. MORBID:UNFAVORABLE::
   a. reputable:favorable
   b. maternal:unfavorable
   c. disputatious:favorable
   d. vigilant:unfavorable
   e. lax:favorable

2. SULLEN:BROOD::
   a. lethargic:cavort
   b. regal:cringe
   c. docile:obey
   d. poised:blunder
   e. despondent:laugh

3. AUTHOR:LITERATE::
   a. cynic:gullible
   b. hothead:prudent
   c. saint:notorious
   d. judge:impartial
   e. doctor:fallible

Answer Key: 1. a 2. c 3. d

Here are more examples, from the Educational Testing Service at *www.ets.org/Media/Tests/GRE/pdf/gre_0809_practice_book.pdf*.

1. STYGIAN:DARK::
   a. abysmal:low
   b. cogent:contentious
   c. fortuitous:accidental
   d. fortuitous:accidental
   e. cataclysmic:doomed

2. WORSHIP:SACRIFICE::
   a. generation:pyre
   b. burial:mortuary
   c. weapon:centurion
   d. massacre:invasion
   e. prediction:augury

3. EVANESCENT:DISAPPEAR::
   a. transparent:penetrate
   b. onerous:struggle

565

c. feckless:succeed
   d. illusory:exist
   e. pliant:yield
4. UPBRAID:REPROACH::
   a. dote:like
   b. lag:stray
   c. vex:please
   d. earn:desire
   e. recast:explain

Answer Key: 1. a 2. e 3. e 4. a

Guide to the Law School Admission Test (LSAT)

The **LSAT** consists of tests of reading comprehension, as well as tests of the ability to perceive and analyze complex relationships. Moreover, these tests purport to assess one's ability to think critically, detect fallacies, identify analogical relationships, reason logically, and evaluate the evidence adduced in support of the conclusions of an argument.

Some of what is classified under "Logical Reasoning" involves constructing hypothetical deductive arguments in conjunction with recombinations of the provided information. This type of problem was analyzed in Section 2.4 in this book.

The following examples are from the Logical Reasoning Practice Tests, available online at *http://www.testprepreview.com*.

> A chess tournament is occurring in the local community school, and the players at all four of the tables are engaged in their fourth game against their respective opponents.
>
> The players with white pieces are: David, Gerry, Lenny and Terry
>
> The players with black pieces are: Don, Mike, Ritchie and Stephen
>
> The scores are 3:0, 2.5:0.5, 2:1, 1.5:1.5
>
> [note: tied games result in a score of 0.5 points for each player]
>
> Lenny is playing at the table to the right of Stephen, who has lost all of his games until now.
>
> Gerry is playing against Mike.
>
> At least one game at table 1 has resulted in a tie.
>
> Ritchie, who is not in the lead over his opponent, has not been in a tied game.
>
> The player who is using the white pieces at table 4 is Terry; however, the current score at table 4 is not 2:1.
>
> Don is leading his match after his last three games.

The following four questions refer to the chess tournament scenario just described:

1. What table is Stephen playing at, and what is the score at that table?
   a. Table 1, 2.5:1.5
   b. Table 1, 3:0
   c. Table 2, 3:0
   d. Table 2, 2.5:1.5
   e. Table 3, 2:1

2. Whose score is highest?
   a. Mike's
   b. Stephen's
   c. Ritchie's
   d. David's
   e. Lenny's

3. Which player has black pieces and is tied?
   a. Mike
   b. David
   c. Ritchie
   d. Don
   e. Terry

4. Who is the winning player at table 4?
   a. Don
   b. Terry
   c. David
   d. Gerry
   e. Ritchie

Answer Key: 1. c 2. d 3. a 4. a

Many of the Analytical Reasoning subtests entail determining which statement would strengthen (render more probable) or weaken (render less probable) an inductive argument. Others involve selecting, from a set of possible choices, a conclusion that can be validly deduced from the premises embedded in a short paragraph. Still others require identification of those tacit assumptions underlying a given argument's conclusion, while others require interpretation of conclusions provided.

Some examples (from the same source):

1. My family doctor said that he would be performing a blood test on me when I visit him today. I know I will feel pain today.

   On which one of the following assumptions does the above argument depend?
   a. The use of a needle always causes pain in the patient.
   b. The doctor will have a hard time finding the patient's vein.
   c. In the past, this patient has experienced pain at the family doctor's.
   d. The needle will leave a bruise.
   e. The doctor will have to try different needles to perform the test.

2. Never again will you have to pay high prices for imported spring water. It is now bottled locally and inexpensively. You'll never taste the difference; however, if you're likely to be embarrassed to serve domestic spring water, simply serve it in a lead crystal decanter.

    What is the assumption made by this ad?
    a. It's not hard to tell domestic water from imported water based on its flavor.
    b. The majority of spring water is bottled at its source.
    c. Restrictions on importing and customs duties make the price of imported water higher.
    d. Spring water tastes best when it's served from a decanter.
    e. Some people purchase imported spring water instead of domestic as a status symbol.

3. Estelle states: When I went fishing the other day, every fish that I caught was a salmon, and every salmon I saw I caught.

    Of the statements listed below, which one can be concluded from the observations of Estelle?
    a. Salmon was the only fish that Estelle saw while she was fishing.
    b. While Estelle was fishing, no fish other than salmon were caught by her.
    c. In the area where Estelle fished, there were no fish other than salmon.
    d. All of the fish that Estelle saw she caught.
    e. Estelle did not see any fish other than salmon while she was fishing.

4. While traveling to Japan, a low-ranking U.S. ambassador asked a Japanese official why Japanese people were so inscrutable. Looking calm and friendly, the official responded in a gentle voice that he much preferred to think upon his race as inscrutable than as wanting in perspicacity, like Americans.

    Of the following statements, which best describes the Japanese official's comment?
    a. All people are inscrutable, not just the Japanese.
    b. Most Americans don't understand Japanese culture.
    c. What a person lacks in perception may be a result of the carelessness of the observer, instead of the obscurity within the object being observed.
    d. The Japanese distrust American ambassadors.
    e. If the East and West are ever to understand each other, there will need to be a much better cultural understanding.

Answer Key: 1. a   2. e   3. b   4. c

The following examples are from *www.lsac.org/LSAT/lsat-prep-materials.asp*.

Passage for Question 1

A medical clinic has a staff of five doctors—Drs. Albert, Burns, Calogero, Defeo, and Evans. The national medical society sponsors exactly five conferences, which the clinic's doctors attend, subject to the following constraints:

If Dr. Albert attends a conference, then Dr. Defeo does not attend it.

If Dr. Burns attends a conference, then either Dr. Calogero or Dr. Defeo, but not both, attends it.

If Dr. Calogero attends a conference, then Dr. Evans does not attend it.

If Dr. Evans attends a conference, then either Dr. Albert or Dr. Burns, but not both, attends it.

Question 1

If Dr. Burns attends one of the conferences, then which one of the following could be a complete and accurate list of the other members of the clinic who also attend that conference?
a. Drs. Albert and Defeo
b. Drs. Albert and Evans
c. Drs. Calogero and Defeo
d. Dr. Defeo
e. Dr. Evans

Explanation for Question 1

This question requires you to determine, from the conditions given, which doctors can attend the same conferences. The question tells us that "Dr. Burns attends one of the conferences," and we are asked to choose the response that could be a list of only those doctors who attend the conference with Dr. Burns. Since we are asked what could be a "complete and accurate list" of those doctors who attend the conference with Dr. Burns, we can eliminate as incorrect those responses that are either inaccurate (that is, cannot be true) or incomplete (that is, do not include everyone who must accompany one or more of the doctors going to the conference). This can be determined easily without the use of a diagram.

Response (a) states that, along with Dr. Burns, Drs. Albert and Defeo also attend the conference. But the first condition tells us that "if Dr. Albert attends a conference, then Dr. Defeo does not attend it." So, Drs. Burns, Albert, and Defeo cannot all attend the same conference. Response (a), then, is incorrect.

Response (b) is incorrect for a similar reason. The fourth condition tells us what must be true if Dr. Evans attends a conference, namely, that "either Dr. Albert or Dr. Burns, but not both, attends it." Since we know that Dr. Burns attends the conference, we know that it cannot be true that both Drs. Albert and Evans also attend that conference.

Response (c) is also incorrect. The second condition tells us what must be true if Dr. Burns attends a conference. Since we know that Dr. Burns does attend the conference, we also know that "either Dr. Calogero or Dr. Defeo, but not both, attends it."

Responses (d) and (e) must be evaluated slightly differently. No condition rules out Dr. Burns's and Dr. Defeo's going to the same conference—response (d)—and no condition forbids Dr. Evans's going with Dr. Burns to a conference—response (e). But recall that the question asks for what could be a "complete and accurate list" of the doctors who attend the conference with Dr. Burns. We know from the second condition that at least one other person must accompany Dr. Burns, and that among those who accompany Dr. Burns is either Dr. Calogero or Dr. Defeo. Since the conditions do not require anyone to accompany Dr. Defeo, it is possible that Dr. Defeo is the only person to accompany Dr. Burns. Thus, response (d) is an accurate response, in that it is possible that Drs. Burns and Defeo attend the same conference, and it is a complete response, in that Drs. Burns and Defeo could be the only doctors of the five to attend the conference. So response (d) is correct.

Response (e) is incorrect because we know that if Dr. Burns goes, someone other than Dr. Evans must also go. Response (e) then is incomplete. It fails to list at least one doctor who we know must also accompany Dr. Burns.

This is a question of "moderate difficulty"; 60 percent of those who took the LSAT on which it appeared answered it correctly. The most common error was selecting response (b) (chosen by 17 percent).

Passage for Question 2

A law firm has exactly nine partners: Fox, Glassen, Hae, Inman, Jacoby, Kohn, Lopez, Malloy, and Nassar. Their salary structure must meet the following conditions:
Kohn's salary is greater than both Inman's and Lopez's.
Lopez's salary is greater than Nassar's.
Inman's salary is greater than Fox's.
Fox's salary is greater than Malloy's.
Malloy's salary is greater than Glassen's.
Glassen's salary is greater than Jacoby's.
Jacoby's salary is greater than Hae's.

Question 2

If Malloy and Nassar earn the same salary, what is the minimum number of partners that must have lower salaries than Lopez?
    a. 3
    b. 4
    c. 5
    d. 6
    e. 7

Explanation for Question 2

As with many problems involving relative rank or order, the test taker should attempt to diagram the various relationships given in the stimulus.

In what follows, each partner's name is abbreviated by its first letter, and the symbol ">" indicates that the person whose initial appears to the left of the sign has a greater salary than that of the person whose initial is to the right of the sign. So, for instance, "K > L" means "Kohn's salary is greater than Lopez's."

The conditions indicate the following eight relative orderings of salary:
1. K > I
2. K > L
3. L > N
4. I > F
5. F > M
6. M > G
7. G > J
8. J > H

It should be obvious that if person A's salary is greater than person B's and if person B's salary is greater than person C's, then person A's salary is also greater than person C's. Using this principle, we may combine and condense several of the above orderings into two separate "chains" of relative order:

9. K > I > F > M > G > J > H (This combines [1], and [4] through [8].)
10. K > L > N (This combines [2] and [3].)

We are now in a position to determine the correct response for question 8. The test taker is asked to determine the minimum number of partners whose salaries must be lower than that of Lopez, if Malloy and Nassar earn the same salary. Assuming that Malloy's and Nassar's salaries are equal allows us to infer from chains (9) and (10) the following chain of relative ordering:

(11)   K > L > (N,M) > G > J > H.

Chain (11) shows that since M and N have the same salary, anyone whose salary is less than M's also has a salary that is less than N's, and therefore also less than L's. So, at least Malloy, Nassar, Glassen, Jacoby, and Hae must have lower salaries than Lopez. This shows that response options (a) and (b) are both incorrect. If we can now show that no partner other than these five must have lower salaries than Lopez, then we will have shown that (c) is the correct response.

To see that there could be fewer than six partners with lower salaries than Lopez, one need merely look at (9) and (10) above to see that as long as Inman and Fox have lower salaries than Kohn, they could have salaries equal to or higher than Lopez's. This allows us to construct the following possible complete chain of relative order:

(12) K > I > F > L > (N,M) > G > J > H.

In this possible case no more than five partners have lower salaries than Lopez, and since there must be at least five such partners, five is the minimum number of such partners. (c) is therefore the correct response. This is considered an item of "middle difficulty."

Questions of this type are essentially mathematical in nature, involving the use of such relations as "greater than" and "less than." This and the previous test item come under the heading of "analytical reasoning," but the type of analyses required is not very sophisticated. No semantic or linguistic analysis is involved, no analysis of the meaning of concepts is involved, no causal analysis is called for, and the logical operations tend to be mechanically combinatorial in nature. Such items are easier to design and score, but require little in the way of deep understanding of abstract concepts or original thought processes, and do not even assess the types of analyses or reasoning performed by scientists and philosophers, let alone people in most other professions.

Guide to the Miller Analogies Test (MAT)

The Miller's Analogies Test or MAT is a test comprised of 120 analogies (only 100 of which are scored) which are administered during a 60-minute time period. The MAT web site provides the following description: "The MAT tests high-level mental ability by requiring those taking the test to solve problems stated as analogies."

Psychologists who study human intelligence and reasoning have found that a person's performance on an analogies test represents one of the best measures of their verbal comprehension and analytical thinking. Since analytical thinking is critical for success in both graduate school and professional life, performance on the MAT can provide insight into a candidate's aptitude in these areas.

The test contains analogies with content from various academic subjects. MAT scores help graduate schools identify candidates whose knowledge and abilities go beyond the mere memorization and recitation of information.

You will find some similarities between what was covered in the analogies section of the GRE, above, with what is found on the MAT.

Examples (from Kaplan Miller Analogies Test, 3rd Edition):

BOLOGNA:COLD CUT::PARFAIT: (a. banana, b. dessert, c. pastrami, d. entree)

As bologna is a type of cold cut, and a parfait is a type of dessert, the correct response is b.

However, in

RENEGADE:INTROVERT::LOYAL: (a. gregarious, b. steadfast, c. treacherous, d. withdrawn)

the relationship sought is between "renegade" and "loyal," which are antonyms; hence, the antonymous relationship between "introvert" and the choices offered is with "gregarious."

Finally, the absent word need not be the last in the sentence. It could be any word.

1. FRAGILE: (a. brittle, b. spoiled, c. sturdy, d. malleable) ::BREAK:MOLD

Answer: d.

2. (a. etymology, b. ontology, c. pedagogy, d. philosophy): ENTOMOLOGY::EDUCATION:INSECTS

Answer: c.

Guide to the Graduate Management Admission Test (GMAT)

The **GMAT** consists of an analytical writing assignment, a quantitative section, and a verbal section. The first two sections are not within our purview here. The verbal section consists of sentence correction, which is grammatical-linguistic in character; reading comprehension, which is self-explanatory; and critical reasoning. It is with critical reasoning that we are concerned.

There are several different types of test items. These include:
1. Assumption questions
2. Strengthen or Weaken questions
3. Inference questions

In assumption questions, the reader is asked to state what tacit premise, or what missing conclusion, is being assumed in a given passage. This was addressed in the section on enthymemes.

In "Strengthen or Weaken" questions, just as in the LSAT discussed above, the task is to determine which statement would render an inductive argument more or less probable.

Other items assess the ability to detect fallacies. This was addressed in Chapter 4.

Finally, items are provided that test one's ability to identify as inductive or deductive the arguments presented, and to identify the methods of inference. These include Mills Methods and parallel argument forms.

Examples (from the Kaplan GMAT Test):
Directions: Select the best answer for each question.

1. In Los Angeles, a political candidate who buys saturation radio advertising will get maximum name recognition.

   The statement above logically conveys which of the following?
   a. Radio advertising is the most important factor in political campaigns in Los Angeles.
   b. Maximum name recognition in Los Angeles will help a candidate win a higher percentage of votes cast in the city.
   c. Saturation radio advertising reaches every geographically distinct sector of the voting population in Los Angeles.
   d. For maximum name recognition a candidate need not spend on media channels other than radio advertising.
   e. A candidate's record of achievement in the Los Angeles area will do little to affect his or her name recognition there.

   Answer: d. From the statement given, it follows that such advertising is sufficient for maximum name recognition without spending on other media. We cannot infer from that premise any of the other statements provided.

2. The extent to which a society is really free can be gauged by its attitude toward artistic expression. Freedom of expression can easily be violated in even the most outwardly democratic of societies. When a government arts council withholds funding from a dance performance that its members deem "obscene," the voices of a few bureaucrats have in fact censored the work of the choreographer, thereby committing the real obscenity of repression.

   Which of the following, if true, would most seriously weaken the argument above?
   a. Members of government arts councils are screened to insure that their beliefs reflect those of the majority.
   b. The term "obscenity" has several different definitions that should not be used interchangeably for rhetorical effect.
   c. Failing to provide financial support for a performance is not the same as actually preventing or inhibiting it.
   d. The council's decision could be reversed if the performance were altered to conform to public standards of appropriateness.
   e. The definition of obscenity is something on which most members of society can agree.

   Answer: c. The passage equates the withholding of funding with censorship. (c) denies that equation, which destroys the argument. (a) is irrelevant—the definition offered by the author of the passage is not contingent upon what anyone thinks. (b) challenges the use of the word "obscenity," but the term under consideration is "censorship." (d) misses the point entirely, and reaffirms the author's position. (e) also invokes majority opinion, but this is beside the point.

3. Ronald is a runner on the track team and is a great hurdler. All runners on the track team are either sprinters or long-distance runners, but a few long-distance runners do not run the sprint because they are not fast enough. Hurdlers never run long distance because they lack the necessary endurance. Therefore, Ronald must be fast.

   For the conclusion drawn above to be logically correct, which of the following must be true?
   a. Sprinters are faster than hurdlers.
   b. All runners on the track team who run hurdles also run long distance.
   c. Hurdling requires more endurance than running long distance.
   d. All sprinters are fast.
   e. Every runner on the track team who is fast is a sprinter.

Answer: d. We are given that Ronald is a hurdler, and that all runners are either sprinters or long-distance runners. No hurdlers are long-distance runners. Therefore Ronald must be a sprinter, and if all sprinters are fast, Ronald must be fast.

(a) makes no sense because we know all hurdlers are sprinters; (b), (c), and (e) all contradict propositions in the argument.

IQ Tests

Various tests of intelligence employ deductive inference and/or analogical reasoning as part of their content.

The Stanford-Binet Intelligence Scales, among many other such tests, includes analogy items among its subtests, as well as opposite analogies and reasoning items.

The Concept Mastery Test, written by Lewis Terman to assess extremely high levels of cognitive ability, includes 75 analogy problems.

The Cattell Test of Crystallized Intelligence used by Mensa contains figure analogies and reasoning items. And the Cattell test of fluid intelligence (the Culture Fair, Scales 2 and 3), while nonverbal, relies upon the ability to infer the correct response to problems based upon perceived patterns of relationships. In sum, reasoning, of one sort or another, lies at the core of aptitude and intelligence testing.

Solutions to Selected Exercises

chapter 5

SECTION 5.3 *Exercises on p. 170*

1. S = historians;
   P = extremely gifted writers whose works read like first-rate novels.
   Form: Particular affirmative.

5. S = members of families that are rich and famous;
   P = persons of either wealth or distinction.
   Form: Particular negative.

10. S = people who have not themselves done creative work in the arts;
    P = responsible critics on whose judgment we can rely.
    Form: Universal negative.

SECTION 5.4 *Exercises on p. 175*

1. Quality: affirmative; quantity: particular; subject and predicate terms both undistributed.
5. Quality: negative; quantity: universal; subject and predicate terms both distributed.
10. Quality: affirmative; quantity: universal; subject term distributed, predicate term undistributed.

SECTION 5.5 *Exercises on p. 180*

1. If we assume that (a) is true, then:
   (b), which is its contrary, is false, and
   (c), which is its subaltern, is true, and
   (d), which is its contradictory, is false.
   If we assume that (a) is false, then:
   (b), which is its contrary, is undetermined, and
   (c), which is its subaltern, is undetermined, and
   (d), which is its contradictory, is true.

SECTION 5.6 *Exercises on pp. 186–188*

A. p. 186

1. No reckless drivers who pay no attention to traffic regulations are people who are considerate of others. Equivalent.
5. Some elderly persons who are incapable of doing an honest day's work are professional wrestlers. Equivalent.

B. p. 187

1. Some college athletes are not nonprofessionals. Equivalent.
5. No objects suitable for boat anchors are objects that weigh less than fifteen pounds. Equivalent.

C. p. 187

1. All nonpessimists are nonjournalists. Equivalent.
5. Some residents are not citizens. Equivalent.

D. p. 187

1. False
5. Undetermined
10. False

E. p. 187

1. False
5. Undetermined
10. False

573

F. p. 188

    1. Undetermined
    5. False
   10. Undetermined
   15. True

G. p. 188

    1. Undetermined
    5. Undetermined
   10. True
   15. Undetermined

SECTION 5.7 *Exercises on p. 197*

E. p. 197

Step (1) to step (2) is invalid: (1) asserts the falsehood of an **I** proposition; (2) asserts the truth of its corresponding **O** proposition. In the traditional interpretation, corresponding **I** and **O** propositions are subcontraries and cannot both be false. Therefore, if the **I** proposition in (1) is false, the **O** proposition in (2) would have to be true, in *that* interpretation. But because both **I** and **O** propositions do have existential import, both *can* be false (in the Boolean interpretation) if the subject class is empty. The subject class *is* empty in this case, because there are no mermaids. Hence the inference from the falsehood of (1) to the truth of (2) is invalid. Corresponding **I** and **O** propositions are not subcontraries in the Boolean interpretation, but the inference from (1) to (2) assumes that they are.

SECTION 5.8 *Exercises on p. 203*

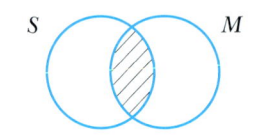

5. $SM = 0$

10. $MP = 0$

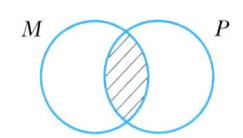

15. $PB \neq 0$

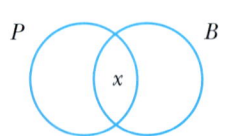

20. $P\overline{M} = 0$

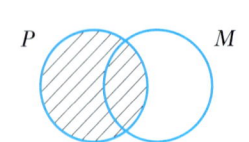

Solutions to Selected Exercises **575**

chapter 6

SECTION 6.1 *Exercises on pp. 209–210*

5.
- STEP 1: The conclusion is: Some conservatives are not advocates of high tariff rates.
- STEP 2: Major term: advocates of high tariff rates.
- STEP 3: Major premise: All advocates of high tariff rates are Republicans.
- STEP 4: Minor premise: Some Republicans are not conservatives.
- STEP 5: This syllogism, written in standard form:
  All advocates of high tariff rates are Republicans.
  Some Republicans are not conservatives.
  Therefore some conservatives are not advocates of high tariff rates.
- STEP 6: The three propositions of this syllogism are, in order: **A**, **O**, **O**. The middle term, "Republicans," is the predicate term of the major premise and the subject term of the minor premise, so the syllogism is in the *fourth* figure. Thus its mood and figure are **AOO–4**.

10.
- STEP 1: The conclusion is: No sports cars are automobiles designed for family use.
- STEP 2: Major term: Automobiles designed for family use.
- STEP 3: Major premise: All automobiles designed for family use are vehicles intended to be driven at moderate speeds.
- STEP 4: Minor premise: No sports cars are vehicles intended to be driven at moderate speeds.
- STEP 5: This syllogism, written in standard form:
  All automobiles designed for family use are vehicles intended to be driven at moderate speeds.
  No sports cars are vehicles intended to be driven at moderate speeds.
  Therefore no sports cars are automobiles designed for family use.
- STEP 6: The three propositions of this syllogism are, in order, **A**, **E**, **E**. The middle term, "vehicles intended to be driven at moderate speeds," is the predicate term of both the major and the minor premise, so the syllogism is in the *second* figure. Thus its mood and figure are **AEE–2**.

SECTION 6.2 *Exercises on p. 213*

5. One possible refuting analogy is this: All unicorns are mammals, so some mammals are not animals, because no animals are unicorns.

10. One possible refuting analogy is this: All square circles are circles, and all square circles are squares; therefore some circles are squares.

SECTION 6.3 *Exercises on pp. 223–224*

A. p. 222

5. No *P* is *M*.
Some *M* is *S*.
∴ Some *S* is not *P*.

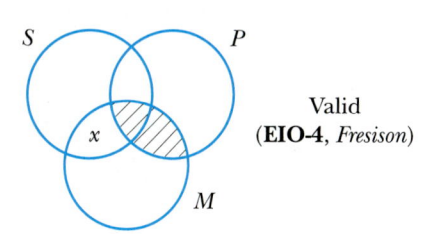

Valid
(**EIO-4**, *Fresison*)

10. Some *P* is *M*.
All *M* is *S*.
∴ Some *S* is *P*.

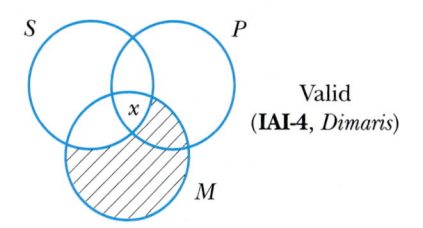

Valid
(**IAI-4**, *Dimaris*)

576 Solutions to Selected Exercises

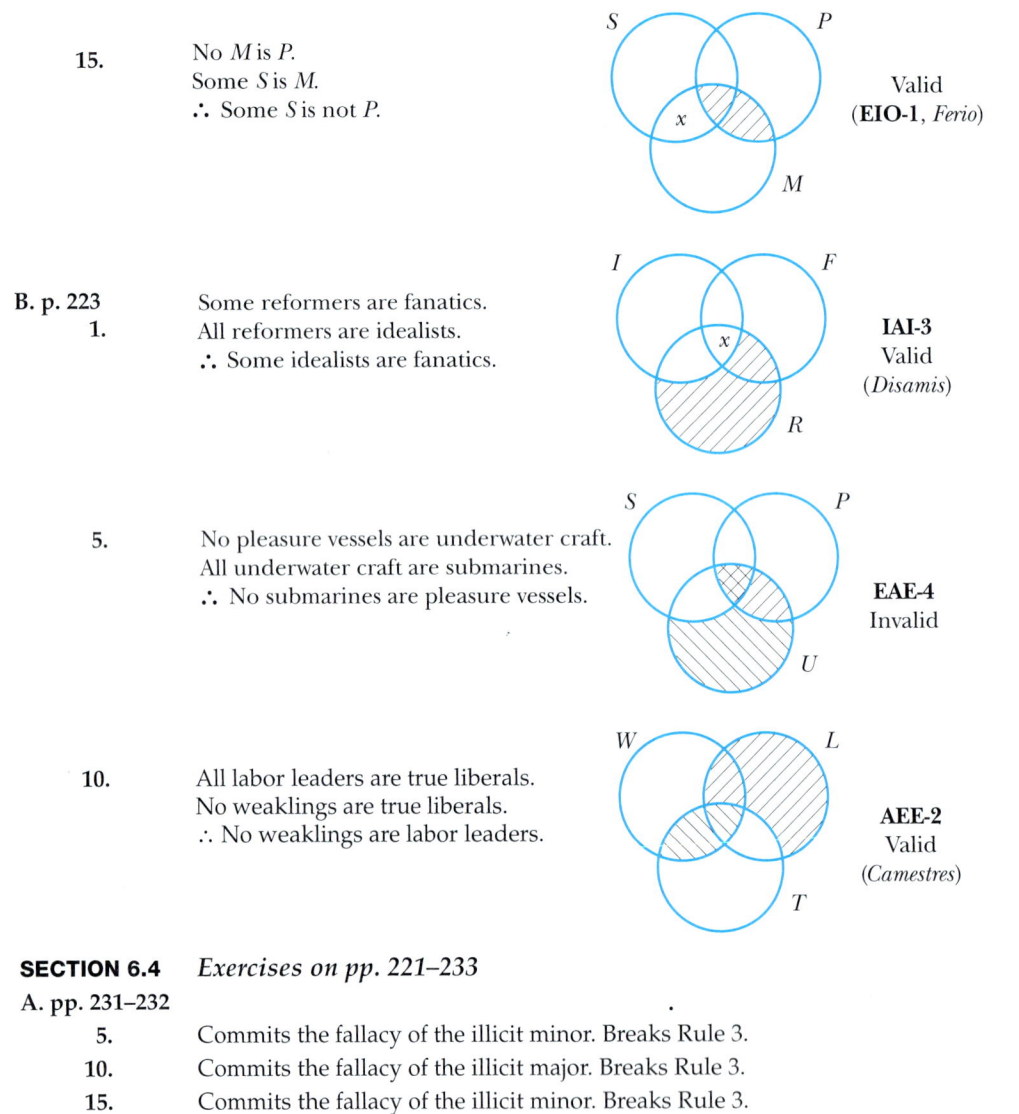

15.   No *M* is *P*.
Some *S* is *M*.
∴ Some *S* is not *P*.

Valid
(**EIO-1**, *Ferio*)

B. p. 223

1.   Some reformers are fanatics.
All reformers are idealists.
∴ Some idealists are fanatics.

IAI-3
Valid
(*Disamis*)

5.   No pleasure vessels are underwater craft.
All underwater craft are submarines.
∴ No submarines are pleasure vessels.

EAE-4
Invalid

10.   All labor leaders are true liberals.
No weaklings are true liberals.
∴ No weaklings are labor leaders.

AEE-2
Valid
(*Camestres*)

SECTION 6.4   *Exercises on pp. 221–233*

A. pp. 231–232
- 5. Commits the fallacy of the illicit minor. Breaks Rule 3.
- 10. Commits the fallacy of the illicit major. Breaks Rule 3.
- 15. Commits the fallacy of the illicit minor. Breaks Rule 3.

B. pp. 232–233
- 5. Commits the existential fallacy. Breaks Rule 6.
- 10. Commits the fallacy of the illicit minor. Breaks Rule 3.

C. pp. 233–234
- 5. Commits the fallacy of the illicit minor. Breaks Rule 3.
- 10. Commits the fallacy of four terms. (There is an equivocation on the term "people who like it," which has a very different meaning in the conclusion from the one it has in the premise.) Breaks Rule 1.

CHAPTER 6 APPENDIX   *Exercises on p. 242*

5.   Plainly this is possible in the first figure, where **AII–1**, which is valid, has only one term distributed, and that term only once. It also is possible in the third figure, where **AII–3** (as well as **IAI–3**) are valid and also have only one term distributed, and distributed only once. It also is possible in the fourth figure, where **IAI–4**, which is valid, has only one term distributed, and distributed only once. But where the middle term is the predicate term of both premises, in the second figure, it is not possible. Consider: To avoid breaking Rule 2, which requires that the middle term be distributed in at least one premise, one of the premises in this figure must be negative. But then, by Rule 5, the conclusion

would have to be negative and would distribute its predicate. Thus, if only one term can be distributed only once, in the second figure that would have to be in the conclusion; but if the distributed term can be distributed only once, that would break Rule 3, because if it is distributed in the conclusion it must be distributed in the premises.

10. None. If the middle term were distributed in both premises, then, in the first figure, the minor premise would have to be negative, whence (by Rule 5) the conclusion would have to be negative, so by Rule 3 the major premise would have to be negative, in violation of Rule 4. In the second figure, both premises would have to be negative, in violation of Rule 4. In the third figure both premises would have to be universal, so the minor premise would have to be negative by Rule 3, and by Rule 5 the conclusion would be negative—so by Rule 3 the major premise would also have to be negative, in violation of Rule 4. In the fourth figure the major premise would have to be negative. Therefore (by Rule 5) the conclusion would have to be negative (**E** or **O**) and it would distribute its major term, which means (by Rule 3) that the major premise would also have to distribute its major term and would therefore be universal (an **E** proposition). The minor premise also must be universal, since it distributes the middle term, and by Rule 4 it cannot be negative, so it must be the **A** proposition *All M is S*. Now Rule 6 precludes the possibility of an **O** proposition in the conclusion, and Rule 3 precludes the possibility of an **E**.

chapter 7

SECTION 7.2 *Exercises on pp. 248–249*

5. Where   E = explosives
            F = flammable things (note that "flammable" and "inflammable" are *synonyms!*)
            S = safe things
This syllogism translates into standard form thus:
        All *E* is *F*.
        No *F* is *S*.
        Therefore no *S* is *E*.
Exhibited in a Venn diagram, this syllogism (in *Camenes*) is shown to be valid.

10. Where   O = objects over six feet long
            D = difficult things to store
            U = useful things
This syllogism translates into standard form thus:
        All *O* is *D*.
        No *D* is *U*.
        Therefore no *U* is *O*.
Exhibited in a Venn diagram, this syllogism (in *Camenes*) is shown to be valid.

SECTION 7.3 *Exercises on pp. 257–258*

5. All Junkos are the best things that money can buy.
10. No people who face the sun are people who see their own shadows.
15. No candidates of the Old Guard are persons supported by the Young Turks.
(Or: No Young Turks are supporters of candidates of the Old Guard.)
20. All people who love well are people who pray well.
25. All soft answers are things that turn away wrath.

SECTION 7.4 *Exercises on p. 260*

A. p. 260

5. All cases in which she gives her opinion are cases in which she is asked to give her opinion.
10. No times when people do not discuss questions freely are times when people are most likely to settle questions rightly.

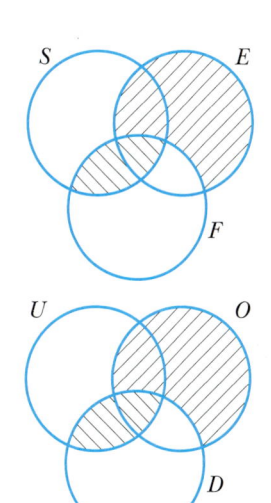

578 Solutions to Selected Exercises

B. pp. 261–263

5. No syllogisms having two negative premises are valid syllogisms.
Some valid syllogisms are not unsound arguments.
∴ Some unsound arguments are syllogisms having two negative premises.

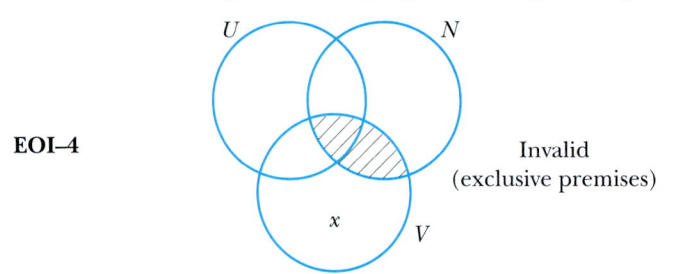

EOI–4   Invalid (exclusive premises)

10. No persons who are truly objective are persons likely to be mistaken.
All persons likely to be mistaken are persons who ignore the facts.
∴ No persons who ignore the facts are persons who are truly objective.

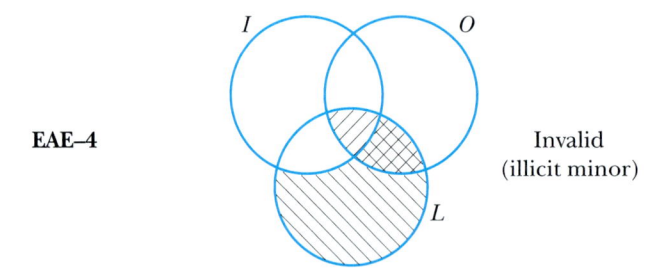

EAE–4   Invalid (illicit minor)

15. All things interesting to engineers are approximations.
No approximations are irrationals.
∴ No irrationals are things interesting to engineers.

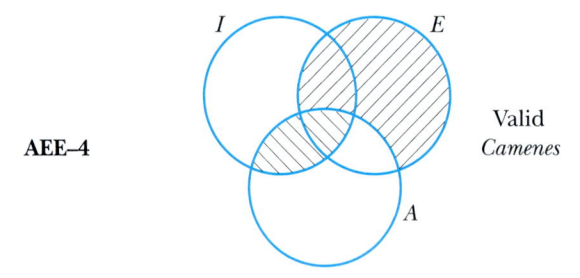

AEE–4   Valid *Camenes*

20. No times when Bill goes to work are times when Bill wears a sweater.
This morning was a time when Bill wore a sweater.
∴ This morning was not a time when Bill went to work.

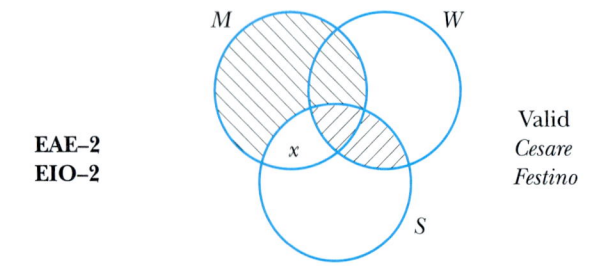

EAE–2
EIO–2   Valid *Cesare Festino*

25. All valid syllogisms are syllogisms that distribute their middle terms in at least one premise.
This syllogism is a syllogism that distributes its middle term in at least one premise.
∴ This syllogism is a valid syllogism.

AAA–2
AII–2

Invalid
(undistributed middle)

30. All situations in which much money is involved are situations in which competition is stiff.
This situation is a situation in which much money is involved.
∴ This situation is a situation in which competition is stiff.

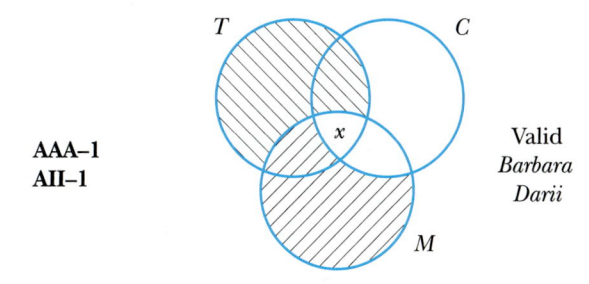

AAA–1
AII–1

Valid
Barbara
Darii

35. All invalid syllogisms are syllogisms that commit an illicit process.
This syllogism is not a syllogism that commits an illicit process.
∴ This syllogism is not an invalid syllogism.

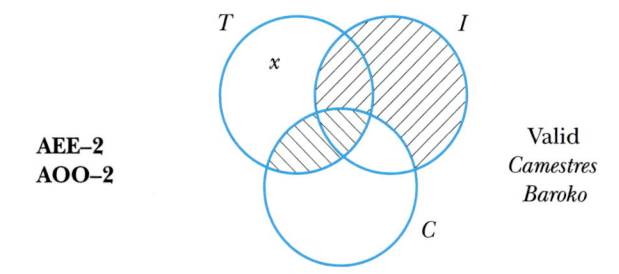

AEE–2
AOO–2

Valid
Camestres
Baroko

SECTION 7.5 *Exercises on pp. 267–269*

5. a. Unstated conclusion: Those persons who are vicious competitors you do not hate.
b. Standard-form translation:
All persons whom you respect are persons whom you do not hate.
All persons who are vicious competitors are persons whom you respect.
∴ All persons who are vicious competitors are persons whom you do not hate.
c. Third-order enthymeme.
d. Valid (in *Barbara*).

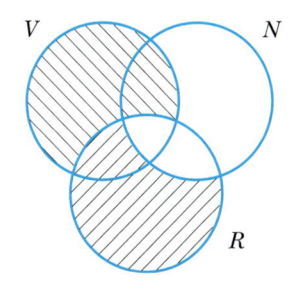

580 Solutions to Selected Exercises

10.   a. Unstated premise: All lies, misstatements, and omissions that are not the result of ignorance are the result of malevolence.
b. Standard-form translation:
All lies, misstatements, and omissions that are not the result of ignorance are lies, misstatements, and omissions that are the result of malevolence.
All lies, misstatements, and omissions in Carter's book are lies, misstatements, and omissions that are not the result of ignorance.
∴ All lies, misstatements, and omissions in Carter's book are lies, misstatements, and omissions that are the result of malevolence.
c. First-order enthymeme.
d. Valid (in *Barbara*).

NOTE: The author of the passage intends to present a valid disjunctive syllogism in the form of an enthymeme. The assumed disjunctive premise is disputable, of course.

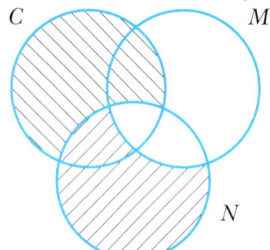

15.   a. Unstated premise: Species that tend to increase at a greater rate than their means of subsistence are occasionally subject to a severe struggle for existence.
b. Standard-form translation:
All species that tend to increase at a greater rate than their means of subsistence are species that are occasionally subject to a severe struggle for existence.
Man is a species that tends to increase at a greater rate than his means of subsistence.
∴ Man is a species that is occasionally subject to a severe struggle for existence.
c. First-order enthymeme.
d. Valid (in *Barbara* or *Darii*).

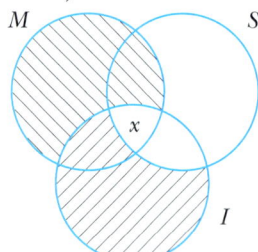

20.   a. Unstated premise: All that betters the condition of the vast majority of the people is desirable.
b. Standard-form translation:
All things that better the condition of the vast majority of the people are things that are desirable.
All productivity is a thing that betters the condition of the vast majority of the people.
∴ All productivity is a thing that is desirable.
c. First-order enthymeme.
d. Valid (in *Barbara*).

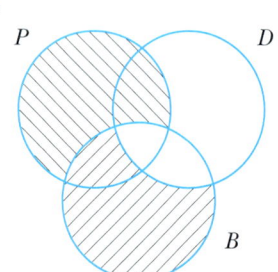

25. a. Unstated premise: The man who says that all things come to pass by necessity cannot criticize those who, by his own admission, do what they do by necessity.
b. Standard-form translation:
All people who are admitted to do what they do by necessity by the man who says that all things come to pass by necessity are people who cannot be criticized by the man who says that all things come to pass by necessity.
All people who deny that all things come to pass by necessity are people who are admitted to do what they do by necessity by the man who says that all things come to pass by necessity.
∴ All people who deny that all things come to pass by necessity are people who cannot be criticized by the man who says that all things come to pass by necessity.
c. First-order enthymeme.
d. Valid (in *Barbara*).

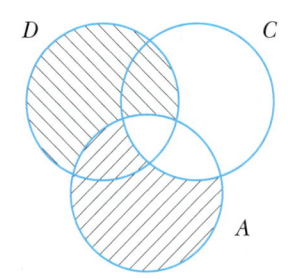

SECTION 7.6 Exercises on pp. 271–272

A. pp. 271–272

5. (1′) All interesting poems are poems that are popular among people of real taste.
(4′) No affected poems are poems that are popular among people of real taste.
(2′) All modern poems are affected poems.
(5′) All poems on the subject of soap bubbles are modern poems.
(3′) All poems of yours are poems on the subject of soap bubbles.
∴ No poems of yours are interesting poems.

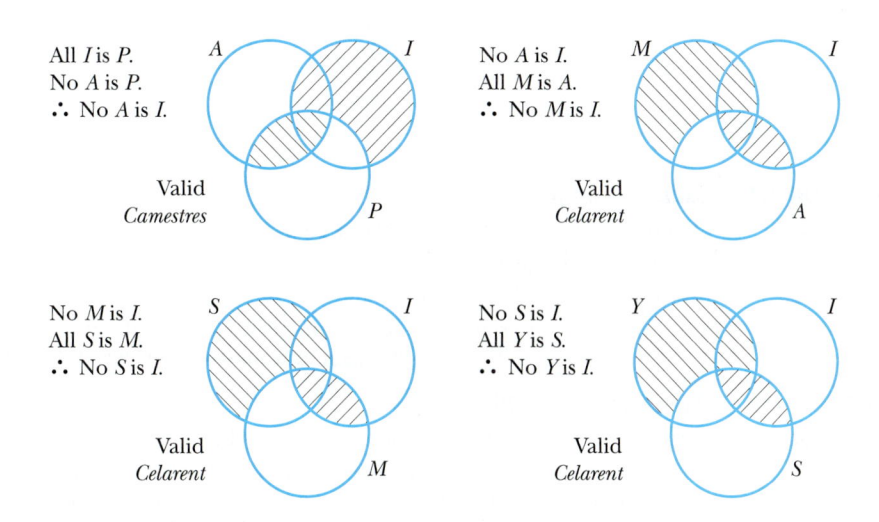

All *I* is *P*.
No *A* is *P*.
∴ No *A* is *I*.

Valid
Camestres

No *A* is *I*.
All *M* is *A*.
∴ No *M* is *I*.

Valid
Celarent

No *M* is *I*.
All *S* is *M*.
∴ No *S* is *I*.

Valid
Celarent

No *S* is *I*.
All *Y* is *S*.
∴ No *Y* is *I*.

Valid
Celarent

Valid

B. p. 272

1. (1′) All those who read *The Times* are those who are well educated.
(3′) No creatures who cannot read are those who are well educated.
(2′) All hedgehogs are creatures who cannot read.
∴ No hedgehogs are those who read *The Times*.

All T is W.
No C is W.
∴ No C is T.

Valid
Camestres

No C is T.
All H is C.
∴ No H is T.

Valid
Celarent

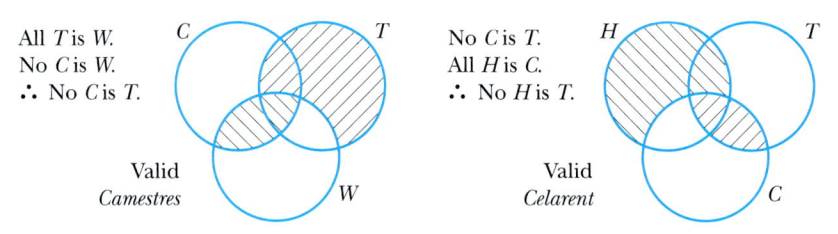

Valid

5. (2′) These sorites are examples not arranged in regular order, like the examples I am used to.
(4′) No examples not arranged in regular order, like the examples I am used to, are examples I can understand.
(1′) All examples I do not grumble at are examples I can understand.
(5′) All examples that do not give me a headache are examples I do not grumble at.
(3′) All easy examples are examples that do not give me a headache.
∴ These sorites are not easy examples.

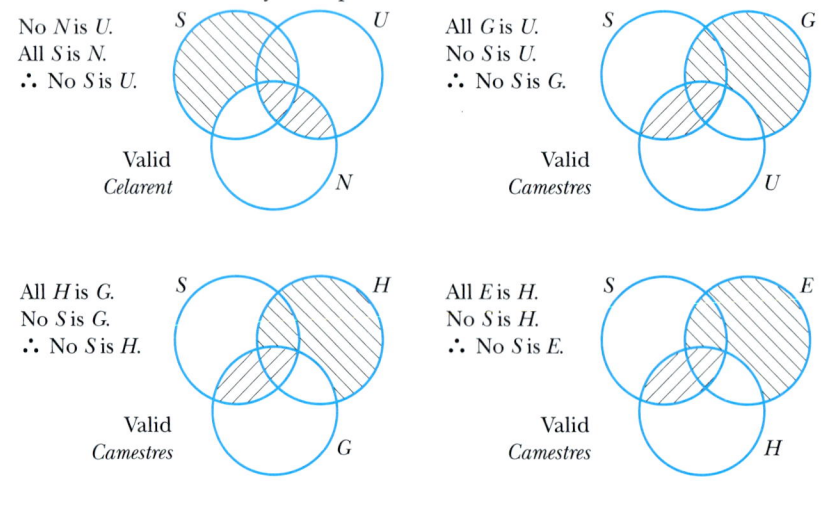

No N is U.
All S is N.
∴ No S is U.

Valid
Celarent

All G is U.
No S is U.
∴ No S is G.

Valid
Camestres

All H is G.
No S is G.
∴ No S is H.

Valid
Camestres

All E is H.
No S is H.
∴ No S is E.

Valid
Camestres

Valid

SECTION 7.7 *Exercises on pp. 277–278*
5. Mixed hypothetical syllogism, *modus ponens*. Valid.
10. Disjunctive syllogism. Valid.
15. Mixed hypothetical syllogism, *modus tollens*. Valid.
20. Two arguments are present here: The first is a pure hypothetical syllogism, the second is a mixed hypothetical syllogism of form *modus tollens*. Both are valid.

SECTION 7.8 *Exercises on pp. 283–285*
5. Very easy to go between the horns here. Plausible to grasp by either horn.
10. Impossible to go between the horns. It is plausible to grasp it by either horn, arguing either (a) that when desiring to preserve we may be motivated simply by inertia and seek to rest in the status quo, even while admitting that a change would not be worse and might even be better—but just "not worth the trouble of changing" or (b) that when desiring to change we may be motivated simply by boredom with the status quo, and seek a change even while admitting that a change might not be better and might even be worse—but "let's have a little variety." These are psychological rather than political or moral considerations, but the original dilemma appears to be itself psychological. The usual rebutting counterdilemma could be used here: When desiring to preserve, we do not wish to bring about something better; when desiring to

change, we do not wish to prevent a change to the worse. It is a question, however, how plausible this is.

15. There were in theory a number of ways to go between the horns here: Between defiance and obedience to the Court decision there are many degrees of partial compliance that fall short of full obedience but do not constitute outright defiance. Either horn could be grasped, at least in theory: An emergency situation in the international sphere might prevent defiance from being followed by impeachment; and it is logically possible that the evidence produced by obedience to the order might not have been sufficient to persuade the Congress to impeach.

20. This is a rather informal version of Pascal's argument, which has been much discussed for more than three hundred years. If it is interpreted as having the disjunctive premise that either God exists or God does not exist, then it is obviously impossible to go between the horns. But each of the horns can be grasped to refute the given argument. It might be argued that if you live a life of conspicuous virtue even though you are not a believer, you will be condemned to spend eternity in the flames of Hell. Or it might be argued that if you live as a believer you will suffer the loss of all those earthly pleasures that you might otherwise have enjoyed, and that that is a very grave penalty indeed.

chapter 8

SECTION 8.2 *Exercises on pp. 297–331*
A. pp. 297–299

1. True.   **5.** True.   **10.** True.   **15.** False.   **20.** True.   **25.** False.

B. p. 299

1. True.   **5.** False.   **10.** True.   **15.** True.   **20.** False.   **25.** False.

C. pp. 299–300

1. $I \bullet \sim L$ **5.** $\sim I \bullet \sim L$ **10.** $\sim(E \vee J)$
15. $\sim I \vee L$ **20.** $(I \bullet E) \vee \sim(J \bullet S)$ **25.** $(L \bullet E) \bullet (S \bullet J)$

SECTION 8.3 *Exercises on p. 308*
A. p. 308

1. True.   **5.** False.   **10.** True.   **15.** False.   **20.** False.   **25.** True.

B. p. 308

1. True.   **5.** False.   **10.** False.   **15.** True.   **20.** False.   **25.** True.

C. pp. 340–341

1. $A \supset (B \supset C)$ **5.** $(A \bullet B) \supset C$ **10.** $\sim[A \supset (B \bullet C)]$
15. $B \supset (A \vee C)$ **20.** $B \vee C$ **25.** $(\sim C \bullet \sim D) \supset (\sim B \vee A)$

SECTION 8.4 *Exercises on pp. 344–346*
A. p. 313

e. 10 is the specific form of *e*.
o. 3 has *o* as a substitution instance, and 24 is the specific form of *o*.

SECTION 8.7 *Exercises on pp. 355–357*
A. p. 322

1.

p	q	$p \supset q$	$\sim q$	$\sim p$	$\sim q \supset \sim p$
T	T	T	F	F	T
T	F	F	T	F	F
F	T	T	F	T	T
F	F	T	T	T	T

Valid

Solutions to Selected Exercises

5.

p	q	p ⊃ q
T	T	T
T	F	F
F	T	T
F	F	T

Invalid (shown by second row)

10.

p	q	p • q
T	T	T
T	F	F
F	T	F
F	F	F

Valid

15.

p	q	r	q ⊃ r	p ⊃ (q ⊃ r)	p ⊃ r	q ⊃ (p ⊃ r)	p ∨ q	(p ∨ q) ⊃ r
T	T	T	T	T	T	T	T	T
T	T	F	F	F	F	F	T	F
T	F	T	T	T	T	T	T	T
T	F	F	T	T	F	T	T	F
F	T	T	T	T	T	T	T	T
F	T	F	F	T	T	T	T	F
F	F	T	T	T	T	T	F	T
F	F	F	T	T	T	T	F	T

Invalid (shown by fourth and sixth rows)

20.

p	q	r	s	$p \cdot q$	$p \supset q$	$(p \cdot q) \supset r$	$r \supset s$	$p \supset (r \supset s)$	$(p \supset q) \cdot [(p \cdot q) \supset r]$	$p \supset s$
T	T	T	T	T	T	T	T	T	T	T
T	T	T	F	T	T	T	F	F	T	F
T	T	F	T	T	T	F	T	T	F	T
T	T	F	F	T	T	F	T	T	F	F
T	F	T	T	F	F	T	T	T	F	T
T	F	T	F	F	F	T	F	F	F	F
T	F	F	T	F	F	T	T	T	F	T
T	F	F	F	F	F	T	T	T	F	F
F	T	T	T	F	T	T	T	T	T	T
F	T	T	F	F	T	T	F	T	T	T
F	T	F	T	F	T	T	T	T	T	T
F	T	F	F	F	T	T	T	T	T	T
F	F	T	T	F	T	T	T	T	T	T
F	F	T	F	F	T	T	F	T	T	T
F	F	F	T	F	T	T	T	T	T	T
F	F	F	F	F	T	T	T	T	T	T

Valid

B. p. 322

1.

$(A \vee B) \supset (A \cdot B)$ $(p \vee q) \supset (p \cdot q)$
$A \vee B$ has the specific $p \vee q$
$\therefore A \cdot B$ form $\therefore p \cdot q$

p	q	$p \vee q$	$p \cdot q$	$(p \vee q) \supset (p \cdot q)$
T	T	T	T	T
T	F	T	F	F
F	T	T	F	F
F	F	F	F	T

Valid

Solutions to Selected Exercises

5. $(I \lor J) \supset (I \cdot J)$ has the specific $(p \lor q) \supset (p \cdot q)$
    $\sim(I \lor J)$ form $\sim(p \lor q)$
    $\therefore \sim(I \cdot J)$ $\therefore \sim(p \cdot q)$

p	q	p ∨ q	p · q	(p ∨ q) ⊃ (p · q)	∼(p ∨ q)	∼(p · q)
T	T	T	T	T	F	F
T	F	T	F	F	F	T
F	T	T	F	F	F	T
F	F	F	F	T	T	T

Valid (Note: Fallacy of denying the antecedent is *not* committed here!)

10. $U \supset (V \lor W)$ has the specific $p \supset (q \lor r)$
    $(V \cdot W) \supset \sim U$ form $(q \cdot r) \supset \sim p$
    $\therefore \sim U$ $\therefore \sim p$

p	q	r	q ∨ r	p ⊃ (q ∨ r)	q · r	∼p	(q · r) ⊃ ∼p
T	T	T	T	T	T	F	F
T	T	F	T	T	F	F	T
T	F	T	T	T	F	F	T
T	F	F	F	F	F	F	T
F	T	T	T	T	T	T	T
F	T	F	T	T	F	T	T
F	F	T	T	T	F	T	T
F	F	F	F	T	F	T	T

Invalid (shown by second and third rows)

C. p. 322

1. $A \supset (B \cdot C)$
$\sim B$ has the specific $p \supset (q \cdot r)$
$\therefore \sim A$ form $\sim q$
                                                   $\therefore \sim p$

p	q	r	q • r	p ⊃ (q • r)	~q	~p
T	T	T	T	T	F	F
T	T	F	F	F	F	F
T	F	T	F	F	T	F
T	F	F	F	F	T	F
F	T	T	T	T	F	T
F	T	F	F	T	F	T
F	F	T	F	T	T	T
F	F	F	F	T	T	T

Valid

5. $M \supset (N \supset O)$
N has the specific $p \supset (q \supset r)$
$\therefore O \supset M$ form q
                                                   $\therefore r \supset p$

p	q	r	q ⊃ r	p ⊃ (q ⊃ r)	r ⊃ p
T	T	T	T	T	T
T	T	F	F	F	T
T	F	T	T	T	T
T	F	F	T	T	T
F	T	T	T	T	F
F	T	F	F	T	T
F	F	T	T	T	F
F	F	F	T	T	T

Invalid (shown by fifth row)

10. $G \supset (I \cdot D)$ $p \supset (q \cdot r)$
     $(I \vee D) \supset B$ has the specific $(q \vee r) \supset s$
     $\therefore G \supset B$ form $\therefore p \supset s$

p	q	r	s	q•r	p⊃(q•r)	q∨r	(q∨r)⊃s	p⊃s
T	T	T	T	T	T	T	T	T
T	T	T	F	T	T	T	F	F
T	T	F	T	F	F	T	T	T
T	T	F	F	F	F	T	F	F
T	F	T	T	F	F	T	T	T
T	F	T	F	F	F	T	F	F
T	F	F	T	F	F	F	T	T
T	F	F	F	F	F	F	T	F
F	T	T	T	T	T	T	T	T
F	T	T	F	T	T	T	F	T
F	T	F	T	F	T	T	T	T
F	T	F	F	F	T	T	F	T
F	F	T	T	F	T	T	T	T
F	F	T	F	F	T	T	F	T
F	F	F	T	F	T	F	T	T
F	F	F	F	F	T	F	T	T

Valid

SECTION 8.8 Exercises on pp. 328

A. pp. 361–362
 1. *c* is the specific form of 1.
 5. *c* has 5 as a substitution instance, and *i* is the specific form of 5.
 10. *e* has 10 as a substitution instance.

B. p. 329

1.

p	q	p ⊃ q	p ⊃ (p ⊃ q)	[p ⊃ (p ⊃ q)] ⊃ q
T	T	T	T	T
T	F	F	F	T
F	T	T	T	T
F	F	T	T	F

Contingent

5.

p	q	~q	q • ~q	p ⊃ (q • ~q)	p ⊃ [p ⊃ (q • ~q)]
T	T	F	F	F	F
T	F	T	F	F	F
F	T	F	F	T	T
F	F	T	F	T	T

Contingent

10.

p	q	r	s	p ⊃ q	r ⊃ s	(p ⊃ q) • (r ⊃ s)	q ∨ s	[(p ⊃ q) • (r ⊃ s)] • (q ∨ s)	p ∨ r	{[(p ⊃ q) • (r ⊃ s)] • (q ∨ s)} ⊃ (p ∨ r)
T	T	T	T	T	T	T	T	T	T	T
T	T	T	F	T	F	F	T	F	T	T
T	T	F	T	T	T	T	T	T	T	T
T	T	F	F	T	T	T	T	T	T	T
T	F	T	T	F	T	F	T	F	T	T
T	F	T	F	F	F	F	F	F	T	T
T	F	F	T	F	T	F	T	F	T	T
T	F	F	F	F	T	F	F	F	T	T
F	T	T	T	T	T	T	T	T	T	T
F	T	T	F	T	F	F	T	F	T	T
F	T	F	T	T	T	T	T	T	F	F
F	T	F	F	T	T	T	T	T	F	F
F	F	T	T	T	T	T	T	T	T	T
F	F	T	F	T	F	F	F	F	T	T
F	F	F	T	T	T	T	T	T	F	F
F	F	F	F	T	T	T	F	F	F	T

Contingent

C. p. 329

1.

p	q	p ⊃ q	~q	~p	~q ⊃ ~p	(p ⊃ q) ≡ (~q ⊃ ~p)
T	T	T	F	F	T	T
T	F	F	T	F	F	T
F	T	T	F	T	T	T
F	F	T	T	T	T	T

Tautology

5.

p	q	$p \vee q$	$p \cdot (p \vee q)$	$p \equiv [p \cdot (p \vee q)]$
T	T	T	T	T
T	F	T	T	T
F	T	T	F	T
F	F	F	F	T

Tautology

10.

p	q	$p \supset q$	$p \vee q$	$(p \vee q) \equiv q$	$(p \supset q) \equiv [(p \vee q) \equiv q]$
T	T	T	T	T	T
T	F	F	T	F	T
F	T	T	T	T	T
F	F	T	F	T	T

Tautology

15.

p	q	r	$q \vee r$	$p \cdot (q \vee r)$	$p \cdot q$	$p \cdot r$	$(p \cdot q) \vee (p \cdot r)$	$[p \cdot (q \vee r)] \equiv [(p \cdot q) \vee (p \cdot r)]$
T	T	T	T	T	T	T	T	T
T	T	F	T	T	T	F	T	T
T	F	T	T	T	F	T	T	T
T	F	F	F	F	F	F	F	T
F	T	T	T	F	F	F	F	T
F	T	F	T	F	F	F	F	T
F	F	T	T	F	F	F	F	T
F	F	F	F	F	F	F	F	T

Tautology

592 Solutions to Selected Exercises

20.

p	q	p ⊃ q	q ⊃ p	(p ⊃ q) • (q ⊃ p)	p • q	~p	~q	~p • ~q	(p • q) ∨ (~p • ~q)	[(p ⊃ q) • (q ⊃ p)] ≡ [(p • q) ∨ (~p • ~q)]
T	T	T	T	T	T	F	F	F	T	T
T	F	F	T	F	F	F	T	F	F	T
F	T	T	F	F	F	T	F	F	F	T
F	F	T	T	T	F	T	T	T	T	T

Tautology

chapter 9

SECTION 9.2 *Exercises on p. 344*
  1. Absorption (Abs.)
 10. Hypothetical Syllogism (H.S.)
 20. Hypothetical Syllogism (H.S.)
  5. Constructive Dilemma (C.D.)
 15. Conjunction (Conj.)

SECTION 9.3 *Exercises on pp. 347–348*
 1. 3. 1, Simp.
    4. 3, Add.
    5. 2, 4, M.P.
    6. 3, 5, Conj.
 5. 5. 2, 4, M.P.
    6. 1, 5, Conj.
    7. 3, 4, D.S.
    8. 6, 7, C.D.
 10. 6. 4, 5, Conj.
    7. 3, 6, M.P.
    8. 7, 1, H.S.
    9. 2, 8, Conj.
    10. 9, 4, C.D.

SECTION 9.4 *Exercises on pp. 349–350*
 5. 1. M ∨ N
    2. ~M • ~O
    ∴ N
    3. ~M   2, Simp.
    4. N   1, 3, D.S.
 15. 1. (P ⊃ Q) • (R ⊃ S)
    2. (P ∨ R) • (Q ∨ R)
    ∴ Q ∨ S
    3. P ∨ R   2, Simp.
    4. Q ∨ S   1, 3, C.D.
 25. 1. (W • X) ⊃ (Y • Z)
    2. ~[(W • X) • (Y • Z)]
    ∴ ~(W • X)
    3. (W • X) ⊃ [(W • X)
       • (Y • Z)]   1, Abs.
    4. ~(W • X)   3, 2, M.T.

 10. 1. A ⊃ B
    2. (A • B) ⊃ C
    ∴ A ⊃ C
    3. A ⊃ (A • B) 1, Abs.
    4. A ⊃ C   3, 2, H.S.
 20. 1. (~H ∨ I) ∨ J
    2. ~(~H ∨ I)
    ∴ J ∨ ~H
    3. J   1, 2, D.S.
    4. J ∨ ~H   3, Add.
 30. 1. Q ⊃ (R ∨ S)
    2. (T • U) ⊃ R
    3. (R ∨ S) ⊃ (T • U)
    ∴ Q ⊃ R
    4. Q ⊃ (T • U) 1, 3, H.S.
    5. Q ⊃ R   4, 2, H.S.

SECTION 9.5 *Exercises on pp. 387–393*
A. pp. 351–352
 5. 1. N ⊃ [(N • O) ⊃ P]
    2. N • O
    ∴ P
    3. N   2, Simp.
    4. (N • O) ⊃ P 1, 3, M.P.
    5. P   4, 2, M.P.

 10. 1. E ∨ ~F
    2. F ∨ (E ∨ G)
    3. ~E
    ∴ G
    4. ~F   1, 3, D.S.
    5. E ∨ G   2, 4, D.S.
    6. G   5, 3, D.S.

15. 1. $(Z \cdot A) \supset B$
    2. $B \supset A$
    3. $(B \cdot A) \supset (A \cdot B)$
       $\therefore (Z \cdot A) \supset (A \cdot B)$
    4. $B \supset (B \cdot A)$ 2, Abs.
    5. $B \supset (A \cdot B)$ 4, 3, H.S.
    6. $(Z \cdot A) \supset (A \cdot B)$ 1, 5, H.S.

B. pp. 353–354

5. 1. $(Q \supset R) \cdot (S \supset T)$
   2. $(U \supset V) \cdot (W \supset X)$
   3. $Q \vee U$
      $\therefore R \vee V$
   4. $Q \supset R$ 1, Simp.
   5. $U \supset V$ 2, Simp.
   6. $(Q \supset R) \cdot (U \supset V)$ 4, 5, Conj.
   7. $R \vee V$ 6, 3, C.D.

10. 1. $(N \vee O) \supset P$
    2. $(P \vee Q) \supset R$
    3. $Q \vee N$
    4. $\sim Q$
       $\therefore R$
    5. N 3, 4, D.S.
    6. $N \vee O$ 5, Add.
    7. P 1, 6, M.P.
    8. $P \vee Q$ 7, Add.
    9. R 2, 8, M.P.

C. pp. 354–356

5. 1. $C \supset R$
   2. $(C \cdot R) \supset B$
   3. $(C \supset B) \supset \sim S$
   4. $S \vee M$
      $\therefore M$
   5. $C \supset (C \cdot R)$ 1, Abs.
   6. $C \supset B$ 5, 2, H.S.
   7. $\sim S$ 3, 6, M.P.
   8. M 4, 7, D.S.

10. 1. $O \supset \sim M$
    2. O
    3. $B \supset \sim N$
    4. B
    5. $(\sim M \cdot \sim N) \supset F$
    6. $(B \cdot F) \supset G$
       $\therefore G$
    7. $\sim M$ 1, 2, M.P.
    8. $\sim N$ 3, 4, M.P.
    9. $\sim M \cdot \sim N$ 7, 8, Conj.
    10. F 5, 9, M.P.
    11. $B \cdot F$ 4, 10, Conj.
    12. G 6, 11, M.P.

SECTION 9.6   *Exercises on pp. 363–634*

5. Material Equivalence (Equiv.)
10. Association (Assoc.)
15. Distribution (Dist.)
20. De Morgan's Theorem (De M.)

SECTION 9.8   *Exercises on pp. 370–383*

A. pp. 370–371

5. 3. 2, Dist.
   4. 3, Com.
   5. 4, Simp.
   6. 5, Taut.
   7. 1, Assoc.
   8. 7, 6, D.S.
   9. 8, Impl

10. 3. 2, Trans.
    4. 3, Exp.
    5. 1, D.N.
    6. 5, Com.
    7. 6, Dist.
    8. 7, Com.
    9. 4, 8, C.D.
    10. 9, Com.
    11. 10, D.N.
    12. 11, De M.

594 Solutions to Selected Exercises

B. pp. 372–374

5.
1. $\sim K \vee (L \supset M)$
   $\therefore (K \cdot L) \supset M$
2. $K \supset (L \supset M)$ 1, Impl.
3. $(K \cdot L) \supset M$ 2, Exp.

10.
1. $Z \supset A$
2. $\sim A \vee B$
   $\therefore Z \supset B$
3. $A \supset B$ 2, Impl.
4. $Z \supset B$ 1, 3, H.S.

15.
1. $(O \vee P) \supset (Q \vee R)$
2. $P \vee O$
   $\therefore Q \vee R$
3. $O \vee P$ 2, Com.
4. $Q \vee R$ 1, 3, M.P.

20.
1. $I \supset [J \vee (K \vee L)]$
2. $\sim[(J \vee K) \vee L]$
   $\therefore \sim I$
3. $\sim[J \vee (K \vee L)]$ 2, Assoc.
4. $\sim I$ 1, 3, M.T.

25.
1. $A \vee B$
2. $C \vee D$
   $\therefore [(A \vee B) \cdot C] \vee [(A \vee B) \cdot D]$
3. $(A \vee B) \cdot (C \vee D)$ 1, 2, Conj.
4. $[(A \vee B) \cdot C] \vee [(A \vee B) \cdot D]$ 3, Dist.

30.
1. $\sim[(B \supset \sim C) \cdot (\sim C \supset B)]$
2. $(D \cdot E) \supset (B \equiv \sim C)$
   $\therefore \sim(D \cdot E)$
3. $\sim(B \equiv \sim C)$ 1, Equiv.
4. $\sim(D \cdot E)$ 2, 3, M.T.

C. p. 375

5.
1. $[(K \vee L) \vee M] \vee N$
   $\therefore (N \vee K) \vee (L \vee M)$
2. $[K \vee (L \vee M)] \vee N$ 1, Assoc.
3. $N \vee [K \vee (L \vee M)]$ 2, Com.
4. $(N \vee K) \vee (L \vee M)$ 3, Assoc.

10.
1. $(Z \vee A) \vee B$
2. $\sim A$
   $\therefore Z \vee B$
3. $(A \vee Z) \vee B$ 1, Com.
4. $A \vee (Z \vee B)$ 3, Assoc.
5. $Z \vee B$ 4, 2, D.S.

15.
1. $[R \supset (S \supset T)] \cdot [(R \cdot T) \supset U]$
2. $R \cdot (S \vee T)$
   $\therefore T \vee U$
3. $(R \cdot S) \vee (R \cdot T)$ 2, Dist.
4. $[(R \cdot S) \supset T] \cdot [(R \cdot T) \supset U]$ 1, Exp.
5. $T \vee U$ 4, 3, C.D.

D. p. 377

5.
1. $K \supset L$
   $\therefore K \supset (L \vee M)$
2. $\sim K \vee L$ 1, Impl.
3. $(\sim K \vee L) \vee M$ 2, Add.
4. $\sim K \vee (L \vee M)$ 3, Assoc.
5. $K \supset (L \vee M)$ 4, Impl.

10.
1. $Z \supset A$
2. $Z \vee A$
   $\therefore A$
3. $A \vee Z$ 2, Com.
4. $\sim\sim A \vee Z$ 3, D.N.
5. $\sim A \supset Z$ 4, Impl.
6. $\sim A \supset A$ 5, 1, H.S.
7. $\sim\sim A \vee A$ 6, Impl.
8. $A \vee A$ 7, D.N.
9. A 8, Taut.

E. p. 380

1.
1. $A \supset \sim B$
2. $\sim(C \cdot \sim A)$
   $\therefore C \supset \sim B$
3. $\sim C \vee \sim\sim A$ 2, De M.
4. $C \supset \sim\sim A$ 3, Impl.
5. $C \supset A$ 4, D.N.
6. $C \supset \sim B$ 5, 1, H.S.

5.
1. $[(M \cdot N) \cdot O] \supset P$
2. $Q \supset [(O \cdot M) \cdot N]$
   $\therefore \sim Q \vee P$
3. $[O \cdot (M \cdot N)] \supset P$ 1, Com.
4. $[(O \cdot M) \cdot N] \supset P$ 3, Assoc.
5. $Q \supset P$ 2, 4, H.S.
6. $\sim Q \vee P$ 5, Impl.

Solutions to Selected Exercises 595

10.
1. $[H \lor (I \lor J)] \supset (K \supset J)$
2. $L \supset [I \lor (J \lor H)]$
   $\therefore (L \bullet K) \supset J$
3. $[(I \lor J) \lor H] \supset (K \supset J)$ — 1, Com.
4. $[I \lor (J \lor H)] \supset (K \supset J)$ — 3, Assoc.
5. $L \supset (K \supset J)$ — 2, 4 H.S.
6. $(L \bullet K) \supset J$ — 5, Exp.

15.
1. $(Z \supset Z) \supset (A \supset A)$
2. $(A \supset A) \supset (Z \supset Z)$
   $\therefore A \supset A$
3. $[(Z \supset Z) \supset (A \supset A)] \lor \sim A$ — 1, Add.
4. $\sim A \lor [(Z \supset Z) \supset (A \supset A)]$ — 3, Com.
5. $A \supset [(Z \supset Z) \supset (A \supset A)]$ — 4, Impl.
6. $A \supset \{A \bullet [(Z \supset Z) \supset (A \supset A)]\}$ — 5, Abs.
7. $\sim A \lor \{A \bullet [(Z \supset Z) \supset (A \supset A)]\}$ — 6, Impl.
8. $(\sim A \lor A) \bullet \{\sim A \lor [(Z \supset Z) \supset (A \supset A)]\}$ — 7, Dist.
9. $\sim A \lor A$ — 8, Simp.
10. $A \supset A$ — 9, Impl.

20.
1. $(R \lor S) \supset (T \bullet U)$
2. $\sim R \supset (V \supset \sim V)$
3. $\sim T$
   $\therefore \sim V$
4. $\sim T \lor \sim U$ — 3, Add.
5. $\sim(T \bullet U)$ — 4, De M.
6. $\sim(R \lor S)$ — 1, 5, M.T.
7. $\sim R \bullet \sim S$ — 6, De M.
8. $\sim R$ — 7, Simp.
9. $V \supset \sim V$ — 2, 8, M.P.
10. $\sim V \lor \sim V$ — 9, Impl.
11. $\sim V$ — 10, Taut.

F. pp. 380–383

1.
1. $\sim N \lor A$
2. N
   $\therefore A$
3. $N \supset A$ — 1, Impl.
4. A — 3, 2, M.P.

5.
1. $R \supset A$
   $\therefore R \supset (A \lor W)$
2. $\sim R \lor A$ — 1, Impl.
3. $(\sim R \lor A) \lor W$ — 2, Add.
4. $\sim R \lor (A \lor W)$ — 3, Assoc.
5. $R \supset (A \lor W)$ — 4, Impl.

10.
1. $(G \bullet S) \supset D$
2. $(S \supset D) \supset P$
3. G
   $\therefore P$
4. $G \supset (S \supset D)$ — 1, Exp.
5. $S \supset D$ — 4, 3, M.P.
6. P — 2, 5, M.P.

15.
1. $M \supset \sim C$
2. $\sim C \supset \sim A$
3. $D \lor A$
   $\therefore \sim M \lor D$
4. $M \supset \sim A$ — 1, 2, H.S.
5. $A \lor D$ — 3, Com.
6. $\sim\sim A \lor D$ — 5, D.N.
7. $\sim A \supset D$ — 6, Impl.
8. $M \supset D$ — 4, 7, H.S.
9. $\sim M \lor D$ — 8, Impl.

20.
1. $P \supset \sim M$
2. $C \supset M$
3. $\sim L \lor C$
4. $(\sim P \supset \sim E) \bullet (\sim E \supset \sim C)$
5. $P \lor \sim P$
   $\therefore \sim L$
6. $(\sim E \supset \sim C) \bullet (\sim P \supset \sim E)$ — 4, Com.
7. $\sim P \supset \sim E$ — 4, Simp.
8. $\sim E \supset \sim C$ — 6, Simp.
9. $\sim P \supset \sim C$ — 7, 8, H.S.
10. $\sim M \supset \sim C$ — 2, Trans.

596 Solutions to Selected Exercises

11.	$P \supset \sim C$	1, 10, H.S.
12.	$(P \supset \sim C) \cdot (\sim P \supset \sim C)$	11, 9, Conj.
13.	$\sim C \vee \sim C$	12, 5, C.D.
14.	$\sim C$	13, Taut.
15.	$C \vee \sim L$	3, Com.
16.	$\sim L$	15, 14, D.S.

G. p. 383

5. 1. $(H \vee \sim H) \supset G$
    $\therefore G$

2.	$[(H \vee \sim H) \supset G] \vee \sim H$	1, Add.
3.	$\sim H \vee [(H \vee \sim H) \supset G]$	2, Com.
4.	$H \supset [(H \vee \sim H) \supset G]$	3, Impl.
5.	$H \supset \{H \cdot [(H \vee \sim H) \supset G]\}$	4, Abs.
6.	$\sim H \vee \{H \cdot [(H \vee \sim H) \supset G]\}$	5, Impl.
7.	$(\sim H \vee H) \cdot \{\sim H \vee [(H \vee \sim H) \supset G]\}$	6, Dist.
8.	$\sim H \vee H$	7, Simp.
9.	$H \vee \sim H$	8, Com.
10.	G	1, 9, M.P.

SECTION 9.9 *Exercises on pp. 385–386*

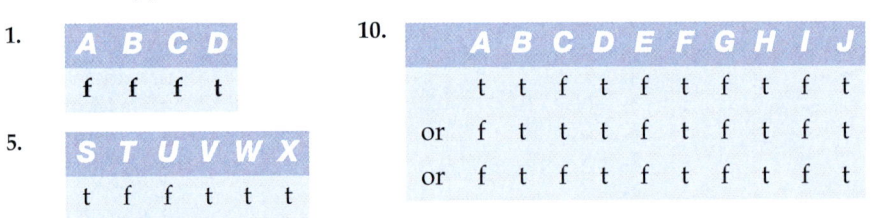

1.

A	B	C	D
f	f	f	t

5.

S	T	U	V	W	X
t	f	f	t	t	t

10.

	A	B	C	D	E	F	G	H	I	J
	t	t	f	t	f	t	f	t	f	t
or	f	t	t	t	f	t	f	t	f	t
or	f	t	f	t	f	t	f	t	f	t

or any of thirteen other truth-value assignments.

SECTION 9.10 *Exercises on pp. 389–392*

A. p. 389

1. 1. $(A \supset B) \cdot (C \supset D)$
    $\therefore (A \cdot C) \supset (B \vee D)$

2.	$A \supset B$	1, Simp.
3.	$\sim A \vee B$	2, Impl.
4.	$(\sim A \vee B) \vee D$	3, Add.
5.	$\sim A \vee (B \vee D)$	4, Assoc.
6.	$[\sim A \vee (B \vee D)] \vee \sim C$	5, Add.
7.	$\sim C \vee [\sim A \vee (B \vee D)]$	6, Com.
8.	$(\sim C \vee \sim A) \vee (B \vee D)$	7, Assoc.
9.	$(\sim A \vee \sim C) \vee (B \vee D)$	8, Com.
10.	$\sim (A \cdot C) \vee (B \vee D)$	9, De M.
11.	$(A \cdot C) \supset (B \vee D)$	10, Impl.

5.

X	Y	Z	A	B	C
t	f	t	f	t	f

10.

	A	B	C	D	E	F	G
	f	f	t	t	f	t	t
or	f	f	t	f	f	t	t
or	f	f	f	t	f	t	t
or	f	f	f	f	f	t	t

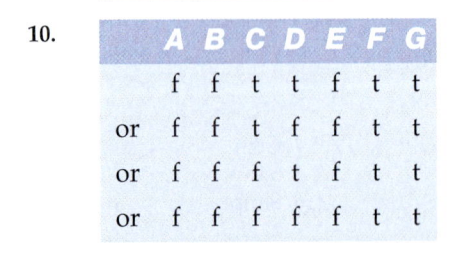

Solutions to Selected Exercises **597**

B. pp. 389–391

1. 1. $C \supset (M \supset D)$
    2. $D \supset V$
    3. $(D \supset A) \cdot \sim A$
       $\therefore M \supset \sim C$
    4. $D \supset A$ 3, Simp.
    5. $\sim A \cdot (D \supset A)$ 3, Com.
    6. $\sim A$ 5, Simp.
    7. $\sim D$ 4, 6, M.T.
    8. $(C \cdot M) \supset D$ 1, Exp.
    9. $\sim(C \cdot M)$ 8, 7, M.T.
    10. $\sim C \vee \sim M$ 9, De M.
    11. $\sim M \vee \sim C$ 10, Com.
    12. $M \supset \sim C$ 11, Impl.

5. $(I \cdot S) \supset (G \cdot P)$
   $[(S \cdot \sim I) \supset A] \cdot (A \supset P)$
   $I \supset S$
   $\therefore P$

	I	S	G	P	A
proved invalid by	f	f	t	f	f
or	f	f	f	f	f

10. $(H \supset A) \cdot (F \supset C)$
    $A \supset (F \cdot E)$
    $(O \supset C) \cdot (O \supset M)$
    $P \supset (M \supset D)$
    $P \cdot (D \supset G)$
    $\therefore H \supset G$

	H	A	C	F	E	O	M	P	D	G
proved invalid by	t	t	t	t	t	f	f	t	f	f

15. 1. $(J \vee A) \supset [(S \vee K) \supset (\sim I \cdot Y)]$
    2. $(\sim I \vee \sim M) \supset E$
       $\therefore J \supset (S \supset E)$
    3. $\sim(J \vee A) \vee [(S \vee K) \supset (\sim I \cdot Y)]$ 1, Impl.
    4. $[(S \vee K) \supset (\sim I \cdot Y)] \vee \sim(J \vee A)$ 3, Com.
    5. $[(S \vee K) \supset (\sim I \cdot Y)] \vee (\sim J \cdot \sim A)$ 4, De M.
    6. $\{[(S \vee K) \supset (\sim I \cdot Y)] \vee \sim J\} \cdot$
       $\{[(S \vee K) \supset (\sim I \cdot Y)] \vee \sim A\}$ 5, Dist.
    7. $[(S \vee K) \supset (\sim I \cdot Y)] \vee \sim J$ 6, Simp.
    8. $[\sim(S \vee K) \vee (\sim I \cdot Y)] \vee \sim J$ 7, Impl.
    9. $\sim(S \vee K) \vee [(\sim I \cdot Y) \vee \sim J]$ 8, Assoc.
    10. $[(\sim I \cdot Y) \vee \sim J] \vee \sim(S \vee K)$ 9, Com.
    11. $[(\sim I \cdot Y) \vee \sim J] \vee (\sim S \cdot \sim K)$ 10, De M.
    12. $\{[(\sim I \cdot Y) \vee \sim J] \vee \sim S\} \cdot$
        $\{[(\sim I \cdot Y) \vee \sim J] \vee \sim K\}$ 11, Dist.
    13. $[(\sim I \cdot Y) \vee \sim J] \vee \sim S$ 12, Simp.
    14. $(\sim I \cdot Y) \vee (\sim J \vee \sim S)$ 13, Assoc.
    15. $(\sim J \vee \sim S) \vee (\sim I \cdot Y)$ 14, Com.
    16. $[(\sim J \vee \sim S) \vee \sim I] \cdot [(\sim J \vee \sim S) \vee Y]$ 15, Dist.
    17. $(\sim J \vee \sim S) \vee \sim I$ 16, Simp.
    18. $[(\sim J \vee \sim S) \vee \sim I] \vee \sim M$ 17, Add.
    19. $(\sim J \vee \sim S) \vee (\sim I \vee \sim M)$ 18, Assoc.
    20. $\sim(J \cdot S) \vee (\sim I \vee \sim M)$ 19, De M.
    21. $(J \cdot S) \supset (\sim I \vee \sim M)$ 20, Impl.
    22. $(J \cdot S) \supset E$ 21, 2, H.S.
    23. $J \supset (S \supset E)$ 22, Exp.

598 Solutions to Selected Exercises

C. p. 392

5. 1. $(R \vee \sim R) \supset W$
    $\therefore W$
    2. $[(R \vee \sim R) \supset W] \vee \sim R$ 1, Add.
    3. $\sim R \vee [(R \vee \sim R) \supset W]$ 2, Com.
    4. $R \supset [(R \vee \sim R) \supset W]$ 3, Impl.
    5. $R \supset \{R \bullet [(R \vee \sim R) \supset W]\}$ 4, Abs.
    6. $\sim R \vee \{R \bullet [(R \vee \sim R) \supset W]\}$ 5, Impl.
    7. $(\sim R \vee R) \bullet \{\sim R \vee [(R \vee \sim R) \supset W]\}$ 6, Dist.
    8. $\sim R \vee R$ 7, Simp.
    9. $R \vee \sim R$ 8, Com.
    10. W 1, 9, M.P.

SECTION 9.11 *Exercises on p. 394*

A. p. 394

5. 1. $D \supset (Z \supset Y)$
    2. $Z \supset (Y \supset \sim Z)$
    $\therefore \sim D \vee \sim Z$
    !3. $\sim(\sim D \vee \sim Z)$ I.P. (Indirect Proof)
    !4. $\sim\sim D \bullet \sim\sim Z$ 3, De M.
    !5. $D \bullet \sim\sim Z$ 4, D.N.
    !6. $D \bullet Z$ 5, D.N.
    !7. D 6, Simp.
    !8. $Z \supset Y$ 1, 7, M.P.
    !9. $Z \bullet D$ 6, Com.
    !10. Z 9, Simp.
    !11. $Y \supset \sim Z$ 2, 10, M.P.
    !12. Y 8, 10, M.P.
    !13. $\sim Z$ 11, 12, M.P.
    !14. $Z \bullet \sim Z$ 10, 13, Conj.

chapter 10

SECTION 10.4 *Exercises on pp. 411–413*

A. p. 411

5. $(\exists x)(Dx \bullet \sim Rx)$
10. $(x)(Cx \supset \sim Fx)$
15. $(x)(Vx \supset Cx)$
20. $(x)(Cx \equiv Hx)$

B. p. 412

5. $[(\exists x)(Gx \bullet \sim Sx)] \bullet [(\exists x)(Dx \bullet \sim Bx)]$
10. $(x)(\sim Bx \supset \sim Wx)$

C. p. 413

1. $(\exists x)(Ax \bullet \sim Bx)$
5. $(\exists x)(Ix \bullet \sim Jx)$
10. $(\exists x)(Sx \bullet \sim Tx)$

SECTION 10.5 *Exercises on pp. 423–424*

A. p. 423

5. 1. $(x)(Mx \supset Nx)$
    2. $(\exists x)(Mx \bullet Ox)$
    $\therefore (\exists x)(Ox \bullet Nx)$
    3. $Ma \bullet Oa$ 2, **E.I.**
    4. $Ma \supset Na$ 1, **U.I.**
    5. Ma 3, Simp.
    6. Na 4, 5, M.P.
    7. $Oa \bullet Ma$ 3, Com.
    8. Oa 7, Simp.
    9. $Oa \bullet Na$ 8, 6, Conj.
    10. $(\exists x)(Ox \bullet Nx)$ 9, **E.G.**

10. 1. $(x)(Bx \supset \sim Cx)$
    2. $(\exists x)(Cx \cdot Dx)$
        $\therefore (\exists x)(Dx \cdot \sim Bx)$
    3. $Ca \cdot Da$ 2, **E.I.**
    4. $Ba \supset \sim Ca$ 1, **U.I.**
    5. Ca 3, Simp.
    6. $\sim\sim Ca$ 5, D.N.
    7. $\sim Ba$ 4, 6, M.T.
    8. $Da \cdot Ca$ 3, Com.
    9. Da 8, Simp.
    10. $Da \cdot \sim Ba$ 9, 7, Conj.
    11. $(\exists x)(Dx \cdot \sim Bx)$ 10, **E.G.**

B. p. 424

1. 1. $(x)(Ax \supset \sim Bx)$
   2. Bc
       $\therefore \sim Ac$
   3. $Ac \supset \sim Bc$ 1, **U.I.**
   4. $\sim\sim Bc$ 2, D.N.
   5. $\sim Ac$ 3, 4, M.T.

5. 1. $(x)(Mx \supset Nx)$
   2. $(\exists x)(Ox \cdot Mx)$
       $\therefore (\exists x)(Ox \cdot Nx)$
   3. $Oa \cdot Ma$ 2, **E.I.**
   4. $Ma \supset Na$ 1, **U.I.**
   5. Oa 3, Simp.
   6. $Ma \cdot Oa$ 3, Com.
   7. Ma 6, Simp.
   8. Na 4, 7, M.P.
   9. $Oa \cdot Na$ 5, 8, Conj.
   10. $(\exists x)(Ox \cdot Nx)$ 9, E.G.

10. 1. $(x)(Ax \supset Rx)$
    2. $\sim Rs$
        $\therefore \sim As$
    3. $As \supset Rs$ 1, **U.I.**
    4. $\sim As$ 3, 2, M.T.

SECTION 10.6 *Exercises on p. 427*

A. pp. 427–428

5. $(\exists x)(Mx \cdot Nx)$ logically $(Ma \cdot Na) \vee (Mb \cdot Nb)$
   $(\exists x)(Mx \cdot Ox)$ equivalent $(Ma \cdot Oa) \vee (Mb \cdot Ob)$
   $\therefore (x)(Ox \supset Nx)$ in $\boxed{a, b}$ to $\therefore (Oa \supset Na) \cdot (Ob \cdot Nb)$

	Ma	Mb	Na	Nb	Oa	Ob
proved invalid by	t	t	t	f	t	t

or any of several other truth-value assignments.

10. $(\exists x)(Bx \cdot \sim Cx)$ logically $(Ba \cdot \sim Ca) \vee (Bb \cdot \sim Cb)$
    $(x)(Dx \supset \sim Cx)$ equivalent $(Da \supset \sim Ca) \cdot (Db \supset \sim Cb)$
    $\therefore (x)(Dx \supset Bx)$ in $\boxed{a, b}$ to $\therefore (Da \supset Ba) \cdot (Db \supset Bb)$

	Ba	Bb	Ca	Cb	Da	Db
proved invalid by	f	t	f	f	t	t

B. pp. 428–429

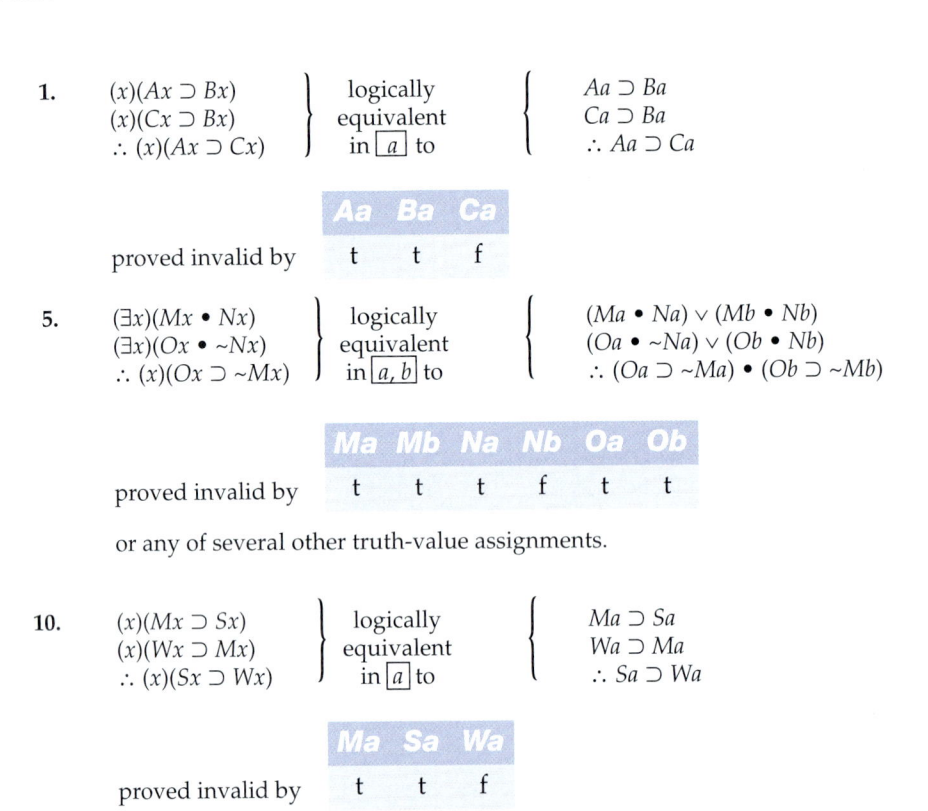

1. $(x)(Ax \supset Bx)$
   $(x)(Cx \supset Bx)$
   $\therefore (x)(Ax \supset Cx)$

   logically equivalent in $\boxed{a}$ to

   $Aa \supset Ba$
   $Ca \supset Ba$
   $\therefore Aa \supset Ca$

	Aa	Ba	Ca
proved invalid by	t	t	f

5. $(\exists x)(Mx \cdot Nx)$
   $(\exists x)(Ox \cdot \sim Nx)$
   $\therefore (x)(Ox \supset \sim Mx)$

   logically equivalent in $\boxed{a, b}$ to

   $(Ma \cdot Na) \vee (Mb \cdot Nb)$
   $(Oa \cdot \sim Na) \vee (Ob \cdot Nb)$
   $\therefore (Oa \supset \sim Ma) \cdot (Ob \supset \sim Mb)$

	Ma	Mb	Na	Nb	Oa	Ob
proved invalid by	t	t	t	f	t	t

   or any of several other truth-value assignments.

10. $(x)(Mx \supset Sx)$
    $(x)(Wx \supset Mx)$
    $\therefore (x)(Sx \supset Wx)$

    logically equivalent in $\boxed{a}$ to

    $Ma \supset Sa$
    $Wa \supset Ma$
    $\therefore Sa \supset Wa$

	Ma	Sa	Wa
proved invalid by	t	t	f

SECTION 10.7 *Exercises on p. 432*

A. pp. 432–433

5. $(x)Gx \supset (Wx \equiv Lx)$
10. $(x)\{Ax \supset [(Bx \supset Wx) \cdot (Px \supset Sx)]\}$

B. pp. 433–434

1. 1. $(x)[(Ax \vee Bx) \supset (Cx \cdot Dx)]$
      $\therefore (x)(Bx \supset Cx)$
   2. $(Ay \vee By) \supset (Cy \cdot Dy)$ 1, U.I.
   3. $\sim(Ay \vee By) \vee (Cy \cdot Dy)$ 2, Impl.
   4. $[\sim(Ay \vee By) \vee Cy] \cdot [\sim(Ay \vee By) \vee Dy]$ 3, Dist.
   5. $\sim(Ay \vee By) \vee Cy$ 4, Simp.
   6. $Cy \vee \sim(Ay \vee By)$ 5, Com.
   7. $Cy \vee (\sim Ay \cdot \sim By)$ 6, De M.
   8. $(Cy \vee \sim Ay) \cdot (Cy \vee \sim By)$ 7, Dist.
   9. $(Cy \vee \sim By) \cdot (Cy \vee \sim Ay)$ 8, Com.
   10. $Cy \vee \sim By$ 9, Simp.
   11. $\sim By \vee Cy$ 10, Com.
   12. $By \supset Cy$ 11, Impl.
   13. $(x)(Bx \supset Cx)$ 12, U.G.

5. $(\exists x)(Sx \cdot Tx)$
   $(\exists x)(Ux \cdot \sim Sx)$
   $(\exists x)(Vx \cdot \sim Tx)$
   $\therefore (\exists x)(Ux \cdot Vx)$

   logically equivalent in $\boxed{a, b, c}$ to

   $(Sa \cdot Ta) \vee (Sb \cdot Tb) \vee (Sc \cdot Tc)$
   $(Ua \cdot \sim Sa) \vee (Ub \cdot \sim Sb) \vee (Uc \cdot \sim Sc)$
   $(Va \cdot \sim Ta) \vee (Vb \cdot \sim Tb) \vee (Vc \cdot \sim Tc)$
   $\therefore (Ua \cdot Va) \vee (Ub \cdot \sim Vb) \vee (Uc \cdot Vc)$

Solutions to Selected Exercises **601**

proved invalid by

Sa	Sb	Sc	Ta	Tb	Tc	Ua	Ub	Uc	Va	Vb	Vc
t	f	t	t	t	f	f	t	f	t	f	t

or any of several other truth-value assignments.

10.

$(x)[(Sx \vee Tx) \supset \sim(Ux \vee Vx)]$

$(\exists x)(Sx \bullet \sim Wx)$
$(\exists x)(Tx \bullet \sim Xx)$
$(x)(\sim Wx \supset Xx)$
$\therefore (\exists x)(Ux \bullet \sim Vx)$

logically equivalent in $\boxed{a, b}$ to

$[(Sa \vee Ta) \supset \sim(Ua \vee Va)] \bullet$
$[(Sb \vee Tb) \supset \sim(Ub \vee Vb)]$
$(Sa \bullet \sim Wa) \vee (Sb \bullet \sim Wb)$
$(Ta \bullet \sim Xa) \vee (Tb \bullet \sim Xb)$
$(\sim Wa \supset Xa) \bullet (\sim Wb \supset Xb)$
$\therefore (Ua \bullet \sim Va) \vee (Ub \bullet \sim Vb)$

and proved invalid by

Sa	Sb	Ta	Tb	Ua	Ub	Va	Vb	Wa	Wb	Xa	Xb
t	t	t	t	f	f	f	f	f	f	t	f

or any of several other truth-value assignments.

C. p. 434

1.
1. $(x)[(Ax \vee Bx) \supset Cx]$
2. $(x)(Vx \supset Ax)$
   $\therefore (x)(Vx \supset Cx)$
3. $(Ay \vee By) \supset Cy$ 1, **U.I.**
4. $Vy \supset Ay$ 2, **U.I.**
5. $\sim Vy \vee Ay$ 4, Impl.
6. $(\sim Vy \vee Ay) \vee By$ 5, Add.
7. $\sim Vy \vee (Ay \vee By)$ 6, Assoc.
8. $Vy \supset (Ay \vee By)$ 7, Impl.
9. $Vy \supset Cy$ 8, 3, H.S.
10. $(x)(Vx \supset Cx)$ 9, **U.G.**

5.
$(x)\{[Ex \bullet (Ix \vee Tx)] \supset \sim Sx\}$
$(\exists x)(Ex \bullet Ix)$
$(\exists x)(Ex \bullet Tx)$
$\therefore (x)(Ex \supset \sim Sx)$

This argument is logically equivalent in $a, \boxed{b \text{ to}}$

$\{[Ea \bullet (Ia \vee Ta)] \supset \sim Sa\} \bullet \{[Eb \bullet (Ib \vee Tb)] \supset \sim Sb\}$
$(Ea \bullet Ia) \vee (Eb \bullet Ib)$
$(Ea \bullet Ta) \vee (Eb \bullet Tb)$
$\therefore (Ea \supset \sim Sa) \bullet (Eb \supset \sim Sb)$

which is proved invalid by

Ea	Eb	Ia	Ib	Ta	Tb	Sa	Sb
t	t	t	f	t	f	f	t

or

Ea	Eb	Ia	Ib	Ta	Tb	Sa	Sb
t	t	f	t	f	t	t	f

10.
1. $(x)[Bx \supset (Ix \supset Wx)]$
2. $(x)[Bx \supset (Wx \supset Ix)]$
   $\therefore (x)\{Bx \supset [(Ix \lor Wx) \supset (Ix \bullet Wx)]\}$
3. $By \supset (Iy \supset Wy)$ — 1, **U.I.**
4. $By \supset (Wy \supset Iy)$ — 2, **U.I.**
5. $[By \supset (Iy \supset Wy)] \bullet [By \supset (Wy \supset Iy)]$ — 3, 4, Conj.
6. $[\sim\!By \lor (Iy \supset Wy)] \bullet [\sim\!By \lor (Wy \supset Iy)]$ — 5, Impl.
7. $\sim\!By \lor [(Iy \supset Wy) \bullet (Wy \supset Iy)]$ — 6, Dist.
8. $\sim\!By \lor (Iy \equiv Wy)$ — 7, Equiv.
9. $\sim\!By \lor [(Iy \bullet Wy) \lor (\sim\!Iy \bullet \sim\!Wy)]$ — 8, Equiv.
10. $\sim\!By \lor [(\sim\!Iy \bullet \sim\!Wy) \lor (Iy \bullet Wy)]$ — 9, Com.
11. $\sim\!By \lor [\sim\!(Iy \lor Wy) \lor (Iy \bullet Wy)]$ — 10, De M.
12. $By \supset [(Iy \lor Wy) \supset (Iy \bullet Wy)]$ — 11, Impl.
13. $(x)\{Bx \supset [(Ix \lor Wx) \supset (Ix \bullet Wx)]\}$ — 12, **U.G.**

D. pp. 435–436

1.
1. $(x)[(Cx \bullet \sim\!Tx) \supset Px]$
2. $(x)(Ox \supset Cx)$
3. $(\exists x)(Ox \bullet \sim\!Px)$
   $\therefore (\exists x)(Tx)$
4. $Oa \bullet \sim\!Pa$ — 3, **E.I.**
5. $Oa \supset Ca$ — 2, **U.I.**
6. $(Ca \bullet \sim\!Ta) \supset Pa$ — 1, **U.I.**
7. Oa — 4, Simp.
8. Ca — 5, 7, M.P.
9. $\sim\!Pa \bullet Oa$ — 4, Com.
10. $\sim\!Pa$ — 9, Simp.
11. $Ca \supset (\sim\!Ta \supset Pa)$ — 6, Exp.
12. $\sim\!Ta \supset Pa$ — 11, 8, M.P.
13. $\sim\!\sim\!Ta$ — 12, 10, M.T.
14. Ta — 13, D.N.
15. $(\exists x)(Tx)$ — 14, **E.G.**

5.
$(\exists x)(Dx \bullet Ax)$
$(x)[Ax \supset (Jx \lor Cx)]$
$(x)(Dx \supset \sim\!Cx)$
$(x)[(Jx \bullet Ix) \supset \sim\!Px]$
$(\exists x)(Dx \bullet Ix)$
$\therefore (\exists x)(Dx \bullet \sim\!Px)$

This argument is logically equivalent in a, b to

$(Da \bullet Aa) \lor (Db \bullet Ab)$
$[Aa \supset (Ja \lor Ca)] \bullet [Ab \supset (Jb \lor Cb)]$
$(Da \supset \sim\!Ca) \bullet (Db \supset \sim\!Cb)$
$[(Ja \bullet Ia) \supset \sim\!Pa] \bullet [(Jb \bullet Ib) \supset \sim\!Pb]$
$(Da \bullet Ia) \lor (Db \bullet Ib)$
$\therefore (Da \bullet \sim\!Pa) \lor (Db \bullet \sim\!Pb)$

proved invalid by

Da	Db	Aa	Ab	Ja	Jb	Ca	Cb	Ia	Ib	Pa	Pb
t	t	t	f	t	f	f	f	t	t	t	t

or

Da	Db	Aa	Ab	Ja	Jb	Ca	Cb	Ia	Ib	Pa	Pb
t	t	f	t	f	t	f	f	t	f	t	t

10.
1. $(\exists x)(Cx \bullet Rx)$
2. $(x)[Rx \supset (Sx \lor Bx)]$
3. $(x)[Bx \supset (Dx \lor Px)]$
4. $(x)(Px \supset Lx)$

5. $(x)(Dx \supset Hx)$
6. $(x)(\sim Hx)$
7. $(x)\{[(Cx \bullet Rx) \bullet Fx] \supset Ax\}$
8. $(x)(Rx \supset Fx)$
9. $(x)[Cx \supset \sim(Lx \bullet Ax)]$
    $\therefore (\exists x)(Cx \bullet Sx)$
10. $Ca \bullet Ra$ 1, **E.I.**
11. $Ra \bullet Ca$ 10, Com.
12. Ra 11, Simp.
13. $Ra \supset Fa$ 8, **U.I.**
14. Fa 13, 12, M.P.
15. $(Ca \bullet Ra) \bullet Fa$ 10, 14, Conj.
16. $[(Ca \bullet Ra) \bullet Fa] \supset Aa$ 7, **U.I.**
17. Aa 16, 15, M.P.
18. $Ca \supset \sim(La \bullet Aa)$ 9, **U.I.**
19. Ca 10, Simp.
20. $\sim(La \bullet Aa)$ 18, 19, M.P.
21. $\sim La \vee \sim Aa$ 20, De M.
22. $\sim Aa \vee \sim La$ 21, Com.
23. $Aa \supset \sim La$ 22, Impl.
24. $\sim La$ 23, 17, M.P.
25. $Pa \supset La$ 4, **U.I.**
26. $\sim Pa$ 25, 24, M.T.
27. $Da \supset Ha$ 5, **U.I.**
28. $\sim Ha$ 6, **U.I.**
29. $\sim Da$ 27, 28, M.T.
30. $\sim Da \bullet \sim Pa$ 29, 26, Conj.
31. $\sim(Da \vee Pa)$ 30, De M.
32. $Ba \supset (Da \vee Pa)$ 3, **U.I.**
33. $\sim Ba$ 32, 31, M.T.
34. $Ra \supset (Sa \vee Ba)$ 2, **U.I.**
35. $Sa \vee Ba$ 34, 12, M.P.
36. $Ba \vee Sa$ 35, Com.
37. Sa 36, 33, D.S.
38. $Ca \bullet Sa$ 19, 37, Conj.
39. $(\exists x)(Cx \bullet Sx)$ 38, **E.G.**

15. 1. $(x)(Ox \supset Sx)$
    2. $(x)(Lx \supset Tx)$
       $\therefore (x)[(Ox \vee Lx) \supset (Sx \vee Tx)]$
    3. $Oy \supset Sy$ 1, **U.I.**
    4. $Ly \supset Ty$ 2, **U.I.**
    5. $\sim Oy \vee Sy$ 3, Impl.
    6. $(\sim Oy \vee Sy) \vee Ty$ 5, Add.
    7. $\sim Oy \vee (Sy \vee Ty)$ 6, Assoc.
    8. $(Sy \vee Ty) \vee \sim Oy$ 7, Com.
    9. $\sim Ly \vee Ty$ 4, Impl.
    10. $(\sim Ly \vee Ty) \vee Sy$ 9, Add.
    11. $\sim Ly \vee (Ty \vee Sy)$ 10, Assoc.
    12. $\sim Ly \vee (Sy \vee Ty)$ 11, Com.
    13. $(Sy \vee Ty) \vee \sim Ly$ 12, Com.
    14. $[(Sy \vee Ty) \vee \sim Oy] \bullet [(Sy \vee Ty) \vee \sim Ly]$ 8, 13, Conj.
    15. $(Sy \vee Ty) \vee (\sim Oy \bullet \sim Ly)$ 14, Dist.
    16. $(\sim Oy \bullet \sim Ly) \vee (Sy \vee Ty)$ 15, Com.
    17. $\sim(Oy \vee Ly) \vee (Sy \vee Ty)$ 16, De M.
    18. $(Oy \vee Ly) \supset (Sy \vee Ty)$ 17, Impl.
    19. $(x)[(Ox \vee Lx) \supset (Sx \vee Tx)]$ 18, **U.G.**

chapter 14

SECTION 14.2 *Exercises on pp. 545–553*

A. pp. 546–547

5. $\frac{1}{4} \times \frac{1}{3} \times \frac{1}{2} \times \frac{1}{1} = \frac{1}{24}$

   The component events here are not independent, but in this case each success (in reaching the right house) *increases* rather than decreases the probability of the next success, because the number of available houses is fixed. After three men reach the correct house, the fourth (having to go to a different house) *must* succeed!

10. The probability that all four students will identify the same tire may be calculated in two different ways—just as the solution to Exercise 6 in this same set may be reached in two different ways.

    Suppose that the first student, A, names the front left tire. The probability of his doing so, *after having done so*, is 1. Now the probability of the second student, B, naming that tire is ¼, there being four tires all (from B's point of view) equipossibly the one that A had named. The same is true of student C, and of student D. Therefore, regardless of which tire A does happen to name (front left, or any other), the probability that all four students will name the same tire is $1 \times \frac{1}{4} \times \frac{1}{4} \times \frac{1}{4} = \frac{1}{64}$ or .016.

    The same result could be achieved by first specifying a particular tire (say, the front left tire) and asking: What is the probability of all four students naming that specified tire? This would be $\frac{1}{4} \times \frac{1}{4} \times \frac{1}{4} \times \frac{1}{4} = .004$. But the condition specified in the problem, that all four name the *same* tire, would be satisfied if all named the front left, or if all named the front right, or if all named the rear left, or if all named the rear right tire. So, if we were to approach the problem in this way, we also would need to inquire as to the probability of *either* the one or the other of these four outcomes—a calculation requiring the addition theorem, explained in Section 14.2B, for alternative outcomes. Because the four successful outcomes are mutually exclusive, we can simply sum the four probabilities: .004 + .004 + .004 + .004 = .016. The two ways of approaching the problem must yield exactly the same result, of course.

    This dual analysis applies likewise to the three patients arriving at a building with five entrances, in Exercise 6. One may calculate $1 \times \frac{1}{5} \times \frac{1}{5} = \frac{1}{25}$; or (using the addition theorem discussed in Section 14.2B,) one may calculate $\frac{1}{5} \times \frac{1}{5} \times \frac{1}{5} = \frac{1}{125}$ and then add $\frac{1}{125} + \frac{1}{125} + \frac{1}{125} + \frac{1}{125} + \frac{1}{125} = \frac{1}{25}$.

B. pp. 552–553

1. Probability of losing with a 2, a 3, or a 12 is $\frac{4}{36}$ or $\frac{1}{9}$
   Probability of throwing a 4, and then a 7 before another 4, is $\frac{3}{36} \times \frac{6}{9} = \frac{1}{18}$
   Probability of throwing a 10, and then a 7 before another 10, is likewise $\frac{1}{18}$
   Probability of throwing a 5, and then a 7 before another 5, is $\frac{4}{36} \times \frac{6}{10} = \frac{1}{15}$
   Probability of throwing a 9, and then a 7 before another 9, is likewise $\frac{1}{15}$
   Probability of throwing a 6, and then a 7 before another 6, is $\frac{5}{36} \times \frac{6}{11} = \frac{5}{66}$
   Probability of throwing an 8, and then a 7 before another 8, is likewise $\frac{5}{66}$
   Sum of the probabilities of the exclusive ways of the shooter's losing is $\frac{251}{495}$
   So the shooter's chance of winning is $1 - \frac{251}{495} = \frac{244}{495}$ or .493.

5. Yes. You lose the bet only if you throw a 2, or a 3, or a 4, or a 5, on *both* rolls of the die. On each throw, the chance of getting one of those four numbers is $\frac{4}{6}$ or $\frac{2}{3}$. The chance of losing the bet is therefore $\frac{2}{3} \times \frac{2}{3}$, or $\frac{4}{9}$. Your chance of winning the bet, therefore, is $1 - \frac{4}{9} = \frac{5}{9} = .556$.

10. **Challenge to the Reader**
    This problem, which has been the focus of some controversy, may be analyzed in two different ways.

    First analysis:
    a. There are 28 possible pairs in the abbreviated deck consisting of four kings and four aces. Of these 28 possible pairs, only seven (equipossible)

pairs contain the ace of spades. Of these seven pairs, three contain two aces. If we know that the pair drawn contains the ace of spades, the probability that this pair contains two aces is ³⁄₇.

b. However, if we know only that one of the cards in the pair is an ace, we know only that the pair drawn is one of the 22 (equipossible) pairs that contain at least one ace. Of these 22 pairs, six contain two aces. Therefore, if we know only that the pair contains an ace, the probability that the pair drawn contains two aces is ⁶⁄₂₂ or ³⁄₁₁.

In this first analysis, the probabilities in the two cases are different.

Second analysis:

a. If one of the cards of the pair drawn is known to be the ace of spades, there are seven other possible cards with which the pair may be completed. Of these seven, three are aces. Therefore, if we know that one of the cards drawn is the ace of spades, the probability that this pair contains two aces is ³⁄₇.

b. If we know only that one of the cards drawn is an ace, we know that it is either the ace of spades, or the ace of hearts, or the ace of diamonds, or the ace of clubs. If it is the ace of spades, the analysis immediately preceding applies, and the probability that this pair contains two aces is again ³⁄₇.

If the ace is the ace of hearts, the same analysis applies; as it does if the card drawn is the ace of diamonds, or the ace of clubs. Therefore, even if we know only that an ace is one of the cards drawn, the probability that the pair contains two aces remains ³⁄₇.

In this second analysis, the probabilities in the two cases are the same. Which of these two analyses do you believe to be correct? Why?

SECTION 14.3 *Exercises on pp. 558–559*

1. a. $3.82
   b. $19,100,000.00
      But note: This was a *very* unusual set of circumstances!

5. This problem requires only a straightforward use of the product theorem. The probability of selecting, at random, just those two cows out of four is the probability of selecting one of that pair on the first choosing (½), times the probability of selecting the other one of that pair on the second choosing, where the first already had been selected (⅓). So the calculation is ½ × ⅓ = ⅙.

10. The calculation of the bettor's chances of winning on the "Don't Pass-Bar 3" line is the probability of the player's losing when the game is played according to the normal rules, *with the provision that he does not lose if he gets a 3 on the first roll*. The probability of a 3 on the first roll is ²⁄₃₆ or .056. The probability of the player losing on the normal rules is .507, as was shown in Section 14.2B. Therefore the probability of the player losing, barring the loss on a first-roll 3, is .507 − .056 = .451. Because this is the probability of the player's losing if he cannot lose by getting a 3 on the first roll, it is the probability of the bettor winning on the "Don't Pass-Bar 3" line. So the expected value of a $100 bet on the "Don't Pass-Bar 3" line is .451 × $200 = $90.20.

    Note that this bet, which the house will gladly accept, is substantially less favorable to the bettor than simply betting on the pass line—that is, simply betting on the player to win. The expected value of such a $100 wager (i.e., on the player to win according to normal rules) is .493 × $200 = $98.60.

Photo Credits

Cover Figure Vladimir Caplinskij/Shutterstock

Chapter 1 Figure 01-01 © Bettmann/CORBIS All Rights Reserved. Figure 01-02 Classic Image/Alamy Images. Figure CO01 M.C. Escher, Dutch, 1898-1972. Sun and Moon, 1948. Woodcut in four colors, 252 × 277 (10 × 10.875″). M.C. Escher Heirs, Cordon Art, Baarn, Holland.

Chapter 2 Figure 02-01 INTERFOTO/Alamy Images. Figure 02-02 INTERFOTO/Alamy Images. Figure CO02 M.C. Escher, Dutch, 1898-1972. Sun and Moon, 1948. Woodcut in four colors, 252 × 277 (10 × 10.875″). M.C. Escher Heirs, Cordon Art, Baarn, Holland.

Chapter 3 Figure 03-01 © Bettmann/CORBIS All Rights Reserved. Figure CO03 M.C. Escher, Dutch, 1898-1972. Sun and Moon, 1948. Woodcut in four colors, 252 × 277 (10 × 10.875″). M.C. Escher Heirs, Cordon Art, Baarn, Holland.

Chapter 4 Figure 04-01 Private Collection/The Bridgeman Art Library. Figure CO04 M.C. Escher, Dutch, 1898-1972. Sun and Moon, 1948. Woodcut in four colors, 252 × 277 (10 × 10.875″). M.C. Escher Heirs, Cordon Art, Baarn, Holland.

Chapter 5 Figure 05-01 © Bettmann/CORBIS All Rights Reserved. Figure 05-11 © Topham/The Image Works. Figure 05-12 Topham/The Image Works. Figure CO05 Images.com.

Chapter 6 Figure CO06 Images.com.

Chapter 7 Figure 07-01 © Bettmann/CORBIS All Rights Reserved.

Chapter 8 Figure 08-01 © Bettmann/CORBIS All Rights Reserved. Figure 08-01a Photodisc/Getty Images. Figure 08-02 Photodisc/Getty Images. Figure 08-03 Photodisc/Getty Images. Figure 08-04 Photodisc/Getty Images. Figure CO08 Images.com.

Chapter 9 Figure 09-01 © Bettmann/CORBIS All Rights Reserved. Figure 09-02 Alfred Eisenstaedt/Time & Life Pictures/Getty Images/Time LifePictures. Figure CO09 Images.com.

Chapter 10 Figure 10-01 Pictorial Press Ltd/Alamy Images. Figure 10-02 © Hulton-Deutsch Collection/CORBIS All Rights Reserved. Figure 10-03 © Bettmann/CORBIS All Rights Reserved. Figure CO10 Images.com.

Chapter 11 Figure 11-02 KING FEATURES SYNDICATE. Figure CO11 M.C. Escher's "Other World" © 2009 The M.C. Escher Company-Holland. All rights reserved. *www.mcescher.com*.

Chapter 12 Figure 12-01 © Bettmann/CORBIS All Rights Reserved. Figure CO12 M.C. Escher's "Other World" © 2009 The M.C. Escher Company-Holland. All rights reserved. *www.mcescher.com*.

Chapter 13 Figure CO13 M.C. Escher's "Other World" © 2009 The M.C. Escher Company-Holland. All rights reserved. *www.mcescher.com*.

Chapter 14 Figure CO14 M.C. Escher's "Other World" © 2009 The M.C. Escher Company-Holland. All rights reserved. *www.mcescher.com*.

Glossary/Index

A

Abelard, Peter, 29, 35

Absorption: A rule of inference; one of nine elementary valid argument forms. If p implies q, absorption permits the inference that p implies *both* p and q. Symbolized as: $p \supset q$, therefore $p \supset (p \cdot q)$, 341

Abusive *ad hominem* argument, 114–15

Accent: An informal fallacy, committed when a term or phrase has a meaning in the conclusion of an argument different from its meaning in one of the premises, the difference arising chiefly from a change in emphasis given to the words used, 142–45

Accident: An informal fallacy, committed when a generalization is applied to individual cases that it does not properly govern, 134. *See also* Converse accident

Adams, Henry, 45

Ad baculum (appeal to force): An informal fallacy in which an inappropriate appeal to force is used to support the truth of some conclusion, 117–18

Addition (Add.): A rule of logical inference, one of nine elementary valid argument forms. Given any proposition p, addition permits the inference that p or q. Also called "logical addition," 342–44

Addition theorem: In the calculus of probability, a theorem used to determine the probability of a complex event consisting of one or more alternative occurrences of simple events whose probabilities are known, 547–51; for exclusive alternatives, 549

Addresses on War (Sumner), 74

Ad hoc: A term with several meanings, used to characterize hypotheses. It may mean only that the hypothesis was constructed after the facts it purports to explain; it may mean that the hypothesis is merely descriptive. Most commonly, "*ad hoc*" is used pejoratively, describing a hypothesis that serves to explain only the facts it was invented to explain and has no other testable consequences, 527

Ad hominem (argument against the person): An informal fallacy in which the object of attack is not the merits of some position, but the person who takes that position, 114–17; abusive, 114–15; circumstantial, 115–17

Ad ignorantiam (argument from ignorance): An informal fallacy in which a conclusion is supported by an illegitimate appeal to ignorance, as when it is supposed that something is likely to be true because we cannot prove that it is false, 126–29

Ad misericordiam (appeal to pity): An informal fallacy in which the support given for some conclusion is an inappropriate appeal to the mercy or altruism of the audience, 110–11

Ad populum (appeal to emotion): An informal fallacy in which the support given for some conclusion is an inappropriate appeal to popular belief, 108–10

Advancement of Learning, The (Bacon), 476

Adventure of Silver Blaze, The (Doyle), 277

Adventures of Huckleberry Fin, The (Twain), 139

Ad verecundiam (appeal to inappropriate authority): An informal fallacy in which an appeal is made to authority, which is fallacious both because the authority appealed to has no special claim to expertness on the matter in question, and because even legitimate authorities are often wrong, 129–30

Affirmative conclusion from a negative premise, Fallacy of, 227

Affirmative singular proposition: A proposition in which it is asserted that a particular individual has some specified attribute, 399

Affirming the consequent: A formal fallacy, so named because the categorical premise in the argument affirms the consequent rather than the antecedent of the conditional premise. Symbolized as: $p \supset q$, q, therefore p, 274, 320

Affluent Society, The (Galbraith), 45

Against the Logicians (Sextus Empiricus), 277

Against the Physicists (Sextus Empiricus), 284

Age of Reason, The (Paine), 69

Agre, Peter, 535–36

Agreement, Method of: A pattern of inductive inference in which it is concluded that, if two or more instances of a phenomenon have only one circumstance in common, that one common circumstance is the cause (or effect) of the phenomenon under investigation, 476–81

Aim and Structure of Physical Theory, The (Duhem), 100

Alev, Imam Fatih, 122

Alford, Henry, 151

Alice's Adventures in Wonderland (Carroll), 389, 463

Alroy, Daniel, 46

Alternative occurrences: In probability theory, a complex event that consists of the occurrence of any one of two or more simple component events. (*e.g.*, the complex event of getting either a spade or a club in the random drawing of a playing card), 543, 547–51

Ambiguity: Uncertainty of meaning, often leading to disputes or to mistakes when the same word or phrase has two (or more) distinct meanings, and the context does not make clear which meaning is intended, 75–79, 103

Ambiguity, Fallacy of: Any fallacy caused by a shift in or confusion of meanings within an argument. Also known as a "*sophism*," 140–54; accent, 142–45; amphiboly, 142; composition, 145–46; division, 146–47; equivocation, 140–42

Ambiguous middle: A formal fallacy, so called because the mistake in some syllogism arises from a shift, within the argument, in the meaning of the middle term, 225

American Journal of Medical Science (Smith), 70

American Notebooks (Hawthorne), 75

Amiel, Henri-Frédéric, 68

Amiel's Journal (Amiel), 68

Glossary/Index 609

Amphiboly: A kind of ambiguity arising from the loose, awkward, or mistaken way in which words are combined, leading to alternative possible meanings of a statement. Also, the name of a fallacy when an argument incorporates an amphibolous statement that is true as used in one occurrence, but false as used in another occurrence of the statement in that argument, 142

Analects, The (Confucius), 100

Analogical argument: A kind of inductive argument in which, from the fact that two entities are alike in some respect(s), it is concluded that they are also alike in some other respect(s), 445–52; criteria for appraising, 452–57

Analogy: A parallel drawn between two (or more) entities by indicating one or more respects in which they are similar, 445, 447; argument by, 445–52; characteristics of argument by, 448; nonargumentative, 446–47

Analytical definition: *See* Definition by genus and difference

Analyzing arguments, 34–62

Anatomy of an Illness (Cousins), 284

Anecdotes of Samuel Johnson (Piozzi), 149

Animals without Backbones (Buchsbaum), 460

Annabel Lee (Poe), 67

Annals (Tacitus), 73

Antecedent: In a conditional statement ("*if* . . . *then* . . ."), the component that immediately follows the "if." Sometimes called the *implicans* or the *protasis*, 274–75, 300–03

Apodosis: The consequent in a hypothetical proposition, 301

Apology (Plato), 284

Apparently verbal but really genuine disputes, 76

Appeal to inappropriate authority: A fallacy in which a conclusion is accepted as true simply because an expert has said that it is true. This is a fallacy whether or not the expert's area of expertise is relevant to the conclusion. Also known as "argument *ad verecundiam*," 129–30

Appeal to force: A fallacy in which the argument relies upon an open or veiled threat of force. Also known as "argument *ad baculum*," 117–18

Appeal to pity: A fallacy in which the argument relies on generosity, altruism, or mercy, rather than on reason. Also known as "argument *ad misericordiam*," 110–11

Appeal to the populace: An informal fallacy in which the support given for some conclusion is an appeal to popular belief. Also known as "argument *ad populum*," 110, 154

***A priori* theory of probability:** A theory in which the probability ascribed to a simple event is a fraction between 0 and 1, of which the denominator is the number of equipossible outcomes, and the numerator is the number of outcomes in which the event in question occurs. Thus on the *a priori* theory, the probability of drawing a spade at random from a deck of playing cards is 13/52, 540–41, 542, 557

Aquinas, Thomas, 9, 42, 261, 277

Argument: Any group of propositions of which one is claimed to follow from the others, which are regarded as providing support or grounds for the truth of that one, 5–9; by analogy, 445–52; analysis of, 34–62; complex, 5, 49–54; conclusion of, 5–6; deductive, 24–27, 164; diagramming, 38–48; explanations and, 18–24; inductive, 24–27, 444–45; interwoven, 43, 49; invalid, 27–32, 383–85; in ordinary language, 278; paraphrasing, 34–38, 44; premise of, 5–6; recognizing, 11–17; sound, 32; syllogistic, 245–49; valid, 27–32

Argument against the person: A fallacy in which the argument relies upon an attack against the person taking a position. This fallacy is also known as "argument *ad hominem*," 114

Argument form: An array of symbols exhibiting logical structure; it contains no statements but it contains statement variables. These variables are arranged in such a way that when statements are consistently substituted for the statement variables, the result is an argument, 310–23; common invalid, 320–21; common valid, 316–20; truth tables and, 314–16; "valid" and "invalid," precise meaning of, 314

Argument from ignorance: A fallacy in which a proposition is held to be true just because it has not been proven false, or false because it has not been proven true. Also known as "argument *ad ignorantiam*," 126

Aringarosa, Bishop, 112

Aristotelian logic: The traditional account of syllogistic reasoning, in which certain interpretations of categorical propositions are presupposed. Often contrasted with the modern symbolic, or Boolean, interpretation of categorical propositions, 164

Aristotle, 7–8, 29, 74, 102, 118, 142–43, 150, 152, 164, 165, 169, 191, 262, 269, 277, 400, 514,

Armour, Stacy, 505

Arnauld, Antoine, 29

Arnett, Cliff, 115

Arnold, Matthew, 101

Art of Scientific Discovery, The (Gore), 494

Association (Assoc.): An expression of logical equivalence; a rule of inference that permits the valid regrouping of simple propositions. According to it, $[p \lor (q \lor r)]$ may be replaced by $[(p \lor q) \lor r]$ and vice versa, and $[p \cdot (q \cdot r)]$ may be replaced by $[(p \cdot q) \cdot r]$ and vice versa, 369

Astell, Mary, 449

Asyllogistic argument: An argument in which one or more of the component propositions is of a form more complicated than the form of the **A**, **E**, **I**, and **O** propositions of the categorical syllogism, and whose analysis therefore requires logical tools more powerful than those provided by Aristotelian logic, 255, 432

Attitude, agreement/disagreement in, 72–73

Authority, appeal to. *See* Ad verecundiam

Autocrat of the Breakfast-Table, The (Holmes), 69

Ayer, Alfred J., 263, 462

B

Bacon, Francis, 10, 29, 69

Baird, Donna D., 480

Baker, Howard, 118

Baker Motley, Constance, 114

Baranovsky, Anatole M., 152

Barbara: The traditional name of one of the 15 valid standard-form categorical syllogisms. A syllogism in the form *Barbara* has the mood and figure **AAA–1**; that is to say, all three of its propositions are **A** propositions, and it is in the first figure because the middle term is the subject of the major premise and the predicate of the minor premise, 211, 240, 259

Baroko: The traditional name of one of the 15 valid standard-form categorical syllogisms. A syllogism in the form *Baroko* has the mood and figure **AOO–2**; that is to say, the minor premise and conclusion are **O** propositions, the major premise is an **A** proposition, and it is in the second figure because the middle term is the predicate of both the major and the minor premise, 236, 241, 256

Barr, Stephen M., 466

Basic Laws of Arithmetic, The (Frege), 451

Bawer, Bruce, 122

Begging the question: An informal fallacy in which the conclusion of an argument is stated or assumed in any one of the premises. Also known as "circular argument" and *petitio principii*, 136–38

Belief, agreement/disagreement in, 71–73

Bellow, Saul, 130

Belvedere (Escher), 162

Bentham, Jeremy, 8

Berkeley, George, 261

Bernstein, Anya, 45

Berra, Yogi, 388

Bettelheim, Bruno, 17

Biconditional statement or proposition: A compound statement or proposition that asserts that its two component statements have the same truth value, and therefore are materially equivalent. So named because, since the two component statements are either both true or both false, they must imply one another. A biconditional statement form is symbolized "$p \equiv q$," which may be read as "p if and only if q," 331, 404, 431

Bierce, Ambrose, 97, 99, 101, 149

Bin Laden, Osama, 115, 467

Blair, Tony, 45

Bodin, Jean, 21

Bokardo: The traditional name of one of the 15 valid standard-form categorical syllogisms. A syllogism in the form *Bokardo* has the mood and figure **OAO–3**; that is to say, its major premise and conclusion are **O** propositions, its minor premise is an **A** proposition, and it is in the third figure because the middle term is the subject of both the minor and the major premise, 242

Bonaparte, Napoleon, 412

Boole, George, 29, 189, 191

Boolean interpretation: The modern interpretation of categorical propositions, adopted in this book and named after the English logician George Boole (1815–1864). In the Boolean interpretation, often contrasted with the Aristotelian interpretation, universal propositions (**A** and **E** propositions) do not have existential import, 189, 194–95, 250–51

Boolean logic, 29

Boolean square of opposition, 198

Brahe, Tycho, 523

Breyer, Stephen, 86

Bright, John, 70

Brill, Steve, 465, 466

Brinkley, Alan, 44

Broad, C.D., 278

Broder, David, 112

Brooks, David, 36, 412

Brooks, John, 261

Browne, Malcolm W., 451

Bruce, Thomas, 450

Bruggemann, Edward, 152

Bruno, Giordano, 533

Buchsbaum, Ralph, 460

Buckvar, Eric, 449

Burke, Edmund, 68

Bush, George W., 119, 273

Buss, David M., 450

Butler, Joseph, 154

Butler, Samuel, 99

C

Calculus of probability: A branch of mathematics that can be used to compute the probabilities of complex events from the probabilities of their component events, 542–47

Callahan, J.J., 278

Camenes: The traditional name of one of the 15 valid standard-form categorical syllogisms. A syllogism in the form *Camenes* has the mood and figure **AEE–4**; that is to say, its minor premise and conclusion are **E** propositions, its major premise is an **A** proposition, and it is in the fourth figure because the middle term is the predicate of the major premise and the subject of the minor premise, 240

Camestres: The traditional name of one of the 15 valid standard-form categorical syllogisms. A syllogism in the form *Camestres* has the mood and figure **AEE–2**; that is to say, its minor premise and conclusion are **E** propositions, its major premise is an **A** proposition, and it is in the second figure because the middle term is the predicate of both the major and the minor premise, 238, 240, 247

Campbell, C. Arthur, 278

Candlish, Stewart, 276

Cargill, Michelle, 506

Carroll, James, 483

Carroll, Lewis, 67, 141, 388, 463, 483

Case, Anne, 507

Categorical proposition: A proposition that can be analyzed as being about classes, or categories, affirming or denying that one class, S, is included in some other class, P, in whole or in part. Four standard forms of categorical propositions are traditionally distinguished: **A**: Universal affirmative propositions (All S is P); **E**: Universal negative propositions (No S is P); **I**: Particular affirmative propositions (Some S is P); **O**: Particular negative propositions (Some S is not P), 164–203, 252–53, 254–55; existential import and, 189–97; symbolism and diagrams for, 197–203; theory of deduction and, 164–65; translating into standard form, 165–69. *See also* Standard-form categorical proposition

Categorical syllogism: A deductive argument consisting of three categorical propositions that contain exactly three terms, each of which occurs in exactly two of the propositions, 205–43, 422, 429; Venn diagram technique for testing, 213–24. *See also* Disjunctive Syllogism; Hypothetical Syllogism; Syllogistic argument

Causal laws: Descriptive laws asserting a necessary connection between events of two kinds, of which one is the cause and the other the effect, 470

Causal reasoning: Inductive reasoning in which some effect is inferred from what is assumed to be its cause, or some cause is inferred from what is assumed to be its effect, 470–511; causal laws/uniformity of nature and, 473; induction by simple enumeration and, 474–76; limitations of inductive techniques, 501–11; meanings of *cause*, 470–73; Mill's methods and, 476–501

Cause: Either the *necessary* condition for the occurrence of an effect (the sense used when we seek to *eliminate* some thing or event by eliminating its cause), or the *sufficient* condition for the occurrence of an effect, understood as the conjunction of its necessary conditions. The latter meaning is more common, and is the sense of cause used when we wish to *produce* some thing or event, 470–73

Cecil, Robert, 70

Celarent: The traditional name of one of the 15 valid standard-form categorical syllogisms. A syllogism in the form *Celarent* has the mood and figure **EAE–1**; that is to say, its major premise and conclusions are **E** propositions, its minor premise is an **A** proposition, and it is in the first figure because the middle term is the subject of the major premise and the predicate of the minor premise, 209, 236, 238, 240

Ceremonial language: language with special social uses normally having a mix of expressive, directive, and informative functions, 65

Cesare: The traditional name of one of the 15 valid standard-form categorical syllogisms. A syllogism in the form *Cesare* has the mood and figure **EAE–2**; that is to say, its major premise and conclusion are **E** propositions, its minor premise is an **A** proposition, and it is in the second figure because the middle term is the predicate of both the major and the minor premise, 236, 238

Challenger, James, 285

Changing Education (Jones), 267

Character and Opinion in the United States (Santayana), 67

Chase, Salmon P., 70

Chen, Fan, 461

Chinese View of Life, The, 150

Chirac, Jacques, 121

Chrysippus, 3, 29

Churchill, Winston, 102

Circular argument: A fallacious argument in which the conclusion is assumed in one of the premises; begging the question. Also called *petitio principii*, 137, 138

Circular definition: A definition that is faulty because its *definiendum* (what is to be defined) appears in its *definiens* (the defining symbols) and therefore is useless, 137

Circumference of Earth, Eratosthenes' measurement of, 518

Circumstantial ad hominem argument: An informal fallacy in which the *ad hominem* attack against the opponent is based upon special circumstances associated with that person, 116–17

Clarence Darrow for the Defense (Stone), 123

Clarke, S., 140

Class: The collection of all objects that have some specified characteristic in common, 165; complement of a, 181, 182; relative complement of a, 182

Classical logic. *See* Aristotelian logic

Classification: The organization and division of large collections of things into an ordered system of groups and subgroups, often used in the construction of scientific hypotheses, 529–37

Cleveland, Grover, 47

Coelho, Tony, 142

Cohen, Randy, 9

Coke, Edward, 100

Coleridge, Samuel Taylor, 75

Coming Struggle for Power, The (Strachey), 126

Coming Through the Fire (Lincoln), 69

Command: One common form of discourse having a directive function, 15

Common Sense (Paine), 121, 125

Communist Manifesto, The (Marx and Engels), 102

Commutation (Com.): An expression of logical equivalence; a rule of inference that permits the valid reordering of the components of conjunctive or disjunctive statements. According to commutation, $(p \vee q)$ and $(q \vee p)$ may replace one another, as may $(p \cdot q)$ and $(q \cdot p)$, 360

Complement, or **complementary class:** The complement of a class is the collection of all things that do not belong to that class, 182

Complex argumentative passages, 49–54

Complex dilemma: An argument consisting of (a) a disjunction, (b) two conditional premises linked by a conjunction, and (c) a conclusion that is not a single categorical proposition (as in a simple dilemma) but a disjunction, a pair of (usually undesirable) alternatives, 279

Complex question: An informal fallacy in which a question is asked in such a way as to presuppose the truth of some conclusion buried in that question, 134–36

Component: A part of a compound statement that is itself a statement, and is of such a nature that, if replaced in the larger statement by any other statement, the result will be meaningful, 290

Composition: An informal fallacy in which an argument erroneously assigns attributes to a whole (or to a collection) based on the fact that parts of that whole (or members of that collection) have those attributes, 145–46

Compound statement or **compound proposition:** A statement that contains two or more statements as components, 4–5, 289

Conan Doyle, Sir Arthur, 53, 264, 277

Concept of Mind, The (Ryle), 101

Conclusion: In any argument, the proposition to which the other propositions in the argument are claimed to give support, or for which they are given as reasons, 5–6, 49

Conclusion indicator: A word or phrase (such as "therefore" or "thus") appearing in an argument and usually indicating that what follows it is the conclusion of that argument, 11

Concomitant Variation, Method of: A pattern of inductive inference in which it is concluded that, when one phenomenon varies consistently with some other phenomenon in some manner, there is some causal relation between the two phenomena, 495–501

Conditional statement: A hypothetical statement; a compound proposition or statement of the form "If *p* then *q*," 5, 300–10

Confessions of a Young Man (Moore), 74

Confucius, 100

Conjunction (Conj.): A truth-functional connective meaning "and," symbolized by the dot. A statement of the form $p \cdot q$ is true if and only if *p* is true *and q* is true. "Conjunction" ("Conj.") is also the name of a rule of inference, one of nine elementary valid argument forms; it permits statements assumed to be true to be combined in one compound statement. Symbolized as: *p*, *q*, therefore $p \cdot q$, 289–92, 327

Conjunctive proposition, 4

Conjunct: Each one of the component statements connected in a conjunctive statement, 289

Connotation: The intension of a term; the attributes shared by all and only those objects to which the term refers, 87

Connotative definition: A definition that states the conventional connotation, or intension, of the term to be defined; usually a definition by genus and difference, 93

Conquest of Happiness, The (Russell), 74

Consequent: In a hypothetical proposition ("*if* . . . *then*"), the component that immediately follows the "then." Sometimes called the *implicate*, or the *apodosis*, 274, 300–02

Constant. *See* Individual constant

Constitution of the United States, The, 9

Constructive Dilemma (C.D.): A rule of inference; one of nine elementary valid argument forms. Constructive Dilemma permits the inference that if $(p \supset q) \cdot (r \supset s)$ is true, and $p \vee r$ is also true, then $q \vee s$ must be true, 320, 341

Conti, Bruno, 491

Contingent: Being neither tautologous nor self-contradictory. A contingent statement may be true or false; a contingent statement form has some true and some false substitution instances, 326

Contradiction: A statement that is necessarily false; a statement form that cannot have any true substitution instances, 326

Contradictories: Two propositions so related that one is the denial or negation of the other. On the traditional square of opposition, the two pairs of contradictories are indicated by the diagonals of the square: **A** and **E** propositions are the contradictories of **O** and **I**, respectively, 176

Contraposition: A valid form of immediate inference for some, but not for all types of propositions. To form the contrapositive of a given proposition, its subject term is replaced by the complement of its predicate term, and its predicate term is replaced by the complement of its subject term. Thus the contrapositive of the proposition "All humans are mammals" is the proposition "All nonmammals are nonhumans," 183–85; table of, 205–06. *See also* Limitation

Contrapositive: The conclusion of the inference called contraposition, 183, 185

Contraries: Two propositions so related that they cannot both be true, although both may be false. On the traditional square of opposition, corresponding **A** and **E** propositions are contraries; but corresponding **A** and **E** propositions are not contraries in the Boolean interpretation, according to which they might both be true, 177, 178, 405. *See also* Subcontraries

Conventional intension: The commonly accepted intension of a term; the criteria generally agreed upon for deciding, with respect to any object, whether it is part of the extension of that term, 92

Converse: The conclusion of the immediate inference called "conversion," 180

Converse accident: An informal fallacy (sometimes called "hasty generalization") committed when one moves carelessly or too quickly from individual cases to a generalization, 133

Conversion: A valid form of immediate inference for some but not all types of propositions. To form the converse of a proposition the subject and predicate terms are simply interchanged. Thus, "No circles are squares" is the converse of "No squares are circles," and "Some thinkers are athletes" is the converse of "Some athletes are thinkers." The proposition converted is called the "convertend," 180–81. *See also* Limitation

Convertend. *See* Conversion

Cooper, Belinda, 10

Copernicus, Nicolaus, 148

Copula: Any form of the verb "to be" that serves to connect the subject term and the predicate term of a categorical proposition, 171

Coren, Stanley, 499–500, 507

Corresponding propositions. *See* Square of Opposition

Cousins, Norman, 284

Cowell, Alan, 45

Critique of Pure Reason (Kant), 21, 99, 102

Crito (Plato), 47

Croce, Benedetto, 151

Cronan, Sheila, 450

Crucial experiment: An experiment whose outcome is claimed to establish the falsehood of one of two competing and inconsistent scientific hypotheses, 524; confirmation of Copernican hypothesis as, 529

Curl: The symbol for negation, ~; the tilde. It appears immediately before (to the left of) what is negated or denied, 292

Cushman, John H., 462

D

Dacke, Marie, 536

Danish, Steve, 451

Darii: The traditional name of one of the 15 valid standard-form categorical syllogisms. A syllogism in the form *Darii* has the mood and figure **AII–1**; that is to say, its minor premise and conclusion are **I** propositions, its major premise is an **A** proposition, and it is in the third figure because the middle term is the subject of both the major and the minor premise, 241, 244

Darrow, Clarence, 123

Darwin, Charles, 69, 129, 268, 450, 517, 520

Datisi: The traditional name of one of the 15 valid standard-form categorical syllogisms. A syllogism in the form *Datisi* has the mood and figure **AII–3**; that is to say, its minor premise and conclusion are **I** propositions, its major premise is an **A** proposition, and it is in the third figure because the middle term is the subject of both the major and minor premise, 241, 244

Davies, H., 493–94

Davies, Paul, 153

Dawkins, Richard, 11

De Beauvoir, Simone, 267

Debs, Eugene, 70

Decatur, Stephen, 73

Deduction: One of the two major types of argument traditionally distinguished, the other being induction. A deductive argument claims to provide conclusive grounds for its conclusion; if it does so it is valid, if it does not it is invalid, 27; distinction between induction and, 444; formal proof of validity and, 337–40; inconsistency and, 386–92; indirect proof of validity and, 392–94; methods of, 337–96; natural, 338, 364–67; proof of invalidity and, 383–86; refutation by logical analogy and, 310–12; rule of replacement and, 357–63; theory of, 164–65

Defective induction, fallacy of: A fallacy in which the premises are too weak or ineffective to warrant the conclusion, 107, 126–33, 155; 174–75; *ad ignorantiam*, 126–29; *ad verecundiam*, 129–30; false cause, 130–32; hasty generalization, 132–33

Defence of Poetry, The (Shelley), 101

Definiendum: In any definition, the word or symbol being defined, 79, 96. *See also* specific types of definition

Definiens: In any definition, a symbol or group of symbols that is said to have the same meaning as the *definiendum*, 79, 90. *See also* specific types of definition

Definition: An expression in which one word or set of symbols (the *definiens*) is provided, which is claimed to have the same meaning as the *definiendum*, the word or symbol defined, 79–103; circular, 96; denotative, 89–91; disputes and, 85; extension/intension and, 86–89; by genus and difference, 105–15; intensional, 91–93; lexical, 81–82; operational, 93; ostensive, 90; persuasive, 85–86; precising, 82–84; quasi-ostensive, 90; rules for, by genus and difference, 94–98; stipulative, 79–80; synonymous, 103; theoretical, 84–85

Definition by genus and difference: A type of connotative definition of a term that first identifies the larger class ("genus") of which the *definiendum* is a species or subclass, and then identifies the attribute ("difference") that distinguishes the members of that species from members of all other species in that genus, 93, 94–103; rules for, 96–98

De Forster, Piers M., 495

De Mello, Fernando Collor, 67

Democracy in America (de Tocqueville), 53

Demonstrative definition: An ostensive definition; one that refers by gesture to examples of the term being defined, 90

De Morgan, Augustus, 29, 358

De Morgan's Theorem (De M.): An expression of logical equivalence; a rule of inference that permits the valid mutual replacement of the negation of a disjunction by the conjunction of the negations of its disjuncts: $\sim(p \vee q) \stackrel{T}{\equiv} (\sim p \cdot \sim q)$; and that permits the valid mutual replacement of the negation of a conjunction by the disjunction of the negations of its conjuncts: $\sim(p \cdot q) \stackrel{T}{\equiv} (\sim p \vee \sim q)$, 331–32, 360, 368, 411

Denotation: The several objects to which a term may correctly be applied; its extension, 89–91

Denotative definition: A definition that identifies the extension of a term, by (for example) listing the members of the class of objects to which the term refers; the members of that class are thus denoted. An extensional definition, 89–91

Denton, Derek, 485

Denying the antecedent: A formal fallacy, so named because the categorical premise in the argument, $\sim p$, denies the antecedent rather than the consequent of the conditional premise. Symbolized as: $p \supset q$, $\sim p$, therefore $\sim q$, 275, 321

Descartes, René, 10

Descent of Man, The (Darwin), 69, 268

De Tocqueville, Alexis, 53

Detweiler, Elaine, 449

Devil's Dictionary, The (Bierce), 97, 101, 149

Devine, Philip E., 100

Dewey, John, 21, 55, 102, 268, 516

Diagramming arguments, 38–48

Dialectic of Sex: The Case for Feminist Revolution, The (Firestone), 126

Dialogues Concerning Natural Religion (Hume), 53, 125, 267, 286, 460, 462

Diary in Australia (Cecil), 70

Diary of a Young Girl, The (Frank), 452

Dickson, W. J., 508

Dictionary of the English Language (Johnson), 92, 97

Difference, Method of: A pattern of inductive argument in which, when cases in which the phenomenon under investigation occurs and cases in which it does not occur differ in only one circumstance, that circumstance is inferred to be causally connected to the phenomenon under investigation, 482–88

Dilemma: A common form of argument in ordinary discourse in which it is claimed that a choice must be made between two alternatives, both of which are (usually) bad, 278–85; ways of evading or refuting, 280–82

DiLorenzo, Thomas, 144

Dimaris: The traditional name of one of the 15 valid standard-form categorical syllogisms. A syllogism in the form *Dimaris* has the mood and figure **IAI–4**; that is to say, its major premise and conclusion are **I** propositions, its minor premise is an **A** proposition, and it is in the fourth figure because the middle term is the predicate of the major premise and the subject of the minor premise, 241, 244

Dirksen, Everett, 388

Directive discourse, 64–65

Disagreement, in attitude and belief, 72–73

Disamis: The traditional name of one of the 15 valid standard-form categorical syllogisms. A syllogism in the *Disamis* has the mood and figure **IAI–3**; that is to say, its major premise and conclusion are **I** propositions, its minor premise is an **A** proposition, and it is in the third figure because the middle term is the subject of both the major and the minor premise, 241

Disanalogy: In an anological argument, a point of difference between the cases mentioned in the premises and the case mentioned in the conclusion, 454–56

Discourse. *See* Language, functions of

Discourse on Method, A (Descartes), 10

Disjunction: A truth-functional connective meaning "or"; components so connected are called "disjuncts." When disjunction is taken to mean that at least one of the disjuncts is true and that they may both be true, it is called a "weak" or "inclusive" disjunction and symbolized by the wedge, $\vee$. When disjunction is taken to mean that at least one of the disjuncts is true and that at least one of them is false, it is called a "strong" of "exclusive" disjunction, 292–94

Disjunctive proposition, 4

Disjunctive statement: A compound statement whose component statements are connected by a disjunction. In modern symbolic logic the interpretation normally given to "or" is weak (inclusive) disjunction, unless further information is provided in the context, 296–97

Disjunctive statement form: A statement form symbolized as: $p \vee q$; its substitution instances are disjunctive statements, 325

Disjunctive Syllogism (D.S.): A rule of inference; a valid argument form in which one premise is a disjunction, another premise is the denial of one of the two disjuncts, and the conclusion is the truth of the other disjunct. Symbolized as: $p \vee q$, $\sim p$, therefore q, 273–74, 275, 293, 302, 316–19, 369

Disputes: apparently verbal but really genuine, 76; merely verbal, 76; obviously genuine, 75–76

Disraeli, Benjamin, 72

Distribution, as a rule of replacement (Dist): An expression of logical equivalence; a rule of inference that permits, in deductive argument, the mutual replacement of certain specified pairs of symbolic expressions, 361, 369

Distribution: An attribute that describes the relationship between a categorical proposition and each one of its terms, indicating whether or not the proposition makes a statement about every member of the class represented by a given term, 171–75

Division: A fallacy of ambiguity in which an argument erroneously assigns attributes to parts of a whole (or to members of a collection) based on the fact that the whole (or the collection) has those attributes, 146–48

Dollar, Steve, 412

Dot: The symbol for conjunction, $\cdot$, meaning "and," 290

Double negation: An expression of logical equivalence; a rule of inference that permits the valid mutual replacement of any symbol by the negation of the negation of that symbol. Symbolized as $p \stackrel{\mathrm{T}}{=} \sim\sim p$, 330, 358, 361

Douglas, Stephen, 28

Douglass, Frederick, 126

Doyle, A. Conan, 53, 264, 277

Doyle, T., 22

Dubois, W.E.B., 67

Dubos, René, 449

Duhem, Pierre, 100

Duns Scotus, 263

Du Pre, Jacqueline, 99

Dworkin, Ronald, 21, 468

Dyson, Freeman, 268

E

Eddington, Arthur, 461, 520

Education of Henry Adams, The (Adams), 45

Edwards, Jonathan, 153

Einstein, Albert, 519–20, 521, 523, 524

Elementary valid argument: Any one of a set of specified deductive arguments that serves as a rule of inference and that may therefore be used in constructing a formal proof of validity, 339

Emotion, appeal to (argument *ad populum*), 121–25

Emotive language, 64, 71–73, 85

Engels, Friedrich, 41, 102

Enthymeme: An argument that is stated incompletely, the unstated part of it being taken for granted. An enthymeme may be of the first, second, or third order, depending upon whether the unstated proposition is the major premise, the minor premise, or the conclusion of the argument, 17, 264–69

Equivocation: An informal fallacy in which two or more meanings of the same word or phrase have been confused. If used with one of its meanings in one of the propositions of the argument but with a different meaning in another proposition of the argument, a word is said to have been used equivocally, 140–42

Erasmus, Desiderius, 73

Eratosthenes, 516, 518

Escher, M.C., xx, 162, 442

Essay Concerning Civil Government (Locke), 101

Essay in Defence of the Female Sex, An (Astell), 449

Essay on Civil Disobedience, An (Thoreau), 75

Essay on the Extinction of the Soul (Chen), 461

Essays (Bacon), 69

Ethics (Spinoza), 54, 71, 85, 101

Euathlus, 281–82

Euphemisms, 71

Euripides, 70

Example, definitions by, 88, 90–91

Exceptive proposition: A proposition that asserts that all members of some class, with the exception of the members of one of its subclasses, are members of some other class. Exceptive propositions are in reality compound, because they assert both a relation of class inclusion, and a relation of class exclusion. Example: "All persons except employees are eligible" is an exceptive proposition in which it is asserted both that "All nonemployees are eligible," and that "No employees are eligible," 255–56, 431–32

Excluded middle, Principle of: The principle that asserts that any statement is either true or false; sometimes called one of the laws of thought, 334, 335

Glossary/Index **615**

Exclusive disjunction or **strong disjunction:** A logical relation meaning "or," that may connect two component statements. A compound statement asserting exclusive disjunction says that at least one of the disjuncts is true *and* that at least one of the disjuncts is false. It is contrasted with an "inclusive" (or "weak") disjunction, which says that at least one of the disjuncts is true and that they may both be true, 293, 302

Exclusive premises, Fallacy of: The formal fallacy that is committed when both premises in a syllogism are negative propositions (**E** or **O**), 227

Exclusive propositions: Propositions that assert that the predicate applies exclusively to the subject named. Example: "None but generals wear stars" asserts that the predicate, "wearing stars," applies only to generals, 254

Existential fallacy: Any mistake in reasoning that arises from assuming illegitimately that some class has members. It is a formal fallacy when, in a standard-form categorical syllogism, a particular conclusion is inferred from two universal premises, 195, 228

Existential Generalization (E.G.): A rule of inference in the theory of quantification that says that from any true substitution instance of a propositional function we may validly infer the existential quantification of the propositional function, 421, 422

Existential import: An attribute of those propositions that normally assert the existence of objects of some specified kind. Particular propositions (**I** and **O** propositions) always have existential import; thus the proposition "Some dogs are obedient" asserts that there are dogs. Whether universal propositions (**A** and **E** propositions) have existential import is an issue on which the Aristotelian and Boolean interpretations of propositions differ, 190

Existential Instantiation (E.I.): A rule of inference in the theory of quantification that says that we may (with some restrictions) validly infer from the existential quantification of a propositional function the truth of its substitution instance with respect to any individual constant that does not occur earlier in that context, 420, 422

Existential presupposition: In Aristotelian logic, the blanket presupposition that all classes referred to in a proposition have members, 207–15

Existential quantifier: A symbol ($\exists$) in modern quantification theory that indicates that any propositional function immediately following it has some true substitution instance; "($\exists x)Fx$" means "there exists an x such that F is true of it," 403

Expectation value: In probability theory, the value of a wager or an investment; determined by multiplying each of the mutually exclusive possible returns from that wager by the probability of the return, and summing those products, 553–55

Explanation: A group of statements from which some event (or thing) to be explained can logically be inferred and whose acceptance removes or diminishes the problematic character of that event (or thing), 18–24; scientific *vs.* unscientific, 513–16. *See also* Scientific explanation

Exportation (Exp): The name of a rule of inference; an expression of logical equivalence that permits the mutual replacement of statements of the form $(p \cdot q) \supset r$ by statements of the form $p \supset (q \supset r)$, 362

Expressive discourse, 64

Extension: The several objects to which a term may correctly be applied; its denotation, 86–91

Extensional definition. *See* Denotative definition

F

Falcoff, Marc, 267

Fallacies of ambiguity. *See* Ambiguity, fallacy of

Fallacies of defective induction: *See* Defective induction, fallacy of

Fallacies of presumption. *See* Presumption, fallacy of

Fallacies of relevance. *See* Relevance, fallacy of

Fallacy: A type of argument that seems to be correct, but contains a mistake in reasoning. Fallacies may be formal or informal, 105–54; classification of, 106–07. *See also* specific fallacies

Fallacy of accent: A fallacy in which a phrase is used to convey two different meanings within an argument, and the difference is based on changes in emphasis given to words within the phrase, 142, 143

Fallacy of accident: A fallacy in which a generalization is wrongly applied to a particular case, 133, 134

Fallacy of amphiboly: A fallacy in which a loose or awkward combination of words can be interpreted in more than one way; the argument contains a premise based upon one interpretation, while the conclusion relies on a different interpretation, 142

Fallacy of composition: A fallacy in which an inference is mistakenly drawn from the attributes of the parts of a whole to the attributes of the whole, 145

Fallacy of division: A fallacy in which a mistaken inference is drawn from the attributes of a whole to the attributes of the parts of the whole, 146

Fallacy of equivocation: A fallacy in which two or more meanings of a word or phrase are used in different parts of an argument, 140

False cause: A fallacy in which something that is not really the cause is treated as as cause. Also known as *non causa pro causa*, 130–32

Falsity, truth and, 28

Fang, Thome H., 150

Fauci, Anthony, 481

Faust (von Goethe), 412

Female Eunuch, The (Greer), 102

Ferio: The traditional name of one of the 15 valid standard-form categorical syllogisms. A syllogism in the form *Ferio* has the mood and figure **EIO–1**; that is to say, its major premise is an **E** proposition, its minor premise is an **I** proposition, its conclusion is an **O** proposition, and it is in the first figure because the middle term is the subject of both the major and the minor premise, 241–42

Ferison: The traditional name of one of the 15 valid standard-form categorical syllogisms. A syllogism in the form *Ferison* has the mood and figure **EIO–3**; that is to say, its major premise is an **E** proposition, its minor premise is an **I** proposition, its conclusion is an **O** proposition, and it is in the third figure because the middle term is the subject of both the major and the minor premise, 241–42

616 Glossary/Index

Festino: The traditional name of one of the 15 valid standard-form categorical syllogisms. A syllogism in the form *Festino* has the mood and figure **EIO–2**; that is to say, its major premise is and **E** proposition, its minor premise is an **I** proposition, its conclusion is an **O** proposition, and it is in the second figure because the middle term is the predicate of both the major and the minor premise, 208–09, 236, 241, 242, 244, 265

Feyerabend. Paul, 123

Feynman, Richard, 279

Figure: The logical shape of a standard-form syllogism as determined by the position of the middle term in its premises. There are four figures, corresponding to the four possible positions of the middle term. First figure: the middle term is the subject term of the major premise and predicate term of the minor premise; second figure: the middle term is the predicate term of both premises; third figure: the middle term is the subject term of both premises; fourth figure: the middle term is the predicate term of the major premise and the subject term of the minor premise, 207–09, 213, 246

Firestone, Shulamith, 126

First-order enthymeme: An incompletely stated argument in which the proposition that is taken for granted but not stated is the major premise of the syllogism, 265

Force, appeal to (argument *ad baculum*), 117–18

Fordice, Kirk, 464

Formal proof of validity: A sequence of statements each of which is either a premise of a given argument, or follows from the preceding statements of the sequence by one of the rules of inference, where the last statement in the sequence is the conclusion of the argument whose validity is proved; for deductive arguments dependent on inner structure of noncompound propositions, 337–39

Foundations of Ethics (Ross), 23

Foundations of Arithmetic, The (Frege), 278

Four terms, fallacy of: A formal fallacy in which a categorical syllogism contains more than three terms, 225

Frank, Anne, 452

Freedom of Choice Affirmed (Lamont), 100

Freeman, Samuel, 34

Frege, Gottlob, 29, 105, 278, 397–98, 416, 451, 462

Fresison: The traditional name of one of the 15 valid standard-form categorical syllogisms. A syllogism in the form *Fresison* has the mood and figure **EIO–4**; that is to say, its major premise is an **E** proposition, its minor premise is an **I** proposition, its conclusion is an **O** proposition, and it is in the fourth figure because the middle term is the predicate of the major premise and the subject of the minor premise, 241

Fritschler, A. L., 123

Future of an Illusion, The (Freud), 126

G

Gagarin, Yuri, 466

Galbraith, John Kenneth, 45, 276

Galdikas, Birute, 11

Galileo Galilei, 127, 493, 514, 523, 526–29, 538

Gamow, George, 154

Gardner, Howard, 22

Garfield, James A., 74

Gates, Bill, 30

Geach, Peter Thomas, 261

Gell-Mann, Murray, 80

Generalization: In quantification theory, the process of forming a proposition from a propositional function by placing a universal quantifier or an existential quantifier before it, 403

General product theorem: A theorem in the calculus of probability used to determine the probability of the joint occurrence of any number of independent events, 544

Genetic Engineering (Smith), 266

Genus and difference: A technique for constructing connotative definitions, 93–94. *See also* Definition by genus and difference

Geronimus, Arline, 498

Gibbs, Walter, 450

Gilg, Oliver, 537

Ginsburg, Ruth Bader, 86

Gladwell, Malcolm, 45

God Delusion, The (Dawkins), 11

Gödel, Kurt, 29, 339–40

God That Failed, The (Silone), 122

Gonzalez, Pancho, 267

Goode, Miranda, 511

Gore, Al, 143, 448

Gore, G., 494

Gotti, Richard, 102

Grant, Ulysses S., 75

Gratz v. Bollinger, 12

Greer, Germain, 102

Grunberger, R., 122

Grunstra, Ken, 37

Guide for the Perplexed, The (Maimonides), 262, 268

Guilt by association, 115

Guss, David, A., 508

H

Hague, Frank, 70

Hammoud, Alex, 37

Haraway, Donna, 412

Hardy, G.H., 35, 262

Harlan, John, 66, 68

Harman, Gilbert, 278

Harry Potter and the Sorcerer's Stone (Rowling), 412

Hasty generalization: A fallacy in which one moves carelessly from individual cases to generalization. Also known as "converse accident," 132–33, 134

Hawking, Stephen, 526

Hawthorne, Nathaniel, 75
Hayden, Dorothy, 283
Hegel, G.W.F., 67
Henry, Patrick, 108
Hentoff, Nat, 268
Herbert, Bob, 38
Hillary, Sir Edmund, 9
History of Chinese Philosophy, A (Fung Yu-Lan), 99
History of the Peloponnesian War (Thucydides), 125
Hitler, Adolf, 69, 108
Hobbes, Thomas, 99, 102
Holmes, Oliver Wendell, 69
Holt, John, 451
Horseshoe: The symbol for material implication, $\supset$, 303
How Children Fail (Holt), 451
Human Accomplishment (Murray), 9
Hume, David, 53, 125, 137, 263, 267, 443, 444, 460, 462, 470, 473
Hungry Soul: Eating and the Perfecting of Our Nature, The (Kass), 152
Hussein, Saddam, 320
Hutchinsons, Richard, 46
Huxley, Thomas Henry, 450
Hypothetical proposition or **hypothetical statement:** A compound proposition of the form "*if p then q*"; a conditional proposition or statement, 6, 17
Hypothetical Syllogism (H.S.): A syllogism that contains a hypothetical proposition as a premise. If the syllogism contains hypothetical propositions exclusively it is called a "pure" hypothetical syllogism; if the syllogism contains one conditional and one categorical premise, it is called a "mixed" hypothetical syllogism. "Hypothetical syllogism" ("H.S.") is also the name of an elementary valid argument form that permits the conclusion that $p \supset r$, if the premises $p \supset q$ and $q \supset r$ are assumed to be true, 274–75, 276, 316, 317, 319–20, 338

I

Iconic representation: The representation of standard-form categorical propositions, and of arguments constituted by such propositions, by means of spatial inclusions and exclusions, as in the use of Venn diagrams, 202
Identity, Principle of: A principle that asserts that if any statement is true then it is true; sometimes held to be one of the laws of thought, 333, 334
Ignorance, argument from (argument *ad ignorantiam*), 126–29
Ignoratio elenchi: The informal fallacy of irrelevant conclusion, 118–20
Illicit major: Short name for the "Fallacy of Illicit Process of the Major Term," a formal mistake made when the major term of a syllogism is undistributed in the major premise, but is distributed in the conclusion. Such a mistake breaks the rule that if either term is distributed in the conclusion, it must be distributed in the premises, 226
Illicit minor: Short name for the "Fallacy of Illicit Process of the Minor Term," a formal mistake made when the minor term of a syllogism is undistributed in the minor premise, but is distributed in the conclusion, 226
Illicit process, fallacy of: The formal fallacy that is committed when a term that is distributed in the conclusion is not distributed in the corresponding premise, 226
Immediate inference: An inference that is drawn directly from one premise without the mediation of any other premise. Various kinds of immediate inferences may be distinguished, traditionally including *conversion, obversion*, and *contraposition*, 179–88
Implicans: The antecedent of a conditional or hypothetical statement; the protasis, 300
Implicate: The consequent of a conditional or hypothetical statement; the apodosis, 300
Implication: The relation that holds between the antecedent and the consequent of a true conditional or hypothetical statement. Because there are different kinds of hypothetical statements, there are different kinds of implication, including: logical implication, definitional implication, causal implication, decisional implication, and material implication. "Impl." is also the abbreviation for "Material Implication," the name of a rule of inference, the expression of a logical equivalence that permits the mutual replacement of a statement of the form "$p \supset q$" by one of the form "$\sim p \vee q$," 301, 302. *See also* Material implication
Inappropriate authority, appeal to, 129
Inclusive disjunction: A truth-functional connective between two components, called disjuncts; a compound statement asserting inclusive disjunction is true when at least one (that is, one or both) of the disjuncts is true. Normally called simply "disjunction" it is also called "weak disjunction" and is symbolized by the wedge, $\vee$, 292–93. *See also* Exclusive disjunction
Inconsistent: Characterizing any set of propositions that cannot all be true together, or any argument having contradictory premises, 386–92
In Defence of Free Will (Campbell), 278
Independent events: In probability theory, events so related that the occurrence or nonoccurrence of one has no effect upon the occurrence or nonoccurrence of the other, 543
Individual constant: A symbol (by convention, normally a lower case letter, *a* through *w*) used in logical notation to denote an individual, 400
Individual variable: A symbol (by convention normally the lower case *x* or *y*), that serves as a placeholder for an individual constant. The universal quantifier, (*x*), means "for all *x* . . ." The existential quantifier, (*x*) means "there is an *x* such that . . .", 400
Induction: One of the two major types of argument traditionally distinguished, the other being deduction. An inductive argument claims that its premises give only some degree of probability, but not certainty, to its conclusion, 27; argument by analogy and, 445–52; defective, fallacies of, 126–33; distinction between deduction and, 27, 444–45; probability and, 26, 27, 539–59; refutation by logical analogy and, 463–65; by simple enumeration, 474–76
Induction, Principle of: The principle, underlying all inductive argument, that nature is sufficiently regular to permit the discovery of causal laws having general application, 137

Induction by simple enumeration: A type of inductive generalization, much criticized, where the premises are instances in which phenomena of two kinds repeatedly accompany one another in certain circumstances, from which it is concluded that phenomena of those two kinds always accompany one another in such circumstances, 474–76. *See also* Methods of experimental inquiry

Inductive generalization: The process of arriving at general or universal propositions from the particular facts of experience, relying upon the principle of induction, 474. *See also* Methods of experimental inquiry

Inference: A process by which one proposition is arrived at and affirmed on the basis of some other proposition or propositions, 5. *See also* Immediate inference

Inference, Rules of: In deductive logic, the rules that may be used in constructing formal proofs of validity, comprising three groups: a set of elementary valid argument forms, a set of logically equivalent pairs of expressions whose members may be replaced by one another, and a set of rules for quantification, 362, 366, 368–83, 415–22

Informative discourse, 64

Ingersoll, Robert G., 121

Instantiation: In quantification theory, the process of substituting an individual constant for an individual variable, thereby converting a propositional function into a proposition, 403

Institutes (Coke), 100

Instructions on Christian Theology (Smith), 74

Intensional definitions, 91–94; conventional, 92; objective, 92; subjective, 92

Intension of a term: The attributes shared by all and only the objects in the class that term denotes; the connotation of the term, 87, 91–93

Interwoven arguments, 43, 49

Invalid: Not valid; characterizing a deductive argument that fails to provide conclusive grounds for the truth of its conclusion; every deductive argument is either valid or invalid, 24, 30, 31; precise meaning of, argument forms and, 314

Invalidity: proving, 424–29

An Investigation into the Laws of Thought (Boole), 189

Irrelevant conclusion: An informal fallacy committed when the premises of an argument purporting to establish one conclusion are actually directed toward the establishment of some other conclusion. Also called the fallacy of *ignoratio elenchi*, 118–20

J

Jablonski, Nina, 20
Jackson, Jesse, 153
Jacoby, James, 262
Jacoby, Oswald, 262
Jacqueline Du Pre: Her Life, Her Music, Her Legend (Du Pre), 99
James, William, 100, 101
Jefferson, Thomas, 71, 75
John Angus Smith v. United States, 152
Johnson, Lyndon, 44
Johnson, Samuel, 24, 69, 97, 99, 108, 413
Johnston, David Cay, 37

Joint Method of Agreement and Difference: A pattern of inductive inference in which the Method of Agreement and the Method of Difference are used in combination to give the conclusion a higher degree of probability, 488–91

Joint occurrence: In probability theory, a compound event in which two simple events both occur. To calculate the probability of joint occurrence the product theorem is applied, 543–47

Jones, W. Ron, 267

Journal of Blacks in Higher Education, 69

K

Kahn, E.J., Jr., 148
Kakutani, Michiko, 121
Kane, Gordon, 22, 467
Kant, Immanuel, 21, 99, 102, 245, 250, 521
Kass, Leon, 152
Kelly, Edmond, 149
Kelly, Jeffrey, 490
Kennedy, Anthony, 34
Kepler, Johannes, 523
Kettering, Charles, 412
Keynes, J.M., 101
Khrushchev, Nikita, 466
Kim, Jaegwon, 278
Kingsley, Charles, 116
Kinsley, Michael, 464
Kipling, Rudyard, 86
Kirsch, Ken I., 21
Koedt, Anne, 450
Kolbert, Elizabeth, 124
Kousa, Anne, 498
Kraska, Keith, 47
Kristof, Nicholas, 37
Kristol, Irving, 261

L

L'Ami du peuple (Marat), 70
Lamont, Corliss, 100
La Mettrie, J. O., 74
Lander, Eric, 446

Language, Functions of: Various uses of language; the informative, expressive, and directive functions of language are most commonly distinguished; ceremonial and performative functions are also noted, 64–71; agreement/disagreement and, 72; emotive and neutral functions, 71–75; language forms and, 65

Language, Truth, and Logic (Ayer), 263
La Rochefoucauld, François, 99
Larson, Gerald James, 103
Last Sermon in Wittenberg (Luther), 11

Laws of thought: Three tautologies—the principle of identity, the principle of noncontradiction, and the principle of excluded middle—that have sometimes been held to be the fundamental principles of all reasoning, 333–35

Lazear, Jesse W., 483

Lebenthal, Alexandra, 101
Lectures on the Philosophy of History (Hegel), 67
Lectures on the Republic of Plato (Nettleship), 154
Left-Hander Syndrome, The (Coren), 507
Leibniz, Gottfried, 29, 163, 205, 269
Lenin, V.I., 466
Lenzi, J., 23
Lepore, Jill, 10
Leviathan (Hobbes), 99, 102
Levine, Ellen, 450
Levit, Fred, 139
Lewis, Bernard, 22
Lexical definition: A definition that reports a meaning the *definiendum* (the term to be explained) already has, and thus a definition that can usually be judged correct or incorrect, 81–82
Lexical definitions, Rules for: Traditional criteria for appraising definitions, normally applied to lexical definitions by genus and difference, 107–09
Limitation, Conversion by and Contraposition by: The immediate inferences of conversion when applied to **A** propositions, and of contraposition when applied to **E** propositions; the phrase "by limitation" in their names indicates that in these special cases such inferences require the existential assumption of Aristotelian logic to legitimize the inference. Immediate inferences "by limitation" are therefore not valid in the Boolean interpretation of categorical propositions, where that traditional existential assumption is rejected, 181, 184
Lincoln, Abraham, 4, 17, 28, 30, 44, 120, 289, 467
Lincoln, C. Eric, 69
Lipset, Seymour, 36
Locke, John, 101, 130
Lodge, Henry Cabot, 152
Logic: The study of the methods and principles used to distinguish correct from incorrect reasoning, 2; visual, 173–75, 191, 220, 305
Logic: The Theory of Inquiry (Dewey), 102
Logical equivalence: In dealing with truth-functional compound propositions, the relationship that holds between two propositions when the statement of their material equivalences is a tautology. A very strong relation; statements that are logically equivalent must have the same meaning, and may therefore replace one another wherever they occur, 329–33; expressions of, 359
Logic, or The Art of Thinking (Arnauld & Nicole), 29
Lombardi, Vince, 73
Losos, Jonathan, 520, 521
Lovelock, James, 46
Lucretius, 37
Lukacs, John, 508
Luther, Martin, 11, 68

M

Macdonald, Mia, 154
Machiavelli, Niccolò, 53, 68
Madison, James, 284

Maimonides, Moses, 262, 268
Major premise: In a standard-form syllogism, the premise that contains the major term, 206, 207
Major term: The term that occurs as the predicate term of the conclusion in a standard-form syllogism, 206, 207, 243
Malthus, Thomas, 517
Marat, Jean-Paul, 70
Margoliash, Danial, 484
Marshall, Thurgood, 84
Marx, Karl, 41, 102, 105
Material equivalence: A truth-functional relation (symbolized by the three bar sign, ≡) that may connect two statements. Two statements are materially equivalent when they are both true, or when the are both false—that is, when they have the same truth value. Materially equivalent statements always materially imply one another. "Material Equivalence" ("Equiv.") also is the name of a rule of logical inference that permits the mutual replacement of certain pairs of logically equivalent expressions, 326–27, 330, 362. *See also* Logical equivalence
Material implication: A truth-functional relation (symbolized by the horseshoe, ⊃) that may connect two statements. The statement "p materially implies q" is true when either p is false or q is true. Material implication is a weak relation; it does not refer to the meaning of the statements connected, but merely asserts that it is not the case both that p is true and that q is false. "Material Implication" ("Impl.") is the name of a rule of inference, an expression of logical equivalence that permits the mutual replacement of statements of the form "$p \supset q$" by statements of the form "$\sim p \vee q$," 300–10, 318, 320, 332; paradoxes of, 332–33
Mathematician's Apology, A (Hardy), 262
Mayo, Elton, 508
McCann, Stuart J.H., 487
McTaggart, John, 153, 267
Mead, Nicole, 511
Mediate inference: Any inference drawn from more than one premise, 179
Mein Kampf (Hitler), 69
Menand, Louis, 449
Mencken, H.L., 100, 101
Meno (Plato), 125, 285
Merely verbal dispute: A dispute in which the fact that there is no real disagreement between the disputants is obscured by the presence of some ambiguous key term, 76
Mente, Andrew, 500
Merrill, David, 490
Methods of experimental inquiry (Mill's Methods): The five patterns of inductive inference, analyzed and formulated by John Stuart Mill, with which hypotheses are confirmed or disconfirmed, 476–501; Joint Method of Agreement and Difference, 488–91; Method of Agreement, 476–81; Method of Concomitant Variation, 495–500; Method of Difference, 482–88; Method of Residues, 491–95; power of, 501
Metrodorus of Chios, 460
Middle term: In a standard-form syllogism (which must contain exactly three terms), the term that appears in both premises, but does not appear in the conclusion, 206, 207, 225–26

Midsummer Night's Dream, A (Shakespeare), 22

Mill, John Stuart, 29, 68, 149, 476–77, 482, 488, 491, 496, 497, 501, 502–03

Miller, Algernon, 127

Mill's Methods. *See* Methods of experimental inquiry

Minor premise: In a standard-form syllogism, the premise that contains the minor term, 206, 207

Minor term: The term that occurs as the subject term of the conclusion in a standard-form syllogism, 206, 207, 226–27

Missing the point: A fallacy in which the premises support a different conclusion from the one that is proposed. Also known as "irrelevant conclusion" and "*ignoratio elenchi*," 118

Mixed hypothetical syllogism. *See* Hypothetical Syllogism

Modern logic. *See* Symbolic logic

Modern symbolic logic, 164–65

***Modus Ponens* (M.P.):** A mixed hypothetical syllogism in which the first premise is a conditional proposition, the second premise affirms the antecedent of that conditional, and the conclusion affirms the consequent of that conditional. One of the nine elementary valid argument forms; a rule of inference according to which, if the truth of a hypothetical premise is assumed, and the truth of the antecedent of that premise is also assumed, we may conclude that the consequent of that premise is true. Symbolized as: $p \supset q$, p, therefore q, 274, 316, 318, 320, 339–41, 343, 349, 365

***Modus Tollens* (M.T.):** A mixed hypothetical syllogism in which the first premise is a conditional proposition, the second premise denies the consequent of that conditional, and the conclusion denies the antecedent of that conditional. One of the nine elementary valid argument forms; a rule of inference according to which, if the truth of a hypothetical premise is assumed, and the falsity of the consequent of that premise is also assumed, we may conclude that the antecedent of that premise is false. Symbolized as $p \supset q$, $\sim q$, therefore $\sim p$, 275–76, 316, 318–19, 321, 338, 351, 352, 365

Mood: A characterization of categorical syllogisms, determined by the forms of the standard-form categorical propositions it contains. Since there are just four forms of propositions, **A**, **E**, **I** and **O**, and each syllogism contains exactly three such propositions, there are exactly 64 moods, each mood identified by the three letters of its constituent propositions, **AAA**, **AAI**, **AAE**, and so on, to **OOO**, 207, 234

Moore, George, 74, 277

More, Thomas, 68

Mozhi, xviii

Mukamal, Kenneth, 510

Murray, Charles, 9

Muscarello v. U.S., 86

Mussolini, Benito, 74

Mutually exclusive events: Events of such a nature that, if one occurs, the other(s) cannot occur at the same time. Thus, in a coin flip, the outcomes "heads" and "tails" are mutually exclusive events, 548

N

National Commission on Civil Disorders, 68

Natural deduction: A method of proving the validity of a deductive argument by using the rules of inference, 338

Nature of Morality, The (Harman), 278

Necessary and sufficient condition: The conjunction of necessary conditions for the occurrence of a given event, this conjunction being all that is needed to ensure the occurrence of the event. It is the sense in which the word *cause* is used when inferences are drawn both from cause to effect and from effect to cause. In deductive reasoning two statements that are materially equivalent are necessary and sufficient conditions for one another, since they imply one another. Hence the sign for material equivalence ($\equiv$) may be read as "if and only if," 473

Necessary condition: That without which some other entity cannot be. In deductive reasoning, the consequent of a hypothetical proposition is the necessary condition of the antecedent of the proposition. In causal reasoning, the circumstance (or set of circumstances) in whose absence an event under examination cannot occur; its *sine qua non*, 471. *See also* Sufficient condition

Negation: Denial; symbolized by the tilde or curl. $\sim p$ simply means "it is not the case that p," and may be read as "not-p," 292–94

Negative definition: A variety of faulty definition that seeks to explain a term by reporting what it does not mean rather than by explaining what it does mean, 98

Nesbit, Winston, 276

Netanyahu, Benjamin, 449

Nettleship, R.L., 154

Neumann, John von, 29, 413–14

Neutral language, 71–75

New Idea of the Universe, The (Copernicus), 148

New Industrial State, The (Galbraith), 276

Newman, John Henry, 116

New Pathways in Science (Eddington), 461

Newton, Isaac, 514, 518, 519, 521–52, 523, 524

Nicomachean Ethics (Aristotle), 269

Nicole, Pierre, 29

Nissen, Steven, 534

Noble, Kenneth, 139

Nominal definition: A stipulative definition; one that arises from the arbitrary assignment of a meaning to a new term, 79

Non causa pro causa: An informal fallacy, more commonly known as the fallacy of false cause, in which one mistakenly treats as the cause of a thing that which is not really its cause, 130–32

Noncontradiction, Principle of: A principle that asserts that no statement can be both true and false; sometimes called one of the laws of thought, 334–35

Nonexclusive events: In probability theory, events so related that the occurrence of one does not preclude the occurrence of the other or others, 548

Non sequitur: Any argument that commits one of those informal fallacies in which the conclusion simply does not follow from the premises, 119–20

Nonstandard-form propositions, techniques for translating into standard form, 249–57. *See also* Standard-form categorical propositions

Normal-form formula: A formula in which negation signs apply to simple predicates only, 411

Notebooks (Butler), 99

Novum Organum, 29

Numerical coefficient of probability: A number that describes the likelihood, or probability, of the occurrence of an event. Its possible values range from 0 (impossibility) to 1 (certainty), 539

O

Objective intension: The total set of characteristics shared by all the objects in the extension of a term, 92

Obverse. *See* Obversion

Obversion: A valid form of immediate inference for every standard-form categorical proposition. To obvert a proposition we change its quality (from affirmative to negative, or from negative to affirmative) and replace the predicate term with its complement. Thus, applied to the proposition "All dogs are mammals" obversion yields "No dogs are non-mammals," which is called its "obverse." The proposition obverted is called the "obvertend," 182–83; table of, 183

Obvertend. *See* Obversion

O'Connor, Sandra Day, 151

O'Flaherty (Shaw), 69

Olbers, Heinrich, 536

On Agriculture (Webster), 68

On Liberty (Mill), 68

Only One Place of Redress: African Americans, Labor Regulations and the Courts from Reconstruction to the New Deal (Bernstein), 21–22

On War (von Clausewitz), 99

Operational definition: A kind of connotative definition that states that the term to be defined is correctly applied to a given case if and only if the performance of specified operations in that case yields a specified result, 93

Opposition: The logical relation that exists between two contradictories, between two contraries, or in general between any two categorical propositions that differ in quantity, quality, or other respects. These relations are displayed on the square of opposition, 176–80, 261

Organon (Aristotle), 164–65, 237

Origin of Species, The (Darwin), 520

Osborne, Thomas, 475

Ostensive definition: A kind of denotative definition in which the objects denoted by the term being defined are referred to by means of pointing, or with some other gesture; sometimes called a demonstrative definition, 90

Outline of History, The (Wells), 68

Owens, Gwinn, 452

Oxford Commentary on the Sentences of Peter Lombard (Duns Scotus), 263

P

Paine, Thomas, 69, 121, 125

Palmer, C.T., 21

Pan Africa (Dubois), 67

Paradoxes of material implication: Certain counterintuitive consequences of the definition of material implication: Because $p \supset q$ means only that either p is false or q is true, it follows that a false statement materially implies any statement, and it follows that a true statement is materially implied by any statement whatever. These so-called "paradoxes" are resolved when the strict logical meaning of material implication is fully understood, 333

Parameter: An auxiliary symbol or phrase that is introduced in translating statements uniformly, helping to express a syllogism with exactly three terms so that it may be accurately tested, 258–59

Paraphrasing arguments, 34–38, 40

Parsons, Cynthia, 125

Particular affirmative (**I**) propositions, 167–68, 171, 172

Particular negative (**O**) propositions, 168, 171

Particular proposition: A proposition that refers to some but not to all the members of a class. The particular affirmative proposition (traditionally called an **I** proposition) says that "Some *S* is *P*." The particular negative proposition (traditionally called an **O** proposition) says that "Some *S* is not *P*." In both traditional and modern logic, particular propositions are understood to have existential import; in the theory of quantification, they are symbolized using the existential quantifier, 177–78

Pascal, Blaise, 461, 495

Pasteur and Modern Science (Dubos), 449

Patterson, Orlando, 23, 36, 320

Paxson, Christina, 507

Peer, Elizabeth, 149

Peirce, Charles Sanders, 29, 80, 101, 324, 539

Peirce's law: A well-known statement form $[(p \supset q) \supset p] \supset p$, 326

Pelkey, Joan D., 44

Performative utterance: An utterance which, under appropriate circumstances, actually performs the act it appears to report or describe. (*E.g.*, "I apologize for my mistake," said or written appropriately, is a performative utterance because, in saying those words, one does apologize), 72

Persuasive definition: A definition formulated and used to resolve a dispute influencing attitudes or stirring emotions, often relying upon the use of emotive language, 85–86

Petitio principii: The informal fallacy of begging the question; an argument in which the conclusion is assumed in one of the premises, 136–38

Philosophical Investigations (Wittgenstein), 103

Philosophical Studies (McTaggart), 267

Philosophy of the Practical (Croce), 151

Physics (Aristotle), 152, 277

Picture of Dorian Gray, The (Wilde), 21

Piozzi, Mrs., 149

Pirsig, Robert, 260

Pitcher, George, 451

Pity, appeal to (argument *ad misericordiam*), 110–11

Place Among the Nations, A (Netanyahu), 449

Planck, Max, 519

Plato, 47, 99, 101, 123, 125, 154, 277, 284, 285

Plessy v. Ferguson, 68

Poe, Edgar Allan, 67

Poetics (Aristotle), 102

Poisoning the well: An informal fallacy; a variety of abusive *ad hominen* argument. So named because, by attacking the good faith or intellectual honesty of the opponent, it undermines continued rational exchange, 116

Politics (Aristotle), 74

Pope Leo XIII, 75

Porter, L., 10

Post hoc ergo propter hoc: A fallacy in which an event is presumed to have been caused by a closely preceding event. Literally, "After this; therefore, because of this," 131

Pratkanis, Anthony, 102

Precht, Robert, 10

Precising definition: A definition devised to eliminate vagueness by delineating a concept more sharply, 82–84

Prejudice (Mencken), 100

Preliminary hypothesis: A hypothesis, usually partial and tentative, adopted at the outset of any scientific inquiry to give some direction to the collection of evidence, 517

Premise indicator: In an argument, a word or phrase (like "because" and "since") that normally signals that what follows it are statements serving as premises, 12

Premises: In an argument, the propositions upon which inference is based; the propositions that are claimed to provide grounds or reasons for the conclusion, 5–6; not in declarative form, 13–16, 65

Preskill, John, 526

Presumption, fallacy of: Any fallacy in which the conclusion depends on a tacit assumption that is dubious, unwarranted, or false, 134–40; accident and converse accident, 134; begging the question, 136–38; complex question, 134–36

Prince, The (Machiavelli), 53, 68

Principia Mathematica (Russell & Whitehead), 29

Principles of Biology (Spencer), 461

Prior Analytics (Aristotle), 262

Probability, 539–63; *a priori* theory of, 539, 540–42; alternative conceptions of, 539–42; of alternative occurrences, 547–52; expectation value and, 552–55; induction and, 27, 29; of joint occurrences, 543–45; probability calculus, 542–52; relative frequency theory of, 541–42, 559

Probability calculus. *See* Calculus of probability

Proceedings of the American Philosophical Association (Devine), 100

Product theorem: In the calculus of probability, a theorem asserting that the probability of the joint occurrence of multiple independent events is equal to the product of their separate probabilities, 543–45

Proposition: A statement; what is typically asserted using a declarative sentence, and hence always either true or false—although its truth or falsity may be unknown, 2–5; compound, 4, 289; disjunctive, 4; exceptive, 255, 430–31; exclusive, 254; sentences and, 4; simple, 4; unstated, 16–17. *See also* Categorical proposition; Standard-form categorical proposition

Propositional function: In quantification theory, an expression from which a proposition may result either by instantiation or by generalization. A propositional function is *instantiated* when the individual variables within it are replaced by individual constants (*e.g.*, Hx is instantiated as Hs). A propositional function is *generalized* when either the universal or the existential quantifier is introduced to precede it [*e.g.*, Hx is generalized as either $(x)Hx$ or as $(\exists x)Hx$], 401

Protasis: The antecedent in a hypothetical proposition, 300

Proud Tower, The (Tuchman), 244

Proximate cause: In any chain of causes and effects, the event nearest to the event whose explanation is sought. Contrasted with "remote" causes, which are more distant in the causal chain, 472

Punctuation: The parentheses, brackets, and braces used in mathematics and logic to eliminate ambiguity, 294–97

Punishing Criminals (Van den Haag), 151

Pure hypothetical syllogism: A syllogism that contains only hypothetical propositions, 274, 275

Q

Quality: An attribute of every categorical proposition, determined by whether the proposition *affirms* or *denies* some form of class inclusion. Every categorical proposition is either affirmative in quality or negative in quality, 171

Quantification: A method for describing and symbolizing noncompound statements by reference to their inner logical structure; the modern theory used in the analysis of what were traditionally called **A**, **E**, **I** and **O** propositions, 397–98; of **A** proposition, 408; of **E** proposition, 408; of **I** proposition, 408; of **O** proposition, 408

Quantification theory, 397–441; asyllogistic inference and, 429–36; proving invalidity and, 424–29; proving validity and, 415–24; singular propositions and, 399–401; subject-predicate propositions and, 406–14; universal and existential quantifiers, 401–05

Quantity: An attribute of every categorical proposition, determined by whether the proposition refers to *all* members or only to *some* members of the class designated by its subject term. Thus every categorical proposition is either universal in quantity or particular in quantity, 171, 253, 254

Quasi-ostensive definition: A variety of denotative definition that relies upon gesture, in conjunction with a descriptive phrase, 90

Question: An expression in the interrogative mood that does not assert anything, and which therefore does not express a proposition—although in ordinary discourse questions are commonly used to assert the propositions their answers suggest, 13–15

Quotations from Chairman Mao, 100

R

Radical Feminism (Koedt, Levine, and Rapone), 450

Ragosine, Victor E., 100

Rajashekhar, Patre S., 9

Rapone, Anita, 450

Rauscher, Frances H., 479–80

Rawls, John, 17

Reagan, Ronald, 118

Reasoning: The central topic in the study of logic: problems in, 54–62

Recognizing arguments, 11–17; conclusion and premise indicators for, 11–12; in context, 12–13; premises not in declarative form and, 13–16

Red herring fallacy: A fallacy in which attention is deliberately deflected away from the issue under discussion, 111–12

Reducing the number of terms in a syllogism: Eliminating synonyms and the names of complementary classes from a syllogism to ensure that it contains exactly three terms; part of the process of translating a syllogism into standard form to test it for validity, 246–49

Reductio ad absurdum, testing argument validity by, 392–93, 395–96

Reduction to standard form: The translation of syllogistic arguments in any form into the standard form in which they can be tested for validity; also called translation to standard form, 245–46

Reed, Walter, 483

Reference and Generality (Geach), 261

Reflections (Rochefoucauld), 99

Refutation by logical analogy: A method that shows the invalidity of an argument by presenting another argument that has the same form, but whose premises are known to be true and whose conclusion is known to be false, 463–68

Rehnquist, William, 13

Reiss, Diana, 535

Relative frequency theory of probability: The view of probability in which the probability of a simple event is determined as a fraction whose denominator is the total number of members of a class, and whose numerator is the number of members of that class that are found to exhibit a particular attribute that is equivalent to the event in question, 541, 557, 558

Relevance: An essential attribute of a good scientific hypothesis, possessed by a hypothesis when the fact(s) to be explained are deducible from that hypothesis, either alone or from it together with known causal laws. Also, one of the criteria with which arguments by analogy are appraised, 454

Relevance, fallacy of: Any fallacy in which the premises are irrelevant to the conclusion, 107, 108–26, 154; *ad baculum*, 117–18; *ad hominem*, 114–17; *ad misericordiam*, 110–11; *ad populum*, 108–10; *ad verecundiam*, 129–30; *ignoratio elenchi*, 118–20

Remote cause: In any chain of causes and effects, an event distant from the effect for which explanation is sought. Contrasted with "proximate" cause, 472

Replacement, Rule of: The rule that logically equivalent expressions may replace each other wherever they occur. The rule of replacement underlies the 10 expressions of logical equivalence that serve as rules of inference, 357–64

Republic, The (Plato), 85, 101, 123

Residues, Method of: A pattern of inductive inquiry in which, when some portions of any phenomenon under investigation are known to be the effects of certain identified antecedents, we may conclude that the remaining portion of the phenomenon is the effect of the remaining antecedents, 491–95

Retrograde analysis: Reasoning that seeks to explain how things must have developed from what went before, 57

Revenge of Gaia: Earth's Climate Crisis and the Fate of Humanity, The (Lovelock), 46

Reynolds, Glenn, 48

Rhetorical question: An utterance used to make a statement, but which, because it is in interrogative form and is therefore neither true nor false, does not literally assert anything, 13, 15

Rice, Grantland, 72

Ridgeley, Stanley, 36

Robertson, James, 23

Robert's Rules of Order, 135

Robin, Corey, 467

Roethlisberger, F. J., 508

Rohr, Michael D., 468

Roosevelt, Franklin, 44

Roosevelt, Theodore, 86

Ross, Sir W. David, 23

Rothstein, Edward, 467

Rousseau, Jean-Jacques, 70, 74

Rowe, Martin, 154

Rowling, J.K., 412

Rule of replacement. *See* Replacement, rule of

Rules and fallacies for syllogisms: The set of rules with which the validity of standard-form syllogisms may be tested. These rules refer to the number and distribution of terms in a valid syllogism, and to the restrictions imposed by the quality and quantity of the premises, 224–34; flowchart for, 230; overview, 229

Rules of inference: The rules that permit valid inferences from statements assumed as premises. Twenty-three rules of inference are set forth in this book: nine elementary valid argument forms, ten logical equivalences whose members may replace one another, and four rules governing instantiation and generalization in quantified logic, 338, 339, 340, 341–42, 343, 350–51, 353, 357–64

Ruskin, John, 71

Russell, Bertrand, 29, 74, 415–16

Ruth, Babe, 6

Ryle, Gilbert, 101

S

Sanders, Charles V., 485–86

Santayana, George, 67, 121

Sargent, S. Stansfeld, 508

Scalia, Antonin, 448

Schiff, Stacy, 121

Schipper, Edith Watson, 1

Schiraldi, Vincent, 45

Scholl, T.O., 506

Schopenhauer, Arthur, 459

Schrader, Harald, 486

Schuck, Victoria, 284

Schurz, Carl, 73

Schwartz, N., 509

Scientific experiments, examples of, 525–26

Scientific explanation: Any theoretical account of some fact or event, always subject to revision, that exhibits certain essential features: relevance, compatibility with previously well-established hypotheses, predictive power, and simplicity, 513–16; criteria for evaluating, 521–29; unscientific explanation *vs.*, 514–15

Scientific investigation. *See* Scientific method

Scientific method: A set of techniques for solving problems involving the construction of preliminary hypotheses, the formulation of explanatory hypotheses, the deduction of consequences from hypotheses, the testing of the consequences deduced, and the application of the theory thus confirmed to further problems, 516–21; confirmation of Copernican hypothesis and, 529

Scientific Thought (Broad), 278

Scope and Methods of Political Economy (Keynes), 101

Second-order enthymeme: An incompletely stated argument in which the proposition taken for granted but not stated is the minor premise of the syllogism, 265

Second Sex, The (Beauvoir), 267

Second Treatise of Government (Locke), 43

Self-contradictory statement form: A statement form all of whose substitution instances are false, 326

Seligman, Daniel, 285

Seneca, 43

Sentence: A unit of language that expresses a complete thought; a sentence may express a proposition, but is distinct from the proposition it may be used to express, 4; declarative, 65; exclamatory, 65; imperative, 65; interrogative, 13, 65

Sextus Empiricus, 278, 285

Shakespeare, William, 22, 71

Shapin, Steven, 488

Shaw, George Bernard, 69

Shaw, Gordon L., 480

Shelley, Percy Bysse, 101

Shepherd, Andrew, 494

Sheridan, Richard, 81

Sherman, William Tecumseh, 69

Silone, Ignazio, 122

Simple dilemma: An argument designed to push the adversary to choose between two alternatives, the (usually undesirable) conclusion in either case being a single categorical proposition, 279

Simple enumeration. *See* Induction by simple enumeration

Simple predicate: In quantification theory, a propositional function having some true and some false substitution instances, each of which is an affirmative singular proposition, 401, 405

Simple statement: A statement that does not contain any other statement as a component, 289

Simplification (Simp.): One of the nine elementary valid argument forms; it is a rule of inference that permits the separation of conjoined statements. If the conjunction of p and q is given, simplification permits the inference that p. Symbolized as $p \cdot q$, therefore p, 341–42

Sine qua non: A necessary condition for something; meaning literally: "that without which not," 471

Singer, Isaac Bashevis, 262

Singular proposition: A proposition that asserts that a particular individual has (or does not have) some specified attribute, 250–52, 399–401, 437

Six Books of the Commonwealth (Bodin), 21

Slagar, Christina, 411

Sleep Thieves (Coren), 500

Slippery slope: A fallacy in which change in a particular direction is asserted to lead inevitably to further changes (usually undesirable) in the same direction, 131, 132

Smart, J.J.C., 262

Smith, Alan E., 266

Smith, J.P., 74

Smith, Theobald, 70

Smoking and Politics (Fritschler), 123

Snow, Clyde Collins, 114

Social Contract, The (Rousseau), 70, 74

Socrates, 26, 111, 383, 399–400

Solomon, Susan, 495

Solzhenitsyn, Alexander, 130

Some Main Problems of Philosophy (Moore), 277

Sommers, Christina, 124

Soper, Rev. Lord, 450

Sophism: Any of the fallacies of ambiguity, 140–54. *See also* Ambiguity, fallacy of

Sorites: An argument whose conclusion is inferred from its premises by a *chain* of syllogistic inferences in which the conclusion of each inference serves as a premise for the next, and the conclusion of the last syllogism is the conclusion of the entire argument, 269–72

Sound: A deductive argument is said to be sound if it is valid *and* has true premises; a deductive argument is *un*sound if it is not valid, *or* if one or more of its premises is false, 32

Specific form (of a given argument): The argument form from which the given argument results when a different simple statement is *consistently* substituted for each different statement variable in that form, 312

Specific form (of a given statement): The statement form from which the statement results by *consistently* substituting a different simple statement for each different statement variable, 325

Spencer, Herbert, 461

Spergel, David, 532

Spinoza, Baruch, 54, 70, 71, 85, 101

Square of opposition: A diagram in the form of a square in which the four types of categorical propositions (**A**, **E**, **I**, and **O**) are situated at the corners, exhibiting the logical relations (called "oppositions") among these propositions. The traditional square of opposition, which represents the Aristotelian interpretation of these propositions and their relations, differs importantly from the square of opposition as it is used in Boolean, or modern symbolic, logic, according to which some traditional oppositions do not hold, 176–80; Boolean, 198, 204; contradictories in, 176; contraries in, 177; diagram, 178–79; subalternation in, 178; subcontraries in, 177–78

Standard-form categorical propositions: The four categorical propositions, named **A**, **E**, **I**, and **O**—for universal affirmative, universal negative, particular affirmative, and particular negative, respectively, 166–70, 204; contradictories and, 176; contraries and, 177; distribution and, 171–73; existential import in interpretation of, 189–97; general schema of, 171; immediate inferences and, 180–88; opposition and, 176–80; quality and, 170; quantity and, 171; square of opposition and, 176–80; subalternation and, 178; subcontraries and, 177–78; symbolism and diagrams for, 197–203; theory of deduction and, 164–65

Standard-form categorical syllogism: A categorical syllogism in which the premises and conclusions are all standard-form categorical propositions (**A**, **E**, **I** or **O**) and are arranged in a specified order; major premise first, minor premise second, and conclusion last, 205–10; deduction of the 15 valid forms of, 239–42; exposition of 15 valid forms of, 234–38; figure and, 207–09; mood and, 207; rules for/fallacies of, 224–34; syllogistic argument and, V; terms of, 206–07; Venn diagram technique for testing, 213–24

Standard-form translation. *See* Reduction to standard form

Statement: A proposition; what is typically asserted by a declarative sentence, but not the sentence itself. Every statement must be either true or false, although the truth or falsity of a given statement may be unknown, 4

Statement form: A sequence of symbols containing no statements, but containing statement variables connected in such a way that when statements are consistently substituted for the statement variables, the result is a statement, 323–25. *See also* Specific form of a given statement

Statement variable: a place-holder; a letter (by convention any of the lower case letters, beginning with $p, q, \ldots$ etc.) for which a statement may be substituted, 311, 312

Sten, Ray, 464

Stipulative definition: A definition in which a new symbol is introduced to which some meaning is arbitrarily assigned; as opposed to a lexical definition, a stipulative definition cannot be correct or incorrect, 79–80

Stoller, Robert, 102

Stone, Irving, 123

Strachey, John, 126

Strauss, Leo, 283

Straw man fallacy: A fallacy in which an opponent's position is depicted as being more extreme or unreasonable than is justified by what was actually asserted, 113

Strom, Stephanie, 140

Study in Scarlet, A (Doyle), 53

Subalternation: The relation on the square of opposition between a universal proposition (an **A** or an **E** proposition) and its corresponding particular proposition (the **I** or the **O** proposition, respectively). In this relation, the particular proposition (**I** or **O**) is called the "subaltern." "Subalternation" is also the name of an immediate inference from a universal proposition to its corresponding particular proposition; it is a valid inference in the traditional interpretation, but generally it is not a valid inference in modern symbolic logic, 178

Subcontraries: Two propositions so related that they cannot both be false, although they may both be true. On the traditional square of opposition the corresponding **I** and **O** propositions, across the bottom of the square, are subcontraries, but on the modern, Boolean interpretation of these propositions, according to which they can both be false, the **I** and **O** propositions are not subcontraries, 177–78, 405

Subjective connotation. *See* Subjective intension

Subjective intension: The set of all attributes that the speaker believes to be possessed by objects denoted by a given term, 92

Subject-predicate propositions: The traditional categorical propositions, identified as universal affirmative (**A**), universal negative (**E**), Particular affirmative (**I**) and particular negative (**O**), 406–11

Substitution instance: For any argument form, any argument that results from the consistent substitution of statements for statement variables is a substitution instance of the given form. For any statement form, any statement that results from the consistent substitution of statements for statement variables is a substitution instance of the statement form, 311–12, 314

Sufficient condition: A circumstance (or set of circumstances) whose presence ensures the occurrence of a given event. So conceived, the sufficient condition must comprise the conjunction of all the necessary conditions of the event in question, and is normally considered the "cause" of that event, 471–72. *See also* Necessary condition

Summa Theologiae (Aquinas), 261, 277

Sumner, Charles, 74, 151

Superaltern. *See* Subalternation

Suppliant Women, The (Euripides), 70

Surowieki, James, 500

Swift, Jonathan, 149, 412

Syllogism: Any deductive argument in which a conclusion is inferred from two premises, 205; in ordinary language, 245–85; principal kinds, 275–76; rules for testing, 213–19; Venn diagram technique for testing, 213–22. *See also* Categorical syllogism; Disjunctive syllogism; Hypothetical syllogism

Syllogistic argument: Any argument that is either a standard-form categorical syllogism or can be reformulated as a standard-form categorical syllogism without any change of meaning, 245–46; enthymemes and, 264–69; in ordinary language, 245–85; reducing number of terms in, 246–49; sorites and, 269–72; translating nonstandard propositions of, 249–58; uniform translation for, 258–64

Syllogistic rules. *See* Rules and fallacies for syllogisms

Symbolic logic: The name commonly given to the modern treatment of deductive logic, 287–335; argument forms and arguments in, 310–22; conditional statements and, 300–08; conjunction in, 289–92; disjunction in, 292–94; laws of thought and, 333–35; logical equivalence and, 329–33; material equivalence and, 326–27; material implication and, 300–08; negation in, 292; punctuation for, 294–97; statement forms and, 323–25; symbolic language and, 287–88; value of special symbols in, 287–88

Symposium (Plato), 277

Synonymous definition: A definition of a symbol that gives its synonym, another word or phrase or set of symbols having the same meaning as the *definiendum*; one kind of connotative definition, 92

Synonyms: Two words with the same meaning, 92; eliminating, in syllogisms, 92

System of Logic, A (Mill), 476

T

Table Talk (Luther), 68

Tacitus, 73

Tamimi, Rulla M., 535

Tautology: A statement form all of whose substitution instances must be true. "Tautology" ("Taut.") is also the name of an expression of logical equivalence, a rule of inference that permits the mutual replacement of p by $(p \lor p)$, and the mutual replacement of p by $(p \cdot p)$, 325–26, 362

Taylor, John, 37

Teehan, John, 23

Tegmark, Max, 533

Tell, David, 468

Teller, Edward, 130

Testability: An attribute of a scientific (as contrasted with an unscientific) hypothesis; its capacity to be confirmed or disconfirmed, 515–16, 522

Thales of Miletus, 29

Theaetetus (Plato), 99

Theoretical definition: A definition that encapsulates an understanding of the theory in which that term is a key element, 84–85

Thernstrom, Abigail, 9

Third-order enthymeme: An incompletely stated argument in which the proposition that is taken for granted but not stated is the conclusion, 265–66

Thoreau, Henry David, 75

Thorne, Kip, 526

Thornhill, R., 21

Three Dialogues between Hylas and Philonous, in Opposition to Sceptics and Atheists (Berkeley), 261

Through the Looking-Glass (Carroll), 67, 140

Thucydides, 125

Thurow, Lester C., 45

Tieck, Ludwig, 70

Tierney, John, 464–65, 467

Tilde: The symbol for negation, ~, appearing immediately before (to the left of) what is denied, 292

Tolstoi, Leo, 100

Tractatus Theologico-politicus (Spinoza), 70

Traditional square of opposition. *See* Square of opposition

Translation to standard form. *See* Reduction to standard form

Transposition (Trans.): The name of an expression of logical equivalence; a rule of inference that permits the mutual replacement of $(p \supset q)$ and $(\sim q \supset \sim p)$, 361

Treatise of Human Nature, A (Hume), 263

Treatise on the Weight of the Mass of the Air (Pascal), 461, 495

Truth, falsity/validity and, 27–32

Truth-functional component: Any component of a compound statement whose replacement there by any other statement having the same truth value would leave the truth value of the compound statement unchanged, 290

Truth-functional compound statement: A compound statement whose truth value is determined wholly by the truth values of its components, 290

Truth-functional connective: Any logical connective (*e.g.*, conjunction, disjunction, material implication and material equivalence) between the components of a truth-functionally compound statement, 290–91, 326

Truth table: An array on which all possible truth values of compound statements are displayed, through the display of all possible combinations of the truth values of their simple components. A truth table may be used to define truth-functional connectives; it may also be used to test the validity of many deductive arguments; testing arguments on, 291, 293, 314

Truth value: The status of any statement as true or false (**T** or **F**), 290, 296–97, 395

Tuchman, Barbara, 225

Tu quoque: An informal fallacy; a variety of circumstantial *ad hominem* argument, 115, 116

Turing, A.M., 276

Turner, Joseph, 37

Twain, Mark, 139

Twentieth Century Socialism (Kelly), 149

U

Ulrich, Frank, 460

Undistributed middle, Fallacy of: A syllogistic fallacy in which the middle of the syllogism term is not distributed in either premise, 225–26

Uniformity of nature, causal laws and, 473

Uniform translation: Techniques (often requiring the use of auxiliary symbols) making possible the reformulation of a syllogistic argument into standard form, so that it may be accurately tested, 258–63

Unit class: A class with only one member, 250

Universal affirmative (**A**) propositions, 166, 169, 175, 178

Universal Generalization (U.G.): A rule of inference in the theory of quantification that permits the valid inference of a generalized, or universally quantified, expression from an expression that is given as true of any arbitrarily selected individual, 419–20, 422

Universal Instantiation (U.I.): A rule of inference in the theory of quantification that permits the valid inference of any substitution instance of a propositional function from the universal quantification of the propositional function, 416–17, 422

Universal negative (**E**) propositions, 167, 169, 172

Universal proposition: A proposition that refers to all the members of a class. The universal affirmative proposition (traditionally called an **A** proposition) says that "All S is P." The universal negative proposition (traditionally called an **E** proposition) says that "No S is P." In the Aristotelian interpretation universal propositions have existential import; in modern symbolic logic they do not have such import, and are symbolized using the universal quantifier, 172, 178, 191, 194, 250, 474

Universal quantifier: A symbol, (x), in the theory of quantification, used before a propositional function to assert that the predicate following the symbol is true of everything. Thus "$(x) Fx$" means "Given any x, F is true of it," 401–02

Unscientific explanation: An explanation that is asserted dogmatically and regarded as unquestionable, 514–16

Unstated propositions, 16–17

U.S. v. Nixon, 284

Utilitarianism (Mill), 149, 150

Utopia (More), 68

V

Vagueness: The attribute of a term having "borderline cases" regarding which it cannot be determined whether the term should be applied to those cases or not, 82, 83, 84. *See also* Ambiguity

Valid conversions, table of, 181

Validity: A characteristic of any deductive argument whose premises, if they were all true, would provide conclusive grounds for the truth of its conclusion. Such an argument is said to be *valid*. Validity is a formal characteristic; it applies only to arguments, as distinguished from truth, which applies to propositions, 24–34; precise meaning of, argument forms and, 314; indirect proof of, 392–94; proving, quantification theory and, 415–24; truth and, 27–33

Van den Haag, Ernest, 151

Variable. *See* Individual variable; Statement variable

Venn, John, 29, 222

Venn diagrams: Iconic representations of categorical propositions, and of arguments, to display their logical forms using overlapping circles, 166, 201–02, 204, 245

Venn diagram technique: The method of testing the validity of syllogisms using Venn diagrams, 213–24

Verbal definition. *See* Stipulative definition

Verbal dispute. *See* Merely verbal dispute

Vohs, Kathleen, 500, 511

Voltaire, Francois, 5

Von Clausewitz, Carl, 99

Von Goethe, Johann Wolfgang, 412

Von Hirsch, Andrew, 68

Von Liebig, Justus, 449

Von Moltke, Helmuth, 68

Von Treitschke, Heinrich, 70

W

Wansink, Brian, 491

Washington, George, 74

Waterfall (Escher), xxx

Waxman, Stephen, 489

Weaver, Jeff, 465

Webster, Adrian, 536

Webster, Daniel, 68

Webster, Frank, 138

Webster, Noah, 10

Wedekind, Claus, 499

Wedge: The symbol (∨) for weak (inclusive) disjunction; any statement of the form $p \lor q$ is true if p is true, or if q is true, or if both p and q are true, 293

Weeks, Jeffrey, 532

Weinberg, Clarice R., 480

Wei-Shih Lun, Ch'eng, 99

Weiss, Michael J., 533

Wells, H.G., 68

Werzberger, A., 488

Whately, Richard, 136

What Is Art? (Tolstoi), 100

What Is Political Philosophy? (Strauss), 283

While Europe Slept: How Radical Islam Is Destroying The West From Within (Bawer), 122

Whitehead, Alfred North, 29

Whitehead, T. N., 508

Wiebe, Phillip H., 192

Wilcox, Allen J., 480

Wilde, Oscar, 21

Will, George, 69

William of Ockham, 48

Williams, B.A.D., 462

Williams, Maurice, 36

Williamson, Robert C., 508

Wilson, E.O., 466

Wilson, Marjorie, 450

Winning Declarer Play (Hayden), 283

Wittgenstein, Ludwig, 103

Wolfe, Alan, 10

Woods, C. G., 489

Z

Zedong, Mao, 100

Zen and the Art of Motorcycle Maintenance (Pirsig), 260

Logic Overview V—Probability Calculations *(see Chapter 14)*

To calculate the probability of the **joint occurrence** of two or more events:
(A) If the events (say, *a and b*) are *independent*, the probability of their joint occurrence is the simple *product* of their probabilities: *P(a and b) = P(a) x P(b)*.
(B) If the events (say, *a and b and c*) are *not independent*, the probability of their joint occurrence is the probability of the first event, times the probability of the second event if the first occurred, times the probability of the third event if the first and second occurred, and so on: *P(a and b and c) = P(a) x P(b if a) x P(c if both a and b)*.

To calculate the probability of the **alternative occurrence** of two or more events:
(A) If the events (say, *a or b*) are *mutually exclusive*, the probability of at least one of them occurring is the simple *sum* of their probabilities: *P(a or b) = P(a) + P(b)*.
(B) If the events (say, *a or b or c*) are *not mutually exclusive*, the probability of at least one of them occurring may be determined by either:
   (1) analyzing the favorable cases into mutually exclusive events and summing the probabilities of those successful events; or
   (2) determining the probability that no one of the alternative events will occur and subtracting that probability from 1.

Logic Overview VI—Quantification Rules *(see Chapter 10)*

UI:	Universal Instantiation	$(x)(\Phi x)$ $\therefore \Phi v$	(where v is any individual symbol)
UG:	Universal Generalization	Φy $\therefore (x)(\Phi x)$	(where y denotes "any arbitrarily selected individual")
EI:	Existential Instantiation	$(\exists x)(\Phi x)$ $\therefore \Phi v$	[where v is any individual constant (other than y) having no previous occurrence in the context]
EG:	Existential Generalization	Φv $\therefore (\exists x)(\Phi x)$	(where v is any individual symbol)

The Four Truth-Functional Connectives

Truth-Functional Connective	Symbol (Name of Symbol)
And	• (dot)
Or	∨ (wedge)
If . . . then	⊃ (horseshoe)
If and only if	≡ (tribar)